Developments in Geotectonics 23

CONTINENTAL LOWER CRUST

FURTHER TITLES IN THIS SERIES

1. J. AUBOUIN
GEOSYNCLINES

2. R.W. VAN BEMMELEN
GEODYNAMIC MODELS

3. A. SUGIMURA and S. UYEDA
ISLAND ARCS

4. A.R. RITSEMA (Editor)
THE UPPER MANTLE

5. C. LOMNITZ
GLOBAL TECTONICS AND EARTHQUAKE RISK

6. X. LE PICHON, J. FRANCHETEAU and J. BONNIN
PLATE TECTONICS

7. R.W. GIRDLER (Editor)
EAST AFRICAN RIFTS

8. S. MUELLER (Editor)
THE STRUCTURE OF THE EARTH'S CRUST

9. N. PAVONI and R. GREEN (Editors)
RECENT CRUSTAL MOVEMENTS

10. S.W. CAREY
THE EXPANDING EARTH

11. A.M. JOHNSON
STYLES OF FOLDING

12. M.H.P. BOTT (Editor)
SEDIMENTARY BASINS OF CONTINENTAL MARGINS AND CRATONS

13. C.A. WHITTEN, R. GREEN and B.K. MEADE (Editors)
RECENT CRUSTAL MOVEMENTS, 1977

14. M.N. TOKSÖZ, S. UYEDA and J. FRANCHETEAU (Editors)
OCEANIC RIDGES AND ARCS

15. C.E. KEEN (Editor)
CRUSTAL PROPERTIES ACROSS PASSIVE MARGINS

16. P. VYSKOCIL, R. GREEN and H. MÄLZER (Editors)
RECENT CRUSTAL MOVEMENTS, 1979

17. J.H. ILLIES (Editor)
MECHANISM OF GRABEN FORMATION

18. E.V. ARTYUSHKOV
GEODYNAMICS

19. P. MORGAN and B.H. BAKER (Editors)
PROCESSES OF CONTINENTAL RIFTING

20. P. VYSKOCIL, A.M. WASSEF and R. GREEN (Editors)
RECENT CRUSTAL MOVEMENTS, 1982

21. F.-C. WEZEL (Editor)
THE ORIGIN OF ARCS

22. W. MANSPEIZER (Editor)
TRIASSIC-JURASSIC RIFTING
Continental Breakup and the Origin of the Atlantic Ocean
and Passive Margins, Part A and Part B

Developments in Geotectonics 23

CONTINENTAL LOWER CRUST

Edited by

D.M. FOUNTAIN

Department of Geology and Geophysics, University of Wyoming, P.O. Box 3006, Laramie, WY 82071, U.S.A.

R. ARCULUS

Department of Geology and Geophysics, University of New England, Armidale, N.S.W. 2351, Australia

and

R.W. KAY

Department of Geological Sciences, Snee Hall, Cornell University, Ithaca, NY 14853, U.S.A.

ELSEVIER

Amsterdam – London – New York – Tokyo 1992

ELSEVIER SCIENCE PUBLISHERS
Sara Burgerhartstraat 25
P.O. Box 211, 1000 AE Amsterdam, The Netherlands

ISBN: 0 444 88294 4

This book is printed on acid-free paper.

Printed in The Netherlands.

Preface

Intense, multi-disciplinary research over the past decade attempted to further elucidate the structure, properties, composition, and processes of the continental lower crust. This high level of interest was generated by near-simultaneous breakthroughs in several fields. One important development was the rapid and significant improvement of geobarometers and geothermometers, which permitted metamorphic petrologists to clearly demonstrate that deep crustal rocks were exposed at the Earth's surface in high-grade metamorphic terrains and as xenoliths within kimberlites and volcanic pipes. When coupled with new geochronologic techniques (e.g., the ion microprobe, Sm-Nd dating), geobarometric and geothermometric data could be used to deduce pressure-temperature-time paths for deep crustal rocks. Concurrently, seismologists applied seismic reflection profiling techniques to the continents to discover that, in many places, the deep continental crust is highly reflective. The causes of this reflectivity remain unresolved and controversial. Finally, the success of deep continental drilling in the former Soviet Union tantalized Earth scientists with the notion that the deep crust and Moho may, someday, be directly accessible for observation. Amidst these developments, petrologists, geochemists, structural geologists, and geophysicists frequently found themselves in the unusual circumstance of exchanging ideas and data at numerous international conferences dedicated to the continental lower crust.

Because of this high level of interest in continental lower crust, it seemed appropriate to summarize the state-of-the-art of deep crustal research. In this volume, we present a diverse array of in-depth essays about the physical properties, petrology, composition, structure, rheology, and evolution of the continental lower crust. The first portion of the volume emphasizes geophysical investigations of the lower crust, studies that assess the contemporary properties of the crust. The second portion of the volume explores the geological view of the deep crust, a view based on field observation, laboratory analysis, and theoretical considerations. All chapters, geological or geophysical, necessarily draw from both areas of study.

As in any effort of this kind, some conclusions drawn by the authors may soon require modification because of new discoveries that post-date manuscript acceptance. Because of the broad scope of the subject, it was practically impossible to cover all facets of the topic. Nevertheless, we hope the novice will find this a good point of departure to begin investigation of the continental lower crust and the active researcher will find that the synoptic view will open new avenues of research.

The editors would like to express their thanks to the authors for investing so much time and effort in this endeavor and for their willingness to share their ideas with the Earth science community. We give our special thanks to Drs. Henk van der Rijst for launching this project and to Drs. Martin Tanke for his patience in seeing the project through to completion.

DAVID M. FOUNTAIN
RICHARD J. ARCULUS
ROBERT W. KAY

List of contributors

R.J. ARCULUS	*Department of Geology and Geophysics, University of New England, Armidale, N.S.W. 2351, Australia*
R.J. BLAKELY	*U.S. Geological Survey, MS 989, 345 Middlefield Road, Menlo Park, CA 94025, U.S.A.*
K.H. BRODIE	*Department of Geology, University of Manchester, Manchester M13 9PL, U.K.*
D.S. CHAPMAN	*Department of Geology and Geophysics, University of Utah, Salt Lake City, UT 84112, U.S.A.*
N.I. CHRISTENSEN	*Department of Earth and Atmospheric Sciences, Purdue University, West Lafayette, IN 47907, U.S.A.*
G.L. FARMER	*Department of Geological Sciences and Cooperative Institute for Research in Environmental Sciences, University of Colorado, Boulder, CO 80309, U.S.A.*
D.M. FOUNTAIN	*Department of Geology and Geophysics, University of Wyoming, P.O. Box 3006, Laramie, WY 82071-3006, U.S.A*
B.R. FROST	*Department of Geology and Geophysics, University of Wyoming, P.O. Box 3006, Laramie, WY 82071-3006, U.S.A*
K.P. FURLONG	*Department of Geosciences, Pennsylvania State University, University Park, PA 16802, U.S.A.*
W.S. HOLBROOK	*Department of Geology and Geophysics, Woods Hole Oceanographic Institution, Woods Hole, MA 02543, U.S.A.*
A.G. JONES	*Lithosphere and Canadian Shield Division, 1 Observatory Crescent, Geological Survey of Canada, Ottawa, Ont. K1A OY3, Canada*
R.W. KAY	*Department of Geological Sciences, Snee Hall, Cornell University, Ithaca, NY 14853, U.S.A.*
S.M. KAY	*Department of Geological Sciences, Snee Hall, Cornell University, Ithaca, NY 14853, U.S.A.*
R. MEISSNER	*Institut für Geophysik, Olshausenstraße 40–60, Christian-Albrechts-Universität zu Kiel, D-2300 Kiel, Germany*
K. MEZGER	*Department of Geological Sciences, C.C. Little Building, University of Michigan, Ann Arbor, MI 48109, U.S.A.* Present address: *Department of Geochemistry, Max-Planck-Institut für Chemie, Saarstraße 23, Postfach 3060, W-6500 Mainz, Germany*
W.D. MOONEY	*U.S. Geological Survey, MS 977, 345 Middlefield Road, Menlo Park, CA 94025, U.S.A.*
J.A. PERCIVAL	*Continental Geoscience Division, Geological Survey of Canada, 588 Booth Street, Ottawa, Ont. K1A 0E4, Canada*

R.L. RUDNICK — *Research School of Earth Sciences, The Australian National University, GPO Box 4, Canberra, A.C.T. 2601, Australia*
and
Department of Geochemistry, Max-Planck-Institut für Chemie, Saarstraße 23, Postfach 3060, W-6500 Mainz, Germany

E.H. RUTTER — *Department of Geology, University of Manchester, Manchester M13 9PL, U.K.*

M.H. SALISBURY — *Atlantic Geoscience Centre, Geological Survey of Canada, P.O. Box 1006, Dartmouth, N.S. B2Y 4A2, Canada*

P.N. SHIVE — *Department of Geology and Geophysics, University of Wyoming, P.O. Box 3006, Laramie, WY 82071-3006, U.S.A*

S.M. WICKHAM — *Department of Geophysical Sciences, University of Chicago, 5734 South Ellis Avenue, Chicago, IL 60637, U.S.A.*

Contents

Chapter 1

The seismic velocity structure of the deep continental crust

W. STEVEN HOLBROOK, WALTER D. MOONEY and NIKOLAS I. CHRISTENSEN

1. Introduction

Despite its importance to the study of continental crustal evolution, the composition of the lower part of the Earth's crust remains poorly known, largely because of the difficulty of directly sampling significant portions of it. Samples of lower-crustal rocks, which can be obtained from either xenoliths or high-grade metamorphic terrains, provide invaluable information on the composition of the deep crust, but it is difficult to assess their applicability to the lower crust as a whole. Geophysical techniques, however, provide a means of indirectly measuring physical properties across large sections of *in situ* lower crust. The physical properties determined by geophysical field experiments can then be compared to those measured in the laboratory on lower-crustal rock types, thereby placing constraints on the composition of the deep crust. One such physical property, which can be measured with seismological techniques, is seismic velocity.

We can infer the seismic velocity structure of the lower continental crust in two ways: either through seismic experiments on the Earth's surface, in which the velocity of seismic waves propagating through the lower crust is determined, or by measuring the speed of sound in laboratory specimens of rocks believed to have once resided in the lower crust. Each of these methods has advantages and drawbacks: seismic reflection/refraction studies provide information on the geometry and velocity structure of the *in situ* lower crust, but they are relatively expensive, and their resolution is limited by the wavelengths used (0.2–1.0 km). Laboratory studies, on the other hand, allow precise, relatively inexpensive measurements of physical properties, but the extrapolation of these properties to the scale of the *in situ* lower crust is uncertain. Naturally, field and laboratory studies are most effective when combined.

In this paper we summarize the current knowledge of the seismic velocity structure of the lower continental crust, as determined from both field seismic experiments and laboratory measurements. We restrict our summary of field results to wide-angle (refraction) seismic data: although narrow-aperture (reflection) seismology has in recent years produced many exciting insights into the structure of the lower crust, only wide-angle data provide accurate deep velocity information. For reviews of recent seismic reflection results and their bearing on deep crustal structure, see Barazangi and Brown (1986a,b) and Matthews and Smith (1987). Our approach will be to compile lower-crustal velocity data from over 90 wide-angle seismic studies and compare them to laboratory-measured rock velocities in order to constrain lower-crustal composition. We have divided the seismic profiles according to tectonic environment (e.g., shields, rifts, etc.) in order to seek systematic velocity patterns. The theme that emerges from this compilation is that the lower continental crust is characterized by considerable diversity, but a diversity which can be characterized by current seismic techniques.

Historically, the concept of a distinct seismic lower crust stems from the work of Conrad (1925), who analyzed the traveltimes of earthquake waves and discovered a layer above the Moho with a seismic P-wave velocity (V_p) between that of the upper crust (6.0 km/s) and upper mantle (8.0 km/s). [The Moho itself, the level at which seismic P-wave velocity abruptly increases to about 8.0 km/s, had only recently been discovered by Mohorovičić (1910); for a historical review of Moho studies, see Jarchow and Thompson (1989).] This layer subsequently came to be identified with a basaltic layer (velocity about 6.5 km/s) underlying a supposedly ubiquitous Conrad discontinuity (Jeffreys, 1926). Even today, the term "seismic lower crust" is sometimes carelessly used in this context. The extreme diversity of crustal structure encountered in recent high-resolution seismic experiments, however, underscores the need to abandon such simplifications (e.g., Litak and Brown, 1989). In this review, we have adopted a more flexible definition of the seismic lower crust, as described below.

We begin with a brief overview of the source of wide-angle seismic data, the source of our *in situ* velocity information.

2. Wide-angle seismic data

Traditionally, seismic studies of the deep continental crust have been classified into two categories, refraction and reflection, depending on their acquisition parameters. Refraction data are recorded to long shot-to-receiver offsets (e.g., up to 200–300 km) but generally with sparse receiver spacing (on the order of 1–5 km) and shot spacing (20–100 km). Normal reflection data, in contrast, are typically recorded with denser shot (50–500 m) and receiver spacing (25–100 m), but to much smaller offsets (2–10 km). As a result of these contrasting recording geometries, reflection and refraction data have complementary strengths: reflection data provide a structural image of the crust — that is, shapes of major impedance contrasts — while refraction data provide an estimate of the seismic velocity distribution in the crust. In recent years, as the number of deep crustal reflection and refraction experiments has increased greatly, there has been increasing recognition that, in order to best characterize the deep crust at a given location, both types of data must be collected (e.g., Mooney and Brocher, 1987). Moreover, the traditional differences between refraction and reflection data are becoming blurred with progressive increases in the shot and receiver spacing of refraction experiments and the maximum offsets of reflection experiments. Our interest in this paper is in refraction, or wide-angle, data, as these are the data that yield information on velocities within the lower crust. Below we give a brief summary of the acquisition and interpretation of wide-angle seismic data; for a more detailed review of seismic methods, see Mooney (1989).

2.1. Acquisition

Wide-angle seismic data can be acquired both onshore and offshore; the fundamental requirement is that a sufficient number of receivers exists at offsets large enough to record identifiable, post-critical reflected and refracted phases from the crust and upper mantle. The most common method used during land experiments is to record seismic energy from large chemical explosions (shots) with a large number (>100) of portable instruments, spaced at

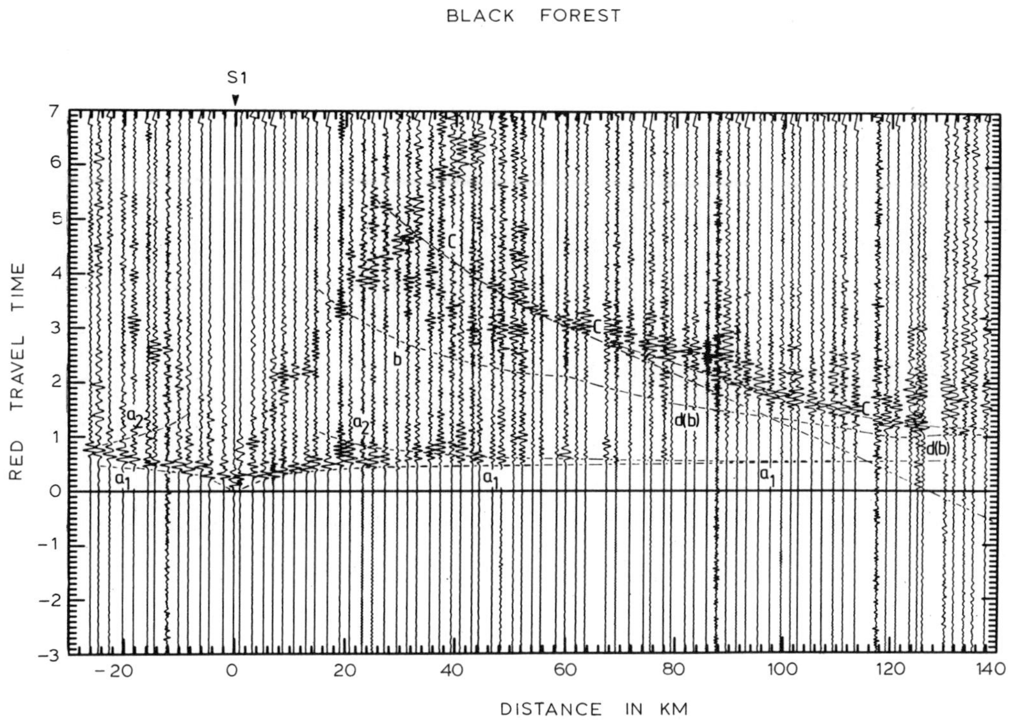

Fig. 1-1. P-wave refraction data from the Black Forest, southwestern Germany, after Gajewski and Prodehl (1987). Vertical-component seismic traces are plotted according to distance from the shotpoint (*S1*). An offset-dependent static shift (Dist/V_r) has been applied to each trace, using a reduction velocity (V_r) of 6.0 km/s; phases with apparent velocity equal to the reduction velocity plot horizontally. Predicted traveltimes of refracted (a_1, $d(b)$) and reflected (a_2, b, c) phases, calculated from the model in Fig. 1-3a, are shown.

regular intervals along a linear profile. The data for a given shot are then plotted as a seismic record section, in which each trace represents energy recorded by a single instrument, plotted as a function of its distance from the shot. A good example of this kind of data, collected in southwestern Germany by Gajewski and Prodehl (1987), is shown in Figure 1-1.

In the marine environment, where it is impractical to deploy large numbers of closely spaced receivers, a relatively new method has been developed to acquire wide-angle seismic data. Large airgun arrays, such as are commonly used in multichannel seismic reflection surveys, are towed behind a ship and detonated at regular intervals (50–100 m). The energy from these airgun pops is recorded on several (5–10) ocean-bottom instruments. In this case, the data from a given instrument are plotted as a seismic record section, in which each trace represents energy from a separate airgun pop, plotted as a function of its distance from the instrument. By the principle of reciprocity, such a receiver gather can be treated in exactly the same manner as a shot gather recorded by many instruments.

Both of the above examples are plots of compressional (P) waves; shear (S) waves, however, can also be recorded in wide-angle experiments. An example of coincident wide-angle compressional- and shear-wave data, recorded in Southwest Germany, is shown in Figure 1-2. Shear waves are transverse waves — that is, the direction of particle motion is perpendicular to the direction of propagation — so that they are best recorded on horizontal-component instruments; however, they are sometimes clearly visible on vertical-component

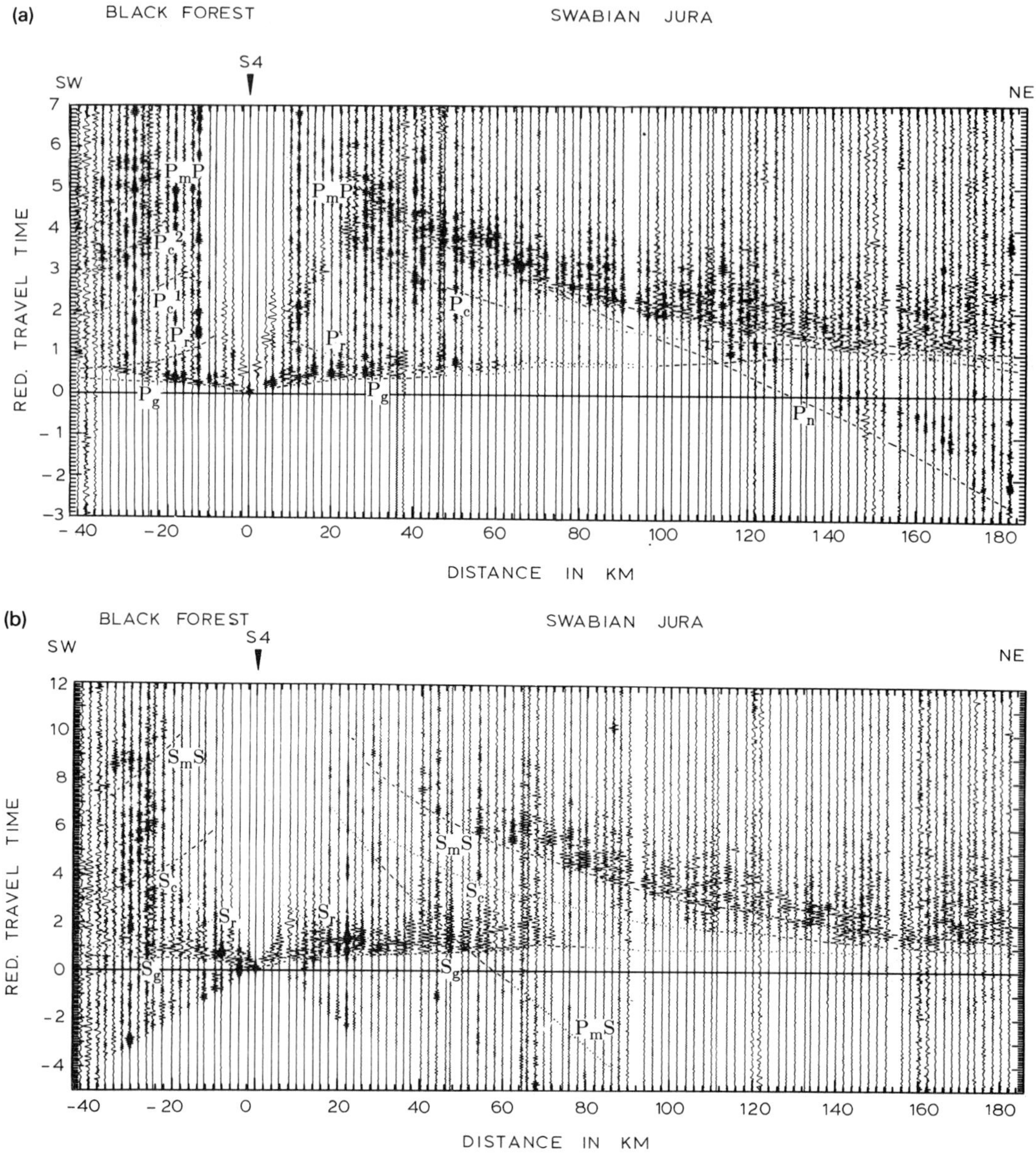

Fig. 1-2. (a) P-wave data (vertical component) from shot U-S4, after Gajewski et al. (1987). Traveltimes of refracted (P_g, P_n) and reflected (P_r, P_c) waves are shown.
(b) S-wave data (horizontal component) from shot U-S4, after Holbrook et al. (1988). Data are plotted with a reduction velocity of 3.46 km/s and a compressed time axis for easier comparison with P-wave record section (Fig. 1-2a).

instruments as well (e.g., Grad and Luosto, 1987). They are easily identifiable on the basis of their slow phase velocities (about $V_p/1.73$). Because shear and compressional waves sense somewhat different physical properties of the rocks through which they propagate, joint P and S interpretations provide more information on crustal composition than P-wave studies alone. Unfortunately, very few interpretations of wide-angle shear waves exist, because of several factors: (1) the difficulty in consistently generating strong shear-wave energy with explosive sources; (2) the paucity of horizontal-component portable seismographs, especially in the U.S.; and (3) the failure of interpreters to routinely look for shear waves in their data.

2.2. Interpretation

Methods for deriving seismic velocity structure from wide-angle seismic data can be divided into one- and two-dimensional techniques. One-dimensional techniques, which assume a laterally uniform velocity structure, include forward traveltime and amplitude modeling (e.g., Fuchs and Müller, 1971), Wiechert-Herglotz inversion (Bullen and Bolt, 1985), and τ_p inversions (e.g., Diebold and Stoffa, 1981). These techniques offer several advantages that two-dimensional methods lack: they are relatively quick and easy to implement, and the latter two provide formal error estimates. One-dimensional methods are limited, however, by their assumption of lateral homogeneity: in regions with significant lateral variability (such as that caused by dipping layers, fault-bounded structures, igneous intrusions, and sedimentary basins) one-dimensional methods yield approximate, or even misleading, results. Therefore, except where there is independent evidence (e.g., geologic, gravimetric, or magnetic) for lateral homogeneity, or where seismic data are unreversed (i.e., only observed in one direction), one-dimensional methods are usually used to provide starting models for two-dimensional modeling.

Two-dimensional interpretation of wide-angle seismic data generally consists of three steps: identification of seismic phases, iterative forward modeling of traveltimes, and calculation of synthetic seismograms. Although these steps proceed roughly in the order they are listed, they are not strictly sequential; rather, they form a continuous feedback loop, where the results of one step often force an earlier step to be repeated. For example, the discovery of an inconsistency in the predicted traveltimes on reversing shots may result in a modified phase correlation; or the comparison of synthetic seismograms to observed amplitudes may necessitate changes in the velocity model which, in turn, require renewed traveltime modeling.

The primary method used for the calculation of traveltimes in a velocity model is the raytracing technique developed by Cerveny et al. (1977). In this method, Snell's Law is used to trace a parcel of energy propagating perpendicular to a wavefront (a ray) through a velocity model consisting of layers or blocks of relatively constant velocity. Traveltimes are calculated by integrating slowness (1/velocity) along the raypath, and amplitudes are determined by applying the Zoeppritz equations (e.g., McCamy et al., 1962) at layer boundaries and calculating the geometrical spreading of adjacent rays (McMechan and Mooney, 1980). Because the raytracing method is based on asymptotic ray theory (i.e., a high-frequency solution to the wave equation), it cannot be used to model such phenomena as diffractions or surface waves. For the applications of crustal seismology, however, these limitations are more than compensated for by the ray method's power in modeling laterally varying structures. Figure 1-3a shows an example of a velocity model derived by raytracing from the Black Forest refraction data of Figure 1-1 (Gajewski and Prodehl, 1987). An idea of the ray coverage provided in this type of experiment can be gained from the ray diagram of Figure 1-3b; the synthetic seismograms calculated from the velocity model (Fig. 1-3c) compare quite favorably with the data (Fig. 1-1).

Workshops on crustal seismology (e.g., Finlayson and Ansorge, 1984; Mooney and Prodehl, 1984) have demonstrated that the greatest uncertainty in forward modeling results from the first, crucial step: phase correlation. Because this is essentially a subjective step, dictated by the experience and judgment of the interpreter, it is difficult to quantify errors in velocity

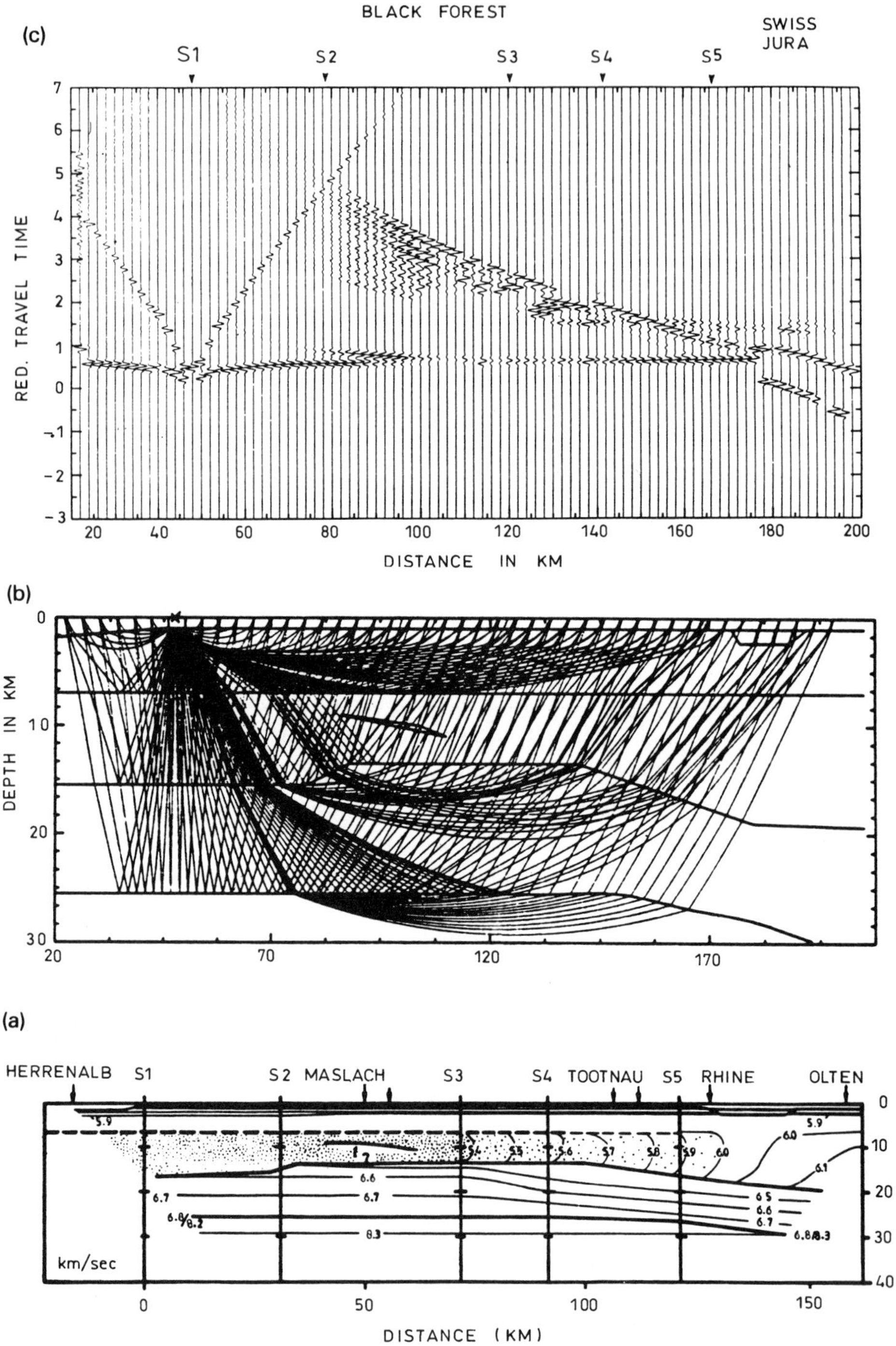

Fig. 1-3. (a) P-wave velocity model for Black Forest (Gajewski and Prodehl, 1987), developed by trial-and-error two-dimensional raytracing. Velocities shown are in km/s.

(b) Raytrace diagram for shot S1 and the P-velocity model of Fig. 1-3a, from Gajewski and Prodehl (1987).

(c) Synthetic seismogram section for Black Forest data (S1), from Gajewski and Prodehl (1987).

models derived by forward modeling. The usual approach is to perturb the velocity model until a significant discrepancy with the observed data is reached; this approach, however, only provides a measure of the error bounds for the velocity model based on a particular phase correlation; still greater differences may result from an alternative phase correlation if the seismic data are sufficiently ambiguous. Future experiments should allow some of these uncertainties to be circumvented: the increased shot and receiver spacing envisioned for these experiments will allow less ambiguous phase correlation and enable increased use of automatic imaging methods, thus reducing the reliance on subjective forward modeling techniques.

2.3. Other methods

Although the type of wide-angle refraction/reflection experiment described above has been the standard means of obtaining velocity information on the deep crust, methods that use different acquisition or interpretive techniques do exist. These include expanding-spread profiles, onshore-offshore profiles, teleseismic receiver functions from earthquakes, and, importantly, common-midpoint reflection profiling. Expanding-spread profiles are commonly collected in two-ship marine experiments where the shooting ship and receiving ship steam away from a common midpoint (Stoffa and Buhl, 1979). Onshore-offshore profiles are landward extensions of marine seismic surveys, in which land seismometers record marine explosions or airgun pops (e.g., Avedik et al., 1984; Gohl et al., 1991). Finally, recordings of teleseismic earthquake waves can be used to supplement wide-angle seismic experiments. In the receiver function method, the teleseismic waveform is modeled to derive a relatively low-resolution one-dimensional crustal structure beneath the recording station (Zandt and Owens, 1986). Because this method is inexpensive and can provide information on the crustal shear-velocity structure, it will probably become a more common complement to wide-angle experiments.

3. Middle- and lower-crustal velocities

The principal goal of this review is to present a summary of continental, middle- and lower-crustal seismic velocities in various tectonic environments, as determined by wide-angle seismic profiling. In choosing velocity models to include in our compilation, we have attempted to strike a balance between selectivity and completeness. Our data base is thus heavily weighted toward studies published in the last decade, when data quality and interpretational techniques improved significantly, though in a few cases we included older or lower-quality data for completeness. In all, our compilation includes results from more than 90 published and unpublished seismic studies (Table 1-1).

In carrying out our compilation, we chose to divide the deep crust into the “middle crust” and the “lower crust”, for two reasons. First, in many regions there is no neat division of upper from lower crust. Rather, the deep crust may comprise two distinct velocity layers, or the crust may be best parceled into thirds, with an upper, middle, and lower crust. Second, we expect that many of the processes which create or modify the lower crust will also affect the middle crust, so that a compilation of the velocity structure immediately above the lowermost crust may prove illuminating. We define the middle and lower crust as follows:

(1) In regions where there is no sharp velocity discontinuity (Conrad) near the middle of the crust, the lower 50% of the crust is defined as the lower crust.

TABLE 1-1

References used in P-velocity compilation

Reference	Environment	Quality	Velocity (km/s)
Asano et al., 1985	arc	4	6.8
Azbel et al., 1989	shield	3	6.7–7.0, 7.3
Bamford et al., 1978	Paleozoic	4	7.0
Banda and Ansorge, 1980	Paleozoic	4	6.7
Banda et al., 1981	Paleozoic	4	6.9
Bartelsen et al., 1982	Paleozoic	3	6.2
Barton and Matthews, 1984	rift, passive margin	4	6.5, 6.8
Beaudoin, 1989	rift	1	6.4
Berry, 1973	shield, platform	3	7.2–7.3
Blümling et al., 1985	forearc	3	5.4
Boland and Ellis, 1989	shield	2	7.4–7.5
Braile et al., 1974	rift	3	6.9
Braile et al., 1982	volcanic	3	6.8
Catchings and Mooney, 1988a	volcanic, rift	1	7.5
Catchings and Mooney, 1988b	volcanic, rift	1	7.4
Catchings and Mooney, 1990	rift	2	6.6, 7.4,
Clowes et al., 1987	platform, Paleozoic	3	6.7, 7.4
Colburn and Mooney, 1986	forearc	1	7.2
Cumming et al., 1979	continent collision	3	6.9
Davydova et al., 1985	platform	3	6.6, 6.8, 7.2
Deichmann et al., 1986	continent collision	3	7.1
Egger et al., 1988	rift	3	6.6–6.9, 7.5
EUGENO-S, 1988	shield, platform, rift	1	6.3–6.9
Finlayson et al., 1980	Paleozoic	4	7.4
Finlayson, 1982	shield	4	7.4
Finlayson and Leven, 1987	platform	3	6.8
Finlayson et al., 1984	platform	4	6.8, 7.1
Fuis et al., 1984	rift	2	7.1
Gajewski and Prodehl, 1987	Paleozoic	1	6.6–6.8
Gajewski et al., 1987	Paleozoic	1	6.4–6.7
Gaulier et al., 1988	passive margin	4	6.4, 6.8, 7.2
Ginzburg et al., 1981	rift	3	6.6
Ginzburg et al., 1985	rift	2	6.5
Goldflam et al., 1977	shield	4	6.8
Grad and Luosto, 1987	shield	1	7.3
Guterch et al., 1983	platform, Paleozoic	4	7.0, 7.3–7.4
Hall and Hajnal, 1973	platform	4	7.0
Hennet et al., 1991	Paleozoic	2	6.8
Holbrook and Mooney, 1987	forearc	1	7.0, 7.2
Holbrook, 1990	rift	1,2	6.6, 7.0–7.2
Howie and Miller, 1988	forearc	2	7.1
Ikami et al., 1985	shield	3	6.9
Iwasaki et al., 1990	arc	1	6.6, 6.8
Jacob et al., 1985	Paleozoic	2	6.8
Jentsch, 1979	shield	4	6.4
Kaila et al., 1981	platform	3	7.0–7.5
Kaila, 1986	volcanic	3	6.9
Kaila et al., 1987	shield	3	6.5
Kaila et al., 1989	shield	2	6.8–6.9
Kan et al., 1986	continent collision	3	6.9–7.0

TABLE 1-1 (continued)

Reference	Environment	Quality	Velocity (km/s)
Kåroly et al., 1986	continent collision, Paleozoic	4	6.8–6.9, 7.2
Keller et al., 1975	rift	4	6.5
Kondorskaya et al., 1981	continent collision, platform	4	6.7, 7.0
LASE Study Group, 1986	passive margin	3	7.2–7.3
Leaver et al., 1984	arc	1	7.0
Lowe and Jacob, 1989	Paleozoic	2	6.9
Luetgert et al., 1987	Paleozoic	1	6.7, 7.0
Luetgert and Mann, 1990	Paleozoic	1	6.8
Luosto and Korhonen, 1986	shield	2	7.2–7.3
Luosto et al., 1989	shield	1	6.9, 7.1
Luosto et al., 1990	shield	1	7.1–7.2
Lutter and Nowack, 1990	Paleozoic	2	6.9
MacGregor-Scott and Walter, 1988	forearc	3	7.2
McCarthy et al., 1989	rift	1	6.6
Mechie et al., 1986	shield	1	6.6–6.8
Mereu et al., 1986	shield	3	6.6, 6.9–7.0
Mooney and Prodehl, 1978	Paleozoic	4	6.7
Mooney et al., 1983	rift	1	7.3
Mooney et al., 1985	shield	1	7.0, 7.3–7.4
Morel–à-l'Huissier et al., 1987	platform	3	7.0–7.2
Mueller et al., 1973	Paleozoic	3	7.1
Mykkeltveit, 1980	Paleozoic	4	7.0
Pakiser and Brune, 1980	arc	4	6.9
Pavlenkova, 1979	platform	4	7.0–7.1, 7.3
Powell and Sinha, 1987	Paleozoic	3	6.7
Prodehl et al., 1984	Paleozoic	4	7.3
Research Group for Seism., 1973	arc	4	6.7–6.8
Sapin et al., 1985	continent collision	4	7.5
Sinno et al., 1986	rift	3	6.7
Sollogub et al., 1973	shield	3	6.8, 7.0
Sparlin et al., 1982	volcanic	3	6.8
Stauber, 1983	rift	3	6.7
Tarkov et al., 1981	shield	4	7.3
Teng et al., 1985	continent collision	3	6.0, 6.7
Trehu and Wheeler, 1987	forearc	3	5.0
Trehu et al., 1989	passive margin	3	7.3–7.4
White et al., 1987	passive margin	3	7.2–7.3
Wilson et al., 1991	rift	1	6.6
Yan and Mechie, 1989	continent collision	2	6.4
Yuan et al., 1986	continent collision	4	6.7, 7.4
Zeis et al., 1990	Paleozoic	2	6.6–6.8
Zelt and Ellis, 1989	platform	1	7.1–7.2
Zucca, 1984	rift	3	6.2

Quality = quality factor (1 = excellent; 2 = good; 3 = fair; 4 = poor).

(2) In regions where there is one velocity layer beneath a Conrad discontinuity, that layer is defined as the lower crust.

(3) If there are two distinct velocity layers beneath a Conrad discontinuity, the middle crust is defined as the layer immediately beneath the Conrad, the lower crust as the layer immediately above the Moho.

TABLE 1-2

Seismic properties of the middle and lower crust

	Shield	Platform	Paleoz	Cont–cont	Volcanic	Rifts	Arcs	Forearcs	Passive
Cross-sectional area surveyed ($\times 10^3$ km^2)									
middle	93	48	47	30	10	34	10	2	0.3
lower	129	46	41	46	16	50	16	3	8
V_p, lower crust (km/s)									
range	6.4–7.5	6.6–7.5	6.2–7.4	6.0–7.5	6.8–7.5	6.2–7.4	6.6–7.0	5.1–7.5	6.4–7.4
mode	6.8, 7.2	6.7, 7.2	6.7	–	6.8, 7.5	6.7, 7.3	6.8, 7.0	7.2	6.5, 7.3
V_p, middle crust (km/s)									
range	6.3–7.0	6.1–7.3	5.4–6.8	6.1–6.6	6.1–6.8	5.5–6.7	6.1–6.6	6.5–6.7	6.8
mode	6.6	6.5	6.4, 6.7	6.2, 6.5	6.5, 6.8	6.2, 6.6	6.1, 6.6	6.6	6.8
V_s, lower crust (km/s)									
range	4.1	4.0	3.6–4.0	–	–	3.3–3.8	–	–	–
mode	4.1	4.0	3.7, 4.0	–	–	3.3	–	–	–
V_s, middle crust (km/s)									
range	4.0	–	3.3–3.7	–	–	3.5	–	–	–
mode	4.0	–	3.7	–	–	3.5	–	–	–
Avg. Poisson ratio									
middle	0.26	–	0.23	–	–	0.29	–	–	–
lower	0.27	0.24	0.26	–	–	0.29	–	–	–
Avg. thickness (km)									
middle	14	15	11	14	11	11	20	5	8
lower	16	11	11	18	20	11	16	6	9
Avg. crustal thickness (km)	45	40	34	55	40	32	38	26	24

Shield = Precambrian shields (e.g., Baltic Shield); Platform = cratonal platforms (e.g., Central U.S. Plains); Paleoz = Paleozoic orogenic areas (e.g., Appalachians, Variscan); Cont–cont = continent–continent collision zones (e.g., Alps, Himalaya); Volcanic = volcanic plateaus (e.g., Columbia plateau, Snake River Plain); Rifts = continental rift zones (e.g., Basin and Range province); Arcs = continental and island arcs (e.g., Japan, Sierra Nevada); Forearcs = forearc regions (e.g., Great Valley, Calif.); Passive = passive continental margins (e.g., U.S. East Coast). Area = cross-sectional area of crust (in units of 1000 km^2) surveyed by P-wave studies. P-wave and S-wave velocities (V_p and V_s) in km/s. Average layer thickness and average crustal thickness in km. Values separated by a comma indicate peaks of bimodal distributions.

(4) Finally, where the velocity structure suggests a natural division of the crust into thirds, the lower crust is defined as deepest third, the middle crust as the middle third.

The flexibility of this set of definitions matches the diversity in the velocity models we examined. For clarity, we will use the term "deep crust" when speaking in a general sense, reserving the terms "middle crust" and "lower crust" for use according to the above definitions. Note that these definitions do not tie the top of the lower crust to a particular seismic velocity, but the base of the lower crust — the Moho — is associated with a seismic velocity (namely, mantle velocities of $\geq$7.6 km/s).

After choosing a velocity model for inclusion in our compilation, we divided its deep crust into blocks of relatively constant velocity, following the lateral and vertical heterogeneity of the model. In layers with strong vertical velocity gradients, it was necessary to approximate the gradient with a constant average velocity. We then categorized the block as belonging to the middle or lower crust and tabulated its P- and S-velocity (V_s), Poisson's ratio (σ), thickness, cross-sectional area (i.e. thickness times horizontal extent), and tectonic environment. In order to provide some measure of the variation in data quality among the seismic studies, we also assigned each velocity determination a subjective quality index, from 1 (highest) to 4 (lowest) (Table 1-1). The quality index was determined by acquisition parameters (e.g., shot and receiver spacing), data quality (e.g., signal-to-noise ratio and presence of identifiable phases), and method of interpretation (e.g., raytracing vs. one-dimensional methods; use of synthetic seismograms). In general, data with a quality factor of 1 had a receiver spacing of 1 km or less, a shot spacing of 30 km or less, good signal-to-noise and identifiable phases at offsets of at least 150–200 km, and were interpreted by two-dimensional raytracing to model both travel times and amplitudes.

We chose to tabulate cross-sectional area of deep crustal units as the best available measure of the amount of crust with a given velocity. Insofar as it is a two-dimensional quantity being used to describe a three-dimensional crust, cross-sectional area is an imperfect parameter; unfortunately, this is an inevitable limitation of two-dimensional seismic surveys. Nevertheless, cross-sectional area is an appropriate measure of the block-like bodies that are common representations of the velocity structure of the deep crust.

The results of the compilation are summarized in Table 1-2. Two important points can be drawn immediately from this table. First, the broad ranges of P-velocity displayed in the deep crust indicate considerable diversity (Table 1-2). The evidence for this diversity has been noted on the basis of both seismic reflection (Smithson, 1978) and refraction (Christensen and Fountain, 1975; Smithson and Brown, 1977; Meissner, 1986) data. P-velocities in the lower crust range from about 6.0 km/s to 7.5 km/s, and in the middle crust from about 5.4 km/s to 7.3 km/s. Second, the tabulated values of cross-sectional area surveyed show that existing deep seismic surveys are biased toward some tectonic environments (e.g., shields, platforms, and rifts) at the expense of others (e.g., arcs, forearcs, and passive margins). Since the distribution of seismic surveys does not necessarily follow the global volumetric abundances of crust in different tectonic provinces, some weighting of the compiled velocity data will be required before conclusions regarding overall (deep) crustal composition can be drawn.

In the following section, we will examine the distribution of deep crustal velocities in more detail and attempt to establish a link to deep crustal composition.

4. Seismic velocities: constraints on composition

The compilations of velocity structure presented above are helpful in providing a crude characterization of the Earth's crust in different tectonic environments, but they are of limited use unless they can be interpreted in terms of lithology. In this section we will discuss the usefulness and limitations of seismic velocities for interpreting the composition of the deep crust.

4.1. Summary of laboratory measurements

The seismic P- and S-wave seismic velocities of the deep crust are primarily determined by five factors: mineralogical composition, confining pressure, temperature, anisotropy, and pore fluid pressure. In order to use seismic velocities to draw inferences about the composition of the deep crust, then, we must first estimate (or assume) the contribution of the other four factors. Lithostatic pressures in the deep crust are easily estimated, but temperatures are less well constrained, and even less is known about the anisotropy and fluid content (and pore pressure) of the deep crust. For the purposes of this paper, we will assume an isotropic, dry crust, with the caveat that this assumption may not hold in all places. There is at present little compelling evidence for the existence of widespread anisotropy or pore pressures high enough to significantly alter seismic velocities in the lower continental crust, although these remain open questions. Hyndman and Klemperer (1989) discuss the implications of the assumption that pore pressures are high enough to reduce seismic velocities in the lower crust.

The effects of pressure and temperature on seismic velocity have been well documented by laboratory measurements (for a compilation of results, see Christensen, 1982; for a review of methods, see Christensen and Wepfer, 1989). As confining pressure increases to about 200 MPa, microcracks close, causing P- and S-velocities to increase rapidly (about 0.5–1.0 km/s/100 MPa); at higher pressures, the increase is slight (about 0.02–0.06 km/s/100 MPa for most rock types; e.g., Birch, 1961; Christensen, 1965, 1966a). As temperature increases, velocities decrease, with typical coefficients of about $2.0–6.0 \times 10^{-4}$ km/s/°C, depending on rock type (e.g., Kern, 1978; Christensen, 1979). Since pressure and temperature both increase with depth in the Earth, their effects on velocity compete. For a typical continental geotherm of about 15°C/km, a crust of uniform lithology will have a constant velocity at mid- and lower-crustal depths for most rock types, while in high heat-flow provinces (25–35°C/km) the temperature effect dominates, causing negative velocity gradients (Christensen, 1979). Because laboratory measurements of temperature coefficients have been made for most major rock types, the chief source of uncertainty in correcting *in situ* velocities for the effect of temperature is estimating the temperature itself (e.g., Lachenbruch and Sass, 1978). An error of 200°C in the temperature estimate for lower crust, for example, results in an error of about 0.1 km/s in the velocity estimate.

Once the effects of pressure and temperature have been accounted for, and assuming isotropy and the absence of high pore pressure, we can consider seismic velocity to be fundamentally a function of mineralogical composition. Because seismic velocity varies from mineral to mineral, ranging from 6.0 km/s for quartz to 8.4 km/s for garnet (e.g., Christensen and Fountain, 1975), it is a useful yardstick of composition. By comparing the velocities determined in field experiments to those measured in laboratory samples, we hope to estimate the gross composition of the deep crust. In order to facilitate that comparison, we have compiled some laboratory-measured physical properties (P- and S-velocity, density,

Poisson's ratio, and P-wave anisotropy) of possible mid- and lower-crustal rock types, spanning compositions from felsic to ultramafic, and metamorphic grades up to granulite and eclogite facies (Table 1-3). In this paper, we will use the term "granulite" strictly to describe metamorphic rocks in granulite facies, whereas we will use the term "amphibolite" in two ways: to describe the rock amphibolite, which consists predominantly of amphiboles, and to describe metamorphic rocks of felsic composition in amphibolite facies (i.e., felsic amphibolite-facies gneiss).

Unfortunately, the correlation between P-wave velocity and composition is sufficiently non-unique that such comparisons are limited to rough estimates of crustal composition. As an extreme example, a P-velocity of 5.8 km/s in the mid-crust could correspond to granite, quartzite, or serpentinite (Table 1-3); in this case, auxiliary geologic or geophysical data, such as shear-wave velocities, drill hole data, potential field data, or detailed knowledge of the geologic history of a given area, would be needed to resolve the ambiguity. In the deep crust, there is a fundamental ambiguity between increasing mafic content and increasing metamorphic grade, both of which can raise seismic velocities: there is significant overlap, for example, between the P-velocities of mafic rocks (gabbro, mafic granulite) and metapelites (Table 1-3).

The non-uniqueness of P-velocity values can be partly overcome, however, by the addition of shear-wave velocity information (e.g., Christensen and Fountain, 1975). The relation between P- and S-velocity for a given rock or mineral can be expressed in terms of Poisson's ratio, which increases with increasing P-velocity or decreasing S-velocity. Poisson's ratio varies from about 0.20 to 0.35 in common rock types and is particularly sensitive to quartz content: while most rock-forming minerals have Poisson's ratios around 0.25–0.30, quartz has a value of 0.08 (Birch, 1961).

By enabling the calculation of Poisson's ratio, coincident compressional- and shear-wave data offer a means of distinguishing between felsic (quartz-rich) and mafic (quartz-poor) rocks that would be indistinguishable on the basis of P-velocity alone. This is demonstrated in Figure 1-4, which shows Poisson's ratio as a function of P-velocity for the rock types in Table 1-3. Although the fields plotted in Figure 1-4 do show some overlap, there is a separation of several rock types whose P-velocities overlap completely. In the middle crust, Poisson's ratio is particularly helpful in distinguishing between quartz-rich rocks such as quartzites and granites from quartz-poor rocks such as serpentinite (Fig. 1-4a). Among lower-crustal rocks, metapelites have discernibly lower Poisson's ratios (0.25–0.29) than mafic (garnet) granulites (0.29–0.33); the ambiguity between increasing mafic content and increasing metamorphic grade is thus largely overcome when shear-wave information is available.

Despite the promise of shear-wave information for resolving some of the ambiguity in linking seismic velocity to lithology, disappointingly few field determinations of shear-wave velocity exist at present: compared to over 90 P-wave studies, we found only eleven published studies of wide-angle shear-waves, geographically restricted to rift zones, shields, and Paleozoic crust (Table 1-4). Because these sparse results are of limited general use in constraining deep-crustal composition, our comparison still suffers from considerable non-uniqueness.

4.2. Results

With that caution in mind, then, we will attempt to draw some petrologic conclusions from our velocity compilation. In order to compare field and laboratory measurements, we have

TABLE 1-3

Laboratory-measured physical properties of possible middle and lower crustal rock types

Rock type	N	ρ	V_p	V_s	σ	P-anis	Ref
Mid-crustal rocks							
1. Serpentinite	16	2.59 ± 0.09	5.50 (5.63) ± 0.55	2.71 (2.79) ± 0.44	0.34 (0.34) ± 0.03	6.9 ± 4.0	6,7,9,20,27
2. Quartzite	9	2.57 ± 0.12	5.52 (5.71) ± 0.57	3.57 (3.60) ± 0.41	0.14 (0.17) ± 0.07	2.9 ± 2.9	2,4,5,19,26,28
3. Granite	27 (26)	2.66 ± 0.05	6.07 (6.22) ± 0.23	3.52 (3.57) ± 0.23	0.24 (0.25) ± 0.04	2.1 ± 1.8	2,4,13,17,21,26,27,28
4. Granodiorite	16	2.69 ± 0.07	6.08 (6.26) ± 0.35	3.43 (3.55) ± 0.21	0.27 (0.26) ± 0.02	2.9 ± 0.9	4,9,13,18,21,26
5. Felsic amph. facies gneiss	18 (15)	2.73 ± 0.07	6.18 (6.27) ± 0.17	3.57 (3.61) ± 0.16	0.25 (0.25) ± 0.03	6.8 ± 3.7	5,13,15,17,21
6. Quartz-mica Schist	8 (6)	2.79 ± 0.07	6.26 (6.37) ± 0.11	3.57 (3.62) ± 0.17	0.26 (0.26) ± 0.04	11.6 ± 7.4	5,13,15,21
7. Metagabbro (greensch.)	14 (9)	2.91 ± 0.09	6.59 (6.76) ± 0.30	3.71 (3.76) ± 0.11	0.27 (0.28) ± 0.01	–	9,16
8. Gabbro	33 (15)	2.94 ± 0.09	6.95 (7.12) ± 0.22	3.74 (3.80) ± 0.18	0.29 (0.30) ± 0.02	2.5 ± 1.5	2,4,9,12,13,16,18,19,22,25
9. Amphibolite	11 (10)	3.05 ± 0.10	7.03 (7.05) ± 0.24	3.77 (3.92) ± 0.21	0.30 (0.28) ± 0.02	5.4 ± 5.3	5,12,13,17,18,19,27
Lower crustal rocks							
1. Quartzite (granulite)	17	2.66 ± 0.12	5.82 (5.97) ± 0.22	3.71 (3.75) ± 0.09	0.15 (0.17) ± 0.05	4.0 ± 0.0	2,4,5,19,26,28
2. Felsic amphib. gneiss	16 (13)	2.73 ± 0.07	6.20 (6.36) ± 0.19	3.54 (3.62) ± 0.16	0.26 (0.26) ± 0.03	7.3 ± 3.2	5,13,15,17,21
3. Felsic granulite	13	2.70 ± 0.06	6.25 (6.44) ± 0.11	3.53 (3.64) ± 0.09	0.27 (0.27) ± 0.02	2.8 ± 2.2	3,10,13,21,24
4. Quartz-mica Schist	8 (6)	2.79 ± 0.07	6.29 (6.47) ± 0.12	3.58 (3.66) ± 0.16	0.26 (0.26) ± 0.04	10.7 ± 6.4	5,13,15,21
5. Intermediate granulite	21	2.79 ± 0.11	6.39 (6.59) ± 0.20	3.58 (3.69) ± 0.14	0.27 (0.27) ± 0.03	2.7 ± 1.5	10,13,21,23,24
6. Anorthosite	15	2.80 ± 0.10	6.81 (7.03) ± 0.31	3.65 (3.72) ± 0.20	0.30 (0.31) ± 0.02	2.9 ± 1.6	2,4,13,15,24,27,29
7. Mafic granulite	38	3.03 ± 0.17	6.86 (7.13) ± 0.27	3.60 (3.90) ± 0.22	0.31 (0.29) ± 0.02	2.6 ± 1.7	10,12,13,14,15,20,23,24
8. Amphibolite	13 (12)	3.05 ± 0.09	7.00 (7.14) ± 0.24	3.81 (4.03) ± 0.22	0.29 (0.27) ± 0.02	5.4 ± 5.2	5,13,15,17,18,19,27
9. Metapelite (granulite)	12 (11)	3.10 ± 0.11	7.17 (7.33) ± 0.36	4.03 (4.11) ± 0.21	0.27 (0.27) ± 0.01	5.5 ± 5.5	14,15,20,21,23
10. Pyroxenite	14	3.27 ± 0.04	7.71 (8.00) ± 0.15	4.23 (4.41) ± 0.27	0.28 (0.28) ± 0.04	2.2 ± 1.7	1,2,4,9,12,15,22,26,27
11. Eclogite	24	3.43 ± 0.09	7.94 (8.11) ± 0.28	4.38 (4.54) ± 0.16	0.28 (0.27) ± 0.02	2.4 ± 0.9	18,23,24,27
12. Dunite/Peridotite	18	3.28 ± 0.04	8.02 (8.28) ± 0.23	4.37 (4.61) ± 0.18	0.29 (0.28) ± 0.02	11.6 ± 4.0	1,2,4,6,18,19,22,26,27

Values shown are averages of highest-pressure measurement from each sample within the ranges 3–5 kbar for the middle crust and 4–10 kbar for the lower crust; ranges shown are one standard deviation. Velocities have been corrected for the effect of temperature assuming a stable cratonic geotherm (15°C/km). N = number of samples (number of samples for which S-velocity was measured in parentheses, if different); ρ = density (gm/cm3); V_p = P-wave velocity (km/s); V_s = S-wave velocity (km/s); σ = Poisson's ratio; P-anis = percent P-wave velocity anisotropy. Ref = references: 1, Babuska, 1972; 2, Birch, 1960, 1961; 3, Bonatti and Seyler, 1987; 4, Bonner and Schock, 1981; 5, Christensen, 1966a; 6, Christensen, 1966b; 7, Christensen, 1972; 8, Christensen, 1977; 9, Christensen, 1978; 10, Christensen and Fountain, 1975; 11, Christensen et al., 1975; 12, Chroston and Evans, 1983; 13, Chroston and Brooks, 1989; 14, Evans, 1980; 15, Fountain, 1976; 16, Fox et al., 1973; 17, Hall and Simmons, 1979; 18, Kanamori and Mizutani, 1965; 19, Kern, 1982; 20, Kern and Schenk, 1985; 21, Kern and Schenk, 1988; 22, Kroenke et al., 1976; 23, Jackson and Arculus, 1984; 24, Manghnani et al., 1974; 25, Nur and Simmons, 1969; 26, Schock et al., 1974; 27, Simmons, 1964; 28, Simmons and Brace, 1965; 29, Wang et al., 1973.

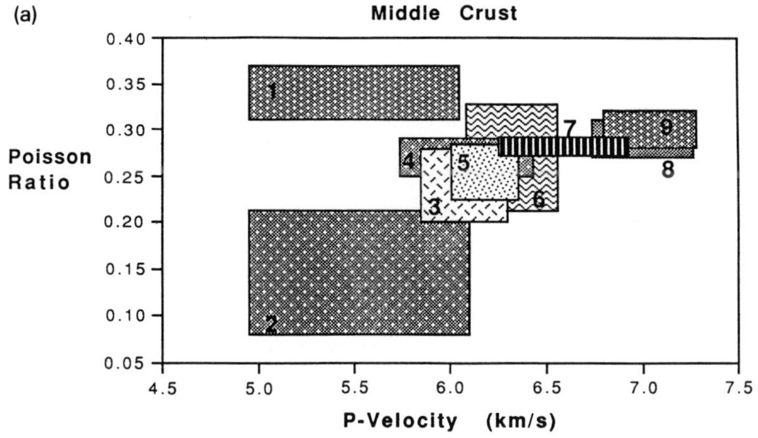

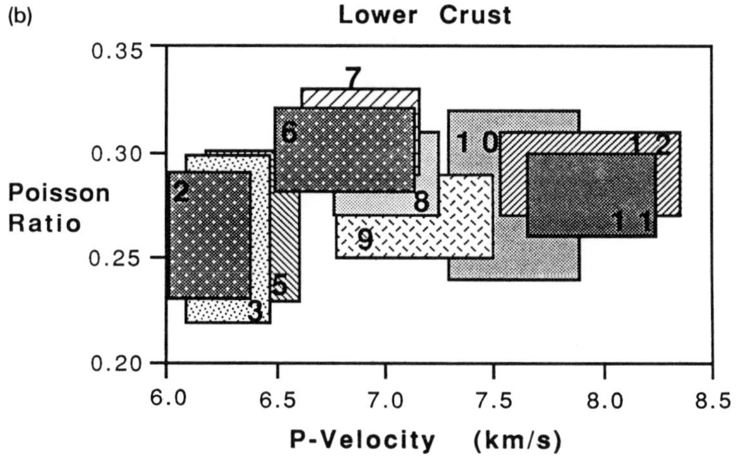

Fig. 1-4. Plots of P-velocity vs. Poisson's ratio for (a) mid-crustal and (b) lower-crustal rock types, from the data of Table 1-3. Width of fields equals two standard deviations; numbers keyed to rock types of Table 1-3. For plot clarity, rock types 1 and 4 (quartzite and schist) are not included in Fig. 1-4b.

TABLE 1-4

References used in S-velocity compilation

Reference	Environment	Quality	S-velocity	σ
Assumpçao and Bamford, 1978	Paleozoic	4	4.0	0.25
Banda et al., 1981	Paleozoic	4	3.9	0.26
Boland and Ellis, 1991	Shield	2	4.2	0.27
Braile et al., 1974	Rift	4	3.8	0.28
Grad and Luosto, 1987	Shield	1	4.1	0.27
Hall and Hajnal, 1973	Shield	4	4.0	0.24
Holbrook et al., 1988	Paleozoic	1	3.6–3.8	0.24–0.29
Keller et al., 1975	Rift	4	3.5	0.29
Luetgert et al., 1987	Paleozoic	1	3.9–4.0	0.25
Luosto et al., 1990	Shield	1	4.0–4.1	0.26–0.27
Tarkov et al., 1981	Shield	4	4.1	0.26

Quality = quality factor; σ = Poisson's ratio.

plotted the laboratory-determined ranges of the velocities and Poisson's ratios for middle- and lower-crustal rocks (Table 1-3) above histograms of the field-measured values in Figures 1-5 to 1-9. All of the lab-measured velocities shown in these plots, as well as in Table 1-3, have been corrected for temperature using published temperature derivatives (e.g., Kern, 1978; Christensen, 1979) and assuming the stable craton geotherm of Lachenbruch and Sass (1978), which predicts a temperature of about 600°C at a depth of 40 km. This correction, which amounts to about 0.2–0.3 km/s for most rock types in the lower crust, is an average estimate, as temperatures vary with tectonic environment; however, this will not affect our conclusions, since the uncertainty in the temperature correction is less than the observed ranges in lab-measured velocity for any given rock type (Table 1-2).

The summary histograms in Figures 1-5 to 1-7 show a deep continental crust with widely varying seismic properties. In the lower crust, 94% of the P-velocity determinations fall between 6.4 km/s and 7.4 km/s (57% between 6.6 km/s and 7.0 km/s), and there is a clear peak at 6.7–6.8 km/s. In the middle crust, 90% of the P-velocities fall between 6.0 km/s and 6.8 km/s, with a strong peak at 6.5–6.6 km/s. Shear-wave velocities in the lower crust show a range of 3.5 km/s to 4.2 km/s, with most values between 3.9 km/s and 4.2 km/s (Fig. 1-7b); in the middle crust, S-velocities range from 3.3 km/s to 4.0 km/s, with most between 3.7 km/s and 4.0 km/s (Fig. 1-7d). The distribution of Poisson's ratio in the lower crust shows a range of 0.24 to 0.29, with most values from 0.24 to 0.27 (Fig. 1-7c); in the middle crust, most values lie between 0.24 and 0.27, with some low (0.15) and high (0.29) determinations (Fig. 1-7a).

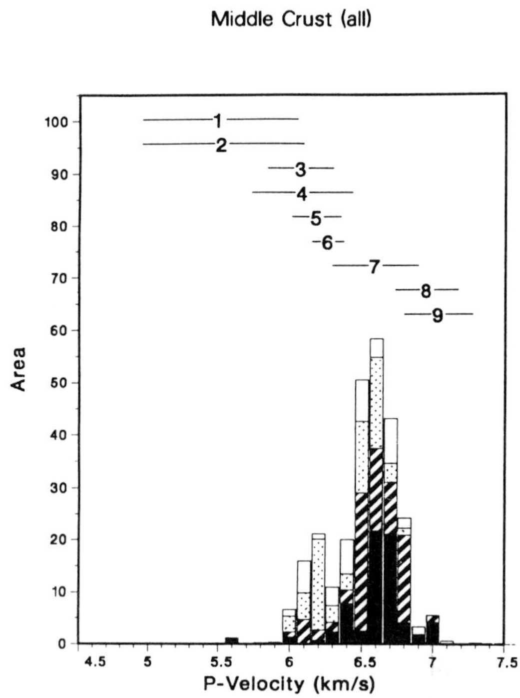

Fig. 1-5. Histogram of cross-sectional area of crust (in km^2 x 10^3) vs. P-wave velocity for the middle crust in all tectonic provinces. Shading indicates quality factor, from highest (1) to lowest (4): 1 = black; 2 = hatchured; 3 = stippled; or 4 = blank. Bars at top show velocity ranges (plus/minus one standard deviation) for possible mid-crustal rock types (Table 1-3): *1* = serpentinite; *2* = quartzite; *3* = granite; *4* = granodiorite; *5* = felsic amphibolite facies gneiss; *6* = schist; *7* = metagabbro; *8* = gabbro; *9* = amphibolite.

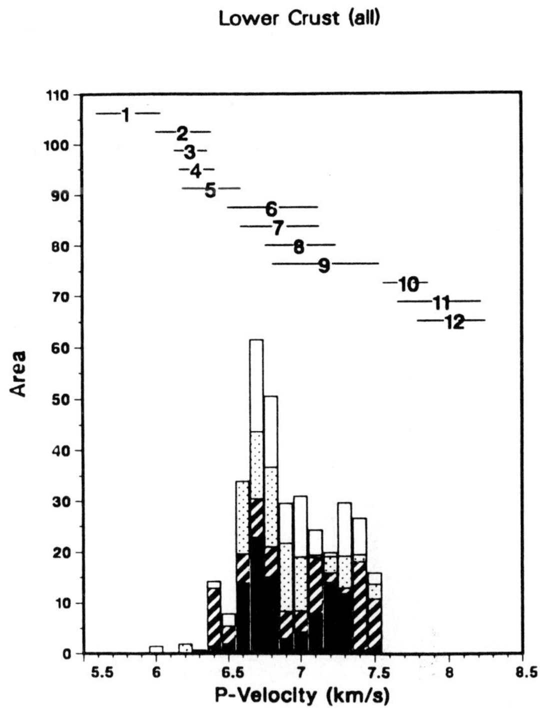

Fig. 1-6. Histogram of cross-sectional area of crust (in $km^2 \times 10^3$) vs. P-wave velocity for the lower crust in all tectonic environments. Shading indicates quality factor (see Fig. 1-5). Bars at top show velocity ranges (plus/minus one standard deviation) for possible lower crustal and upper mantle rock types (Table 2): *1* = quartzite (granulite); *2* = felsic amphibolite facies gneiss; *3* = felsic granulite; *4* = schist; *5* = intermediate granulite; *6* = anorthosite; *7* = mafic granulite; *8* = amphibolite; *9* = felsic and intermediate garnet granulite; *10* = pyroxenite; *11* = eclogite; and *12* = dunite/peridotite.

The large range of P-velocities plotted in Figures 1-5 and 1-6 implies that the deep continental crust must be composed of a diversity of rock types. The span of the lower-crustal velocities, from 6.4 km/s to 7.5 km/s, overlaps the velocity ranges of quartz-mica schist, intermediate and mafic granulites, anorthosite, amphibolite, and metapelites (Fig. 1-6). The lower-crustal velocity distribution has several peaks: there are strong peaks at 6.7–6.8 km/s and 7.2–7.5 km/s, and a subsidiary peak at 7.0 km/s. The low-velocity (6.7–6.8 km/s) peak roughly corresponds to the velocity ranges of anorthosite or mafic granulite, while the high-velocity peak (7.2–7.5 km/s) corresponds to granulite-facies metapelite or pyroxenite. In the middle crust, where the velocity distribution is narrower, the peak at 6.5–6.8 km/s best matches the velocities of greenschist-facies metagabbro, although they also fall within the ranges of quartz-mica schist, gabbro, and amphibolite (Fig. 1-6).

Using the histograms of Figure 1-8 we can draw more detailed conclusions about the average composition of the deep crust as a function of tectonic environment. In each environment, the range of observed velocities in both the middle crust and lower crust requires considerable lateral changes in bulk composition. The strongest manifestation of this is the bimodal distribution of P-velocities in the lower crust of some tectonic environments — shields, platforms, passive margins, and, to a lesser extent, rifts and volcanic plateaus — which indicates strong lateral changes in composition. The high velocities

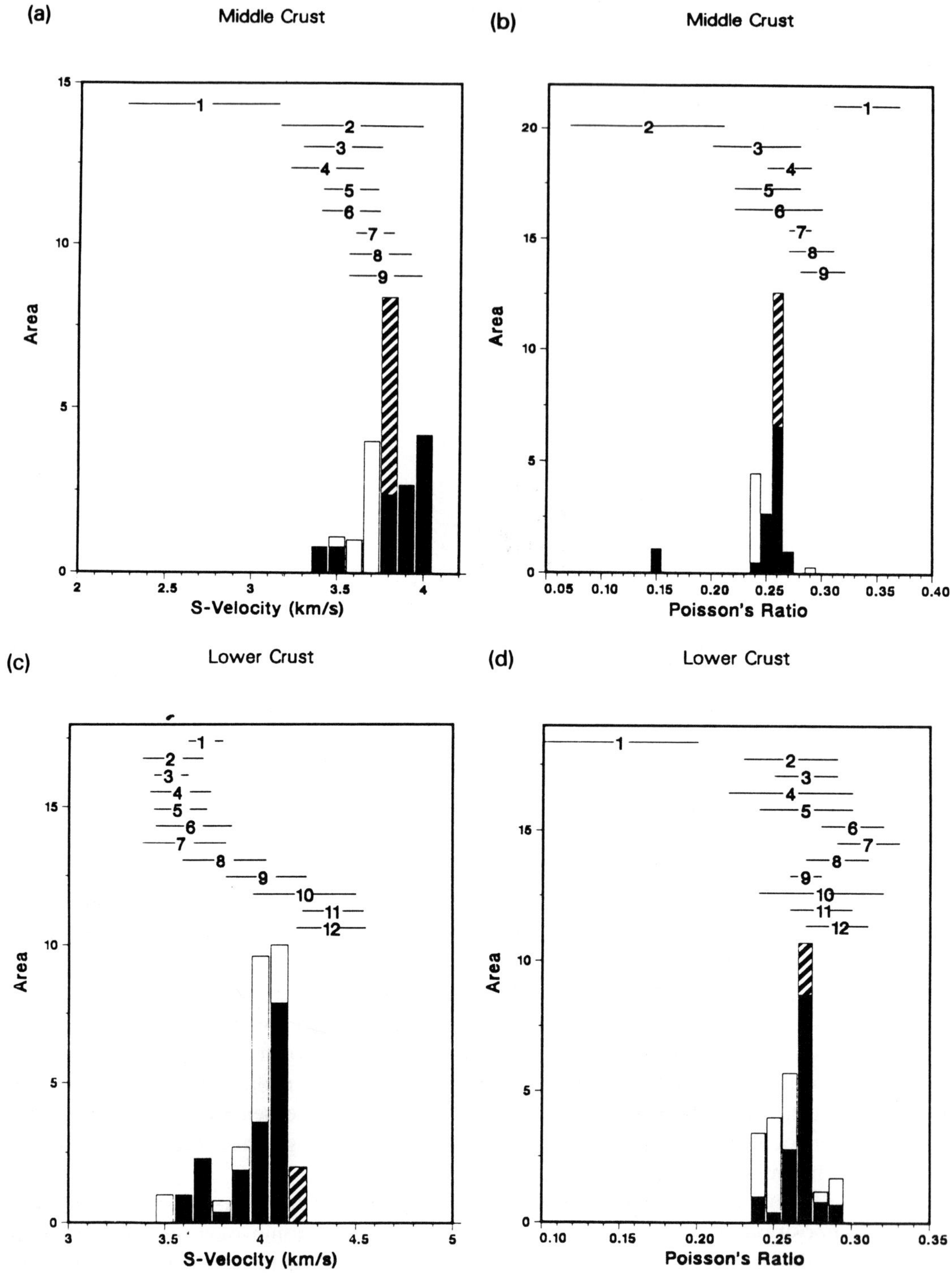

Fig. 1-7. Histograms of cross-sectional area of crust vs. S-wave velocity and Poisson's ratio in the deep crust. (a) S-velocity in the middle crust, (b) Poisson's ratio in the middle crust, (c) S-velocity in the lower crust, (d) Poisson's ratio in the lower crust.

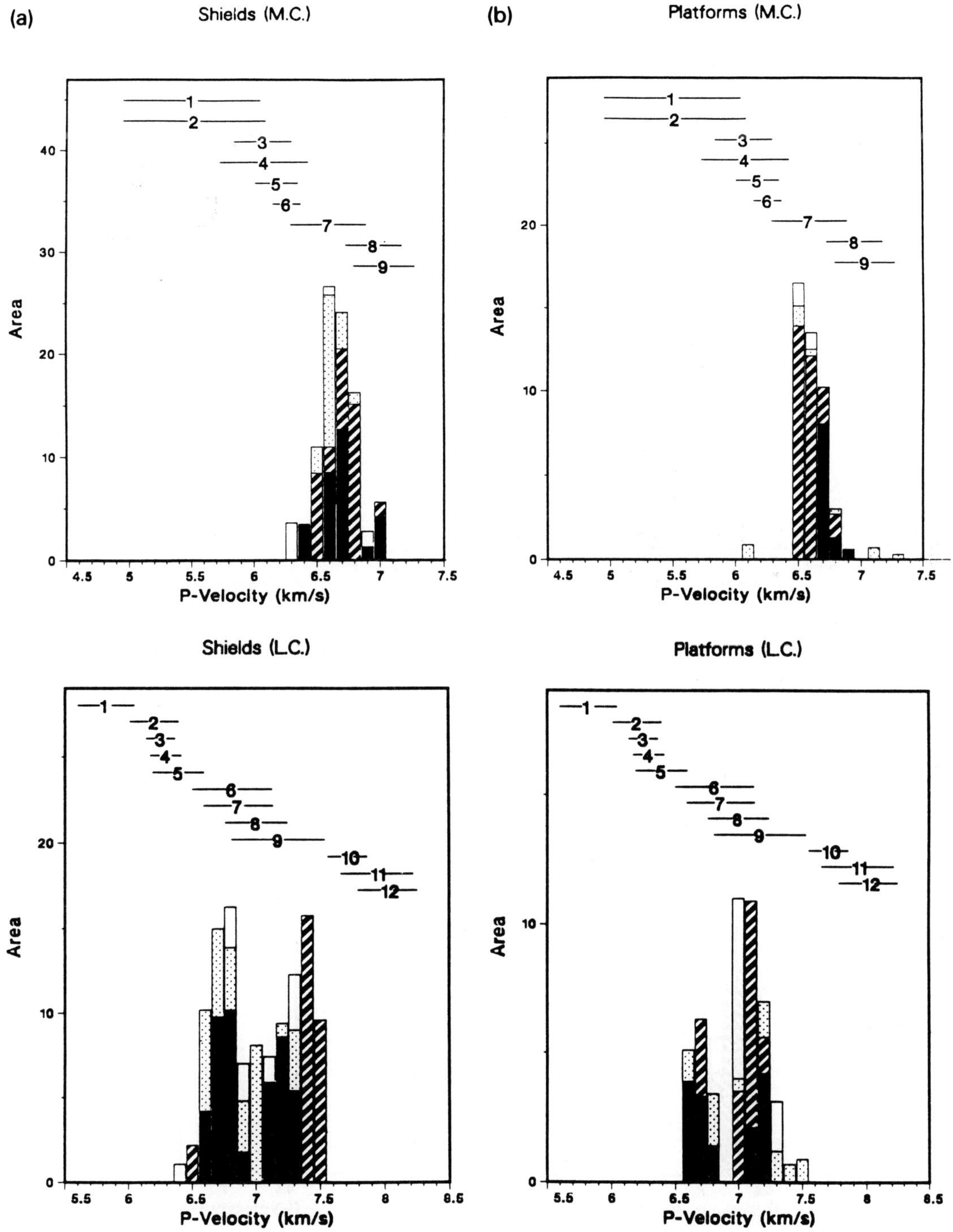

Fig. 1-8. Histograms of cross-sectional area of crust (in $km^2 \times 10^3$) vs. P-wave velocity for the middle crust (M.C.) and lower crust (L.C.), plotted as in Fig. 1-5, for (a) shields, (b) platforms.

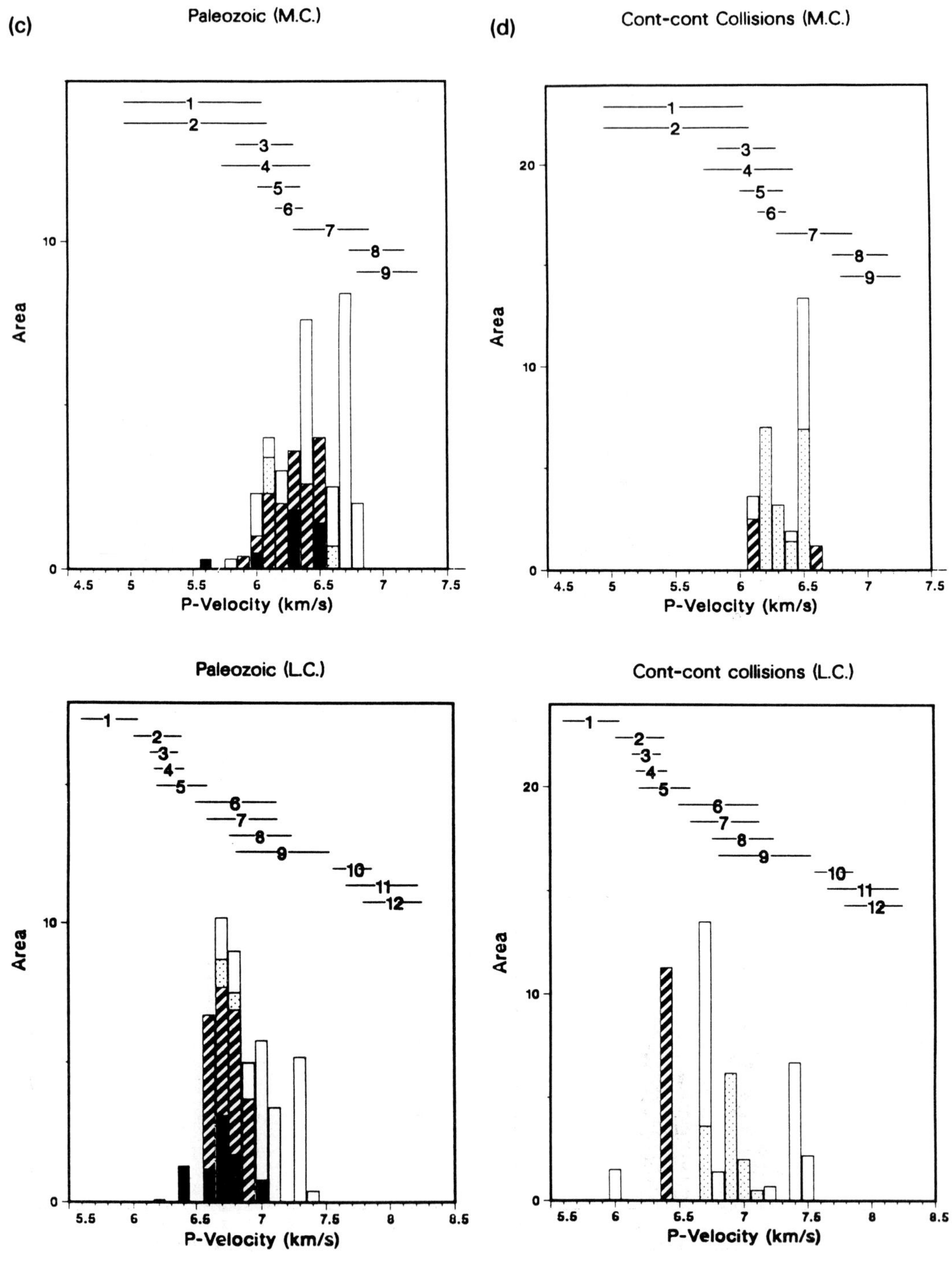

Fig. 1-8 (continued). Histograms of cross-sectional area of crust (in km^2 x 10^3) vs. P-wave velocity for the middle crust (M.C.) and lower crust (L.C.), plotted as in Fig. 1-5, for (c) Paleozoic crust, (d) continent–continent collision zones.

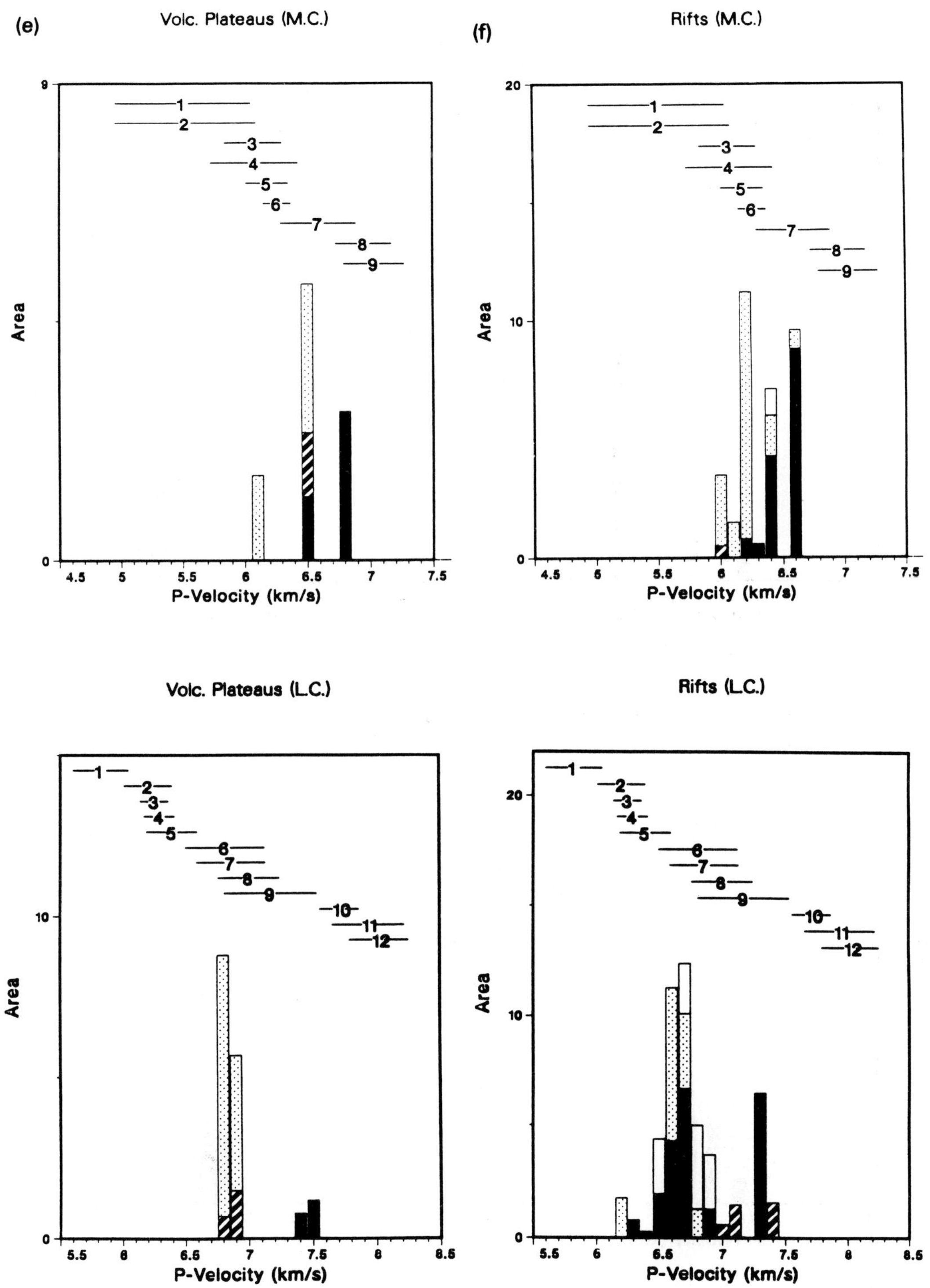

Fig. 1-8 (continued). Histograms of cross-sectional area of crust (in $km^2 \times 10^3$) vs. P-wave velocity for the middle crust (M.C.) and lower crust (L.C.), plotted as in Fig. 1-5, for (e) volcanic plateaus, (f) rifts.

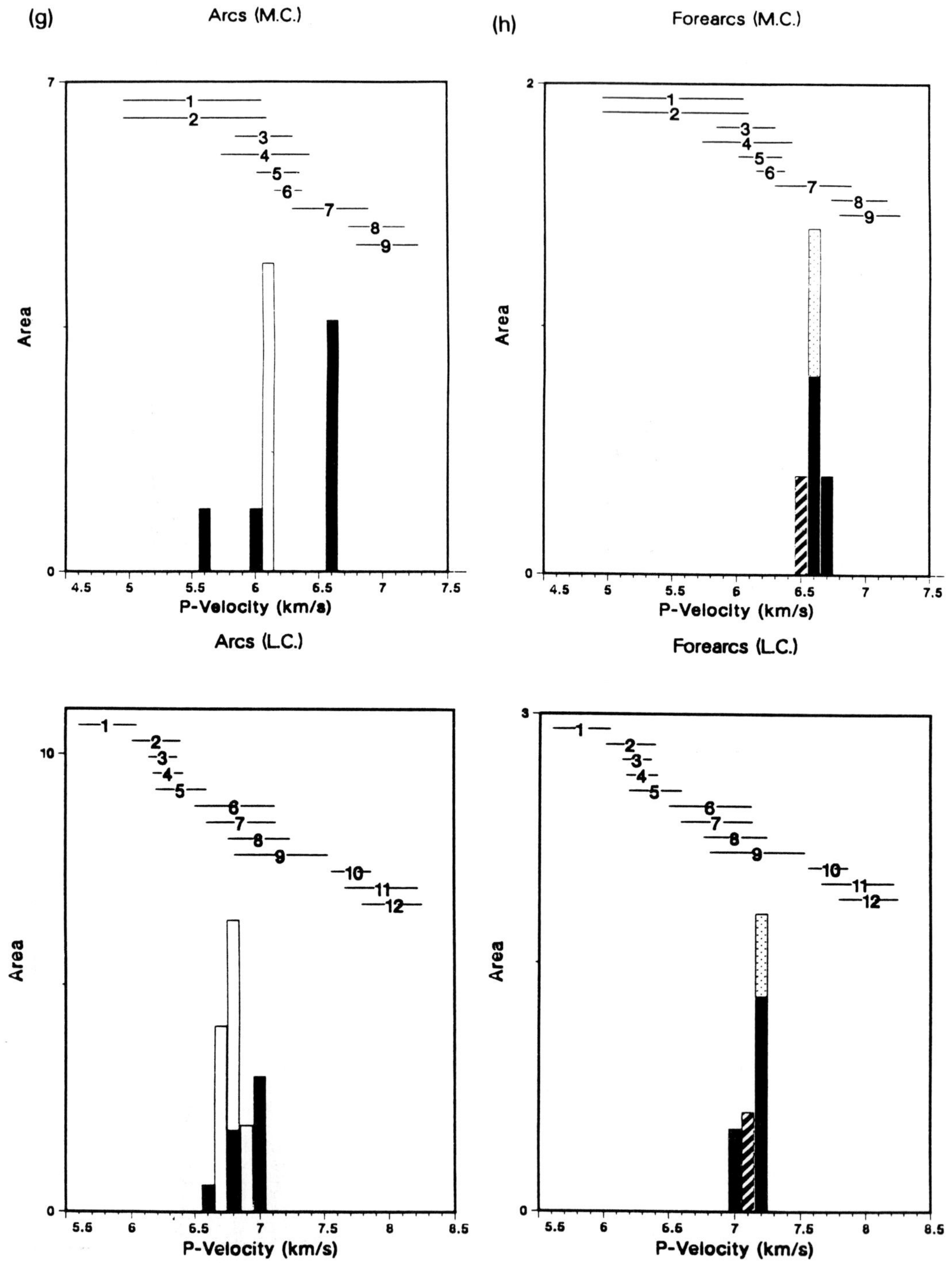

Fig. 1-8 (continued). Histograms of cross-sectional area of crust (in $km^2 \times 10^3$) vs. P-wave velocity for the middle crust (M.C.) and lower crust (L.C.), plotted as in Fig. 1-5, for (g) arcs, (h) forearcs.

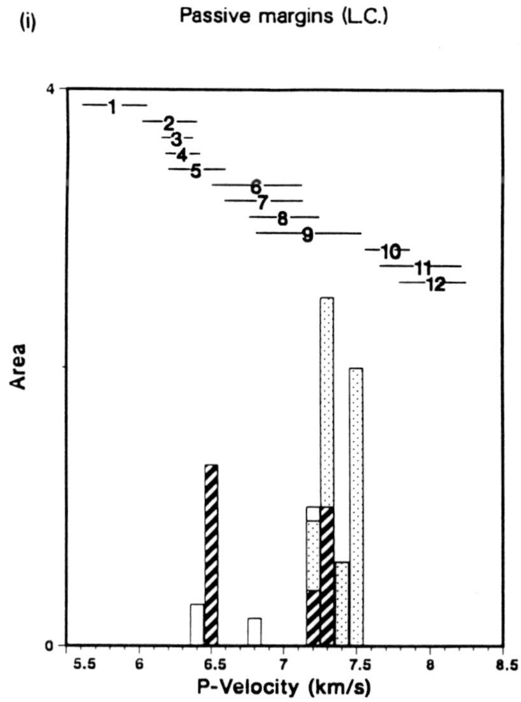

Fig. 1-8 (continued). Histograms of cross-sectional area of crust (in $km^2 \times 10^3$) vs. P-wave velocity for the middle crust (M.C.) and lower crust (L.C.), plotted as in Fig. 1-5, for (i) passive margins.

(7.2–7.5 km/s) in these areas require the presence of metapelite, pyroxenite, or a mixture of mafic and ultramafic rocks. The lower-velocity peaks (6.6–6.8 km/s) are best modeled as mafic granulites, anorthosites, or possibly intermediate granulites. In the middle crust, likely compositions vary from metagabbro (greenschist) in shields, platforms, and some volcanic plateaus, rifts, and continent–continent collision zones, to more felsic compositions (granite, granodiorite, felsic amphibolite-facies gneiss, and schist), in Paleozoic crust and some volcanic plateaus, rifts, and continent–continent collision zones.

These results are summarized in Table 1-5, which shows the rock types best matched by the major peaks in the velocity histograms of Figure 1-8. It is important to keep in mind that this summary represents an average or typical result for each environment — due to the extreme diversity of field-measured velocities, however, these average lithologies may not apply at a particular location. Moreover, a lower crust which can be modeled as having, say, a bulk mafic composition may also contain roughly balanced amounts of felsic and ultramafic rocks. Nevertheless, the comparison of average seismic velocity to laboratory-measured rock velocities provides a useful constraint on the allowable proportions of various rock types in the lower crust.

The broad range of rock types which are possible constituents of the lower crust reflects the non-uniqueness of petrological inferences drawn from P-velocities. Some help can be obtained from the sparse shear-wave (i.e., Poisson's ratio) information available in rifts, Paleozoic regions, and shields (Fig. 1-9). In shields and platforms, the relatively low Poisson's ratios (0.24–0.27) favor less mafic compositions-quartz-mica schists, intermediate granulites, and granulite-facies metapelites. In rifts, the somewhat higher Poisson's ratios favor anorthosites,

TABLE 1-5

Possible bulk composition of middle and lower crust

Middle crust

Environment	serp	qtz	gran	gdior	f.gns	sch	metag	gabb	amph
Shields						□ +	●	□	□
Platforms						□	●		
Paleozoic			□ +	□ +	● +	● +	□	□	
Cont. col.(low)			●	●	●	□			
(high)						□	●		
Volc.(l)			●	●					
(h)						□	●	□	
Rift (l)	□	□ +	●	●	●	□	●	●	●
(h)						● +	● +		
Arcs		□	□	□	□	●			
Forearc					□	●	□		

Lower crust

Environment	qtz	f.gns	fels	sch	int	anor	maf	amp	metap	px	maf+ult
Shields (low)					□	●	●	□	□		
(high)								□	● +	□	●
Platforms (l)					□	●	●	□			
(h)						□	□	●	●		□
Paleozoic					□	●	●	□	□ +		
Cont. col. (l)		□	●	●	●	●	●	●	●		
(h)									□	□	●
Volc. (l)						●	●	●	□		
(h)									□		●
Rift (l)					□	● +	●	□ +			
(h)									●		●
Arcs						●	●	●	●		
Forearc (l)	□										
(h)						●	●	●	●		
Passive (l)					●	□					
(h)								□	●		●

Symbols mark compositions which show good (●) or fair (□) agreement with observed histogram peaks; + denotes compositions favored by available Poisson's ratio data. Low (l) and high (h) refer to separate low- and high-velocity peaks, where appropriate. Compositions in middle crust: serp = serpentinite; qtz = quartzite; gran = granite; gdior = granodiorite; f.gns = felsic amphibolite gneiss; sch = schist; metag = metagabbro; gabb = gabbro; amph = amphibolite. Compositions in lower crust: qtz = quartzite; f.gns. = felsic amphibolite gneiss; fels = felsic granulite; sch = schist; int = intermediate (granodioritic) granulite; anor = anorthosite; maf = mafic granulite; amp = amphibolite; metap = metapelite; px = pyroxenite; maf+ult = mix of mafic and ultramafic rocks.

amphibolites, and intermediate granulites, while in Paleozoic regions, the broad range of Poisson's ratios suggests compositional diversity.

These interpretations are best shown by plotting the field-determined P-velocities and Poisson's ratios from a particular location on the fields of laboratory-measured values, as shown in Figure 1-10 for the Paleozoic crust of southwest Germany (Holbrook et al., 1988), the Ukrainian and Baltic shields (Tarkov et al., 1981; Grad and Luosto, 1987) and the

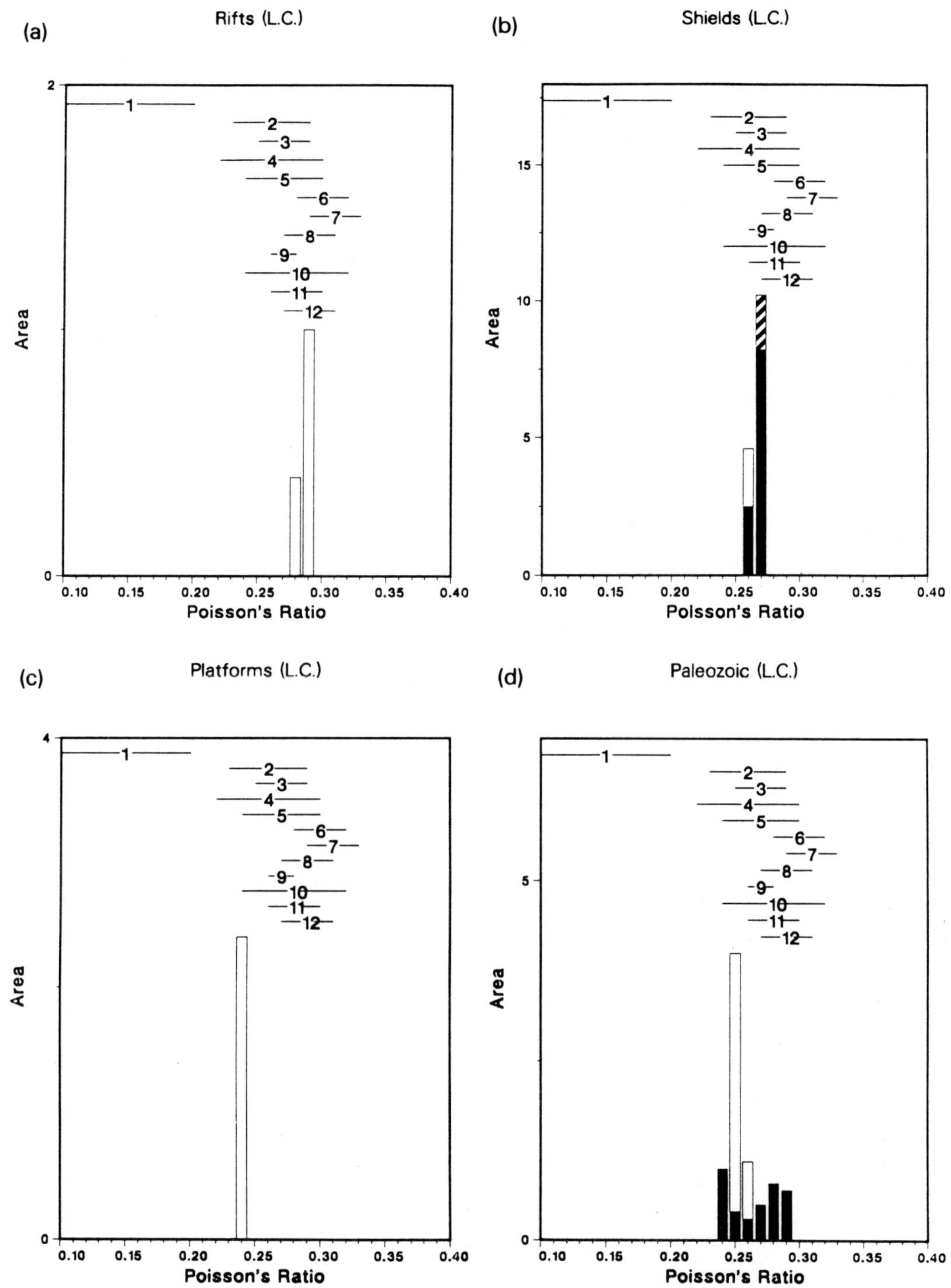

Fig. 1-9. Histograms of cross-sectional area of lower crust vs. Poisson's ratio in (a) rifts, (b) shields, (c) platforms, and (d) Paleozoic crust.

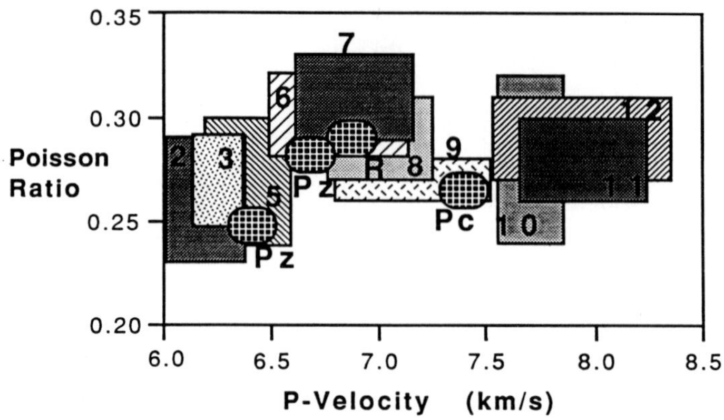

Fig. 1-10. Comparison of field- and laboratory-measured P-velocities and Poisson's ratios. Field-measured data from several areas (cross-hatchured ovals) are plotted on fields of lab-measured data from Fig. 1-4. *R* = rift zone (Basin and Range, Braile et al., 1974); *Pc* = Precambrian shield (Kapuskasing uplift, Boland and Ellis, 1991); *Pz* = Paleozoic crust (Black Forest and Urach areas, Holbrook et al., 1988).

Basin and Range rift zone (Braile et al., 1974). In the lower crust of Precambrian shields, the high P-velocities (7.3 km/s) and intermediate Poisson's ratios (0.26–0.27) fall within the field of metapelites (Fig. 1-10). An alternative interpretation, however, would be a mixture of high-σ mafic granulites and low-σ pyroxenites (bronzite, for example, has a low Poisson's ratio of 0.21; Christensen and Fountain, 1975). The intermediate P-velocity (6.9 km/s) and high Poisson's ratio (0.28–0.29) determined for the Basin and Range are consistent with mafic granulite, anorthosite, or amphibolite, or a mixture of these rock types. In the Paleozoic of Southwest Germany, the measured P-velocities and Poisson's ratios vary from values appropriate for felsic to intermediate granulites (V_p = 6.4 km/s, σ = 0.25) to values indicative of a mix of intermediate and mafic granulites (V_p = 6.7 km/s, σ = 0.28).

Of course, these conclusions must be viewed with the proper caution. So far, very few shear-wave data have been interpreted, and in most existing studies the data are of marginal quality. Even more importantly, we must remember that determinations of lithologies from seismic velocities generally refer to the average composition of an entire lower-crustal block measuring tens of kilometers across and 10–20 km thick. Yet we know from multichannel seismic reflection data and high-grade metamorphic terrains that the deep crust must be heterogeneous on the scale of tens to hundreds of meters (e.g., Fountain and Salisbury, 1981). Thus, a lower crust whose seismic properties are compatible with a bulk composition of intermediate granulite probably contains significant amounts of more felsic and more mafic rocks. Improving the resolution with which the physical properties of the lower crust can be determined remains one of the challenges of deep crustal seismology.

5. Fluids and anisotropy

The conclusions about deep crustal composition we have drawn above depend on the key assumptions that the deep continental crust is isotropic and free of fluids. Because these assumptions are inevitably idealizations of an imperfect Earth, we must ask ourselves (1) how anisotropic is the Earth's crust? and (2) how prevalent are pore fluids in the deep crust? The answers to these questions will undoubtedly change from place to place, and perhaps from experiment to experiment, as seismic source characteristics change.

The primary evidence for fluids in the deep continental crust comes from interpretations of electromagnetic data, which in many places (particularly in Phanerozoic crust) require a zone of high conductance in the lower crust (Hyndman and Hyndman, 1968; Shankland and Ander, 1983; Jones, 1987). Saline fluids in rocks with a few per cent porosity would account for the high conductivity (Hyndman and Shearer, 1989). The effect that such porosity would have on seismic velocities depends on the pore pressure of the fluid: fluids at low pore pressure will have a negligible effect on seismic velocity, while pore pressures increasing to near-lithostatic will have a drastic effect on velocity (Todd and Simmons, 1972; Christensen, 1984). For the low-porosity metamorphic and igneous rocks expected in the lower crust, porosity of several per cent would have to be held open by high pore pressure.

If saturated pore spaces are indeed widespread in the lower crust, the estimates of bulk composition inferred from seismic velocities will be biased toward felsic compositions, as pointed out by Hyndman and Klemperer (1989). Our estimates of composition, which assume a dry crust, could therefore be considered a felsic end-member of the possible bulk composition. Shear-wave data offer a potential means of detecting zones of fluid-filled porosity, because the presence of fluids at high pore pressure increases Poisson's ratio (Christensen, 1984). Fluid-filled cracks have also been proposed as a mechanism for generating lower-crustal reflections (Matthews and Cheadle, 1986; Hyndman, 1988; Hyndman and Shearer, 1989); however, this mechanism is difficult to reconcile with the weak amplitudes of shear-wave reflections corresponding to the lower-crustal P-wave reflections (Holbrook et al., 1987, 1988; Goodwin and McCarthy, 1990).

The existence of seismic anisotropy in the lower continental crust remains an open question. Seismic anisotropy has been measured in the oceanic and continental upper mantle (Raitt et al., 1969; Bamford, 1977; Fuchs, 1983) and in the upper crust (e.g., Crampin et al., 1984), but no conclusive measurements of lower-crustal anisotropy have been made. This is somewhat surprising, because laboratory measurements of P-velocities (Table 1-3) show that several lower-crustal rock types have significant anisotropy, particularly quartz-mica schists (10.7%), felsic amphibolite gneiss (7.3%), granulite-facies metapelite (5.5%), and amphibolite (5.4%). Moreover, the horizontally layered lower crust seen in many multichannel reflection sections (e.g., Brown et al., 1986) should produce transverse isotropy (Crampin, 1989), a particular kind of anisotropy in which the horizontal velocity is constant in all directions, but the horizontal and vertical velocities differ. Such anisotropy in the lower crust would be expected to produce shear-wave splitting of the Moho reflection, S_mS; unfortunately, few wide-angle, three-component seismic studies have been carried out, and existing studies have not detected significant splitting of S_mS (e.g., Gajewski and Prodehl, 1987; Holbrook et al., 1988). Furthermore, shear-wave splitting observed on earthquake records in Japan has been interpreted as the result of anisotropy in the upper, not the lower, crust (Kaneshima and Ando, 1989). On the basis of present data, then, we must conclude at present that either the lower crust is isotropic, or that anisotropy exists in isolated blocks whose size lies beneath the resolution of current experiments. Nevertheless, seismic anisotropy is an important property of crustal samples recovered from outcrops and drill holes, and probably influences crustal reflectivity (e.g., Jones and Nur, 1982, 1984; Christensen and Szymanski, 1988; Christensen, 1989; Fountain et al., 1990).

6. Discussion

6.1. Average continental crustal structure

Our compilation of field-measured seismic velocities allows the construction of some generalized models of deep crustal structure as a function of tectonic environment. Although such models run the risk of oversimplification, they can provide useful characterizations of the seismic structure of the continental crust. Using the data from Table 1-2, we can construct a standard crustal column for each tectonic environment, showing (1) average thicknesses of the crust, middle crust, and lower crust, and (2) modal velocities of the deep crust (Fig. 1-11). These columns show that the crust is thickest beneath continent–continent collision zones, shields, and platforms, while the crust is thinnest beneath passive margins, forearcs, and rifts. In general, a three-layer crust, with velocities increasing with depth, is a useful model; in most tectonic environments, the middle and lower crust together comprise about 2/3 of the total crust.

We have constructed from these crustal columns a generalized cross-section through a hypothetical continent bounded by an active and a passive margin (Fig. 1-12). This cartoon depicts the general velocity structure of the continental crust across some common tectonic environments: clearly shown, for example, is the bimodal distribution of lower-crustal velocities (6.7–6.8 and 7.1–7.3 km/s) and the predominance of mid-crustal velocities in the range 6.4–6.7 km/s.

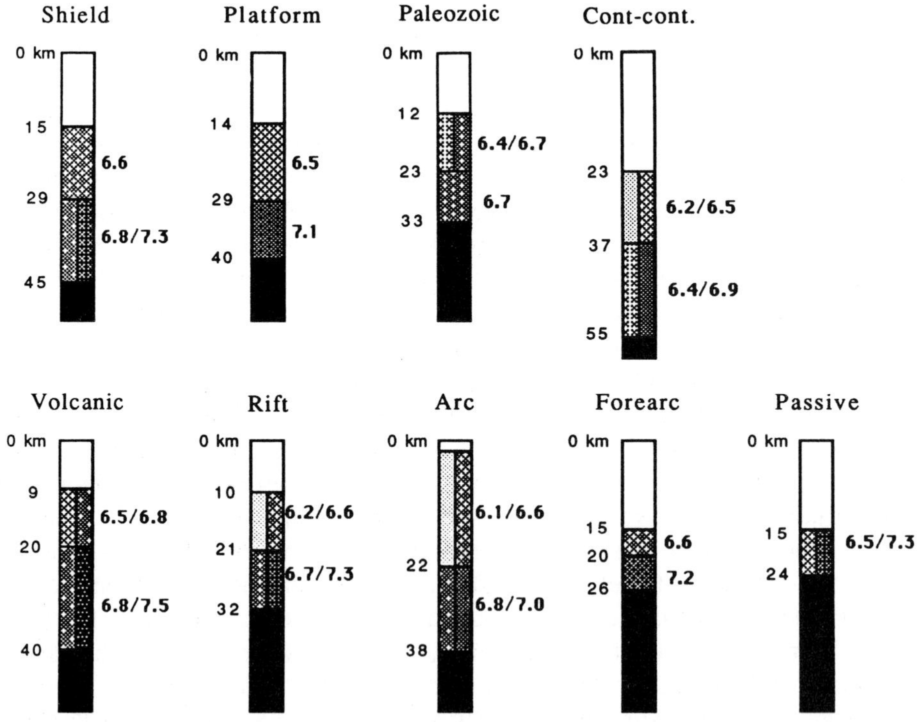

Fig. 1-11. Summary of average crustal columns in various tectonic environments, showing average thickness of crustal layers and modal velocities, determined from histograms of Fig. 1-8. Shading indicates velocity values; split layers indicate bimodal velocity distributions. Black represents upper mantle. The crust is thickest beneath continent–continent collision zones, shields, platforms, and volcanic plateaus, while it is thinnest beneath passive margins, forearcs, and rifts.

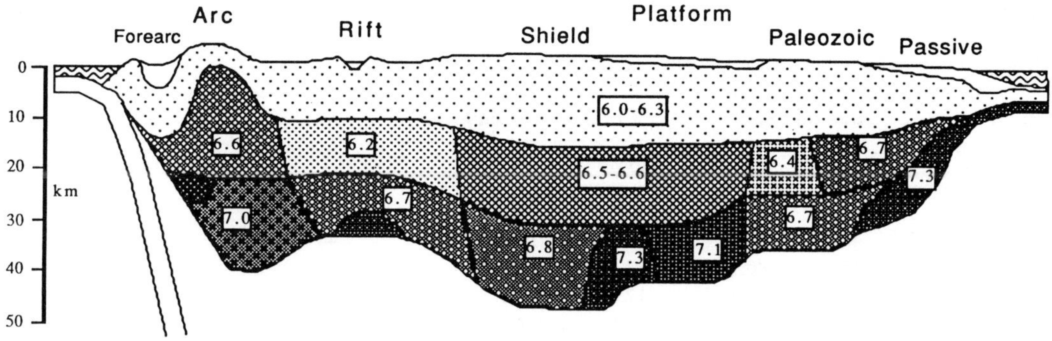

Fig. 1-12. Cartoon cross-section across a hypothetical continent, showing average seismic properties in various tectonic provinces. The velocity in the upper crust is assumed to be 6.0–6.3 km/s everywhere. Note the strong lateral changes in crustal structure from province to province. High-velocity (7.1–7.3 km/s) layers are found beneath rifts, shields, platforms, and passive margins. Mid-crustal velocities are lowest beneath rifts, perhaps due to higher temperatures.

6.2. Bulk lower-crustal composition

The bulk composition of the continental crust has long been under debate, yet is still poorly known. The primary reason for this uncertainty is the difficulty in estimating the composition of the deep crust. Geological estimates of lower-crustal composition derived from studies of Precambrian granulite terrains (e.g., Weaver and Tarney, 1984; Shaw et al., 1986), island arcs (Taylor and McLennan, 1981), and xenoliths (e.g., Dupuy et al., 1979) generally fall in the range 54–61% SiO_2, corresponding to an intermediate (dioritic) composition.

Smithson and his coworkers (Smithson and Brown, 1977; Smithson, 1978; Smithson et al., 1981), using measurements of mean crustal velocity and observations from multichannel reflection data, also inferred a dioritic (59% SiO_2) lower-crustal composition. According to Smithson and Brown (1977), “the lower crust must be distinctly less mafic... than gabbro.” This conclusion was based on (1) their observation that most lower crustal P-velocities are less than 7.0 km/s, corresponding to the range of felsic and intermediate granulites, and (2) the existence of numerous reflections in the lower crust on multichannel reflection profiles, thus implying the presence of more felsic (i.e., lower-velocity) rocks.

Our summary of seismic velocities has an important bearing on such estimates of bulk lower-crustal composition. For this purpose, however, the velocity histograms of Figures 1-5 and 1-6 are somewhat misleading, in that some tectonic provinces have been over- or under-sampled with respect to their global volumetric abundance. In order correct for this sampling bias, we applied a weighting factor to each velocity determination, based on the ratio of the proportion of total crustal area measured in each tectonic environment to the approximate global volumetric abundance of crust in that environment (Table 1-6). The weighting factors thus calculated show that, relative to their global volumetric abundance, shields, volcanic plateaus, and rifts have been oversampled, while platforms, forearcs, and passive margins have been undersampled by seismic experiments.

The resulting corrected velocity histograms, shown in Figure 1-13, show several important differences to the uncorrected histograms (Figs. 1-5 and 1-6), especially in the lower crust (Fig. 1-13b). Particularly obvious is the marked increase in the proportion of velocities in the range 7.0–7.2 km/s; this is primarily due to the increasing contribution of platforms, which have a strong velocity peak in that range (Fig. 1-8b). There has been a relative

TABLE 1-6

Calculation of weighting factors for velocity histogram correction

Tectonic environment [1]	Area surveyed [2]		% of total area surveyed [3]		Vol.% worldwide [4]	Weighting factor [5]	
	lower	middle	lower	middle		lower	middle
Shield	129.0	93.4	35.4	35.1	15	0.4	0.4
Platform	48.4	45.7	13.3	17.2	40	3.0	2.3
Paleozoic	47.1	40.7	12.9	15.3	16	1.2	1.0
Cont. collision	46.0	30.3	12.6	11.2	16	1.3	1.4
Volcanic plat.	16.5	9.6	4.5	3.6	[1]	0.2	0.3
Rift	49.9	34.1	13.7	12.8	1	0.1	0.1
Arc	16.0	9.8	4.4	3.7	4	0.9	1.1
Forearc	3.3	2.2	0.9	0.8	[2]	2.2	2.5
Passive margin	7.9	0.3	2.2	0.1	[5]	2.3	10.0

[1] Tectonic environment;
[2] Total cross-sectional area of middle and lower crust surveyed by wide-angle seismic experiments;
[3] Area surveyed as a percentage of total area surveyed;
[4] Estimated global volumetric abundance, expressed as percent of total continental crust, taken from Condie (1982) except where denoted by brackets;
[5] Weighting factors to correct for bias in distribution of wide-angle seismic surveys, calculated from the ratio of percentage of worldwide volume to percentage of total area surveyed.

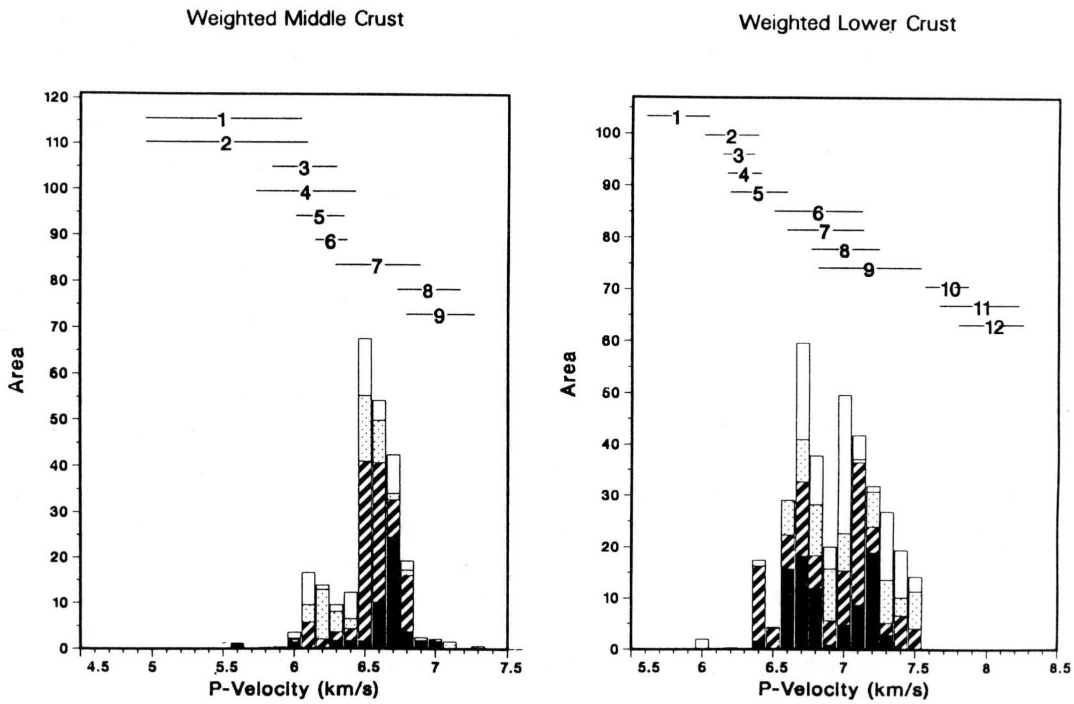

Fig. 1-13. Velocity histograms for the (a) middle crust and (b) lower crust, calculated from the histograms of Figs. 1-5 and 1-6 but using the weighting factors of Table 1-6. These weighted histograms more accurately reflect the average properties of the middle and lower crust, because over- and undersampling of some tectonic provinces has been approximately corrected for. Shading indicates quality factor (see Fig. 1-5). Bars at top show velocity ranges of possible (a) mid-crustal and (b) lower-crustal rocks (see Table 1-3 and Figs. 1-5 and 1-6).

decrease in the proportion of velocities in the range 7.3–7.5 km/s, especially among the better-determined values; this is due to the decreased contribution of shields, rifts, and volcanic plateaus, all of which show high-velocity peaks. As a result, the bimodal nature of the lower-crustal velocity distribution has been enhanced, with 64% of the total area (72% of the best-determined values) falling within the ranges 6.7–6.8 km/s and 7.0–7.2 km/s. This is a remarkable observation, as it implies that about 2/3 of the lower continental crust worldwide has an average P-velocity within those relatively narrow ranges.

If the weighted lower-crustal velocity histogram is approximately correct, it implies that roughly half (53%) of the lower continental crust has a velocity of 7.0 km/s or greater. Smithson's (1978) estimate of the average velocity of the lower crust was therefore low, due to the smaller refraction data base available to him. Moreover, many lower-velocity regions (6.7–6.8 km/s) may consist of predominantly mafic material, since the temperature-corrected velocities of many mafic granulites fall within that range (6.86 $\pm$ 0.27 km/s, Table 1-3). These two observations suggest that the composition of the lower crust in many places is closer to gabbroic than to the dioritic compositions derived by Smithson (1978), or from geologic and geochemical data by Taylor and McLennan (1981), Weaver and Tarney (1984), and Shaw et al. (1986). This conclusion would concur with the review of xenolith evidence (Griffin and O'Reilly, 1987; Rudnick, this volume), which suggests that mafic rocks are far more prevalent in the lower crust than felsic or intermediate rocks. An alternative explanation for the high velocities would be the presence of garnet, which would allow more felsic compositions, if significant amounts of metapelite are present.

There is, of course, some danger in overinterpreting summary diagrams such as those in Figure 1-13: as we have seen, the seismic characteristics of the deep continental crust vary markedly from place to place, even within similar tectonic environments. The most reliable estimates of lower-crustal composition will be those at individual, well-studied locations, where a combination of geophysical and geological data exists.

6.3. Constraints from seismic reflection profiling

The results of deep seismic reflection profiles provide additional constraints on conclusions regarding the composition and physical properties of the crust. Although seismic reflection data provide structural resolution unavailable from refraction methods, they do not provide accurate deep velocity information, due to their limited aperture (usually <10 km). The primary observation from reflection profiles that we address here is the high reflectivity of the lower crust, often with a well-defined top and bottom to the reflective zone (e.g., Mooney and Brocher, 1987). Unfortunately, any comparison of seismic reflection results with wide-angle reflection/refraction results is made difficult by the dissimilar coverage currently available for the two methods. For example, we noted above that wide-angle seismic data have been extensively recorded within shields and platforms, but these regions are sparsely sampled by reflection profiles, which are biased toward continental margins and regions of relatively thin ($\leq$35 km) crust. Despite the differences in coverage, the following conclusions can be drawn.

Although lower crustal reflectivity varies significantly from region to region, it is generally quite high in rifts and areas of extended crust (e.g., McCarthy and Thompson, 1988). The wide-angle seismic results presented here show that these areas have high Poisson's ratio, indicative of a mafic lower crust. This inferred composition is consistent with the suggestion that crustal reflectivity in these areas is in part due to mafic intrusions. Alternatively, ductile shear in extended lower crust would result in the alignment of anisotropic minerals with the higher P-wave axis horizontal; anastomozing shear zones could produce high lower crustal reflectivity without requiring numerous mafic intrusions (Jones and Nur, 1984). The high velocity layers (7.2–7.5 km/s) found beneath some rifts and passive margins likely consist of a mix of mafic and ultramafic rocks. If layered, this zone would be highly reflective, consistent with a clearly defined "reflection Moho" in extended regions (e.g., Klemperer et al., 1986; McCarthy and Thompson, 1988).

The origin of lower crustal reflectivity remains a major unknown. Because deep reflections originate from boundaries separating layers of contrasting seismic impedance (velocity x density), they must hold important clues for improving our knowledge of deep crustal composition. Several origins have been proposed for deep crustal reflections, including compositional layering (igneous or metamorphic layering; Hale and Thompson, 1982; Christensen, 1989), ductile shear zones (Jones and Nur, 1982, 1984), igneous intrusions (Meissner, 1973), lenses of partial melt (Meissner, 1973; Hale and Thompson, 1982), and the presence of fluid-filled cracks (Matthews and Cheadle, 1986). The wide range of lower-crustal velocities compiled above, which implies a large diversity in rock types in the lower crust, suggests that no single cause is responsible for all deep reflections. Although we are unable to determine the causes of deep reflections with our compilation, there is clearly a need to make detailed comparisons of reflection and refraction results at locations where both types of data exist (e.g., Bartelsen et al., 1982; Wever, 1989). Coincident shear-wave data can be particularly illuminating in constraining the origin of deep reflections (e.g., Holbrook et al., 1988).

6.4. Implications for crustal evolution

Models of the growth of continental crust through time vary widely, but most suggest that the bulk of continental crust was produced during the Archean, by about 2.5 Ga (e.g., McLennan and Taylor, 1982; Reymer and Schubert, 1984). The widespread Archean granite-greenstone belts were created by the melting of mafic source rocks (Rudnick and Taylor, 1986), probably in intra-arc (Sleep and Windley, 1982) or intracontinental rift (Kröner, 1984) settings. During the Phanerozoic, in contrast, accretion of continental crust took place primarily by arc magmatism (e.g., Hamilton, 1981, 1988), with contributions from rift and hotspot magmatism. In tackling the problem of crustal evolution from a seismological standpoint, then, we must examine two kinds of tectonic environments: stable cratons and regions of more recent magmatic activity, such as arcs, rifts and volcanic plateaus.

A widely used model for the growth of the continental crust at magmatic arcs during the Phanerozoic is the andesite model of Taylor (1977). This model, which assumes additions to the crust of bulk andesitic composition, uses a mass balance to predict a mafic composition for the lower crust (e.g., Taylor and McLennan, 1981). Since the work of Smithson (1978), many authors (e.g., Kay and Kay, 1986) have pointed out an apparent contradiction between the mafic lower crust required by the andesite model and the intermediate lower crust proposed by Smithson (1978) on the basis of P-wave velocities. As pointed out above, however, our compilation shows that in most tectonic environments, *in situ* lower crustal compressional velocities are indeed compatible with mafic compositions (Table 1-5). The limited seismic information from arcs themselves show velocities of 6.7–6.8 km/s in island arcs (Asano et al., 1985; Iwasaki et al., 1990) and 6.9–7.0 km/s in Andean-type arcs (Pakiser and Brune, 1980; Leaver et al., 1984); these values are all consistent with mafic granulites (6.86 ± 0.27 km/s; Table 1-3). There is thus no need to appeal to alternative explanations such as the presence of fluid-filled cracks (Hyndman and Klemperer, 1989), widespread hydrous minerals, or foundering of mafic rocks into the mantle to reconcile the seismic and geologic evidence.

The bimodal distribution of P-velocities beneath shields and platforms (Fig. 1-8a,b) suggests that two distinct crust-forming processes may have been active during the Precambrian. The similarity between the low-velocity peak (6.6–6.8 km/s) and the values found beneath arcs (Fig. 1-8g) suggests that the lower crust beneath some shields and platforms was formed by arc magmatism. This interpretation is consistent with most models of Precambrian crustal evolution, which generally assert that modern-style subduction zones have existed since the Archean (e.g., Tarney and Windley, 1977; Windley, 1981) or Proterozoic (e.g., Hargraves, 1981). The high-velocity peak (7.1–7.5 km/s), on the other hand, can be explained either as the result of the underplating of mafic and ultramafic magmas (e.g., Furlong and Fountain, 1986) or the presence of high-grade metamorphic rocks of supracrustal origin (metapelites). The interpretation of underplated material at the base of cratons is consistent with models of early Precambrian crustal evolution, which envision continental growth by underplating of magmas from a vigorously convecting mantle (e.g., Kröner, 1981). Alternatively, the high P-velocities could be the result of a prevalence of garnetiferous metapelites, which may have been accreted to the base of the crust in Precambrian orogenic belts. This process must have occurred to some extent during the Archean, as evidenced by the existence of granulite-facies metapelites in some Archean terrains (e.g., Fountain and Salisbury, 1981).

The activity of mafic/ultramafic underplating during the Phanerozoic is indicated by the presence of high seismic velocities (7.1–7.5 km/s) beneath many rifts, volcanic plateaus, and passive margins (Fig. 1-8). This process has been suggested to occur beneath volcanic plateaus and at rifted continental margins where a mantle thermal anomaly exists (White et al., 1987; White and McKenzie, 1989). The high velocities are predicted by thermal and petrologic considerations (Furlong and Fountain, 1986; White and McKenzie, 1989). Thus magmatic underplating appears to supplement arc magmatism as an important means of continental growth during the Phanerozoic.

6.5. Directions for future research

Our knowledge of the evolution of the lower continental crust is severely hampered by the paucity of high-quality data in several key tectonic environments, especially island arcs, Andean-type arcs, and continent–continent collision zones. These regions typically present logistical obstacles to the collection of wide-angle data, but these must be overcome if we are to understand the influence of these areas on crustal evolution.

A further important development will be the increased use of shear-wave information in wide-angle seismic experiments. Crustal seismologists must free themselves of the pervasive bias toward P-waves: even if techniques are developed which greatly improve resolution, P-wave velocity alone will always give an incomplete picture of the lower crust, due to the inescapable overlap of the P-velocities of diverse rock types (e.g., Table 1-2). Shear-wave information helps overcome this ambiguity.

There is a vital need for field and laboratory measurements addressing the two fundamental assumptions of this study: the lack of (1) anisotropy and (2) fluids in deep crust. Studies of *in situ* anisotropy in the Earth have focused on the upper crust and the oceanic and continental upper mantle; very little work has been done, however, on the anisotropy of the lower crust, despite the fact that many metamorphic rock types expected to exist in the lower crust show considerable anisotropy in the laboratory (Table 1-3). The role of fluids in the deep crust is still under debate (e.g., Fyfe et al., 1978; Shankland and Ander, 1983; Hall, 1986; Hyndman and Klemperer, 1989), but little is known about the influence of fluids on seismic velocities at pressures and temperatures appropriate for the lower crust.

7. Conclusions

The summary of deep crustal velocities presented here is, first and foremost, testimony to the diversity of the deep continental crust. The wide range of P-velocities found in the middle crust (6.0–7.1 km/s) and lower crust (6.4–7.5 km/s) implies the existence of bulk compositions varying, from place to place, from felsic to mafic/ultramafic. Still, some generalizations can be made: most mid-crustal P-velocities fall in the range 6.5–6.8 km/s, while most in the lower crust lie in the ranges 6.7–6.8 km/s or 7.0–7.2 km/s. Lower-crustal velocities have a bimodal distribution beneath shields, platforms, passive margins, rifts, and volcanic plateaus, suggesting the influence of either mafic/ultramafic magmatic activity or high-grade metamorphism to raise P-velocities.

Inferences about composition drawn from P-velocities are non-unique, but this ambiguity can be significantly reduced by the existence of complementary S-velocity (and hence Poisson's ratio) information. The sparse shear-wave information available for the lower crust suggests that S-velocities are relatively high beneath shields and platforms (i.e. Poisson's ratio is low), supporting the presence of either quartz-rich metapelites or low-σ pyroxene. The bimodal pattern may be the manifestation of two distinct processes of Precambrian crustal genesis: mafic/ultramafic underplating and arc magmatism. P-velocities beneath most Phanerozoic environments fall in the range 6.6–6.8 km/s, consistent with a predominantly mafic bulk composition; this interpretation is in accord with a model of crustal growth by arc magmatism. The presence of high velocities (7.1–7.5 km/s) beneath Phanerozoic rifts, volcanic plateaus, and passive margins is likely the result of mafic/ultramafic underplating in those regions.

Finally, our compilation shows that high velocities ($\geq$7.0 km/s) are more common in the lower crust than previously thought. In most tectonic environments, P-wave velocities are consistent with a bulk mafic lower crust. This conclusion eliminates the often-quoted paradox between estimates of bulk lower-crustal composition from seismic methods and geologic evidence.

Acknowledgments

We thank David Fountain and Roy Johnson for thorough, constructive reviews. Support for W.S.H. was provided by a Woods Hole Oceanographic Institution Postdoctoral Scholarship.

References

Asano, S. and 10 others, 1985. Crustal structure in the northern part of the Philippine Sea Plate as derived from seismic observations of Hatoyama off Izu Peninsula explosions. J. Phys. Earth, 33: 173–189.

Assumpçao, M. and Bamford, D., 1978. LISPB — V. Studies of crustal shear waves. Geophys. J. R. Astron. Soc., 54: 61–73.

Avedik, F., Berendsen, D., Fucke, H., Goldflam, S., Hirschleber, H., Meissner, R., Sellevoll, M.A. and Weinrebe, W., 1984. Seismic investigations along the Scandinavian "Blue Norma" profile. Ann. Geophys., 2: 571–578.

Azbel, I.Y., Buyanov, A.F., Ionkis, V.T., Sharov, N.V. and Sharova, V.P., 1989. Crustal structure of the Kola Peninsula from inversion of deep seismic sounding data. Tectonophysics, 162: 87–99.

Babuska, V., 1972. Elasticity and anisotropy of dunite and bronzitite. J. Geophys. Res., 77: 6955–6965.

Bamford, D., 1977. Pn velocity anisotropy in a continental upper mantle. Geophys. J. R. Astron. Soc., 49: 29–48.

Bamford, D., Nunn, K., Prodehl, C. and Jacob, B., 1978. LISPB — IV. Crustal structure of northern Britain. Geophys. J. R. Astron. Soc., 54: 43–60.

Banda, E. and Ansorge, J., 1980. Crustal structure under the central and eastern part of the Betic Cordillera. Geophys. J. R. Astron. Soc., 63: 515–532.

Banda, E., Suriñach, E., Aparicio, A., Sierra, J. and Ruiz de la Parte, E., 1981. Crust and upper mantle structure of the central Iberian Meseta (Spain). Geophys. J. R. Astron. Soc., 67: 779–789.

Barazangi, M. and Brown, L. (Editors), 1986a. Reflection Seismology: A Global Perspective. American Geophysical Union, Geodynamics Series 13, Washington, D.C., 311 pp.

Barazangi, M. and Brown, L. (Editors), 1986b. Reflection Seismology: A Global Perspective. American Geophysical Union, Geodynamics Series 14, Washington, D.C., 339 pp.

Bartelsen, H., Lüschen, E., Krey, T., Meissner, R., Schmoll, H. and Walter, C., 1982. The combined seismic reflection-refraction investigation of the Urach geothermal anomaly. In: R. Haenel (Editor), The Urach Geothermal Project. Schweizerbart'sche Verlag, Stuttgart, pp. 247–262.

Barton, P.J. and Matthews, D.H., 1984. Deep structure and geology of the North Sea region interpreted from a seismic refraction profile. Ann. Geophys., 2: 663–668.

Beaudoin, B., 1989. Crustal structure of the Yukon-Tanana terrane along the TACT corridor: Evidence for lower crustal detachment. EOS, Trans. AGU, 70: 1337.

Berry, M.J., 1973. Structure of the crust and upper mantle in Canada. Tectonophysics, 20: 183–201.

Birch, F., 1960. The velocity of compressional waves in rocks to 10 kilobars, Part 1. J. Geophys. Res., 65: 1083–1102.

Birch, F., 1961. The velocity of compressional waves in rocks to 10 kilobars, Part 2. J. Geophys. Res, 66: 2199–2224.

Blümling, P., Mooney, W.D. and Lee, W.H.K., 1985. Crustal structure of the Calaveras Fault zone, central California, from seismic refraction investigations. Bull. Seismol. Soc. Am., 75: 193–209.

Boland, A.V. and Ellis, R.M., 1989. Velocity structure of the Kapuskasing uplift, northern Ontario, from seismic refraction studies. J. Geophys. Res., 94: 7189–7204.

Boland, A.V. and Ellis, R.M., 1991. A geophysical model for the Kapuskasing Uplift from seismic and gravity studies. Can. J. Earth Sci., 28: 342–354.

Bonatti, E. and Seyler, M., 1987. Crustal underplating and evolution in the Red Sea Rift: Uplifted gabbro/gneiss complexes of Zabargad and Brothers Islands. J. Geophys. Res., 92: 12,803–12,821.

Bonner, B.P. and Schock, R.N., 1981. Seismic wave velocity. In: W. R. Judd and R.F. Roy (Editors), Physical Properties of Rocks and Minerals. McGraw-Hill, McGraw-Hill/CINDAS Data Series on Material Properties II-2, pp. 221–256.

Braile, L.W., Smith, R.B., Keller, G.R. and Welch, R.M., 1974. Crustal structure across the Wasatch front from detailed seismic refraction studies. J. Geophys. Res., 79: 2669–2677.

Braile, L.W. and 9 others, 1982. The Yellowstone–Snake River Plain seismic profiling experiment: crustal structure of the eastern Snake River Plain. J. Geophys. Res., 87: 2597–2609.

Brown, L., Barazangi, M., Kaufman, S. and Oliver, J., 1986. The first decade of COCORP: 1974–1984. In: M. Barazangi and L. Brown (Editors), Reflection Seismology: A Global Perspective. American Geophysical Union, Geodynamics Series 13, Washington, D.C., pp. 107–120.

Bullen, K.E. and Bolt, B.A., 1985. An Introduction to the Theory of Seismology. Cambridge University Press, Cambridge, U.K., 499 pp.

Catchings, R.D. and Mooney, W.D., 1988a. Crustal structure of the Columbia Plateau: evidence for continental rifting. J. Geophys. Res., 93: 459–474.

Catchings, R.D. and Mooney, W.D., 1988b. Crustal structure of East Central Oregon: relation between Newberry Volcano and regional crustal structure. J. Geophys. Res., 93: 10,081–10,094.

Catchings, R.D. and Mooney, W.D., 1990. Basin and Range crustal and upper mantle structure along the 40°N parallel, northwest Nevada. J. Geophys. Res., in review.

Cerveny, V., Molotkov, I.A. and Psencik, I., 1977. Ray Method in Seismology. University of Karlova, Prague, 214 pp.

Christensen, N.I., 1965. Compressional wave velocities in metamorphic rocks at pressures to 10 kilobars. J. Geophys. Res., 70: 6147–6164.

Christensen, N.I., 1966a. Shear wave velocities in metamorphic rocks at pressures to 10 kilobars. J. Geophys. Res., 71: 3549–3556.

Christensen, N.I., 1966b. Elasticity of ultrabasic rocks. J. Geophys. Res., 71: 5921–5931.

Christensen, N.I., 1972. The abundance of serpentinites in the oceanic crust. J. Geol., 80: 709–719.

Christensen, N.I., 1977. The geophysical significance of oceanic plagiogranite. Earth Planet. Sci. Lett., 36: 297–300.

Christensen, N.I., 1978. Ophiolites, seismic velocities and oceanic crustal structure. Tectonophysics, 47: 131–157.

Christensen, N.I., 1979. Compressional wave velocities in rocks at high temperature and pressures, critical thermal gradients and crustal low-velocity zones. J. Geophys. Res., 84: 6849–6857.

Christensen, N.I., 1982. Seismic velocities. In: R. S. Carmichael (Editors, Handbook of Physical Properties of Rocks, 2. CRC Press, Boca Raton, Fla., pp. 1–228.

Christensen, N.I., 1984. Pore pressure and oceanic crustal seismic structure. Geophys. J. R. Astron. Soc., 79: 411–423.

Christensen, N.I., 1989. Reflectivity and seismic properties of the deep continental crust. J. Geophys. Res., 94: 17,793–17,804.

Christensen, N.I. and Fountain, D.M., 1975. Constitution of the lower continental crust based on experimental studies of seismic velocities in granulite. Geol. Soc. Am. Bull., 86: 227–236.

Christensen, N.I. and Szymanski, D.L., 1988. Origin of reflections from the Brevard fault zone. J. Geophys. Res, 93: 1087–1102.

Christensen, N.I. and Wepfer, W.W., 1989. Laboratory techniques for determining seismic velocities and attenuations, with applications to the continental lithosphere. In: L. C. Pakiser and W.D. Mooney (Editors), The Geophysical Framework of the Continental United States. Geol. Soc. Am., Mem., 172: 91–102.

Christensen, N.I., Carlson, R.L., Salisbury, M.H. and Fountain, D.M., 1975. Elastic wave velocities in volcanic and plutonic rocks recovered on DSDP Leg 31. In: Initial Reports of the DSDP, 31: 607–610.

Chroston, P.N. and Brooks, S.G., 1989. Lower crustal seismic velocities from Lofoten-Vesteralen, N. Norway. Tectonophysics, 157: 251–269.

Chroston, P.N. and Evans, C.J., 1983. Seismic velocities of granulites from the Seiland petrographic province, N. Norway. J. Geophys., 52: 14–21.

Clowes, R.M., Gens-Lenartowicz, E., Demartin, M. and Saxov, S., 1987. Lithospheric structure in southern Sweden –results from FENNOLORA. Tectonophysics, 142: 1–14.

Colburn, R.H. and Mooney, W.D., 1986. Two-dimensional velocity structure along the synclinal axis of the Great Valley, California. Bull. Seismol. Soc. Am., 76: 1305–1322.

Condie, K.C., 1982. Plate Tectonics and Crustal Evolution. Pergamon Press, Oxford, 310 pp.

Conrad, V., 1925. Laufzeitkurven des Tauernbebens vom 28. November 1923. Mitt. Erdb. Komm. Wiener Akad. Wiss., 59: 1–23.

Crampin, S., Chesnokov, E.M. and Hipkin, R.G., 1984. Seismic anisotropy –the state of the art: II. Geophys. J. R. Astron. Soc., 76: 1–16.

Crampin, S., 1989. Suggestions for a consistent terminology for seismic anisotropy. Geophys. Prospect., 37: 753–770.

Cumming, W.B., Clowes, R.M. and Ellis, R.M., 1979. Crustal structure from a seismic refraction profile across southern British Columbia. Can. J. Earth Sci., 16: 1024–1040.

Davydova, N.I., Pavlenkova, N.E., Tulina, Yu.V. and Zverev, S.M., 1985. Crustal structure of the Barents Sea from seismic data. Tectonophysics, 114: 213–231.

Deichmann, N., Ansorge, J. and Mueller, S., 1986. Crustal structure of the southern Alps beneath the intersection with the European Geotraverse. Tectonophysics, 126: 57–83.

Diebold, J.B. and Stoffa, P.L., 1981. The traveltime equation, tau-p mapping and inversion of common midpoint data. Geophysics, 46: 238–254.

Dupuy, C., Leyreloup, A. and Vernieres, J., 1979. The lower continental crust of the Massif Central (Bornac, France) –with special references to REE, U and Th composition, evolution and heat-flow production. Phys. Chem. Earth, 11: 401–415.

Egger, A., Demartin, M., Ansorge, J., Banda, E. and Maistrello, M., 1988. The gross structure of the crust under Corsica and Sardinia. Tectonophysics, 150: 363–389.

EUGENO-S, W.G., 1988. Crustal structure and tectonic evolution of the transition between the Baltic Shield and the North German Caledonides (the EUGENO-S Project). Tectonophysics, 150: 253–348.

Evans, C.J., 1980. The seismic velocities of the Ox Mountain granulites of Ireland and the implications for the interpretation of the crustal structure of north Britain. Geophys. J. R. Astron. Soc., 63: 417–426.

Finlayson, D.M., Collins, C.D.N. and Denham, D., 1980. Crustal structure under the Lachlan fold belt, southeastern Australia. Phys. Earth Planet. Inter., 21: 321–342.

Finlayson, D.M., 1982. Seismic crustal structure of the Proterozoic North Australian craton between Tennant Creek and Mount Isa. J. Geophys. Res., 87: 10,569–10,578.

Finlayson, D.M. and Ansorge, J., 1984. Workshop proceedings: Interpretation of seismic wave propagation in laterally heterogeneous structures. Report 258, Bureau of Mineral Resources, Geology and Geophysics, Australian Govt. Publishing Service, Canberra, A.C.T.

Finlayson, D.M. and Leven, J.H., 1987. Lithospheric structures and possible processes in eastern Australia from deep seismic investigations. Tectonophysics, 133: 199–215.

Finlayson, D.M., Collins, C.D.N. and Lock, J., 1984. P-wave velocity features of the lithosphere under the Eromanga Basin, eastern Australia, including a prominent mid-crustal (Conrad?) discontinuity. Tectonophysics, 101: 267–291.

Fountain, D.M., 1976. The Ivrea–Verbano and Strona–Ceneri zones, northern Italy: A cross-section of the continental crust –new evidence from seismic velocities of rock samples. Tectonophysics, 33: 145–165.

Fountain, D.M. and Christensen, N.I., 1989. Composition of the continental crust and upper mantle –a review. In: L. C. Pakiser and W.D. Mooney (Editors), Geophysical Framework of the Continental United States. Geol. Soc. Am., Mem., 172: 711–742.

Fountain, D.M. and Salisbury, M.H., 1981. Exposed cross-sections through the continental crust: Implications for crustal structure, petrology and evolution. Earth Planet. Sci. Lett., 56: 263–277.

Fountain, D.M., Salisbury, M.H. and Percival, J., 1990. Seismic structure of the continental crust based on rock velocity measurements from the Kapuskasing Uplift. J. Geophys. Res., 95: 1167–1186.

Fox, P.J., Schreiber, E. and Peterson, J.J., 1973. The geology of the ocean crust: Compressional wave velocities of oceanic rocks. J. Geophys. Res., 78: 5155.

Fuchs, K., 1983. Recently formed elastic anisotropy and petrological models for the continental subcrustal lithosphere in southern Germany. Phys. Earth Planet. Inter., 31: 93–118.

Fuchs, K. and Müller, G., 1971. Computation of synthetic seismograms with the reflectivity method and comparison with observations. Geophys. J. R. Astron. Soc., 23: 417–433.

Fuis, G.S., Mooney, W.D., Healy, J.H., McMechan, G.A. and Lutter, W.J., 1984. A seismic refraction survey of the Imperial Valley region, California. J. Geophys. Res., 89: 1165–1189.

Furlong, K.P. and Fountain, D.M., 1986. Continental crustal underplating: Thermal considerations and seismic-petrologic consequences. J. Geophys. Res., 91: 8285–8294.

Fyfe, W.S., Price, N.J. and Thompson, A.B., 1978. Fluids in the Earth's Crust. Elsevier, Amsterdam-New York, 383 pp.

Gajewski, D. and Prodehl, C., 1987. Seismic refraction investigation of the Black Forest. Tectonophysics, 142: 27–48.

Gajewski, D., Holbrook, W.S. and Prodehl, C., 1987. A three-dimensional crustal model of southwest Germany derived from seismic refraction data. Tectonophysics, 142: 49–70.

Gaulier, J.M., Le Pichon, X., Lyberis, N., Avedik, F., Geli, L., Moretti, I., Deschamps, A. and Hafez, S., 1988. Seismic study of the crust of the northern Red Sea and Gulf of Suez. Tectonophysics, 153: 55–88.

Ginzburg, A., Makris, J., Fuchs, K. and Prodehl, C., 1981. The structure of the crust and upper mantle in the Dead Sea Rift. Tectonophysics, 80: 109–119.

Ginzburg, A., Whitmarsh, R.B., Roberts, D.G., Montadert, L., Camus, A. and F. Avedik, 1985. The deep seismic structure of the northern continental margin of the Bay of Biscay. Ann. Geophys., 3: 499–510.

Gohl, K., Smithson, S.B. and Kristoffersen, Y., 1991. The structure of the Archean crust in SW Greenland from seismic wide-angle data: a preliminary analysis. In: R. Meissner, L. Brown, H.-J. Dürbaum, W. Franke, K. Fuchs and F. Seifert (Editors), Continental Lithosphere: Deep Seismic Reflections. Am. Geophys. Union, Geodynamics Series, 22: 53–57.

Goldflam, S., Hirschleber, H.B. and Janle, P., 1977. A refined crustal model and the isostatic state of the Scandinavian Blue Road area. J. Geophys., 42: 419–428.

Goodwin, E.B. and McCarthy, J., 1990. Composition of the lower crust, West-central Arizona transition zone, from vertical-incidence and wide-angle measurements of P and S. J. Geophys. Res., 95: 20,097–20,109.

Grad, M. and Luosto, U., 1987. Seismic models of the crust of the Baltic Shield along the SVEKA profile in Finland. Ann. Geophys., 5: 639–650.

Griffin, W.L. and O'Reilly, S.Y., 1987. The composition of the lower crust and the nature of the continental Moho — xenolith evidence. In: P.H. Nixon (Editor), Mantle Xenoliths. John Wiley and Sons, New York, N.Y., pp. 413–430.

Guterch, A., Grad, M. and Materzok, R., 1983. Structure of the earth's crust of the Permian Basin in Poland. Acta Geophys. Pol., 31: 121–138.

Hale, L.D. and Thompson, G.A., 1982. The seismic reflection character of the continental Mohorovičić discontinuity. J. Geophys. Res., 87: 4625–4635.

Hall, D.H. and Hajnal, Z., 1973. Deep seismic crustal studies in Manitoba. Bull. Seismol. Soc. Am., 63: 885–910.

Hall, J., 1986. The physical properties of layered rocks in the deep continental crust. In: J.B. Dawson, D.A. Carswell, J. Hall and K.H. Wedepohl (Editors), The Nature of the Lower Continental Crust. Geol. Soc. London, Geol. Soc. Spec. Publ., 24: 51–62.

Hall, J. and Simmons, G., 1979. Seismic velocities of Lewisian metamorphic rocks at pressures to 8 kilobars: Relationship to crustal layering in North Britain. Geophys. J. R. Astron. Soc., 58: 337–347.

Hamilton, W.B., 1981. Crustal evolution by arc magmatism. Philos. Trans. R. Soc. London, 301: 279–291.

Hamilton, W.B., 1988. Plate tectonics and island arcs. Bull. Geol. Soc. Am., 100: 1503–1527.

Hargraves, R.B., 1981. Precambrian tectonic style: A liberal uniformitarian interpretation. In: A. Kröner (Editor), Precambrian Plate Tectonics. Elsevier, Amsterdam, pp. 21–56.

Hennet, C.G., Luetgert, J.H. and Phinney, R.A., 1991. The crustal structure in central Maine from coherency processed refraction data. J. Geophys. Res., 96: 12,023–12,037.

Holbrook, W.S., 1988. Wide-angle seismic studies of crustal structure and composition in Nevada, California and Southwest Germany. Ph.D. thesis, Stanford University.

Holbrook, W.S., 1990. The crustal structure of the northwestern Basin and Range province, Nevada, from wide-angle seismic data. J. Geophys. Res., 95: 21,843–21,869.

Holbrook, W.S. and Mooney, W.D., 1987. The crustal structure of the axis of the Great Valley, California, from seismic refraction measurements. Tectonophysics, 140: 49–63.

Holbrook, W.S., Gajewski, D. and Prodehl, C., 1987. Shear-wave velocity and Poisson's ratio structure of the upper lithosphere in Southwest Germany. Geophys. Res. Lett., 14: 231–234.

Holbrook, W.S., Gajewski, D., Krammer, A. and Prodehl, C., 1988. An interpretation of wide-angle compressional and shear wave data in Southwest Germany: Poisson's ratio and petrological implications. J. Geophys. Res., 93: 12,081–12,106.

Howie, J. and Miller, K., 1988. Interpretation of coincident vertical-incidence and wide-angle seismic profiles across the central California margin. EOS, Trans. Am. Geophys. Union, 69: 1315.

Hyndman, R.D., 1988. Dipping seismic reflectors, electrically conductive zones and trapped water in the crust over a subducting plate. J. Geophys. Res., 93: 13,391–13,405.

Hyndman, R.D. and Hyndman, D.W., 1968. Water saturation and high electrical conductivity in the lower continental crust. Earth Planet. Sci. Lett., 4: 427–432.

Hyndman, R.D. and Klemperer, S.L., 1989. Lower-crustal porosity from electrical measurements and inferences about composition from seismic velocities. Geophys. Res. Lett., 16: 255–258.

Hyndman, R.D. and Shearer, P.M., 1989. Water in the lower continental crust: modelling magnetotelluric and seismic reflection results. Geophys. J. Int., 98: 343–365.

Ikami, A., Ito, K., Shibuya, K. and Kaminuma, K., 1985. Geophysical studies of crustal structure of the Ongul islands and the northern Mizuho Plateau, East Antarctica. Tectonophysics, 114: 371–387.

Iwasaki, T., Hirata, N., Kanazawa, T., Melles, J., Suyehiro, K., Urabe, T., Möller, L., Makris, J. and Shimamura, H., 1990. Crustal and upper mantle structure in the Ryukyu island arc deduced from deep seismic sounding. Geophys. J. Int., 102: 631–651.

Jackson, I. and Arculus, R.J., 1984. Laboratory wave velocity measurements on lower crustal xenoliths from Calcutteroo, South Australia. Tectonophysics, 101: 185–197.

Jacob, A.W.B., Kaminski, W., Murphy, T., Phillips, W.E.A. and Prodehl, C., 1985. A crustal model for a northeast–southwest profile through Ireland. Tectonophysics, 113: 75–103.

Jarchow, C.M. and Thompson, G.A., 1989. The nature of the Mohorovičić discontinuity. Annu. Rev. Earth Planet. Sci., 17: 475–506.

Jeffreys, H., 1926. On near earthquakes. Monthly Not. R. Astron. Soc. Geophys. Suppl., 1: 385.

Jentsch, M., 1979. Reinterpretation of a deep-seismic-sounding profile on the Ukrainian Shield. J. Geophys., 45: 355–372.

Jones, A.G., 1987. MT and reflection: an essential combination. Geophys. J. R. Astron. Soc., 89: 7–18.

Jones, T.D. and Nur, A., 1982. Seismic velocity and anisotropy in mylonites and the reflectivity of deep crustal faults. Geology, 10: 260–263.

Jones, T.D. and Nur, A., 1984. The nature of seismic reflections from deep crustal fault zones. J. Geophys. Res., 89: 3153–3171.

Kaila, K.L., 1986. Tectonic framework of Narmada–Son Lineament –a continental rift system in central India from deep seismic soundings. In: M. Barazangi and L. Brown (Editors), Reflection Seismology: A Global Perspective. American Geophysical Union, Geodynamics Series 13, Washington, D.C., 133–150.

Kaila, K.L., Krishna, V.G. and Mall, D.M., 1981. Crustal structure along Mehmadabad–Billimora profile in the Cambay Basin, India, from deep seismic soundings. Tectonophysics, 76: 99–130.

Kaila, K.L., Murty, P.R.K., Mall, D.M., Dixit, M.M. and Sarker, D., 1987. Deep seismic soundings along Hirnpur–Mandla profile, central India. Geophys. J. R. Astron. Soc., 89: 399–404.

Kaila, K.L., Rao, I.B.P., Rao, P.K., Rao, N.M., Krishna, V.G. and Sridhar, A.R., 1989. DSS studies over Deccan Traps along the Thuadara–Sendhwa–Sindad profile, across Narmada–Son lineament, India. In: R. F. Mereu, S. Müller and D.M. Fountain (Editors), Properties and Processes of Earth's Lower Crust. Am. Geophys. Union, Monogr., 51: 121–125.

Kan, R.-J., Hu, H.-X., Zeng, R.-S., Mooney, W.D. and McEvilly, T.V., 1986. Crustal structure of Yunnan province, People's Republic of China, from seismic refraction profiles. Science, 234: 433–437.

Kanamori, H. and Mizutani, H., 1965. Ultrasonic measurements of elastic constants of rocks uner high pressures. Bull. Earthquake Res. Inst., 43: 173.

Kaneshima, S. and Ando, M., 1989. An analysis of split shear waves observed above crustal and uppermost mantle earthquakes beneath Shikoku, Japan: Implications in effective depth extent of seismic anisotropy. J. Geophys. Res., 94: 14,077–14,092.

Kay, R.W. and Kay, S.M., 1981. The nature of the lower continental crust: Inferences from geophysics, surface geology and crustal xenoliths. Rev. Geophys. Space Phys., 19: 271–297.

Kay, R.W. and S.M., Kay,, 1986. Petrology and geochemistry of the lower continental crust: an overview. In: J. B. Dawson, D.A. Carswell, J. Hall and K.H. Wedepohl (Editors), The Nature of the Lower Continental Crust. Geol. Soc. London, Geol. Soc. Spec. Publ., 24: 147–159.

Káроly, P., István, A., Géza, R. and Géza, V., 1986. Characteristics of the reflecting layers in the earth's crust and upper mantle in Hungary. In: M. Barazangi and L. Brown (Editors), Reflection Seismology: A Global Perspective. American Geophysical Union, Geodynamics Series 13, Washington, D.C., pp. 55–65.

Keller, G.R., Smith, R.B. and Braile, L.W., 1975. Crustal structure along the Great Basin–Colorado Plateau transition from seismic refraction studies. J. Geophys. Res., 80: 1093–1098.

Kern, H., 1978. The effect of high temperature and high confining pressure on compressional wave velocities in quartz-bearing and quartz-free igneous and metamorphic rocks. Tectonophysics, 44: 185–203.

Kern, H., 1982. Elastic-wave velocity in crustal and mantle rocks at high pressure and temperature: The role of the high–low quartz transition and of dehydration reactions. Phys. Earth Planet. Inter., 29: 12–23.

Kern, H. and Schenk, V., 1985. Elastic wave velocities in rocks from a lower crustal section in southern Calabria (Italy). Phys. Earth Planet. Inter., 40: 147–160.

Kern, H. and Schenk, V., 1988. A model of velocity structure beneath Calabria, southern Italy, based on laboratory data. Earth Planet. Sci. Lett., 87: 325–337.

Klemperer, S.L., 1989. Deep seismic reflection profiling and the growth of the continental crust. Tectonophysics, 161: 233–244.

Klemperer, S.L., Hauge, T.A., Hauser, E.C., Oliver, J.E. and Potter, C.J., 1986. The Moho in the northern Basin and Range province, Nevada, along the 40°N seismic-reflection transect. Geol. Soc. Am. Bull., 97: 603–618.

Kondorskaya, N.V. and 10 others, 1981. Joint analysis of seismological data by the USSO and DSS stations in the Caucasus region. Pure Appl. Geophys., 119: 1167–1179.

Kroenke, I.W., Manghnani, M.H., Rai, C.S., Fryer, P. and Ramananantoandro, R., 1976. Elastic properties of selected ophiolite rocks from Papua, New Guinea: Nature and composition of oceanic lower crust and upper mantle. In: G. H. Sutton, M.H. Manghnani and R. Moberly (Editors), The Geophysics of the Pacific Ocean Basin and its Margins. Am. Geophys. Union, Monogr., 19: 407–421.

Kröner, A., 1981. Precambrian plate tectonics. In: A. Kröner (Editor), Precambrian Plate Tectonics. Elsevier, Amsterdam, pp. 57–90.

Kröner, A., 1984. Evolution, growth and stabilization of the Precambrian lithosphere. Phys. Chem. Earth, 15: 69–106.

Lachenbruch, A.H. and Sass, J.H., 1978. Models of an extending lithosphere and heat flow in the Basin and Range province. In: R. B. Smith and G.P. Eaton (Editors), Cenozoic Tectonics and Regional Geophysics of the Western

Cordillera. Geol. Soc. Am., Mem., 152: 209–250.

LASE, S.G., 1986. Deep structure of the U.S. East Coast passive margin from large aperture seismic experiments (LASE). Mar. Pet. Geol., 3: 234–242.

Leaver, D.S., Mooney, W.D. and Kohler, W.M., 1984. A seismic refraction study of the Oregon Cascades. J. Geophys. Res., 89: 3121–3134.

Litak, R.K. and Brown, L.D., 1989. A modern perspective on the Conrad discontinuity. EOS, Trans. AGU, 70, 713.

Lowe, C. and Jacob, A.W.B., 1989. A north–south seismic profile across the Caledonian suture zone in Ireland. Tectonophysics, 168: 297–318.

Luetgert, J.H. and C.E., Mann,, 1990. Avalon terrane in eastern coastal Maine: Seismic refraction–wide-angle reflection data. Geology, 18: 878–881.

Luetgert, J.H., Mann, C.E. and Klemperer, S.L., 1987. Wide-angle deep crustal reflections in the northern Appalachians. Geophys. J. R. Astron. Soc., 89: 183–188.

Luosto, U. and Korhonen, H., 1986. Crustal structure of the Baltic Shield along the SVEKA profile in Finland. Ann. Geophys., 5: 639–650.

Luosto, U., Flüh, E.R., Lund, C.-E. and Working Group, 1989. The crustal structure along the POLAR Profile from seismic refraction investigations. Tectonophysics, 162: 51–85.

Luosto, U., Tiira, T., Korhonen, H., Azbel, I., Burmin, V., Buyanov, A., Kosminskaya, I., Ionkis, V. and Sharov, N., 1990. Crust and upper mantle structure along the DSS Baltic profile in SE Finland. Geophys. J. Int., 101: 89–110.

Lutter, W.J. and Nowack, R.L., 1990. Inversion for crustal structure using reflections from the PASSCAL Ouachita experiment. J. Geophys. Res., 95: 4633–4646.

MacGregor-Scott, N. and Walter, A., 1988. Crustal velocities near Coalinga, California, modeled from a combined earthquake/explosion refraction profile. Bull. Seismol. Soc. Am., 78: 1475–1490.

Manghnani, M.H., Ramananantoandro, R. and Clark, S.P. Jr., 1974. Compressional and shear wave velocities in granulite facies rocks and eclogites to 10 kilobars. J. Geophys. Res., 79: 5427–5446.

Matthews, D.H. and Cheadle, M.J., 1986. Deep reflections from the Caledonides and Variscides west of Britain and comparison with the Himalayas. In: M. Barazangi and L. Brown (Editors), Reflection Seismology: A Global Perspective. American Geophysical Union, Geodynamics Series 13, Washington, D.C., pp. 5–19.

Matthews, D.H. and Smith, C., 1987. Deep seismic reflection profiling of the continental lithosphere. Geophys. J. R. Astron. Society 89, 498 pp.

McCamy, K., Meyer, R.P. and Smith, T.J., 1962. Generally applicable solutions of Zoeppritz' amplitude equations. Bull. Seismol. Soc. Am., 52: 923–955.

McCarthy, J. and Thompson, G.A., 1988. Seismic imaging of extended crust with emphasis on the western United States. Geol. Soc. Am. Bull., 100: 1361–1374.

McCarthy, J., et al., 1989. Unpublished PACE results.

McLennan, S.M. and Taylor, S.R., 1982. Geochemical constraints on the growth of the continental crust. J. Geol., 90: 347–361.

McMechan, G.A. and Mooney, W.D., 1980. Asymptotic ray theory and synthetic seismograms for laterally varying structures: Theory and application to the Imperial Valley, California. Bull. Seismol. Soc. Am., 70: 2021–2035.

Mechie, J., Prodehl, C. and Koptschalitsch, G., 1986. Ray path interpretation of the crustal structure beneath Saudi Arabia. Tectonophysics, 131: 333–352.

Meissner, R., 1973. The 'Moho' as a transition zone. Geophys. Surv., 1: 195–216.

Meissner, R., 1986. The Continental Crust, a Geophysical Approach. Academic Press, Orlando, Fla., 426 pp.

Mereu, R.F. and 11 others, 1986. The 1982 COCRUST experiment across the Ottawa–Bonnechere graben and Grenville Front in Ontario and Quebec. Geophys. J. R. Astron. Soc., 84: 491–514.

Mohorovičić, A., 1910. Das Beben vom 8.X.1909. Jahrb. Meteorol. Obs. Zagreb, 9: Teil 4, Abschn. 1.63.

Mooney, W.D., 1989. Seismic methods for determining earthquake source parameters and lithospheric structure. In: L.C. Pakiser and W.D. Mooney (Editors), The Geophysical Framework of the Continental United States. Geol. Soc. Am., Mem., 172: 11–34.

Mooney, W.D. and Brocher, T.M., 1987. Coincident seismic reflection/refraction studies of the continental lithosphere: a global review. Rev. Geophys., 25: 723–742.

Mooney, W.D. and Prodehl, C., 1978. Crustal structure of the Rhenish Massif and adjacent areas; a reinterpretation of existing seismic-refraction data. J. Geophys., 44: 573–601.

Mooney, W.D. and Prodehl, C. (Editors), 1984. Proceedings of the 1980 workshop of the International Association of Seismology and Physics of the Earth's Interior on the seismic modeling of laterally varying structures: Contributions based on data from the 1978 Saudi Arabian refraction profile. U.S. Geol. Surv., Circ. 937, 158 pp.

Mooney, W.D., Andrews, M.C., Ginzburg, A., Peters, D.A. and Hamilton, R.M., 1983. Crustal structure of the northern Mississippi embayment and a comparison with other continental rift zones. Tectonophysics, 94: 327–348.

Mooney, W.D., Gettings, M.E., Blank, H.R. and Healy, J.H., 1985. Saudi Arabian seismic refraction profile: a traveltime interpretation of crustal and upper mantle structure. Tectonophysics, 111: 173–246.

Morel-á-l'Huissier, P., Green, A.G. and Pike, C.J., 1987. Crustal refraction surveys across the Trans-Hudson orogen/Williston Basin of South Central Canada. J. Geophys. Res., 92: 6403–6420.

Müller, S., Prodehl, C., Mendes, A.S. and Sousa Moreira, V., 1973. Crustal structure in the southwestern part of the Iberian Peninsula. Tectonophysics, 20: 307–318.

Mykkeltveit, S., 1980. A seismic profile in southern Norway. Pure Appl. Geophys., 118: 1310–1325.

Nur, A. and Simmons, G., 1969. The effect of saturation on velocity in low porosity rocks. Earth Planet. Sci. Lett., 7: 183–193.

Pakiser, L.C. and Brune, J.N., 1980. Seismic models of the root of the Sierra Nevada. Science, 210: 1088–1094.

Pavlenkova, N.I., 1979. Generalized geophysical model and dynamic properties of the continental crust. Tectonophysics, 59: 381–390.

Phinney, R.A. and Roy-Chowdury, K., 1989. Reflection seismic studies of crustal structure in the eastern United States. In: L. C. Pakiser and W.D. Mooney (Editors), The Geophysical Framework of the Continental United States. Geol. Soc. Am., Mem., 172: 613–654.

Powell, C.M.R. and Sinha, M.C., 1987. The PUMA experiment west of Lewis. Geophys. J. R. Astron. Soc., 89: 259–264.

Prodehl, C., Schlittenhardt, J. and Stewart, S.W., 1984. Crustal structure of the Appalachian Highlands in Tennessee. Tectonophysics, 109: 61–76.

Raitt, R.W., Shor, G.G., Jr., Francis, T.J.G. and Morris, G.B., 1969. Anisotropy of the Pacific upper mantle. J. Geophys. Res., 74: 3095–3109.

Reiter, E.C., Purdy, G.M., Sawyer, D., Stoffa, P.L., Phillips, J.A., Austin, J.A., Jr. and Toksoz, N., 1988. Ocean bottom hydrophone results from a multichannel seismic survey in the Carolina Trough and Southeast Georgia Embayment [abs.]. EOS, Trans. AGU 69: 1405.

Research Group for Seismology, 1973. Crustal structure of Japan as derived from explosion seismic data. Tectonophysics, 20: 129–135.

Reymer, A. and Schubert, G., 1984. Phanerozoic addition rates to the continental crust and crustal growth. Tectonics, 3: 63–77.

Rudnick, R.L. and Taylor, S.R., 1986. Geochemical constraints on the origin of Archean tonalitic-trondhjemtic rocks and implications for lower crustal composition. In: J. B. Dawson, D.A. Carswell, J. Hall and K.H. Wedepohl (Editors), The Nature of the Lower Continental Crust. Geol. Soc. London, Geol. Soc. Spec. Publ., 24: 179–191.

Sapin, M., Wang, X.-J., Hirn, A. and Z.X., Xu,, 1985. A seismic sounding in the crust of the Lhasa block, Tibet. Ann. Geophys., 3: 637–646.

Schock, R.N., Bonner, B.P. and Louis, H., 1974. Collection of ultrasonic velocity data as a function of pressure for polycrystalline solids. National Tech. Info. Servide, Lawrence Livermore Lab. Tech. Rept., UCRL-51508 Springfield, Va., 262 pp.

Shankland, T.J. and Ander, M.E., 1983. Electrical conductivity, temperatures and fluids in the lower crust. J. Geophys. Res., 88: 9475–9484.

Shaw, D.M., Cramer, J.J., Higgins, M.D. and Truscott, M.G., 1986. Composition of the Canadian Precambrian shield and the continental crust of the earth. In: J. B. Dawson, D.A. Carswell, J. Hall and K.H. Wedepohl (Editors), The Nature of the Lower Continental Crust. Geol. Soc. London, Geol. Soc. Spec. Publ., 24: 275–282.

Simmons, G., 1964. Velocity of shear waves in rocks to 10 kilobars. J. Geophys. Res., 69: 1123–1130.

Simmons, G. and Brace, W.F., 1965. Comparison of static and dynamic measurements of compressibility of rocks. J. Geophys. Res., 70: 5649–5656.

Sinno, Y.A., Daggett, P.H., Keller, G.R., Morgan, P. and Harder, S.H., 1986. Crustal structure of the southern Rio Grande Rift determined from seismic refraction profiling. J. Geophys. Res., 91: 6143–6156.

Sleep, N.H. and Windley, B.F., 1982. Archean plate tectonics: Constraints and inferences. J. Geol., 90: 363–379.

Smithson, S.B., 1978. Modeling continental crust: structural and chemical constraints. Geophys. Res. Lett., 5: 749–752.

Smithson, S.B., 1986. A physical model of the lower crust from North America based on seismic reflection data. In: J. B. Dawson (Editor), The Nature of the Lower Continental Crust. Geol. Soc. London, Geol. Soc. Spec. Publ., 25: 23–34.

Smithson, S.B. and Brown, S.K., 1977. A model for lower continental crust. Earth Planet. Sci. Lett., 35: 134–144.

Smithson, S.B. and Johnson, R., 1989. Crustal structure of the western U.S. based on reflection seismology. In: L. C. Pakiser and W.D. Mooney (Editors), Geophysical Framework of the Continental United States. Geol. Soc. Am., Mem., 172: 577–612.

Smithson, S.B., Johnson, R.A. and Wong, Y.K., 1981. Mean crustal velocity: a critical parameter for interpreting crustal structure and crustal growth. Earth Planet. Sci. Lett., 53: 323–332.

Sollogub, V.B., Litvinenko, I.V., Chekunov, A.V., Ankudinov, S.A., Ivanov, A.A., Kalyuzhnaya, L.T., Kokorina, L.K. and Tripolsky, A.A., 1973. New D.S.S. data on the crustal structure of the Baltic and Ukrainian shields. Tectonophysics, 20: 67–84.

Sparlin, M.A., Braile, L.W. and Smith, R.B., 1982. Crustal structure of the eastern Snake River Plain determined from ray trace modeling of seismic refraction data. J. Geophys, Res., 87: 2619–2633.

Stauber, D.A., 1983. Crustal structure in northern Nevada from seismic refraction data. In: Geothermal Resources Council, Spec. Rep., 13: 319–332.

Stoffa, P.L. and Buhl, P., 1979. Two-ship multichannel seismic experiments for deep crustal studies: Expanded spread and constant offset profiles. J. Geophys. Res., 84: 7645–7660.

Tarkov, A.P., Basula, I.P., Generalov, V.G., Dubyansky, A.I. and Chernykh, V.V., 1981. Composite travel times of seismic waves and general velocity models of the Voronezh Shield crust and upper mantle. Geophys. J. R. Astron. Soc., 67: 137–143.

Tarney, J. and Windley, B.F., 1977. Chemistry, thermal gradients and evolution of the lower continental crust. J. Geol. Soc. London, 134: 153–172.

Taylor, S.R., 1977. Island arc models and the composition of the continental crust. In: M. Talwani and W.C. Pitman (Editors), Am. Geophys. Union, Maurice Ewing Series, 1: 229–242.

Taylor, S.R. and McLennan, S.M., 1981. The composition and evolution of the continental crust: rare earth element evidence from sedimentary rocks. Philos. Trans. R. Soc. London, 301: 381–399.

Teng, J., Xiong, S., Yin, Z., Xu, Z., Wang, X. and Lu, D., 1985. Structure of the crust and upper mantle pattern and velocity distributional characteristics in the northern Himalayan mountain region. J. Phys. Earth, 33: 157–171.

Todd, T. and Simmons, G., 1972. Effect of pore pressure on the velocity of compressional waves in low-porosity rocks. J. Geophys. Res., 77: 3731–3743.

Trehu, A.M. and Wheeler, W.H., IV, 1987. Possible evidence for subducted sedimentary materials beneath central California. Geology, 15: 254–258.

Trehu, A.M., Ballard, A., Dorman, L.M., Gettrust, J.F., Klitgord, K.D. and Schreiner, A., 1989. Structure of the lower crust beneath the Carolina Trough, U.S. Atlantic continental margin. J. Geophys. Res., 94: 10,585–10,600.

Wang, H., Todd, T., Richter, D. and Simmons, G., 1973. Elastic properties of plagioclase aggregates and seismic velocities in the moon. Proc. 4th Lunar Sci. Conf., 4-3: 2663–2671.

Weaver, B.L. and Tarney, J., 1984. Empirical approach to estimating the composition of the continental crust. Nature, 310: 575–577.

Wever, T., 1989. The Conrad discontinuity and the top of the reflective lower crust –do they coincide? Tectonophysics, 157: 39–58.

White, R.S., Spence, G.D., Fowler, S.R., McKenzie, D.P., Westbrook, G.K. and Bowen, A.N., 1987. Magmatism at rifted continental margins. Nature, 330: 439–444.

White, R.S. and McKenzie, D., 1989. Magmatism at rift zones: The generation of volcanic continental margins and flood basalts. J. Geophys. Res., 94: 7685–7730.

Windley, B.F., 1981. Precambrian rocks in the light of the plate-tectonic concept. In: A. Kröner (Editor), Precambrian Plate Tectonics. Elsevier, Amsterdam, pp. 1–20.

Wilson, J.M., McCarthy, J., Johnson, R.A. and Howard, K.A., 1991. An axial view of a metamorphic core complex: Structure of the Whipple and Chemehuevi Mountains, southeastern California. J. Geophys. Res., 96: 12,293–12,311.

Yan, Q.Z. and Mechie, J., 1989. A fine structural section through the crust and lower lithosphere along the axial region of the Alps. Geophys. J., 98: 465–488.

Yuan, X., Wang, S., Li, L. and Zhu, J., 1986. A geophysical investigation of deep structure in China. In: M. Barazangi and L. Brown (Editors), Reflection Seismology: A Global Perspective. American Geophysical Union, Geodynamics Series 13, Washington, D.C., pp. 151–160.

Zandt, G. and Owens, T.J., 1986. Comparison of crustal velocity profiles determined by seismic refraction and teleseismic methods. Tectonophysics, 128: 155–161.

Zeis, S., Gajewski, D. and Prodehl, C., 1990. Crustal structure of southern Germany from seismic refraction data. Tectonophysics, 176: 59–86.

Zelt, C.A. and Ellis, R.M., 1989. Seismic structure of the crust and upper mantle in the Peace River Arch region, Canada. J. Geophys. Res., 94: 5729–5744.

Zucca, J.J., 1984. The crustal structure of the southern Rhinegraben from re-interpretation of seismic refraction data. J. Geophys., 55: 13–22.

Chapter 2

Multi-genetic origin of crustal reflectivity: a review of seismic reflection profiling of the continental lower crust and Moho

W.D. MOONEY and R. MEISSNER

1. Introduction

Seismic reflection profiles provide the highest resolution information on the *in situ* structure and physical properties of the lower continental crust and Moho. The seismic properties most readily obtained from reflection profiles are reflectivity patterns, which correlate with distinct geologic settings. When reflectivity patterns are interpreted with complementary seismic velocity and non-seismic crustal parameters, inferences regarding the composition and evolution of the lower crust can be made.

The study of the lower continental crust with seismic reflection data is a rapidly evolving field; improved data sets continually provide new insights and alter initial concepts based on more limited data. In this chapter we summarize the most important observations and conclusions derived from reflection seismology and relate these to models of the physical properties and evolution of the lower crust. We focus particularly on the structure of the lower crust in diverse tectonic settings, the nature of the Moho, and the multi-genetic origin of crustal reflectivity. The reader should consult the companion papers by Holbrook et al. (this volume) and Jones (this volume) for the related topics of seismic velocities and conductivity in the lower crust.

2. The seismic reflection method

Seismological observations at the beginning of this century led to the detection of the Earth's core in 1909, and the crust–mantle boundary, the Moho (Mohorovičić, 1910). Seismic refraction measurements of potential hydrocarbon traps within sedimentary structures also began at about the same time, but it was not until the end of the 1930's that useful seismic reflection data became available (Sheriff and Geldart, 1982). This relatively late start in the widespread application of reflection seismology was due to the great technical effort needed to enhance near-vertical reflected signals that often are 10,000 times weaker than the seismic ground-roll from near-surface layers. In the 1940's and 1950's new techniques, like common depth point (CDP) stacking, versatile seismic sources, and magnetic tape recording were developed and contributed to the emerging superiority of reflection profiling over refraction seismics for prospecting in sedimentary basins.

2.1. Early deep crustal seismic reflection studies

It was in the 1950's that the first sporadic observations of near-vertical reflections from the lower continental crust were reported (Junger, 1957; Galfi and Stegena, 1957). However, serious doubts about the reality of these deep reflections were expressed by some workers (e.g., Steinhart and Meyer, 1961). These doubts led to an "experimentum crucis" in 1964 near Augsburg, Germany, where deep crustal reflectors were observed at both near-vertical and wide-angle incidence (Meissner, 1967), thus confirming, at least in this area of the Bavarian Molasse Basin, the existence of deep reflectors. At about the same time, similar investigations in Australia by the Bureau of Mineral Resources (BMR) (Moss, 1962; additional work summarized in Moss and Dooley, 1988, and Dooley and Moss, 1988) and in Canada (Kanasewich and Cummings, 1965; Clowes et al., 1968) also gave conclusive evidence for intracrustal and Moho reflections at near-vertical incidence. The lower crust was shown to be highly reflective at near-vertical incidence, and the names "lamellae" or "laminae" for the seismic layering in the crust were coined (Glocke and Meissner, 1976). Seismic lamellae are depicted in Figure 2-1.

2.2. The last 15 years of deep crustal reflection studies

Sporadic seismic reflection campaigns continued in the early 1970's across important geological structures in Germany (Dohr and Meissner, 1975) and elsewhere. However, the year 1975 marks the beginning of a nearly continuously operating seismic crew by COCORP (Consortium for Continental Reflection Profiling) in the U.S. (Oliver et al. 1976; Brown et al., 1986). Since then, a concerted effort to study the continental crust has been made, and similar groups have been formed elsewhere in the USA (U.S. Geological Survey, Univ. of Wyoming, and CALCRUST), Great Britain (BIRPS), Canada (LITHOPROBE and the Geological Survey of Canada), France (ECORS), Australia (ACORP and the BMR), Germany (DEKORP), Switzerland, and elsewhere (summarized in volumes edited by Barazangi and Brown, 1986a,b; Matthews and Smith, 1987, Leven et al., 1990).

Many of these research groups did not attempt to measure lower crustal seismic velocities. The use of spread lengths of about 10 km on land and about 3 km at sea did not permit the determination of seismic velocities to depths greater than about 10 km. The detailed picture available on near-vertical sections was so fascinating, and data were so quickly collected — especially at sea — that crustal velocities were of secondary interest. More recently, considerable effort has gone into the determination of seismic velocities (P-wave, S-wave) and anisotropy in an effort to place tighter constraints on crustal composition and physical state (e.g., Bartelson et al., 1982; Clowes et al., 1987; Mooney and Brocher, 1987; Wenzel et al., 1987; Fisher et al., 1989; Valasek et al., 1989; BABEL Working Group, 1990; Goodwin and McCarthy, 1990; Lueschen et al., 1990).

2.3. Resolution and uncertainties

The acquisition and processing of seismic reflection data has been well advanced by the hydrocarbon exploration industry, as described in such standard texts as Telford et al. (1976), Sheriff and Geldart (1982), Claerbout (1985), Waters (1987), and Yilmaz (1987). However, the application of reflection seismology to the deep continental crust presents special problems not often encountered in the exploration of sedimentary basins. Among the most important

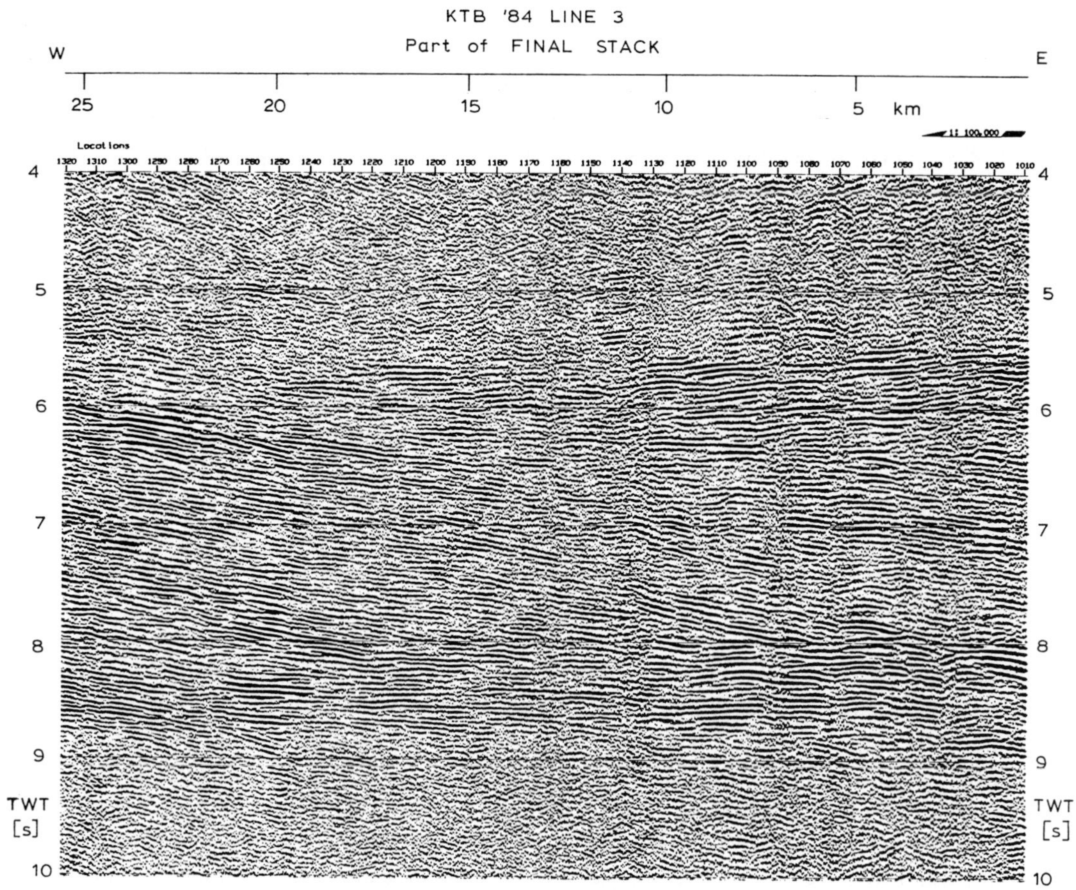

Fig. 2-1. Lower crustal seismic lamellae (or "laminae") as observed on a seismic reflection profile recorded in the Black Forest, SW Germany (Lueschen et al., 1987). Crustal reflectivity increases abruptly at about 5.5 s two-way time (TWT) and terminates at 9 s at the Moho. The high amplitudes of the reflections requires the existence of significant velocity contrasts (±5% or larger) in the lower crust. Lower crustal seismic lamellae have been recorded throughout the world with varying amplitudes, lateral continuity, and dips depending on the tectonic setting.

problems, as summarized by Mooney (1989), are:

(1) low reflection amplitudes due to geometrical spreading, scattering, and seismic attenuation;

(2) large static corrections (time shifts) for land profiles due to variations in elevation and near-surface geology;

(3) an inadequate knowledge of deep crustal velocities (including their lateral variations), which degrades the effectiveness of the common depth point (CDP) stacking method; and

(4) poor data quality where there is high cultural noise, imperfect coupling of vibrators and geophones, or low, uneven-fold data arising from skips or crooked-line recording.

These factors can make the difference between recording coherent deep seismic reflections or incoherent noise (cf., Taner et al., 1970; Johnson and Smithson, 1986; Phinney and Roy-Chowdhury, 1989; Brocher and Hart, 1991).

Smithson and Johnson (1989) point out three other problems that reflection seismologists encounter when working in the crystalline crust: (1) reflection coefficients are smaller for layered crystalline rocks (about 0.1) in comparison with those for layered sedimentary rocks (about 0.3); (2) crystalline rocks are often more deformed and distorted than sedimentary strata; and (3) there is generally a lack of drill hole control to test interpretations. We discuss some cases where upper crustal drill hole control is available in Section 4.1.

The relative vertical resolution of seismic reflections is generally comparable to one-quarter of the wavelength of the seismic signal according to:

$$\lambda = V_p/f$$

where λ = wavelength (km), V_p = compressional wave velocity (km/s), and f = frequency (Hz, cycles/s).

Using a crustal V_p-value of 6.0 km/s and a typical frequency of $f = 25$ Hz, the seismic wavelength is 240 m and the optimal resolution in the seismic section will be equal to $\lambda/4$, or 60 m. High-frequency signals ($f = 50$ Hz) show higher resolution (30 m), and in sediments, with their lower V_p-values, resolution may reach 10–20 m. Layers thinner than one-quarter wavelength will also be resolved, up to approximately one-tenth of a wavelength (Fuchs, 1969; Widess, 1973), but signal amplitudes will be lower and favorable signal-to-noise conditions are necessary. The best resolution of the physical properties of crustal reflectors is obtained with broadband (approximately 8–80 Hz and broader) seismic data.

Horizontal resolution is less than vertical (thickness) resolution, and decreases with depth. The concept of the first Fresnel zone (Sheriff and Geldhart, 1982) describes the minimum length of a well-resolved reflecting horizon. Typical Fresnel zone radii (for 25 Hz data) in the crystalline crust are 1.5 km at 10 km depth, and 3 km at 30 km depth. Gaps that are smaller than a Fresnel zone will not be evident and, therefore, a discontinuous reflection horizon will appear to be more continuous than it actually is. Computer modeling of hypothetical structures provides quantitative evaluation of seismic resolution and uncertainties (Hurich and Smithson, 1987; Reston, 1987; Wenzel et al., 1987; Gibson and Levander, 1988, 1990; Blundell, 1990).

The determination of a reflector's absolute depth, for example, the Moho, depends on a knowledge of the velocity structure of the crust. This information normally comes from wide-angle reflection/refraction studies that generally use lower frequencies (5–10 Hz) and thus have correspondingly lower resolution compared with seismic reflection data. The error in the calculated depth (z) of a reflecting interface is proportional to the accuracy of the measured average velocity (V_p). V_p–z (velocity–depth) curves may easily assume errors of 5–6% (Holbrook et al., this volume).

3. Observations of continental lower crust and Moho

Some 40,000 km of deep crustal reflection data have been collected worldwide during the past 15 years in continental areas. About half of the data were obtained by marine surveys in continental shelf areas. Observations in similar geologic environments, for example, young mountain belts, are consistent from region to region. Significantly, reflectivity patterns do not generally appear to be controlled by data acquisition parameters, and data collected in similar geologic environments using vibroseis, explosives, or airguns usually show similar reflectivity patterns. The primary exceptions to this are that explosive and airgun sources appear to produce superior deep reflections for crustal thickness greater than about 25 km, and explosive sources sometimes produce fewer artifacts in land data than vibroseis sources (e.g., Brocher and Hart, 1991). The similarities in crustal reflectivity for common geologic environments is an important observation that indicates common processes of tectonic evolution for these environments.

3.1. Terminology and overview of important contributions

As an introduction to the following sections, some frequently used expressions in the reflection literature are defined, after which we list some of the most important contributions made by seismic reflection studies.

a. *Laminated lower crust*: a lower crust that is full of densely packed, multiple sets of reflections referred to as seismic lamellae or laminae. These reflections are commonly subhorizontal, but occasionally dip, curve, and overlap, and often show a sudden onset at 4–6 s two-way traveltime (twt) (12–18 km depth) and an abrupt termination at the Moho.

b. *Transparent upper (or lower) crust*: a significant time interval of a seismic section that is relatively free of seismic reflections.

c. *Structureless crust*: reflections are predominantly short (1.5 km or less in length), the strength of reflections (i.e., reflectivity) generally decreases with depth, and the crust shows few reliable reflectivity patterns (see below); may sometimes be caused by data acquisition problems (section 2.3 Resolution and uncertainties).

d. *Diffractive crust*: crust is dominated by long (3–5 km or more) hyperbola-shaped events, possibly generated by strong folds or small-scale inhomogeneities, such as vertical intrusives.

e. *Seismic duplex*: Seismic reflections in a rhomboid shape, similar to cross-sections of structural duplexes, and usually seen in shortened (compressional) environments (Fig. 2-2a).

f. *Seismic wedge*: Diverging, wedge-shaped seismic reflectors generally seen in shortened (compressive) environments; these are more reliably identified in migrated sections (Fig. 2-2b).

We illustrate and discuss the significance of these features within various geologic setting in the forthcoming sections. Some of the most important contributions of crustal reflection studies in the last 15 years are:

(1) The determination of the geometry at depth of prominent fault zones, some of which reach the lower crust. Examples are the Wind River fault in Wyoming (Smithson et al., 1979), the Hunsruck-fault in Europe (DEKORP Res. Group, 1991), and the Flannan fault north of Scotland (Peddy et al., 1986).

(2) The discovery of the seismically laminated lower crust, sometimes accompanied by a relatively transparent upper crust, in areas of extended crust (Allmendinger et al., 1986, 1987; BIRPS and ECORS, 1986; Matthews, 1986).

(3) The recognition of distinct structural developments for different ancient orogens; for example, *thin-skinned* tectonics in the southern Appalachians (Clark et al., 1978; Cook et al., 1981) and the Variscan deformation in Europe (DEKORP Research Group, 1990), compared with *thick-skinned* compressional features in the New England Appalachians (Brown et al., 1983), the Grenville orogenic belt (Green et al., 1988) and the eastern Rhenish Massif (DEKORP Research Group, 1991).

(4) The confirmation of thick crustal roots under young orogenic belts and the absence of crustal roots in most Caledonian and Variscan belts (Meissner et al., 1987).

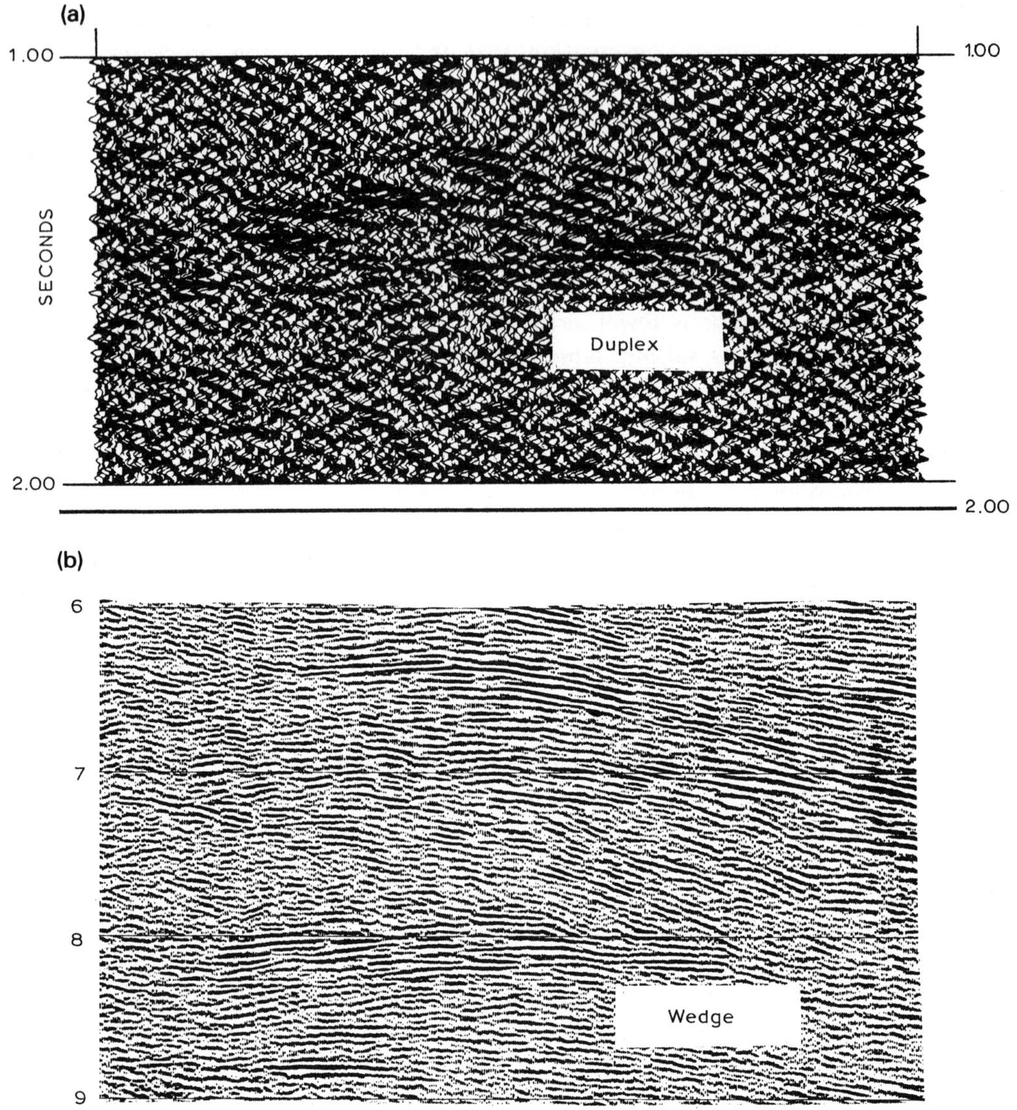

Fig. 2-2. Examples of two common structural features observed in deep seismic reflection data: (A) a seismic duplex (Hubbard et al., 1991) and (B) a seismic wedge (DEKORP Res. Group, 1991). These features are interpretated as resulting from crustal shortening (see sections 2.3.2–2.3.4).

3.2. Precambrian crust

Early seismic reflection profiles in Precambrian crust indicated a generally structureless lower crust, but revealed pronounced structural features in the upper and middle crust that appear to date from Precambrian times (Oliver et al., 1976; Smithson et al., 1980; Gibbs, 1986; Dahl-Jensen, 1987; Smithson and Johnson, 1989). The most significant features are steeply-dipping zones of reflections that often can be correlated to depths greater than 15 km, although data quality generally is significantly better (and therefore interpretation more reliable) above this depth. The major inference is that observed structures are consistent with modern tectonic processes, such as collisional orogens, crustal extension, and basin development, operating in Precambrian time. Furthermore, the fact that many Precambrian structures have persisted through time (Gibbs, 1986) is indicative of the long-term thermal and tectonic stability of the Precambrian crust.

The lower crust imaged in these early reflection profiles is characterized by a decrease in the number of reflections with depth, and the Moho is indicated only by a few weak and discontinuous reflections. These observations led to the suggestion that the lower crust of Precambrian shields is cold and brittle and has lost its reflectivity due to brittle deformation (Meissner, 1986). The lack of a distinct Moho reflection appeared to be consistent with a transitional Moho beneath Precambrian crust, possibly due to a gradual transition from mafic lower crust to an ultramafic upper mantle (Pavlenkova, 1987, 1988).

These concepts of the seismic properties of the Precambrian lower crust have undergone a significant re-evaluation in the light of recently collected profiles that replaced conventional mechanical seismic sources (vibrators) with more powerful sources (large tuned airgun arrays and dynamite shots). Deep penetrating profiles over Precambrian crust have been recorded in central Australia (Fig. 2-3a; Finlayson et al., 1989; Goleby et al., 1990), the Superior Province of Canada and the United States (Figs. 2-3b and 2-4b; Behrendt et al., 1988, 1989, 1990; Cannon et al., 1989, 1991; Geis et al., 1990; Green et al., 1989; Jackson et al., 1990), Africa (Wright and Hall, 1990) and the Baltic shield (Behrens et al., 1989; BABEL Working Group, 1990; Fig. 2-4a). The common characteristics are strong lower crustal reflectivity and Moho reflections that can be traced in a piecewise continuous fashion across all records. These features have previously been ascribed only to Phanerozoic crust (e.g., Meissner, 1986). Precambrian lower crust, like Phanerozoic crust, appears to possesses the full range of seismic reflectivity responses, ranging from transparent to highly reflective laminae. Deciphering the correlation of these Precambrian reflection patterns with tectonic processes is an area of forefront research.

The seismic reflection data of BABEL line 2 in the Gulf of Bothnia shows a strongly reflective lower crust and Moho throughout (BABEL Working Group, 1990; Fig. 2-4a). Compared to the lower crust of extensional regimes (see below), reflector lengths seem to be shorter, and reflectivity continues deeper. The termination of reflectivity at the Moho (at 40–50 km depth) agrees with the depth of refraction/wide-angle reflection Moho events. A Proterozoic compressional belt (suture zone) is interpreted in the middle of the profile (Fig. 2-4a), apparently representing an early version of plate tectonics (BABEL Working Group, 1990).

These recent Precambrian profiles are exciting new additions to our seismic constraints on crustal structure and evolution. Most Precambrian data collected to date are from Proterozoic terranes, and an outstanding question is whether Archean profiles will show similar reflectivity patterns, as suggested by Jackson et al. (1990), or will show a consistently thinner crust and some distinctive reflectivity patterns, as might be expected on the basis of seismic refraction data (Durrheim and Mooney, 1991).

3.3. Ancient orogens

We use the term ancient orogen to refer to Proterozoic to Mesozoic orogens, and therefore, there is some overlap with the previous catagory "Precambrian crust", for example, the Early Proterozoic orogen within the Baltic shield (Fig. 2-4a). In that example, and many other ancient orogens, it is significant that the evidence of a suture zone is still preserved in the lower crust and Moho. While the surface topographic expression of the orogen has long since

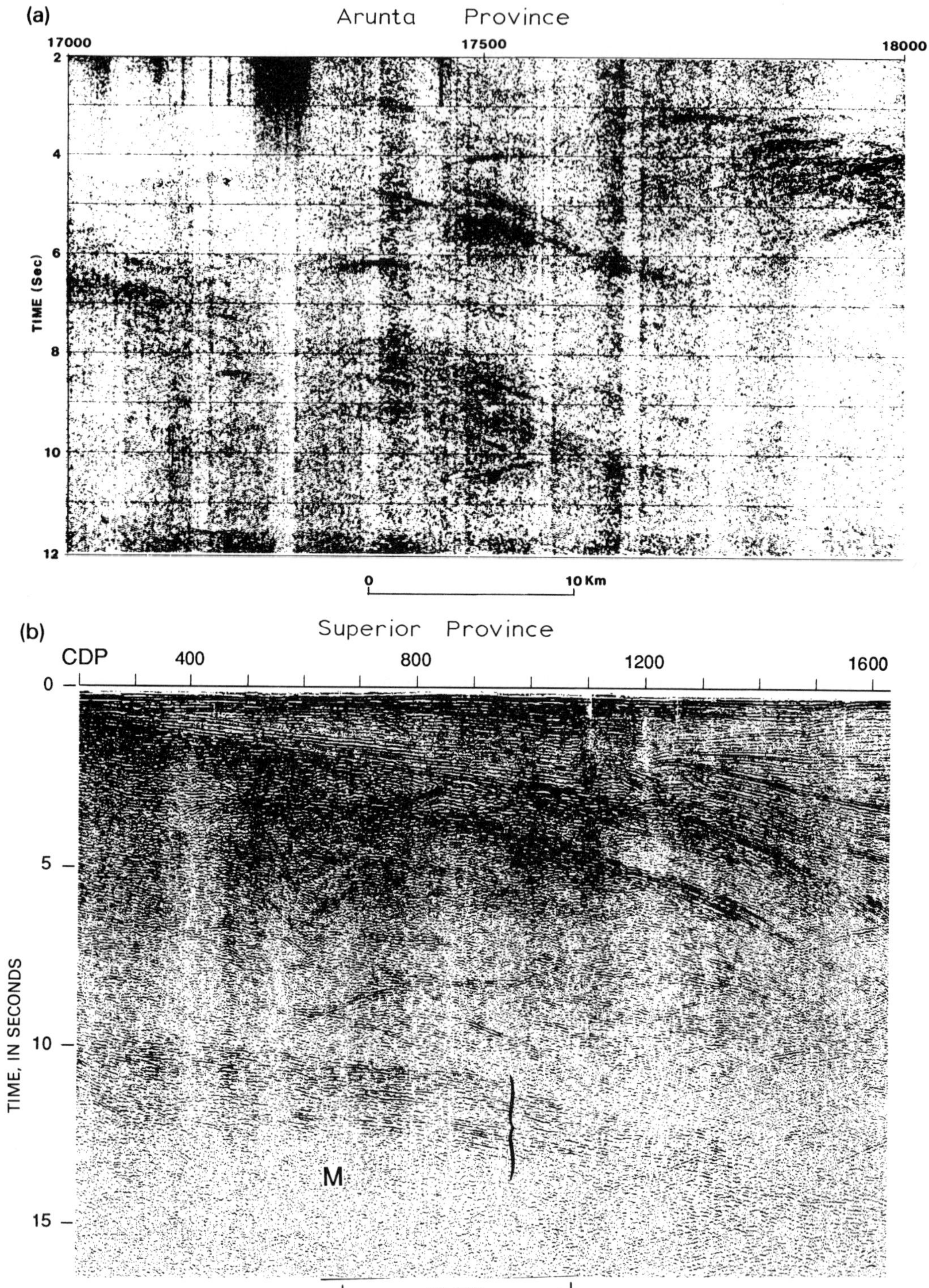

Fig. 2-3. Two examples of seismic reflection data from Precambrian crust: (A) the Arunta Province of central Australia (Goleby et al., 1990) and (B) the Superior Province of central North America (Hutchinson et al., 1991). These profiles were among the first to show strong lower crustal reflectivity in Precambrian crust, and therefore were significant in establishing the fact that a reflective lower crust and Moho (*M*) is not restricted to Phanerozoic crust. The dipping reflections beneath the Arunta block (A) appear to be related to a Precambrian collision that formed the block, and the reflectivity at 10–12 s TWT (above the Moho) beneath Lake Superior (B) may be related to rifting and igneous activity associated with the formation of the mid-continental rift.

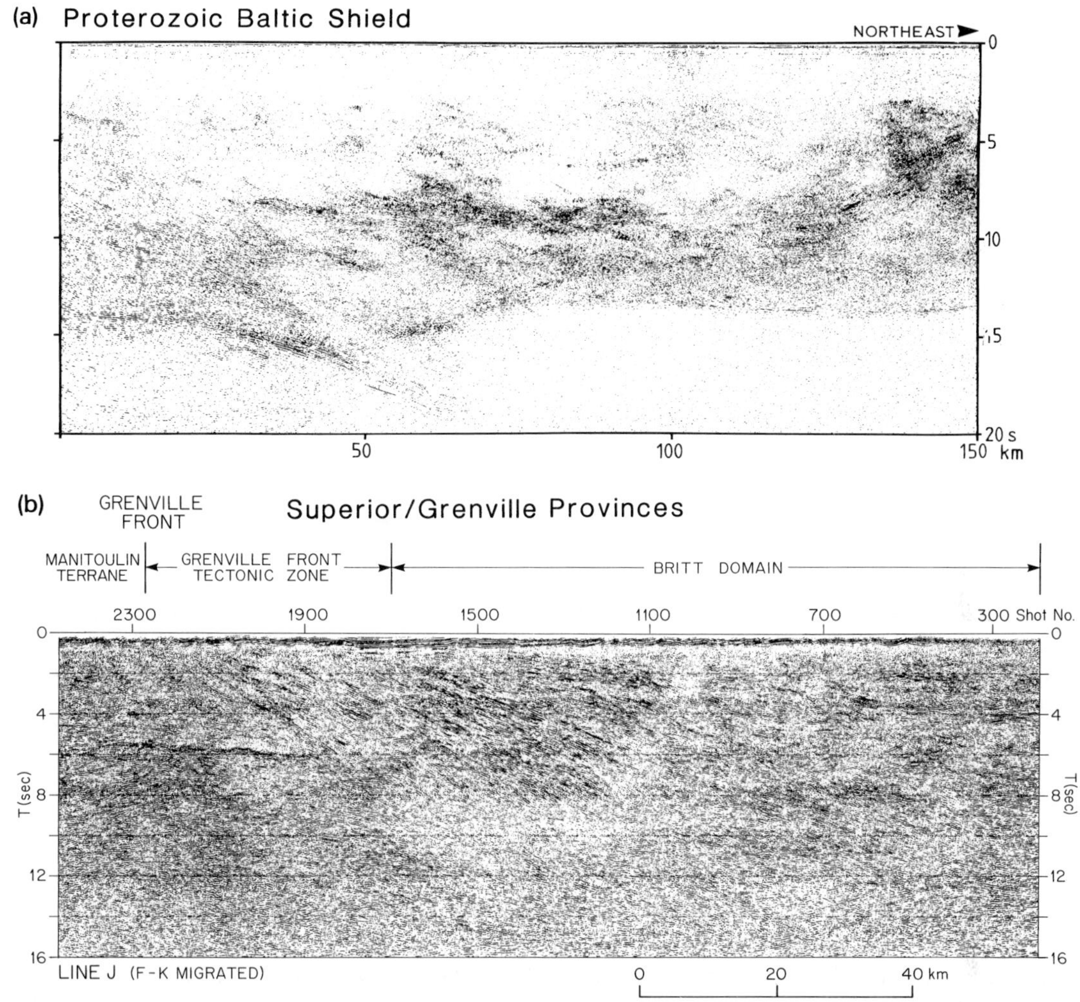

Fig. 2-4. Seismic reflection profiles from (A) the Precambrian crust of the Baltic shield (BABEL Working Group, 1990) and (B) across the Grenville Front beneath Lake Huron of North America (Green et al., 1988, 1989). The Baltic shield data show remarkably high reflectivity in the lower crust (10–15 s TWT) and a clear reflection Moho above a nearly transparent upper mantle. A Precambrian suture is evident at 50 km on the distance scale. Lower crustal reflectivity is less evident across the Grenville Front (B) due to the dominance of the distinctive dipping reflectivity pattern in the upper 8 s of the profiles. Nevertheless, sub-horizontal lower crustal and Moho reflectivity can be traced in a manner similar to other Precambrian reflection sections. The fact that the reflectivity patterns due to ancient tectonic events ("primary" patterns) are preserved beneath Precambrian crust (these examples and Fig. 2-3) attests to the long-term thermal and tectonic stability of the Precambrian crust; specifically, neither ductile flow nor magmatic intrusion from the mantle have occurred on a scale sufficient to disrupt primary lower crustal reflectivity patterns.

been eroded, a remnant of the crustal root is still present. Apparently, lower crustal creep has not eliminated the crustal root, as is sometimes the case elsewhere (see below).

A distinct dipping reflectivity pattern is seen in the upper 8 s of the crust beneath the Grenville Front in Canada (Behrendt et al., 1988; Green et al., 1988, 1989), a late Proterozoic compressional belt. Here, the reflection elements are slightly longer and stronger, especially in the upper and middle crust where evidence for thrusting is observed (Fig. 2-4b). Lower

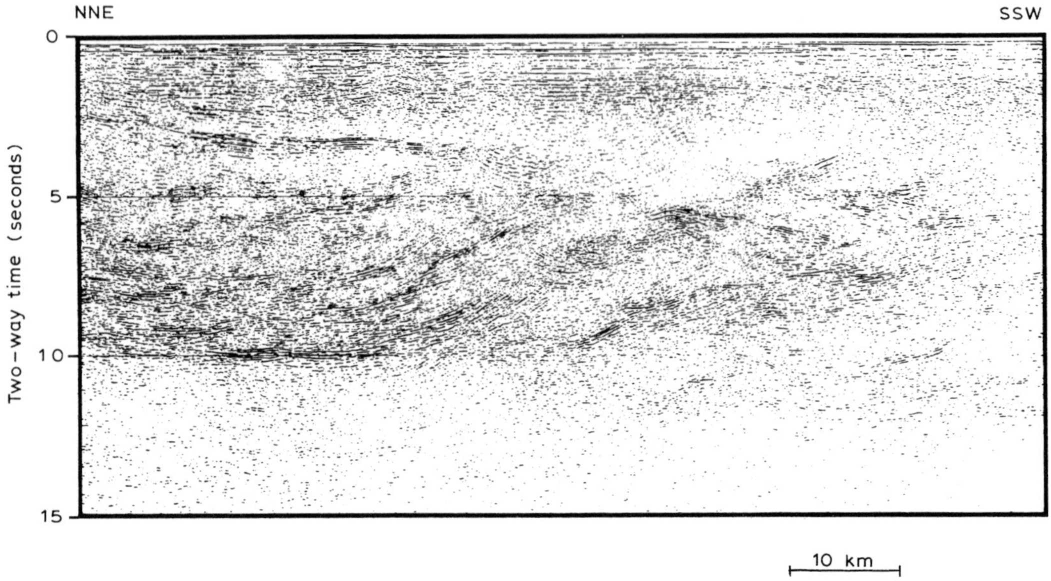

Fig. 2-5. Marine seismic reflection data recorded west of Ireland across the Iapetus suture, which is the result of the closure of the Iapetus Sea in the early Paleozoic Caledonian orogeny (Klemperer, 1989; Matthews and the BIRPS Group, 1990). The suture is marked by a series of prominent northward-dipping lower crustal reflectors that are truncated at the Moho (10 s TWT). The crustal root that presumably was formed during the orogeny (c.f., Fig. 2-4a) has been eliminated by post-orogenic extension and lower crustal ductile flow. We infer, therefore, that the observed reflectivity is multi-genetic in origin, and is due to processes associated with *both* lithospheric collision and later extension.

crustal reflectivity, aside from the dipping pattern, is less well defined but appears to be gently dipping to sub-horizontal, like much of the lower crustal reflectivity beneath the other Precambrian areas (Figs. 2-3 and 2-4a).

Reflection profiles for Phanerozoic orogenic belts show significant differences in comparison with Proterozoic belts. The Caledonian and Variscan orogenic belts of Europe have apparently lost their crustal roots and show well developed, predominantly subhorizontal (rather than dipping) lamellae in the lower crust. These are features that appear to be due to post-orogenic crustal extension. Despite this overprinting by extensional processes, some structures remain that are presumably associated with Variscan crustal shortening. A prominent example is the Iapetus suture that is evidenced in the lower crust by a wedge-like reflectivity pattern (Fig. 2-5; Klemperer and Matthews, 1987; Freeman and Klemperer, 1988; Klemperer, 1988).

The Paleozoic Appalachian orogen of North America has received much attention from reflection seismology since the first surveys reported evidence for thin-skinned megathrust tectonics (Clark et al., 1978; Cook et al., 1981). Later studies in the western Appalachians have confirmed this model and have traced a master decollement separating autochthonous Grenville basement from overlying allochthonous rocks of the Appalachian orogen (Fig. 2-6; Iverson and Smithson, 1983; Brown et al., 1986; Stewart et al., 1986; Spencer et al., 1989). In

---

Fig. 2-6. Three pairs of seismic line drawings and interpretive crustal sections for the Appalachians of eastern North America. Information is limited regarding the lower crust in the top (a) and middle (b) panels due to sparse reflectivity within Grenville basement.

(a) Southern Appalachians profile showing thin-skinned tectonics in the upper crust, and dipping reflections from shear zones beneath the Coastal Plain where multiple thrusts are interpreted (Cook et al., 1981). Grenville basement is appears to be seismically transparent in these data.

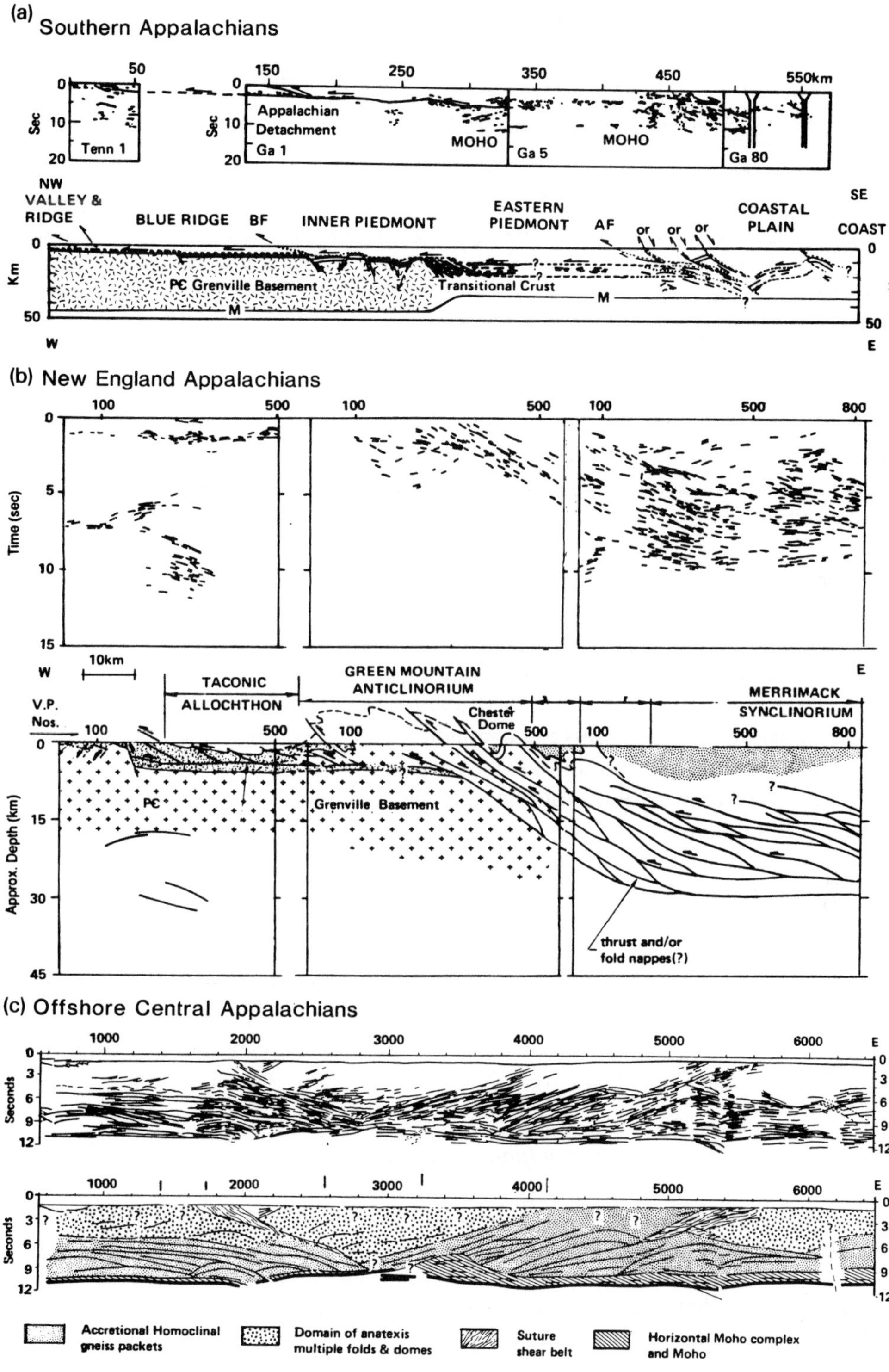

(b) New England, USA, Appalachians showing high reflectivity throughout the lower crust beneath the Merrimack Synclinorium (Brown et al., 1983). This reflectivity rises to the west and ramps over the Grenville basement beneath the Green Mountain Anticlinorium. Sporadic lower crustal reflectivity can be identified within Grenville basement, but coherent Moho reflections are not visible as are seen beneath the Grenville Province in the GLIMPCE data from Lake Huron (Fig. 2-4b) or beneath the Superior Province (Fig. 2-3b).

(c) Highly reflective lower crust is recorded beneath the offshore eastern Appalachians where superimposed orogenic shortening and later extension has deformed the lower crust and strongly enhanced reflectivity (Phinney, 1986).

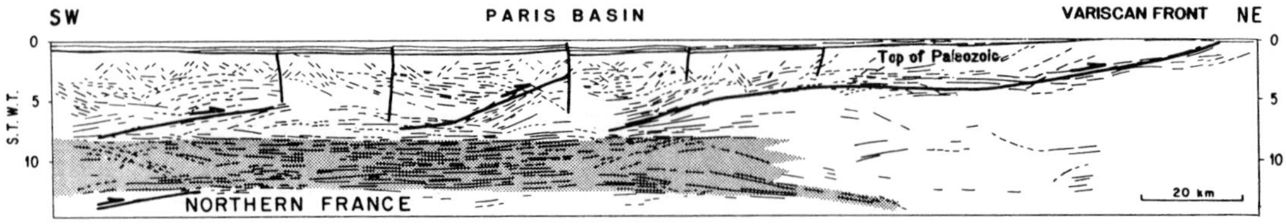

Fig. 2-7. Line drawing of seismic reflection data across the Paris Basin and Variscan Front in France (Bois et al., 1989) showing thin-skin tectonics similar to the Appalachians (Fig. 2-6), but with abundant sub-horizontal reflectivity in the lower crust below the Paris Basin. The Moho is nearly flat and a pre-existing crustal root has been eliminated by post-orogenic extension, as is also seen across the Iapetus Suture (Fig. 2-5) and numerous other regions that have experienced post-collisional extension. In contrast, a crustal root is identified beneath the younger orogens of the Alps and Pyrenees (Fig. 2-8), and the much older Baltic Shield (Fig. 2-4a).

all of these interpretations the lower crust (Grenville basement) is seismically transparent and has, to date, been interpreted as relatively undeformed by the collision. In contrast, the lower crust of the eastern Appalachians (east end of Fig. 2-6a,b) is highly reflective due to extensive deformation and shearing.

The lower crust of the eastern Appalachians, which now underlies the eastern coastal plain and shelf (passive margin) of North America, was moderately extended during the Mesozoic–Tertiary opening of the Atlantic Ocean, and is highly reflective (Fig. 2-6c). The lower crustal reflectivity patterns of the eastern Appalachians have been interpreted to be the result of a tectonic cycle of superimposed shortening and extension (Fig. 2-6c; Phinney, 1986; Hutchinson et al., 1986, 1988; Hall et al., 1989, 1990; Marillier, 1989a,b). The superposition of successive, contrasting tectonic events is a common feature of seismic reflectivity patterns and presents a challenge to the interpreter of these patterns.

The seismic reflectivity patterns of the European North Variscan Deformation Front provide evidence for thin-skinned tectonics that is similar to the Appalachians (Matte and Hirn, 1988; Fig. 2-7). Within the Variscan terranes there are abundant seismic wedges distributed over hundreds of kilometers. However, nearly all seismic profiles show a nearly flat Moho, indicating that the former crustal roots of these orogenic belts are "eroded", and the present-day Moho is apparently a young feature (Meissner et al., 1987; see Recently Extended Crust, Section 3.5). This flat Moho geometry is indicative of lower crustal creep during the post-orogenic development of these shortened orogenic belts. We assume that this crustal creep occurred during the long period of post-orogenic wrench tectonics and extension that affected the Variscan and Caledonian area (Ziegler, 1981, 1986; Bois et al., 1989).

3.4. Young (post-Mesozoic) orogens

Young orogenic belts have crustal roots that are approximately in isostatic equilibrium with their high topography. In general the crustal root is 6–8 times thicker than the mean excess elevation (e.g., Woollard, 1962; Simpson et al., 1986). The crustal root beneath the Alps is about 20 km thick, and that of the Hamalayas about 30–40 km thick.

A representative example of the seismic structure of recent orogens is provided by the Swiss Eastern Alps Traverse (Fig. 2-8). Here, the crustal root, as previously determined by a large number of wide-angle/refraction studies, is also imaged in the seismic reflection section. The lower crust dips toward the center of the root from both sides and shows seismic laminations terminating approximately at the refraction Moho. The reflectivity pattern of the

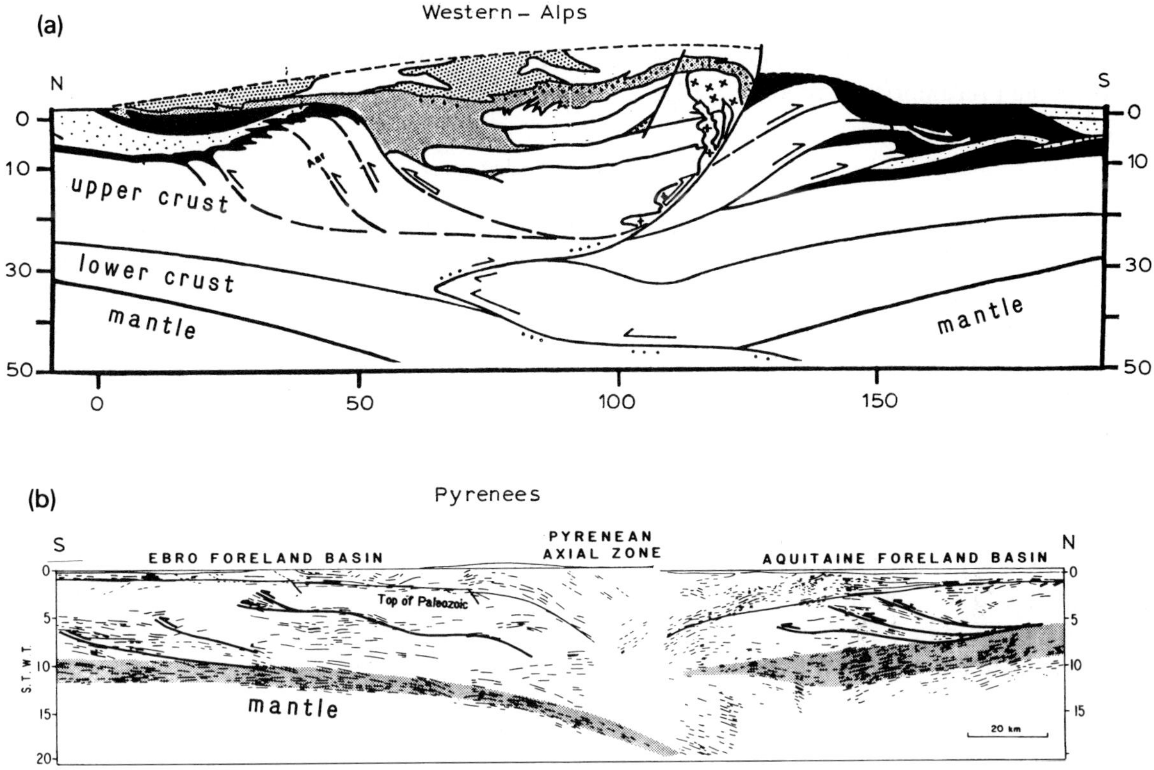

Fig. 2-8. Crustal structure of two young orogens from seismic reflection data.
(A) The Eastern Swiss Alps have a prominent crustal root as evidenced by lower crustal reflectivity (not shown) that dips strongly toward the center of the root (Pfiffner et al., 1988). Significantly, the smaller calculated volume of material within the crustal root as compared with the estimated amount of crustal shortening implies that some crustal material has been delaminated beneath the orogen.
(B) Line drawing of seismic reflection data across the Pyrenees showing northward lithospheric convergence at a moderate dip (Choukroune and the ECORS Team, 1989). Reflectivity can be followed to greater than 20 s TWT (more than 60 km), which suggests that lower crustal delamination also occurred beneath this orogen.

lower crust may be the result of deformation along multiple shear zones. The pronounced crustal shortening suggests that at least some lower crustal material might be delaminated beneath the deepest part of the root (England and Houseman, 1989). The upper and middle crust, on the other hand, show a more complex reflectivity pattern that is suggestive of crustal wedging, with nappe displacements above these wedges (Pfiffner et al., 1988; Schweizerische Arbeitsgruppe fuer Reflexionsseismik, 1988).

A deep seismic profile across the Pyrenees imaged, for the first time, the entire crust of this orogen, and provided the basis of a palinspastic restoration of the collision zone (Fig. 2-8b; ECORS Pyrenees Team, 1988; Choukroune and the ECORS Team, 1989; Roure et al., 1989). The seismic data indicate about 100 km of north–south shortening between the Iberian and European plates and confirm a Moho offset (deeper to the south) of 15 km. The crust within the axial zone of the collision belt has been thickened by at least 15 km by thick-skinned stacking of whole-crustal flakes (Fig. 2-8b). This profile is significant for its confirmation of both a large Moho offset and the thick-skinned deformation of the Iberian crust. Whereas thin-skinned tectonics is commonly observed in the Paleozoic orogens discussed above, there is excellent evidence for thick-skinned crustal shortening and crustal thickening for these late-Cenozoic orogenic belts. Overall, the style of deformation is very similar to the examples from the Precambrian (Section 3.2; Figs. 2-3 and 2-4).

3.5. Recently extended crust

Perhaps no lower crustal seismic reflection data are more remarkable than the profiles recorded in areas where the last tectonic event was extensional, such as much of western Europe, the Basin and Range province of the United States, and many passive margins. In these areas, the lower crust (below a depth of about 10–14 km) is consistently highly reflective, in sharp contrast to the relatively transparent upper crust and upper mantle (e.g., Fig. 2-1). As mentioned above, the reflection Moho is usually nearly flat, because the extensional process eliminates crustal roots below pre-extensional orogenies. Upper crustal faults, where they can be traced into the lower crust, are listric at depth and merge into the reflective lower crust.

Some of the best examples of laminated lower crust and the other reflection patterns described above are from the extended crust around the British Isles (Matthews, 1986; BIRPS and ECORS, 1986; Le Gall, 1990; Fig. 2-9). The reflectivity pattern is variable, sometimes filling the lower crust nearly uniformly, and other times concentrating into two or more narrow zones, or forming a single narrow zone of reflections close to the Moho (Blundell, 1990). The Moho is remarkably flat, and there are several instances of upper mantle reflections, both dipping and flat-lying (Warner and McGeary, 1987).

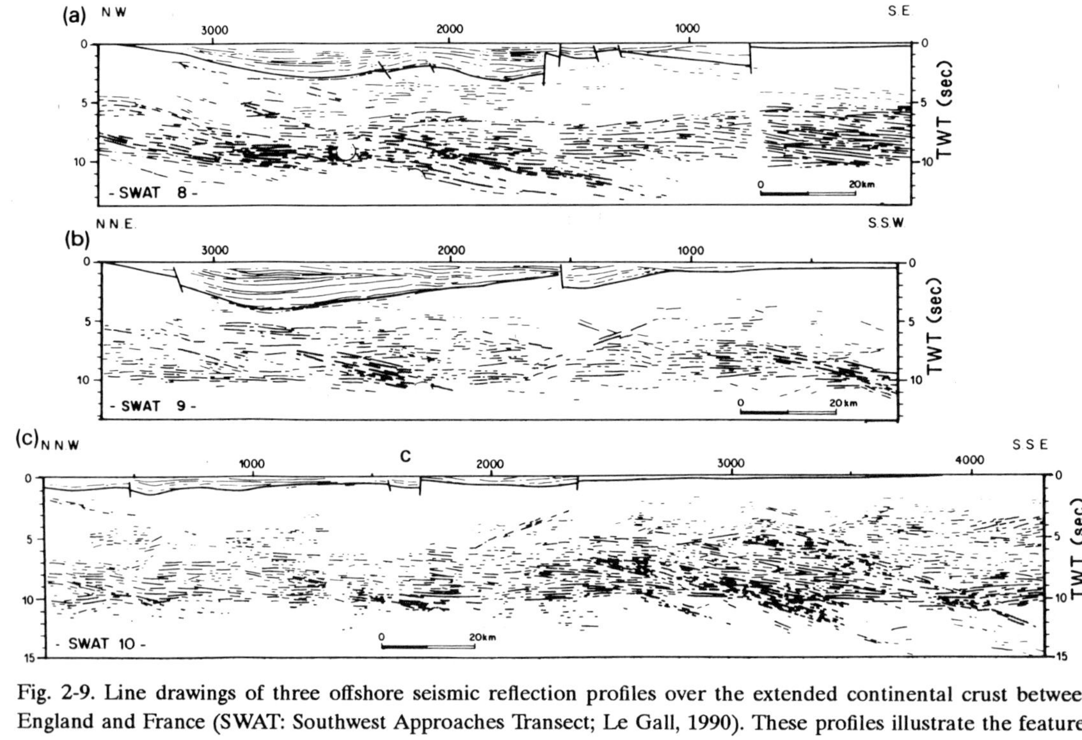

Fig. 2-9. Line drawings of three offshore seismic reflection profiles over the extended continental crust between England and France (SWAT: Southwest Approaches Transect; Le Gall, 1990). These profiles illustrate the features that characterize seismic reflection profiles in extended crust: significant seismic reflectivity concentrated within the lower crust (and in supracrustal sedimentary basins); a predominantly sub-horizontal reflector geometry and a clear reflection Moho; some dipping upper mantle reflections. The reflectivity of the lower crust is enhanced by ductile stretching during extension (see section 2.5).

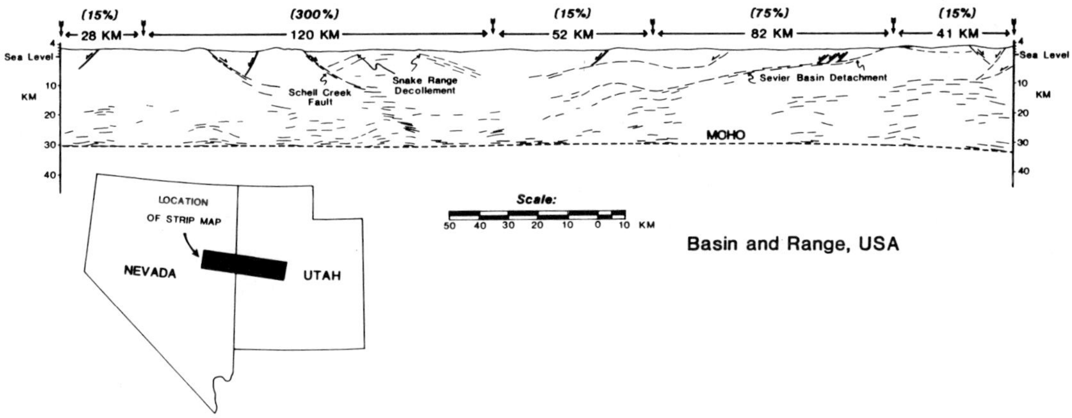

Fig. 2-10. Line drawing of seismic reflection data from the Basin and Range province of the western USA illustrating a flat Moho over a distance of 323 km, and beneath terrains with crustal extension estimated to vary from 15% to 300% (Gans, 1987). The important implication is that lower crustal flow has compensated for the laterally varying amounts of extension, with ductile flow into the areas with the largest amount of extension. The seismic sections from the Basin and Range show high lower crustal reflectivity and are similar to the data illustrated in Figures 1, 9 and 11, which also were recorded over extended crust.

The ECORS Nord de la France line (Fig. 2-7; Cazes et al., 1986) and DEKORP's lines in the Black Forest (Fig. 2-1; Lueschen et al., 1987) also show a pronounced lower crustal lamination. The four lines in the Black Forest are especially interesting because this region lies on the shoulders of the Tertiary Rhinegraben and has experienced recent uplift but little crustal extension. We speculate that both the lower crustal laminae and the uplift of the Rhinegraben were caused by minor extension, together with intrusion of magmas (from the mantle) into the lower crust. Hence, the laminae seem to represent a two-phase process: first, the intrusion of mafic magma and, second, an ordering process (lateral shearing) which makes reflectors near-horizontal and therefore readily observable on seismic reflection profiles. This process is discussed further in Section 5, below.

The Basin and Range province of the western United States underwent Cenozoic extension that generated listric and high-angle normal faults that are imposed on a crust that was shortened during a Mesozoic orogenic event. Similar to the examples cited above, this extended crust shows a reflective lower crust and a remarkably planar reflection Moho (Fig. 2-10; Klemperer et al., 1986). It appears almost certain that the Moho in the Basin and Range, and other recently-extended crust, is a young feature that formed during extension. The processes forming the Moho would include the ductile stretching of recent igneous intrusions and the metamorphism and deformation of lower crustal rocks. Basin bounding normal faults can not be traced through the lower crust and do not appear to offset the Moho. A possible exception is a profile in Arizona which has been interpreted as showing a Moho offset (Hauser et al., 1987), although this interpretation has been disputed (Goodwin and Thompson, 1988).

3.6. The Moho

The Moho is defined as the boundary (or transition zone) where the compressional wave seismic velocity, as measured by seismic refraction data, increases to greater than or equal to 7.6 km/s (Steinhart, 1967). Sub-Moho seismic velocities of 8.0 ± 0.2 km/s are typical, and the upper mantle is generally interpreted to have an ultramafic bulk composition (cf., references in Holbrook et al., this volume, but see also Snyder and Flack, 1990, for a recent discussion of eclogite in the upper mantle). Seismic reflection observations over the past fifteen years have resulted in a major reevaluation of the nature of the Moho (cf. Jarchow and Thompson, 1989). Two observations were particularly critical in this reevaluation.

First, the Moho was expected to provide one of the strongest near-vertical reflections (after correcting for wave spreading losses) because seismic refraction models commonly show large seismic velocity steps (0.5–1.5 km/s) at the Moho. In practice, the reflection Moho has proved to be a highly variable and, at times, elusive feature. Mid-crustal reflections are often of higher amplitude than the Moho, and, as a result, the reflection Moho has come to be defined as the deepest set of reflections before the nearly-complete die out of reflectivity at the top of the upper mantle. In addition, since seismic refraction data established the Moho as a universal feature of the lithosphere, it was expected that Moho reflections would be observed nearly continuously across seismic reflection sections. In fact, the reflection Moho is generally a 1–2-s-wide zone of reflectivity that can only be traced laterally in a piecewise fashion, and less often as a very narrow, nearly continuous horizon.

Second, many crustal models derived from seismic refraction and gravity data showed considerable relief on the Moho across major physiographic features, and it was therefore

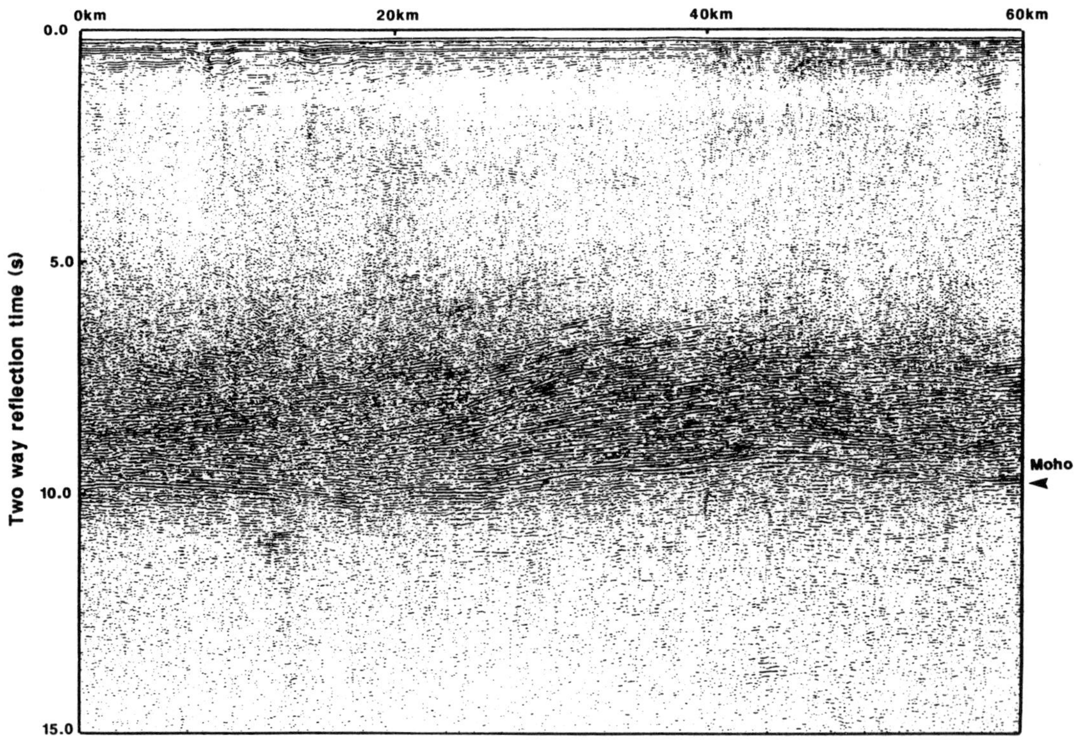

Fig. 2-11. Seismic reflection profile recorded in shallow water offshore of Britain showing a highly reflective lower crust and relatively transparent upper crust and mantle (Warner, 1990). This remarkable seismic section illustrates the 10-km-and-longer continuity of individul reflections, as well as the strength of the reflections (spherical divergence and trace equalization have been applied). Similar seismic sections have been obtained in other regions of extended continental crust. The origin of this reflectivity is discussed in sections 2.4 and 2.5, and in Figs. 2-12, 2-14 and 2-15.

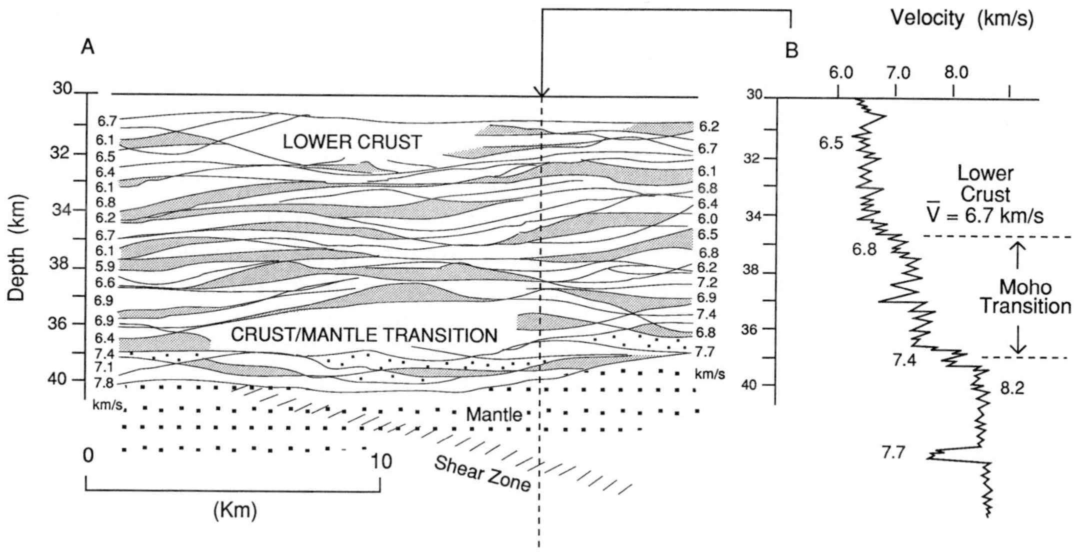

Fig. 2-12. (A) Model for the lower crust and Moho that consists of random, anastomosing, high- and low-velocity lamellae. These lamellae (or laminae) are variously estimated to be 400 ± 300 m thick and show random velocity variations of ±0.05–0.5 km/s with respect to the mean lower crustal velocity (e.g., Wenzel et al., 1987; Sandmeier et al., 1987). Seismic lamellae most probably have their primary origin in lithologic and metamorphic layering, including that caused by igneous intrusions and shear zones. In many cases, the reflectivity of lamellae (e.g., Figs. 2-1 and 2-11) are enhanced by ductile stretching of the lower crust (section 2.5).
(B) Sketch of a representative one-dimensional velocity–depth profile through model (A).

expected that seismic reflection profiles would confirm and perhaps sharpen the definition of this high Moho relief. Unexpectedly, many reflection profiles showed the Moho to be nearly horizontal over large distances, despite the presence of significant togography and/or structure at the surface. Examples include profiles across either rugged topography and/or sedimentary basins, such as the Basin and Range, USA (Fig. 2-10), Variscan Europe (Fig. 2-10), and the continental shelf of the British Isles (Fig. 2-11). Profiles that confirmed high relief, and even offsets and/or overlapping of the Moho (double Mohos) include the Alps and Pyrenees profiles (Fig. 2-8).

Several new insights have been derived from these observations, some of which have been alluded to in the previous sections. The characteristic of the reflection Moho as a gradual die out in reflectivity that can usually only be followed in a piecewise continuous manner is best explained by a model of the crust in which the Moho is a 3–5-km-wide transition zone consisting, like the lower crust, of alternating thin, random high- and low-velocity layers (seismic laminae; Fig. 2-12). These laminae are anastomosing rather than continuous and flat-lying, with the result that there is both constructive and destructive interference that increases and decreases reflection amplitudes (Braile and Chang, 1986). Since the entire lower crust, including the Moho, is laminated with thin, random velocity layers, the Moho may or may not be the highest amplitude reflection, and often can only be identified by the transition to the transparent upper mantle.

The flat Moho reflection geometry was unexpected in several areas where previous seismic reflection results indicated significant Moho relief. As discussed above, the most likely explanation for the lack of relief is that the Moho has been reformed by igneous intrusions and ductile deformation in the lower crust (Klemperer et al., 1986). This explanation implies that the Moho transition zone is often the youngest "layer" within the continental crust, a concept that is contrary to conventional stratigraphic succession (younger over older).

Where older seismic refraction interpretations were incompatible with modern seismic reflection observations, newer coincident seismic refraction data confirm that the Moho is in fact measured at the same depth on the two data sets (e.g., Mooney and Brocher, 1987; Holbrook, 1990; Catchings and Mooney, 1991). In all cases where modern data are available, the seismic refraction and reflection Mohos appear to correspond in depth to within the resolution of the techniques. However, a few discrepancies remain. The Moho depth beneath the Colorado Plateau, U.S., is reported as 42 km from older seismic reflection data and as 50 km from modern seismic reflection data (Hauser and Lundy, 1989). Modern seismic refraction measurements could clarify this apparent discrepancy.

4. Multi-genetic origin of lower crustal reflections

The examples described in the previous sections document the diversity in lower crustal reflectivity patterns for different tectonic settings and age provinces. This diversity is indicative of a similar variabliity in the origin of crustal reflectivity (Smithson, 1986, 1989). In this section we discuss the major hypotheses for the origin of lower crustal reflectivity in the context of these and other examples, and in the context of laboratory measurements of lower crustal properties. We separate this discussion into instances for which there exists direct borehole or outcrop evidence for the origin of seismic reflectors and those instances that lack this direct evidence.

4.1. Origin of reflectivity: concepts with direct borehole or outcrop evidence

In several instances, the origin of upper crustal reflectivity has been established either by drilling into a reflection horizon or by following it to outcrop. Since these upper crustal examples are also relevant to the lower crust, we discuss these well-determined cases first.

Reflectivity caused by igneous intrusives has been documented by drill-hole and field mapping studies. A seismic reflection profile across the Siljan meteorite impact structure, Sweden, revealed a series of discrete, near-horizontal, upper crustal reflections (Juhlin and Pedersen, 1987). Subsequent deep drilling established that these reflections were due to dolerite sheets (V_p = 6.6 km/s) within granitic country rock (V_p = 6.0 km/s) (Juhlin, 1988). The individual sheets ranged in thickness from about 4 to 60 m. Similar thin igneous sheets are possible in the lower crust and would cause high reflectivity.

The reflectivity of a high-grade metamorphic terrane in the southern Appalachians, U.S. (Coruh et al., 1987) was investigated with laboratory measurements on cores from a borehole (Christensen 1989a). The lithologies in the study area are significant in that they were once part of the middle crust and are typical of many high-grade metamorphic terranes throughout the world. Computer modelling of compressional-wave velocity and density data determined in the laboratory demonstrate that fine-scale layering of these metamorphic rocks can produce reflection amplitudes typical of lower crustal events. The results reported by Christensen (1989a) emphasize three additional observations: (1) anisotropy in metamorphic rocks can *decrease* vertical-incidence impedance contrasts at lithologic contacts and hence, reflection amplitudes; (2) lower crustal seismic bright spots, which are commonly attributed to high pore

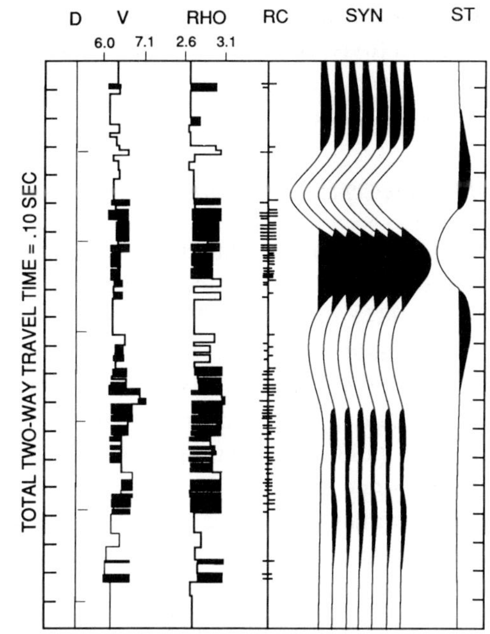

A CORES AT 600 MPs

Fig. 2-13. Reflection originating from a zone of alternating high- and low-velocities that demonstrates that multiple thin layers can produce significant reflections (Christensen, 1989). These thin layers may have multiple causes, such as layering of lithology, metamorphic grade, or seismic anisotropy. Seismic "bright spots" need not be due to thick crustal layers. Key: *D* = depth scale in 50 meter intervals; *V* = compressional-wave velocity in km/s; *RHO* = density in gm/cc; *RC* = reflection coefficient; *SYN* = synthetic seismogram; *ST* = standard wavelet corresponding to a reflection coefficient of 0.1, is shown for amplitude comparison.

pressure or partial melting (e.g., Brown et al., 1980, 1986), can also be induced by fine tuning of thin metamorphic layers (Fig. 2-13); and (3) the seismic properties of laminated, thin layers *within* geologic units may be more important for the generation of deep crustal near-vertical reflections than the contacts between major mapped geologic units. The latter conclusion is contrary to the common notion that seismic reflection data provide an approximate cross section of deep-crustal geology.

In a few instances, seismic reflections from fault zones have been traced to outcrops. The high reflectivity observed for some deeply penetrating fault zones (e.g., Smithson et al., 1979; Brewer et al., 1983) can be caused either by the juxtaposition of different rock types across the fault or by the physical characteristics within the fault zone itself. The latter, crustal reflections from mylonitic shear zones, have been investigated by several workers. Laboratory measurements and computer modelling demonstrate that preferred mineral orientations and the presence of retrograde mineral assemblages can result in lower seismic velocities (for propagation directions normal to mylonite foliation within fault zones) (Jones and Nur, 1984; Fountain et al., 1984). The correlation of high upper crustal reflectivity with mylonite zones was established by tracing this reflectivity to the outcrop of mylonites (Hurich et al., 1985; McDonough and Fountain, 1988; Christensen and Szymanski, 1988; Wang et al., 1989; Calvert and Clowes, 1990). The structural and petrologic characteristics of these mylonite zones indicate that they are representative of mid- to lower crustal mylonite zones. These studies demonstrate that fault zone reflectivity does not result simply from reflections from the top and bottom of the fault, but that reflections with the highest amplitudes can originate from within the fault zone. These reflections are caused by internal fault zone structure and a complex interaction of lithologic layering. These layers may often have varying amounts of seismic anisotropy resulting from ductile strain.

4.2. Origin of reflectivity: concepts without direct borehole or outcrop evidence

Upper crustal reflections that are similar in character to those beneath the Siljan structure that were confirmed to be due to dolerite sheets (see above) were recorded in southwestern Arizona, USA (Hauser et al., 1987) and may also be due to igneous sheets that are mapped nearby. Although no drill-hole confirmation is available, these reflectors have been successfully modeled by Goodwin et al. (1989) as thin, high-velocity layers based on the phase (positive polarity) and amplitudes observed on three-component reflection data. The modeling results of Goodwin et al. (1989) are incompatible with an origin of the reflections from water-saturated shear zones, since these would be associated with a negative reflection polarity. These and other similar studies emphasize the importance of igneous intrusions as one origin for crustal reflectivity (Dohr and Meissner, 1975; Oliver et al., 1976; McCarthy and Thompson, 1988; Warner, 1990).

The possible origin of crustal reflections by fluid layers in the deep crust has found support from geologic (e.g., Fyfe et al., 1978) and geoelectrical conductivity data (e.g., Shankland and Ander, 1983). The evidence for deep crustal fluids is summarized by Wickham (this volume) and Jones (this volume). More recently, a number of workers have suggested that lower crustal reflectivity is related to layers with stratified porosity contrasts (Matthews, 1986; Gough, 1986; Jones, 1987; Klemperer, 1987; Hyndman, 1988; Hyndman and Shearer, 1989). Hyndman and Shearer (1989) argue that reasonable physical conditions permit the existence of 1–4% pore water with a salinity close to that of sea-water in the lower crust. They note the common coincidence of high lower-crustal electrical conductivity with high seismic reflectivity, and suggest that this correspondence is indicative of a common origin. In their model, layers with porosity contrasts occur primarily in greenschist and lower amphibolite facies rocks where resistivites of 20–30 Ω m and high reflectivity are observed. They correlate low reflectivity to the absence of pore water; examples include much of the upper crust (where pore water has escaped by flow through fractures), and the lower crust of shields (presumed to be composed of dry granulite facies rocks). Hyndman and Shearer (1989) cite the BIRPS data from the continental shelf around the British Isles as an example of high lower crustal reflectivity that may be associated with high porosity.

An alternate view of the origin of the high lower-crustal reflectivity is expressed by Warner (1990). He summarizes the geologic and geophysical evidence for and against layered lower crustal porosity, and concludes that there are two factors that are difficult to reconcile with a fluid origin for crustal reflectivity: (1) large quantities (2–4%) of pore fluids needed, and (2) long fluid residence times in the lower crust seem unreasonable given the prevailing physical conditions. He concludes that igneous intrusions and underplating of the lower crust with mantle-derived mafic rocks provide a more suitable general explanation for lower crustal reflectivity. Non-fluid causes of high crustal conductivity are summarized in Jones (this volume).

In the model of Hyndman and Shearer (1989), layered lower crustal porosity contrasts result in layers with high and low seismic velocities. However, we note that either high porosity and moderate pore pressure or, in the case of lower (2–4%) porosity, very high pore pressure is needed to produce a significant reduction in seismic velocity in the lower crust (Christensen, 1989b). Geologic evidence would appear to contradict the hypothesis of high porosity within the high-grade metamorphic and igneous rocks that probably are common within the lower crust. Furthermore, very high pore pressures would be difficult to maintain for long times. Therefore, large velocity contrasts associated with layers with stratified porosity contrasts seem physically implausible.

An additional piece of evidence that weighs against abundant stratified porosity contrasts are estimates of the seismic attenuation (inverse Q) in the lower crust. Qualitatively, the recording of high-frequency (60 Hz and higher) lower crustal and Moho reflections argues for high Q. Quantitative Q estimates have large error bars, but nearly all indicate lower crustal compressional-wave Q values that are moderately high (400–1000; Hwang and Mooney, 1986; Hobbs et al., 1990; Benz et al., 1990). These observations would appear to preclude porosity layers of significant cumulative thickness (greater than several kilometers), assuming that high-porosity layers would have low Q values (10–50).

The recent seismic reflection profiles reported from Proterozoic shields provide a test of the hypothesized correlation between high crustal reflectivity and conductivity. While the lower crustal reflectivity of recently-reported shield profiles is comparable to that reported in Phanerozoic areas (Section 3.2), the conductivity of shields is commonly (but not always) several orders of magnitude lower than the regions cited by Hyndman and Shearer (1989). Thus, there does not appear to be a widespread correlation between lower crustal reflectivity and conductivity in Proterozoic crust. Furthermore, some of the reflectivity patterns in Proterozoic crust appear to date from ancient collisions; a fluid origin for this reflectivity would require either very long residence times, or fluid recharge along pre-existing shear zones, neither of which seems plausible. Therefore, it is unlikely that lower crustal fluids contribute to the reflectivity of Proterozoic crust.

The high lower crustal reflectivity observed in passive margins, in the Basin and Range of the United States, and in Variscan Europe led to the suggestion that reflectivity is related to the process of extension (Phinney and Jurdy, 1979; Matthews and Cheadle, 1986; Smithson, 1986; Allmendinger et al., 1987; McCarthy and Thompson, 1988; Reston, 1988). Although the proposed mechanisms differ in detail, they all incorporate ductile flow in the lower crust to align minerals and stretch lower crustal bodies into a subhorizontal geometry. The resultant structure is reflective due to compositional, metamorphic, and mineralogical layering. The laboratory and borehole data summarized in the previous section support these ideas. The additional possibility that high strain zones are overpressured with fluids (Matthews and Cheadle, 1986) is a special and probably rare case, as discussed above. We expand on the evidence for the enhancement of lower crustal reflectivity by ductile flow in Section 5, below.

In a number of studies, realistic models of the lower crust have been constructed from geologic cross-sections, and velocities and densities have been estimated for these cross-sections based on laboratory and/or borehole measurements (Hale and Thompson, 1982; Fountain et al., 1984; McDonough and Fountain, 1987; Sandmeier et al., 1987; Christensen, 1989a; Fountain and Christensen, 1989). These studies demonstrate that compositional layering by common crustal rocks is an important cause of lower crustal reflectivity.

In summary, several causes of crustal reflectivity have been directly confirmed by drill-hole or outcrop evidence. These causes are igneous intrusions, lithologic and metamorphic layering, and reflectivity from mylonite zones. Additional evidence shows that compositional layering by common rock types contributes to lower crustal reflectivity, and that lower crustal ductile flow enhances reflectivity by aligning minerals and stretching lower crustal bodies into a subhorizontal geometry. For the reasons outlined above, stratified layers with porosity contrasts are likely to be the exception, rather than the rule, as a source of crustal reflectivity.

5. Reflectivity and viscosity in the lower crust

The previous sections presented evidence that lower crustal reflectivity is multi-genetic, and illustrated that the tectonic history of the crust is strongly correlated with the observed reflectivity patterns. In this section we summarize the evidence that ductile deformation in the lower crust acts to enhance lower crustal reflectivity. This ductile deformation may be ancient, as in Precambrian suture zones, or young, as in recently extended crust. We seek to explain the widespread occurrence of a reflective lower crust by considering three lines of evidence for the existence (and nature) of a lower crustal ordering process that enhances crustal reflectivity. The three primary constraints are: (1) the geometry of lower crustal reflections; (2) the evidence for a global stress system; and (3) the rheology of the lower crust. We consider each of these in turn.

5.1. Geometry of lower crustal reflections

Crustal reflectors (or closely-spaced reflector segments) must have a minimum length in order to achieve constructive interference at the surface. This length is about 5 km in the lower crust, provided the interface is planar and not curved. In warm and young lower crusts, mostly rather long, planar, and possibly slightly anastomosing reflectors are present in the lower crust (Section 3.4; Figs. 2-1, 2-9, and 2-11). The widespread, regional appearance of the laminated lower crust and the lack of a consistent correlation with higher seismic velocities (greater than 6.8 km/s) appears to rule out an origin of reflectivity solely from extensive magmatic intrusions. For example, the lower crust of volcanic areas like the Eifel, Germany, do *not* show a significant increase in seismic lamination, as would be expected if igneous intrusion were the primary cause (DEKORP Res. Group, 1991). Rifted and unrifted areas in the North Sea have the same seismic signatures (Klemperer, 1988). Furthermore, magmatic intrusions should produce many diffractions, as are recorded across the Mid-Continental Rift system (Brown et al., 1982), rather than the planar reflections that are commonly observed elsewhere. While magmatic intrusions certainly play a role in causing lower crustal reflections (cf., section 4.1), a way to explain the wide-spread presence of lower crustal lamellae is an ordering process due to regional or global stress system (Zoback et al., 1990). This ordering process would include an ordering of magmatic intrusions.

5.2. Stress-strain evidence for major displacements in the lower crust

Several observations argue for major displacements in a ductile lower crust. The first observation is the existence of large regional stress systems in the upper crust of the continents, as determined by a number of measurements in North America (Zoback and Zoback, 1989), Europe and other parts of the world (Zoback et al., 1990). These large-scale stress systems are related to plate boundary forces that must also affect the lower crust, and can apparently be transferred over distances of more than 1000 km.

The second observation relates to movements of the upper crust relative to the lower crust and subcrustal lithosphere. The mapping of crustal detachment faults that often become listric in the middle or lower crust is among the outstanding achievements of crustal reflection studies (Cook et al, 1981; Allmendinger et al., 1983; Matthews, 1986). The presence of a flat Moho beneath an upper crust that has been extended by an amount that varies locally (ranging from 15 to 300% in the Basin and Range, U.S.A.) implies that brittle upper crustal deformation is decoupled from a more uniform deformation in the middle and lower crust. Another relevant observation is that of tectonic rotation of upper crustal blocks (Matte, 1986; Nur et al., 1989). These block rotations can be modeled by stress concentrations in the rigid upper crust above a weak, low viscosity lower crust (Kusznir, 1982; Kusznir and Park, 1987; Strehlau and Meissner, 1987).

5.3. Rheological control of seismic reflectors

There is a growing body of evidence that the lower continental crust must be a rheologically weak layer between the upper, rigid crust and the subcrustal lithosphere (Sibson, 1982; Meissner and Strehlau, 1982; Kirby and Kronenberg, 1987). Maximum rock strength in the upper crust as calculated from Byerlee's general relationship (Byerlee, 1987) increases with depth (z), but is drastically reduced by the onset of creep, generally at the base of the upper crust between 8 and 15 km depth. Creep may give rise to regional scale flow of material within the lower crust beneath the rigid mid-crust, thereby providing isostatic compensation for domal core complexes (exposed mid-crustal rocks) and maintaining a flat Moho beneath these complexes (Gans, 1987; Block and Royden, 1990). Figure 2-14 shows a calculated viscosity–depth curve for Phanerozoic areas with moderate heat flow. This curve is idealized, and abrupt changes can only be real in the case of an equally abrupt change of rock composition.

The minima in viscosity–depth curves are of particular importance to studies of seismic reflectivity. In the case of young crust these minima coincide with the seismic lamellae of the lower crust, especially for the broad viscosity-minimum of the lowermost crust (Meissner and Kusznir, 1987). Figure 2-14 shows a conceptual picture of this correlation between crustal viscosity and reflectivity. This correlation suggests that lamellae are enhanced by ductile shear in the lower crust. Mooney and Brocher (1987) have shown that the onset of lower crust lamination does not correlate globally with any specific intracrustal velocity boundaries. Thus, crustal velocities (and by inference, composition) seem to be secondary to crustal viscosity for the formation of seismic lamellae.

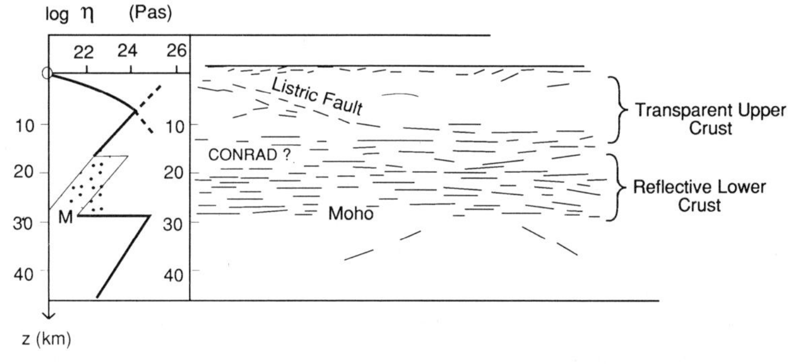

Fig. 2-14. Conceptual picture of the relationship between crustal viscosity and reflectivity that emphasizes the enhancement of lower crustal lamellae (e.g., Figs. 2-1 and 2-11) by ductile shear. Key: M = Moho. See discussion in section 2.5.

While global plate stresses may be important in accounting for the nearly ubiquitous observation of lower crustal reflectivity, stresses at collisional and extensional boundaries are clearly equally important, given that nearly all continental crust has experienced some convergent and/or extensional tectonism at some point. Thus, lower crustal reflectivity may be classified as *primary* if it is associated with early crust forming processes, or *secondary* if it is associated with later ductile deformation or metamorphism.

5.4. Seismically transparent lower crust and mixed reflectivity patterns

We have argued that the global stress system will enhance lower crustal reflectivity nearly everywhere by inducing lower crustal ductile flow that produces a subhorizontal ordering. One may therefore ask why planar seismic lamellae are not present everywhere in the lower crust, that is, what is the origin of seismically transparent lower crust, where it occurs? We first note that crustal stress does not necessarily imply crustal strain. The fact that some Precambrian sutures preserve their reflectivity patterns (Section 3.2) indicates that the crust cooled and has not undergone lower crustal ductile deformation for almost 2 Ga. Much of the observed reflectivity is therefore primary, that is, it is related to deformation and shear that occurred during the suturing of micro-plates to form the Precambrian shields. Second, depending on composition and mineralogy, the lower crust may be more or less likely to form thin metamorphic layers and mylonitic shear zones that enhance reflectivity. Third, either vertically oriented igneous intrusions, or deep faulting could disrupt lamallae after they have formed. A final factor that has already been mentioned (section 2.3) are data aquisition problems that can degrade the quality of deep reflection data, at times to the point that there are few coherent reflectors.

Mixed reflectivity patterns were discussed above (section 3.5) and are indicative of a rheologically stratified crust. Some pre-Cenozoic shortened orogenic belts display both upper crustal wedging and interfingering patterns, and lower crustal lamellae that indicate ductile stretching. Examples include the Basin and Range province (Allmendinger et al., 1983), the eastern Appalachians (Keen et al., 1986; Marillier, 1989a,b; Hall et al., 1989, 1990), and the Variscan crust of western Europe (DEKORP Res. Group, 1990). A Mesozoic age for Variscan lower crustal laminations can be estimated by noting that they have obliterated Paleozoic thrusts and faults at depth, but have themselves been offset by Tertiary deformation (Bois et al., 1989). The determination of the chronology of seismic reflectivity patterns provides a major constraint on the structural evolution of the crust (e.g., Klemperer et al., 1990).

6. Conclusions

Seismic reflection profiles collected over the past 15 years range in tectonic setting from stable Precambrian shields to actively extending areas and young collision zones. These profiles have provided a variety of new insights into the structure and evolution of the continental lower crust and Moho. Here we summarize some of the key insights; the appropriate primary references are cited in the previous sections. We arrange this discussion into three principle topics: the structural development of the lower crust in various tectonic settings, the nature of the Moho, and the origin of lower crustal reflectivity.

Prominent insights regarding the structural development of the crust include the demonstration of the mechanics of crustal shortening in either a thick- or thin-skinned fashion. The southern Appalachians (Fig. 2-6a) and parts of the Variscan front (Fig. 2-7) illustrate thin-skinned thrusting that extends for 100 km and more. Thick-skinned tectonics, which involves deformation of the lower crust, is seen beneath the New England Appalachians (Fig. 2-6b) and is implied by the depth extent of the Iapetus suture (Fig. 2-5). Seismic profiles across the Alps and the Pyrenees (Fig. 2-8) both show significant crustal shortening, with the lower crust deforming and thickening beneath the high topography. Significantly, the lower crust may delaminate beneath some collision belts, thereby creating a less mafic orogenic crust.

Recent profiles within Precambrian crust have identified lower crustal reflectivity patterns that are consistent with modern tectonic processes, including collisional orogenies (Figs. 2-3 and 2-4). The data permit the extension of plate tectonic models into the Early Proterozoic with greater confidence. The fact that ancient reflectivity patterns remain "frozen in" the lower crust attests to the tectonic and thermal stability of the Precambrian crust over the past 1–2 Ga. Expanded deep reflection profiling of Archean crust remains a scientific frontier.

Perhaps the most well known seismic reflection profiles are those recorded over extended continental crust that show striking sub-horizontal lamellae throughout the lower crust (Figs. 2-1, 2-9 and 2-11). Well-studied examples include the crust around the British Isles, the Basin and Range, USA, and much of western Europe. Normal faults generally become listric with depth, and anastomosing shear zones appear to predominate in the lower crust (Fig. 2-11). The stretching of the lower crust has eliminated pre-extensional crustal roots beneath these terranes, and the flat Moho relief, despite laterally varying amounts of extension (15–300%) is indicative of lower crustal flow to equalize crustal thickness.

Mixed reflectivity patterns are also common, and are indicative of successive tectonic episodes that have superposed their structural signatures on the crust. For example, the eastern Appalachians of North American and Variscan Europe preserve evidence for Paleozoic crustal shortening in the upper crust and Mesozoic ductile stretching in the lower crust. The superposition of crustal reflectivity patterns can lead to interpretational ambiguities, but also provides a rich source of information regarding the tectonic history of the crust. The dating of diverse seismic reflectivity patterns, generally by geometrical relationships, is an important goal for seismic reflection studies.

Seismic reflection profiles show great diversity in the structure of the Moho. Recently recorded airgun and explosion source profiles have provided the best images to date of the Moho beneath Precambrian shields. In these profiles the lower crust and Moho are moderately-to-highly reflective, with horizontal to gently dipping reflections predominating, especially away from ancient collision zones (Figs. 2-3 and 2-4). In young orogenic zones, the Moho dips steeply downward from either side of the mountain belt (e.g., the Alps and the Pyrenees; Fig. 2-8), and the Mohos of opposing convergent lithopsheric plates are offset, with either a Moho step, or a "double Moho" beneath a limited portion of the mountain belt.

The fine-scale properties of the Moho have been investigated as well, and seismic reflection data indicate that the Moho is commonly a 3–5-km-wide transition that, like the lower crust above it, consists of anastomosing thin layers with randomly alternating high and low velocities (Fig. 2-12). Despite the significant velocity increase (0.5–1.5 km/s) from the lower crust to the mantle, the Moho reflection is often not the strongest event in a seismic section because the fine structure of reflective layers is more important than the bulk velocity change. Thus, middle or lower crustal laminations may produce stronger reflections than the crust/mantle transition.

The Moho is a young, newly formed feature in areas of highly extended crust, and thus the principle of stratigraphic succession, younger over older, is violated in the crystalline continental crust and Moho. As mentioned above, the nearly flat geometry of the Moho beneath highly variably-extended crust implies that there is lateral flow in the lower crust to preserve a nearly uniform crustal thickness.

Various hypotheses have been advanced for the origin of lower crustal reflectivity, and we have argued that this reflectivity is multi-genetic, a point that is unambiguously demonstrated by the correlation of reflection profiles with boreholes and by tracing reflection horizons to outcrop. Causes of crustal reflectivity that have been demonstrated by such direct evidence

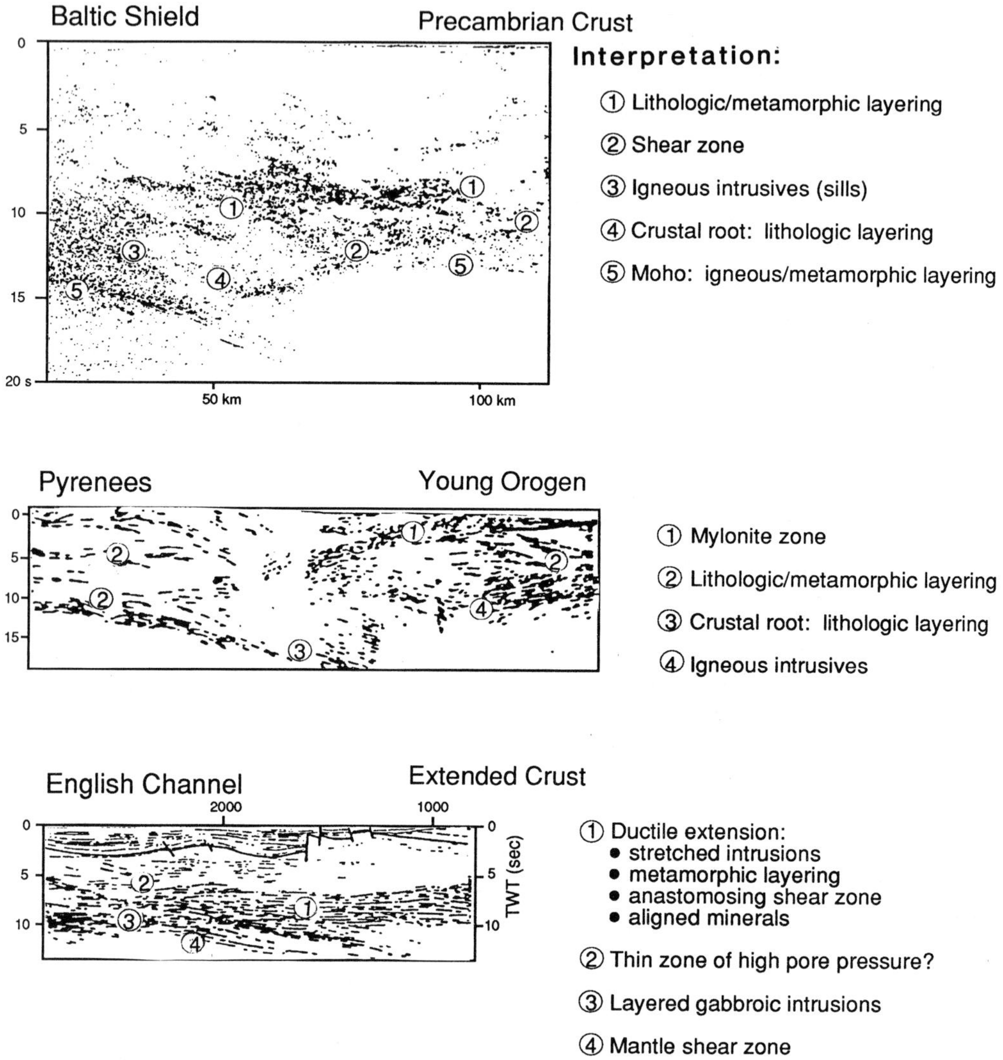

Fig. 2-15. Examples of reflective lower crust from three diverse geologic settings that emphasize the multi-genetic origin of crustal reflectivity. Interpretations of selected reflective zones (numbers within circles) are based on the evidence presented in sections 2.4 and 2.5. However, these interpretations are only working hypotheses, there being no drill-hole confirmation available for these deep-crustal examples. The most common causes of crustal reflectivity are indicated in the top and middle examples (lithologic/metamorphic layering, shear zones, etc.), and we refer to these as "primary" causes; extended crust nearly always shows enhanced lower crustal reflectivity due to ductile flow, which we term a "secondary" cause. See text for discussion.

includes igneous intrusions (sills), fine-scaled metamorphic layering, mylonite zones, and lithologic layering.

Recent reflection profiles in Precambrian shields reveal strong crustal reflectivity, including sub-horizontal laminae in the lower crust, thereby upsetting an apparent correlation between lower crustal lamination and crustal extension. Many of the Precambrian reflectivity patterns appear to preserve structures that date from ancient collisional events, which implies that this reflectivity is *not* due to post-orogenic ductile shear, igneous intrusions, or fluids. Rather, reflectivity is due to primary lithologic and metamorphic layering, and Precambrian shear zones that were formed during the compressional orogenies that formed the Precambrian crust (Fig. 2-15). Similar mechanisms can account for reflective lower crust in young orogens.

In extended crust we favor a model that is depth-dependent, with ductile shear enhancing reflectivity that has its primary origin in lithologic and metamorphic layering in the lower crust, and igneous layering within the 3–5-km-thick Moho transition zone (Fig. 2-15). Layered zones of high pore pressure (up to several kilometers thick and due to metamorphic dewatering of the lower crust) may sometimes be present at the top of the lower crust, as indicated by low seismic refraction velocities and high electrical conductivity.

The general observation of lower crustal reflectivity leads to the suggestion that a common process acts to promote reflectivity. We favor a model whereby regional and global plate stresses act to enhance reflectivity by inducing lower crustal ductile flow that produces subhorizontal lamination. This ordering process would enhance reflectivity that has its primary cause in igneous intrusions, or compositional, metamorphic, or mineralogic layering (Fig. 2-15). This ordering process requires elevated temperatures to promote ductile flow, and the dipping reflection patterns in Precambrian terranes appear to be primary patterns associated with Proterozoic micro-plate collisions, rather than secondary reflectivity associated with ductile flow. Thus, the Precambrian crust appears to possess considerable long-term thermal and tectonic stability.

Deep seismic reflection exploration of the crystalline crust has yielded an abundance of insights into the continental lower crust and Moho. Future insights will come from expanded geographic coverage, particularly in Proterozoic and Archean crust, the application to shear-wave profiling, and from continued correlation with related geologic/geophysical studies, and borehole and laboratory studies of the physical properties of rocks.

Acknowledgments

Discussions with numerous colleagues over the past several years have shaped and clarified our ideas. We acknowledge support from the USGS Deep Continental Studies Program (WDM), the Continental Dynamics Program of the US NSF, and the German Research Society (D.F.G.) which has support DEKORP and related activties (RM). Review comments by B. Beaudoin, R. D. Catchings, N. I. Christensen, and D. M. Fountain are greatly appreciated, as is the assistance with figure preparation by T. Williams. S.L. Klemperer and A.G. Green kindly provided copies of reflection data for Figure 2-4.

References

Allmendinger, R.W., and 6 others, 1983. Cenozoic and Mesozoicstructures of the eastern Basin and Range province, Utah, from COCORP seismic-reflection data. Geology, 11: 532–536.

Allmendinger, R., and 6 others, 1986. Phanerozoic tectonics ofthe Basin and Range–Colorado Plateau transition from COCORP data and geologic data, a review. In: M. Barazangi and L.D. Brown (Editors), Reflection Seismology: The Continental Crust. Am. Geophys. Union, Geodyn. Ser., 14: 257–267.

Allmendinger, R.W., and 7 others, 1987. Overview of the COCORP40° transect, western United States; The fabric of an orogenic belt. Geol. Soc. Am. Bull., 98: 308–319.

BABEL Working Group, 1990. Evidence for early Proterozoic platetectonics from seismic reflection profiles in the Baltic shield. Nature, 348: 34–38.

Barazangi, M. and Brown, L.D., 1986a. Reflection Seismology; A Global Perspective. Am. Geophys. Union, Geodyn. Ser., 13: 311 pp.

Barazangi, M. and Brown, L.D. (Editors), 1986b. Reflection Seismology; The Continental Crust. Am. Geophys. Union, Geodyn. Ser., 14: 339 pp.

Bartelsen, H., Lueschen, E., Krey, Th., Meissner, R., Scholl, J. and Walter, Ch., 1982. The combined seismic reflection-refraction investigation of the Urach Geothermal Anomaly. In: R. Haenel (Editor), The Urach Geothermal Project. Schweizerbart, Stuttgart, pp. 247–262.

Behrendt, J.C., Green, A.G., Cannon, W.F., Hutchinson, D.R., Lee, M.W. Milkereit, B., Agena, W.F. and Spencer, C., 1988. Crustal structure of the Midcontinent Rift System: results from GLIMPCE deep seismic reflection profiles. Geology, 16: 81–85.

Behrendt, J.C., Green, A.G., Lee, M.W., Hutchinson, D.R., Cannon, W.F., Milkereit, B., Agena, W.F. and Spencer, C., 1989. Crustal extension in the midcontinent rift system — results from Glimpce deep seismic reflection profiles over Lakes Superior and Michigan. In: R.F. Mereu, S. Mueller and D.M. Fountain (Editors), Properties and Processes of Earth's Lower Crust. Am. Geophys. Union, Geophys. Monogr., 51: 81–89.

Behrendt, J.C., Hutchinson, D.R., Lee, M., Thornber, C.R., Trehu, A., Cannon, W. and Green, A., 1990. GLIMPCE, seismic reflection evidence of deep-crustal and upper-mantle intrusions and magmatic underplating associated with the Midcontinent Rift System of North America. In: J.H. Leven, D.M. Finlayson, C. Wright, J.C. Dooley and B.L.N. Kennett (Editors), Seismic Probing of Continents and Their Margins. Tectonophysics, 173: 595–615.

Behrens, K., Goldflam, S., Heikkinen, P., Hirschleber, H., Lindguist, G. and Lund, C.E., 1989. Reflection seismic measurements across the Granulite Belt of the POLAR Profile in the northern Baltic Shield, North Finland. Tectonophysics, 162: 101–111.

Benz, H.M., Smith, R.B. and Mooney, W.M., 1990. Crustal structure of the northwestern Basin and Range Province from the 1986 Program for Array Seismic Studies of the Continental Lithosphere seismic experiment. J. Geophys. Res., 95: 21,823–21,842.

BIRPS and ECORS, 1986. Deep seismic reflection profiling between England, France and Ireland. J. Geol. Soc. London, 143: 45–52.

Block, L.B. and Royden, L.H., 1990. Core complex geometriesand regional scale flow in the lower crust. Tectonics, 9: 557–567.

Blundell, D.J., 1990. Seismic images of continental lithosphere. J. Geol. Soc. London, 147: 895–913.

Bois, C. Cazes M., Damotte, B., Galdeano, A., Hirn, A., Mascle,A., Matte, P., Raoult, J.F. and Torreilles, G., 1986. Deep seismic profiling of the crust in northern France: The ECORS project. In: M. Barazangi and L.D. Brown (Editors), Reflection Seismology: A Global Perspective. Am. Geophys. Union, Geodyn. Ser., 13: 21–30.

Bois, C., Pinet, B. and Roure, F., 1989. Dating lower crustalfeatures in France and adjacent areas from deep seismic profiles. In: R.F. Mereu, S. Mueller and D.M. Fountain (Editors), Properties and Processes of Earth's Lower Crust. Am. Geophys. Union, Geophys. Monogr., 51: 17–31.

Braile, L. and Chiang, C.S., 1986. The continental Mohorovicic discontinuity; Results from near-vertical and wide-angle seismic refraction studies. In: M. Barazangi and L.D. Brown (Editors), Reflection Seismology: A Global Perspective. Am. Geophys. Union, Geodyn. Ser., 13: 257–272.

Brewer, J.A., Matthews, D.H., Warner, M.R., Hall, J.R., Smythe, D.K. and Whittington, R.J., 1983. BIRPS deep seismic reflection studies of the British Caledonides. Nature, 305: 206–210.

Brocher, T.M. and Hart, P., 1991. Comparison of vibroseis and explosive source methods for deep crustal seismic reflection profiling in the Basin and Range Province. J. Geophys. Res., 96: 18,197–18,213.

Brown, L.D., Chapin, C.E., Sanford, A.R., Kaufman, S. and Oliver, J., 1980. Deep structure of the Rio Grande Rift from seismic reflection profiling. J. Geophys. Res., 85: 4773–4800.

Brown, L.D., Jensen, L., Oliver, J., Kaufman, S. and Steiner, D., 1982. Rift structure beneath the Michigan Basin from COCORP profiling. Geology, 10: 645–649.

Brown, L.D., Ando, C., Klemperer, S., Oliver, J., Kaufman, S., Czuchra, B., Walsh, T. and Isachsen, Y.W., 1983. Adirondack–Appalachian crustal structure: the COCORP northeast traverse. Geol. Soc. Am. Bull., 94: 1173–1184.

Brown, L.D., Barazangi, M., Kaufman, S. and Oliver, J., 1986. The first decade of COCORP; 1974–1984. In: M. Barazangi and L.D. Brown (Editors), Reflection Seismology: A Global Perspective. Am. Geophys. Union, Geodyn. Ser., 13: 107–120.

Calvert, A.J. and Clowes, R.M., 1990. Deep high-amplitude reflections from a major shear zone above the subducting Juan de Fuca plate. Geology, 18: 1091–1094.

Cannon, W.F., and eleven others, 1989. The North American midcontinent rift beneath Lake Superior from GLIMPCE seismic reflection profiling. Tectonics, 8: 305–332.

Cannon, W.F., Lee, M.W., Hinze, W.J., Schulz, K.J. and Green, A.G., 1991. Deep crustal structure of the Precambrian basement beneath northern Lake Michigan, midcontinent North America. Geology, 19: 207–210.

Catchings, R.C. and Mooney, W.D., 1991. Basin and Range crustal and upper mantle structure, northwest to central Nevada. J. Geophys. Res., 96: 6,247–6,267.

Cazes, M., Mascle, A. and Scientific Party, 1986. Large Variscan overthrusts beneath the Paris Basin. Nature, 323: 144–147.

Choukroune, P. and the ECORS Pyrenees Team, 1989. The ECORS Pyrenean deep seismic profile. Reflection data and the overall structure of an orogenic belt. Tectonics, 8: 23–39.

Choukroune, P., 1990. Main results of the ECORS Pyrenees profile In: J.H. Leven, D.M. Finlayson, C. Wright, J.C. Dooley and B.L.N. Kennett (Editors), Seismic Probing of Continents and Their Margins. Tectonophysics, 173: 411–423.

Christensen, N.I. and Szymanski, D.I., 1988. Origin of reflections from the Breward fault zone. J. Geophys. Res., 93: 1087–1102.

Christensen, N.I., 1989a. Reflectivity and seismic properties of the deep continental crust. J. Geophys. Res., 94: 17,793–17,804.

Christensen, N.I., 1989b. Pore pressure, seismic velocities, and crustal structure. In: L.C. Pakiser and W.D. Mooney (Editors), Geophysical Framework of the Continental United States. Geol. Soc. Am. Mem., 172: 783–798.

Claerbout, J.F., 1985. Imaging the Earth's Interior. Blackwell Scientific Publications, Oxford, 398 pp.

Clark, H., Costain, J. and Glover, L., III, 1978. Structure and seismic reflection studies on the Brevard ductile deformation zone near Rosman, North Carolina. Am. J. Sci., 278: 419–441.

Clowes, R.M., Kanasewich, E.R. and Cumming, G.L., 1968. Deep crustal seismic reflections at near vertical incidence. Geophysics, 33: 441–451.

Clowes, R.M., Brandon, M., Green, A., Yorath, C. and Sutherland-Brown, A., 1987. LITHOPROBE — Southern Vancouver Island: Cenozoic subduction complex imaged by deep seismic reflections. Can. J. Earth Sci., 24: 31–51.

Cook, F.A., Brown, L.D., Kaufman, S., Oliver, J.E. and Petersen, 1981. COCORP seismic profiling of the Appalachian orogen beneath the coastal plain of Georgia. Geol. Soc. Am. Bull., Part 1, 92: 738–748.

Coruh, C., Costain, J.K., Hatcher, R.D., Jr., Pratt, T.L., Williams, R.T. and Phinney, R.A., 1987. Results from regional vibroseis profiling: Appalachian ultradeep core hole site study. Geophys. J. R. Astron. Soc., 89: 147–155.

Dahl-Jensen, T., Dyrelius, D., Juhlin, C., Palm, H. and Pedersen, L.B., 1987. Deep reflection seismics in the Precambrian of Sweden. R. Astron. Soc. Geophys. J., 89: 371–378.

DEKORP Res. Group, represented by Meissner, R., Wever, Th. and Sadowiak, P., 1990b. Reflectivity patterns in the Variscan mountain belts and adjacent areas: an attempt for pattern recognition and correlation to tectonic units. In: J.H. Leven, D.M. Finlayson, C. Wright, J.C. Dooley and B.L.N. Kennett (Editors), Seismic Probing of Continents and their Margins. Tectonophysics, 173: 361–378.

DEKORP Res. Group, 1991. Results of the DEKORP 1 (BELCORP-DEKORP) deep seismic reflection studies in the western part of the Rhenish Massif. Geophys. J. Int., 106: 203–227.

Dooley, J.C. and Moss, F.J., 1988. Deep crustal reflections in Australia 1957–1973 - II. Crustal models. Geophys. J. R. Astron. Soc., 93: 239–249.

Dohr, G.P. and Meissner, R., 1975. Deep crustal reflections in Europe. Geophysics, 40: 25–39.

Durrheim, R.J. and Mooney, W.D., 1991. Archean and Proterozoic crustal evolution: evidence from crustal seismology. Geology, 19: x–x.

ECORS Pyrenees Team, 1988. The ECORS deep reflection seismic survey across the Pyrenees. Nature, 331: 508–511.

England, P. and Houseman, G., 1989. Extension during continental convergence with applications to the Tibetan Plateau. J. Geophys. Res., 94: 15,561–15,575.

Finlayson, S.M., Leven, J.H. and Wake-Dyster, K.D., 1989. Large-scale lenticles in the lower crust under an intra-continental basin in eastern Australia. In: R.F. Mereu, S. Mueller and D.M. Fountain (Editors), Properties and Processes of Earth's Lower Crust. Am. Geophys. Union, Geophys. Monogr., 51: 3–16.

Fountain, D.M. and Christensen, N.I., 1989. Composition of the continent crust and upper mantle — a review. In: L.C. Pakiser and W.D. Mooney (Editors), Geophysical Framework of the Continental United States. Geol. Soc. Am. Mem., 172: 711–742.

Fountain, D.M., Hurich, C.A. and Smithson, S.B., 1984. Seismic reflectivity of mylonite zones in the crust. Geology, 12: 195–198.

Freeman, B., Klemperer, S.L. and Hobbs, R.W., 1988. The deep structure of northern England and the Iapetus Suture zone from BIRPS deep seismic reflection profiles. J. Geol. Soc. London, 145: 727–740.

Fuchs, K., 1969. On the properties of deep crustal reflections. J. Geophys., 35: 133–149.

Fyfe, W., Price, R. and Thompson, A., 1978. Fluids in the Earth's Crust. Elsevier, Amsterdam, 383 pp.

Gans, P.G., 1987. An open-system, two-layer crustal stretching model for the eastern Great Basin. Tectonics, 6: 1–12.

Geis, W.T., Cook, F.A., Green, A.G., Milkereit, B., Percival, J.A. and West, G.F., 1990. Thin thrust sheet formation of the Kapuskasing structural zone revealed by LITHOPROBE seismic reflection data. Geology, 18: 513–516.

Gibbs, A.K., 1986. Seismic reflection profiles of Precambrian crust: a qualitative assessment. In: M. Barazangi and L.D. Brown (Editors), Reflection Seismology: The Continental Crust. Am. Geophys. Union, Geodyn. Ser., 14: 95–106.

Gibson, B.S. and Levander, A.R., 1988. Modeling and processing of scattered waves in seismic reflection surveys. Geophysics, 53: 466–478.

Gibson, B.S. and Levander, A.R., 1990. Apparent layering in common-midpoint stacked images of two-dimensionally heterogeneous targets. Geophysics, 55: 1466–1477.

Glocke, A. and Meissner, R., 1976. Near-vertical reflections recorded at the wide-angle profile in the Rhenish Massif. In: P. Giese, C. Prodehl and A. Stein (Editors), Explosion Seismology in Central Europe. Springer-Verlag, Berlin, pp. 252–256.

Goleby, B.R., Kennett, B.L.., Wright, C., Shaw, R.D. and Lambeck, K., 1990. Seismic reflection profiling in the Proterozoic Arunta Block, central Australia: processing for testing models of tectonic evolution. In: J.H. Leven, D.M. Finlayson, C. Wright, J.C. Dooley and B.L.N. Kennett (Editors), Seismic Probing of Continents and their Margins. Tectonophysics, 173: 257–268.

Goodwin, E.B. and McCarthy, J., 1990. Composition of the lower crust in west central Arizona from three-component seismic data. J. Geophys. Res., 95: 20,097–20,109.

Goodwin, E.B. and Thompson, G.A., 1988. The seismically reflective crust beneath highly extended terranes: Evidence for its origin in extension. Geol. Soc. Am. Bull., 100: 1616–1626.

Goodwin, E.B., Thompson, G.A. and Okaya, D.A., 1989. Seismic identification of basement reflectors: the Bagdad reflection sequence in the Basin and Range Province–Colorado Plateau Transition Zone, Arizona. Tectonics, 8: 821–831.

Gough, D.I., 1986. Seismic reflectors, conductivity, water and stress in the continental crust. Nature, 323: 143–144.

Green, A. G., and nine others, 1988. Crustal structure of the Grenville front and adjacent terranes. Geology, 16: 788–792.

Green, A., Cannon, W., Milkereit, B., Hutchinson, D., Davidson, A., Behrendt, J., Spencer, C., Lee, M., Morel-a-Hussier, P. and Agena, W., 1989. A "Glimpce" of the deep crust beneath the Great Lakes. In: R.F. Mereu, S. Mueller and D.M. Fountain (Editors), Properties and Processes of Earth's Lower Crust. Am. Geophys. Union, Geophys. Monogr., 51: 65–80.

Hale, L.D. and Thompson, G.A., 1982. The seismic reflection character of the continental Mohorovicic discontinuity. J. Geophys. Res., 87: 4525–4635.

Hall, J., Quinlan, G., Wright, J., Keen, C. and Marillier, F., 1989. Styles of deformation observed in deep seismic reflection profiles of the Appalachian Caledonide system. In: R.F. Mereu, S. Mueller and D.M. Fountain (Editors), Properties and Processes of Earth's Lower Crust. Am. Geophys. Union, Geophys. Monogr., 51: 33–43.

Hall, J., Quinlan, G., Marillier, F. and Keen, C., 1990. Dipping shear zones and the base of the crust in the Appalachians, offshore Canada. In: J.H. Leven, D.M. Finlayson, C. Wright, J.C. Dooley and B.L.N. Kennett

(Editors), Seismic Probing of Continents and their Margins. Tectonophysics, 173: 581–593.

Hauser, E.C. and Lundy, J., 1989. COCORP deep reflections: Moho at 50 km (16 s) beneath the Colorado Plateau. J. Geophys. Res., 94: 7071–7081.

Hauser, E.C., Gephart, J., Latham, T., Oliver, J., Kaufman, S., Brown, L. and Lucchitta, I., 1987. COCORP Arizona transect: Strong crustal reflections and offset Moho beneath the transition zone. Geology, 15: 1103–1106.

Holbrook, W.S., 1990. The crustal structure of the northwestern Basin and Range Province, Nevada, from wide-angle seismic data. J. Geophys. Res., 95: 21,843–21,869.

Hubbard, S.S., Coruh, C. and Costain, J.K., 1991. Paleozoic and Grenvillian structures in the southern Appalachians: extended interpretation of seismic reflection data. Tectonics, 10: 141–170.

Hurich, C.A., Smithson, S.B., Fountain, D.M. and Humphreys, M.C., 1985. Seismic evidence of mylonite reflectivity and deep structure in the Kettle Dome metamorphic core complex, Washington. Geology, 13: 577–580.

Hurich, A.A. and Smithson, S.B., 1987. Compositional variation and the origin of deep crustal reflections. Earth Planet. Sci. Lett., 85: 416–426.

Hutchinson, D.R., Grow, J.A., Klitgord, K.D. and Detrick, R.S., 1986. Moho reflections from the Long Island platform, eastern United States. In: M. Barazangi and L.D. Brown (Editors), Reflection Seismology: The Continental Crust. Am. Geophys. Union, Geodyn. Ser., 14: 95–106.

Hutchinson, D.R., Klitgord, K.D., Lee, M.W. and Trehu, A.M., 1988. U.S. Geological Survey deep seismic reflection profile across the Gulf of Maine. Geol. Soc. Am. Bull., 100: 172–184.

Hyndman, R.D., 1988. Dipping seismic reflectors, electrical conductive zones, and trapped water in the crust over a subducting plate. J. Geophys. Res., 93: 13,391–13,405.

Hyndman, R.D. and Shearer, P.M., 1989. Water in the lower continental crust: modelling magnetotelluric and seismic reflection results. Geophys. J., 98: 343–365.

Hwang, L.J. and Mooney, W.D., 1986. Velocity and Q structureof the central Valley, California, based on synthetic seismogram modelling of seismic refraction data. Bull. Seismol. Soc. Am., 76: 1053–1067.

Iverson, W.P. and Smithson, S.B., 1983. Reprocessing and reinterpreting of COCORP southern Appalachian profiles. Earth Planet. Sci. Lett., 62: 75–90.

Jackson, S. L., Sutcliffe, R.H., Ludden, J.N., Hubert, C., Green, A.G., Milkereit, B., Mayrand, L., West, G.F. and Verpaelst, P., 1990. Southern Abitibi greenstone belt: Archean crustal structure from seismic-reflection profiles. Geology, 18: 1086–1090.

Jarchow, C.M. and Thompson, G.A., 1989. The nature of the Mohorovicic discontinuity. Annu. Rev. Earth Planet. Sci., 17: 475–506.

Johnson, R.A. and Smithson, S.B., 1986. Processing crustal reflection data. In: M. Barazangi and L.D. Brown (Editors), Relection Seismology: A Global Perspective. Am. Geophys. Union, Geodyn. Ser., 13: 197–208.

Jones, A.G., 1987. MT and seismic reflection: an essential combination. Geophys. J. R. Astron. Soc., 89: 7–18.

Jones, A.G., 1992. Electrical conductivity of the lower continental crust. In: D.M. Fountaim, R. Arculus and R.M. Kay (Editors), Continental Lower Crust. Elsevier, Amsterdam, this volume.

Jones, T.D. and Nur, A., 1984. The nature of seismic reflections from deep crustal fault zones. J. Geophys. Res., 89: 3153–3171.

Juhlin, C., 1988. Interpretation of seismic reflectors in the Grarberg-1 well. In: A. Boden and N.N. Eriksson (Editors), Deep Drilling in Crystalline Bedrock. Springer-Verlag, Berlin, pp. 113–121.

Juhlin, C. and Pedersen, L.B., 1987. Reflection seismic investigations of the Siljan Impact structure, Sweden. J. Geophys., Res., 92: 14,113–14,122.

Junger, A., 1957. Deep basement reflections in Big Horn County, Montana. Geophysics, 16: 449–505.

Kanasewich, E.R. and Cumming, G.L., 1965. Near vertical-incidence seismic reflections from the "Conrad" discontinuity. J. Geophys. Res., 70: 3441–3446.

Keen, C.E., and 10 others, 1986. A deep seismic reflection profile across the northern Appalachians. Geology, 14: 141–145.

Kirby, S. and Kronenberg, A., 1987. Rheology of the lithosphere: selected topics. Rev. Geophys. Space Phys., 25: 1219–1244.

Klemperer, S.L., 1988. Crustal thinning and the nature of extension in the northern North Sea from deep seismic reflection profiling. Tectonics, 7: 803–821.

Klemperer, S.L., 1989. Seismic reflection evidence for the location of the Iaspetus suture west of Ireland. J. Geol. Soc. London, 146: 409–412.

Klemperer, S.L. and Matthews, D.H., 1987. Iapetus suture located beneath the North Sea by BIRPS deep seismic reflection profiling. Geology, 15: 195–198.

Klemperer, S.L., Hauge, T.A., Hauser, E.C., Oliver, J.E. and Potter, C.J., 1986. The Moho in the northern Basin and Range province, Nevada, along the COCORP 40°N seismic-reflection transect. Geol. Soc. Am. Bull., 97: 603–618.

Klemperer, S.L., Hobbs, R. W. and Freeman, B., 1990. Dating the source of lower crustal reflectivity using BIRPS deep seismic profiles across the Iapetus suture. In: J.H. Leven, D.M. Finlayson, C. Wright, J.C. Dooley and B.L.N. Kennett (Editors), Seismic Probing of Continents and their Margins. Tectonophysics, 173: 445–454.

Kohlstadt, D.L. and Goetz, C., 1974. Low stress high-temperature creep in olivine single crystals. J. Geophys. Res., 79: 2045–2051.

Kusznir, N.J., 1982. Lithosphere response to externally and internally derived stresses: a viscoelastic stress guide with amplification. Geophys. J. R. Astron. Soc., 70: 399–414.

Kusznir, N.J. and Park, R.G., 1987. The extensional strength of the continental lithosphere: its dependence on geothermal gradiet, and crustal composition and thickness. In: M.P. Coward, J.F. Dewey and P.L. Hancock (Editors), Continental Extensional Tectonics. Geol. Soc. London, Spec. Publ., 28: 35–52.

Le Gall, B., 1990. Evidence of an imbricate crustal thrust belt in the southern British Variscides: Contributions of Southwestern Approaches Traverse (SWAT) deep seismic reflection profiling recorded through the English Channel and the Celtic Sea. Tectonics, 9: 283–302.

Leven, J.H., Finlayson, D.M., Wright, C., Dooley, J.C. and Kennett, B.L.N. (Editors), 1990. Probing of Continents and their Margins. Tectonophysics, 173: 641 pp.

Lueschen, E., and fourteen others, 1987. Near-vertical and wide-angle seismic surveys in the Black Forest, SW Germany. J. Geophys., 62: 1–30.

Lueschen, E., Nolte, B., Fuchs, K., 1990. Shear-wave evidence for an anisotropic lower crust beneath the Black Forest, southwest Germany. In: J.H. Leven, D.M. Finlayson, C. Wright, J.C. Dooley and B.L.N. Kennett (Editors), Seismic Probing of Continents and Their Margins. Tectonophysics, 173: 483–493.

Marillier, F., Keen, Ch., Stockmal, G., Quinlan, G., Williams, H., Colman-Sadd, S. and O'Brien, S., 1989a. Crustal structure and surface zonations of the Canadian Appalachians: implications of deep seismic reflection data. Can. J. Earth Sci., 26: 305–321.

Marillier, F., Keen, C. and Stockmal, G., 1989b. Laterally persistent seismic characteristics of the lower crust: examples from the northern Appalachians. In: R.F. Mereu, S. Mueller and D.M. Fountain (Editors), Properties and Processes of Earth's Lower Crust. Am. Geophys. Union, Geophys. Monogr., 51: 45–52.

Matte, P., 1986. Tectonics and plate tectonics model for the Variscan Belt of Europe. Tectonophysics, 126: 329–374.

Matte, P. and Hirn, A., 1988. Seismic signature and tectonic cross section of the Variscan crust in western France. Tectonics, 7: 157–180.

Matthews, D.H., 1986. Seismic reflections from the lower crust around Britain. In: J.B. Dawson, D.A. Carswell, J. Hall and K.H. Wedepohl (Editors), Nature of the Lower Continental Crust. Geol. Soc. London, Spec. Publ., 24: 11–21.

Matthews, D.H. and Cheadle, M.J., 1986. Deep reflections from Caledonides and the Variscides of Britain and comparison with the Himalayas. In: M. Barazangi and L.D. Brown (Editors), Reflection Seismology: A Global Perspective. Am. Geophys. Union, Geodyn. Ser., 13: 5–19.

Matthews, D.H. and Smith, C. (Editors), 1987. Deep seismic reflection profiling of the continental lithosphere. Geophys. J. R. Astron. Soc., 89, 447.

Matthews, D.H., and the BIRPS Group, 1990. Progress in BIRPS deep seismic reflection profiling around the British Isles. Tectonophysics, 173: 387–396.

McCarthy, J. and Thompson, G.A., 1988. Seismic imaging of extended crust with emphasis on the western United States. Geol. Soc. Am. Bull., 100: 1361–1374.

McDonough, D.T. and Fountain, D.M., 1988. Reflection characteristics of a mylonite zone based on compressional wave velocities of rock samples. Geophys. J. R. Astron. Soc., 93: 547–558.

Meissner, R., 1986. The Continental Crust: a Geophysical Approach. Academic Press, London, 426 pp.

Meissner, R. and Kusznir, N., 1987. Crustal viscosity and the reflectivity of the lower crust. Ann. Geophys., 5B: 365–374.

Meissner, R. and Strehlau, J., 1982. Limits of stresses in continental crusts and their relation to the depth-frequency distribution of shallow earthquakes. Tectonics, 1: 73–89.

Meissner, R., Wever, Th. and Flueh, E.R., 1987. The Moho in Europe — implications for crustal development. Ann. Geophys., 5: 357–364.

Mohorovičić, A., 1910. Das Beben vom 8.X.1909. Jahrbuch Meteorol. Obs. Zagreb, 9, Teil 4, Abschn. 1–63.

Mooney, W.D., 1989. Seismic methods for determining earthquake source parameters and lithospheric structure. In: L.C. Pakiser and W.D. Mooney (Editors), Geophysical Framework of the Continental United States. Geol. Soc. Am. Mem., 172: 11–34.

Mooney, W.D. and Brocher, T.M., 1987. Coincident seismic reflection and refraction measurements of the continental lithosphere: a global review. Rev. Geophys., 25: 723–742.

Moss, F.J., 1962. Amadeus Basin (southern margin) seismic survey, Northern Territory and Queensland, 1965. Bur. Mineral. Res. Aust., Record 1966/28.

Moss, F.J. and Dooley, J.C., 1988. Deep crustal reflection recordings in Australia 1957–1973 - I. Data acquisition and presentation. Geophys. J. R. Astron. Soc., 93: 229–237.

Nicolas, A., Hirn, A., Nicolich, R., Polino, R. and ECORS-CROP Working Group, 1990. Lithospheric wedging in the western Alps inferred from the ECORS-CROP traverse. Geology, 18: 587–590.

Nur, A., Ron, H. and Scotti, O., 1986. Fault mechanics and the kinematics of block rotations. Geology, 14: 746–749.

Oliver, J.E., Dobrin, M., Kaufman, S., Meyer, R. and Phinney, R., 1976. Continuous seismic reflection profiling of the deep basement, Hardeman County Texas. Geol. Soc. Am. Bull., 87: 1537–1546.

Pavlenkova, N., 1987. Properties of middle and lower crust in platform areas. Ann. Geophys., 5B: 6, 651–666.

Pavlenkova, N., 1988. The nature of seismic velocities in the continental lithosphere. Tectonophysics, 154: 211–225.

Peddy, C.P., Brown, L.D. and Klemperer, S.L., 1986. Interpreting the deep structure of rifts with synthetic seismic sections. In: M. Barazangi and L.D. Brown (Editors), Reflection Seismology: A Global Perspective. Am. Geophys. Union, Geodyn. Ser., 13: 301–311.

Pfiffner, O.A., Frei, W., Finckh, P. and Valasek, P., 1988. Deep seismic reflection profiling in the Swiss Alps: Explosion seismology results for line NFP 20-East. Geology, 16: 987–990.

Phinney, R.A., 1986. A seismic cross section of the New England Appalachians: The orogen exposed. In: M. Barazangi and L.D. Brown (Editors), Reflection Seismology: The Continental Crust. Am. Geophys. Union, Geodyn. Ser., 14: 95–106.

Phinney, R.A. and Jurdy, D.M., 1979. Seismic imaging of deep crust. Geophysics, 44: 1637–1660.

Phinney, R.A. and Roy-Chowdhury, K., 1989. Reflection seismic studies of crustal structure in the eastern United States. In: L.C. Pakiser and W.D. Mooney (Editors), Geophysical Framework of the Continental United States. Geol. Soc. Am. Mem., 172: 613–653.

Pratt, T., Culotta, R., Hauser, E., Nelson, D., Brown, L., Kaufman, S., Oliver, J. and Hinze, 1989. Major Proterozoic basement features of the eastern midcontinent of North America revealed by recent COCORP profiling. Geology, 17: 505–509.

Reston, T.J., 1987. Spatial interference, reflection character and the structure of the lower crust under extension. Results from 2-D seismic modelling. Ann. Geophys. Ser. B, xx: 339–348.

Reston, T.J., 1988. Evidence for shear zones in the lower crust offshore Britain. Tectonics, 7: 929–945.

Roure, F., and nine others,, 1989. ECORS deep seismic data and balanced cross sections: geometric constraints on the evolution of the Pyrenees. Tectonics, 8: 41–50.

Sandmeier, K.J., Waelde, W. and Wenzel, F., 1987. Physical properties and structure of the lower crust revealed by one- and two-dimensional modelling. Geophys. J. R. Astron. Soc., 89: 339–344.

Schweizerische Arbeitsgruppe fur Reflexionsseismik, 1988. Vorlaufige Ergebnisse der Alpentraversen des NFP-20: Geologische Tiefenstruktur de Schweiz. Bull. Ver. Schweiz. Pet. Geol. Lug., 1–30.

Shankland, T.J. and Ander, M.E., 1983. Electrical conductivity, temperatures, and fluids in the lower crust. J. Geophys. Res., 88: 9475–9484.

Sheriff, R.E. and Geldhart, L.P., 1982. Exploration Seismology; Vol. 1, History, Theory, and Data Acquisition. Cambridge University Press, 253 pp.

Sibson, R.H., 1982. Fault zone models, heat flow, and the depth distribution of earthquakes in the continental crust of the United States. Bull. Seismol. Soc. Am., 72: 151–163.

Simpson, R.W., Jachens, R.C., Blakely, R.J. and Saltus, R.W., 1986. A new isostatic residual gravity map of the conterminous United States with a discussion on the significance of isostatic residual anomalies. J. Geophys. Res., 91: 8348–8372.

Smithson, S.B., 1986. A physical model of the lower crust from North America based on seismic reflection data. In: J.B. Dawson, D.A. Carswell, J. Hall and K.H. Wedepohl (Editors), The Nature of the Lower Continental Crust.

Geol. Soc. London, Spec. Publ., 24: 23–34.

Smithson, S.B., 1989. Contrasting types of lower crust. In: R.F. Mereu, S. Mueller and D.M. Fountain (Editors), Properties and Processes of Earth's Lower Crust. Am. Geophys. Union, Geophys. Monogr., 51: 53–63.

Smithson, S.B. and Johnson, R.A., 1989. Crustal structure of the western U.S. based on reflection seismology. In: L.C. Pakiser and W.D. Mooney (Editors), Geophysical Framework of the Continental United States. Geol. Soc. Am. Mem., 172: 577–612.

Smithson, S.B., Brewer, J.A., Kaufman, S., Oliver, J.E. and Hurich, C., 1979. Structure of the Laramide Wind River uplift, Wyoming, from COCORP deep reflection data and gravity data. J. Geophys. Res., 84: 5955–5972.

Smithson, S.B., Brewer, J.A., Kaufman, S., Oliver, J.E. and Zawislak, R.L., 1980. Complex Archean lower crustal structure revealed by COCORP crustal reflection profiling in the Wind River Range, Wyoming. Earth Planet. Sci. Lett., 46: 295–305.

Snyder, D.B. and Flack, C.A., 1990. A Caledonian age for reflectors within the mantle lithosphere north and west of Scotland. Tectonics, 9: 903–922.

Spencer, C., Green, A., Morel-al-Huissier, Luetgert, J. Stewart, D., Unger, J. and Phillips, 1989. The extension of Grenville basement beneath the northern Appalachians: results from the Quebec–Maine seismic reflection and refraction surveys. Tectonics, 8: 677–696.

Steinhart, J.S., 1967. Mohorovicic discontinuity. In: S.K. Runcorn (Editor), International Dictionary of Geophysics. Pergamon, Oxford, 2: 991–994.

Stewart, D.B., Unger, J.D., Phillips, J.D., Goldsmith, R., Poole, W.H. Spencer, C.P., Green, A.G., Loiselle, M.C. and St. Julien, P., 1986. The Quebec–Western Maine seismic reflection profile; Setting and first year results. In: M. Barazangi and L.D. Brown (Editors), Reflection Seismology: The Continental Crust. Am. Geophys. Union, Geodyn. Ser., 14: 189–199.

Strehlau, J. and Meissner, R., 1987. Estimation of crustal viscosities and their role in geodynamic processes. In: K. Fuchs and C. Froidevaux (Editors), The Composition, Structure and Dynamics of the Lithosphere–Asthenosphere System. Am. Geophys. Union, Geodyn. Ser., 16: 69–87.

Taner, M.T., Cook, E.F. and Neidell, N.S., 1970. Limitations of the reflection seismic method: Lessons from computer simulations. Geophysics, 35: 55–73.

Telford, W.M., Geldart, L.P., Sheriff, R.E. and Keys, D.A., 1976. Applied Geophysics. Cambridge University Press, Cambridge, 860 pp.

Valasek, P.A., Snoke, A.W., Hurich, C.A. and Smithson, S.B., 1989. Nature and origin of seismic reflection fabric, Ruby–East Humboldt metamorphic core complex, Nevada. Tectonics, 8: 391–415.

Wang, C.Y., Okaya, D.A., Ruppert, C., Davis, G.A., Guo, T-S., Zhong, Z. and Wenk, H-R., 1989. Seismic reflectivity of the Whipple Mountain shear zone in southern California. J. Geophys. Res., 94: 2989–3005.

Warner, M., 1990. Basalts, water, or shear zones in the lower continental crust? In: J.H. Leven, D.M. Finlayson, C. Wright, J.C. Dooley and B.L.N. Kennett (Editors), Seismic Probing of Continents and their Margins. Tectonophysics, 173: 163–174.

Warner, M. and McGeary, S., 1987. Seismic reflection coefficients from mantle fault zones. Geophys. J. R. Astron. Soc., 89: 223–230.

Waters, K.H., 1987. Reflection Seismology; A Tool for Energy Resource Exploration. John Wiley and Sons, New York, NY, 537 pp.

Wenzel, F., Sandmeier, K.J., Walde, W., 1987. Properties of the lower crust from modeling refraction and reflection data. J. Geophys. Res., 92: 11,575–11,583.

Wickham, S.M., 1992. Fluids in the deep crust - petrological and isotopic evidence. In: D.M. Fountain, R. Arculus and R.M. Kay (Editors), Lower Continental Crust. Elsevier, Amsterdam, this volume.

Widess, M. B., 1973. How thin is a thin bed? Geophysics, 38: 1176–1180.

Woollard, G.P., 1962. The relation of gravity anomalies to surface elevation, crustal structure, and geology. Univ. Wis., Geophys. Polar Res. Center, Res. Rep. 62-9, 356 pp.

Wright, J.A. and Hall, J., 1990. Deep seismic profiling in the Nosop Basin, Botswana: cratons, mobile belts and sedimentary basins. In: J.H. Leven, D.M. Finlayson, C. Wright, J.C. Dooley and B.L.N. Kennett (Editors), Seismic Probing of Continents and their Margins. Tectonophysics, 173: 333–343

Yilmaz, O., 1987. Seismic data processing. Soc. Explor. Geophys., Invest. Geophys., 2, 526.

Ziegler, P.A., 1981. Evolution of sedimentary basins in Northwest Europe. In: L.V. Illing and G.D. Hobson (Editors), Petroleum Geology of the Continental Shelf of North-West Europe. Institute of Petroleum, Heyden and Sons, London, pp. 3–39.

Ziegler, P., 1986. Geodynamic model for the Paleozoic crustal consolidation of Western and Central Europe. Tectonophysics, 126: 302–328.

Zoback, M.L. and Zoback, M.D., 1989. Tectonic stress field of the conterminous United States. In: L.C. Pakiser and W.D. Mooney (Editors), The Geophysical Framework of the Continental United States. Geol. Soc. Am. Mem., 172: 523–539.

Zoback, M.L., and twenty-eight others, 1989. Global patterns of tectonic stress. Nature, 341: 291–298.

Chapter 3

Electrical conductivity of the continental lower crust

ALAN G. JONES

1. Introduction

The continental middle to lower crust (CLC) remains one of the most enigmatic parts of the Earth about which comparatively little is known. Remote-sensing geophysical and geochemical data illustrate our gaps in complete knowledge of the state and composition of the CLC. Generally, it is thought that the CLC, when compared to the upper crust, is somewhat more uniform in its properties. However, studies undertaken in the last decade have revealed that the CLC can be as heterogeneous as the upper crust, but that some regions of surprising homogeneity exist. In this chapter, I will describe attempts to image one particular physical parameter of the CLC, namely its electrical conductivity σ.

Electromagnetic (EM) sounding methods, as with most geophysical methods, are sensitive to the structure that exists *today*, and as such are dependent on the *current* state and composition of the CLC, rather than its state at formation or as a consequence of tectonic or metamorphic activity as are petrological studies on exhumed samples from the CLC. All EM methods give responses that are volumetric averages of the Earth's conductivities sensed by the diffusive EM fields, and as such are in the same class of geophysical techniques as seismic surface wave studies, potential field methods, and geothermal investigations. These are distinct from seismic reflection and refraction methods which deal with non-diffusive waves. However, with EM methods the governing field equations (§2) and wide range of the physical parameter being sensed (electrical conductivity, §1.1) ensure a far greater resolving power to anomalies than with the other diffusive geophysical techniques. EM and seismic methods are the only geophysical techniques for which probing of the deep crust is assured; with all others there are inherent screening effects and depth ambiguities.

Imaging the electrical conductivity structure beneath the surface has a number of uses, classified mainly by the depth of investigation:

(1) Identifying zones of mineralization, which are of economic importance.

(2) Detecting fluids in the deep crust, such as those above downgoing slabs, e.g., the Juan de Fuca plate (Kurtz et al., 1986a, 1990; EMSLAB, 1989).

(3) Resolving one physical property of crustal zones. As expressed by Dohr et al. (1989): "... *magnetotellurics works like a broad paint brush, colouring the layers bounded by the seismic reflections*".

Geological Survey of Canada Contribution No. 17492.

(4) Determining the structure of the mantle, in particular the depth to the "electrical asthenosphere", which is a zone of enhanced conductivity in the upper mantle below the lithosphere (e.g., Jones, 1982). This depth often correlates with the depth to a seismic low velocity zone.

Of these uses, (2) and (3) are relevent to the electrical conductivity of the CLC. Haak and Hutton (1986) have reviewed this topic, and I intend this chapter to complement their work by adding new results and interpretations, only repeating certain of their points for completeness. After a discussion of the physical parameter that EM methods are sensing, I will review EM methods appropriate for determining the conductivity of the CLC and their problems and limitations. A summary of recent (1980s) EM results pertinent to the topic follows with specific details from two geographic areas: the Kapuskasing structure in northern Ontario and the Valhalla complex in southeastern British Columbia. Finally, I examine the proposed causes of the observed enhanced electrical conductivity of the CLC focussing on the two currently most popular: saline fluids and grain-boundary films of carbon.

1.1. The parameter being sensed: electrical conductivity (σ)

The physical parameter that is being imaged by EM methods is electrical conductivity, σ, measured in Siemens per metre (S/m). The effects of magnetic permeability variations (e.g., Kao and Orr, 1982) are not considered as they are of little consequence for the CLC. For historical reasons it is more common to discuss the reciprocal of conductivity, which is electrical resistivity, ρ. Usually, as will be described in §2.4.3, it is not the conductivity or thickness of a zone of enhanced conductivity that is resolved from the observations, but the product of these two. This conductivity-thickness product, $S = \sigma h$ where h is the thickness of the layer, is called the *conductance* and is measured in units of Siemens (S). Its complementary product, which is usually poorly resolved, is the resistivity-thickness of the zone, $T = \rho h$, or *resistance* in units of ohms (Ω).

Electrical conductivity is sensitive to very small changes in minor constituents of the rock, and hence is a complementary parameter to other geophysical techniques, such as potential field methods and most seismic methods, which are sensitive to the bulk properties of the medium. As the conductivity of most rock matrices is very low, the conductivity of a rock unit is generally a function of the interconnection of a minor constituent, such as fluids, partial melts, or highly conducting minerals like graphite, that provide ionic or electronic pathways. Rarely is the rock so competent that the resistivity of the rock matrix is measured, although such highly resistive blocks have been found in the upper crust (e.g., Bailey et al., 1989; Kurtz et al., 1989; Beamish, 1990).

The well known Archie's Law (Archie, 1942), which was initially proposed to describe the conductivity of saturated sediments, is often an appropriate first-order model for the total conductivity of a medium:

$$\sigma_m = \sigma_f \eta^m \tag{1}$$

where σ_m and σ_f are the conductivities of the bulk medium and of the fluid respectively, η is the porosity, and the exponent m has a value between 1 and 2, with 2 being shown empirically to be valid for a wide range of rocks to mid-crustal depths (Brace et al., 1965; Brace and Orange, 1968). Note that the rock matrix conductivity, σ_r, is assumed to be sufficiently high to be of little effect (see §4.6 for a more complete discussion of pore geometry considerations). The electrical resistivity of zones within the Earth's crust can vary by over eight orders of

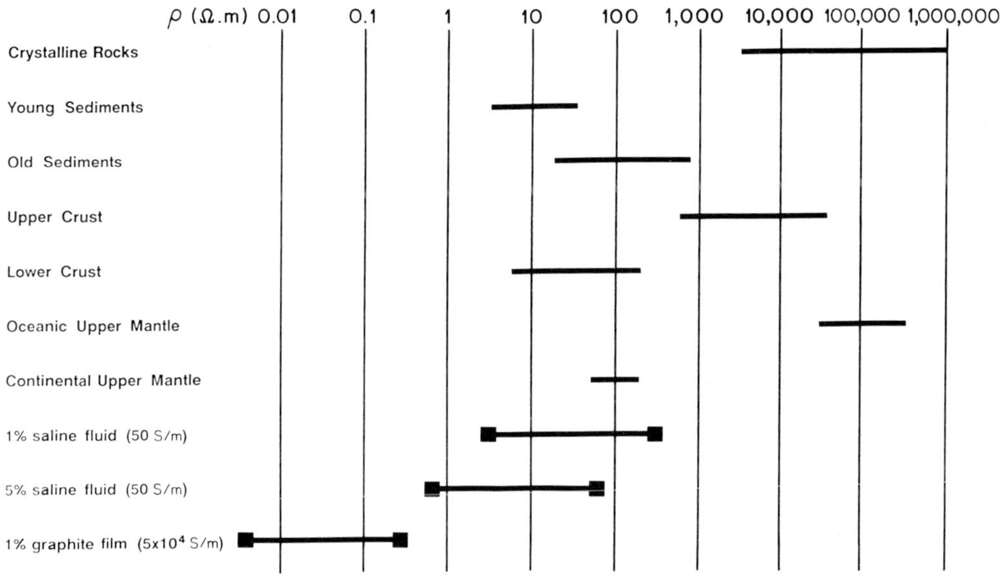

Fig. 3-1. Range of electrical resistivity of earth materials: dry crystalline rocks; "young" porous brine-saturated sediments; "old" less porous sediments; typical upper crust; typical lower crusts; oceanic upper mantle resistivity; continental upper mantle resistivity; fluid-saturated resistive rock of 1% porosity (fluid of 50 S/m conductivity): the upper and lower bounds are for Archie's Law exponents of 2 and 1 respectively; fluid-saturated resistive rock of 5% porosity; thin graphite film (graphite of 5 x 10^4 S/m conductivity) (modified from Haak and Hutton, 1986).

magnitude (Fig. 3-1, redrawn from Haak and Hutton, 1986), which is the widest range of any of the physical parameters that can be remotely sensed from the surface of the Earth.

2. Review of methods

Imaging the electrical conductivity structure at CLC depths requires either natural-source or deep-probing controlled-source EM techniques. I will review these briefly; for more complete discussions see Berdichevsky and Zhdanov (1984), Vozoff (1986), Nabighian (1987, 1991), and Keller (1989a).

Natural-source EM methods have the advantages that:

(1) penetration to all depths is assured by the skin depth phenomenon;

(2) no transmitter is required; and

(3) the mathematics of the assumed uniform source-field are relatively tractable, with the consequence that interpretation techniques are more advanced than controlled-source methods.

In comparison, controlled-source EM methods have the advantages that:

(1) the source is controllable in space, so it can be configured to excite optimally the structure of interest;

(2) the source is controllable in time (or frequency) and is thus repeatable thereby enabling signal enhancement techniques; and

(3) in the transition zone between the near field (where the source geometry dominates the responses) and far field (where the fields approach those of a uniform source) there is greater resolving power.

Obviously, given the advantages and disadvantages of both techniques one should not be exclusive, but instead choose the method, or combination of methods, most appropriate and with the highest chance of success for solving the particular problem at hand. However, for probing to great depths within the continental crust, logistical considerations usually exclude controlled-source methods.

The governing field equations can be written in terms of potential functions. However, in comparison to gravity, magnetics and geothermal formulations, the potential functions are *vector* not scalar potentials; this enables the construction of model uniqueness theorems (§2.4.3), whereas scalar potential methods are inherently non-unique. The skin depth property of EM fields (Eq. 5 below) ensures penetration to all depths, although the resolution is strongly affected by the conductance of the material overlying the depth of interest.

2.1. Magnetotelluric method

The most commonly used technique for determining the conductivity distribution within the CLC is the magnetotelluric (MT) method. MT is a natural-source technique utilizing the time-varying electromagnetic fields due to electric storms (for frequencies above 8 Hz) and solar activity (for periods longer than 0.125 s). A good collation of fifty-five papers on the MT method, detailing its historical development and the state-of-the-art up to the early 1980s, is given in the Society of Exploration Geophysicists book edited by Vozoff (1986).

The external magnetic fields penetrate into the ground and induce electric (also called telluric) fields and secondary magnetic fields, and components of the total electric and magnetic fields are measured on the Earth's surface. In the case of the magnetic field, $\boldsymbol{H}$, all three components are measured. These are h_x, h_y, and h_z, where x usually denotes either north (geographic or geomagnetic) or along strike (for a 2D body), y denotes either east or perpendicular to strike, and z denotes vertically downwards. For the telluric field, E, only the two horizontal components, E_x and E_y, are measured. The fields are measured in the time domain and are transformed into the frequency domain where cross-spectra are computed, and from these spectra the MT response function estimates are derived. The amplitude and phase relationship at a particular frequency, ω, between $\boldsymbol{E}(\omega)$ and $\boldsymbol{H}(\omega)$ are indicative of the conductivity distribution below, and the horizontal field components are related by the complex MT impedance tensor:

$$\begin{bmatrix} E_x \\ E_y \end{bmatrix} = \begin{bmatrix} Z_{xx} & Z_{xy} \\ Z_{yx} & Z_{yy} \end{bmatrix} \begin{bmatrix} H_x \\ H_y \end{bmatrix} \tag{2}$$

(dependence on ω assumed). The principal impedances Z_{xy} and Z_{yx} are converted to apparent resistivities (ρ_a) and phases (ϕ) using:

$$\rho_{a,xy}(\omega) = (1/\omega\mu)|Z_{xy}(\omega)|^2 \tag{3}$$

and

$$\phi_{xy}(\omega) = \tan^{-1}[\text{Imag } Z_{xy}(\omega)/\text{Real } Z_{xy}(\omega)] \tag{4}$$

[similarly for $\rho_{a\,yx}(\omega)$ and $\phi_{yx}(\omega)$], where μ is the permeability (usually taken as the free space value, $\mu_0 = 4\pi \times 10^{-7}$ H/m). The MT apparent resistivity represents the scaled total impedance of the layers and zones of varying conductivity in the Earth and may be modelled by equivalent series and parallel electrical circuits. Apparent resistivity and phase are not independent quantities, but are related to each other: it can be shown that for responses from all one-dimensional (1D) Earth models, and almost all two-dimensional (2D) Earth models, the two form a Hilbert transform pair (Weidelt, 1972; Fischer and Schnegg, 1980). However, because of the unknown constant in the Hilbert transform integral, although the phase curve may be predicted from the apparent resistivity curve, only the *shape* of the apparent resistivity curve can be predicted from the phase curve; there is an unknown multiplicative scaling constant on the apparent resistivities. For responses from three-dimensional (3D) Earth models the relationship has yet to be proven analytically, but empirically the two appear to obey Hilbert transformation.

For a uniform Earth, the apparent resistivities are the true resistivity of the half-space, and the phases are 45° (actually the ϕ_{yx} phases would be $-135°$, i.e., $Z_{yx} = -Z_{xy}$, because of the sign convention adopted). For regions of high electrical conductivity there is little induced E-field, and so the apparent resistivities are reduced and the phases are high (close to 90° for ϕ_{xy}, $-90°$ for ϕ_{yx}). Conversely, for regions of low conductivity the apparent resistivities are high and the phases are low (close to 0° or $-180°$). In contrast to DC resistivity, penetration to all depths is assured because of the skin depth phenomenon of EM fields. The skin depth, $\delta(T)$, in metres, given by:

$$\delta(T) = [2\rho_a(T)/\omega\mu]^{1/2} \approx 503\,[\rho_a(T)T]^{1/2} \tag{5}$$

where T is the period in seconds ($T = 2\pi/\omega$), can be a reasonable measure of the depth of investigation of the MT method. Spies (1989) considers that the *maximum* depth of investigation is given by $1.5\,\delta(T)$, whereas for Niblett-Bostick transform the depth is given by $(1/2^{1/2})\delta(T)$ (Niblett and Sayn-Wittgenstein, 1960; Bostick, 1977; Jones, 1983a). For Schmucker's ρ^*-z^* transform (Schmucker, 1970), z^* is $(1/2)\delta(T)$ and is thought to be the "centre of gravity of the in-phase induced current system" by Weidelt (1972).

In a one-dimensional (1D) environment when the conductivity varies with depth alone, $\sigma(z)$, the MT impedance tensor reduces to the simple form:

$$\mathbf{Z}_{1D} = \begin{bmatrix} 0 & Z_{xy} \\ -Z_{xy} & 0 \end{bmatrix} \tag{6}$$

The negative sign for the lower off-diagonal element indicates that the sign convention is such that the phases of this element are in the third rather than the first quadrant. Such a 1D description of the conductivity distribution may be valid at sufficiently short periods for sites over a sedimentary basin, or at long periods when the EM fields are sampling the continental upper mantle. The MT method has advanced to the stage at which 1D MT data are well understood and conductivity-depth profiles can be determined using a variety of inversion methods (e.g., Parker, 1980; Parker and Whaler, 1981; Constable et al., 1987; Oldenburg, 1992), each of which has a fundamentally different philosophy. MT apparent resistivity and phase curves for two three-layered 1D earth models are illustrated in Figure 3-2. Note that the phase curve for the model with a monotonically decreasing resistivity-depth function (solid line) is always above 45°, and conversely for the monotonically increasing resistivity-depth model (hatched line).

In a two-dimensional (2D) model of the Earth, when the axes of the co-ordinate frame are aligned parallel (x) and perpendicular (y) to strike, the impedance tensor becomes:

$$\mathbf{Z}_{2D} = \begin{bmatrix} 0 & Z_{xy} \\ Z_{yx} & 0 \end{bmatrix} \tag{7}$$

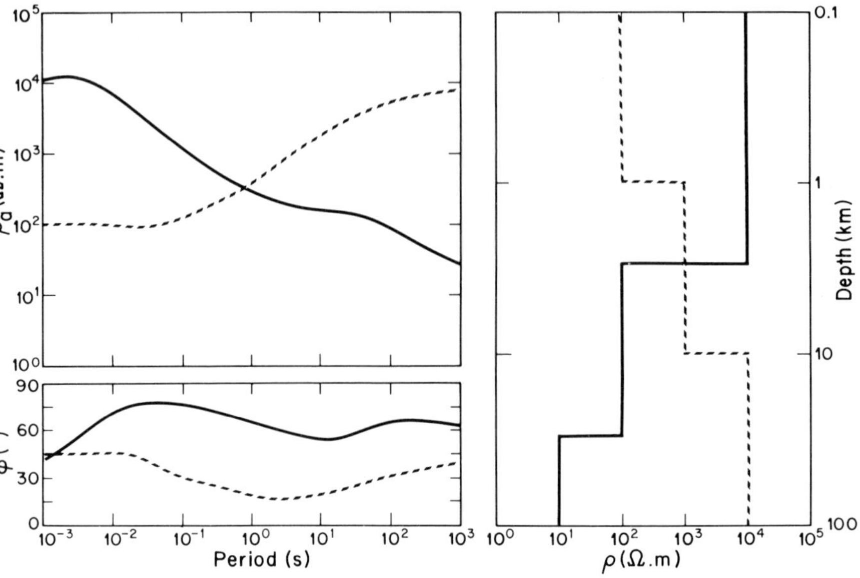

Fig. 3-2. One-dimensional apparent resistivity (top left) and phase (bottom left) curves for the two 1D three-layer models illustrated (right).

One approach for handling 2D data is to find a suitable approximate 1D response that may be inverted to illustrate, perhaps in only a qualitative sense, the structure beneath the recording site (e.g., Jones and Hutton, 1979a). However, it is becoming common to undertake more thorough trial-and-error forward 2D modelling of the data (e.g., Kurtz et al., 1986a, 1990; Wannamaker et al., 1989b; Gupta and Jones, 1990; Jones and Craven, 1990), and recent advances in 2D inversion methods are very promising (deGroot-Hedlin and Constable, 1990; Smith and Booker, 1991; Oldenburg, 1992).

In 2D, Maxwell's equations separate into two modes: the *TE-mode* (transverse electric) describes the field components (E_x, H_y and H_z) observed when the currents are flowing *along* (parallel to) the structure, and the *TM-mode* (transverse magnetic) relates the field components (H_x, E_y and E_z) when the currents are *crossing* (perpendicular to) the structure. The TE-mode is also often called E-polarization or E-parallel, and the TM-mode B-polarization or E-perpendicular. An instructive 2D body to consider is the fault model with two infinite quarter spaces of differing resistivity juxtaposed (Fig. 3-3). The TE-mode responses (solid lines) are continuous with lateral distance with the apparent resistivities at a given frequency going from ρ_1 to ρ_2 smoothly. In contrast, because physics requires continuity of electric current J, the electric field is discontinuous across the boundary (J is given by $J = \sigma E$, so if σ is discontinuous then so must be E to ensure continuity of J). Accordingly, the TM-mode apparent resistivity curve (dashed line) is discontinuous; this leads to a fundamentally higher resolution for the TM-mode responses to lateral variation in conductivity than for the TE ones.

In a fully three-dimensional (3D) Earth model, where $\mathbf{Z}$ takes on the general form as expressed in Eq. 2 above, MT data are not as well understood. Due to prohibitively high computational costs, full 3D modelling is usually undertaken only for representative structures, not to model actual field data. Approximate 3D solutions, such as "thin sheets", are computationally much faster and are adequate for modelling continent–ocean boundaries where 0.3 Ω·m sea-water is juxtaposed against 10^3–10^4 Ω·m land (e.g., Weaver, 1982).

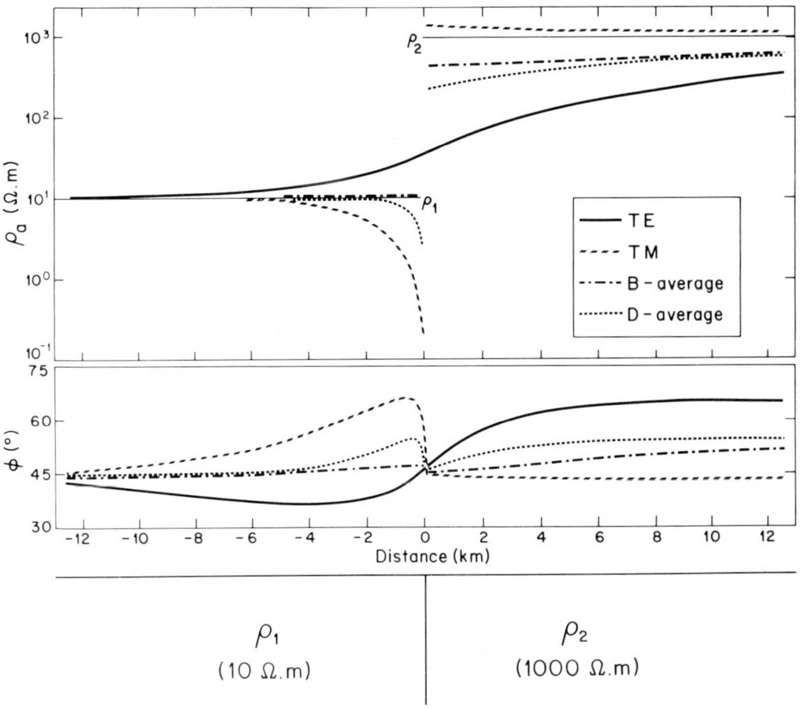

Fig. 3-3. Apparent resistivity (top) and phase (middle) curves for the fault model illustrated (bottom) at 1 s. The TE-mode responses (solid curves) are continuous across the structure, whereas the TM-mode (dashed curves) apparent resistivity curve is discontinuous. Also shown are the Berdichevsky averages (B-average; dashed-dot curves) and the determinant averages (D-average; dotted curves). Note that on the conductive side of the discontinuity the Berdichevsky average gives virtually the correct resistivity (ρ_1) and phase (45°) for valid 1D interpretation right up to the contact. On the resistive side a 1D interpretation of the TM data will yield the most correct model.

Advances have been made to reduce the dimensionality of the data, from 3D to 2D or even 1D, and still yield a valid model. In certain cases, the TM-mode responses from a 3D situation can be interpreted in a 2D manner to give a reasonably correct conductivity distribution (Jones, 1983b; Wannamaker et al., 1984).

The use of two invariant forms of the impedance estimates is becoming more prevalent in MT studies. These two forms, called the "Berdichevsky average", Z_B, and the "determinant average", Z_D, are defined by:

$$Z_B = (Z_{xy} - Z_{yx})/2 \tag{8}$$

and

$$Z_D = (Z_{xx}Z_{yy} - Z_{xy}Z_{yx})^{1/2} \tag{9}$$

(Berdichevsky and Dmitriev, 1976). The negative sign in Z_B is again an expression of the coordinate convention used with the phase of Z_{yx} in the third quadrant. The Berdichevsky and determinant averages for the fault model are shown on Figure 3-3. Various authors have illustrated that under certain conditions, even with 3D data, a 1D interpretation of Z_B or Z_D can give a reasonable indication of the conductivity distribution beneath the recording site (Ranganayaki, 1984; Ingham, 1988; Park and Livelybrooks, 1989). Note that for 1D and 2D

Earth models Z_B and Z_D are the arithmetic and geometric means respectively of Z_{xy} and Z_{yx}. The determinant phase, ϕ_D, has the particular appeal of being unaffected by galvanic static shift distortions of the MT impedance tensor (see §2.4.1 for a description of static shift and its implications for MT interpretations).

2.2. Geomagnetic Depth Sounding (GDS) and Horizontal Spatial Gradient (HSG)

Two methods are used for sensing the Earth's conductivity structure in the CLC by recording the magnetic field components alone. In the Geomagnetic Depth Sounding (GDS) method, the two transfer response functions T_x and T_y which relate the vertical magnetic field component to the two horizontal magnetic field components:

$$H_z = \begin{bmatrix} T_x & T_y \end{bmatrix} \begin{bmatrix} H_x \\ H_y \end{bmatrix} \tag{10}$$

(dependence on frequency assumed) are interpreted. They are often displayed as "induction arrows" (Schmucker, 1970), and the real arrows generally point towards zones of enhanced conductivity (Jones, 1986). This method is excellent as a mapping tool for locating anomalous structures, and array studies by Gough and others have mapped continental-scale anomalies in many parts of the world (Gough, 1981, 1989). A full description of the analysis and interpretation techniques used for magnetometer array studies can be found in Gough and Ingham (1983).

The Horizontal Spatial Gradient (HSG) method differs from both MT and GDS in that the external source fields are assumed to be non-uniform. Over large-scale laterally homogeneous regions, the spatial gradients of the horizontal components of the magnetic field induce a vertical magnetic component, and the ratio of the two can be interpreted in terms of conductivity-depth structure (Schmucker, 1970; Kuckes, 1973a,b; Jones, 1980). This ratio, called Schmucker's C-function (Schmucker, 1970), is given by:

$$C(\omega,k) = H_z(\omega) / \left[\frac{\partial}{\partial x}H_x(\omega) + \frac{\partial}{\partial y}H_y(\omega)\right] \tag{11}$$

where k is the wavenumber representative of the non-uniform source field, and $C(\omega,k)$ is related to the MT impedance for a layered earth model $Z_{1D}(\omega,k)$ by:

$$C(\omega,k) = (1/i\omega\mu)Z_{1D}(\omega,k) \tag{12}$$

Thus, $C(\omega,k)$ can be interpreted using the same advanced techniques available for $Z_{1D}(\omega)$. One advantage of the HSG method over MT is that, because only magnetic fields are measured, galvanic static shift problems associated with the telluric fields (§2.4.1) are avoided. The HSG method is not more routinely used because of the requirements for: (a) a large laterally homogeneous region; (b) a sufficient gradient in the source fields; and (c) synoptic observations of the magnetic fields at at least five locations and preferably many more.

2.3. Controlled-source EM methods

Very few controlled-source techniques are capable of resolving structure at CLC depths due to the requirement for large source energy and large source–receiver separations. Ward (1983), Nabighian (1987), Keller (1989a), and recently Boerner (1992), have reviewed the methods and results obtained from controlled-source EM surveys. A number of large-scale experiments have been carried out using devices developed for other purposes, such as power transmission lines (van Zijl, 1969; Blohm et al., 1977; Lienert, 1979; Lienert and Bennet, 1977; Towle, 1980), decommissioned telephone lines (Constable et al., 1984) and the Kola peninsula magneto-hydrodynamic (MHD) generator (Velikhov et al., 1986). In these experiments, EM fields penetrated and sampled the CLC. Conventional deep-probing controlled-source techniques, e.g., LOTEM (Long Offset Transient ElectroMagnetic system: Keller et al., 1984; Strack, 1984), CSAMT (Controlled Source Audio MagnetoTellurics: Boerner et al., 1990) and UTEM (University of Toronto ElectroMagnetic system: Kurtz et al., 1989; Bailey et al., 1989), usually do not penetrate into the CLC, although the two UTEM studies are important because they sample an upthrust lower crustal block (see §3.5). However, using MEGASOURCE (Keller et al., 1984), a very high-current electric bipole source, Keller, Skokan and colleagues believe that they are able to map the CLC and the Moho (Skokan, 1990). Boerner (1992) discusses problems and limitations of controlled-source methods for deep probing of the crust.

2.4. *Problems and limitations of the MT method*

The periods at which MT data sample CLC depths are usually in the range 10–100 s (frequencies of 0.1–0.01 Hz). Over highly resistive upper crust, periodicities of 1 s and above may be important, and over thick sedimentary basins periodicities of 30 s or greater may be required. At these periods, the sources are mainly due to solar plasma ejected by the Sun that enters the Earth's magnetosphere and induces magnetic fields in the Earth's ionosphere. Usually, there are excellent signal levels at periods greater than 10 s. Between 0.1 and 10 s there is a minimum in the telluric field spectrum and a maximum in the natural noise (due to microseismic activity, mainly wind noise but also ocean effects) and cultural noise spectra, with resultant low signal-to-noise ratios. This is the so-called MT "dead-band" and until the late 1970s it was difficult to obtain good quality MT response estimates in this band. During the 1980s there have been many significant advances in the instrumentation used to acquire MT data, in the acquisition procedures used (e.g., remote-reference acquisition; Gamble et al., 1979), and in the processing methods employed to estimate the MT impedance elements from the time series (see Jones et al., 1989) to the extent that *modern* MT data have errors of only a few percent over the whole period range of observation. This is compared to the typical one quarter of an order of magnitude errors for data collected and processed during the 1970s. Precise data, which are required if one wants to resolve certain features (e.g., Cavaliere and Jones, 1984), demand highly sophisticated modelling methods for their interpretation.

Equally important is that the data are now wide-band: a modern MT system typically covers six orders of magnitude in frequency (10^3 Hz–10^3 s) compared to two or three orders of magnitude previously (10–10^3 s). Although the data in the range 10–100 s are the most sensitive to the CLC, high frequency data are important for determining the upper crustal structure in order to understand its effects on the longer period data.

Perhaps the most pressing current problem is the effects that local near-surface inhomogeneities have on the MT impedance estimates. These effects have been termed "static shifts" but should more generally be referred to as "static distortions". It is necessary to correct the MT responses for these distortions prior to any attempt to derive a conductivity model. Intimately coupled with determination of the static distortions is determination of the appropriate strike angle of the coordinate frame into which to rotate the data for 2D modelling. Finally, MT data are limited in their resolving power, and this is discussed with particular reference to the CLC below.

2.4.1. Static distortions

Advances are being made in understanding the effects that near-surface inhomogeneities have on MT responses. A spectacular example of this phenomenon is illustrated by Poll et al. (1987) in which a decrease of more than an order of magnitude in the electric field amplitude perpendicular to a fault was observed over 50 m. This is a problem which expresses our current difficulty in dealing with the very wide range of scales in a typical crustal-sounding MT experiment, from the metres-scale of the locations of the telluric electrodes themselves to the hundreds of kilometres-scale of regional tectonic features. To address this problem approximate techniques have been developed which deal with local features as if the EM fields are at the galvanic limit. The only observable EM effects in the data of the inhomogeneities are due to electric charges bound to the surfaces of them. The effects of these charges are predominantly on the electric field components, but at sufficiently high frequencies the magnetic field components are also affected (Zhang et al., 1987; Groom, 1988; Groom and Bailey, 1991).

In its simplest form for data from either 1D or 2D (in the correct strike direction) Earth models, this problem manifests itself as a shift of the apparent resistivities by a frequency independent multiplicative constant without affecting the phases, and is termed "static shift" (e.g., Jones, 1988a; Sternberg et al., 1988). An example is given in Figure 3-4 for data from two MT sites on a thick sedimentary basin that were separated by only 50 m. Three of the four ρ_a curves are at the same level, and all four phases are similar, but one of the ρ_a curves is shifted downwards by local conditions. A 1D interpretation of the shifted ρ_{yx} curve (dashed line, Fig. 3-4) would lead to an erroneous model with the interfaces closer to the surface by a factor that is the square root of the multiplicative shift constant, and layer resistivities too small by a factor equal to the multiplicative shift constant.

Interpretations of MT data possibly affected by static shifts can be found in the published literature. For example, Figure 3-5 is an interpretation of MT data from a profile of four sites

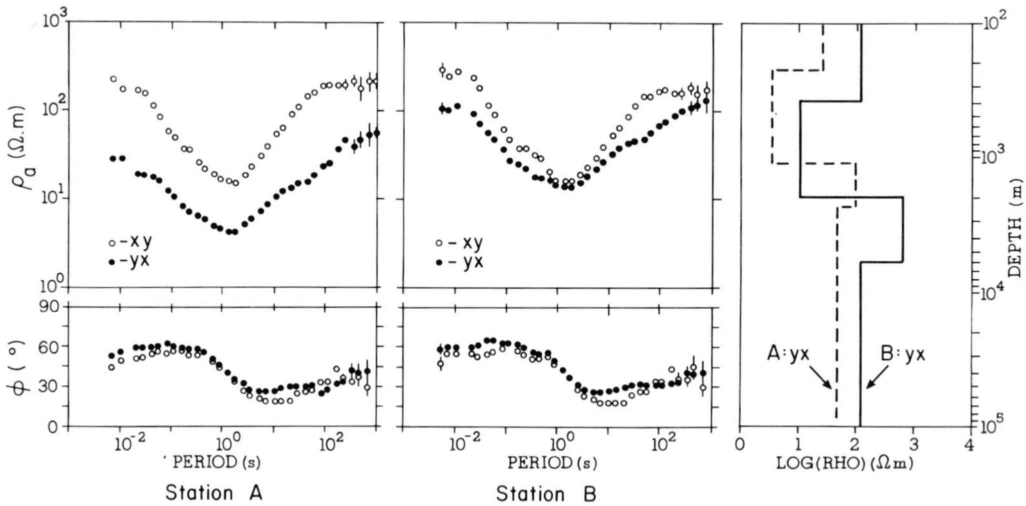

Fig. 3-4. Apparent resistivity and phase curves from two MT sites that shared a common electrode, i.e., were centred 50 m apart. Note that one ρ_a curve (ρ_{yx} at station A) is static shifted downwards by half an order of magnitude compared to the other three, but that all four phase curves are similar. The 1D model of the static shifted data (A: *yx*; dashed line) is in error compared to the correct model (B: *yx*; full line). with layer interfaces and resistivities underestimated.

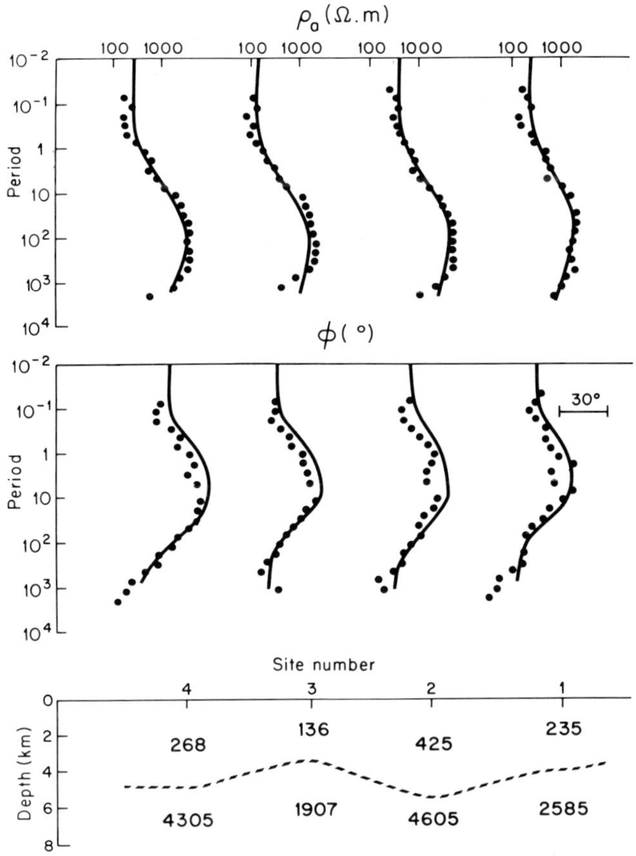

Fig. 3-5. Apparent resistivity and phase curves, and their 1D inversions, for four locations in northern Australia (redrawn from Cull, 1982). Note that the four phase curves are virtually identical, and that the four apparent resistivity curves have the same shape but are at different base levels. Static shift of the resistivity curves is an alternative explanation for the implied lateral variation in basement topography.

in the McArthur Basin of northern Australia (Cull, 1982). Note that the resistivity curves are similar in shape, and the phase curves are virtually equal to one another, but that the levels of the resistivity curves are shifted by a factor of nearly 5. The pseudo 2D model, obtained by stitching together the 1D models from each site, shows purported basement topography beneath site 3. However, the sediment and basement resistivities beneath this site are both smaller by a factor of approximately 2 than at site 4, and the depth to basement is smaller by a factor of approximately $2^{1/2}$. Accordingly, whereas one might expect the three Earth parameters ρ_1 (resistivity of the sedimentary layer), h_1 (thickness of the sedimentary layer) and ρ_2 (resistivity of the basement) to vary with distance, it is suspicious that they vary in a manner consistent with a static shift of 2. Without *a priori* knowledge, it is impossible to be certain of the correct level of the MT apparent resistivity curves, and thus perhaps the data from *all* of the sites are affected by varying amounts. The model may be correct, but there is sufficient doubt to warrant independent assessment of the depths to basement. Many other examples abound in the older MT literature of suspect interpretations due to possible static shift problems, and recently a variety of methods have been advanced to estimate the shift factor (Jones, 1988a; Sternberg et al., 1988; Craven et al., 1990; Pellerin and Hohmann, 1990).

Others have noted that, whereas the phases from a number of sites all lie within a narrow range, the apparent resistivity curves all have the same shape but are displaced by up to two orders of magnitude (e.g., Kurtz et al., 1986a, 1990; Jones, 1988a; Vanyan et al., 1989; Beamish, 1990; Volbers et al., 1992), which is also a manifestation of static shift. Sternberg et al. (1988) illustrated that, for their MT data from 70 sites, the static shifts on the apparent resistivity curves had a standard deviation of about one-quarter to one-third of an order of magnitude. Vanyan et al.'s (1989) compilation for almost 400 MT sites on the eastern Siberian shield covering a very large area (1500 × 1000 km) exhibits a statistical standard deviation of the static shift factors of half an order of magnitude.

In its more general form these distortions have been addressed by modelling their effects as 3D galvanic bodies over a 2D regional earth:

$$\boldsymbol{Z} = \boldsymbol{C}\boldsymbol{Z}_{2D} = \begin{bmatrix} a & b \\ c & d \end{bmatrix} \boldsymbol{Z}_{2D} \tag{13}$$

where a, b, c and d are real and frequency-independent. Methods have been developed to determine the 2D strike direction, and to decompose the MT tensor observed, $\boldsymbol{Z}$, into component tensors that describe the distortions and the regional 2D conductivity distribution separately (see §2.4.2). No methods currently exist for dealing with the completely general case of a 3D distorting tensor $\boldsymbol{C}$ over a region with a 3D conductivity distribution. However, because the phase of the determinant of a complex matrix is unaffected by multiplication with a real matrix, the determinant phase of $\boldsymbol{Z}$ (Eq. 9) is a true estimate of the determinant phase of the regional structure. Accordingly, a pseudo-section plot of this phase with distance along the abscissa and period down the ordinate can give a good image of the first-order conductivity features beneath the profile (e.g., Jones et al., 1988). One note of caution, however, is that the estimation of Z_D is unstable in the presence of noise and the condition number of $\boldsymbol{Z}$ should be determined to indicate if the matrix is singular.

2.4.2. Strike determination

If the conductivity structure of the Earth near a site can be reasonably approximated by a 2D model, at least for a range of frequencies, then it is necessary to determine the appropriate strike direction, θ, so as to obtain a tensor that best approximates the form of Eq. 7 in some manner. The observations in co-ordinate frame θ, $\boldsymbol{Z}_\theta$, are related to the 2D impedance tensor by:

$$\boldsymbol{Z}_\theta = \boldsymbol{R}\boldsymbol{Z}_{2D}\boldsymbol{R}^{t} \tag{14}$$

where $\boldsymbol{R}$ is the Cartesian rotation matrix and the superscript t denotes transpose. This angle can either be obtained from the MT data or can be assumed from *a priori* knowledge of the region.

Quite a number of methods were developed through the 1970s for deriving the strike angle from the MT data. The majority of workers used a form of Swift's (1967) algorithm that leads to an analytical solution for an angle which maximized the power in the off-diagonal components (or equivalently minimized the power in the diagonal components) of $\boldsymbol{Z}$ at each frequency. Generally, this was applied independently at each frequency and at each site, so that sites very close together (in an inductive scale length sense; Eq. 5) had data rotated to angles that could differ by up to 90°. However, local static 3D distortions and noise contributions can cause meaningless angles to be determined from Swift's formula; this leads to frequency- and site-dependent angles that result in estimates of $\mathbf{Z}_{2D}$ which are not rotated to the correct axes with possible mode-mixing of the TM and TE responses.

Accordingly, new methods have been advanced to determine the correct strike angle of the 2D regional structure in the presence of local distortions (Bahr, 1985, 1988, 1991; Zhang et al., 1986). These methods suffer in the presence of noise as the regional strike angle is the least stable distortion parameter that can be extracted from the data. Groom and Bailey's (1989, 1991) mathematical decomposition gives the parameters required as well as the confidence with which to believe those estimates, and the Groom-Bailey approach is now being used routinely by many workers in the field for deriving the appropriate θ and the distortion parameters.

Alternatively, regional geology/tectonics can indicate the appropriate strike angle to choose; for example, the strike of the coast (Vancouver Island: Kurtz et al., 1986a, 1990; Oregon/Washington (EMSLAB): Wannamaker et al., 1989a,b), the strike of accreted terrains (southeastern B.C.: Jones et al., 1988), or the strike of major faults. No matter how the strike angle θ is determined however, it is important that the data from the profile all be rotated into a *consistent* reference frame to permit valid 2D modelling.

2.4.3. Model resolution and uniqueness

The current state-of-the-art for MT data modelling is that 1D inversions, 2D trial-and-error forward model fitting, and 3D thin-sheet modelling are routine, and 2D inversions and full 3D model studies for generic structures are becoming more used. Full 3D inversions for arbitrarily-shaped realistic bodies of widely differing scales is our ultimate goal and, given advances at the current rate, should be attainable by the turn of the century.

One question that must be asked is just how well resolved and how unique are our conductivity models of the CLC? In the majority of studies, especially the older ones, 1D inversions were used to estimate the conductivity of the CLC. For perfect 1D MT data at all frequencies *only one* $\sigma(z)$ model exists that will satisfy the observations (Bailey, 1970; Weidelt, 1972). In theory therefore, MT data are distinct from other geophysical responses of diffusive phenomena (gravity, magnetics, geothermal) in that there is no inherent non-uniqueness. The realization of perfect data is obviously unrealistic, but the existence of the uniqueness theorem gives us a rationale for striving to obtain the highest precision data possible.

Consider a typical crust of 40 km thickness with a conducting CLC beginning at a depth of 20 km. Assuming that there are no sedimentary sequences, we may adopt a resistivity of 10^4 $\Omega{\cdot}$m for the upper crust, and a value of 10^3 $\Omega{\cdot}$m for the continental upper mantle. The response to a single-layer CLC of 100 $\Omega{\cdot}$m material is illustrated in Figure 3-6 (solid curves). A singular value decomposition (SVD) analysis of this model (e.g., Edwards et al., 1981; Jones, 1982), for an observation range of 10^3 Hz -10^3 s, indicates that the parameters are resolved in the following order: the thickness of the upper crust (h_1), the resistivity of the upper crust (ρ_1), the conductance of the lower crust ($S_2 = \sigma_2 h_2$), and the resistivity of the upper mantle (ρ_3). Least well resolved is the resistance of the lower crust ($T_2 = \rho_2 h_2$). This indicates that the correct (orthogonal) parameterization of the CLC of this model is not one in terms of layer thickness and resistivity, but one with a Dar Zarrouk parameterization (Maillet, 1947; Koefoed, 1979) in terms of S_2 and T_2. This phenomenon, of sensitivity to $S(z)$ rather than $\sigma(z)$, has been known for a decade. In the COPROD project (COmparison of PRofiles from One-dimensional Data) a number of workers derived a wide variety of 1D models that fit the MT data from a site in southern Scotland (the NEW (Newcastleton) data of

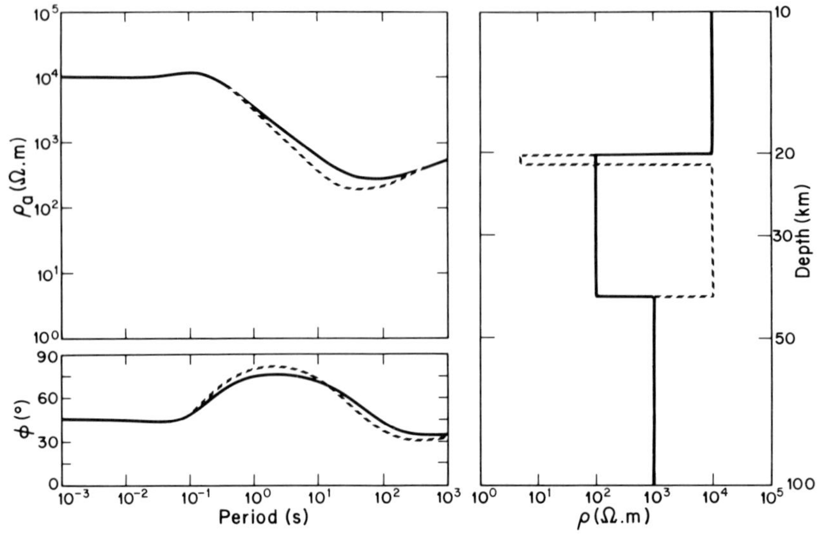

Fig. 3-6. Apparent resistivity and phase curves for two models of the continental lower crust of total conductance 200 S: in one model the conductance is distributed in a single 100 Ω·m layer (solid curves) whereas in the other model the conductance is distributed in a thin top layer of 1 km thick and 5 Ω·m underlain by a resistive lowermost lower crust (dashed curves).

Jones and Hutton, 1979a,b), but it can be shown that the conductances determined from all of the models fell within a narrow range (Weidelt, 1985). Some inversion algorithms favour thin zones — without constraints the natural tendency to minimize misfit is for infinitely-thin zones (D^+: Parker, 1980) — whereas others tend towards thicker zones, e.g., the smooth inversions of Constable et al. (1987) and Smith and Booker (1988).

The conductance of the lower crust in this model is 200 S (a reasonable average for the CLC, see §3.4), which could also be distributed as a thin layer (layer 2) of 1 km thickness and 5 Ω·m overlying a 19-km-thick layer (layer 3) of much higher resistivity (taken as 10^4 Ω·m). For such a model the parameters that are well-resolved are the thickness of the upper crust (h_1), the conductance of the conducting zone (S_2), the resistivity of the upper crust (ρ_1) and the resistivity of the upper mantle (ρ_4). For MT data with 5% errors the thickness of the resistive part of the lower crust (h_3) is marginally resolved, whereas the resistance of the conducting zone (T_2) and the resistivity of the resistive part of the lower crust (ρ_3) are unresolved. The responses of this two-layer CLC are also illustrated on Figure 3-6 (dashed curves), and the difference between the two responses is a few percent. Discriminating between these two end-member models for the distribution of the conductance in the CLC — either a thick zone of moderate resistivity (100s Ω·m), or a thin zone of low resistivity (10s Ω·m) overlying a resistive lowermost lower crust — depends on resolving the resistive part of the CLC. This requires MT data with 2% errors or better and a region which is 1D electrically to within these errors. Thus, many of the zones classified by Jones (1981b) as "*intermediate*" and by Haak and Hutton (1986) as "*normal*", of resistivity of 100–300 Ω·m and thickness some 20 km could, in fact, be 10–30 Ω·m and only 2 km thick for example.

As a general rule-of-thumb it is not possible to resolve conducting zones of conductance *less* than the conductance of all the zones above it. Figure 3-7 shows the responses of two models of the lower crust, one with a single 1-km-thick zone of conductance 200 S at 20 km depth (solid curves) and the other with an additional zone of enhanced conductivity at 30

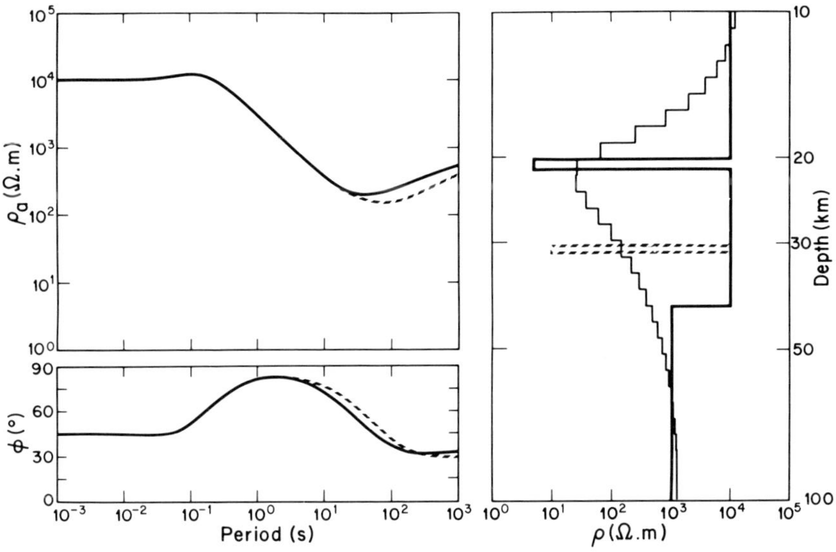

Fig. 3-7. Apparent resistivity and phase curves for two models of the continental lower crust: one model (solid curves) is of a two-layer CLC with the top layer being 1 km thick of 5 Ω·m (thin layer model of Fig. 3-4), and the other model is a four-layer CLC with a second 5 Ω·m conducting zone of 100 S at a depth of 30 km (dashed curves). A smooth inversion of the theoretical data with 5% errors of the four-layer model (thin line) illustrates that the lower conducting zone cannot be detected.

km with half the conductance (100 S, or 10 Ω·m) of the upper one (dashed curves). Note that there is little difference in the responses between the two models. A smooth inversion (Constable et al., 1987) of the MT data with 5% errors from the two-conducting zone model of the CLC is also illustrated in Figure 3-7 (thin line), and the lower conducting zone cannot be detected. This shielding effect could be important for certain geological symmetries, such as anticlines and synclines.

This aspect, of screening by overlying conductive structures, has an important consequence for resolving the conductivity structure of the CLC beneath sedimentary basins. Figure 3-8 compares the MT responses that would be observed on a 2-km-thick basin with fill of 5 Ω·m sediments (dashed curves), i.e., total conductance of 400 S, to the responses that would be observed on resistive upper crust with no sedimentary cover (solid curves) for the thin-layer CLC example of Figure 3-6. For no sedimentary cover, the existence of the conducting layer within the CLC is indicated by the drop in the ρ_a curve at periods greater than 0.1 s, and its conductance is indicated by the level of the minimum at 30 s. For the ρ_a curve on the basin neither of these features is readily apparent in the curves. Smooth inversion (Constable et al., 1987) of the MT data, with 5% errors, on the basin illustrates that the conducting zone within the CLC cannot be detected (thin line). For MT data with 2% errors the existence of the zone can be inferred, and with 1%, or better, data its conductance and the depth to its centre can be marginally resolved.

In 2D studies, because of the two polarizations of the EM fields, it is possible with high-quality data to discriminate between thin zones and thicker zones even beneath sedimentary basins, as shown, for example, by Jones and Craven (1990) in their modelling study of the North American Central Plains conductivity anomaly (NACP, §3.3.4). However, few MT studies of the lower crust are of sufficient quality or are interpreted two-dimensionally. The

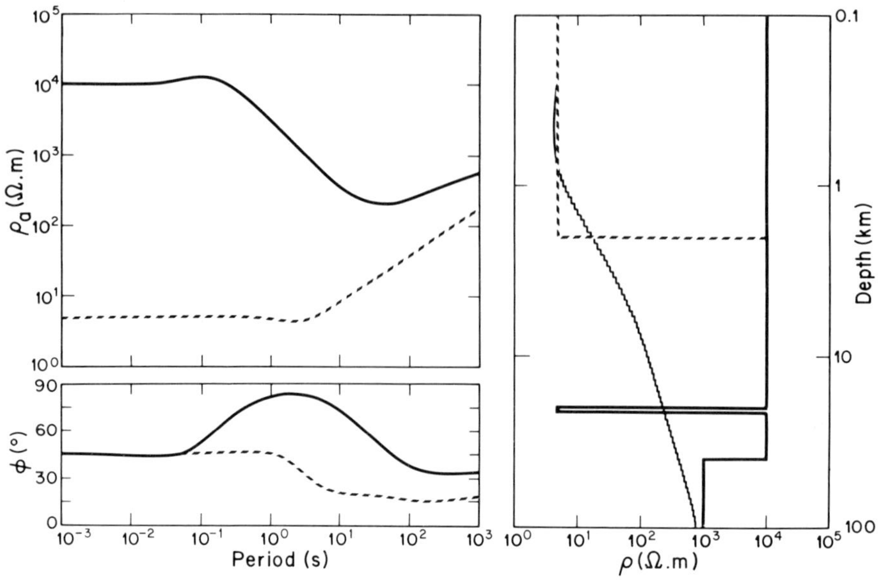

Fig. 3-8. Apparent resistivity and phase curves for a model of the continental lower crust of 1 km thickness, 5 Ω·m beneath an exposed upper crust (solid curves) and beneath a 3-km-thick sedimentary basin with fill of 10 Ω·m (dashed curves). A smooth inversion of the theoretical data with 5% errors of the basin model (thin line) illustrates that the conducting zone in the lower crust cannot be detected.

shielding effect is still important and conductors beneath other conductors may be masked (see fig. 1 of Jones, 1987). The inherent sensitivity to lateral variations of conductivity is greater for TM data than for TE data.

The MT data that can be acquired today are of sufficient precision (1% is possible) to resolve theoretically the details of the conductivity structure of the CLC beneath regions of little sedimentary cover. The important limiting factor is now the geological noise, rather than instrumentation noise or cultural noise, which is an expression of our modelling limitations.

3. Summary of results

Interpretations of the conductivity distribution within the CLC abound in the published literature. However, many models were derived from responses that were of insufficient precision for the quoted resolution due to older technologies used for data acquisition and processing (see Jones et al., 1989). Accordingly, I will concentrate on the more recent results and refer to older results when they are significant or historic.

I have grouped the specific results into three broad categories; (1) shields, (2) rifts, and (3) continental margins, though there is obviously much overlap between these. Another feature that warrants consideration in a general treatise on crustal EM studies is sedimentary basins. However, in this paper we are concerned with the CLC beneath the basins, not the structure of the basins themselves. It should be noted that although thick sedimentary sequences can mask the features of the CLC that we are attempting to elucidate, strongly anomalous structures within the CLC can create secondary EM fields that can be observed on the surface of the Earth, e.g., the NACP anomaly (see Trans-Hudson Orogen, §3.3.4) and the anomaly in the CLC beneath the Flathead basin in southeastern British Columbia (Gupta and Jones, 1990).

The generic results for the resistivity of the CLC from statistical compilations are also considered (§3.4), as are the results from two special areas: the Kapuskasing Uplift in Ontario (§3.5) and the Valhalla complex in British Columbia (§3.6). The former is important as it is an exposure of rocks from deep levels within the crust, whereas the latter is unusual in exhibiting high conductivity but no seismic reflectivity in the lower crust.

In the following I will not discuss an electrical analogue to the seismic Moho. In my opinion there is not a single EM study which shows unequivocally a change in electrical conductivity at the base of the crust as defined seismically by the depth of the Moho discontinuity. Indeed, studies on resistive windows, which would enable the greatest chances of resolving an electrical Moho, suggest that there is no dramatic change in conductivity across it (Jones, 1982; Beamish, 1990).

3.1. Shields

Electromagnetic investigations have been carried out on all of the Earth's shields and platforms, and a review of the work up to the mid-1970s was given by Kovtun (1976). Three shield regions are chosen for discussion; namely the Baltic, Canadian and Siberian shields.

3.1.1. Baltic shield

By far the most extensively studied shield region using modern equipment, analysis and interpretation techniques is the Baltic shield of northern Europe. Results and models for predominantly the Finnish and Soviet portions of the shield were recently collated by Hjelt and Vanyan (1989).

The first results from the Baltic — or Fennoscandian — shield (Jones, 1982, 1984a; Jones et al., 1983) from long-period (>100 s) MT and HSG measurements indicated that the CLC beneath the northern parts of the shield was moderately resistive — of the order of some hundreds of Ω·m — whereas the CLC beneath southern Finland appeared to be more conductive — some tens of Ω·m. The dividing line between these two was conjectured to be the Ladoga–Bothnian Bay zone (LBBZ), which is discussed in greater detail below (§3.3.5). The LBBZ, described as a 1.9–1.87 Ga Svecofennidic schist, separates Archean basement (2.8–2.6 Ga) to the northeast from a Svecokarelian (1.9–1.8 Ga) complex to the southwest. Although the actual resistivity values of the CLCs on either side of the LBBZ may be in question, there is little doubt that there must be an increase in the conductance of the CLC of approximately an order of magnitude to the south of the LBBZ compared to the north (about 1000 S to the south compared to 100 S to the north). This difference is also evident in the vertical magnetic fields observed by the Scandinavian International Magnetospheric Study (IMS) array (Küppers et al., 1979; Jones et al., 1983).

The shield has been further studied using MT methods by Swedish and Finnish groups with Soviet and Hungarian participation. Rasmussen et al. (1987) conducted an extensive regional MT survey of forty sites along a 1250 km profile following the N–S Fennolora (Fennoscandian Long Range profile) seismic refraction profile in Sweden, but the stations are generally too widely spaced (separation typically 40 km) to be certain of the results. The three most northerly sites show a zone of increased conductivity in the CLC, with estimates of conductance in the range 100–1000 S. This wide variation may be explained by a downward static shift of the data at one anomalous site giving a regional average value for the

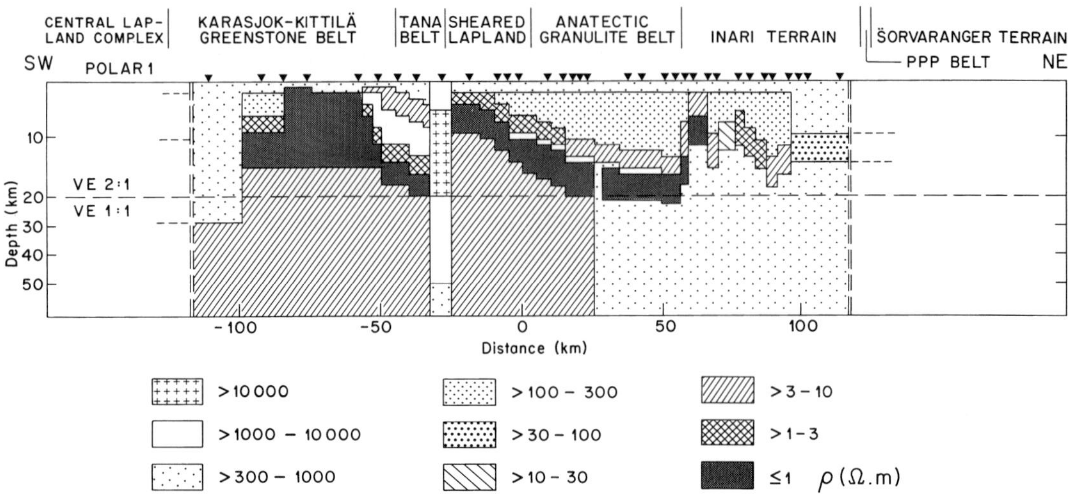

Fig. 3-9. A 2D model of the POLAR profile data from northeastern Finland (redrawn from Korja et al., 1989). The locations of the MT sites are indicated by the inverted triangles.

conductance of the CLC of 100 S. In the south there does not appear to be a zone of markedly high conductance in the CLC.

An extensive MT study was performed along the POLAR profile, the northernmost segment of the European Geotraverse (EGT), which strikes northeast–southwest in northeastern Finland (Korja et al., 1989). MT measurements were made at 40 sites along 300 km of the transect, with a shorter parallel transect of ten sites to the southeast. After attempting to correct the data for static distortion effects, the authors derived a 2D model which fitted the MT data (Fig. 3-9) with greatest weight being given to the phase information. (The resistivity shading scheme used is consistent on all 2D models illustrated in this chapter to facilitate comparison.) Note that the northeastward-dipping conducting zones within the granulite belt appear to stop at mid-crustal levels, ≈20 km. Within the CLC the conductance decreases from SW to NE; the actual conductances are not well resolved and the boundary between the high- (resistivity about 5 Ω·m) and low- (resistivity about 100 Ω·m) conductivity CLC regions is not well defined. Features in the upper and middle crust correlate well with other geophysical information (seismic reflection, refraction and gravity). In the lower crust, however, there is less coincidence.

An MT study in southern Sweden across the Mylonite Zone by Rasmussen (1988) led to two possible families of models to explain the observations: either (1) a 2D model with inductive scale lengths that are large compared to the profile length so that lateral variations in resistivity are modelled to occur off the ends of the profile (more than one profile length away), or (2) a 1D model incorporating a transversely anisotropic layer between 12 km and 33 km of resistivities of 400 Ω·m and 17000 Ω·m in the two directions (Fig. 3-10). Moho is thought to be around 40 km in this region. Rasmussen (1988) expresses a preference for the latter interpretation because it explains better the observations over the whole frequency range of observation. This is discussed further in §4.5.

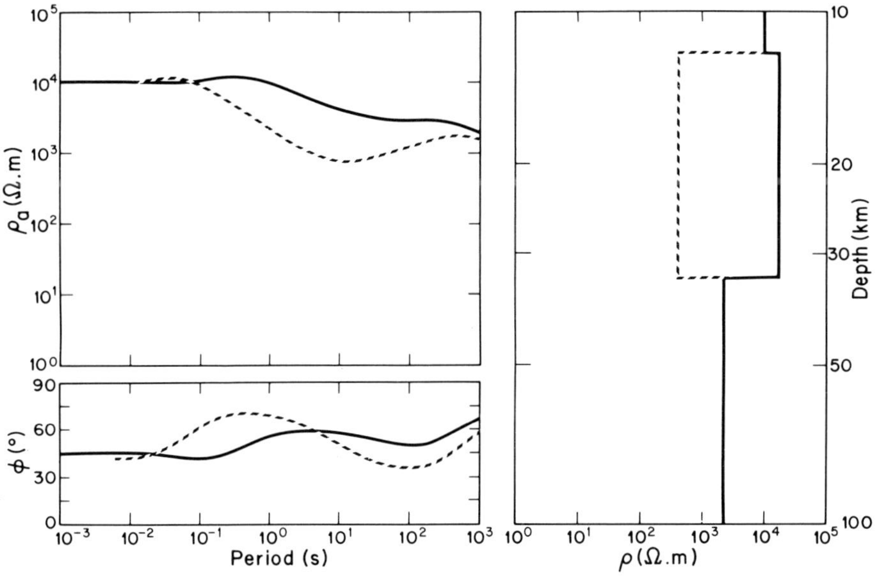

Fig. 3-10. A 1D model with an anisotropic middle and lower crust of Rasmussen (1988) and the theoretical responses.

3.1.2. Canadian shield

Studies on the Canadian shield up to the early 1980s were reviewed by Jones (1981b, 1984b) who proposed that the shield's CLC could be zoned into three general types (1981b, fig. 4) with the central part (beneath Alberta and Saskatchewan) highly resistive (10^3–10^4 Ω·m), an edge of intermediate resistivity (100–300 Ω·m), and some zones of anomalously low resistivity (10–50 Ω·m). Given the screening effects of the sediments (§2.4.3) and our knowledge of the sensitivity to the conductance of the CLC rather than its intrinsic conductivity (§2.4.3), this zonation should be treated with caution. However, Keller's (1989b) recent map of the zonation of the United States shows some similarities to Jones' map and, while some details may be in error, there may be a general validity to the zonation.

The early works of greatest reliability were the deep-probing controlled-source measurements of Connerney et al. (1980) in the Adirondacks and those of Duncan et al. (1980) in northern Ontario. In both areas a marked decrease in electrical resistivity was observed at mid-crustal depths; 23 ± 2 km to the top of a zone of 20 ± 10 Ω·m for the Adirondacks and 17–29 km to a zone of 270 ± 90 Ω·m for northern Ontario. The differences in resistivity are not considered significant: for both studies the resistivity of the lower half-space (the conducting zone) was the least-resolved parameter (see Table 4 of Duncan et al., 1980). HSG measurements in the Adirondacks by Connerney and Kuckes (1980) indicated that the conducting zone has a conductance of ≈400 S.

In contrast to Scandinavian work, deep-probing EM studies in Canada during the 1980s were more target-oriented, such as sedimentary basins (Prince Edward Island: Jones and Garland, 1986; Williston Basin: Jones, 1988a; Fredericton basin: Jones, unpubl.), earthquake zones (Buchbinder et al., 1983; Kurtz and Gupta, 1992), tectonic features (Trans-Hudson Orogen: see §3.3.4; Wopmay Orogen: Camfield et al., 1989; Kapuskasing Uplift Structure: see §3.5), the Abitibi greenstone belt (Chouteau et al., 1991), and plutons as possible sites for the disposal of radioactive waste (Kurtz et al., 1986b).

The MT study by Kurtz et al. (1986b) on the East Bull Lake pluton in northern Ontario showed a highly resistive CLC (2,500 Ω·m) underlain by an 800-Ω·m mantle at a depth of 200 km. Given that this mantle resistivity is high compared to global averages of about 100 Ω·m (Schmucker, 1985), the apparent resistivity data are possibly static shifted upwards at the longer periods by an order of magnitude. Correcting for such a shift would imply that the resistivity of the CLC here is about 250 Ω·m, which is consistent with Duncan et al.'s (1980) estimate, although it would be much shallower (6 km compared to 17–29 km for Duncan et al., 1980). The longest-period apparent resistivity values of Kurtz et al.'s (1986b) data are 3,000 Ω·m at 10^3 s, which is an order of magnitude greater than Vanyan's global average curve (Vanyan et al., 1980; Vanyan and Cox, 1983), and this would also be consistent with a static shift of that order. Obviously, the distorting effects of the resistive pluton, as illustrated by the Nelson batholith in British Columbia (Jones et al., 1988), need to be taken fully into account.

MT studies of the earthquake-prone Miramichi region of Canada's maritimes (New Brunswick) by Kurtz and Gupta (1991) showed a highly resistive crust ($> 10^5$ Ω·m) down to 20 km then a transition to 1,000s Ω·m underlain by a zone of 100s Ω·m at a depth of ≈30 km.

3.1.3. *Siberian shield*

The Siberian shield has been much studied by U.S.S.R. groups, and recently Vanyan et al. (1989) gave a review of the interpretations from more than 2500 MT soundings made in its eastern part since the late 1960s. They showed that the responses could be grouped into four type areas, each with its own conductivity-profile. Types 1 and 2 curves, which are representative of most of the area studied, had a conducting zone in the CLC of about 350–650 S with their centres between 30 and 40 km in a crust of approximately 45–50 km thickness. Type 3 curves are insensitive to a CLC conducting layer because of the screening by sediments of 500 S total conductance on the surface (see §2.4.3). Type 4 curves are indicative of a highly conducting zone of resistivity <0.5 Ω·m, at a depth of 3–5 km which also screens the CLC. Although the MT data were not acquired by modern instrumentation and were not corrected for static shifts, the existence of a zone of enhanced conductivity in the CLC beneath most of the eastern Siberian shield cannot be doubted due to the vast quantity of data available.

3.1.4. *Shields: Conclusions*

It is apparent that, in detail, the shields are almost as heterogeneous in the conductivity of their lower crusts as younger or active regions. Although statistically (§3.4) the conductance of the CLC beneath shield regions is low — of the order of 40 S — it is still far higher than would be expected for dry rock (<2 S for an CLC of 20 km and resistivity of 10^4 Ω·m, §4). Also, there is evidence that the conductance of the CLC can be an order of magnitude greater, such as beneath the eastern part of the Siberian shield and the southern part of the Baltic shield in Finland. What is causing the enhanced conductivity of these old regions? This question is discussed further in section §4.

3.2. *Rifts*

Because of their intriguing nature quite a few studies of continental rifts have been undertaken. One particular problem with imaging the conductivity-distribution of the CLC beneath rifts is that the sediments filling the rift, or the rift structure itself, are often highly conducting and mask the deeper structure. In a comparison of interpretations from four rifts, Jiracek et al. (1979) concluded that a ubiquitous feature of rifts is a zone of enhanced conductivity with $\rho < 50$ Ω·m at 10–30 km depth beneath them.

3.2.1. East African Rift

Two-dimensional model studies of MT and GDS data from the East African Rift (Banks and Ottey, 1974; Rooney and Hutton, 1977; Banks and Beamish, 1979) suggest that much of the crust within the Gregory Rift is highly conducting (10 Ω·m) from the surface down to >35 km depth, with crustal thickness in the region being about 36 km (Henry et al., 1990). Although these MT data are not of sufficient bandwidth (30–3000 s period range) or quality for resolution of fine structure within the rift, the conclusion that much of the crust beneath the rift is conducting is not easily amenable to other interpretation. A recent interpretation of the KRISP85 (Kenya Rift International Seismic Project 1985) seismic refraction study in the same region (Henry et al., 1990) presented a cartoon (their fig. 4) to explain the observations. One obvious interpretation for the enhanced conductivity is magma that permeates the whole of the crust beneath the Rift Valley. A more likely explanation is that the uppermost part of the crust is conducting because of the valley sediments, of thickness ≈3 km, below which there are zones of interconnected partial melt, as suggested by Rooney and Hutton (1977) and Banks and Beamish (1979). Any resistive layer between these two conducting zones would not be detectable in the data (Rooney and Hutton, 1977).

3.2.2. Rio Grande Rift

The Rio Grande Rift has been the subject of EM studies for three decades (Schmucker, 1964, 1970; Swift, 1967; see historical review in Keshet and Hermance, 1986), and recently MT measurements have been made across it by Jiracek and his colleagues (Jiracek et al., 1979, 1983, 1987). Jiracek's initial studies followed COCORP's* seismic reflection profiling of the rift in 1975 and 1976 (Oliver and Kaufman, 1976) that imaged a seismic reflection bright spot at 18–22 km depth beneath the rift. Jiracek's data are reasonably broad-band (0.1–1000 s), and are of good quality. Considerable along-strike variation in the conductivity distribution is evident from the MT data along two east–west profiles separated by ≈30 km.

On the northern line, through Bernardo, a thick zone of low resistivity, 10 Ω·m, which is thicker and closer to the surface to the west, was modelled in the middle to lower crust (Fig. 3-11a). This was interpreted as due to fluid trapped beneath an impermeable cap (Jiracek et al., 1983). In contrast, on the southern line through Socorro no such zone was present (Fig. 3-11b). Given the enhanced earthquake activity in the region of Socorro compared to further north, Jiracek et al. (1987) interpreted the lack of a zone of enhanced conductivity as due to magmatic fracturing of the impermeable cap which led to loss of the trapped fluid. One puzzling aspect is that where there is active magma injection to upper crustal levels on the southern profile, the crust is relatively resistive (400 Ω·m). Given that the geothermal gradient and heat flow are higher in the Socorro area (Reiter et al., 1978), this observation appears to contradict laboratory results that partially molten rocks are more conducting (Waff, 1974; Sato and Ida, 1984). Could this be due to lack of interconnectivity of the magma? **.

* COCORP: COnsortium for COntinental Reflection Profiling based at Cornell University.

** *Note added in proof:* A recent study across the Socorro magma body by Hermance and Neumann (1991) challenges Jiracek et al.'s (1987 model (Fig. 3-11b). Hermance and Neumann (1991) suggest that there is indeed a conducting zone, of 10–30 Ω·m, beneath the rift at intermediate depths (15–20 km).

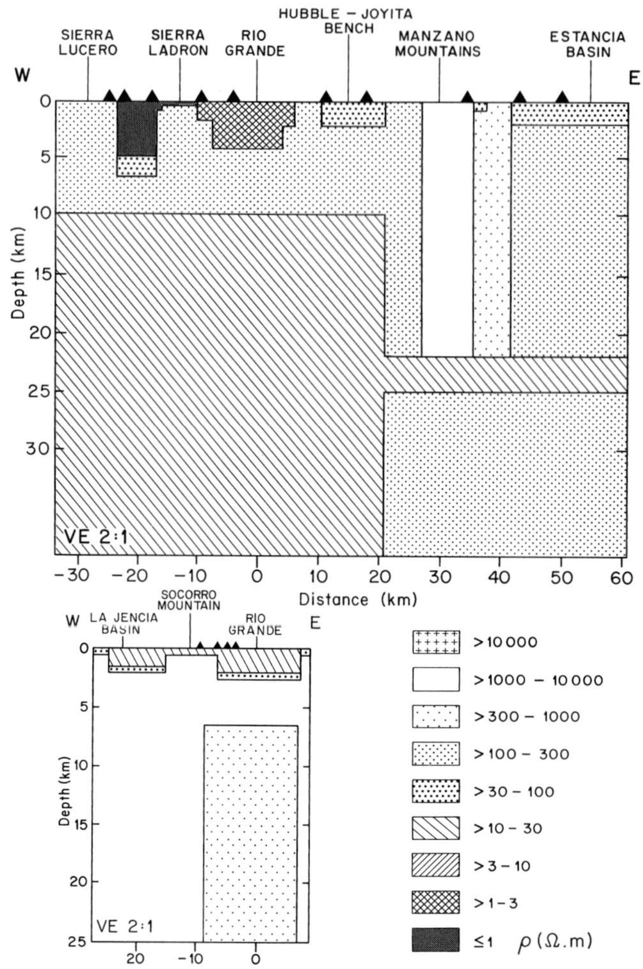

Fig. 3-11. Two 2D models of the Rio Grande Rift from profiles 30 km apart (redrawn from Jiracek et al., 1983). Note that whereas the lower crust beneath the northern profile is interpreted to be of very low resistivity (3–10 Ω·m), the lower crust beneath the southern profile is thought to be of much higher resistivity (>100 Ω·m).

Approximately 250 km to the south (32°N latitude), Keshet and Hermance (1986) interpret their observations in terms of a conducting zone in the CLC, with ρ = 1–10 Ω·m beginning at ≈20 km depth. Recent results by Jiracek and his colleagues from an E–W profile 160 km north of the Bernardo line are indicative of a strongly conducting zone, with ρ = 1 Ω·m, at depths below ≈15 km (G.R. Jiracek, pers. commun., 1991).

3.2.3. Rhinegraben

The Rhinegraben has been the focus of much EM work since the early 1970s (see the review by Schmucker and Tezkan, 1988), and initial interpretations were of a zone of enhanced conductivity in the lower crust and upper mantle (Reitmayr, 1975). However, the highly attenuating effects that the conductive sediments can have on the observations were not initially appreciated. Interpretations of the GDS and MT responses at periods shorter than 1000 s in terms of bodies of enhanced conductivity in the deep crust or mantle have

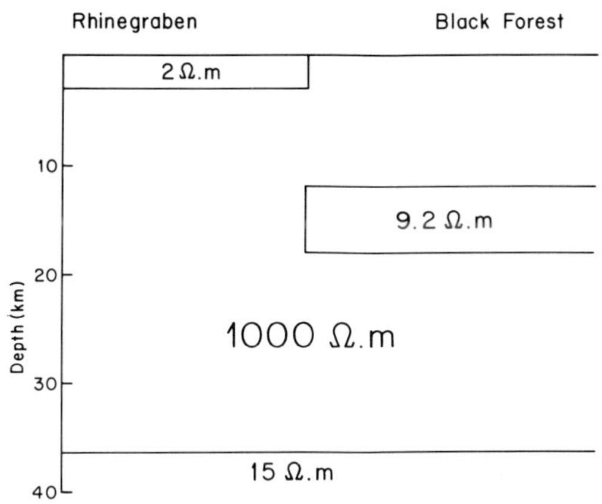

Fig. 3-12. A 2D model of the Rhinegraben with a zone of enhanced conductivity at a depth of 12 km beneath the Black Forest which is absent beneath the Rhinegraben (redrawn from Wilhelm et al., 1989).

been shown to be in error, and conductive sedimentary fill can explain these shorter period responses (Dupis and Thera, 1982; Jones, 1983b). At longer periods, however, there are features in the MT and GDS data that require explanation, and Richards et al. (1981; reported in Fuchs et al., 1987) interpret their MT data as indicating a zone of slightly enhanced conductivity between 20 and 40 km depth. Limited 3D thin-sheet modelling of the Rhinegraben area by Kaikkonen et al. (1985) was interpreted in terms of current channelling, but the modelling algorithm did not allow for poloidal current flow so that currents were confined to the surface thin-sheet and thus were insensitive to lateral variation in conductivity beneath the surficial layer.

Recently, MT and LOTEM measurements were made in the Black Forest as a component of the multidisciplinary studies for the KTB * drill site (Wilhelm et al., 1989; Strack et al., 1990). The MT data were interpreted, using a 2D model (Fig. 3-12), as indicative of a conductive zone about 12–18 km beneath the eastern flank of the Rhinegraben, with no conductive zone, apart from the sedimentary fill, beneath the Rhinegraben itself (Berktold et al., 1985; Tezkan and Schmucker, 1985: both reported in Fuchs et al., 1987; Schmucker and Tezkan, 1988; Wilhelm et al., 1989). The LOTEM data, which were inverted for 1D structure, revealed a zone of enhanced conductivity at depths of 7–9 km, at least 500 m thick (Strack et al., 1990). As the shortest period of the MT data was 10 s, it is possible that the zone imaged by the LOTEM results was not resolvable from the MT data, as suggested by Wilhelm et al. (1989). However, an alternative explanation is that the MT data are static shifted (§2.4.1): Tezkan's model comprises of a layer of 1000 Ω·m material overlying 9.2 Ω·m material at a depth of 12 km, whereas the LOTEM results indicate that the uppermost part of the crust is about 400 Ω·m with a conductive zone of about 4 Ω·m material (Wilhelm et al., 1989; Strack et al., 1990). Applying a static shift factor of 2.5 to Tezkan's model, so that the MT upper crustal resistivity agrees with LOTEM, gives a depth of $\approx$7.5 km (i.e., $12/2.5^{1/2}$) to a zone of 3.7 Ω·m (i.e., 9.2/2.5), which agrees well with the LOTEM results. The LOTEM results are insensitive to the thickness of this conducting zone, but impose a lower bound of at least 500 m (Strack et al., 1990). However, if the MT model (Tezkan and Schmucker, 1985) is scaled with a shift of 2.5,

* KTB (Kontinentales Tief Bohrung): the German Continental Deep Drilling Program.

then the base of the zone is at 11.4 km ($18/2.5^{1/2}$). Thus, Tezkan's model, when corrected for static shift with a factor of 2.5, has a zone of enhanced conductivity between 7.5 and 11.4 km, which coincides with a pronounced low-velocity layer between depths of 7–14 km, above a laminated lower crust (Gajewski and Prodehl, 1987; Wilhelm et al., 1989). Alternatively, the LOTEM data could be affected by static shift (K.-M. Strack, pers. commun., 1991), or both the MT and the LOTEM could, but the correlation of the depth to the LOTEM anomaly and the depth to the top of the seismic low-velocity zone suggests that the LOTEM data are reliable. Also, the magnitude of any static shifts of the LOTEM data would be in terms of percent rather than factors of 2.5 as for the MT. The conductive zone has been interpreted in terms of either fluids or graphitic metasediments (Wilhelm et al., 1989), with the fluid explanation being supported by the geothermal results.

3.2.4. Baikal rift

Generally, the lower crust in the Baikal region exhibits a zone of increased conductivity in the middle to lower crust (Popov, 1987, and references therein), and this zone broadly correlates with a seismic low-velocity zone. Popov (1987) illustrates that the conductance of this zone increases from 200–300 S to 800 S as the rift is approached from either flank, which implies a four-fold decrease in resistivity beneath the rift for a layer of constant thickness. The depth to this zone is considered to rise from 25 km on the flanks to 14 km beneath the rift itself. Generalized 1D electrical and seismic structures for the platform region and for the rift zone are illustrated in Figure 3-13. Note that for the lower crust the models indicate a $V_p \approx 7.0–7.5$ km/s and $\rho \approx 200$ Ω·m, whereas beneath the rift these parameters are 6.5–7.0 km/s and 30 Ω·m, respectively.

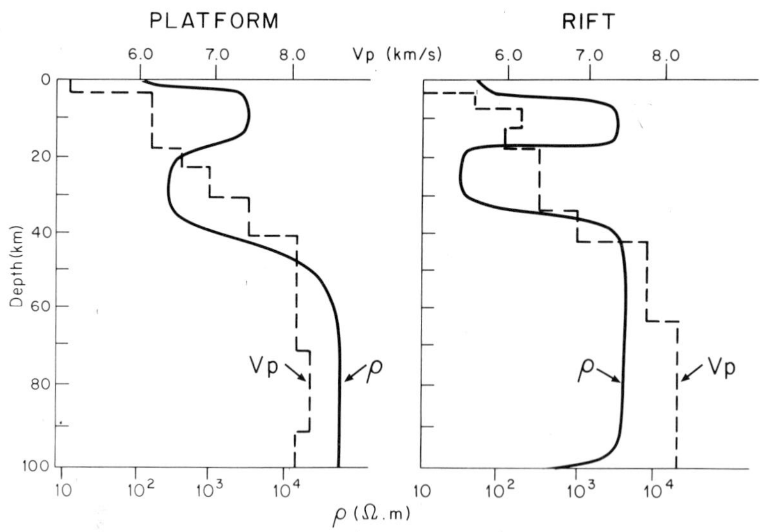

Fig. 3-13. Generalized electrical and seismic 1D models for the flank and the rift in the Baikal (redrawn from Popov, 1987).

3.2.5. Midcontinent Rift

Although the thrust of this section of the chapter is tectonically active rift zones, for comparison I have included studies over an ancient rift. The 1.1-Ga Keweenawan Midcontinent Rift (MCR) in North America has not been extensively studied by EM methods, and the only recently published results are those of Young and Wunderman (Wunderman et al., 1985; Wunderman, 1986; Wunderman and Young, 1987; Young et al., 1989; Adams and Young, 1989). One significant result in the MT data over the MCR is the asymmetry of conductivity structure beneath each of the flanks. Beneath the western flank, MT surveys imaged an easterly-dipping conducting zone, of $\rho < 10\ \Omega{\cdot}\text{m}$, to a depth of ≈ 5 km at the rift. This zone, when traced updip, is thought to correspond to rocks of the Animikie basin, with the upper contact representing a decollement zone (Wunderman and Young, 1987). The zone was also seen on proprietary commercial MT data on the western flank of the MCR in Iowa (C.T. Young, pers. commun., 1991). COCORP data from Kansas image a prominent east dipping feature in the basement to the west of the MCR which has been interpreted as "*a low angle detachment similar to those seen in the Basin and Range*" (Serpa et al., 1984). In contrast, the eastern flank does not exhibit such a conductive feature. Beneath the MCR itself, the MT data illustrated in Wunderman et al. (1985, site MR) indicate that the crust is resistive, with resistivity greater than 10,000 $\Omega{\cdot}$m down to a depth of some 30 km.

3.2.6. Rifts: Conclusions

It is apparent that Jiracek et al.'s (1979) general conclusions regarding modern rifts have to be modified in the light of more recent investigations. No single conductivity model can be proposed for rift zones, and certainly variations of structure along strike should be kept in mind. In most locations over modern rifts the lower crust generally has a high conductivity (low resistivity), which is most probably associated with an interconnected fluid phase (either partial melt or saline fluids). Intriguingly, in at least the central portion of the Rio Grande Rift there is the suggestion that a partially molten zone is resistive — an interpretation that is at odds with laboratory studies (§4.4).

3.3. Modern and ancient continental margins

Imaging the conductivity structure in the vicinity of modern continental margins is difficult because of the dominant coast-effect. The large contrast in electrical conductivity between land (typically $\rho = 1{,}000$–$10{,}000\ \Omega{\cdot}\text{m}$) and sea water ($\rho = 0.3\ \Omega{\cdot}\text{m}$) results in electric current redistributions at coastlines. In the TE-mode (current flow parallel to the coast) there is virtually a line of current flowing offshore, whereas in the TM-mode (current flow perpendicular to the coast) the currents flowing in the sea water deepen significantly on entering the land. The observed MT responses are therefore dominated by the geometry of the ocean–continent boundary, and conductivity structures beneath margins are usually of second-order on the responses. Thus, in studies close to margins it is very important to obtain the highest possible accuracy and precision in the responses in order to resolve structure.

As suggested initially by Law and Riddihough (1971), EM methods are one possible geophysical technique for delineating the locations of ancient margins interior to the continents. One cause for conductive anomalies at ancient plate margins, proposed by Drury and Niblett (1980), is the presence of subducted and obducted sediments saturated with saline brines.

Whilst this explanation may not be correct ubiquitously (e.g., Trans-Hudson Orogen, §3.3.4), the existence of conductive zones as identifiable markers in the CLC of ancient tectonic boundaries is an important observation. An alternative explanation is that the zones are associated with conductive minerals — possibly graphite — in metamorphosed and fractured rocks in the basement (Camfield and Gough, 1977).

3.3.1. West coast of North America

The most recent, and most precise, EM investigations on a continental margin have been carried out along the west coast of North America. These studies were aimed at imaging the Juan de Fuca (JdF) plate as it dips beneath the mainland of Oregon, Washington and British Columbia. The first of these studies, by Kurtz et al. (1986a, 1990; Green et al., 1987), was undertaken as part of LITHOPROBE's* Vancouver Island transect investigations. Kurtz et al. (1986a, 1990) made high quality measurements at twenty-seven locations coincident with a SE–NW seismic reflection profile. Their interpretation yielded a model (Fig. 3-14) that has a dipping conductive zone, of $\rho = 30\ \Omega\cdot m$, which correlates spatially with a strong reflection zone. The conductivity of the zone is consistent with a porosity of $\approx 1.6\%$ infilled with saline fluid (Kurtz et al., 1986a; Hyndman, 1988). Initial interpretation of the coincident reflecting

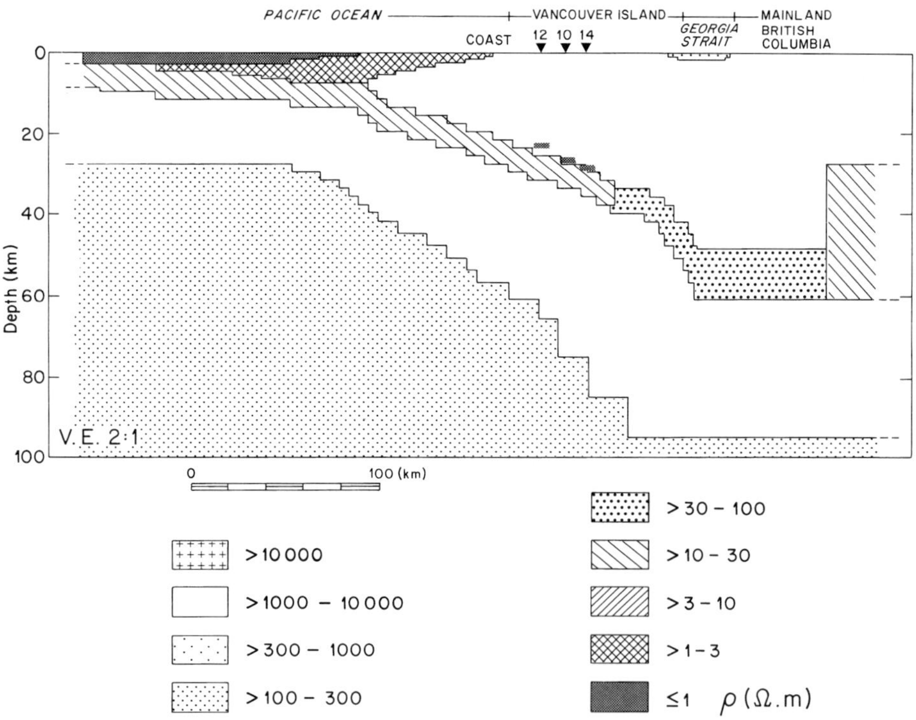

Fig. 3-14. A 2D resistivity model of the crust and upper mantle beneath Vancouver Island and the nearby mainland to account for the observed responses (redrawn from Kurtz et al., 1986).

* LITHOPROBE is Canada's national, collaborative, multidisciplinary earth science research program designed to answer fundamental questions on the nature and evolution of the lithosphere beneath Canada and its surrounding oceans.

and conductive zone suggested that it represents the top of the downgoing Juan de Fuca plate, but more recent studies, including the locating of earthquakes and further offshore seismic experiments, infer that the top of the plate is below this zone. It has been suggested that the anomalous zone may be due to sediments derived from the accretionary wedge, with fluids causing the high reflectivities and conductivities. Alternatively, a recent interpretation of the strength of seismic reflections concludes that the reflection coefficients of up to 0.2 are too high to be caused by fluids alone and that the reflections are explained by shear zones (Calvert and Clowes, 1990a,b).

Just to the south of Vancouver Island, onshore and offshore of Oregon and Washington, the largest EM experiment to date was carried out in 1985. This experiment, named EMSLAB*, involved 24 collaborating institutes supplying over 100 instruments (EMSLAB, 1989). The main results of the EMSLAB experiment are discussed in the special issue devoted to the study (Booker and Chave, 1989). Essentially, the same electrical structure was found beneath Oregon as beneath Vancouver Island. However, whereas Vancouver Island's upper crust is resistive (some thousands of Ω·m), the Coast Range of Oregon is relatively conductive (some hundreds of Ω·m) with total conductance of the same order as that of the zone associated with the downgoing plate. Accordingly, whereas for Vancouver Island the presence of the conducting zone was a first-order feature of the data (phase differences of 15°), essentially the same zone beneath Oregon was a second-order feature (phase differences of 2°) that demanded sophisticated processing, modelling and interpretation (see Booker and Chave, 1989, for a review).

3.3.2. East coast of North America

The east coast of North America is distinct from the west coast in that it is a passive margin rather than an active one. Along four widely-spaced transects, highly conducting lower crust appears to extend from the continental edge inland over 600 km beneath much of the Appalachians (Fig. 3-15; see review by Greenhouse and Bailey, 1981). Some specific details of this interpretation may be suspect because the models were derived from GDS data alone, but the general result of a highly conducting lower crust is probably correct. MT measurements in South Carolina (Young et al., 1986) show a conducting zone beneath the survey area, with the indication of a sharp rise in the depth to the layer from 20 km to some 5 km closer to the coast. The conducting layer is modelled to be of around 100 Ω·m. These results have to be followed up by more precise EM studies of the CLC beneath this and other passive margins.

3.3.3. Iapetus suture

A prominent and well known conductivity anomaly is the "Eskdalemuir anomaly" in the Southern Uplands of Scotland. This anomaly has been studied in Scotland using MT and GDS techniques since the mid-1960s (Jain, 1964; Edwards et al., 1971; Jones and Hutton, 1979a,b; Ingham and Hutton, 1982; Beamish and Smythe, 1986; Sule and Hutton, 1986), and most recently in Ireland by Whelan et al. (1990). The conductive anomaly is related to the Iapetus suture formed by the closure of the Iapetus — or Proto-Atlantic — ocean originally proposed to have existed by Wilson (1966) mainly from consideration of faunal realms. The Iapetus

* EMSLAB: ElectroMagnetic Study of the Lithosphere and Asthenosphere Beneath the Juan de Fuca plate.

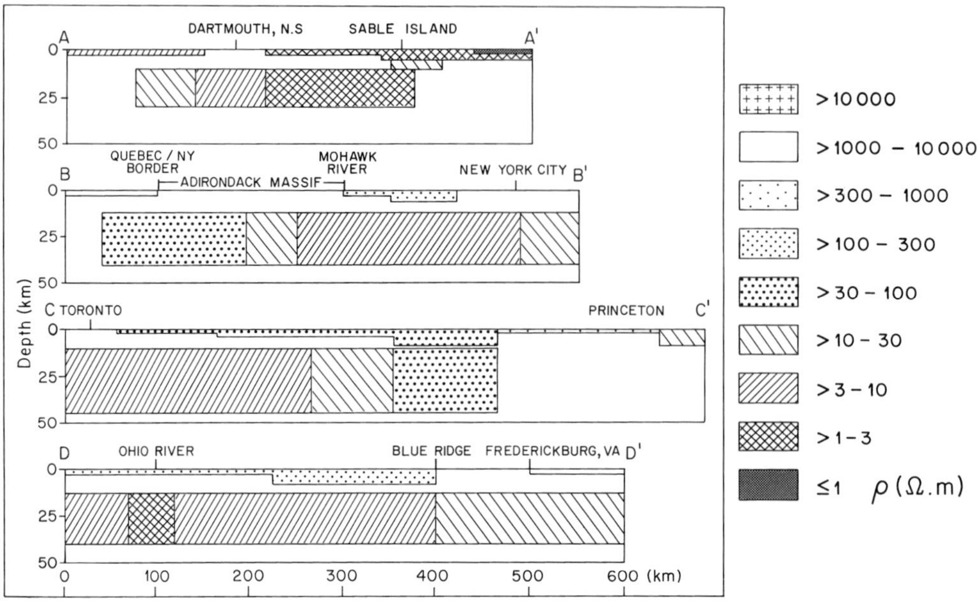

Fig. 3-15. 2D resistivity models for four profiles across the Appalachians (redrawn from Greenhouse and Bailey, 1981).

suture (Fig. 3-16) is believed to be represented in Britain by the Solway Line (Phillips et al., 1976; McKerrow and Soper, 1989a) and in Ireland by the Navan–Silvermines Fault (Phillips et al., 1976; McKerrow and Soper, 1989a), although the latter interpretation is debated (Murphy and Hutton, 1986; Hutton and Murphy, 1987; Harper and Murphy, 1989; McKerrow and Soper, 1989b). A variety of tectonic models have been proposed for the closure of the Iapetus Ocean, including:

(1) northward-dipping subduction beneath the Midland Valley of Scotland along a line now marked by the Southern Uplands Fault (Garson and Plant, 1973);

(2) southward-dipping subduction beneath northern England (Fitton and Hughes, 1970);

(3) two subduction zones (Dewey, 1969), with a 14–16° oblique angle of convergence between them (Phillips et al., 1976); and

(4) northward subduction initially located close to the present Southern Uplands Fault, but which migrated southwards with time, with subduction of continental crust from eastern Avalonia during the final stages (Legget et al., 1983; McKerrow and Soper, 1989a, and references therein). Northward subduction appears to be supported by BIRPS reflection data in the Irish Sea (Brewer et al., 1983; Beamish and Smythe, 1986) and in the North Sea (Klemperer and Matthews, 1987). These data imaged a reflective horizon dipping at 25–40° from 10 km below northern England to a depth of 25 km beneath the Southern Uplands. There appears to be a correlation between this seismic image and a northward-dipping conductive structure determined from three MT sites (Beamish and Smythe, 1986). However, updip projection of this reflecting horizon would intersect the surface in the Lake District, and thus McKerrow and Soper (1989a) consider it not to be the plate boundary but probably an intra-crustal shear zone.

Hutton and her colleagues have recorded MT data at over thirty locations in the Southern Uplands and have imaged several conducting zones in the upper and lower crust. The locations of these conducting zones are depicted in Figure 3-16 together with zones recently

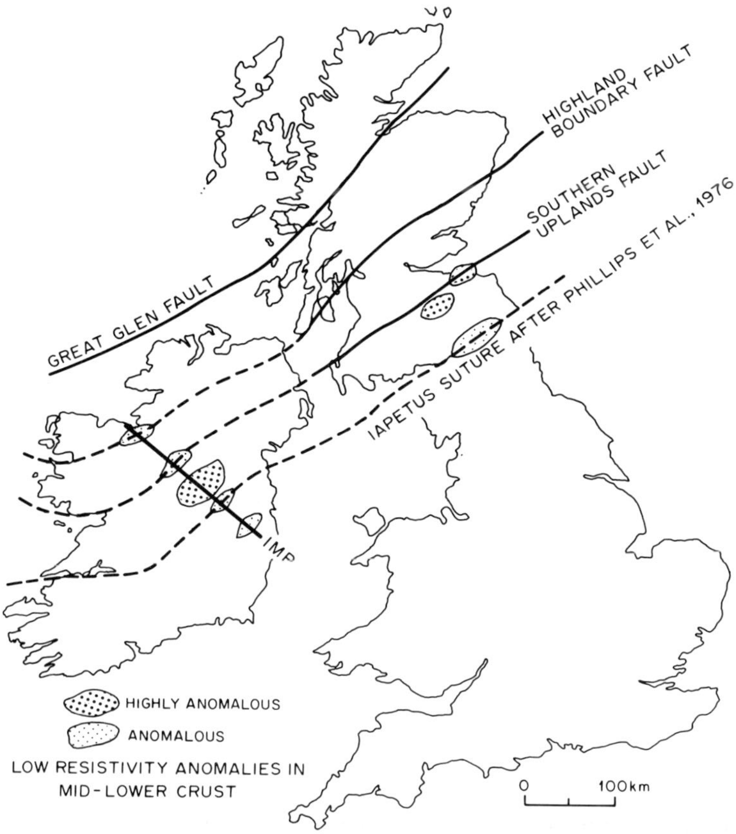

Fig. 3-16. Ireland and Britain showing the location of the two major tectonic features, the Southern Uplands fault and the Iapetus Suture, and the location of lower crustal conductivity zones (redrawn from Whelan et al., 1990).

identified in Ireland (Whelan et al., 1990 and references therein). The anomalous zone in the Northumberland Trough is the same as that identified by Beamish and Smythe (1986). The more major conductive features, labelled as "highly anomalous" zones on Figure 3-16, are all close to the surface (4–6 km in Scotland and 4 km in Ireland) and all exhibit the same geometry. The locations of these major zones also are in qualitative agreement with the induction arrows for both Scotland and Ireland (Edwards et al., 1971; Hutton and Jones, 1980). Interestingly, the trace of these zones is not parallel to either the proposed Iapetus Suture (Solway Line — Navan-Silvermines Fault) or the Southern Uplands Fault, but is oblique to them both at an angle of $\approx 5°$. The geometry of these zones appears to support Phillips et al.'s (1976) hypothesis of oblique subduction along two zones.

3.3.4. Trans-Hudson Orogen

The enigmatic North American Central Plains conductivity anomaly (NACP, Fig. 3-17) has been mapped since the late 1960s using GDS array studies (cross-hatched shading) by Gough and others (see Alabi et al., 1975 and references therein; Handa and Camfield, 1984; Gupta et al., 1985) and was postulated to be an EM expression of a Proterozoic plate boundary by Camfield and Gough (1977). In regional terms their hypothesis linked a postulated

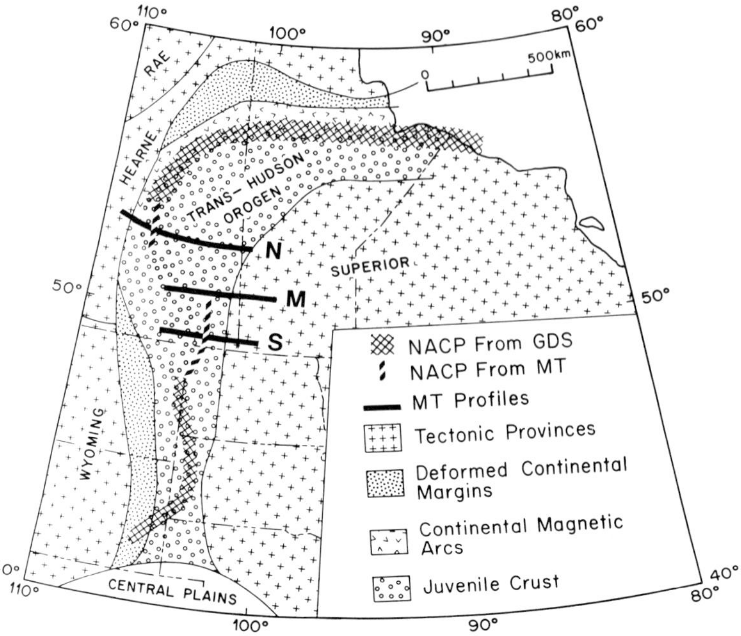

Fig. 3-17. Trace of the NACP from GDS and MT studies and Hoffman's (1988) tectonic map of North America (redrawn from Jones and Craven, 1990).

Proterozoic subduction zone in southeastern Wyoming (Hills et al., 1975) with a proposed suture in the Canadian shield (Gibb and Walcott, 1971), although the spatial correlations were not coincident. The NACP (Fig. 3-17) lies within an Early Proterozoic assemblage of juvenile crust between the Superior and Churchill (now Hearne and Rae) cratons termed the Trans-Hudson Orogen (Hoffman, 1981, 1988).

Recent MT studies in Canada over the NACP (Jones and Savage, 1986; Jones, 1988b; Jones and Craven, 1990; Rankin and Pascal, 1990) have imaged the structure with greater resolution than possible with the GDS data. They have shown that between 49° and 50° latitude the NACP is some 75 km east of the location indicated by the GDS studies, and that the NACP is not a continuous feature but exhibits offsets (Fig. 3-17; dashed shading of NACP); a feature also identified by Thomas et al. (1987) from the horizontal gravity gradient map of North America. Two-dimensional modelling of the MT responses (Jones and Craven, 1990) shows that the top of the structure is at a depth of about 9–10 km and that it is a zone of high conductivity, of $\rho < 1\ \Omega\cdot$m, with a truncated asymmetric anticlinal geometry and predominantly westward dip (Jones and Craven, 1990). A single-body model proposed by Jones and Craven (1990) to explain the observations is illustrated in Figure 3-18. Two-dimensional smooth inversions of the data from line S in Figure 3-17 essentially substantiate the main features of this model (deGroot-Hedlin and Constable, 1990; J.R. Booker, pers. commun., 1991). A body of approximately equal shape with a positive density contrast of 270 kg/m^3 was shown to explain the locally observed gravity high coincident with the NACP (Jones and Craven, 1990). Furthermore, the trace of the NACP is coincident with a zone of anomalously high heat flow and with a magnetic quiet zone.

The high conductivities of the NACP cannot be explained in terms of fluids of 50 S/m conductivity (see §4.1.4) in a porous medium, as this would require implausibly high porosities

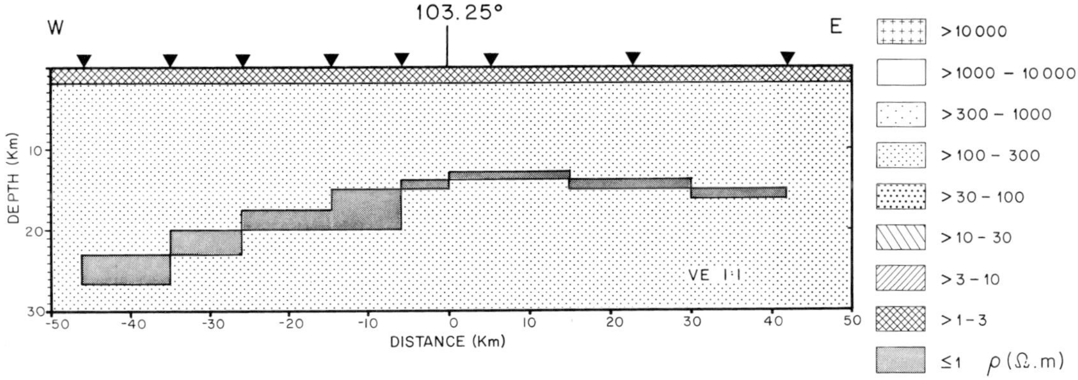

Fig. 3-18. A single body 2D model derived to explain the MT observations along profile S of Fig. 3-17 (redrawn from Jones and Craven, 1990).

of >10%. Shortly after its discovery, Camfield et al. (1971) suggested that the NACP structure might be associated with conductive minerals such as graphite in schistose rocks. Camfield and Gough (1977) discussed the spatial correlation of the ends of the NACP anomaly with exposures of graphite in southeastern Wyoming and in northern Saskatchewan, although these correlations must be revised in the light of more recent studies in northern Canada. Resolution of other more subtle variations in the conductivity of the lower crust beneath the Trans-Hudson Orogen beneath the Phanerozoic cover is not possible due to the attenuating effect of the thick conducting sediments (3 km of 3 Ω·m material, i.e., S = 1000 S, see §2.4.3).

3.3.5. Ladoga–Bothnian Bay–Skellefteå zone

By far the most interesting results from the Fennolora MT profile (§3.1.1) of Rasmussen et al. (1987) are from the Skellefteå region in northern Sweden, which is a mining district believed to be a relic of a Proterozoic island arc that separates Archean basement to the north from Svecokarelian rocks to the south. A significant EM anomaly (Storavan anomaly) in this region was first mapped by Jones (1981a) using the coarsely-spaced IMS magnetometer array. Various tectonic models have been presented for the region, including a northeast-dipping subduction zone. A high station density (average of 20 km separation compared to 40 km for the rest of the Fennolora profile, §3.1.1) ensured greater resolution of the Skellefteå structure, and the proposed 2D model to explain the observations is illustrated in Figure 3-19. Note that the conducting zone, of conductance 2000–4000 S, dips to the northeast with its base at the base of the crust. This model is similar to Jones and Craven's (1990) model for the NACP (Fig. 3-18), which is also associated with a Proterozoic collision zone of similar age (1.9–1.8 Ga). Very recent reflection profiling in the Bothnian Bay (the BABEL* experiment, BABEL Working Group 1990) shows enhanced reflectivity in the zone of increased conductivity mapped by Rasmussen et al. (1987).

Certain workers have postulated that the Skellefteå zone is an extension of the Ladoga–Bothnian Bay Zone (LBBZ, §3.1.1) into Sweden. The Finnish group has studied the LBBZ with magnetometer arrays, MT profiles around Oulu, and a 550 km profile (called SVEKA) of 62 MT sites in central Finland. The array studies (Pajunpää, 1984, 1986, 1987) confirmed

* BAltic and Bothnian Echoes from the Lithosphere.

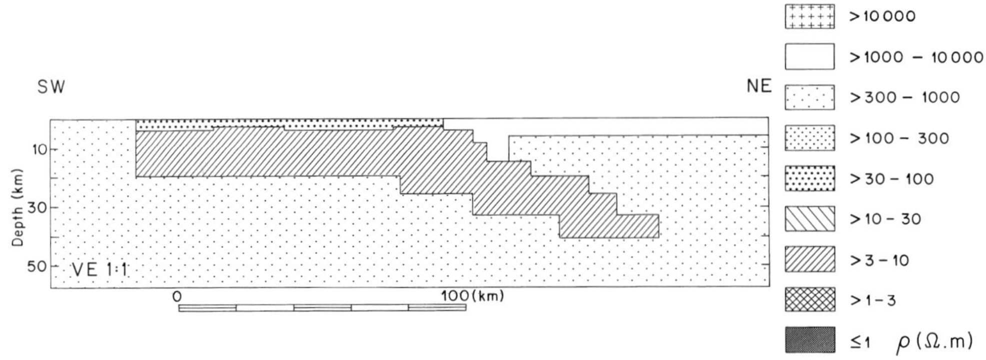

Fig. 3-19. A 2D model of the Skellefteå area in northern Sweden (redrawn from Rasmussen et al., 1987).

the existence of an anomaly around Oulu previously identified by Küppers et al. (1979), and revealed an important anomaly in southeastern Finland (the "Outokumpu" anomaly), which is along strike from the Ladoga conductivity anomaly to the northeast of Leningrad in the southeastern part of the shield (Rokityansky et al., 1981). The Ladoga anomaly was modelled as a thin sheet of anomalously high conductivity with a depth to top less than 10 km and with a dip to the north of total horizontal extent of the order of 40 km. This model was obtained from GDS measurements, which are sensitive to strong anomalies in Earth structure rather than the actual host earth structure itself. The body was modelled with a total longitudinal conductance (conductivity times thickness times length extent) of the order of 2×10^8 S·m, which for the proposed length extent of 45 km implies a conductance of the order of 4,500 S. MT measurements (Rokityansky, 1983) essentially confirmed the interpretation, although there were differences in some minor details. Interpretation of densely-spaced GDS and MT studies in the region of Oulu (Pajunpää et al., 1983; Korja et al., 1986) required a highly conducting feature (some 0.5 Ω·m) very close to the surface (4–7 km), with a slightly less conductive deeper block (5 Ω·m, 12 km) to the southwest. This anomaly was thought to be due to Archean brines. An alternative model of the TM-mode only data was presented by Golubev and Varentsov (1989), who used an unconstrained inversion program, in which the conducting zone was 4 km thick, of resistivity greater than 10 Ω·m, and total conductance less than 500 S. This exemplifies the need for both objective 2D inversion and for interpretation of all the data available, not just a limited subset.

Although interpretation of the SVEKA MT data has yet to be completed, initial results (Korja, 1990) indicate the presence of a conducting zone beneath the LBBZ which may be linked with the Ladoga and Storavan anomalies (Skellefteå zone). Such an elongated feature from northeast of Leningrad to northern Sweden was originally suggested by Rokityansky (1983), which he termed the "Transscandinavian anomaly". Comparing Rokityansky et al.'s (1981) model with that of Rasmussen et al. (1987) (Fig. 3-19), there is excellent agreement with the approximate shape of the top of the dipping portion and with the vertical conductance (of the order of 4,000 S).

3.3.6. Continental margins: Conclusions

It is apparent that at various geological times and locations the process of subduction has emplaced bodies of enhanced conductivity within the middle and lower crust which can survive for half the age of the earth. These bodies can be imaged by high-quality MT studies, and possibly by deep-probing controlled-source methods. Also, the seismic impedance contrast between these bodies and the material above them is such that there is a reflecting horizon associated with them. What is the nature of these bodies, and what causes their enhanced conductivity? Is this a ubiquitous feature of subduction? If so then the locations of all ancient active margins can be mapped using EM methods. If not, then what are the differences between the processes?

Beneath the passive margin of eastern North America there is a zone of enhanced conductivity also. What is the nature of this anomaly? Is it a ubiquitous feature of all passive margins? Certainly both active and passive margins, in particular the east coast of North America, warrant further studies to elucidate these points.

3.4 "Generic" results

A number of compilations have been made of the resistivity of the CLC (e.g., Jones, 1981b; Shankland and Ander, 1983; Haak and Hutton, 1986; Ádám, 1987; Hjelt, 1988; Keller, 1989b; Schwarz, 1990) without regard for the quality of data or interpretations and, as such, these compilations could be misleading. However, as discussed in §2.4.3, one parameter that is reasonably well resolved in MT studies and that is reasonably robust (i.e., will not change greatly after re-measuring or re-interpretation) is the total conductance S of the CLC. If an MT apparent resistivity curve is static shifted by a factor of c, the estimate of the conductance of the CLC is reduced by $c^{1/2}$. Given that static shifts display a standard deviation of one-quarter to one-half an order of magnitude (Sternberg et al., 1985), statistically 95% of the estimates of S_{CLC} will be correct to within factors of 2–3, and will be both underestimates and overestimates.

Hyndman and Shearer (1989) normalized the results of the resistivities and thicknesses of conductive zones in the CLC presented in compilations of Shankland and Ander (1983) and Haak and Hutton (1986) for a 10-km-thick zone, and also classified them into Phanerozoic and Precambrian areas. Representing their histogram in terms of conductances (Fig. 3-20) it is apparent that there is a statistical difference between the average conductance for Phanerozoic and for Precambrian areas, with the mean for the former of 400 S and the mean for the latter of 20 S.

Jones (1981b) had previously compared seismic and electromagnetic CLC values, and had surmised that the Canadian Shield may be zoned with a central resistive "core" and a more

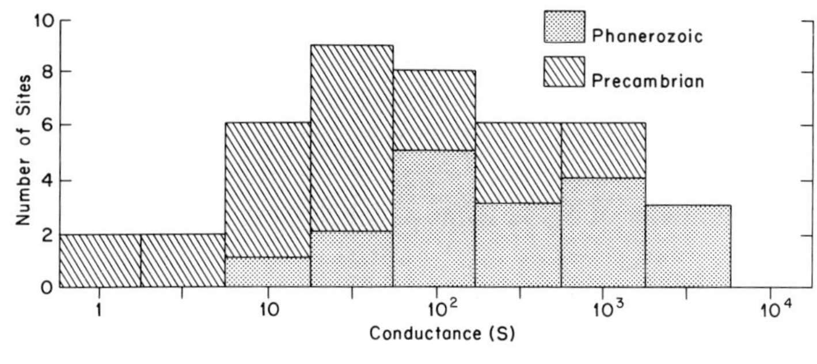

Fig. 3-20. The conductance of the continental lower crust compiled from a variety of studies and classified into Phanerozoic and Precambrian terrains (modified from Hyndman and Shearer, 1989).

conducting outer edge. This model must be modified in the light of more recent investigations, but nevertheless the lateral variation in resistivity of the CLC, even within an old shield region, is apparent.

3.5. Kapuskasing structure

The Kapuskasing uplift in northern Ontario, Canada, is of particular importance because it represents deep crustal material that has been exposed on the surface by tectonic trusting along a fault with ramp-and-flat geometry.

Originally discovered in the late-1940s by its very large gravity anomaly (Garland, 1950), it has recently been the subject of intense geoscientific study as one of the LITHOPROBE targets. The 500-km-long structure strikes NE, lies between Lake Superior and James Bay, and consists of high-grade (granulite facies) metamorphic rocks. The latest tectonic models of the region, based initially on geobarometry and geothermometry studies and gravity modelling, included a lower crustal block upthrust from 30 km depth in the Proterozoic along a major thrust fault with attendant crustal thickening (Percival and Card, 1983, 1985). A short pilot reflection study appeared to confirm this model (Cook, 1985).

Accordingly, it was considered an excellent opportunity to address the question as to whether the conductivity of the lower crust is *positionally-dependent* or *compositionally-dependent*, i.e., what happens to the conductivity when CLC rocks are brought to the surface. The initial EM experiment was a GDS array study by Woods and Allard (1986), and the results indicated that the uplifted lower crustal section was even more resistive than the surrounding upper crust. This demonstrated that the location of the rocks in the crust is important, and was interpreted by Woods and Allard (1986) as indicating that intrinsic conditions at depth rather than mineralogical composition controlled lower crustal conductivity.

An MT study by Kurtz et al. (1988) over one part of the Kapuskasing structure showed remarkable lateral homogeneity. A virtually 1D model consisting of a 40,000-Ω·m upper crust to 15 km depth overlays a 4,000-Ω·m mid-crust to 25 km. This is underlain by a 100-Ω·m lower crustal zone to 35 km and a 900-Ω·m upper mantle (to 250 km). The model was not completely 1D; the conducting zone in the lower crust was laterally bounded with a width of 30–60 km and a strike of N65°E (Fig. 3-21). However, this model generates vertical magnetic field components that should have been observed by the GDS array study of Woods and Allard (1986). As these components were not observed, Kurtz et al. (1988) proposed a model that included electrical anisotropy (see §4.5). Perhaps of significance is that the electrical strike of N65°E is similar to a number of trends on the magnetic anomaly map of the region and to diabase dyke swarms of various ages (Kurtz et al., 1988).

A similar virtual-1D conductivity structure was interpreted by Mareschal et al. (1988) and Mareschal (1990) for MT data from a profile further north over a different part of the Kapuskasing structure. Controlled-source studies across the master thrust fault (Bailey et al., 1989; Kurtz et al., 1989) confirmed the highly resistive nature and lateral homogeneity of the upthrust lower crustal material, and located some conductive features — where conductive here refers to zones of resistivity of 5,000 Ω·m — which were interpreted as due to brines in rocks with 0.5% porosity.

Seismic refraction and reflection studies led to modifications of Percival and Card's original tectonic model with the conclusion that, in fact, the lowermost crust is not exposed but that deep crustal levels (to 30 km) are brought to the surface along ramp-and-flat structures (Boland et al., 1988; Boland and Ellis, 1989; Percival et al., 1989; Geis et al., 1990), and that

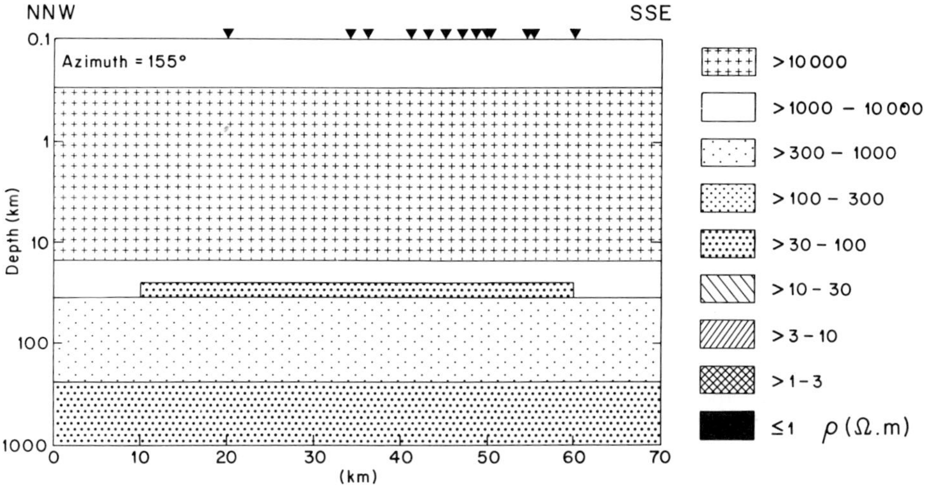

Fig. 3-21. A two-dimensional model of MT data observed across the Kapuskasing uplift structure in northern Ontario (redrawn from Kurtz et al., 1988).

there exists a decollement in the mid-crust (≈20 km in Percival et al., 1989 and ≈12 km in Geis et al., 1990). The lowermost crust is not actually uplifted and exposed, but upper lower crustal levels are. Thus, the conducting regions in the lowermost lower crust (Fig. 3-21) detected by Kurtz et al. (1988) and Mareschal et al. (1988; Mareschal, 1990) remain hidden from inspection.

The proposed crustal thickening was verified by the seismic studies, and it is of interest to note that the 100 Ω·m lower crustal block in Kurtz et al.'s model (Fig. 3-21) spatially correlates with the depression in the Moho, although the orientations of the two features are not equivalent (N45°E for the Moho depression, N65°E for the conductivity anomaly).

3.6. Valhalla complex

The Valhalla gneiss complex is a metamorphic core complex located in southeastern British Columbia which has been the subject of multidisciplinary geoscientific studies as part of the LITHOPROBE Southern Cordillera transect. Seismic reflection studies (Cook et al., 1987, 1988; Eaton and Cook, 1990) over the complex imaged a shear zone in the uppermost crust that was interpreted as bounding the complex on its western edge and being structurally related to the Slocan Lake Fault, which bounds the complex to the east. Beneath this shear zone is a major reflector, the Valhalla reflector, which is associated with a shear zone exposed to the north.

Other than these features, the seismic data show moderate reflectivity for the upper 8-s two-way travel time (TWT), after which there are few reflections to about 12 s where a weak band of events is probably associated with the Moho. This latter TWT, equivalent to ≈36 km, corresponds with a crustal thickness estimate of 35 km from a seismic refraction experiment (Cumming et al., 1979). There can be no question that the seismic energy was sufficient for sampling the lower crust; Moho reflections, although weak, were recorded.

Data from two MT stations on the Valhalla complex were reasonably 1D for periods that sampled the crust, and the conductivity model that matches the data has a highly resistive uppermost layer of ≈1,000 Ω·m to a depth of ≈9 km below which there is a less resistive zone

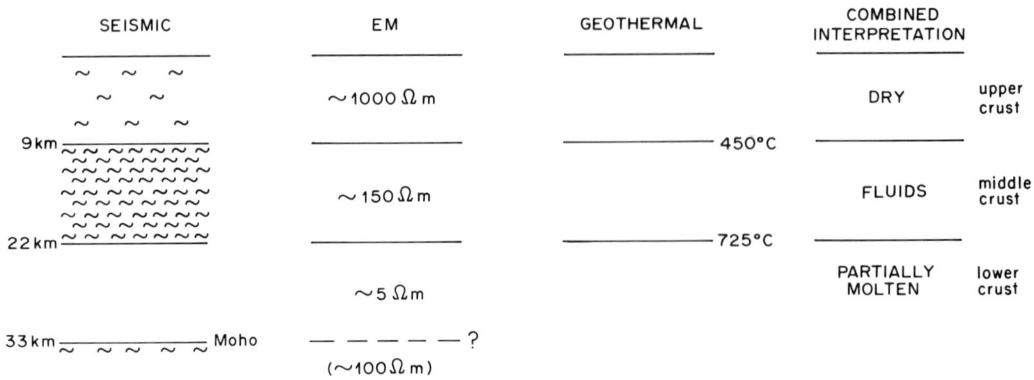

Fig. 3-22. A combined interpretation for the Valhalla gneiss complex showing a cold, highly resistive, virtually unreflective uppermost crust to ≈10 km, a warm, somewhat conductive, reflective mid-crust to ≈22 km, underlain by a hot, highly conducting, unreflective lower crust to the Moho at ≈33 km.

of 150 Ω·m down to 19 km, underlain by a highly conducting lower crust of 5 Ω·m (Jones et al., 1987). That all four ρ_a curves from the two stations lie on or close to one another is an indication that there are no severe static shift problems associated with these data. The resolution of this model is good with standard errors on the layer parameters of the order of 5–10%. Lewis (1990) and Lewis et al. (1992) recently presented a geothermal model for the complex in which he concluded, from analyses of the heat generation, that temperatures of 450°C are reached at a depth of 12 km and 730°C at 21 km. In summary, the upper crust (depths <10 km) is moderately reflective, highly resistive (thousands of Ω·m), and cold (<450°C). The mid-crust (10–22 km) is reflective, moderately conductive (hundreds of Ω·m), and warm (450–730°C). The lower crust (>22 km) is non-reflective, highly conductive (≈5 Ω·m), and hot (>730°C). This crustal stratification into three zones is shown in Figure 3-22. Lewis's (Lewis, 1990; Lewis et al., 1992) interpretation of this stratification involves a "dry" upper crust, a "wet" mid-crust, and a partially molten lower crust.

3.7. Results: Conclusions

It is apparent from the above information that it is not possible to construct a generic model of the conductivity of the CLC. Detailed investigations illustrate that there are many anomalous conductivity features within the CLC, and given that EM data are sensitive to high-conductivity anomalies rather than featureless zones, any global compilations are prone to bias and should be treated cautiously — every locale must be considered unique unless otherwise shown differently.

Notwithstanding these remarks, there is the strong indication that at many localities the electrical conductivity increases sharply at mid-crustal depths (15–20 km). Whether these zones of enhanced conductance are typically thin and highly conducting, or thick and moderately conducting, is unknown at this time. Also, there is the suggestion that the CLC becomes more resistive with age (§3.4), and that perhaps shields that are sufficiently large may have a central, cold resistive core surrounded by an edge displaying more transitional geophysical parameters (Jones, 1981; Keller, 1989b). Whether there is an age dependency or a function of proximity to the edge, where processes such as fluid infiltration may be controlling the electrical conductivity, should be determined by EM studies on shields of varying sizes.

4. Causes of enhanced electrical conductivity in the continental lower crust

The high electrical conductivity of the lower crust observed in many parts of the world is an enigma that needs to be addressed. During the 1960s and 1970s, any structure within the CLC of resistivity less than thousands of Ω·m was considered "*anomalous*" because it was known from laboratory measurements that dry rocks at the appropriate temperatures and pressures have resistivities in excess of 10^4 Ω·m (Brace, 1971). More recent work on the resistivities of dry rocks that are possible lower crustal material yield values in excess of 10^4 Ω·m at mid-crustal temperatures (400°C) decreasing to 10^2 Ω·m for silicic rocks and to 10^3 Ω·m for mafic rocks at 800°C temperature (Kariya and Shankland, 1983).

Recent compilations of lower crustal resistivities from various parts of the globe (§3.4) illustrate that *resistive* CLCs (resistivities in excess of some thousands of Ω·m) are the exception rather than the rule, and that resistivities of hundreds of Ω·m, with the top of the zones in the mid-crust, are more common. Thus, according to Haak and Hutton (1986) we should now consider that the CLC is *normally* within the range 10^2–10^3 Ω·m, and that values above and below this range can be classified as anomalously high or anomalously low respectively. Such a classification was also adopted by Keller (1989b) in his zonal subdivision of the continental United States. As stated in §2.4.3, it is virtually impossible with 1D models to differentiate between a thin zone of high conductivity and a thicker zone of lower conductivity. Thus, many of the zones interpreted to be of 300 Ω·m and 20 km thickness *could*, in fact, be 30 Ω·m and only 2 km thick, for example. Accordingly, for the present the CLCs should be classified by their conductance values *not* their interpreted resistivities. This has the additional attraction that static shifts of the apparent resistivity curves only bias the estimates of the conductances of the CLC by the square root of their multiplicative factors, whereas they bias the estimates of the resistivities of the zones by the factors themselves. Thus, whereas the definitions of the CLCs into "*normal*" and "*anomalously high*" based on the resistivities is suspect, a categorization in terms of S_{CLC} (Fig. 3-20) may have some meaning, as shown (indirectly) by Hyndman and Shearer (1989). Obviously, one first-order question that needs to be addressed by the EM community is the thickness of the conductive zone within the CLC. MT methods are not likely to provide the answer and specifically-designed controlled-source EM experiments must be undertaken.

What are the causes of the observed enhanced conductivities? Many have been proposed during the last two decades but consideration will be given here to the four most likely candidates: (1) saline fluids; (2) carbon grain-boundary films; (3) conducting minerals; and (4) partial melts.

During the mid- to late-1970s, it was common to interpret enhanced conductivity as due to hydrated minerals, especially serpentine or serpentinized rocks (e.g., van Zijl, 1977; Connerney et al., 1980; Jones, 1981b; de Beer et al., 1982) as laboratory studies had shown that serpentinized rocks exhibited high electrical conductivity (Stesky and Brace 1978). However, in an experiment in which the outgasses were carefully controlled, Olhoeft (1981) showed that at lower crustal conditions a hydrated hornblende schist with structural water had an electrical resistivity very similar to that for dry granite (Fig. 3-23). This illustrated that structural water appears to have no significant effect on electrical properties, contrary to the comments by Lee et al. (1983). For this reason, compositional contrasts, as advocated most recently by Baker (1990) to explain the conductivity of the CLC, are not considered here.

Duba (1976) questioned whether interpretations of laboratory measurements on rock samples were meaningful given that it is impossible to recreate the *P-T-t* (pressure-temperature-

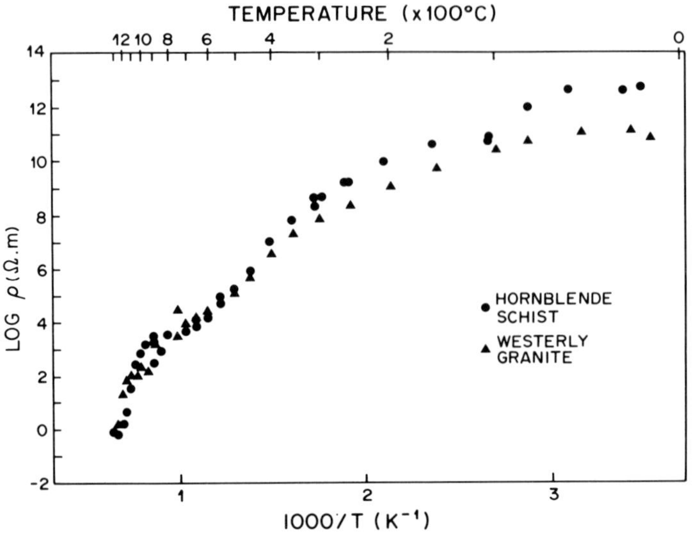

Fig. 3-23. Log electrical resistivity versus reciprocal temperature for Westerly Granite and hornblende schist. The two curves are essentially identical at lower crustal temperatures even though the hornblende has considerably more structural water than the granite (redrawn from Olhoeft, 1981).

time) conditions and environment of the CLC. In particular, simply wetting rocks is inappropriate because of the chemical reactions that take place, and, most importantly, the time element can never be reproduced. Given our understanding of the "ductile" nature of the CLC on geological time scales, this latter point is critical. It has been shown that holding a sample at certain *P-T* conditions for even different short lengths of time (1500 h instead of 500 h) can lead to an increase in the conductivity of albite by three orders of magnitude at critical temperatures just below the onset of partial melting (Piwinskii and Duba 1974; Duba 1976).

4.1. Saline fluids

Saline fluids have been proposed as an explanation of lower crustal conducting anomalies since such features were first discovered (Hyndman and Hyndman, 1968; Brace, 1971; Parkomenko et al., 1972). The dramatic effect that even small amounts of saline fluids have on the electrical conductivity is well illustrated in Figure 3-24, and many workers have appealed to the presence of aqueous fluids with a high ionic content to explain the conducting CLC (Shankland and Ander, 1983; Lee et al., 1983; Haak and Hutton, 1986; Jones, 1987; Vanyan and Shilovski, 1989). A model proposed to explain both the observed enhanced electrical conductivity and the increased seismic reflectivity presented by Jones (1987) consists of a thin zone of interconnected fluids trapped beneath an impermeable zone, which is at a temperature of the order of 400°C (Fig. 3-25). At greater depths, where the temperatures are in excess of about 600°C, the resistivity increases because the fluids are no longer interconnected. This model is based upon the fluid circulation model of Etheridge et al. (1983, 1984) shown in Figure 3-26. Field evidence for the growth of an impermeable zone sealed by mineral deposition at a temperature of 370°C is given in Fournier (1991). Jones' model (1987) was examined by Hyndman (1988), who considered the petrological consequences in terms of metamorphic facies. Hyndman (1988) showed that the temperature range 400–700°C covers greenschist to amphibolite facies conditions, for which petrological studies indicate

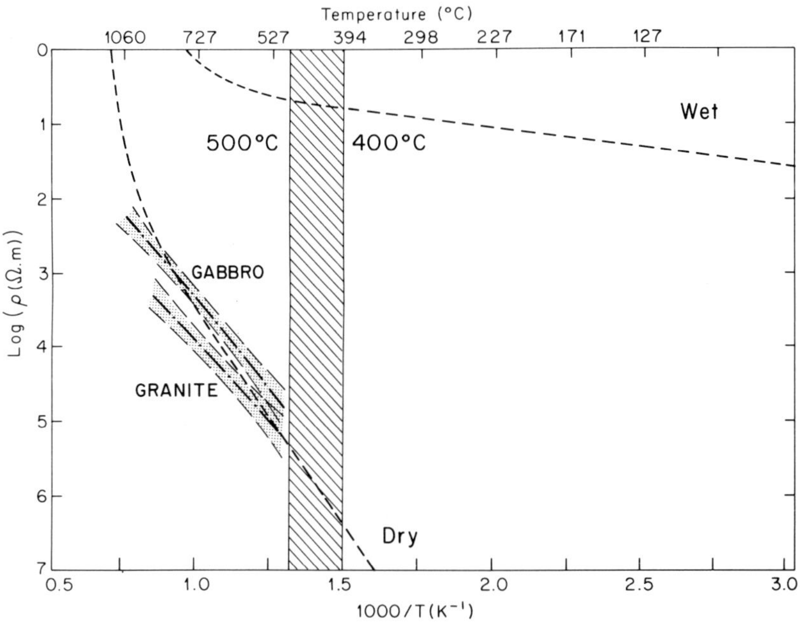

Fig. 3-24. Log electrical resistivity versus reciprocal temperature for dry granite and gabbro compared to their wet counterparts. (redrawn from Shankland and Ander, 1983). Note that in the temperature range believed to be appropriate for the upper part of the lower crust (400–500°C; shaded region) there is five orders of magnitude difference in resistivity.

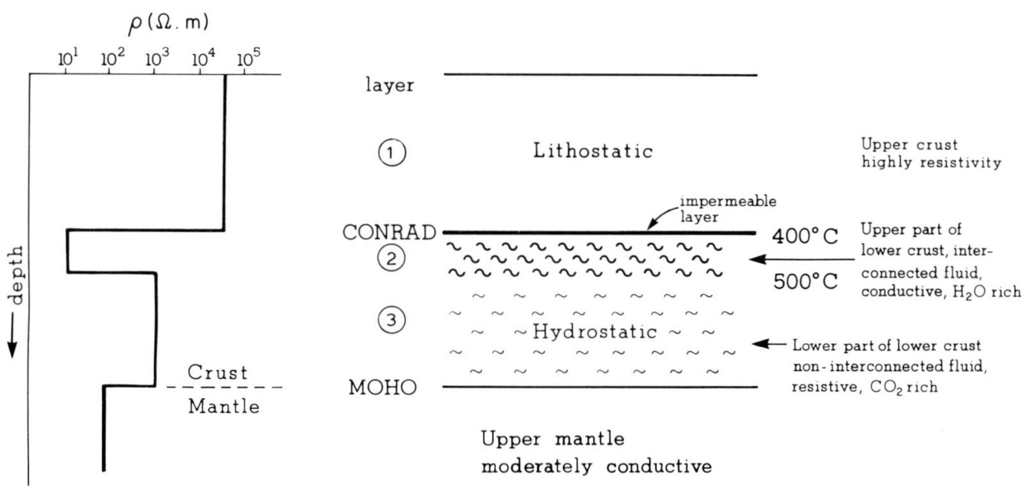

Fig. 3-25. A model for the continental crust with a zone of interconnected H_2O-rich fluids trapped beneath an impermeable layer (the seismic "Conrad" discontinuity?) at mid-crustal depths at a temperature of 400°C. At deeper depths within the lower crust the temperature is greater, the fluids are CO_2-rich, and the resistivity is greater (modified from Jones, 1987). The difference in pressure regime, with the upper crust at hydrostatic pressure and the lower crust at lithostatic, is possibly an explanation for the highly resistive nature of the upper crust.

chemical equilibrium with free saline water at close to lithostatic pressure.

A number of questions arise when a "wet" lower crust is postulated: (1) Do the fluids exist? (2) What are the fluids made of? (2) Do they stay there? (3) Do they move? and (4) What is the likely conductivity of the saline fluids?

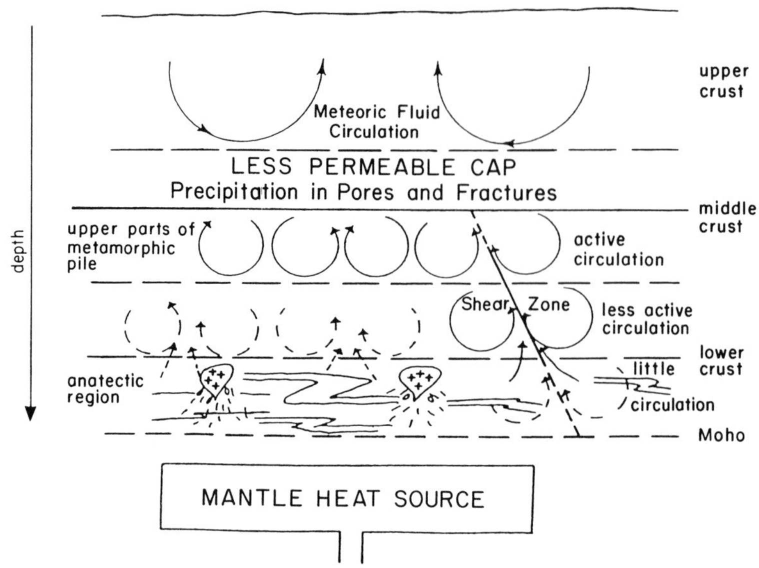

Fig. 3-26. A model of the fluid circulation within the continental crust with the development of an impermeable cap at mid-crustal depths (redrawn from Etheridge et al., 1983).

4.1.1. Existence and types of fluids

The existence of fluids within the deep continental crust is a topic that provokes intense debate within much of the geoscientific community, and recent workshops have focussed on the general themes of crustal fluids and fluid movements (see Bridgwater, 1989; Fountain and Boriani, 1990; Langseth and Moore, 1991; Torgersen, 1991).

In certain geographical areas there is general acceptance that fluids are present at least locally. This is particularly so in active subduction zones where it is well-known that at depths less than 40 km there is a large amount of free water available from expulsion of pore waters and from CH_4-H_2O fluids produced by diagenetic and low-grade metamorphic reactions (Peacock, 1990). At greater depths, H_2O and CO_2 are released by metamorphic reactions in the subducting oceanic crust. It is estimated that the transport rate of H_2O down subduction zones is of the order of 10^{15} g/yr (Fyfe et al., 1987). These fluids may have been detected using the MT method in two studies of the subducting Juan de Fuca plate (§3.3.1).

Is there evidence for free fluids in stable regimes?

There appears to be acceptance by some geologists and geochemists since the mid-1970s that fluids can exist in the CLC and that they account for a variety of metamorphic and petrological processes. It has been proposed that in certain cases magma generation in the deep crust by crustal anatexis requires the ingress of external fluid (Wickham and Taylor, 1987). In his review of the evidence for fluid flow during regional metamorphism, Rumble (1989) emphasized the advances that are currently taking place in metamorphic petrology due to an understanding of the dynamic physical processes responsible for metamorphism such as heat flow, fluid flow and tectonic advection. As noted by Rumble (1989), compared to grain-boundary permeability for which fluid–rock ratios of 1:1 to 4:1 are required, fracture permeability requires fluid–rock ratios of 330:1 for quartz to 600:1 for graphite. The presence of hydrothermal graphite in a sillimanite zone in New Hampshire is evidence for pervasive fluid flow throughout the 2000 km^2 of the studied zone (Rumble et al., 1986), and Rumble (1989) tentatively concluded that there is significant intergranular permeability at depths of 20–25 km.

One enigma is the apparent paucity of data for inferring the existence of fluids in the CLC (Yardley, 1986a). This is discussed further below (§4.1.5).

What types of fluids are there at various crustal depths?

Studies of fluids in the C-O-H system on metamorphic rocks from the Grenville Orogen and the Alps are interpreted as suggesting a three-layer stratification of the crust: a methane-rich upper crust, a H_2O-dominant mid-crust, and a CO_2-dominant lower crust (Frey et al., 1980; Touret, 1987; Skippen, 1988). This is consistent with Frost et al.'s (1989a) hypothesis that magmas may form conduits by which mantle CO_2 is transported into the lower crust while H_2O is moved to higher levels from the lower crust. The temperatures inferred from microprobe analyses of the Alpine rocks are 450–550°C for the mid-crust, and 600–800°C for the lower crust (Frey et al., 1980), which is consistent with CO_2 being less soluble in silicate melts than H_2O leading to CO_2-saturated melts crystallizing at higher temperatures than those that are H_2O-saturated (Frost et al., 1989a). H_2O-rich fluids cannot exist in the deep crust at temperatures in excess of 650°C because silicate melts would replace them (Newton, 1990), and beyond even 500°C wollastonite crystals begin to grow by which process CO_2, released by the metamorphic reactions, begins to dilute the H_2O (Rumble, 1989). That crustal fluids at high temperatures are H_2O-poor and CO_2-rich was also interpreted from fluid inclusion studies of the Laramie Anorthosite Complex (Wyoming) which crystallized at around 950°C (Frost and Touret, 1989). These mid-crustal temperatures of 450°C are close to those proposed by Jones (1987) for the thin crustal conductor, and the fluid consequences are included in the model of Figure 3-25.

4.1.2. Retention/replacement of fluids

One objection to the proposal that saline fluids are the cause of high conductivity in the CLC is, in the absence of a recharge mechanism, the apparently unreasonably long retention times required. Etheridge et al. (1984) suggested that there exists an impermeable cap in the mid-crust beneath beneath which fluids are trapped. Such an impermeable cap is considered to be untenable because the pressure at the top of a sealed lithostatic reservoir will soon exceed the local lithostatic pressure and hydraulic fracturing will break the cap (Yardley, 1986b). Subsequently, the trapped fluids will escape and migrate upwards (McCaig, 1988). This then raises the question of the rate of fluid migration. A recent theoretical study (Bailey, 1990) showed that the retention time is critically dependent on crustal temperature. For 1% initial porosity, fluids at 30 km depth would take 100 years and 3×10^8 years to rise to the surface for crustal temperatures of 800°C and 400°C, respectively. Assuming an average crustal temperature of 400°C yields retention times of 0.1–1 Ga. Degassing rates of ^4He from depths of 40 km are also consistent with a 1-Ga-time scale (Torgersen, 1989). Also, once the cap is broken will all the fluid escape or will the cap re-seal after loss of sufficient fluid to reduce the pressure to lithostatic as discussed by Fournier (1991)?

Alternatively, the fluids could be continuously recharged from the mantle, as proposed by Haak and Hutton (1986). This poses the question of why the upper crust is so resistive if fluid is passing up through it. Gough (1986) proposed that the electrical state of the crust depends on the different stress regimes. In his (Gough, 1986) generalized model of the continental crust, the upper crust is electrically resistive, seismically transparent and brittle. In contrast the lower crust is conductive, reflective and ductile. Gough suggested that saline fluid permeating the whole crust is under compressive stress in the upper crust, and thus is in separated cavities, whereas in the lower crust the fluid forms interconnected films on the crystal surfaces.

4.1.3. Large-scale fluid circulation

Fluid-driven mass transport in the crust is a necessary prerequisite for emplacement of ores in faults and fractures, and fluids from the surface are often interpreted to convect to great depths (10–20 km) within the continental crust (see McCaig, 1989, and review by Torgersen, 1990). Seismic pumping is thought to be a mechanism by which the fluids are brought to the surface along wrench faults (Sibson et al., 1975).

Crustal fluid circulation models have been presented by Etheridge et al. (1983, 1984) in which there is depth-dependent circulation rates beneath an impermeable cap that is sealed by precipitation in pores and fractures (Fig. 3-26). This model has not been generally accepted (Wood and Walter, 1986; Valley, 1986; Yardley, 1986b), but other evidence exists to support large-scale fluid circulation. For example, ^{4}He degassing fluxes suggests that the whole crust degasses on a 1-Ga time scale. This is possibly the first observational support for a grain-boundary diffusion process that operates on crustal/continental scales (Torgersen, 1989, 1990). Furthermore, studies of Lewisian gneisses in NW Scotland were interpreted to indicate that fluid circulation along cracks had been active for over 2 Ga (Hay et al., 1988). These studies support the concept of rapid, ubiquitous and large-scale mass transport by fluids in the Earth's crust (Torgersen 1989, 1990).

Obviously, fluid movement will be at two rates. It will be very slow, of the time scales given by Bailey's (1990) calculations, in a relatively unfractured rock matrix. Once fluids enter a fracture/fault system, however, they will move rapidly. Such systems can exist at the frictional-quasiplastic ("brittle–ductile") transition, above which fluid pressures are hydrostatic but below which are lithostatic (McCaig, 1988) (Fig. 3-25). This is an alternative explanation to Gough's (1986) for the high resistivity of the upper crust: fluids cannot exist in the upper crust for long geological times and are quickly expelled.

4.1.4. Fluid conductivity

Pore geometry (§4.6) is obviously an important factor when considering "effective porosity". As important, however, for interpretation of EM data is an accurate assessment of the fluid conductivity, σ_f. Hyndman and Shearer (1989), using data from Quist and Marshall (1968) and Ucok (1979), illustrated the effect of increasing salinity on fluid resistivity at 500°C and 4 kbar (Fig. 3-27), and adopted a fluid conductivity of 50 S/m (resistivity of 0.02 Ω·m), equivalent to an NaCl concentration of 5%, for their calculations. As stated by Hyndman and Shearer (1989), the values from Figure 3-27 should be considered *maximum* values of resistivity because of the necessary extrapolation from 4 kbar to 6–10 kbar pressure appropriate for the CLC.

However, fluid inclusion studies from Alpine thrust faults show typically 20–25 wt.% dissolved salts (Grant et al., 1991), from deep-penetrating shear zones up to 40 wt.% NaCl (S. Tempest, reported in McCaig, 1989), and from the Laramie Anorthosite Complex indicate 50–60 wt.% NaCl (Frost and Touret, 1989). The occurrence of highly saline waters at depth beneath the Canadian Shield has long been known from mines (Lane, 1914), and borehole

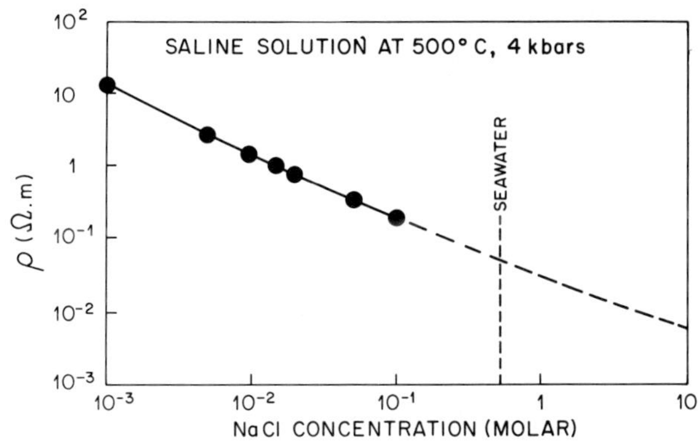

Fig. 3-27. Resistivity as a function of NaCl concentration at 500°C and 4 kbar pressure (redrawn from Hyndman and Shearer, 1989).

studies show increasing salinity of the groundwater with depth, with salt concentrations in excess of 30% by weight from depths of 1.5 km (Frape and Fritz, 1987). These borehole data are interpreted by Frape et al. (1984) to indicate the existence of a common source brine at great depth beneath the whole Canadian shield, with total salt concentrations of ≈40 wt.% in the parent brines (Gascoyne et al., 1987). The origin of the salinity is still debated, but theories are either for an allochthonous origin, such as sedimentary basin brines or modified Paleozoic seawater, or an autochthonous origin, such as leaching of fluid inclusions, grain boundary salts or weathered mineral phases (see Frape and Fritz, 1987).

From Figure 3-27 it is apparent that an order of magnitude increase in salinity concentration produces approximately an order of magnitude decrease in the electrical resistivity of the solution. This, in turn, will reduce the thickness required to produce observed conductances by an order of magnitude. *

4.1.5. Petrological arguments against free fluids

The interpretation of the observed enhanced electrical conductivity being due to acqueous fluids is not accepted by some petrologists who argue for a dry lower crust. Their thesis, summarized succinctly by Yardley (1986a), is that the rocks which predominate within the CLC are granulite facies assemblages. Such rocks are only stable in the presence of fluid at the extreme metamorphic formation temperatures of 700–800°C, whereas the temperature at the top of the zone of enhanced conductivity in the mid-crust is believed to be around 400°C. Water penetrating granulites at these lower temperatures are consumed by hydration reactions to form lower-grade hydrous minerals until the granulite facies assemblage is entirely retrograded. These retrograde reactions should be detected in the CLC rocks exhumed. Accordingly, areas undergoing active metamorphism or plutonism will be full of fluids, but large-scale fluid flow should not occur in stable regions (B.R. Frost, pers. commun., 1990). Thus, the fluid explanation is acceptable for areas of active subduction or rifting, but from these petrological arguments it is thought to be invalid for the CLC beneath stable regions.

It is interesting to note the results of Hulsebosch and Frost (1989), which suggest that large quantities of saline brines penetrating into the deep crust down zones of ductile deformation

* *Note added in proof:* Recent work by Nesbitt (1992) shows that the resistivity decrease asymptotically approaches $\approx 10^{-2}$ with increasing salt concentration, rather than the simple increase shown in Fig. 3-27.

do not produce major chemical modifications. If this is so, then perhaps it is not surprising that the petrology of rocks from the CLC brought to the surface does not indicate the existence of saline brines.

Sanders (1991) proposed that the petrological evidence for a dry predominantely granulitic lower crust, and the proposal that fluids are the cause for the enhanced electrical conductivity, can be resolved by considering the lower crust to consist of "kilometre-scale granulite-facies lenses (mega-augen) separated from each other by a system of anastomosing, gently-dipping brine-soaked ductile shear zones". The brine in the shear zones is postulated to be stable, and that retrogression into the mega-augen fails because the hydration reactions on the margins form an impermeable seal preventing further ingress of water. Sanders (1991) details the observations that are consistent with this model, but concludes that the model raises many questions which can only be satisfactorily addressed when the interaction of brine and water-deficient rock is better understood.

4.2. Grain-boundary films

The most recent explanation for observed high conductivities concerns films of carbon on grain boundaries (Duba et al., 1988; Frost et al., 1989b). Duba and Shankland (1982) proposed that grain-boundary carbon was a possible cause of a high-conductivity layer in the upper mantle and Jödicke (1985) had suggested that carbon on the grain boundaries was the cause of the low resistivity of black shales. Fluid-inclusion and Raman studies on rocks from the 1.4 Ga Laramie Anorthosite Complex indicate that the rocks crystallized at pressures of 3 kbar and temperatures of 950–1030°C in the presence of graphite and a nearly-pure CO_2-CO vapour phase (Frost et al., 1989b). Graphite films of $\approx$ 1,000 Å thickness were observed using Auger spectrometry on the grain boundaries, and Frost et al., (1989b) suggest that an interconnected carbon film may explain the high conductivities of the CLC.

This phenomenon appears to be depth-dependent: at depths where the temperatures exceed rock crystallization temperatures, CO_2 is excluded as a candidate because it is an insulating liquid and a non-polar gas, and thus has little or no effect on electrical properties (Olhoeft, 1981; Kay and Kay, 1981). Equally, rocks crystallizing at shallow depths may not reach graphite saturation (Frost et al., 1989b). Also, as shown by Frost et al. (1989b) there is a grain-size dependence of the expected conductivity that is given approximately by $\log[\sigma] = -4 - \log[x]$, where σ is in S/m, and x is the grain size in metres. This converts to $\rho = 10^4 x$, where ρ is the resistivity in Ω·m. Small-grain rocks (<1 cm) are conductive (ρ <100 Ω·m), whereas large-grain rocks (>10 cm) are resistive (ρ >1000 Ω·m). Such a grain-size dependence was previously noted by Kariya and Shankland (1983) in their statistical study of the conductivity of dry lower crustal rocks. In the ductile conditions of the lower crust one would expect that large-sized grains would predominate.

Given the petrological arguments advanced against fluids as a cause of enhanced conductivity in stable regions, EM practitioners are beginning to appeal to the explanation that grain-boundary films of graphite are responsible, e.g., Mareschal (1990), Haak et al. (1991) and Jödicke (1991). For some zones of enhanced conductivity for which there exists careful corroborating evidence this explanation may be acceptable. However, certain questions must be addressed if we are not to repeat the same mistake that we made with serpentine in the 1970s (see §4) by interpreting ubiquitously such regionally-widespread zones in the CLC as due to interconnected grain-boundary thin films over many hundreds of kilometres.

4.2.1. Wetting angle paradox and carbon mobility

Studies of the dihedral wetting angle θ in the presence of C-O-H fluids indicate that the angle will be too great for wetting by pure CO_2 or pure H_2O (Watson and Brenan, 1987; Yund and Farver, 1990; Lee et al., 1991). This has the consequence that, in the absence of existing fractures, fluid transport in such systems is limited to transient events during which the fluid pressure exceeds the confining pressure and crack propagation occurs. However, introduction of only small amounts of solutes (e.g., NaCl) to H_2O causes a substantial reduction in θ with consequent wetting along all grain boundaries and possible pervasive fluid transport. If this is the case, why is graphite observed by Frost et al. (1989b) along the grain boundaries and not just at the grain junctions? Perhaps the fluids present are not either pure CO_2 or pure H_2O, but a mixture of the two. However, as discussed previously (§4.1.1) fluid stratification into a H_2O-rich middle crust and a CO_2-rich lower crust is indicated, with little mixing likely, and Frost et al. (1989b) remarked that their rocks crystallized in the presence of a nearly-pure CO_2-CO vapour phase.

Laboratory studies of the mobility of carbon in olivine (Tsong et al., 1985) and in dunite (Watson, 1986) grain boundaries illustrated that carbon is sufficiently immobile as to preclude development of a continuous carbon film by diffusion from an initially localized source. With diffusivities of the order of 10^{-10} cm^2/s, carbon would move only a few metres in 200 million years (Watson, 1986), from which one must conclude that regional continuity over hundreds of kilometres of a carbon film derived initially from localized sources of carbon cannot be achieved. Thus, the existence of a CO- and/or CO_2-rich fluid that had previously infiltrated the region in question must be postulated to explain the existence of a hypothetical carbon film.

4.2.2. Untestable hypothesis?

One unfortunate aspect of Frost et al.'s (1989b) proposal is that the hypothesis is possibly untestable. It is argued that when lower crustal rocks are tectonically exhumed, decompression destroys the continuity of the fine graphite film along the grain boundaries, leading to an increase in the resistivity of the rock. In addition, laboratory studies are thought to be of little consequence because the discontinuous film cannot be reconnected under laboratory conditions, thus making it impossible to observe the high conductivity either in the field or in the laboratory.

As a possible example of this conundrum, measurements on graphitic pelites from northern Canada, which were of such high graphite content that they were black in colour, gave values in excess of 10^5 Ω·m (Camfield et al., 1989). This is explained by Frost et al. (1989b) as destruction by fracturing of the grain-boundary films, and that the relatively coarse individual grains seen are unimportant for conductivity-enhancement as they are not interconnected.

4.2.3. Black shales

Jödicke (1985) suggested that carbon films on grain boundaries were the source of the low resistivity (2–3 Ω·m) of black shales. Duba et al. (1988) undertook laboratory studies on Carboniferous rock samples abundant in black shales, the colour of which is due to 5–8% organic carbon, extracted from a 5425-m-deep borehole in northern Germany. Rock analyses showed that there was no graphite present in the samples, and that the organic component consisted of ≈95% carbon. At high frequencies strong dependence of the electrical impedance with bedding direction and between dry and wet samples was observed. At frequencies of interest for deep studies (frequencies lower than 17 Hz in their experiments) the data were nearly frequency-independent. Their experiments showed that the conduction mechanism was controlled by an interconnected phase, and they concluded that they had confirmed Jödicke's earlier suggestion that the low resistivity is caused by a thin film of carbon on the grain boundaries.

When the samples were heated to 417°C in the absence of distilled water, their resistivities were observed to be frequency dependent and were very high ($> 10^6$ Ω·m at 17 Hz). When saturated with distilled water they were observed to be frequency-independent and much lower (10 Ω·m) (Duba et al., 1988). After cooling back to room temperature it was observed that the carbon had oxidized and the interconnectivity was obviously broken. Pyrite, also an excellent conductor with a resistivity of 5×10^{-5} Ω·m, is stable to 743°C but it did not contribute to the low resistivities because there was no continuous contacts between the pyrite spheres.

Stanley (1989) suggested that metamorphosed black shales subducted to depths in excess of 20 km may be responsible for conductivity anomalies beneath the Carpathians and beneath the Cascade mountains of northwestern U.S.A. Given the observation by Duba et al. (1988) regarding the oxidation of the carbon film at temperatures in excess of 400°C, it appears unlikely that this explanation is valid for the whole depth extent (to 20 km) postulated for the Cascades conductor which lies coincident with the Cascades volcanic belt.

Jödicke et al. (1983) interpreted their thin conducting zones in the mid-crust beneath the Rhenish Massif as due to graphite-bearing metamorphic black shales of Precambrian age. Fluids were excluded because of the lack of evidence for large amounts of free water in the middle and lower crust from xenolith studies.

4.3. Sulphides/metallic ores/graphite

EM methods have been used by the exploration industry to locate highly conducting mineralized zones for over half a century. Many long linear zones of enhanced conductivity in the CLC (see Haak and Hutton, 1986, their Table 2) may be associated with mineralized zones. For the longest of these, the 3000-km NACP anomaly (§3.3.4), Camfield and Gough (1977) discussed a spatial correlation of its ends with known exposures of graphite in both southeastern Wyoming and northern Saskatchewan, and concluded that the structure is probably comprised of graphite sheets in highly metamorphosed and folded basement rocks. Although this correlation must now be revised, for at least the northern segment, the high conductivity of the NACP (>1 S/m) appears to preclude fluids as an explanation (Jones and Craven, 1990).

Olhoeft (1981) suggested that sulphides might be the cause of enhanced conductivity of certain zones. Conducting zones located by commercial MT studies in northern Montana and southeastern British Columbia were intepreted as due to sulphides in belt rocks; an interpretation that was confirmed by drilling.

4.4. Partial melt

Partial melt has been shown in laboratory studies to increase the conductivity of rock by orders of magnitude (Waff, 1974; Sato and Ida, 1984). Tyburczy and Waff (1983) showed that conductivities are far less dependent on pressure than on temperature, and even at pressures as high as 25 kbar the conductivity approaches 70% of the zero pressure values.

4.5. Anisotropy?

Seismic anisotropy of the continental upper crust and upper mantle has been known for over a decade (Crampin et al., 1984), and evidence for electrical anisotropy in orthogonal horizontal directions of the oceanic upper mantle was recently presented by Chave et al. (1990).

Previous studies on land have also proposed transverse anisotropy to explain some observations (e.g., Jones and Garland, 1984; Kurtz et al., 1986b), but the anisotropy was confined to the upper crust and was "macro anisotropy" rather than intrinsic anisotropy. Two-dimensional models were found that satisfied the data but they included zones that had resistivities that, when averaged over a lateral distance of the order of the appropriate inductive scale length, differed in the two horizontal directions (see discussion of Cull, 1985, below).

Rasmussen (1988) concluded that the CLC of southern Sweden is electrically anisotropic (Fig. 3-10), with an anisotropy factor, given by the ratio of the two horizontal resistivities of the layer, of 42.5, or 4,250% (§3.1.1). Such an anisotropic model of the CLC has also recently been proposed by Kurtz et al. (1988) for the Kapuskasing structure in the Canadian shield (§3.5) and by Kellet et al. (1992) for the Abitibi belt. Given our generalized rheological model of the continental crust with a "ductile" lower crust that cannot withstand long-term stress, it is difficult to understand how a layer deep within the crust could maintain intrinsic anisotropy over long geological times. Is this proposed electrical anisotropy evidence for 1D intrinsic micro-anisotropy within the CLC with a preferential fluid-filled crack orientations? Or is it evidence for 2D macro-anisotropy with a conductive CLC (of ρ presumably $\approx$400 Ω·m) intersected by numerous resistive dykes (of ρ presumably far greater than 17,000 Ω·m)? Such a model was given as an alternative interpretation by Cull (1985) for MT data recorded on Precambrian terrain in Australia (Fig. 3-28), although Cull (1985) preferred the 2D contact model (model A in Fig. 3-28) to the dyke model (model B).

In Sweden there are no surface geological observations to support the "dyke" model, and Rasmussen (1988) preferred intrinsic micro-anisotropy as the explanation. This concept, of possible anisotropic CLCs beneath at least three Precambrian shield regions — the Baltic, Canadian and Australian shields — needs to be tested further.

In this regard, it is of import to note that the rheology of the lower crust beneath shield regions is such that deep earthquakes do occur (Chen, 1989), albeit rarely; the most recent on the Canadian shield was the Saguenay earthquake just north of Quebec City on November

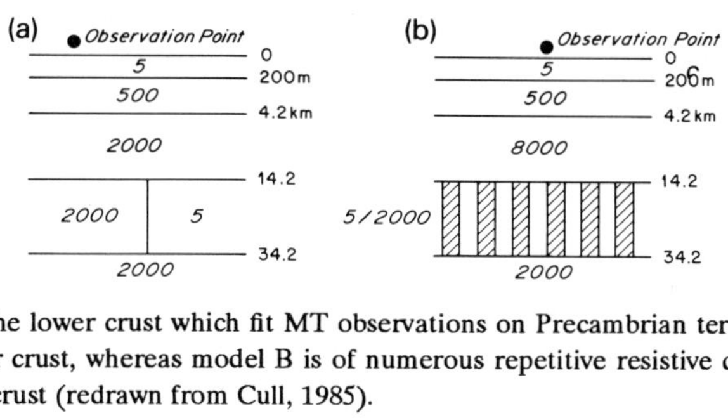

Fig. 3-28. Two models of the lower crust which fit MT observations on Precambrian terrain in Australia. Model A is of a contact in the lower crust, whereas model B is of numerous repetitive resistive dykes (2000 Ω·m) cutting a conducting (5 Ω·m) lower crust (redrawn from Cull, 1985).

25th, 1988, measuring 5.9 on the m_b scale located at a depth of 29 km (North et al., 1989) in a region of 39 km crustal thickness (Berry and Fuchs, 1973). This unusually large depth for eastern North America was confirmed by an aftershock study which located many events deeper than 25 km (North et al., 1989). Most of these deeper intracratonic earthquakes appear to be close to the crust–mantle boundary (Chen, 1989). Thus, our rheological model of the continental lower crust beneath shield regions may have to be modified somewhat to permit sufficient stress build up for earthquakes to occur and possibly for the existence of intrinsic electrical anisotropy in two horizontal directions.

4.6. Pore geometry considerations

Whatever is responsible for the enhanced electrical conductivity of the CLC — fluids, carbon films, partial melt — a mathematical model of the effect can be described by considering possible pore geometries. As stated in the Introduction (§1.1), Archie's Law (Eq. 1) has been used frequently for estimating the porosity from electrical conductivity estimates and assumed pore fluid conductivity. An exponent of 2 has been shown to be appropriate where the porosity is due largely to high aspect-ratio microcracks, whereas 1 is appropriate for interconnected low aspect-ratio cracks (Evans et al., 1982). Departures from Archie's Law have been noted (Lee et al., 1983, and references therein), particularly at very low or high porosities, and Hermance (1979) proposed a modified form for low porosities:

$$\sigma_A = \sigma_r + (\sigma_f - \sigma_r)\eta^2 \tag{15}$$

where σ_r and σ_f are the rock matrix and fluid conductivities, respectively, and η is the porosity. Note that in Eq. (15), in contrast to Eq. (1), the rock matrix conductivity is considered. Evans et al. (1982) suggest using Hoening's (1979) formulation modified for cigar-shaped cracks, and showed that, for pores with aspect ratio greater than 0.01, there is a critical porosity above which the effective conductivity increases rapidly to approach Archie's Law with unit exponent, and below which the effective conductivity is negligible.

More detailed consideration of possible pore geometries leads to a wide range of conductivities for a given porosity (Schmeling, 1986), with the highest conductivity values being obtained for interconnected thin films that wet the grain boundaries. These conditions are described by the Hashin-Shtrikman upper bound (Hashin and Shtrikman, 1963):

$$\sigma_{HS+} = \sigma_f + (1 - \eta)[1/(\sigma_r - \sigma_f) + \eta/3\sigma_f]^{-1} \tag{16}$$

which approximates to:

$$\sigma_{HS+} \approx (2/3)\sigma_f\eta \tag{17}$$

at low porosities and $\sigma_r \ll \sigma_f$. Of historical note is that the Hashin-Shtrikman upper bound is identical to one developed by Maxwell (1892) for widely dispersed spheres embedded in a fluid. Waff (1974) considered the problem of arbitrary-sized spheres and generally confirmed the Hashin-Shtrikman formula with the modification that:

$$\sigma_W \approx (2/3)\sigma_f\eta + (1 - \eta)\sigma_r \tag{18}$$

at very low porosities. This formula describes a parallel electric circuit with ionic or electronic conduction along fluid pathways and electronic conduction through the rock matrix itself.

For fluid-filled tubes along grain edges, Grant and West (1965) showed that an effective conductivity is given by:

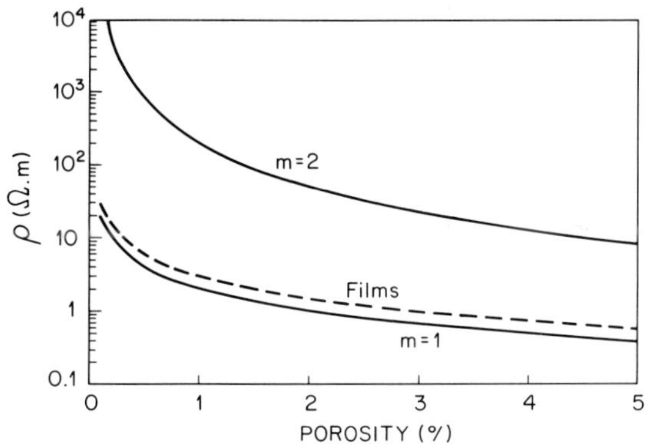

Fig. 3-29. Resistivity vs. porosity curves for fluid of 50 S/m conductivity in a resistive host matrix with Archie's Law exponents of 2 and 1 and for thin film geometry (calculated from the equations in §4.6).

$$\sigma_{\mathrm{GW}} \approx (1/3)\sigma_{\mathrm{f}}\eta + (1-\eta)\sigma_{\mathrm{r}} \tag{19}$$

Hyndman and Shearer (1989) have recently undertaken a comparison of the resistivity of a medium for various pore geometries for a host-rock resistivity of 10^4 Ω·m and fluid resistivity of 0.02 Ω·m (equivalent to an NaCl concentration of ≈1 M (molar), or 5%, at 500°C; Quist and Marshall 1968). Using these parameters in the above equations (Fig. 3-29) it is clear that pore geometry is important for conclusions regarding possible porosities. For example, Kurtz et al. (1986a) assumed an Archie's Law equation for their 30-Ω·m layer with a fluid of 20 wt.% NaCl of 0.008 Ω·m resistivity for a calculated porosity of 1.6%. However, a thin connected film of the same resistivity would only require a porosity of 0.04% according to Eq. 17.

Assuming a resistivity for the saline fluid of 0.002 Ω·m and 100 S for the CLC (§3.4) leads to a relationship between films of porosity η and thickness of the conducting zone h (in m) of:

$$\eta = 3/10h \tag{20}$$

that for a zone of 6-km thickness (i.e., a resistivity of 60 Ω·m), requires a porosity of only 0.005%. For more conducting zones of 1000 S, this value rises to 0.05%.

4.7. Causes: Conclusions

One should not fall into the trap of searching for a single cause for the enhanced conductivity of the CLC — it is clear that there may be many explanations, and that different effects are operating in different tectonic environments and at different depths. For currently-active subduction zones there are large quantities of free water available from expulsion of pore fluids and from CH_4-H_2O fluids produced by diagenetic and low-grade metamorphic reactions at depths less than 40 km (Peacock, 1990). It is expected that after the end of metamorphic activity the fluids can be maintained for more than 70 Ma (Thompson and Connolly, 1990) or even much longer (Bailey, 1990). For the Early Proterozoic subduction zone within the Trans-Hudson Orogen, the values of conductivity are too high (>3 S/m) to be explained by fluids alone; for fluid of 50 S/m conductivity the porosity would have to be implausibly high (>10%). Explanations involving conducting minerals, such as graphite sheets, must be appealed to for this type of structure. For zones where the temperature exceeds 730°C, such as the lower crust beneath the Valhalla Gneiss Complex in British

Columbia (§3.6, Fig. 3-22), partial melting may be the cause of the very high conductivities observed.

Given our observations that the conductance beneath old shield regions is two or three orders of magnitude larger than would be expected for dry rocks, we still do not have a satisfactory explanation free of considerable objections. A small percentage (less than 1%) of saline-filled porosity is currently favoured by the electromagnetic community. Is it reasonable to expect free fluids? Many petrologists see no evidence for free water within the lower crust in their data. Accordingly, other explanations — such as carbon films on grain-boundaries — are being advanced. The required porosities could be very low if the fluids are highly saline with conductivities of 500 S/m rather than 50 S/m. Would the effects of saline fluids in thin zones, of such very low porosities, in the mid-crust be observed in exhumed rocks from the CLC, especially given Hulsebosch and Frost's (1989) remarks concerning the lack of chemical modification of rocks by deeply-penetrating saline brines? Is it reasonable to expect interconnectivity at such low porosities? Is the model proposed by Sanders (1991) an explanation of this apparent dichotomy between petrologists and EM geophysicists?

Alternatively, is it reasonable to expect interconnected thin carbon films over many thousands of kilometres, such as the Siberian shield (§3.1.3), especially given the observed immobility of carbon (Tsong et al., 1985; Watson, 1986). Also, can we further expect these films to remain stable over very long geological times?

It would narrow the field of possible causes considerably if we could determine precisely the thicknesses of the various conducting zones observed in the CLC. As discussed in §2.4.3, the deep-probing EM methods used are virtually all detecting the conductance, or conductivity-thickness product, of the zone of enhanced conductivity and not either parameter independently. Stacked multiple thin zones, rather than a single thick zone, may be more consistent with the seismic reflectivity patterns observed for the CLC, but we must try to obtain evidence of this from the EM data themselves.

5. Conclusions

Mapping the lower continental crust is only in its infancy, and this activity will engage us for many years to come. As envisioned by Oliver (1991), by the year 2100 all the continents will be criss-crossed by deep seismic reflection profiles with complimentary information from other geophysics that will give us a far more comprehensive understanding of the continental crust than we have today. As a component of such a multidisciplinary geoscientific program, imaging the electrical conductivity can add much to understanding the state and composition of the crust. The EM results from Kapuskasing (§3.5) and beneath the Valhalla complex (§3.6) are giving key information, and EM methods may be the only method for detecting crustal-scale fluids and fluid transport systems of the type proposed by Torgersen (1990).

One enigma that must be addressed is "*what is the cause, or causes, of the zones of enhanced electrical conductivity beneath stable regions?*" Interpretations of such zones beneath active or young (<100 Ma) regions in terms of fluids or partial melt are generally acceptable. Linear anomalies of enhanced conductivity associated with ancient subduction zones can be interpreted as due to conducting minerals. But what of the CLCs beneath stable regions such as the Canadian, Siberian and Baltic shields which have conductances one to two orders of magnitude greater than would be expected for dry granulite facies assemblages? Compositional contrasts appear to be excluded. Less than 1% of an interconnected saline fluid is sufficient to account for the observations — but there have been petrological arguments advanced against such a hypothesis which must be addressed. Alternatively, is it reasonable to expect an interconnected grain boundary film of carbon beneath extensive stable regions? Are there other conduction mechanisms operating that we have yet to discover?

Such questions can only be answered by the reference to other geophysical, geological and geochemical data. Statistical correlations between EM results and heat flow (Ádám, 1976, 1980), seismic refraction (Jones, 1981b; Hyndman and Klemperer, 1989) and seismic reflection (Jones, 1987), and between heat flow and seismic reflection results (Klemperer, 1987), must be treated with caution, but are the necessary first steps.

Perhaps it should be re-emphasized that the whole of the CLC may not be conducting, but a thin zone of high conductivity at mid-crustal depths may be the cause for our EM observations. Obviously, the prime task of the EM community in the near-term is to resolve the actual depth-distribution of conductivity within the CLC rather than just its conductance; this will involve specifically-designed experiments at well-chosen sites. To aid our interpretation of the conductivity structure determined, the geological and geochemical community must come to a consensus on the role of saline fluids in the deep crust of stable regions and either permit or exclude them. I predict that by the beginning of the 21st century we will have a far clearer picture of the electrical conductivity of the lower continental crust and its interpretation.

Acknowledgements

I wish to acknowledge with deep gratitude the invitation of the editors, in particular David Fountain, to contribute this chapter to their book, and their fortitude and patience whilst waiting for its completion. My colleagues responded abundantly to my requests for reprints and preprints of their work, and Steve Constable kindly provided the code for his OCCAM inversion scheme. Finally, I thank Alan Green, David Fountain, Ron Frost, David Boerner and Ron Kurtz for comments on an earlier version. Their contributions, especially Alan Green's, were substantial; the errors remain mine alone, however.

References

Ádám, A., 1976. Quantitative connections between regional heat flow and the depth of conductive layers in the earth's crust and upper mantle. Acta Geod. Geophys. Montanist., Acad. Sci. Hung., 11: 503–509.

Ádám, A., 1980. Connection between the electric conductivity increase due to phase transition and heat flow. J. Geomagn. Geoelectr., 32(Suppl I): SI115–SI124.

Ádám, A., 1987. Are there two types of conductivity anomaly (CA) caused by fluid in the lower crust. Phys. Earth Planet. Inter., 45: 209–215.

Ádám, A., Nagy, Z., Nemesi, L. and Varga, G., 1990. Crustal conductivity anomalies in the Pannonian Basin. Acta Geod. Geophys. Montanist. Hung., 25: 279–290.

Adams, D.C. and Young, C.T., 1989. Magnetotelluric investigation of the contact between the Duluth Complex and Animikie Basin in northeastern Minnesota. Paper presented at Inst. on Lake Superior Geology, Duluth, Minnesota, extended abstracts, pp. 1–2.

Alabi, A.O., Camfield, P.A. and Gough, D.I., 1975. The North American Central Plains anomaly. Geophys. J. R. Astron. Soc., 43: 815–834.

Archie, G.E., 1942. The electrical resistivity log as an aid in determining some reservoir characteristics. Trans. Am. Inst. Min. Metall. Eng., 146: 54–67.

AGU Chapman Conference, 1990. Crustal-Scale Fluid Transport: Magnitude and Mechanisms. Snowbird, Utah, 4–8 June.

BABEL Working Group, 1990. Evidence for early Proterozoic plate tectonics from seismic reflection profiles in the Baltic Shield. Nature, 348: 34–38.

Bahr, K., 1985. Magnetotellurische Messung des Elektrischen Widerstandes der Erdkruste und des Oberen Mantels in Gebieten mit Lokalen und Regionalen Leit-fähigkeits-anomalien. Ph.D. thesis, Univ. Göttingen.

Bahr, K., 1988. Interpretation of the magnetotelluric impedance tensor: regional induction and local telluric distortion. J. Geophys., 62: 119–127.

Bahr, K., 1991. Geological noise in magnetotelluric data: a classification of distortion types. Phys. Earth Planet. Inter., 66: 24–38.

Bailey, R.C., 1970. Inversion of the geomagnetic induction problem. Proc. R. Soc. London, Ser. A, 315: 185–194.

Bailey, R.C., 1990. Trapping of acqueous fluids in the deep crust. Geophys. Res. Lett., 17: 1129–1132.

Bailey, R.C., Craven, J.A., Macnae, J.C. and Polzer, B.D., 1989. Imaging of deep fluids in Archean crust. Nature, 340: 136–138.

Baker, A.J., 1990. Fluid–rock interactions in the Ivrea Zone and the origin of high lower crustal conductivities. In: M.H. Salisbury and D.M. Fountain (Editors), Exposed Cross-Sections of the Continental Crust. Kluwer, Dordrecht, pp. 437–452.

Banks, R.J. and Beamish, D., 1979. Melting in the crust and upper mantle beneath the Kenya Rift: evidence from Geomagnetic Deep Sounding experiments. J. Geol. Soc. London, 136: 225–233.

Banks, R.J. and Ottey, P., 1974. Geomagnetic deep sounding in and around the Kenya rift valley. Geophys. J. R. Astron. Soc., 36: 321–335.

Beamish, D., 1990. A granite window to the lower electrical crust and upper mantle. Terra Nova, 2: 314–319.

Beamish, D. and Smythe, D.K., 1986. Geophysical images of the deep crust: the Iapetus suture. J. Geol. Soc. London, 143: 489–497.

Berdichevsky, M.N. and Dmitriev, V.I., 1976. Distortion of magnetic and electric fields by near-surface lateral inhomogeneities. Acta Geod. Geophys. Montanist., Acad. Sci. Hung., 11: 447–483.

Berdichevsky, M.N. and Zhdanov, M.S., 1984. Advanced Theory of Deep Electromagnetic Sounding. Elsevier, Amsterdam, 408 pp.

Berktold, A., Musmann, G., Tezkan, B. and Wohlenberg, J., 1985. Electrical conductivity studies, Schwartzwald working group. In: Abstracts 2nd Int. Symp. on Observation of the Continental Crust through Drilling. Alfred-Wegener-Stiftung, Bonn, p. 70.

Berry, M. and Fuchs, K., 1973. Crustal structure of the Superior and Grenville provinces of the northeastern Canadian shield. Bull. Seismol. Soc. Am., 63: 1393–1432.

Blohm, E.K., Worzyk, P. and Scriba, H., 1977. Geoelectric deep soundings in southern Africa using the Cabora Bassa power line. J. Geophys., 43: 665–679.

Boerner, D.E., 1992. Deep controlled source electromagnetic sounding: theory, results and correlation with natural source results. Surv. Geophys., in press.

Boerner, D.E., Spencer, C.P., Wright, J.A., Carroll, P. and Reed, L.E., 1990. Developing new methods at an old mine. Eng. Min. J., 191: 25–29.

Boland, A.V. and Ellis, R.M., 1989. Velocity structure of the Kapuskasing uplift, northern Ontario, from seismic refraction studies. J. Geophys. Res., 94: 7189–7204.

Boland, A.V., Ellis, R.M., Northey, D.J., West, G.F., Green, A.G., Forsyth, D.A., Mereu, R.F., Meyer, R.P., Morel-á-l'Hussier, P., Buchbinder, G.R., Asudeh, I. and Haddon, W., 1988. Seismic delineation of upthrust Archean crust in Kapuskasing, northern Ontario. Nature, 335: 711–713.

Booker, J.R. and Chave, A.D., 1989. Introduction to the Special Section on the EMSLAB-Juan de Fuca Experiment. J. Geophys. Res., 94: 14,093–14,098.

Bostick, F.X., 1977. A simple almost exact method of MT analysis. In: Workshop on Electrical Methods in Geothermal Exploration. U.S. Geol. Surv., Contract No. 14080001-8-359.

Brace, W.F., 1971. Resistivity of saturated crustal rocks to 40 km based on laboratory measurements. In: J.G. Heacock (Editor), Structure and Physical Properties of the Earth's Crust. AGU Geophys. Monogr. Ser., 14: 206–210.

Brace, W.F. and Orange, A.S., 1968. Further studies of the effects of pressure on electrical resistivity of rocks. J. Geophys. Res., 73: 5407–5420.

Brace, W.F., Orange, A.S. and Madden, T.R., 1965. The effect of pressure on the electrical resistivity of water-saturated crystalline rocks. J. Geophys. Res., 70: 5669–5678.

Brewer, J.A., Matthews, D.H., Warner, M.R., Hall, J., Smythe, D.K. and Whittington, R.J., 1983. BIRPS deep seismic reflection studies of the British Caledonides. Nature, 305: 206–210.

Bridgwater, D., 1989. Fluid Movements –Element Transport and the Composition of the Deep Crust. Kluwer, Dordrecht, 416 pp.

Buchbinder, G.G.R., Kurtz R.D. and Lambert, A., 1983. A review of time-dependent geophysical parameters in the Charlevoix Region, Quebec. Earthquake Predict. Res., 2: 149–166.

Calvert, A.J. and Clowes, R.M., 1990a. Deep, high amplitude reflections from a major shear zone above the subducting Juan de Fuca plate. Contributed paper at "LITHOPROBE Cordilleran Workshop", Calgary, Alberta, 3–4 March. Cordilleran Workshop Proc., 2: 156–165.

Calvert, A.J. and Clowes, R.M., 1990b. Deep, high-amplitude reflections from a major shear zone above the subducting Juan de Fuca plate. Geology, 18: 1091–1094.

Camfield, P.A. and Gough, D.I., 1977. A possible Proterozoic plate boundary in North America. Can. J. Earth Sci., 14: 1229–1238.

Camfield, P.A., Gough, D.I. and Porath, H., 1971. Magnetometer array studies in the northwestern United States and southwestern Canada. Geophys. J. R. Astron. Soc.,22: 201–222.

Camfield, P.A., Gupta, J.C., Jones, A.G., Kurtz, R.D., Krentz, D.H., Ostrowski, J.A. and Craven, J.A., 1989. Electromagnetic sounding and crustal electrical conductivity structure in the region of the Wopmay Orogen, Northwest Territories, Canada. Can. J. Earth Sci., 26: 2385–2395.

Cavaliere, T. and Jones, A.G., 1984. On the identification of a transition zone in electrical conductivity between the lithosphere and asthenosphere: a plea for more precise phase data. J. Geophys., 55: 23–30.

Chave, A.D., Weertman, B.R. and Booker, J.R., 1990. Application of a galvanic static distortion removal technique to EMSLAB seafloor magnetotelluric data reveals a 2D magnetotelluric impedance tensor aligned with the relative motion of the Juan de Fuca plate. EOS, 71: 1623.

Chen, W.-P., 1989. A brief update on the focal depths of intracontinental earthquakes and their correlations with heat flow and tectonic age. Seismol. Res. Lett., 59: 263–272.

Chouteau, M., Thevenin, J. and Kurtz, R.D., 1991. Determination of the crustal structure in the central part of the Abitibi Greenstone Belt (Canada) by the magnetotelluric method. Phys. Earth Planet Inter., submitted.

Connerney, J.E.P. and Kuckes, A.F., 1980. Gradient analysis of geomagnetic fluctuations in the Adirondacks. J. Geophys. Res., 85: 2615–2624.

Connerney, J.E.P., Nekut, A. and Kuckes, A.F., 1980. Deep crustal electrical conductivity in the Adirondacks. J. Geophys. Res., 85: 2603–2614.

Constable, S.C., McElhinny, M.W. and McFadden, P.L., 1984. Deep Schlumberger sounding and the crustal resistivity structure of central Australia. Geophys. J. R. Astron. Soc., 79: 893–910.

Constable, S.C., Parker, R.L. and Constable, C.G., 1987. Occam's inversion: a practical algorithm for generating smooth models from electromagnetic sounding data. Geophysics, 52: 289–300.

Cook, F.A., 1985. Geometry of the Kapuskasing structure from a LITHOPROBE pilot reflection survey. Geology, 13: 368–371.

Cook, F.A., Simony, P.S., Coflin, K.C., Green, A.G., Milkereit, B., Price, R.A., Parrish, R.R., Patenaude, C., Gordy, P.L. and Brown, R.L., 1987. LITHOPROBE southern Canadian Cordilleran transect: Rocky Mountain thrust belt to Valhalla gneiss complex. Geophysics, 89: 91–98.

Cook, F.A., Green, A.G., Simony, P.S., Price, R.A., Parrish, R.R., Milkereit, B., Gordy, P.L., Brown, K.C., Coflin, K.C. and Patenaude, C., 1988. LITHOPROBE seismic reflection structure of the southeastern Canadian Cordillera: Initial results. Tectonics, 7: 157–180.

Crampin, S., Chesnokov, E.M. and Hipkin, R.G., 1984. Seismic anisotropy –the state of the art, II. Geophys. J. R. Astron. Soc., 76: 1–16.

Craven, J.C., Jones, A.G., Boerner, D.E., Groom, R.W. and Kurtz, R.D., 1990. The correction of static shift in magnetotelluric data from the LITHOPROBE Southern Cordilleran transect. Soc. Explor. Geophys., Extended Abstracts with Biographies, pp. 561–564.

Cull, J.P., 1982. Magnetotelluric profiles in the McArthur Basin of northern Australia. BMR J. Aust. Geol. Geophys., 7: 275–286.

Cull, J.P., 1985. Magnetotelluric soundings over a Precambrian contact in Australia. Geophys. J. R. Astron. Soc., 80: 661–675.

Cumming, W.B., Clowes, R.M. and Ellis, R.M., 1979. Crustal structure from a seismic refraction profile across southern British Columbia. Can. J. Earth Sci., 16: 1024–1040.

de Beer, J.H., van Zijl, J.S.V. and Gough, D.I., 1982. The Southern Cape conductive belt (South Africa): its composition, origin, and tectonic significance. Tectonophysics, 83: 205–225.

deGroot-Hedlin, C. and Constable, S., 1990. Occam's inversion to generate smooth two-dimensional models from magnetotelluric data. Geophysics, 55: 1613–1624.

Dewey, J.F., 1969. Evolution of the Appalachian-Caledonian orogen. Nature, 241: 124–129.

Dohr, G., Dürschner, H. and Edelmann, H.A.K., 1989. Exploration geophysics in Germany. First Break, 7: 153–171.

Drury, M.J., Niblett, E.R., 1980. Buried ocean crust and continental crust geomagnetic induction anomalies: a possible association. Can. J. Earth Sci., 17: 961–967.

Duba, A., 1976. Are laboratory electrical conductivity data relevant to the earth?. Acta Geod. Geophys. Montanist., Acad. Sci. Hung., 11: 485–495.

Duba, A.L. and Shankland, T.J., 1982. Free carbon and electrical conductivity in the Earth's mantle. Geophys. Res. Lett., 9: 1271–1274.

Duba, A., Huenges, E., Nover, G., Will, G. and Jödicke, H., 1988. Impedance of black shale from Münsterland 1 borehole: an anomalous good conductor? Geophys. J., 94: 413–419.

Duncan, P.M., Hwang, A., Edwards, R.N., Bailey, R.C. and Garland, G.D.. 1980. The development and applications of a wide band electromagnetic sounding system using a pseudo-noise source. Geophysics, 45: 1276–1296.

Dupis, A. and Thera, A.L., 1982. Natural electromagnetism in the Rhine Graben. Geophys. J. R. Astron. Soc., 68: 545–557.

Eaton, D.W.S. and Cook, F.A., 1990. Crustal structure of the Valhalla complex, British Columbia, from LITHO-PROBE seismic-reflection and potential-field data. Can. J. Earth Sci., 27: 1048–1060.

Edwards, R.N., Law, L.K. and White, A., 1971. Geomagnetic variations in the British Isles and their relation to electrical currents in the ocean and shallow seas. Philos. Trans. R. Soc. London, 270: 289–323.

Edwards, R.N., Bailey, R.C. and Garland, G.D., 1981. Conductivity anomalies: lower crust or asthenosphere?. Phys. Earth Planet. Inter., 25: 263–272.

EMSLAB, 1989. Special Issue of Journal of Geophysical Research devoted to the EMSLAB experiment (14 papers). J. Geophys. Res., 94: 14,093–14,283.

Etheridge, M.A., Wall, V.J. and Vernon, R.H., 1983. The role of the fluid phase during regional metamorphism and deformation. J. Metamorph. Geol., 1: 205–226.

Etheridge, M.A., Wall, V.J., Cox, S.F. and Vernon, R.H., 1984. High fluid pressures during regional metamorphism and deformation: implications for mass transport and deformation mechanisms. J. Geophys. Res., 89: 4344–4358.

Evans, C.J., Chroston, P.N. and Toussaint-Jackson, J.E., 1982. A comparison of laboratory measured electrical conductivity in rocks with theoretical conductivity based on derived pore aspect ratio spectra. Geophys. J. R. Astron. Soc., 71: 247–260.

Fischer, G. and Schnegg, P.-A., 1980. The dispersion relations of the magnetotelluric response and their incidence on the inversion problem. Geophys. J. R. Astron. Soc., 62: 661–673.

Fitton, J.G. and Hughes, D.J., 1970. Volcanism and plate tectonics in the British Ordovician. Earth Planet. Sci. Lett., 8: 223–228.

Fountain, D.M. and Boriani, A., 1990. The Nature of the Lower Continental Crust. Tectonophysics, 182: 1–192.

Fournier, R.O., 1991. The transition from hydrostatic to greater than hydrostatic fluid pressure in presently active continental hydrothermal systems in crystalline rock. Geophys. Res. Lett., 18: 955–958.

Frape, S.K. and Fritz, P., 1987. Geochemical trends for groundwaters from the Canadian Shield. In: P. Fritz and S.K. Frape (Editors), Saline Water and Gases in Crystalline Rocks. Geol. Assoc. Can., Spec. Pap., 33: 19–38.

Frape, S.K., Fritz, P. and McNutt, R.H., 1984. The role of water–rock interation in the chemical evolution of groundwaters from the Canadian Shield. Geochim. Cosmochim. Acta, 48: 1617–162.

Frey, M., Bucher, K., Frank, E. and Mullis, J., 1980. Alpine metamorphism along the Geotraverse Basel–Chiasso –a review. Eclogae Geol. Helv., 73: 527–546.

Frost, B.R. and Touret, J.L.R., 1989. Magmatic CO_2 and saline melts from the Sybille Monzosyenite, Laramie Anorthosite Complex, Wyoming. Contrib. Mineral. Petrol., 103: 178–186.

Frost, B.R., Fyfe, W.S., Tazaki, K. and Chan, T., 1989a. Grain-boundary graphite in rocks and implications for high electrical conductivity in the lower crust. Nature, 340: 134–136.

Frost, B.R., Frost, C. and Touret, J.L.R., 1989b. Magmas as a source of heat and fluids in granulite metamorphism. In: D. Bridgewater (Editor), Fluid Movements –Element Transport and the Composition of the Deep Crust. Kluwer, Dordrecht, pp. 1–18.

Fuchs, K., Bonjer, K.-P., Gajewski, D., Lüschen, E., Prodehl, C., Sandmeier, K.-J., Wenzel, F. and Wilhelm, H., 1987. Crustal evolution of the Rhinegraben area. 1. Exploring the lower crust in the Rhinegraben rift by unified geophysical experiments. Tectonophysics, 141: 261–275.

Fyfe, W.S., Price, N.J. and Thompson, A.B., 1978. Fluids in the Earth's Crust. Elsevier, Amsterdam, 383 pp.

Fyfe, W.S., Kerrich, R., Hicock, S.R. and Colloza, F.C., 1987. LITHOPROBE: faults and fluids. Geosci. Can., 14: 75–81.

Gajewski, D. and Prodehl, C., 1987. Crustal evolution of the Rhinegraben area II. Seismic refraction investigation of the Black Forest. Tectonophysics, 142: 27–48.

Gamble, T.D., Goubau, W.M. and Clarke, J., 1979. Magnetotellurics with a remote reference. Geophysics, 44: 53–68.

Garland, G.D., 1950. Interpretation of gravimetric and magnetic anomalies on traverses in the Canadian shield. Publ. Dom. Observ. Canada, vol. 16, 57 pp.

Garson, M.S. and Plant, J., 1973. Alpine type ultramafic rocks and episodic mountain building in the Scottish Highlands. Nature, 242: 34–38.

Gascoyne, M., Davison, C.C., Ross, J.D. and Pearson, R., 1987. Saline groundwaters and brines in plutons in the Canadian Shield. In: P. Fritz and S.K. Frape (Editors), Saline Water and Gases in Crystalline Rocks. Geol. Assoc. Can., Spec. Pap., 33: 19–38.

Geis, W.T., Cook, F.A., Green, A.G., Milkereit, B., Percival, J.A. and West, G.F., 1990. Thin thrust sheet formation of the Kapuskasing structural zone revealed by LITHOPROBE seismic reflection data. Geology, 18: 513–516.

Gibb, R.A. and Walcott, R.I., 1971. A Precambrian suture in the Canadian Shield. Earth Planet. Sci. Lett., 10: 417–422.

Golubev, N.B. and Varentsov, I.M., 1989. Additional notes on the geoelectric structure of the OULU I. In: S.-E. Hjelt and L.L. Vanyan (Editors), Geoelectric Models of the Baltic Shield. Dept. Geophysics, Univ. Oulu Rep. No. 16, pp. 132–133.

Gough, D.I., 1981. Magnetometer arrays and geodynamics. In: Evolution of the Earth. AGU Geodyn. Ser., 5: 87–95.

Gough, D.I., 1983. Electromagnetic geophysics and global tectonics. J. Geophys. Res., 88: 3367–3377.

Gough, D.I., 1986. Seismic reflectors, conductivity, water and stress in the continental crust. Nature, 323: 143–144.

Gough, D.I., 1989. Magnetometer array studies, earth structure, and tectonic processes. Rev. Geophys., 27: 141–157.

Gough, D.I. and Ingham, M.R., 1983. Interpretation methods for magnetometer arrays. Rev. Geophys., 21: 805–827.

Grant, F.S. and West, G.F., 1965. Interpretation Theory in Applied Geophysics. McGraw-Hill, New York, N.Y., 584 pp. .

Green, A.G., Milkereit, B., Mayrand, L., Spencer, C., Kurtz, R.D. and Clowes, R.M., 1987. LITHOPROBE seismic reflection profiling across Vancouver Island. Geophys. J. R. Astron. Soc., 89: 85–90.

Greenhouse, J.P. and Bailey, R.C., 1981. A review of geomagnetic variation measurements in the eastern United States: implications for continental tectonics. Can. J. Earth Sci., 18: 1268–1289.

Groom, R.W., 1988. The effects of inhomogeneities on magnetotellurics. Ph.D. thesis, Univ. Toronto.

Groom, R.W. and Bailey, R.C., 1989. Decomposition of magnetotelluric impedance tensor in the presence of local three-dimensional galvanic distortion. J. Geophys. Res., 94: 1913–1925.

Groom, R.W. and Bailey, R.C., 1991. Analytical investigations of the effects of near-surface three-dimensional galvanic scatterers on MT tensor decomposition. Geophysics, 56: 496–518.

Gupta, J.C. and Jones, A.G., 1990. Conductivity structure of the Flahhead basin in south-eastern British Columbia, Canada. Can. J. Earth Sci., August issue.

Gupta, J.C., Kurtz, R.D., Camfield, P.A. and Niblett, E.R., 1985. A geomagnetic induction anomaly from IMS data near Hudson Bay, and its relation to crustal electrical conductivity in central North America. Geophys. J. R. Astron. Soc., 81: 33–46.

Haak, V. and Hutton, V.R.S., 1986. Electrical resistivity in continental lower crust. In: J.B. Dawson, D.A. Carswell, J. Hall and K.H. Wedepohl (Editors), The Nature of the Lower Continental Crust. Geol. Soc. London, Spec. Publ., 24: 35–49.

Haak, V., Stoll, J. and Winter, H., 1991. Why is the electrical resistivity around the KTB hole so low?. Phys. Earth Planet. Inter., 66: 12–23.

Handa, S. and Camfield, P.A., 1984. Crustal electrical conductivity in north-central Saskatchewan: the North American Central Plains anomaly and its relation to a Proterozoic plate margin. Can. J. Earth Sci., 21: 533–543.

Harper, D.A.T. and Murphy, F.C., 1989. The Iapetus suture in the British Isles –comment on its position in eastern Ireland. Geol. Mag., 126: 723–724.

Hashin, Z. and Shtrikman, S., 1963. A variational approach to the theory of the elastic behaviour of multiphase materials. J. Mech. Phys. Solids, 11: 12–140.

Hay, S.J., Hall, J., Simmons, G. and Russell, M.J., 1988. Sealed microcracks in the Lewisian of NW Scotland: a record of 2 billion years of fluid circulation. J. Geol. Soc. London, 145: 819–830.

Henry, W.J., Mechie, J., Maguire, P.K.H., Khan, M.A., Prodehl, C., Keller, G.R. and Patel, J., 1990. A seismic investigation of the Kenya Rift Valley. Geophys. J. Int., 100: 107–130.

Hermance, J.F., 1979. The electrical conductivity of materials containing partial melts: a simple model from Archie's Law. Geophys. Res. Lett., 6: 613–616.

Hermance, J.F. and Pedersen, J., 1980. Deep structure of the Rio Grande rift: A magnetotelluric interpretation. J. Geophys. Res., 85: 3899–3912.

Hermance, J.F. and Neumann, G.A., 1991. The Rio Grande rift: electromagnetic constraints on the Socorro magma body. Phys. Earth Planet. Inter., 66: 101–117.

Hills, F.A., Houston,R.S. and Subbarayudu, G.V., 1975. Possible Proterozoic plate boundary in southern Wyoming. Geol. Soc. Am. (Abstracts), 7: 614.

Hjelt, S.-E., 1988. Regional EM studies in the 80s. Surv. Geophys., 9: 349–387.

Hjelt, S.-E. and Vanyan, L.L., 1989. Geoelectric Models of the Baltic Shield. Dept. Geophysics, Univ. Oulu Rep. No. 16.

Hoening, A., 1979. Electrical conductivities of a cracked solid. Pure Appl. Geophys., 117: 690–709.

Hoffman, P., 1981. Autopsy of Athapuscow aulacogen: a failed arm affected by three collisions. In: F.H.A. Campbel (Editor), Proterozoic Basins of Canada. Geol. Surv. Can., Pap., 81-10: 97–102.

Hoffman, P., 1988. United plates of America, the birth of a craton: early Proterozoic assembly and growth of Proto-Laurentia. Annu. Rev. Earth Planet. Sci., 16: 543–603.

Hulsebosch, T.P. and Frost, B.R., 1989. Mineral changes, element mobility, and fluids associated with deep shearing in the Mount Helen structural belt, Wyoming, U.S.A. In: D. Bridgewater (Editor), Fluid Movements –Element Transport and the Composition of the Deep Crust. Kluwer, Dordrecht, pp. 139–150.

Hutton, D.H.W. and Murphy, F.C., 1987. The Silurian of the Southern Uplands and Ireland as a succesor basin to the end-Ordovician closure of Iapetus. J. Geol. Soc., 144: 765–772.

Hutton, V.R.S. and Jones, A.G., 1980. Magnetovariational and magnetotelluric investigations in S. Scotland. In: U. Schmucker (Editor), Electromagnetic Induction in the Earth and Moon. AEPS-9, Centr. Acad. Publ. Japan, Tokio and D. Reidel Publ. Co., Dordrecht, Suppl. Issue J. Geomagn. Geoelectr., pp.141–150.

Hyndman, R.D., 1988. Dipping seismic reflectors, electrically conductive zones, and trapped water in the crust over a subducting plate. J. Geophys. Res., 93: 13,391–13,405.

Hyndman, R.D. and Hyndman, D.W., 1968. Water saturation and high electrical conductivity in the lower crust. Earth Planet. Sci. Lett., 4: 427–432.

Hyndman, R.D. and Klemperer, S.L., 1989. Lower-crustal porosity from electrical measurements and inferences about composition from seismic velocities. Geophys. Res. Lett., 16: 255–258.

Hyndman, R.D. and Shearer, P.M., 1989. Water in the lower continental crust: modelling magnetotelluric and seismic reflection results. Geophys. J. Int., 98: 343–365.

Ingham, M.R., 1988. The use of invariant impedances in magnetotelluric interpretation. Geophys. J. R. Astron. Soc., 92: 165–169.

Ingham, M.R. and Hutton, V.R.S., 1982. Crustal and upper mantle electrical conductivity structure in southern Scotland. Geophys. J. R. Astron. Soc., 68: 579–594.

Jain, S., 1964. Electrical conductivity of the crust and upper mantle at Eskdalemuir, southern Scotland. Nature, 203: 631–632.

Jiracek, G.R., Ander, M.E. and Holcombe, H.T., 1979. Magnetotelluric soundings of crustal conductive zones in major continental rifts. In: Robert E. Riecker (Editor), Rio Grande Rift: Tectonics and Magmatism. Am. Geophys. Union, Washington DC, pp. 209–222.

Jiracek, G.R., Gustafson, E. and Mitchell, P.S., 1983. Magnetotelluric results opposing magma origin of crustal conductors in the Rio Grande rift. Tectonophysics, 94: 299–326.

Jiracek, G.R., Rodi, W.L. and Vanyan, L.L., 1987. Implications of magnetotelluric modeling for the deep crustal environment in the Rio Grande rift. Phys. Earth Planet. Inter., 45: 179–192.

Jödicke, H., 1985. A large self-potential anomaly at the SE flank of the Stavelot-Venn anticline originating from meta-anthracite bearing black shales at the Salm/Revin boundary. Neues Jahrb. Geol. Paläeontol. Abh., 171: 387–402.

Jödicke, H., 1991. Interpretation of conductivity models. Surv. Geophys., submitted

Jödicke, H., Untiedt, J., Olgemann, W., Schulte, L. and Wagenitz, V., 1983. Electrical conductivity structure of the crust and upper mantle beneath the Rhenish Massif. In: K. Fuchs (Editor), Plateau Uplift. Springer-Verlag, Berlin-Heidelberg, pp. 288–302.

Jones, A.G., 1980. Geomagnetic induction studies in Scandinavia-I. Determination of the inductive response function from the magnetometer data. J. Geophys., 48: 181–194.

Jones, A.G., 1981a. Geomagnetic induction studies in Scandinavia-II. Geomagnetic depth sounding, induction vectors and coast effect. J. Geophys., 50: 23–36.

Jones, A.G., 1981b. On a type classification of lower crustal layers under Precambrian regions. J. Geophys., 49: 226–233.

Jones, A.G., 1982. On the electrical crust–mantle structure in Fennoscandia: no Moho and the asthenosphere revealed?. Geophys. J. R. Astron. Soc., 68: 371–388.

Jones, A.G., 1983a. On the equivalence of the "Niblett" and "Bostick" transformations in the magnetotelluric method. J. Geophys., 53: 72–73.

Jones, A.G., 1983b. The problem of "current channelling": a critical review. Geophys. Surv., 6: 79–122.

Jones, A.G., 1984a. The electrical structure of the lithosphere and asthenosphere beneath the Fennoscandian shield. J. Geomagn. Geoelectr., 35: 811–827.

Jones, A.G., 1984b. Electromagnetic investigations in Eastern Canada — a concise review. In: S.-E. Hjelt (Editor), The Development of the Deep Geoelectric Model of the Baltic Shield. Dept. of Geophysics, Oulu University, pp. 166–178.

Jones, A.G., 1986. Parkinsons' pointers' potential perfidy!. Geophys. J. R. Astron. Soc., 87: 1215–1224.

Jones, A.G., 1987. MT and reflection: an essential combination. Geophys. J. R. Astron. Soc., 89: 7–18.

Jones, A.G., 1988a. Static shift of magnetotelluric data and its removal in a sedimentary basin environment. Geophysics, 53: 967–978.

Jones, A.G., 1988b. Discussion of "A magnetotelluric investigation under the Williston Basin" by J.M. Maidens and K.V. Paulson. Can. J. Earth Sci., 25: 1132–1139.

Jones, A.G. and Craven, J.A., 1990. The North American Central Plains conductivity anomaly and its correlation with gravity, magnetics, seismic, and heat flow data in the Province of Saskatchewan. Phys. Earth Planet. Inter., 60: 169–194.

Jones, A.G. and Garland, G.D., 1984. A magnetotelluric survey of Prince Edward Island: evidence for an electrically isotropic zone. Contributed paper at Seventh Workshop on Electromagnetic Induction in the Earth and Moon. Ile-Ife, Nigeria, 15–24 August.

Jones, A.G. and Garland, G.D., 1986. Prelimenary interpretation of the upper crustal structure beneath Prince Edward Island. Ann. Geophys., 4B: 157–164.

Jones, A.G. and Hutton, R., 1979a. A multi-station magnetotelluric study in southern Scotland — I. Fieldwork, data analysis and results. Geophys. J. R. Astron. Soc., 56: 329–349.

Jones, A.G. and Hutton, R., 1979b. A multi-station magnetotelluric study in southern Scotland — II. Monte-Carlo inversion of the data and its geophysical and tectonic implications. Geophys. J. R. Astron. Soc., 56: 351–368.

Jones, A.G. and Savage, P.J., 1986. North American Central Plains conductivity anomaly goes east. Geophys. Res. Lett., 13: 685–688.

Jones, A.G., Olafsdottir, B. and Tiikkainen, J., 1983. Geomagnetic induction studies in Scandinavia — III. Magnetotelluric observations. J. Geophys., 54: 35–50.

Jones, A.G., Kurtz, R.D., Oldenburg, D.W., Boerner, D.E. and Ellis, R., 1988. Magnetotelluric observations along the LITHOPROBE southeastern Canadian Cordilleran transect. Geophys. Res. Lett., 15: 677–680.

Jones, A.G., Chave, A.D., Egbert, G., Auld, D. and Bahr, K., 1989. A comparison of techniques for magnetotelluric response function estimation. J. Geophys. Res., 94: 14,201–14,213.

Kaikkonen, P., Vanyan, L.L., Martanus, E.R. and Okulessky, B.A., 1985. Contribution of the surficial effects on the low frequency magnetotelluric anomaly at the Rhinegraben area. Phys. Earth Planet. Inter., 37: 223–227.

Kao, D. and Orr, D., 1982. Magnetotelluric response of a uniformly stratified earth containing a magnetized layer. Geophys. J. R. Astron. Soc., 70: 339–347.

Kariya, K.A. and Shankland, T.J., 1983. Electrical conductivity of dry lower crustal rocks. Geophysics, 48: 52–61.

Kay, R.W. and Kay, S.M., 1981. The nature of the lower continental crust: Inferences from geophysics, surface geology, and crustal xenoliths. Rev. Geophys. Space Phys., 19: 271–297.

Keller, G.V., 1989a. Electrical structure of the crust and upper mantle beneath the United States: Part 1, Methods for determining the conductivity profile. In: L.C. Pakiser and W.D. Mooney (Editors), Geophysical Framework of the Continental United States. Geol. Soc. Am. Mem., 172: 71–89.

Keller, G.V., 1989b. Electrical structure of the crust and upper mantle beneath the United States: Part 2, Survey of data and interpretation. In: L.C. Pakiser and W.D. Mooney (Editors), Geophysical Framework of the Continental United States. Geol. Soc. Am. Mem., 172: 425–446.

Keller, G.V., Pritchard, J.I., Jacobson, J.J. and Harthill, N., 1984. Megasource time-domain electromagnetic sounding methods. Geophysics, 49: 993–1009.

Kellet, R.L., Mareschal, M. and Kurz, R.D., 1992. A model of lower crustal electrical anisotropy for the Pontiac Subprovince of the Canadian Shield. Geophys. J. Int., in press.

Keshet, Y. and Hermance, J.F., 1986. A new regional electrical model for the southern section of the Rio Grande rift and the adjacent Basin and Range and Great Plains. J. Geophys. Res., 91: 6359–6366.

Klemperer, S.L., 1987. A relation between continental heat-flow and the seismic reflectivity of the lower crust. J. Geophys., 61: 1–11.

Klemperer, S.L. and Matthews, D.H., 1987. Iapetus suture located beneath the North Sea by BIRPS deep seismic reflection profiling. Geology, 15: 195–198.

Koefoed, O., 1979. Geosounding Principles, 1. Resistivity sounding measurements. Elsevier, Amsterdam, 276 pp. .

Korja, T., 1990. Electrical conductivity variations in central Fennoscandian Shield, Finland. Contributed paper at 10th Workshop on Electromagnetic Induction in the Earth. Ensenada, Mexico, 22–29 August.

Korja, T., Zhang, P. and Pajunpää, K., 1986. Magnetovariational and magnetotelluric studies of the Oulu anomaly on the Baltic Shield in Finland. J. Geophys., 59: 32–41.

Korja, T., Hjelt, S.-E., Kaikonnen, P., Koivukoski, K., Rasmussen, T.M. and Roberts, R.G., 1989. The geoelectric model of the POLAR profile. Tectonophysics, 162: 113–133.

Kovtun, A.A., 1976. Induction studies in stable shield and platform areas. Acta Geod. Geophys. Montanist., Acad. Sci. Hung., 11: 333–346.

Kuckes, A.F., 1973a. Relations betwenn electrical conductivity of a mantle and fluctuating magnetic fields. Geophys. J. R. Astron. Soc., 32: 119–131.

Kuckes, A.F., 1973b. Correspondence between the magnetotelluric and field penetration depth analysis for measuring electrical conductivity. Geophys. J. R. Astron. Soc., 32: 381–385.

Küppers, F., Untiedt, J., Baumjohann, W., Lange, K. and Jones, A.G., 1979. A two-dimensional magnetometer array for ground-based observations of auroral zone electric currents during the International Magnetospheric Study (IMS). J. Geophys., 46: 429–450.

Kurtz, R.D. and Gupta, J.C., 1992. Conductivity studies in the Miramichi, New Brunswick, earthquake zone. Can. J. Earth Sci., July issue.

Kurtz, R.D., DeLaurier, J.M. and Gupta, J.C., 1986a. A magnetotelluric sounding across Vancouver Island sees the subducting Juan de Fuca plate. Nature, 321: 596–599.

Kurtz, R.D., Ostrowski, J.A. and Niblett, E.R., 1986b. A magnetotelluric survey over the East Bull Lake Gabbro–Anorthosite Complex. J. Geophys. Res., 91: 7403–7416.

Kurtz, R.D., Niblett, E.R., Craven, J.A., Stevens, R.A. and Macnae, J.C., 1988. Electromagnetic studies over the Kapuskasing structural zone. In: G.F. West (Editor), Proceedings of the Kapuskasing LITHOPROBE Workshop. Toronto, 16–17 Feb., pp. 169–175.

Kurtz, R.D., Macnae, J.C. and West, G.F., 1989. A controlled-source, time-domain electromagnetic survey over an upthrust section of Archean crust in the Kapuskasing Structural Zone. Geophys. J. Int., 99: 195–203.

Kurtz, R.D., DeLaurier, J.M. and Gupta, J.C., 1990. The electrical conductivity distribution beneath Vancouver Island: a region of active plate subduction. J. Geophys. Res., in press

Lane, A.C., 1914. Mine water composition, an index to the course of ore-bearing currents. Econ. Geol., 9: 239–262.

Law, L.K. and Riddihough, R.P., 1971. A geographical relation between geomagnetic variation anomalies and tectonics. Can. J. Earth Sci., 8: 1094–1106.

Lee, C.D., Vine, F.J. and Ross, R.G., 1983. Electrical conductivity models for the continental crust based on laboratory measurements on high-grade metamorphic rocks. Geophys. J. R. Astron. Soc.,72: 353–371.

Lee, V.W., Mackwell, S.J. and Brantley, S.L., 1991. The effect of fluid chemistry on wetting textures in novaculite. J. Geophys. Res., 96: 10,023–10,037.

Legget, J.K., McKerrow, W.S. and Soper, N.J., 1983. A model for the crustal evolution of southern Scotland. Tectonics, 2: 18–210.

Lewis, T., 1990. Crustal temperatures along the LITHOPROBE southern Cordilleran transect. LITHOPROBE Cordilleran Workshop, Calgary, Alberta, 3–4 March. Cordilleran Workshop Proc., 2: 171–173.

Lewis, T.J., Bentkowski, W.H. and Hyndman, R.D., 1992. Crustal temperatures near the Lithoprobe Southern Cordilleran Transect. Can. J. Earth Sci., June issue.

Lienert, B.R., 1979. Crustal electrical conductivities along the eastern flank of the Sierra Nevadas. Geophysics, 44: 1830–1845.

Lienert, B.R. and Bennet, D.J., 1977. High electrical condictivities in the lower crust of the northwestern Basin and Range: An application of inverse theory to a controlled-source deep-magnetic-sounding experiment. In: J.G. Heacock (Editor), The Earth's Crust. AGU Geophys. Monogr. Ser., 20: 531–552.

Maillet, R., 1947. The fundamental equations of electrical prospecting. Geophysics, 12: 529–556.

Maxwell, J.C., 1892. A Treatise on Electricity and Magnetism, 3rd ed. (2 vol). Clarendon Press, Oxford.

Mareschal, M., 1990. Electrical conductivity: the story of an elusive parameter, and how it possibly relates to the Kapuskasing Uplift (LITHOPROBE Canada). In: M.H. Salisbury and D.M. Fountain (Editors), Exposed Cross-Sections of the Continental Crust. Kluwer, Dordrecht, pp. 453–468.

Mareschal, M., Chakridi, R. and Chouteau, M., 1988. A magnetotelluric survey across the Groundhog River block: progress report on the pseudo 1-D interpretation. In: G.F. West (Editor), Proceedings of the Kapuskasing LITHOPROBE Workshop. Toronto, 16–17 Feb., pp. 77–82.

McCaig, A.M., 1988. Deep fluid circulation in fault zones. Geology, 16: 867–870.

McCaig, A.M., 1989. Fluid flow through fault zones. Nature, 340: 600.

McKerrow, W.S. and Soper, N.J., 1989a. The Iapetus suture in the British Isles. Geol. Mag., 126: 1–8.

McKerrow, W.S. and Soper, N.J., 1989b. The Iapetus suture in the British Isles –reply. Geol. Mag., 126: 724.

Murphy, F.C. and Hutton, D.H.W., 1986. Is the Southern Uplands of Scotland really an accretionary prism?. Geology, 14: 354–357.

Nabighian, M., 1987. Electromagnetic Methods in Applied Geophysics. Vol. 1, Theory. Soc. Explor. Geophys., Tulsa, Okla.

Nabighian, M., 1991. Electromagnetic Methods in Applied Geophysics, Vol. 2. Applications. Soc. Explor. Geophys., Tulsa, Okla., 2 parts.

Nesbitt, B.E., 1992. Electrical resistivities of crustal fluids. Geophys. Res. Lett., in press.

Newton, R.C., 1990. Fluids and shear zones in the deep crust. Tectonophysics, 182: 21–37.

Niblett, E.R. and Sayn-Wittgenstein, C., 1960. Variation of the electrical conductivity with depth by the magnetotelluric method. Geophysics, 25: 998–1008.

North, R.G., Wetmiller, R.J., Adams, J., Anglin, F.M., Hasegawa, H.S., Lamontagne, M., Berger, R.D., Seeber, L. and Armbruster, J., 1989. Preliminary results from the November 25, 1988 Saguenay (Quebec) earthquake. Seism. Res. Lett., 60: 89–93.

Oldenburg, D.W., 1992. Inversion of electromagnetic data: an overview of new techniques. Geophys. Surv., in press.

Olhoeft, G.R., 1981. Electrical properties of granite with implications for the lower crust. J. Geophys. Res., 86: 931–936.

Oliver, J., 1991. Solid Earth science during the 21st century. EOS, 72: 124–126.

Oliver, J. and Kaufman, S., 1976. Profiling the Rio Grande rift. Geotimes, 21: 20–23.

Pajunpää, K., 1984. Magnetometer array studies in Finland –determination of single station transfer functions. J. Geophys., 55: 153–160.

Pajunpää, K., 1986. Magnetometer array studies in southeastern Finland on the Baltic Shield. J. Geophys., 59: 23–31.

Pajunpää, K., 1987. Conductivity anomalies in the Baltic Shield in Finland. Geophys. J. R. Astron. Soc., 91: 657–666.

Pajunpää, K., Heikka, J. and Korja, T., 1983. Magnetometer array studies in Finland. J. Geomagn. Geoelectr., 35: 543–553.

Park, S.K. and D. Livelybrooks: 1989. Quantitative interpretation of rotationally invariant parameters in magnetotellurics. Geophysics, 54: 1483–1490.

Parker, R.L., 1980. The inverse problem of electromagnetic induction: existence and construction of solutions based on incomplete data. J. Geophys. Res., 85: 4421–4425.

Parker, R.L. and Whaler, K.A., 1981. Numerical methods for establishing solutions to the inverse problem of electromagnetic induction. J. Geophys. Res., 86: 9574–9584.

Parkomenko, E.I., Stafankevich, Z.V. and Vysokova, I.M., 1972. On the role of moisture in the electrical conductivity of crystalline rocks at high pressure and temperatures. Izv. Phys. Solid Earth, 8: 323–332.

Peacock, S.M., 1990. Fluid processes in subduction zones. Science, 248: 329–337.

Pellerin, L. and Hohmann, G.W., 1990. Transient electromagnetic inversion: a remedy for magnetotelluric static shifts. Geophysics, 55: 1242–1250.

Percival, J.A. and Card, K.D., 1983. Archean crust as revealed in the Kapuskasing Uplift, Superior Province, Canada. Geology, 11: 323–326.

Percival, J.A. and Card, K.D., 1985. Structure and evolution of Archean crust in central Superior Province, Canada. Spec. Pap., Geol. Assoc. Can., 28: 19–192.

Percival, J.A., Green, A.G., Milkereit, B., Cook, F.A., Geis, W. and West, G.F., 1989. Seismic reflection profiles across deep continental crust exposed in the Kapuskasing uplift structure. Nature, 342: 416–420.

Phillips, W.E.A., Stillman, C.J. and Murphy, T., 1976. A Caledonian plate tectonic model. J. Geol. Soc. London, 132: 576–609.

Piwinskii, A.J. and Duba, A., 1974. High temperature electrical conductivity of albite. Geophys. Res. Lett., 1: 209–211.

Poll, H.E., Weaver, J.T. and Jones, A.G., 1987. Calculations of voltage differences for magnetotelluric modelling of a region with near-surface inhomogeneities. Phys. Earth Planet. Inter., 53: 287–297.

Popov, A.M., 1987. Causes of increased electrical conductivity in the crust (by example of the Baikal region). Geol. Geofiz. [Sov. Geol. Geophys.], 28: 56–65.

Quist, A.S. and Marshall, W.L., 1968. Electrical conductances of aqueous sodium chloride solutions from 0° to 800°C and effective pressures to 4000 bars. J. Phys. Chem., 72: 684–703.

Ranganayaki, R.P., 1984. An interpretive analysis of magnetotelluric data. Geophysics, 49: 1730–1748.

Rankin, D. and Pascal, F., 1990. A gap in the North American Central Plains conductivity anomaly. Phys. Earth Planet. Inter., 60: 132–137.

Rasmussen, T.M., 1988. Magnetotellurics in southwestern Sweden: evidence for electrical anisotropy in the lower crust?. J. Geophys. Res., 93: 7897–7907.

Rasmussen, T.M., Roberts, R.G. and Pedersen, L.B., 1987. Magnetotellurics along the Fennoscandian Long Range profile. Geophys. J. R. Astron. Soc., 89: 799–820.

Reiter, M., Eggleston, R.E., Broadwell, B.R. and Minier, J., 1986. Estimates of terrestrial heat flow from deep petroleum tests along the Rio Grande rift in central and southern New Mexico. J. Geophys. Res., 91: 6225–6245.

Reitmayr, G., 1975. An anomaly of the upper mantle below the Rhinegraben, studied by the inductive response of natural electromagnetic fields. J. Geophys., 41: 651–658.

Richards, M.L., Schmucker, U., Steveling, E. and Watermann, J.. 1981. Erdmagnetische und magnetotellurische Sondierungen im Gebeit des mitteleuropäischen Riftsystems. Inst. Geophys. Univ. Göttingen, Forschungsber., T81–111.

Rokityansky, I.I., 1983. Geoelectromagnetic studies of the Baltic and Ukranian shield: review of some results. In: S.-E. Hjelt (Editor), The Development of the Deep Geoelectric Model of the Baltic Shield. Dept. of Geophysics, Oulu University, pp. 110–150.

Rokityansky, I.I., Kulik, S.N. and Rokityanskaya, D.A., 1981. The Ladoga electric conductivity anomaly. Geophys. J. (Geofiz Zh.), 3: 301–304.

Rooney, D. and Hutton, V.R.S., 1977. A magnetotelluric and magnetovariational study of the Gregory Rift Valley, Kenya. Geophys. J. R. Astron. Soc., 51: 91–119.

Rumble, D., 1989. Evidences for fluid flow during regional metamorphism. Eur. J. Mineral., 1: 731–737.

Rumble, D., Duke, E.F. and Hoering, T.C., 1986. Hydrothermal graphite in New Hampshire: evidence for carbon mobility during regional metamorphism. Geology, 14: 452–455.

Sanders, I.S., 1991. Exhumed lower crust in NW Ireland, and a model for crustal conductivity. J. Geol. Soc., London, 148: 131–135.

Sato, H. and Ida, Y., 1984. Low frequency electrical impedance of partially molten gabbro: the effect of melt geometry on electrical properties. Tectonophysics, 107: 105–134.

Schmeling, H., 1986. Numerical models on the influence of partial melt on elastic, anelastic and electrical properties of rocks. Part II: Electrical conductivity. Phys. Earth Planet. Inter., 43: 123–136.

Schmucker, U., 1964. Anomalies of geomagnetic variations in the southwestern United States. J. Geomagn. Geoelectr., 15: 192–221.

Schmucker, U., 1970. Anomalies of geomagnetic variations in the southwestern United States. Bull. Scripps Inst. Oceanogr., Univ. Calif. Press, 13.

Schmucker, H., 1985. Depth distribution of the electrical conductivity. In: K.-H. Hellwege and O. Madelung (Editors), Landolt-Börnstein, Vol. 2. Geophysics of the Solid Earth, the Moon and the Planets. Springer, Berlin, pp. 370–397.

Schmucker, U. and Tezkan, B., 1988. 20 Jahre elektromagnetische Tiefenforschung im Rheingraben –eine Zusammenfassung mit Ausblick auf neuere Ergebnisse. In: V. Haak and J. Homilius (Editors), Protokoll über das 12. Kolloquium "Elektromagnetische Tiefenforschung". Königstein im Taunus, 1–3 March, pp. 17–34.

Schwarz, G., 1990. Electrical conductivity of the Earth's crust and upper mantle. Surv. Geophys., 11: 133–161.

Serpa, L., Setzer, T., Farmer, H., Brown, L., Oliver, J., Kaufman, S., Sharp, J. and Steeples, D., 1984. Structure of the southern Keweenawan Rift from COCORP surveys across the midcontinent geophysical anomaly in northeastern Kansas. Tectonics, 3: 367–384.

Shankland, T.J. and Ander, M.E., 1983. Electrical conductivity, temperatures, and fluids in the lower crust. J. Geophys. Res., 88: 9475–9484.

Sibson, R.H., Moore, J.McM. and Rankin, A.H., 1975. Seismic pumping — a hydrothermal fluid transport mechanism. J. Geol. Soc. London, 131: 653–659.

Skippen, G., 1988. Studies of fluids in the crust. Logan Club seminar, Geol. Surv. Can., Ottawa, April 11th.

Skokan, C.K. and Andersen, H.T., 1991. Deep long-offset transient electromagnetic surveys for crustal studies in the U.S.A.. Phys. Earth Planet. Inter., 66: 39–50.

Smith, J.T. and Booker, J.R., 1988. Magnetotelluric inversion for minimum structure. Geophysics, 53: 1565–1576.

Smith, J.T. and Booker, J.R., 1991. Rapid inversion of two and three-dimensional magnetotelluric data. J. Geophys. Res., 96: 3905–3922.

Spies, B.R., 1989. Depth of investigation in electromagnetic sounding methods. Geophysics, 54: 872–888.

Stanley, W.D., 1989. Comparison of geoelectric/tectonic models for suture zones in the western U.S.A. and eastern Europe: are black shales a possible source of high conductivities?. Phys. Earth Planet. Inter., 53: 228–238.

Sternberg, B.K., Washburne, J.C. and Pellerin, L., 1988. Correction for the static shift in magnetotellurics using transient electromagnetic soundings. Geophysics, 53: 1459–1468.

Stesky, R.M. and Brace, W.F., 1978. Electrical conductivity of serpentinized rocks to 6 kilobars. J. Geophys. Res., 41: 529–547.

Strack, K.-M., 1984. The deep transient electromagnetic sounding technique: First field test in Australia. Explor. Geophys., 15: 251–259.

Strack, K.-M., Lüschen, E. and Kötz, A.W., 1990. Long-offset transient electromagnetic (LOTEM) depth soundings applied to crustal studies in the Black Forest and Swabian Alb, Federal Republic of Germany. Geophysics, 55: 834–842.

Sule, P.O. and Hutton, V.R.S., 1986. A broad-band magnetotelluric study in southeastern Scotland. Data acquisition, analysis and one-dimensional modelling. Ann. Geophys., 4(B): 145–156.

Swift, C.M., 1967. A magnetotelluric investigation of an electrical conductivity anomaly in the south-western United States. Ph.D. thesis, M.I.T., Cambridge, Mass.

Tezkan, B. and Schmucker, U., 1985. Erdmagnetische und tellurische Untersuchung im Schwartzwald. Bericht zum Unterantrag der Univ. Karlsruhe zum BMFT Förderungs-vorhaben RG8314, Inst. für Geophysik, Univ. Göttingen.

Thomas, M.D., Sharpton, V.L. and Grieve, R.A.F., 1987. Gravity patterns and Precambrian structure in the North American Central Plains. Geology, 15: 489–492.

Thompson, A.B. and Connolly, J.A.D., 1990. Metamorphic fluids and anomalous porosity in the lower crust. Tectonophysics, 182: 47–55.

Torgersen, T., 1989. Terrestrial helium degassing fluxes and the atmospheric helium budget: Implications with respect to the degassing processes of continental crust. Chem. Geol. (Isotope Geosci. Section), 79: 1–14.

Torgersen, T., 1990. Crustal scale fluid transport: magnitude and mechanisms. EOS, 71: 1,4–5.

Torgersen, T., 1991. Crustal fluid flow: continuous or episodic. EOS, 72: 18–19.

Touret, J., 1987. Fluid distribution in the continental lithosphere In: A. Kröner (Editor), Proterozoic Lithospheric Evolution. Am. Geophys. Union, Geodynamics Series, 17: 27–33.

Towle, J.N., 1980. Observations of direct current concentration on the eastern Sierran Front: Evidence for shallow crustal conductors on the eastern Sierran Front and beneath the Coso Range. J. Geophys. Res., 85: 2484–2490.

Tsong, I.S.T., Knipping, U., Loxton, C.M., Magee, C.W. and Arnold, G.W., 1985. Carbon on surfaces of magnesium oxide and olivine single crystals. Diffusion from the bulk or surface contamination?. Phys. Chem. Minerals, 12: 261–270.

Tyburczy, J.A. and Waff, H.S., 1983. Electrical conductivity of molten basalt and andersite to 25 kilobars pressure: geophysical significance and implications for charge transport and melt structure. J. Geophys. Res., 88: 2413–2430.

Ucok, H., 1979. Temperature dependence of electrical resistivity of aqueous salt solutions and solution saturated rocks. Ph.D. thesis, Univ. Southern California, Los Angeles, 154 pp.

Valley, J.W., 1986. Stable isotope geochemistry of metamorphic rocks. In: J.W. Valley, H.P. Taylor. and J.R. O'Neil (Editors), Stable Isotopes at High Temperature Geological Processes. Mineral. Soc. Am. Rev. Mineral., 16: 445–489.

Vanyan, L.L. and Cox, C.S., 1983. Comparison of deep conductivities beneath continents and oceans. J. Geomagn. Geoelectr., 35: 805–809.

Vanyan, L.L. and Shilovsky, A., 1989. Fluids in the lower crust inferred from electromagnetic data. In: R.F. Mereu, S. Mueller and D.M. Fountain (Editors),Properties and Processes of Earth's Lower Crust. AGU Geophys. Monogr., 51: 243–246.

Vanyan, L.L., Berdichevsky, M.N., Vasin, N.D., Okulyessky, B.A. and Shilovsky, P.P., 1980. On the normal geoelectric profile. Izv. Earth Phys., 16: 131–133.

Vanyan, L.L., Shilovsky, A.P., Okulessky, B.A., Semenov, V.Y. and Sidelnikova, T.A., 1989. electrical conductivity of the crust of the Siberian platform. Phys. Earth Planet. Inter., 54: 163–166.

van Zijl, J.S.V., 1969. A deep Schlumber sounding to investigate the electrcal structures of the crust and upper mantle in South Africa. Geophysics, 34: 450–462.

van Zijl, J.S.V., 1977. Electrical studies of the deep crust in various tectonic provinces of southern Africa. In: J.G. Heacock (Editor), The Earth's Crust: Its Nature and Physical Properties. AGU, Geophys. Monogr. Ser., 20: 470–500.

Velikhov, Y.P., Zhamaletdinov, A.A., Belkov, I.V., Gorbunov, G.I., Hjelt, S.-E., Lisin, A.S., Vanyan, L.L., Zhdanov, M.S., Demidova, T.A., Korja, T., Kirillov, S.K., Kuksa, Y.I., Poltanov, A.Y., Tokarev, A.D. and Yevstigneyev, V.V., 1986. Electromagnetic studies on the Kola peninsula and in northern Finland by means of a powerful controlled source. J. Geodyn., 5: 237–256.

Volbers, R., Jödicke, H. and Untiedt, J., 1992. Magnetotelluric study of the earth's crust along the deep seismic reflection profile DEKORP 2-N. Geol. Rundsch., in press.

Vozoff, K., ed., 1986. Magnetotelluric Methods. Soc. Explor. Geophys. Reprint Ser. No. 5, Publ. by Soc. Expl. Geophys., Tulsa, Okla., ISBN 0-931830-36-2.

Waff, H., 1974. Theoretical considerations of electrical conductivity in a partially molten mantle with implications for geothermometry. J. Geophys. Res., 79: 4003–4010.

Wannamaker, P.E., Hohmann, G.W. and Ward, S.H., 1984. Magnetotelluric responses of three-dimensional bodies in layered earths. Geophysics, 49: 1517–1533.

Wannamaker, P.E., Booker, J.R., Filloux, J.H., Jones, A.G., Jiracek, G.R., Chave, A.D., Waff, H.S., Young, C.T., Stodt, J.A., Martinez, M., Law, L.K., Yukitake, T., Segawa, J.S., White, A. and Green, A.W., 1989a. Magnetotelluric observations across the Juan de Fuca subduction system in the EMSLAB project. J. Geophys. Res., 94: 14,111–14,125.

Wannamaker, P.E., Booker, J.R., Jones, A.G., Chave, A.D., Filloux, J.H., Waff, H.S. and Law, L.K., 1989b. Resistivity cross-section through the Juan de Fuca subduction system and its tectonic implications. J. Geophys. Res., 94: 14,127–14,144.

Ward, S., 1983. Controlled source electrical methods for deep exploration. Geophys. Surv., 6: 137–152.

Watson, E.B., 1986. Immobility of reduced carbon along grain boundaries in dunite. Geophys. Res. Lett., 13: 529–532.

Watson, E.B. and Brenan, J.M., 1987. Fluids in the lithosphere, 1. Experimentally-determined wetting characteristics of CO_2-H_2O fluids and their implications for fluid transport, host-rock physical properties, and fluid inclusion formation. Earth Planet. Sci. Lett., 85: 497–515.

Weaver, J.T., 1982. Regional induction in Scotland: an example of three-dimensional numerical modelling using the thin sheet approximation. Phys. Earth Planet. Inter., 28: 161–180.

Weidelt, P., 1972. The inverse problem of geomagnetic induction. Z. Geophys., 38: 257–289.

Weidelt, P., 1985. Construction of conductance bounds from magneotelluric impedances. J. Geophys., 57: 191–206.

Whelan, J.P., Brown, C., Hutton, V.R.S. and Dawes, G.J.K., 1990. A geoelectric section across Ireland from magnetotelluric soundings. Phys. Earth Planet. Inter., 60: 138–146.

Wickham, S.M. and Taylor, H.P.J., 1987. Stable isotope constraints on the origin and depth of penetration of hydrothermal fluids associated with Hercynian regional metamorphism and crustal anatexis in the Pyrenees. Contrib. Mineral. Petrol., 95: 255–268.

Wilhelm, H., Berktold, A., Bonjer, K.-P., Jäger, K., Stiefel, A. and Strack, K.-M., 1989. Heat flow, electrical conductivity and seismicity in the Black Forest crust, SW Germany. In: R.F. Mereu, S. Mueller and D.M. Fountain (Editors), Properties and Processes of Earth's Lower Crust. Am. Geophys. Union, Geophys. Monogr., 51: 215–232.

Wilson, J.T., 1966. Did the Atlantic close and then re-open?. Nature, 211: 676–681.

Wood, B.J. and Walther, J.V., 1986. Fluid flow during metamorphism and its implications for fluid–rock ratios. In: J.V. Walther and B.J. Wood (Editors), Fluid–Rock Interactions during Metamorphism. Springer-Verlag, Berlin-Heidelberg-New York, pp. 109–131.

Woods, D.V. and Allard, M., 1986. Reconnaissance electromagnetic induction study of the Kapuskasing Structural Zone: implications for lower crustal conductivity. Phys. Earth Planet. Inter., 42: 135–142.

Wunderman, R.L., 1986. Structure of the Keweenaw Rift and adjacent crust from a magnetotelluric survey in central Minnesota–Wisconsin. Paper presented at 32nd Inst. on Lake Superior Geology, Wisconsin Rapids, Wisconsin, extended abstract pp. 91–92.

Wunderman, R.L. and Young, C.T., 1987. Evidence for widespread basement decollement and related crustal asymmetry associated with the western limb of the Midcontinent Rift. Paper presented at 33rd Inst. on Lake Superior Geology, Wawa, Ontario, extended abstract pp. 85–86.

Wunderman, R.L., Young, C.T. and Rogers, J.R., 1985. A magnetotelluric traverse across the exposed midcontinent rift and adjacent basement, central Minnesota–Wisconsin –prelimenary data. EOS, 66: 874.

Wyllie, P.J., 1971. A discussion of water in the crust. In: J.G. Heacock (Editor), The Structure and Physical Properties of the Earth's Crust. AGU Geophys. Monogr. Ser., 14: 257–260.

Yardley, B.W.D., 1986a. Is there water in the deep continental crust? . Nature, 323: 111.

Yardley, B.W.D., 1986b. Fluid migration and veining in the Connemara schists, Ireland. In: J.V. Walther and B.J. Woods (Editors),Fluid–Rock Interactions during Metamorphism. Springer-Verlag, New York, pp. 109–131.

Young, C.T., Kitchen, M.R., Rogers, J.C., Mareschal, J.C. and Brandwein, S., 1986. Magnetotelluric soundings in the Charleston, South Carolina, area. In: Proceedings, Third U.S. National Conference on Earthquake Engineering, Earthquake Engineering Res. Instit., pp. 83–91.

Young, C.T., Adams, D., Loukili, A. and Wunderman, R., 1989. Electrical conductivity models of the Lake Superior region (midcontinent rift). EOS, 70: 274.

Yund, R.A. and Farver, J.R., 1990. Oxygen grain boundary diffusion in quartz aggregates: role of microstructure and fluid composition. EOS, 71: 1664.

Zhang, P., Roberts, R.G. and Pedersen, L.B., 1987. Magnetotelluric strike rules. Geophysics, 52: 267–278.

Chapter 4

Magnetic properties of the lower continental crust

PETER N. SHIVE, RICHARD J. BLAKELY, B. RONALD FROST
and DAVID M. FOUNTAIN

1. Introduction

Our knowledge of the magnetic properties of the deep continental crust results from two different kinds of studies. First, aeromagnetic and satellite magnetic measurements above the Earth's surface can, subject to considerable interpretive ambiguities imposed by potential field theory, yield information about the distribution of magnetic sources at depth. Second, laboratory measurements of magnetic properties of rocks that were once in the deep crust provide more detailed, albeit localized, information. Both types of studies are well represented in the literature, but they do not always agree on the more important questions of deep crustal magnetization.

In this chapter we first review studies of long-wavelength magnetic anomalies, which are generally regarded as due to sources in the deep crust. We then discuss laboratory magnetic and petrologic studies of rocks believed to represent the lower crust. Although our primary focus is on the continental crust, brief comparative discussion of the mantle and the oceanic crust is also in order. Our goal is to provide the best available answers to the following questions:

(1) What is the intensity of magnetization of the deep crust?

(2) What is the relative importance of induced and remanent magnetization?

(3) What are the important magnetic minerals?

(4) How does magnetization vary with temperature and depth?

(5) How does magnetization vary with metamorphic grade?

Finally, we use the answers to these questions to discuss the constraints placed on structural and petrologic models of the deep continental crust by magnetic studies to date.

2. Long-wavelength magnetic anomalies

Aeromagnetic and satellite magnetometer surveys commonly show anomalies whose wavelengths are comparable to or greater than the distance from the survey level to the lower crust. Studies of these long-wavelength magnetic anomalies are remarkably consistent in one respect. Highly magnetic sources appear to exist deep in the crustal section at depths above the Curie-temperature isotherm. Table 4-1, expanded and updated from Mayhew et al. (1985a), shows various estimates of magnetization and depth-extent of crustal magnetic sources as inferred from long-wavelength magnetic anomalies. All of these studies found relatively large

TABLE 4-1

Magnetization and depth extent of magnetic crust inferred from long-wavelength magnetic anomalies

Location	Data	Method	M (A/m)	Extent (km)	Cause	Reference
Ukranian and Baltic shields	A	2-D FM	>4	40–50	undulations on bottom of lower-crustal layer	Krutikhovskaya and Pashkevich (1977, 1979)
British Columbia	A	2-D FM w/pp	3–5	30–40	highly magnetic rocks, deep Curie-T isotherm related to subduction	Coles and Currie (1977)
Northwest Territories	A	LSI	4	16	Precambrian volcanic rocks	Coles (1976)
British Colombia	A	2-D FM	5	24	effect of western boundary of Canadian shield	Coles (1976)
Manitoba and Ontario	A	C w/FA	5	40	structure within lower-crustal layer	Hall (1974)
Northwestern Germany	A	FAGD	2	various	relief on top of lower-crustal layer	Hahn et al. (1976)
Conterminous U.S.	S	EL/CT	2–6	20–55	lower-crustal layer bottom relief and lateral magnetization variations	Schnetzler (1985)
Southeastern U.S.	S	3-D FM	2	45	compositional variations in lower-crustal layer	Ruder and Alexander (1986)
Utah	A	2-D FM	6	20–35	Curie-T isotherm undulations	Shuey et al. (1973)
Kentucky	A, S	3-D FM w/G and SE	4	40	mafic intrusive complex extending extending throughout crust	Mayhew et al. (1982, 1985b)
Fennoscandia	Gr	LSI	2	35	Curie-T isothem undulations	Elming and Törne (1976)
Sweden	A	2-D FM	3–5	10–23	anorogenic granitic complex	Riddihough (1972)
Central African Republic	S, A	3-D FM	3	35	mafic plutons	Regan and Marsh (1982)
Japanese Volcanic Arc	A	2-D FM	5	30	Cretaceous volcanic-plutonic belt	Segawa and Oshima (1975
Broken Ridge	S	3-D FM	6	17	high VRM in extra thick crust	Johnson (1985)
Alpha Ridge	S	3-D FM	3	30	continental crust surrounded by oceanic crust	Taylor (1983)
Western Pacific	Sh	2-D FM	5	>15	crustal thickness undulations	LaBrecque et al. (1985)
Pacific and Atlantic basins	Sh, S	2-D FM	7	1	seafloor-spreading, TRM decay, CRM growth, geomagnetic reversals	Raymond and LaBrecque (1987)
Lord Howe Rise	S	3-D FM w/SE	2	2	altered lower crust	Frey (1985)
Aleutian arc	S	3-D FM	4	7	subducting slab	Clark et al. (1985)

Abbreviations: A: aeromagnetic; C: correlation of crustal structure; EL/CT: equivalent layer divided by crustal thickness; FA: filtered anomalies; FAGD: Fourier analysis of gridded data; FM: forward models; G: gravity; Gr: ground magnetic; LSI: least-squares inverse; M: magnetization; pp: physical properties measurement; S: satellite magnetic; SE: seismic; Sh: shipboard magnetic; T: temperature; 2-D: two-dimensional; 3-D: three-dimensional.

magnetizations, ranging from 2 to 10 A/m and typically about 4 A/m, to exist in the deep crust. Moreover, these large magnetizations appear to extend either to the bottom of the crust or to the Curie-temperature isotherm, whichever is shallower. Various deep-crustal models have been proposed to explain long-wavelength magnetic anomalies: (1) undulations of the crust-mantle interface; (2) undulations of an interface between mid- and lower-crustal layers (Hall et al., 1985); (3) undulations in the Curie-temperature isotherm (Mayhew, 1985); and (4) lateral compositional variations (e.g., Arkani-Hamed and Strangway, 1985). Deciding on the favorable model in any particular area is exceedingly difficult because of the inherent nonuniqueness of magnetic data.

The techniques and assumptions that underlie these determinations vary widely from study to study (Table 4-1). Crustal models usually are developed by trial-and-error forward calculations constrained by independent information, such as magnetic measurements of exposed rocks (e.g., Krutikhovskaya and Pashkevich, 1977, 1979; Coles and Currie, 1977), seismically determined crustal structure (Hall, 1974; Mayhew et al., 1982, 1985b; Frey, 1985; Schnetzler, 1985), gravity anomalies (Mayhew et al., 1982, 1985b), or heat-flow measurements (Shuey et al., 1973). Several models have been constructed with nonlinear least-square methods (Elming and Törne, 1976; Coles, 1976), and several studies have used Fourier analysis of gridded magnetic anomalies (Hahn et al., 1976; Connard et al., 1983; Blakely, 1988).

As discussed subsequently, the values of magnetization reported in Table 4-1 are several times larger than magnetizations actually measured in rocks believed to have originated in the deep crust. Reconciliation of this discrepancy between laboratory-determined and anomaly-inferred magnetizations of deep crustal rocks is one of the most significant geophysical challenges of the decade.

2.1. Big sources versus deep sources

Models of magnetic sources constructed from magnetic anomalies are nonunique; any magnetic anomaly has an infinite variety of mathematically possible sources. This fact is of particular importance to interpretation of long-wavelength anomalies; without geologic or geophysical information to the contrary, a long-wavelength anomaly does not necessarily require a deep source. This can be demonstrated by example. Consider a magnetic anomaly $h(x,y)$ observed on a horizontal surface at height $z = 0$, with z-axis positive down. Let the crustal magnetization $m(x,y)$ vary only in the horizontal dimensions and be confined between depths z_1 and z_2 where $z_1 < z_2$ In the following, we represent the Fourier transform of a two-dimensional function $f(x,y)$ by the notation $F\{f(x,y)\}$, where:

$$F\{f(x,y)\} = \int_{-\infty}^{\infty}\int_{-\infty}^{\infty} f(x,y)\ \exp\{-i(k_x x + k_y y)\}\ \mathrm{d}x\mathrm{d}y \tag{1}$$

The independent variables of $F\{f(x,y)\}$ are wave numbers k_x and k_y which are inversely proportional to wavelengths λ_x and λ_y in the x and y directions, respectively; i.e., $k_x = 2\pi/\lambda_x$ and $k_y = 2\pi/\lambda_y$. The Fourier transform of $h(x,y)$ is given by:

$$F\{h(x,y)\} = \theta(k_x,k_y) \cdot F\{m(x,y)\} \cdot [\exp\{-kz_1\} - \exp\{-kz_2\}] \tag{2}$$

where

$$k = \left[\frac{2\pi}{\lambda_x^2 + \lambda_y^2}\right]^{1/2} \qquad (3)$$

(Blakely and Connard, 1989). The factor $\theta(k_x,k_y)$ in equation (2) is a function of k_x and k_y that depends only on the direction of magnetization and direction of the regional field (Schouten, 1971). The assumptions implicit in equation (2), although appropriate for aeromagnetic surveys, are not strictly valid for global-scale satellite studies. Nevertheless, the arguments that follow could also be derived using spherical coordinates and thereby be made applicable to global data.

Now consider a new magnetization $m'(x,y)$ confined within a horizontal layer of identical thickness located a distance above the layer in equation (2). The anomaly caused by $m'(x,y)$ is given by:

$$Ff\ \{h'(x,y)\} = \theta(k_x,k_y) \cdot F\ \{m'(x,y)\}\left[\exp\{-k(z_1 - \Delta z)\} - \exp\{-k(z_2 - \Delta z)\}\right] \qquad (4)$$

The magnetization $m'(x,y)$ that produces the same magnetic anomaly as $m(x,y)$ is derived by requiring $h'(x,y) = h(x,y)$ and combining equations (2) and (4):

$$F\ \{m'(x,y)\} = F\ \{m(x,y)\}\exp\{-k\Delta z\} \qquad (5)$$

Equation (5) demonstrates that any magnetization confined to a deep crustal layer can be transformed to another magnetization within a shallower layer with no effect on the magnetic anomaly. The transformation is simply a filtering operation defined in the Fourier domain by $e^{-k\Delta z}$; this filter is identical to upward continuation (Blakely and Connard, 1989), a well-known smoothing operation. Hence, the distribution of magnetization within the shallow layer will be broader and smoother than the magnetization of the deep layer, but the anomalies produced by the two layers will be identical. For example, an anomaly that can be explained by abrupt compositional variations near the crust–mantle interface may be explained equally well by smooth relief at the top of a lower-crustal layer (Fig. 4-1).

2.2. Satellite magnetic surveys

Satellite magnetic surveys provide the only globally consistent information about crustal magnetization. This significant advantage over aeromagnetic data is offset by the necessity to orbit at high altitudes, where the anomaly signal is much smaller and diurnal noise is much greater than at the Earth's surface. The POGO mission resulted in the first consistent map of crustal magnetic fields on a global scale (Regan et al., 1975). The POGO satellite orbited within an altitude range of 400 to 1510 km that permits a resolution of crustal magnetic anomalies of only 250–500 km across (Langel and Thorning, 1982). The Magsat mission, flown between October 1979 and June 1980, orbited between altitudes of 200–550 km, and crustal magnetic maps produced with Magsat data resolve magnetic features with lateral dimensions of 150–300 km (Langel et al., 1982). Although the Magsat mission has greatly expanded our knowledge of crustal magnetic fields, the resolving power of Magsat is still on the order of the dimensions of many geologic provinces. Both the National Aeronautics and Space Administration and the European Space Agency are planning future orbital missions at even lower altitudes, which will greatly benefit crustal investigations.

Magnetic sources within the Earth are located primarily in two regions: in the crust above the Curie-temperature isotherm, caused primarily by spontaneous magnetization of

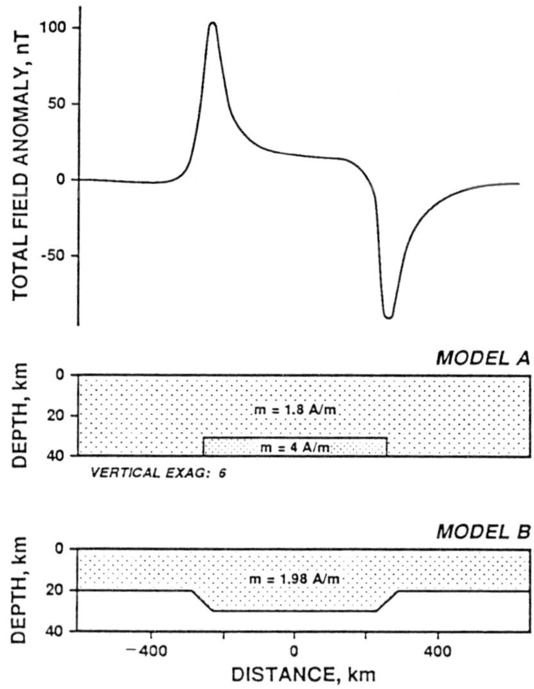

Fig. 4-1. Two crustal models for long-wavelength magnetic anomalies. The same magnetic anomaly is produced by both models A and B. Model A represents a highly magnetic body at the base of the crust. Model B represents a downwarp of the Curie-temperature isotherm. In both models, flight elevation is 1 km above the surface, profile is directed north–south, inclination of magnetization and regional field is 45°, and declination of magnetization and regional field is 0°.

titanomagnetite (see below), and in the outer core, caused by magnetohydrodynamic effects. The region between the Curie-temperature isotherm (or crust–mantle interface, whichever is shallower) and the core–mantle interface is thought to be relatively nonmagnetic except for minor magnetic fields caused by electrical currents. Because the shape and amplitude of magnetic anomalies measured above the Earth depend critically on the depth to their respective sources, it would seem that distinguishing between crustal and core source regions would be straightforward. On the contrary, one of the most difficult problems in deriving crustal magnetization from satellite magnetic data is the difficulty in distinguishing long-wavelength anomalies caused by sources within the crust from anomalies caused by the Earth's core.

The most common techniques to construct crustal magnetic maps from satellite data involve spherical harmonic analysis. Gauss showed as early as 1839 that the magnetic field due to sources internal to the Earth can be represented at any point (r, ϕ, θ) outside the Earth by a spherical harmonic expansion of the form:

$$\boldsymbol{B} = \nabla V \tag{6}$$

where

$$V = a \sum_{n=1}^{N} \sum_{m=0}^{n} \left(\frac{a}{r}\right)^{(n+1)} [g_n^m \cos m\phi + h_n^m \sin m\phi]\, P_n^m(\cos\theta) \tag{7}$$

a is the mean radius of the Earth, and P_n^m is a normalized form of associated Legendre polynomials. Specific terms of the spherical harmonic expansion are identified by their degree n and order m. The coefficients g_n^m and h_n^m are usually calculated by least-squares analysis of observed field values. Langel and Estes (1982), for example, analyzed 26,500 Magsat measurements taken on 14 quiet days of its nine-month mission to develop a 23-degree representation of the field known as MGST(10/81). They then used the spherical harmonic coefficients to estimate a power spectrum of the global magnetic field given by:

$$R_n = (n+1) \sum_{m=0}^{N} \left[(g_n^m)^2 + h_n^m)^2 \right] \tag{8}$$

(Lowes, 1966). The spectrum thus determined has a definite change in slope between degrees 13 and 14, in agreement with earlier studies (e.g., Cain et al., 1974). Langel and Estes (1982) interpreted the steep part of the spectrum ($n < 14$) to be caused by sources within the core and the flatter part of the spectrum ($n > 14$) to be dominated by crustal sources.

It would seem, therefore, that a crustal magnetic map could be constructed from satellite data by subtracting a 13-degree spherical harmonic expansion derived from the same data. Cain et al. (1984), for example, selected 40,000 Magsat measurements taken during magnetically quiet periods and constructed a spherical harmonic expansion of degree and order 29. They produced a crustal magnetic map by simply dropping the first 13 degrees. However, the tesserae that result from a 29-degree expansion have dimensions of 6° by 6° (approximately 440,000 km^2 at the equator), too large to properly represent many short-wavelength features.

Langel et al. (1982) used an alternative procedure that improved the resolution. They also selected data from magnetically quiet periods. They subtracted a 13-degree spherical harmonic expansion derived from a subset of Magsat data from these data. They also modeled the effects of external magnetic fields, caused by magnetospheric currents, with a simple spherical harmonic expansion and subtracted these effects from the data. The resulting residual values were assigned to their respective 2° by 2° tesserae (approximately 49,000 km^2 at the equator) between latitudes 50°N and 50°S, and an average value was calculated for each tessera. This 2°-average map has been widely used to understand the global crustal magnetic field.

Unfortunately, subtraction of a 13-degree spherical harmonic expansion does not eliminate long-wavelength contributions from the core (Carle and Harrison, 1982). Satellite analyses require the use of total-field data, which are calculated by squaring and summing the three vector components of equation (7); i.e.:

$$F = \left[\left(\frac{\partial V}{\partial x}\right)^2 + \left(\frac{\partial V}{\partial y}\right)^2 + \left(\frac{\partial V}{\partial z}\right)^2 \right]^{1/2} \tag{9}$$

Equation (7) shows that a spherical harmonic expansion of the vector components is the sum of sinusoidal terms. When squared to represent the total field, each sinusoidal term becomes a combination of lower- and higher-degree terms. For example, consider just the lowest-degree term in equation (7) with $r = a$:

$$V = g_1^0 P_1^0(\cos\theta) \tag{10}$$

$$= g_1^0 \cos\theta \tag{11}$$

This simple potential has a single zonal wavelength of 360°, but the total field calculated from this term:

$$F = g_1^0 \left[\tfrac{3}{2} \cos 2\theta + \tfrac{5}{2}\right]^{1/2} \tag{12}$$

has wavelengths of both 180° and infinity. Likewise, each sinusoidal term in the magnetic potential becomes a combination of lower- and higher-degree terms in the total-field representation; subtracting a low-degree expansion will not eliminate all of these contributions from the crustal magnetic field.

This result has profound implications for anomalies caused by deep crustal sources. It is commonly assumed that subtraction of a 13-degree spherical harmonic expansion from satellite data (or any other magnetic survey) produces a magnetic field caused primarily by crustal sources. Carle and Harrison (1982) showed that these residual fields also contain long wavelength components. Harrison et al. (1986) concluded that these long-wavelength components must originate in the core. They used the highest degree ($n > 13$) terms of a 29-degree field model (Cain et al., 1984) to compute a global equivalent surface layer of magnetization and found two surprising results. First, the calculated magnetization showed no latitudinal dependence, contrary to what would be expected if induced or viscous remanent magnetization were important. Second, vertically summed magnetizations from the continents were not significantly different from the oceans, which implies that crustal thickness and average magnetization are inversely proportional worldwide. These two results are highly unlikely if the high-degree expansion represents only crustal magnetization, and Harrison et al. (1986) concluded that significant contributions from the core must be contained in harmonics above $n = 13$. Arkani-Hamed and Strangway (1985) and Harrison et al. (1986) independently concluded that the crustal portion of the magnetic field is limited to degrees 19 and greater. When these higher-degree terms are examined, continental areas have significantly greater anomalies than oceanic regions, as expected.

2.3. Aeromagnetic surveys

The application of airborne magnetometers to geologic problems began in 1944 when J.R. Balsley and Homer Jensen of the U.S. Geological Survey used a stabilized fluxgate magnetometer to measure total field in the first geophysical survey near Boyertown, Pennsylvania (Jensen, 1961). Early aeromagnetic surveys and many modern surveys were designed to support tectonic or mineralogic studies and, therefore, focused on upper crustal geologic sources. Various other studies, however, focused on the depth-extent of magnetic sources and the nature of deep crustal magnetization (Table 4-1). The techniques employed in deep crustal studies can be classed in four general groups:

(1) comparison of long-wavelength aeromagnetic anomalies with independent information about deep crustal structure or heat flow (Hall, 1974; Krutikhovskaya and Pashkevich, 1979);

(2) forward models based on assumed geologic cross sections (Shuey et al., 1973; Krutikhovskaya and Pashkevich, 1977; Coles and Currie, 1977);

(3) inverse methods based on simplifying assumptions about crustal magnetization (Elming and Törne, 1976; Coles, 1976); and

(4) Fourier analysis of magnetic anomalies (Hahn et al., 1976; Connard et al., 1983; Blakely, 1988).

Recent compilations of magnetic surveys into continent-scale data bases have rejuvenated interest in long-wavelength magnetic anomalies. Maps of Canada (Morley et al., 1984), the United States (USGS and SEG, 1982; Bond and Zietz, 1987), and North America (Committee for the Magnetic Anomaly Map of North America, 1987) provide unprecedented opportunities for correlation with regional geologic, tectonic, and thermal characteristics of the crust.

A few studies have attempted to estimate the depth and configuration of Curie-temperature isotherms using either the shape of isolated magnetic anomalies (e.g., Bhattacharyya and Leu, 1977; Byerly and Stolt, 1977) or the statistical properties of patterns of magnetic anomalies (e.g., Shuey et al., 1977; Connard et al., 1983). These studies are done typically in areas of elevated heat flow, such as the Basin and Range province (Blakely, 1988), the Cascade Range (Connard et al., 1983), and the Japanese volcanic arc (Okubo et al., 1985). Consequently, they tend to find relatively shallow Curie-temperature isotherms, with depths in the range of 10–20 km. Mayhew (1985) developed a technique to estimate the depth to the Curie-temperature isotherm from satellite data and found good agreement with isotherms estimated from aeromagnetic data in the Pacific Northwest (Connard et al., 1983).

Estimation of depth to Curie-temperature from magnetic anomalies is a risky proposition at best. The problems are both mathematical and geological. First, anomalies caused by deep magnetic sources are long in wavelength and low in amplitude relative to anomalies caused by shallower sources. Distinguishing deep-source anomalies from shallow-source anomalies is problematic, and the problem is compounded because sources of long-wavelength noise are common and difficult to detect, as discussed subsequently. Moreover, all techniques to estimate basal depth of magnetic sources involve gross oversimplification of geology. The method of Connard et al. (1983), for example, assumes that magnetization is uncorrelated with itself (i.e., varies randomly in horizontal directions), which may be unrealistic in many geologic environments.

Most of the discussion thus far has focused on the importance of deep crustal rocks to long-wavelength magnetic anomalies. Recent regional compilations, however, emphasize a point made earlier by Hall (1974); some long-wavelength anomalies correlate with exposed upper crustal sources. This implies (1) that upper crustal magnetization may be an important contributor to long-wavelength anomalies and/or (2) that deep crustal structure may be reflected in surface geology. Patterns of anomalies with sources in the upper crust are often consistent over geologic terrains and provinces. When viewed at higher altitudes, these patterns may appear as individual long-wavelength anomalies. Hildenbrand (1985), for example, proposed that continent-scale crustal discontinuities are reflected in digital aeromagnetic data from the central United States. Blakely and Griscom (1988) continued the North American magnetic map to higher elevations and found excellent correlations with Precambrian and Paleozoic geologic provinces. A positive anomaly appears in the upward-continued data over the Grenville and Appalachian provinces, for example; the anomaly has lateral dimensions similar to typical anomalies seen in Magsat data, and models of its crustal source have edges that correspond with province boundaries.

Aeromagnetic data are susceptible to various sources of errors due to navigational or altimetric mislocation, improper regional field removal, and external magnetic fields. Some of these errors can appear in magnetic profiles as long-wavelength magnetic anomalies and, therefore, may have detrimental effects on models of deep crustal sources. Perhaps the most serious long-wavelength error can occur in compilations of regional aeromagnetic surveys. If regional fields are not removed properly and consistently from individual surveys, wavelengths comparable to the dimensions of the surveys may be generated in the compilation that are unrelated to crustal sources.

Long-wavelength errors in aeromagnetic data become especially pronounced when data are continued upward. A tenet of potential theory states that if the magnetic potential is known accurately and entirely over a surface that encloses all magnetic sources it can be calculated at any point above that surface, a principal that underlies the process of upward continuation. It should be possible, therefore, to continue regional magnetic maps to satellite altitudes and thereby duplicate observations actually made at the higher altitude. Various studies have attempted this comparison, sometimes with unsettling results. Schnetzler et al. (1985), for example, continued the composite magnetic map of the continental United States (USGS and SEG, 1982) to 320 km altitude in order to compare with equivalent-source anomalies derived from Magsat data (Mayhew and Galliher, 1982). The upward-continued aeromagnetic data include a very large positive anomaly over Washington, Idaho and Montana, and a large negative anomaly centered over Mississippi and Alabama, neither of which are present in Magsat crustal maps.

The cause of these long-wavelength, high-amplitude anomalies can be seen in colored versions of the U.S. aeromagnetic compilation (USGS and SEG, 1982). This map clearly includes long-wavelength components, apparently caused by inconsistent removal of regional fields. Schnetzler et al. (1985) found that agreement between upward-continued U.S. data and satellite data was greatly improved if a polynomial surface was removed from the aeromagnetic compilation prior to upward continuation. The long-wavelength errors present in the U.S. map were removed in compiling the North American map (R. Godson, pers. commun., 1990). Hence, although large discrepancies in amplitude were noted, Blakely and Griscom (1988) found a nearly one-to-one correspondence between upward-continued versions of the North American magnetic map and Magsat anomalies.

A great deal of evidence from studies of both aeromagnetic and satellite data indicates that deep regions of the continental crust are highly magnetic. These results should be tempered with a few cautions. First, maps of crustal magnetic anomalies are usually derived by subtracting a low-degree (usually $n < 13$) spherical harmonic representation of the core field from measured data. This subtraction, however, does not remove all wavelengths longer than the spherical harmonic representation, and long-wavelength signal from the core remains in the residual crustal maps. Second, the determination of magnetization from magnetic anomalies is an inherently nonunique procedure. Without geologic or geophysical evidence to the contrary, any deep crustal model for a magnetic anomaly can be replaced with a broader, shallower source.

3. Magnetic properties of lower crustal rocks

As noted above, magnetic properties of the lower crust may be inferred from analysis of long-wavelength magnetic anomalies. Another constraint is provided by laboratory measurement of magnetic properties of rocks thought to represent the lower crust. Such rocks may be from high-grade metamorphic terrains, they may be xenoliths, or they may be from the lower portions of exposed cross sections of continental crust. In this section we summarize the results of such studies, emphasizing laboratory measurements on lower crustal samples and the additional constraints provided by petrological investigations.

3.1. Types of samples studied

Magnetic properties of high-grade metamorphic terrains were studied long before there was a particular concern about lower crustal magnetization. These studies were primarily designed to provide information relevant to problems in geophysical prospecting, or to evaluate the paleomagnetic potential of metamorphic rocks. Although not ideally designed for our purposes, these studies provide analysis of a large number of samples. A partial review includes papers by Balsley and Buddington (1958), Hall (1968), Larson et al. (1969), Powell (1970), Chernyuk (1971), Aravamadhu (1974), Lidiak (1974), Coles and Currie (1977), Ishihara (1981), Krutikhovskaya and Pashkevich (1977), Dunlop (1979), Gasparini et al. (1979), Hall et al. (1979), Krutikhovskaya et al. (1979), Mutton and Shaw (1979), Piper (1979), Subrahmanyan and Verma (1981), Ramaswamy and Ramanathan (1982), and Wagner (1984).

Laboratory studies of xenoliths and lower crustal cross sections are aimed more specifically at lower crustal magnetization. Because of their relatively rapid rise to the surface, xenoliths from the lower crust and upper mantle may show less retrograde alteration than tectonically uplifted terrains (Wasilewski et al., 1979; Wasilewski and Mayhew, 1982; Haggerty and Toft, 1985; Toft and Haggerty, 1988). On the other hand, they may represent atypical compositions (Frost and Shive, 1989) and systematic relationships of physical properties in the parent crustal sections are obscured by mechanical scrambling during ascent. Nevertheless, information from this source is particularly valuable because many samples come from the lowermost crustal depths, which are probably under-represented in other sample suites. It is unfortunate, therefore, that systematic magnetic studies have been done on only a very few available xenolith collections.

The most detailed magnetic information about lower crustal rocks comes from systematic sampling of the lower portions of cross sections representing rocks that were once deep in the continental crust, but which have been subsequently uplifted and exposed (Fountain and Salisbury, 1981). The advantages of this type of study are that the relative volumes of and relationships between rock units can be deciphered through geologic mapping and that the relative depth of samples in the crust is known. There are several disadvantages. These cross sections do not represent the deepest crustal layers (e.g., Fountain, 1985; Bohlen and Mezger, 1989; Fountain et al., 1990). Retrograde metamorphism may destroy the mineral assemblages that were present when the sections were actually deep in the crust, although geobarometric/geothermometric studies can evaluate this possibility and provide an absolute depth calibration. Subsolidus exsolution of oxides may produce magnetic changes that can only be understood if the temperature-oxygen fugacity (T-f_{O_2}) history of the cross section can be constrained. Of the dozen or so well-recognized lower-crustal cross sections, four have been the subject of systematic magnetic studies. Wasilewski and Fountain (1982) did preliminary work on Ivrea zone samples and this was recently extended by Wasilewski and Warner (1988) and by Belluso et al. (1990). Schlinger (1985) carried out a comprehensive magnetic study of the Lofoten–Vesterålen terrain of Norway. Williams et al. (1985/86) studied the Pikwitonei–Cross Lake subprovinces in Central Manitoba, and Shive and Fountain (1988) studied the Kapuskasing uplift of Ontario.

Taken as a group, the lower crustal samples are fairly well distributed in space and time. The Canadian crustal cross sections are representative of Archean rocks of the Superior Province (see Percival et al., this volume; Mezger, this volume). The Lofoten–Vesterålen terrain includes Archean and Proterozoic rocks (see Griffin et al., 1978); metamorphism of the Ivrea section occurred during the Paleozoic (see Percival et al., this volume). The more numerous studies of high-grade metamorphic terrains noted above represent most of the Earth's major shield areas and many smaller crystalline uplifts. Xenolith studies are based on a total of fewer than 30 samples, but these samples are distributed worldwide.

3.2. Magnetic susceptibility (induced magnetization) of lower crustal rocks — laboratory data

Room-temperature initial susceptibility is the most commonly measured magnetic property for lower crustal rocks. About half of the tens of thousands of such determinations show insignificant amounts of ferromagnetic minerals and thus nearly or completely paramagnetic behavior. Of the remaining determinations, most have induced magnetizations less than 1 A/m (10^{-3} emu/cm^3). The most strongly magnetized samples are usually associated with serpentinized ultramafic rocks, iron-formation, or gabbro.

Lower-crustal samples do not show a Hopkinson effect (Nagata, 1961), the enhancement of susceptibility at elevated temperatures just below the Curie-temperature. Rather, susceptibility is quite insensitive to temperature, varying only about $\pm 10\%$ from room temperature to about 560°C, at which point it drops rapidly to zero (Fig. 4-2; Schlinger, 1985; Williams et al., 1985/86; Shive and Fountain, 1988). Thus room temperature measurements are sufficient to characterize the induced magnetization of a unit at depth in the crust.

An approximate relationship between susceptibility and magnetite content has been well established from field (e.g., Carmichael, 1982), laboratory (e.g., Dunlop, 1986) and theoretical (e.g., Stacey and Banerjee, 1974) studies. It is:

$$k = 0.2 \times 4\pi f \tag{13}$$

where k is the susceptibility (in SI units) and f is the volume fraction of magnetite. Thus, measured susceptibilities allow us to estimate the amount of magnetite in many lower-crustal

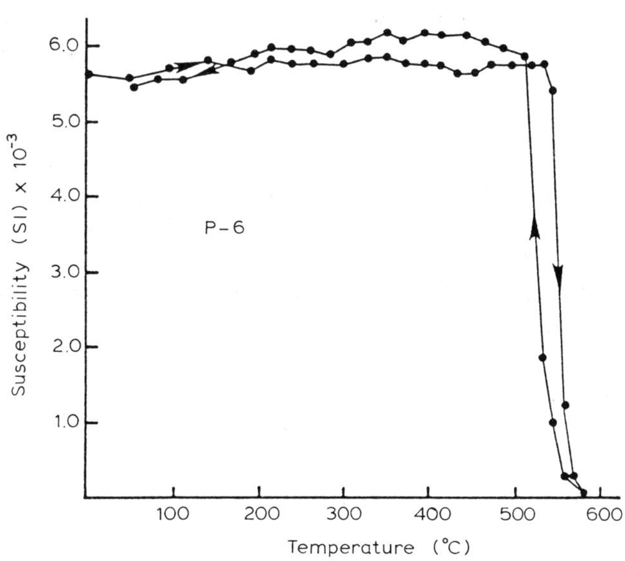

Fig. 4-2. Susceptibility versus temperature for a sample from the Pikwitonei granulite domain (Williams et al., 1985/86). The sample contains magnetite, and shows behavior that is typical for strongly magnetic samples from other lower crustal terrains.

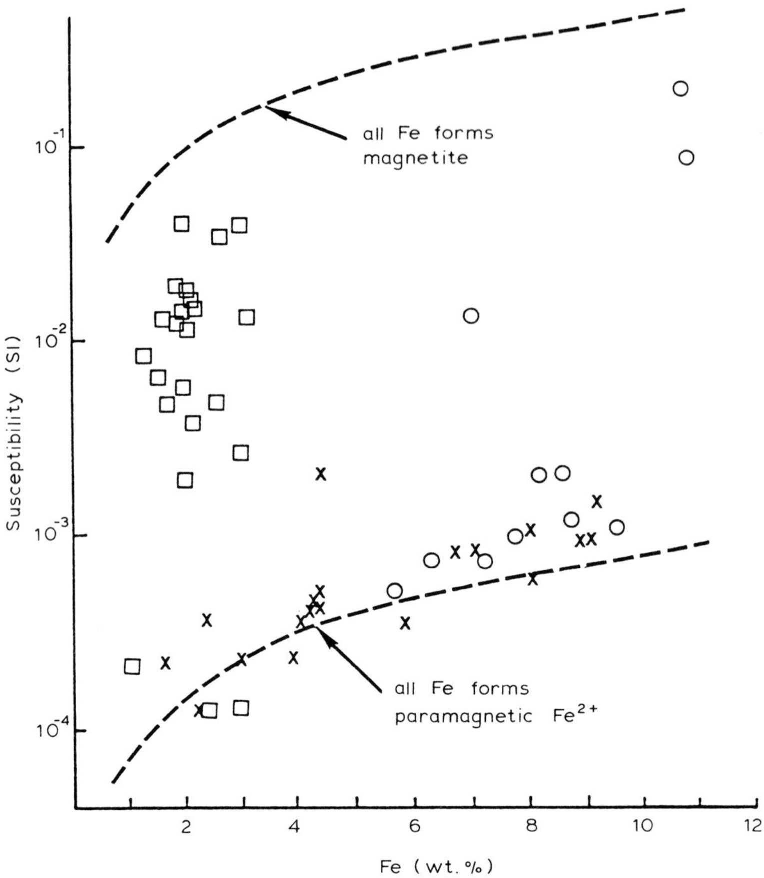

Fig. 4-3. Susceptibility versus weight per cent iron for samples from the Pikwitonei granulite domain (Williams et al., 1985/86). Circles: mafic gneisses and intrusives; squares: silicic gneisses and intrusives; crosses: supracrustal rocks. The upper dotted line shows is the theoretical susceptibility for the case where all iron is in the form of magnetite; the lower dotted line shows the theoretical susceptibility for the case where all iron is in the paramagnetic state. Magnetite may be formed more efficiently in rocks with small amounts of iron.

samples assuming that magnetite is the dominant magnetic phase (see below). For example, susceptibilities of samples from lower-crustal cross sections have an average value of about 10^{-2} SI, which corresponds to a magnetite content of less than 0.5% by volume.

In general, magnetization increases with iron content, but there are some striking exceptions. For example, Figure 4-3 shows that silicic rocks from the Pikwitonei terrain are about an order of magnitude more magnetic than mafic rocks (Williams et al., 1985/86). This is because the amount of iron in the form of magnetite is more important than total iron. Silicic rocks may have significant amounts iron in the form of magnetite, while mafic rocks may have the bulk of their iron in the form of ferromagnesian silicates and/or relatively nonmagnetic oxides and sulfides.

Locally, magnetization intensities can be correlated with metamorphic grade. The conventional wisdom is that susceptibility should increase with increasing metamorphic grade, and this is the case in the Lofoten–Vesterålen (Schlinger, 1985) and Lewisian (Powell, 1970) terrains. On the other hand, susceptibility is not significantly dependent on metamorphic grade in samples from any of the other three lower crustal terrains studied to date (e.g., Fig.

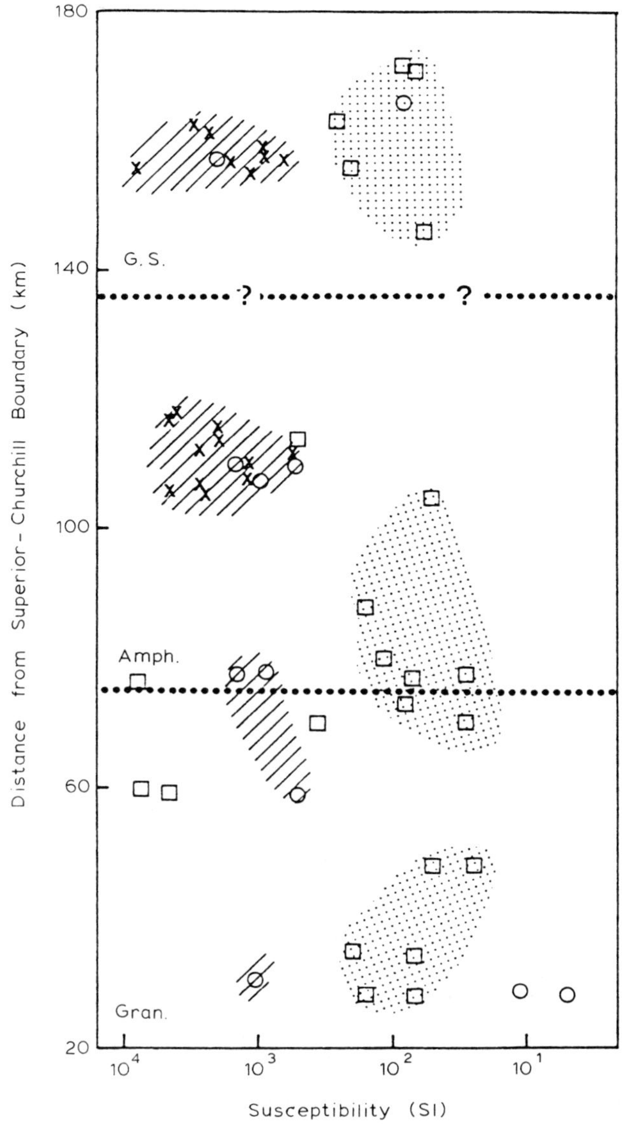

Fig. 4-4. Susceptibility versus metamorphic grade, from greenschist (G.S.) to granulite (Gran.) facies, for samples from the Pikwitonei terrain (Williams et al., 1985/86). Symbols are the same as in Fig. 4-3.

4-4). Serpentinites are a special case. Although serpentinites can be strongly magnetic (Saad, 1969), their parent peridotites are generally paramagnetic or weakly magnetic (Wasilewski et al., 1979). Prograde metamorphism of serpentinite toward the granulite facies tends to increase the the amount of Cr and Al in the spinel and thus the intensity of magnetization (Evans and Frost, 1975; Shive et al., 1988).

3.3. Magnetic mineralogy of lower crustal rocks — laboratory observations

Identification of magnetic minerals in a rock is complicated by the fact that significant magnetization may be carried by a very small portion (<1% by volume) of the sample.

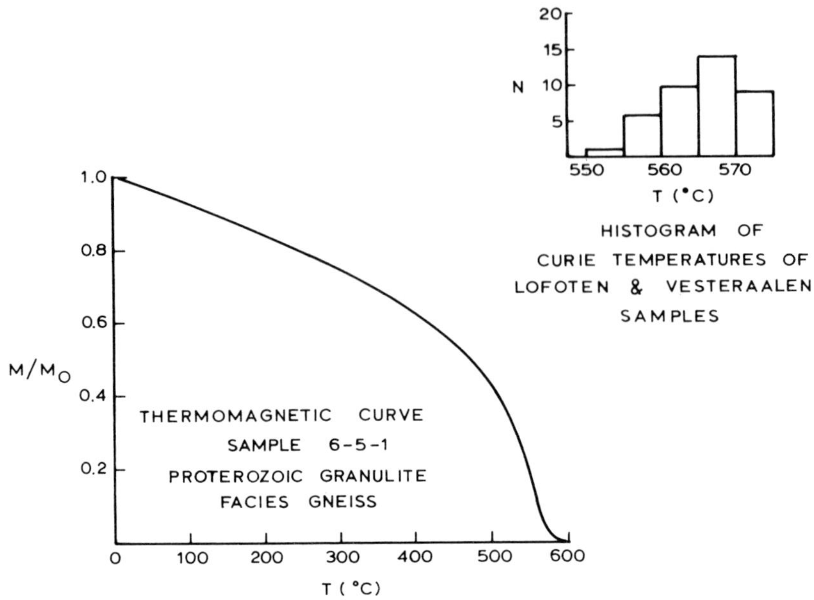

Fig. 4-5. Typical thermomagnetic curve and histogram of 40 Curie-temperatures for samples from the Lofoten–Vesterålen terrain (Schlinger, 1985). This behavior is typical for strongly magnetic samples from other lower crustal cross sections and from many high-grade metamorphic terrains.

Furthermore, magnetic grains may be submicron in size. Because of this, the most reliable identifications result from a combination of petrologic and thermomagnetic analysis (e.g., Larson et al., 1969; Lidiak, 1974; Hall et al., 1979; Piper, 1979; Shive and Morel, 1980; Schlinger, 1985; Williams et al, 1985/86; Shive and Fountain, 1988; Shive et al., 1988). The overwhelming majority of all studies of lower crustal rocks show that magnetite (Fe_3O_4) with at most a few percent of ulvöspinel (Fe_2TiO_4) in solid solution is the dominant magnetic mineral in lower crustal rocks. Curie-temperatures generally fall between 550 and 580°C (Fig. 4-5). Some samples have minor amounts of pyrrhotite ($Fe_{1-x}S$) often coexisting with magnetite.

Minerals in the hematite-ilmenite (Fe_2O_3-$FeTiO_3$) solid solution series are less common but may produce strong lower crustal magnetic anomalies. Hematite and ilmenohematites with up to 50 mole percent ilmenite are so weakly magnetic (Stacey and Banerjee, 1974) that they cannot produce significant anomalies. In upper crustal rocks, hemoilmenite rich in titanium has strong magnetization and can produce anomalies such as that over the Allard Lake hemoilmenite deposits in Quebec (Balsley and Buddington, 1958; Hargraves, 1959). However, their Curie-temperatures are so low ($T_c < 220$°C) that they will not be magnetic under deep crustal conditions.

Native metal alloys (mostly Fe-Ni) are sometimes observed in serpentinites and xenoliths, and must be considered as possible sources of deep crustal (and upper mantle) magnetization (Haggerty, 1978, 1979; Haggerty and Toft, 1985; Toft and Haggerty, 1988). Xenoliths studied to date do not contain metals in magnetically significant amounts. The xenoliths studied by Wasilewski et al. (1979) and Wasilewski and Mayhew (1982) contained no metals, and the small quantity of metals in the kimberlite xenoliths studied by Haggerty and Toft (1985) could be produced by the special reducing conditions that accompany kimberlite emplacement (Frost and Shive, 1989). Serpentinite may contain Ni-Fe alloys such as awaruite (Ni_3Fe)

(Dick, 1974; Leinert and Wasilewski, 1979) but metamorphosed serpentinites contain far greater amounts of magnetite than metal alloys (Shive et al., 1988). Moreover, metal-bearing serpentinite is thought to be uncommon in the deep crust and upper mantle because FeNi alloys are found only in serpentinites that have been metamorphosed at grades lower than mid-greenschist (Frost, 1985).

Consequently, we conclude that magnetite is the key magnetic mineral in lower crustal samples studied to date. It is the only significant magnetic mineral in samples from lower crustal cross sections, virtually all high-grade metamorphic terrains, and many magnetic xenoliths. It does not follow directly, however, that magnetite is the dominant magnetic mineral in the lower crust. It is possible that the spinel phase in the lower crust is titanomagnetite and that magnetite found in exposed rocks is a retrograde phase formed during uplift and cooling. We must next consider the temperature and f_{O_2} constraints on spinel composition in the crust.

3.4. Effect of cooling on magnetite composition

During cooling, the Fe-Ti oxides in all plutonic rocks undergo significant changes in chemistry through inter-oxide and intra-oxide re-equilibration (Buddington and Lindsley, 1964; Frost et al., 1988). The most important effect of this re-equilibration is that Ti is purged from spinel. As a result, the spinel phase found in all plutonic rocks is nearly pure magnetite. It may be laced by crystallographically oriented plates of ilmenite or even of ulvöspinel [see Haggerty (1976) for crystallographic details], but the spinel itself is low in Ti and will behave magnetically like pure magnetite, even if the primary spinel phase was enriched in Ti.

An important question is whether Ti was purged from magnetite at temperatures above the Curie-temperature. If so, the magnetic properties measured in the laboratory do replicate the properties that the rock had when it was in the lower crust. The amount of titanium in magnetite in deep crustal rocks depends on the ambient temperature and f_{O_2} (Buddington and Lindsley, 1964; Andersen and Lindsley, 1988). Frost and Chacko (1989) and Frost et al. (1989) noted that many granulites and mafic plutonic rocks cool along the graphite-saturated surface (Fig. 4-6); for CO_2-bearing rocks this is the lowest f_{O_2} likely to be encountered during cooling. Other rocks cool along the hematite-ilmenite solvus, with oxygen fugacities around 3 log units above the fayalite-magnetite-quartz buffer (Frost, 1988); very few rocks are more oxidized. With these limits to oxygen fugacity in common rocks, Figure 4-6 may be used to determine the composition of a titanomagnetite phase coexisting with a hemoilmenite phase, a typical assemblage in deep crustal rocks. The figure shows that as the rock cools, titanomagnetites will become progressively more magnetite-rich while hemoilmenites become more ilmenite-rich. The assemblage will become magnetic when it cools to a temperature of about 520–550°C, at which point the titanomagnetite will have about 4–8% ulvöspinel in solid solution. Data from Frost and Chacko (1989, fig. 6) indicates that the re-equilibration process closes at temperatures of 400–500°C. This is lower than the Curie-temperature for Ti-bearing magnetites (see Fig. 4-6), but even if the magnetite is saturated with Ti for the given temperature, it will have a Curie-temperature above 500°C.

Thus, unless the lower crust is far more reducing than FMQ, the major magnetic phase will be magnetite with less than 5–10% ulvöspinel in solid solution. Coexisting rhombohedral phases will have compositions within a few mole percent of pure ilmenite and will be nonmagnetic because their Curie-temperatures are 220°C or lower. Pyrrhotite may make a minor contribution to magnetization in some rocks, but this will vanish at temperatures above 320°, the Curie-temperature of pyrrhotite. Metal alloys could contribute to magnetization of

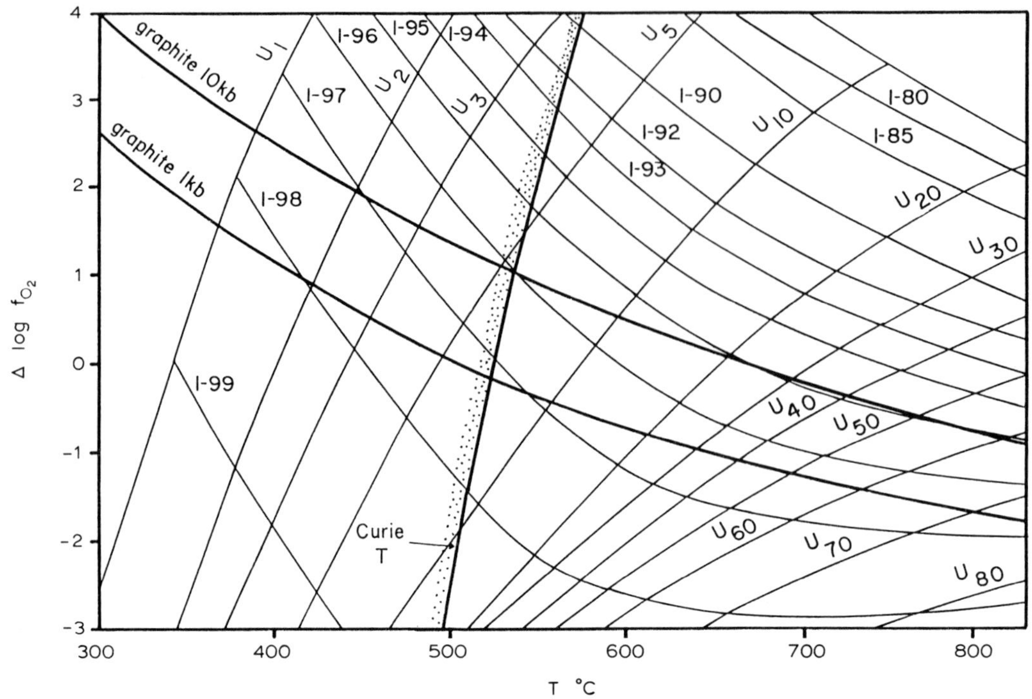

Fig. 4-6. $\Delta \log f_{O_2}$ (with respect to the fayalite-magnetite-quartz buffer assemblages) versus temperature showing the composition of coexisting Fe-Ti oxides (light lines) as calculated from data in Andersen and Lindsley (1988). I = mole percent ilmenite in rhombic phase; U = mole percent ulvöspinel in spinel. Heavy line with stippling shows the Curie-temperature for magnetite coexisting with ilmenite as a function of temperature and oxygen fugacity. Stippling represents the uncertainty as propagated from error in the Fe-Ti oxide thermometer. Also shown are the positions of graphite saturation surfaces for 1 and 10 kbar. Many titanomagnetites appear to cool near graphite saturation; thus these curves mark likely typical oxygen fugacities for two-oxide assemblages.

lower crustal rocks under very reduced conditions. Further studies of xenoliths are needed before the role of alloys in the deep crust can be fully evaluated.

3.5. Effect of metamorphism on magnetite stability

The relationship between degree of metamorphism and magnetization is still unclear. There are several reasons for this. First, there is still considerable uncertainty about geochemical conditions in the deep crust and thus about the metamorphic reactions that might stabilize magnetite or other magnetic minerals at the expense of silicates and nonmagnetic oxides. Second, crustal cross sections representing the lowermost crust are unavailable for sampling and observations (Fountain et al., 1990). Third, it is difficult to separate effects of metamorphism from composition, and key experiments, such as study of isochemical rock units of differing metamorphic grade, are just beginning. However, much about magnetite stability can be inferred from simple petrologic principles. Two types of reactions will likely affect magnetite abundance: (1) buffer type equilibria, and (2) oxygen-conserved equilibria. These are discussed in detail below.

3.5.1. Buffer type equilibria

Redox reactions (i.e., buffer type) are those that involve reactions between magnetite and ferrous iron in silicates during metamorphism. In a very broad manner this can be considered a variant of the fayalite-magnetite-quartz (FMQ) buffer except that in most plutonic rocks such reactions involve pyroxene or the participation of a fluid phase (i.e., H_2O or CO_2). The free oxygen content of geologic fluids is vanishingly low. For example, at 600°C in a rock containing hematite and magnetite (a relatively oxidizing condition for the crust) there will be 10^{-20} moles of free oxygen in each cubic centimeter of fluid, hardly enough to drive any oxidation or reduction reaction. Thus, although buffer type equilibria can buffer the fluid over large ranges of oxygen fugacity and have important effects on the composition of the fluid phase, under most conditions only very small amounts of magnetite will be produced or consumed by these reactions (Frost, 1988).

3.5.2. Oxygen-conserved oxide-silicate equilibria

Because oxygen is present in such small amounts in metamorphic fluids, significant amounts of magnetite may be produced or consumed by metamorphic reactions only if oxygen is conserved. There are two types of oxygen-conserved reactions. One involves the participation of an oxide-silicate reaction with other oxygen-dependent equilibria, such as an oxygen buffer or equilibria involving other elements (C, H, S) with multiple valence states. The other type of oxygen-conserved equilibrium is the reaction between magnetite and a silicate that contains ferric iron.

One important oxygen-conserved reaction occurs when a buffering reaction crosses the hematite-magnetite buffer. In such a situation the following reaction should occur:

$$\underset{\text{in silicates}}{FeO} + \underset{\text{hematite}}{Fe_2O_3} = \underset{\text{magnetite}}{Fe_3O_4} \tag{14}$$

Reaction (14) could lead to a 67% increase in the amount of Fe-oxide in the rock. Such a reaction is likely to account for the major increase in oxide content during diagenesis of iron-formations (Han, 1982). This reaction may also be important in rocks containing coexisting ilmenite and magnetite. Because the amount of Fe_2O_3 dissolved in ilmenite is a function of oxygen fugacity, a continuous version of reaction (14) will occur in any rock containing two Fe-Ti oxides that is subjected to oxidation or reduction during metamorphism or magmatic cooling.

Other oxygen-conserved equilibria will involve the participation of elements, in addition to iron, that can occur in multiple oxidation states. In geologic environments the most important of these are carbon, sulfur, and hydrogen. The oxidation of graphite to CO_2 is a well known reduction process in geological situations (French, 1966; Ohmoto and Kerrick, 1977; Frost, 1979a). Magnetite and graphite are not commonly found together because they are incompatible in quartz-bearing, Ti-free rocks (Frost, 1979b). The presence or absence of sulfur as a major fluid species also plays an important role in the abundance of magnetite in rocks. The outward migration of sulfur during metamorphism of an ore body can remove magnetite from the surrounding rock by altering it to sulfides (Nesbitt, 1986). It is possible, although not well documented, that magnetite can be produced during metamorphism of sulfide-bearing (but graphite-absent) rocks by reactions such as:

$$\underset{\text{pyrite}}{3\,FeS_2} + \underset{\text{in silicates}}{6\,FeO} + 6\,H_2O = \underset{\text{magnetite}}{3\,Fe_3O_4} + 6\,H_2S \tag{15}$$

Unlike reactions with sulfur or carbon, dissociation of H_2O is not likely to be an efficient means for oxidation or reduction. An aqueous fluid in equilibrium with the FMQ buffer (a relatively reducing condition for crustal rocks) at 600°C and 500 MPa will have a mole fraction of H_2 of only 0.0026. Disassociation of huge amounts of H_2O (and consequent escape of H_2) would be needed in such conditions to produce even minor amounts of magnetite (see also Candela, 1986).

Another type of oxide-silicate reaction that is oxygen-conserving involves the reaction of magnetite to a silicate containing ferric iron. Ferric iron can be an important constituent in epidote, chlorite, biotite, hornblende, clinopyroxene or garnet. In reactions between such minerals and biotite there need not be oxidation or reduction and, therefore, the progress of the reaction is not inhibited by the availability of oxygen. It is postulated that reactions like this are the cause of the disappearance of magnetite during prograde metamorphism of mafic rocks (Frost, 1988).

It is important to note, however, that the reactions involving the solution of ferric iron into silicates are multivariant. This means that, in addition to temperature and pressure, the extent to which these reactions progress is dependent on many compositional parameters. For example magnetite may be absent in one rock because the ferric iron can be accommodated entirely in hornblende and yet it may be present in significant amounts in another rock at the same grade that has a lower abundance of hornblende (or perhaps with the same amount of hornblende and with higher ferric iron content).

3.6. Magnetite stability in crustal rocks

The abundance of magnetite in crustal rocks is controlled by the type of reactions discussed above. As such, it is complexly dependent on rock composition, as well as on pressure and temperature conditions of metamorphism. Below, we summarize the current understanding about the petrological controls on the occurrence of magnetite within rocks that may constitute the lower crust. A discussion of magnetite occurrence in some igneous rocks is included so as to provide a background for discussing the changes associated with metamorphism.

3.6.1. Granitic rocks

The factors that control magnetite abundance in granitic rocks are still poorly known. It is well known that magnetite is absent from many granitic rocks. Indeed, the oxide mineralogy of granitic rocks has become the basis for classification of granites into magnetite series versus ilmenite series (Ishihara, 1979; Takahashi et al., 1980). It is clear that the magnetite series granitoids are more oxidized than the ilmenite series granitoids. Not only are the Fe-Ti oxides more abundant in the magnetite series rocks (Ishihara, 1981), but the ilmenite in these rocks is much richer in ferric iron (Czamanske et al., 1981). The unanswered question is what has produced this oxidation. Although there is no clear chemical distinction between ilmenite-series and magnetite-series granitoids, alkaline and subalkaline granitoids tend to be magnetite bearing, while calc-alkalic plutons may be either magnetite-bearing or magnetite-free (Ishihara, 1981).

The three suggested causes for the differing oxygen fugacity between the magnetite-series and ilmenite-series granitoids are: (1) reduction due to incorporation of graphitic metasedimentary rocks (Ishihara, 1979); (2) oxidation of magnetite-series granites by dissociation of

H_2O during second boiling (Czamanske et al., 1981) and (3) reaction of Ti-magnetite with K-feldspar to produce biotite (Diamond and Frost, 1988). It is possible that examples of all three types of processes can be found in nature [although process (2) has a major mass-balance problem; see Candela (1986)].

Because granitic rocks are made up mostly of quartz and feldspar, the effect of metamorphism on the abundance of magnetite will not be as strong as it is in other rock types; the small amount of amphibole or biotite produced during metamorphism of granitic rocks is not enough to dissolve significant amounts of ferric iron. Thus one might suspect that the magnetic signature of granitic gneisses is mostly inherited from the parent igneous rock.

3.6.2. Mafic igneous and meta-igneous rocks

Mafic igneous rocks can be strongly magnetic, both because of the occurrence of magmatic Fe-Ti oxides (Haggerty, 1976), and because of the presence of sub-microscopic magnetite exsolved out of the feldspars (Geissman et al., 1988) or pyroxenes (Schlinger and Veblen, 1989). Three distinct types of magnetite-rich intrusions can be recognized: Duke Island-type intrusions, evolved tholeiitic intrusions, and anorthosite complexes.

Duke-Island type intrusions are calc-alkalic intrusions in which magnetite appears early in the crystallization history, while the silicates are still very magnesian (X_{Mg} of the pyroxenes is generally greater than 0.5). The Duke Island intrusion (Irvine, 1965, 1974) typifies such mafic intrusions. Similar intrusions include the Bear Mountain Complex (Snoke et al., 1981), Lake Owens Complex (Patchen, 1987), and Somerset Dam (Mathison, 1975). Magnetite is abundant in such intrusions both in ultramafic rocks and in gabbros, with some units containing up to 15% magnetite.

Tholeiitic layered intrusions commonly contain abundant magnetite in their upper, evolved portions. Although primitive tholeiitic magmas do not crystallize abundant magnetite, such magmas emplaced at depth evolve toward magnetite saturation and the evolved melts from these magmas may crystallize significant amounts of magnetite. In the Skaergaard, for example, magnetite crystallized at the top of the lower zone and persisted throughout the rest of the crystallization history (Wager and Brown, 1968). Magnetite abundance is variable over this thickness of nearly 2000 m. The average abundance ranges from 1 to 8 volume percent, but local horizons may contain as much as 36 volume percent magnetite (Vincent and Phillips, 1954). Magnetite is also present in the Upper Zone of the Bushveld where, over a thickness of at least 1500 m, it has an average abundance of 15 volume percent and locally forms seams of nearly pure magnetite (Willemse, 1969). Oxide-rich horizons are also found in abundances up to 5% in a 50-m-thick horizon of the Muskox intrusion (Irvine and Smith, 1969). Magnetite occurs in abundances between 2 and 12% in the upper 200 m of the Dufek intrusion (Himmelberg and Ford, 1977), where it produces a distinct aeromagnetic anomaly (Behrendt et al., 1980).

Anorthosite complexes, a third type of mafic igneous rock association, commonly contain abundant magnetite. Although anorthosites themselves do not contain abundant magnetite, they are usually associated with magnetite-bearing gabbros and massive Fe-Ti oxide bodies (Ashwal, 1982). For example, the upper portion of the Bjerkem-Sogndal massif, Norway, contains about 5% magnetite (Duchesne, 1972) and the Strong Creek Gabbro of the Laramie Anorthosite Complex, Wyoming, contains between 5 and 10% magnetite (Kling, 1986).

Some metabasites are strongly magnetic (Wasilewski and Mayhew, 1982; Wasilewski and Warner, 1988), whereas others are not (Williams et al., 1985/86; Shive and Fountain, 1988).

Similarly, in some localities it appears that magnetite abundance increases in the transition from amphibolite to granulite facies (Schlinger, 1985), while in other areas the results are more equivocal (Williams et al., 1985/86; Shive and Fountain, 1988; Wasilewski and Warner, 1988). These apparently conflicting observations are a reflection of the multivariate nature of the equilibria that control the solution of ferric iron into silicates (see above). Wet chemical analyses tabulated by Deer et al. (1966) indicate that garnet and hornblende accommodate more ferric iron than does augite. Thus, one would expect that magnetite would be produced during isochemical dehydration of amphibolite to two-pyroxene granulite and that magnetite would be consumed in granulites as garnet becomes progressively more abundant with increasing pressure. In general this is true, for garnet-free granulites do seem to be more magnetic than amphibolites or garnet granulites (cf. Schlinger et al., 1992). In detail it is not so simple; there are amphibolites and garnet granulites that are highly magnetic and there are mafic granulites that are almost paramagnetic. These apparent contradictions reflect the dependency of magnetite abundance on compositional parameters alluded to above.

An important question to be answered here is whether magnetite is stable in high-pressure metamorphosed mafic rocks. Brown (1974) proposed that the reaction that produces crossite in blueschist consumes ferric iron from the oxides. This would suggest that magnetite abundance (and hence magnetic susceptibility) should be lower in blueschists than in lower pressure metabasites. There has been, unfortunately, no adequate study of magnetic susceptibility in these rock types. The petrologic data available indicate that, while magnetite is absent from some blueschist terrains (Roeske, 1986), it is present in others (Brown, 1986). There is clear evidence that magnetite is consumed during reactions that produce eclogites (Austrheim and Griffin, 1985; Pognante, 1985; Smulikowski and Smulikowski, 1985). Further, magnetite is commonly formed from eclogites in retrograde reactions (Pinet and Smith, 1985; Sautter, 1985). Indeed, although a few eclogites are noted to contain magnetite (Pinet and Smith, 1985) it is distinctly missing in many others (Mørk, 1985; Sautter, 1985). This implication, that eclogites should be paramagnetic, is confirmed by the limited (i.e., two samples) amount of magnetic susceptibility data available (Wasilewski et al., 1979).

3.6.3. Ultramafic rocks

The stability of magnetite in metamorphosed serpentinites has been studied extensively by Evans and Frost (1975) and by Frost (1985). They showed that magnetite is the stable oxide in serpentinites but that the spinel phase contains sufficient Cr_2O_3 to make it non-magnetic at metamorphic grades above those of serpentine stability. Magnetic susceptibility measurements are consistent with these simple phase equilibria studies indicating that, whereas peridotites are paramagnetic (Wasilewski et al., 1979), serpentinites are strongly magnetic (see references listed in Shive et al., 1988).

The occurrence of stable iron alloys in serpentinites is restricted to peridotites containing both olivine and serpentine (Frost, 1985). In prograde assemblages this would represent metamorphic conditions associated with the first appearance of olivine (roughly upper greenschist conditions). In retrograde rocks this would constitute any rock in which relict olivine is still present with serpentine. Shive et al. (1988) made detailed study of the change of magnetic susceptibility in metaperidotites as a function of metamorphic grade in the Malenco Serpentinite. They showed that even in rocks containing relatively abundant iron alloys the magnetic signature was dominated by magnetite. Furthermore, their results indicate that even though magnetite is not the stable oxide phase in high-grade metaserpentinites, significant amounts of magnetite remains in these rocks to the highest grades, perhaps as relicts in olivine.

3.6.4. Metasedimentary rocks

The magnetic susceptibility of metasedimentary rocks is dependant on the composition of the protolith. Metamorphosed carbonates and graphitic metapelites are weakly magnetic, whereas metamorphosed greywackes or graphite-free pelitic schists may have a substantial magnetic susceptibility. The magnetite may be distinctly restricted in metamorphic grade. For example Speer (1981) described a contact aureole in graphite-free pelitic rocks where magnetite was produced at low grades by the breakdown of epidote and then was consumed at higher grades by reactions producing cordierite and biotite.

The one metasedimentary rock type that would have a significant magnetic susceptibility will be iron-formations. Iron-formations traditionally are considered in terms of a four-fold classification: oxide, silicate, carbonate, or sulfide, depending on the major Fe-bearing mineral (James, 1954). Magnetite can be a major constituent in any of these types of iron-formations. In the oxide iron-formation it may be present in amounts up to 60%, although 30–40% is more typical (James, 1966). The typical Proterozoic iron-formation usually contains interlayered oxide, silicate, and carbonate horizons. They are generally magnetite rich. For example, the Biwabik Iron-Formation, a typical middle Proterozoic iron-formation, may contain up to 20 or 30% magnetite (Gundersen and Schwartz, 1962). The factors controlling magnetite stability in iron-formation were outlined by Frost (1979b, 1988), who showed that even those iron-formations that originally contained hematite +quartz with siderite or Fe-silicate should undergo relative reduction during metamorphism with the production of magnetite by reaction (14). However, iron-formations do not form large bodies (i.e., more than 100 km on a side) and it is not likely that there could be a sufficient amount of iron-formation in the deep crust to produce significant magnetic anomalies.

3.7. Remanent magnetization in lower crustal rocks

Total magnetization in the lower crust is the vector sum of induced and remanent magnetization. Thus it is also important to consider the remanent magnetization of lower crustal rocks. Two types of remanence must be considered, stable and viscous. Some rocks may have a stable, primary remanence acquired when the rock cooled to its present temperature. This magnetization will be acquired in a direction parallel to the Earth's magnetic field during cooling, which may be quite different from its present direction because of the possibility of field reversals, continental drift and/or tectonic disruption of the rock. Stable remanence is most effectively carried by single-domain grains, which for magnetite are smaller than one micron (Stacey and Banerjee, 1974). Primary remanence in larger, multi-domain size grains is less stable, and may undergo viscous decay with time. If so, the original remanence will be replaced by a viscous remanent magnetization (VRM) parallel to the present field direction. This type of remanence would thus always produce a total magnetization greater than the induced magnetization.

First consider stable remanence. The natural remanent magnetization (NRM) of exposed rocks gives a rough measure of their capacity to carry stable remanence in the upper crust. Most magnetic studies of lower crustal rocks include measurements of NRM. These studies usually show that the NRM intensities of lower crustal rocks are comparable to or less than the induced magnetization intensities. For example, several thousand samples from the Ukranian shield (Krutikhovskaya and Pashkevich, 1977) show Koenigsberger ratios (Q_n: the ratio of remanent to induced magnetization) of about 1.0. Lofoten–Vesterålen samples (Schlinger, 1985) have Q_n values that average about 1/3 to 1/2. On the other hand, there are a number of examples of lower crustal rocks such as layered mafic complexes and anorthosites that have (at least locally) NRM intensities that are much stronger than induced intensities (e.g., Balsley and Buddington, 1958; Shive et al., 1990).

Stable remanence will be much less important under deep crustal conditions than under upper crustal conditions. It will be attenuated by elevated temperatures in two ways. First, its intensity is proportional to the intensity of saturation magnetization, which decreases with temperature. The saturation magnetization of magnetite at 400°C is only 63% of its room temperature value; at 500°C it has dropped to 43% (Pauthenet and Bochirol, 1951), and it disappears altogether at the Curie-temperature of 575°C. Even more important, however, is the fact that viscous decay is thermally activated (Shimizu,1960). Stable remanence carried by both single- and multi-domain grains will decay exponentially with time, although at different rates. There is considerable debate about the most appropriate time constant for a given grain size (e.g., Smith, 1984; Dunlop, 1984), but it is clear that a considerable portion of stable remanence is likely to decay over time periods of several million years when exposed to lower crustal temperatures (Pulliah et al., 1975).

Shive (1989a) pointed out that stable remanence in most lower crustal bodies is likely to be acquired in zones of alternating polarities because the cooling time is greater than the length of a typical polarity interval. This subdivision dramatically reduces the long-wavelength power of the resulting anomaly, because the lower crustal bodies with alternating polarities of remanence cannot produce significant anomalies at the surface of the Earth, even if their remanent intensities are greater than their induced intensities.

Stable remanence in deep crustal rocks cannot be responsible for the high magnetization required by anomaly analyses (Table 4-1). For it to do so, the following conditions must all be met: (1) an extended rock body must have a Koenigsberger ratio of 2–3; (2) the remanence must be acquired rapidly, or during a long polarity interval; and (3) the remanence must resist viscous demagnetization during the time since it was first acquired. Satisfaction of all these conditions on a scale as worldwide as implied by the data in Table 4-1 is not probable.

We then turn to a consideration of VRM acquired in the present field direction. Could such remanence, acquired during the Brunhes Normal Polarity-Chron, provide intensities 2–3 times as great as induced magnetization? Although a definitive answer is not yet available, there have been experimental and theoretical studies relevant to the question of lower crustal VRM. Treloar et al. (1986) measured VRM acquisition in lower crustal samples at lower crustal temperatures. They found that acquired VRM intensities were not as great as induced intensities on a laboratory time scale, but it is not possible to extrapolate this data to predict VRM intensities acquired during the Brunhes (0–680,000 years B.P.). The most that can be said is that their experiment did not substantiate the acquisition of strong VRM in deep crustal samples.

Shive (1989a) showed from theoretical considerations that viscous remanence in multi-domain grains cannot have intensities more than about 25% as great as induced magnetization in the same grains. VRM could be significantly more intense than induced magnetization in single- and pseudo-single-domain grains (Wörm, 1989). If this mechanism were to account for a significant VRM, however, the major portion of the magnetization of deep crustal rocks would be carried by grains much smaller than one micron. This is unlikely because the average size of Fe-Ti oxides in lower crustal rocks is around 100 microns (e.g., Schlinger, 1985). Although the magnetite is typically subdivided by ilmenite lamellae, the susceptibility versus temperature behavior of these samples indicates that significant amounts of magnetization is not carried by very fine grains (Shive, 1989b). We conclude that remanent magnetization in deep crustal rocks studied to date is not likely to be as intense as the induced magnetization, which means that the total magnetization of these rocks is not significantly greater than their induced magnetization.

4. Nature of magnetic bodies in the lower crust

At the beginning of this chapter we noted that a variety of deep crustal models (undulations in the Curie-temperature isotherm, lateral compositional variations, etc.) could be invoked to explain long-wavelength magnetic anomalies, and furthermore that it is difficult to specify which model applies to any particular area because of the inherent nonuniqueness in potential field data. Although this is certainly true, it should not obscure the essential point, which is that whatever the correct model may be, it requires sources that have magnetizations of about 5 A/m. This is a moderately strong magnetization. In the upper crust of the Earth magnetizations of that intensity are generally restricted to bodies of metamorphic and basic igneous rocks that contain 5% magnetite or more by volume.

On the basis of both petrologic and observational evidence discussed above, we favor the view that magnetite is also the dominant carrier of magnetization in the deep crust. Although we certainly can not rule out the possibility that some other strongly magnetic phase (such as iron alloys) may be important in the deep crust and upper mantle, it is less likely. Such phases have not been observed in sufficient quantity in deep crustal samples to produce a magnetization of 5 A/m, and the lower crust is probably not reducing enough to stabilize metals over regional distances. We will now consider possible deep sources for long-wavelength magnetic anomalies, assuming that the bulk of the magnetization is carried by magnetite. The amount of magnetite required to produce a magnetization of about 5 A/m depends on the type of magnetization. A strong and coherent component of remanent magnetization could relieve the necessity for the rocks to contain 5% magnetite by volume, as would be required were the anomalies produced by induced magnetization. As noted above, the remanence in the hot, generally coarse-grained rocks of the lower crust is probably neither strong nor coherent.

4.1. Sources of magnetization in the lower crust

Long-wavelength magnetic anomalies may be produced by rock bodies that contain 5% or more magnetite by volume and located above the 600°C isotherm in the crust. In the following discussion we will consider what some of these rock bodies may be. This discussion will be strongly influenced by the magnetic behavior of rock units that have been observed and studied, and by known magnetic anomalies that can be confidently associated with some rock body or structure in the upper crust. Keep in mind, however, that arguments based on familiarity may be deceptive because the deep crust (below about 25 km depth) may contain rock types that have not yet been seen and studied.

The most likely sources for long-wavelength magnetic anomalies are mafic and ultramafic rocks of various types. In particular, layered intrusions often have high magnetite contents, and the crystallization of evolved mafic magma at the base of the crust can lead to the development of sizable magnetic anomalies (Wasilewski and Mayhew, 1982; Mayhew et al., 1982). Continental rifts, where erupted flood basalts tend to include units that are significantly enriched in iron over magnesium (Basaltic Volcanism Study Group, 1981), are one tectonic environment in which this is likely to happen.

A good example of a magnetic anomaly that is probably a product of such mafic plutonism is the one over the Midcontinent Rift, which is characterized by a narrow gravity and coincident magnetic anomaly that extends from Lake Superior, south through Minnesota, Wisconsin, Iowa, Nebraska, and Kansas, a distance of over 1000 km. Thiel (1956) correlated the gravity anomaly in the Lake Superior area with rift-related volcanic and intrusive rocks of the 1.2–1.0 Ga Keweenawan Supergroup, but the sources of the anomalies are buried beneath Paleozoic strata along most of its length. The anomalies are thought to represent a narrow (40–85 km wide) rift active during the Middle Proterozoic (King and Zietz, 1971; van Schmus and Hinze, 1985). McSwiggin et al. (1987) produced detailed gravity and magnetic models of the rift along four profiles in Minnesota and Wisconsin. They concluded that significantly magnetic units exist within a central block of the rift and that the central block extends along most of length of the rift. These rocks extend to 20 km depth, have total magnetizations of up to 4.9 A/m, and have Koenigsberger ratios typically greater than one. The magnetic rocks are interpreted to be mafic volcanic and intrusive rocks of the Keweenawan Supergroup.

The mafic roots of calc-alkalic plutons may also be a major source of magnetic anomalies. Not only are observed mafic plutons of calc-alkalic affinity, such as Duke Island (see above), notably magnetic, but some magnetic anomalies are considered to be associated with calc-alkalic plutonism. These include the anomalies in the southern Coast Mountains of British Columbia (Coles et al., 1976) as well as the Early Proterozoic orogenic belts of the Canadian Shield and the Alaska Border Ranges. The Thelon orogen, an example of an Early Proterozoic orogenic belt, resulted from a collision between the Archean Slave and Rae provinces. It is characterized by a belt of magnetic anomalies, typically 80 km wide and >1000 nT amplitude, that can be traced for 2250 km. The belt of anomalies is related to 2.02–1.91 Ga granitic to dioritic plutons exposed along 1000 km of the anomaly and interpreted to be both a precollisional magmatic arc and postcollisional anatectic batholith (Hoffman, 1988).

A final example of this type of occurrence is marked by a narrow discontinuous aeromagnetic and marine magnetic anomaly that extends along 1000 km of the suture zone in southern Alaska that joins the Chugach terrain to the south with the Peninsular, Wrangellia, and other exotic terrains to the north (Plafker et al., 1977). It is related in some places to exposures of mafic and ultramafic plutonic bodies of Jurassic age. A typical section consists of residual harzburgite, 2–3 km of ultramafic cumulate rocks, and 4–7 km of gabbroic rocks. Burns (1985) inferred that the entire 1000-km-long magnetic anomaly is caused by a belt of plutonic rocks that she informally called the Border Ranges ultramafic and mafic complex, and interpreted the complex as the plutonic core of an intraoceanic island arc. The mineralogy and composition of the gabbroic rocks of this complex are consistent with crystal fractionation of andesite from basaltic melt at pressures of 5–12 kbar. The largest exposure of the complex is a gabbro body located roughly 160 km east of Anchorage. The aeromagnetic anomaly over this body has an amplitude in excess of 2000 nT. Burns (1982) constructed cross-sectional models of this gabbro body from magnetic profiles that demonstrate that the plutonic body extends 3–8 km below the surface and has magnetizations ranging from 1.5 to 6 A/m.

Serpentinite and its metamorphic equivalents often contain significant amounts of magnetite and can produce strong magnetic anomalies. Magnetic anomalies outline most serpentinite belts, both those from ophiolites (e.g., Vine et al., 1973; Zietz, 1982) and those in ancient suture zones such as the Thompson Nickel Belt (Morley et al., 1967) and the Zermatt zone of the Alps (Fischer et al., 1979). Long-wavelength magnetic anomalies may outline areas where the upper mantle beneath ancient oceanic crust has been serpentinized through deep circulation of ocean water (Arkani-Hamed, 1988). In addition, areas where serpentinite is undergoing active subduction should show strong anomalies, because in such low heat flow regimes strong magnetization in serpentinites can survive to depths of more than 45 km (Shive et al., 1988). Subducted serpentinite may well contribute to the strong positive anomaly over the Middle-American Trench observed from Magsat data by Vasicek et al. (1988).

A linear magnetic anomaly with typical amplitudes of 1000 nT that extends 700 km along the length of the Great Valley (Cady, 1975; Blake et al., 1975) is possibly associated with serpentinite. The anomaly overlies the Great Valley sequence, a thick sequence of Late Jurassic to Late Cretaceous marine clastic rocks. These sedimentary rocks are not sufficiently magnetic to cause the anomaly, but the anomaly is related spatially and geologically with a parallel anomaly over Coast Range ophiolite to the west, a belt of serpentinite and less abundant mafic rocks of Jurassic age. Various authors (Griscom, 1966; Bailey et al., 1970; Blake et al., 1975; Cady, 1975) have inferred that the buried source of the Great Valley anomaly also is oceanic crust obducted from the west during Jurassic time. Griscom and Jachens (in press) constructed crustal models of the southern Great Valley in the Coalinga area based on gravity and magnetic data and constrained by seismic refraction and reflection data and geologic information. They concluded that the source of the anomaly is a thick west-dipping slab of ophiolite extending to 20 km depth; the bottom of the magnetic part of the ophiolite is controlled in part by the depth to the Curie-temperature isotherm. They determined magnetizations ranging from 3.5 to 5.5 A/m, somewhat greater than measured magnetizations of the Coast Ranges ophiolite. Deep drill holes in the central Great Valley recovered gabbro, diabase, serpentinite and mafic metaigneous rocks (Thompson and Talwani, 1964), all possible ophiolitic components, but magnetic properties have not been determined for these samples.

4.2. Depth of magnetic layers

If magnetite is the dominant magnetic mineral in the crust, the bottom of the magnetic layer will occur at the depth at which the magnetization of magnetite drops to insignificant values. The high temperature susceptibility studies discussed above indicate that induced magnetization disappears at temperatures near 550°C. Remanence, if important, should disappear at or below 550°C. The Curie-temperature of magnetite increases at the rate of 1.85°C/kbar (Schult, 1970), and remanent and induced magnetization may be similarly affected by pressure. Taking this into account, it is reasonable that rocks should lose their magnetization at a depth corresponding approximately to the 600°C isotherm.

The depth corresponding to the 600°C isotherm depends on the geothermal conditions of the particular crustal section. Frost and Shive (1986) computed the depth to the 600°C isotherm based on measured geothermal gradients for 10 different provinces for which geothermal conditions were fairly well characterized (see Chapman and Furlong, this volume). It ranges from a depth of 20 km in the Basin and Range province to 74 km for the Sierra Nevada. The depths lie above the Moho for three provinces (Basin and Range, Eastern Australia, Central Australia), so the lower crust should not be magnetic in these areas. The 600°C isotherm lies in the mantle in the other seven areas. Wasilewski et al. (1979)

contend that the Moho is a magnetic boundary, because mantle peridotites should contain nonmagnetic spinels rather than magnetite. If so, the base of the crust will define the bottom of the magnetic layer. If the upper mantle has been serpentinized, it could be magnetic to depths greater than 70 km in some low heat flow regions. Of course, if metal alloys occur in significant amounts in the upper mantle, ferromagnetic behavior could extend to depths as great as 95 km (Toft and Haggerty, 1988).

5. Concluding remarks

The most important goal of the magnetic modeling, rock magnetic analysis and petrologic studies discussed above is to provide accurate magnetic models of the deep continental crust, models which may be used to draw inferences about the composition and structure of the lower crust. We still fall considerably short of that goal. Magnetic modeling tells us that the deep crust contains significant volumes of highly (~5 A/m) magnetic material. Rock magnetic analysis shows that magnetite is the most likely source for this magnetization, and that rocks containing about 5% magnetite by volume could provide the required magnetization if they occur above the 600°C isotherm. Petrologic analysis indicates that low-Ti magnetite should be stable in the lower crust in mafic and ultramafic rock units similar to those in the upper crust that produce large magnetic anomalies. Although these rock units do not constitute significant portions of lower-crustal cross sections studied to date, there is no reason why they may not exist in sufficiently large volumes at mid- and lower crustal levels to produce long wavelength magnetic anomalies.

The discrepancy between magnetization intensities of lower crustal rocks and the source requirements for the lower crust from anomaly analyses is the most perplexing problem in studies of lower crustal magnetization. Perhaps lower crustal rocks studied to date are not representative of the lower crust. Perhaps metal alloys are more common than we think. Perhaps the magnetization resides in serpentinite, which can be strongly magnetic. Although the weight of geologic evidence favors a largely serpentinite-free lower crust and upper mantle (Frost and Shive, 1986), some authors (Jordan and Frazer, 1975; Arkani-Hamed, 1988) believe that serpentinite may be volumetrically significant in the lower crust, particularly under old oceanic crust where access of H_2O to the upper mantle is possible.

References

Andersen, D.J. and Lindsley, D.H., 1988. Internally consistent solution models for Fe-Mg-Mn-Ti oxides: Fe-ti oxides. Am. Mineral., 73: 714–726.

Aravamadhu, P.S., 1974. Certain aspects of magnetic studies of charnockites at Waltair, India - field investigations. Can. J. Earth. Sci., 11: 503–509.

Arkani-Hamed, J., 1988. Remanent magnetization of the oceanic upper mantle. Geophys. Res. Lett., 15: 48–51.

Arkani-Hamed, J. and Strangway, D.W., 1985. Lateral variations of apparent magnetic susceptibility of lithosphere deduced from Magsat data. J. Geophys. Res., 90: 2655–2664.

Ashwal, L.D., 1982. Mineralogy of mafic and Fe-Ti oxide-rich differentiates of the Marcy anorthosite massif, Adirondacks, New York. Am. Mineral., 67: 14-27.

Austrheim, H. and Griffin, W.L., 1985. Shear deformation and eclogite formation within granulite-facies anorthosites of the Bergen Arcs, Western Norway. Chem. Geol., 50: 267–281.

Bailey, E.H., Blake, M.C., Jr., and Jones, D.L., 1970. On-land Mesozoic oceanic crust in California. U.S. Geol. Surv., Prof. Pap., 700-C: C70-C81.

Balsley, J.R. and Buddington, A.F., 1958. Iron-titanium oxide minerals, rocks, and aeromagnetic anomalies of the Adirondack area, New York. Econ. Geol., 53: 777–805.

Basaltic Volcanism Study Group, 1981. Basaltic Volcanism on the Terrestrial Planets. Pergamon Press, Oxford-New York, 1286 pp.

Behrendt, J.C., Drewry, D.J., Jankowski, E. and Grim, M.S., 1980. Aeromagnetic and radio echo ice-sounding measurements show much greater area of the Dufek Intrusion, Antarctica. Science, 209: 1014–1017.

Belluso, E., Biino, G. and Lanza, R., 1990. New data on the rock magnetism in the Ivrea-Verbano Zone (northern Italy) and its relations to the magnetic anomaly. Tectonophysics, 182: 79-89.

Bhattacharyya, B.K. and Leu, L.K., 1977. Spectral analysis of gravity and magnetic anomalies due to rectangular prismatic bodies. Geophysics, 42: 41–50.

Blake, M.C., Jr., Zietz, I. and Daniels, D.L., 1975. Aeromagnetic and generalized geologic map of parts of central California. U.S. Geol. Surv., Geophys. Inv. Map GP-918, scale 1:1,000,000

Blakely, R.J., 1988. Curie temperature isotherm analysis and tectonic implications of aeromagnetic data from Nevada. J. Geophys. Res., 93: 11,817–11,832.

Blakely, R.J. and Connard, G.G., 1989. Crustal studies using magnetic data, In: W.D. Mooney and L.C. Pakiser (Editors), Geophysical Framework of the Continental United States. Geol. Soc. Am. Mem., 172: 45-60.

Blakely, R.J. and Griscom, A., 1988. Geologic insights from upward continuation and boundary analysis of North American magnetic and gravity maps (abstract). Geol. Soc. Am., Abstr. Programs, 20: A325.

Bohlen, S.R. and Mezger, K., 1989. Origin of granulite terrains and the formation of lowermost continental crust. Science, 244: 326–329.

Bond, K.R. and Zietz, I., 1987. Composite magnetic anomaly map of the conterminous United States West of 96° longitude. U.S. Geol. Surv., Geophys. Inv. Map GP-977, 2 sheets, scale 1:2,500,000.

Brown, E.H., 1974. Comparison of the mineralogy and phase relations of blueschists from the North Cascades, Washington, and greenschists from Otago, New Zealand. Geol. Soc. Am. Bull., 85: 334–344.

Brown, E.H., 1986. Geology of the Shuksan Suite, North Cascades, Washington. In: B.W. Evans and E.H. Brown (Editors), Blueschists and Eclogites. Geol. Soc. Am. Mem., 164: 143–154.

Buddington, A.F. and Lindsley, D.H., 1964. Iron-titanium oxide minerals and synthetic equivalents. J. Petrol., 5: 310–357.

Burns, L.E., 1982. Gravity and aeromagnetic modelling of a large gabbroic body near the Border Ranges fault, southern Alaska. U.S. Geol. Surv., Open-File Rep. 82-460, 66 pp.

Burns, L.E., 1985. The Border Ranges ultramafic and mafic complex, south-central Alaska: cumulate fractionates of island arc volcanics. Can. J. Earth Sci., 22: 1020-1038.

Byerly, P.E. and Stolt, R.H., 1977. An attempt to define the Curie point isotherm in northern and central Arizona. Geophysics, 42: 1394–1400.

Cady, J.W., 1975. Magnetic and gravity anomalies in the Great Valley and western Sierra Nevada metamorphic belt, California. Geol. Soc. Am., Spec. Pap. 168, 56 pp.

Cain, J.C., Davis, W.M. and Regan, R.D., 1974. An $N = 22$ model of the geomagnetic field (abstract). EOS, Trans. Am. Geophys. Union, 56: 1108.

Cain, J.C., Schmitz, D.R. and Muth, L., 1984. Small-scale features in the Earth's magnetic field observed by Magsat. J. Geophys. Res., 89: 1070-1076.

Candela, P.A., 1986. The evolution of aqueous vapor from silicate melts: Effect on oxygen fugacity. Geochim. Cosmochim. Acta, 50: 1205–1211.

Carle, H.M. and Harrison, C.G.A., 1982. A problem in representing the core magnetic field of the Earth using spherical harmonics. Geophys. Res. Lett., 9: 265–268.

Carmichael, R.S., 1982. Magnetic properties of minerals and rocks, In: R.S. Carmichael (Editor), Handbook of Physical Properties of Rocks. CRC Press, Inc., Boca Raton, FL, pp. 229–287.

Chernyuk, M.V., 1971. Distribution of the magnetic susceptibility of intrusive rocks. Izv. Phys. Solid Earth (Engl. trans.), 3: 225–229.

Clark, S.C., Frey, H. and Thomas, H.H., 1985. Satellite magnetic anomalies over subduction zones: The Aleutian Arc anomaly. Geophys. Res. Lett., 12: 41–44.

Coles, R.L., 1976. A flexible interactive magnetic anomaly interpretation technique using multiple rectangular prisms. Geoexploration, 14: 125–141.

Coles, R.L. and Currie, R.G., 1977. Magnetic anomalies and rock magnetizations in the southern Coast Mountains, British Columbia: possible relation to subduction. Can. J. Earth Sci., 14: 1753–1770.

Coles, R.L., Haines, G.V. and Hanneford, W., 1976. Large scale magnetic anomalies over western Canada and the Arctic: a discussion. Can. J. Earth Sci., 13: 790-802.

Committee for the Magnetic Anomaly Map of North America, 1987. Magnetic anomaly map of North America. Geol. Soc. Am., Boulder, CO, 4 sheets, scale 1:5,000,000.

Connard, G., Couch, R. and Gemperle, M., 1983. Analysis of aeromagnetic measurements from the Cascade Range in central Oregon. Geophysics, 48: 376–390.

Czamanske, G.K., Ishihara, S. and Atkin, S.A., 1981. Chemistry of rock-forming minerals of the Cretaceous-Paleocene batholith in southwestern Japan and implications for magma genesis. J. Geophys. Res., 86: 10,341-10,469.

Deer, W.A., Howie, R.A. and Zussman, J., 1966. An Introduction to the Rock-Forming Minerals. J. Wiley and Sons, New York, NY, 528 pp.

Diamond, J.L. and B.R. Frost, 1988. Mineralogical controls on rock magnetism in the Sierra San Pedro Martir Pluton, Baja California, Mexico (abstract). EOS, Trans. Am. Geophys. Union, 69: 1506.

Dick, H.J.B., 1974. Terrestrial nickel-iron from the Josephine Peridotite, its geologic occurrence, association and origin. Earth Planet. Sci. Lett., 24: 291–298.

Duchesne, J-C., 1972. Iron-titanium oxide minerals in the Bjerkem-Sogndal massif, southwestern Norway. J. Petrol., 13: 57–81.

Dunlop, D.J., 1979. A regional paleomagnetic study of Archean rocks from the Superior geotraverse area, Northwestern Ontario. Can. J. Earth Sci., 16: 1906–1919.

Dunlop, D.J., 1984. A method of determining demagnetizing factor from multidomain hysteresis. J. Geophys. Res., 89: 553–558.

Dunlop, D., 1986. Hysteresis properties of magnetite and their dependence on particle size: a test of pseudo-single-domain remanence models. J. Geophys. Res., 91: 9569–9584.

Elming, S.-A. and Törne, A., 1976. The blue road geotraverse: A magnetic ground survey and the interpretation of magnetic anomalies. Geol. Foeren. Stockholm Forh., 98: 264–270.

Evans, B.W. and Frost, B.R., 1975. Chrome-spinel in progressive metamorphism –A preliminary analysis. Geochim. Cosmochim. Acta, 39: 959–972.

Fischer, G., Schnegg, P.A. and Sesiano, J., 1979. A new geomagnetic survey of Switzerland. Materiquix pour la Geologie de la Suisse "Geophysique", 19: 43 pp.

Fountain, D.M., 1985. Geological and geophysical nature of the lower continental crust as revealed by exposed cross sections of the continental crust. In: J.S. Noller, S.H. Kirby and J.E. Nielsen-Pike (Editors), Geophysics and Petrology of the Deep Crust and Upper Mantle. U.S. Geol. Surv., Circ., 956: 25-26.

Fountain, D.M. and Salisbury, M.H., 1981. Exposed cross-sections through the continental crust: implications for crustal structure, petrology and evolution. Earth Planet. Sci. Lett., 56: 263–277.

Fountain, D.M., Percival, J. and Salisbury, M.H., 1990. Exposed cross sections of the continental crust - synopsis. In: M.H. Salisbury and D.M. Fountain (Editors), Exposed Cross-Sections of the Continental Crust. Kluwer Academic Publ., Dordrecht, pp. 653-662.

French, B.M., 1966. Some geological implications of equilibrium between graphite and a C-H-O gas phase at high temperatures and pressures. Rev. Geophys., 4: 223–253.

Frey, H., 1985. Magsat and POGO magnetic anomalies over the Lord Howe Rise: evidence against a simple continental crustal structure. J. Geophys. Res., 90: 2631–2639.

Frost, B.R., 1979a. Mineral equilibria involving mixed-volatiles in a C-O-H fluid phase: The stabilities of graphite and siderite. Am. J. Sci., 279: 1033–1059.

Frost, B.R., 1979b. Metamorphism of iron-formation: Paragenesis in the system Fe-Si-C-O-H. Econ. Geol., 74: 775–785.

Frost, B.R., 1985. On the stability of sulfides, oxides, and native metals in serpentinite. J. Petrol., 26: 31–63.

Frost, B.R., 1988. A review of graphite-sulfide-oxide-silicate equilibria in metamorphic rocks. Recond. Soc. Ital. Miner. Petrol., 43: 25–40

Frost, B.R. and Chacko, T., 1989. The granulite uncertainty principle: Limitations on thermobarometry in granulites. J. Geol., 97: 435–450.

Frost, B.R. and Shive, P.N., 1986. Magnetic mineralogy of the lower continental crust. J. Geophys. Res., 91: 6513–6522.

Frost, B.R. and Shive, P.N., 1989. Comment on "Limiting depth of magnetization in cratonic lithosphere". Geophys. Res. Lett., 16: 477–479.

Frost, B.R., Fyfe, W.S., Tazaki, K. and Chan, T., 1989. Grain-boundary graphite in rocks and implications for high electrical conductivity in the lower crustal. Nature, 340: 134-136.

Frost, B.R., Lindsley, D.H. and Andersen, D.J., 1988. Fe-Ti oxide–silicate equilibria: Assemblages with fayalitic olivine. Am. Mineral., 73: 727–740.

Gasparini, P., Mantovani, M.S.M., Corrado, G. and Rapolla, A., 1979. Depth of Curie temperature in continental shields: a compositional boundary? Nature, 278: 845–846.

Geissman, J.W., Harlan, S.S. and Brearley, A.J., 1988. The physical isolation and identification of carriers of geologically stable remanent magnetization: Paleomagnetic and rock magnetic microanalysis and electron microscopy. Geophys. Res. Lett., 15: 479–482.

Griffin, W.L., Taylor, P.N., Hakkinen, J.W., Heier, K.S., Iden, I.K., Krogh, E.J., Malm, O., Olsen, K.I., Ormaasen, D.E. and Tveten, E., 1978. Archaean and Proterozoic crustal evolution in Lofoten-Vesterålen, N Norway. J. Geol. Soc. London, 135: 629-647.

Griscom, A., 1966. Magnetic data and regional structure in northern California. In: E.H. Bailey (Editor), Geology of Northern California. Calif. Div. Mines Geol. Bull., 190: 407-417.

Griscom, A. and Jachens, R.C., in press. Tectonic implications of gravity and magnetic models along east-west seismic profiles across the Great Valley near Coalinga. U.S. Geol. Surv., Prof. Pap.

Gundersen, J.N. and Schwartz, G.M., 1962. The geology of the metamorphosed Biwabik iron-formation, Eastern Mesabi District, Minnesota. Minn. Geol. Surv. Bull. 43, 103 pp.

Haggerty, S.E., 1976. Opaque mineral oxides in terrestrial igneous rocks. Rev. Mineral., 3: 101-300.

Haggerty, S.E., 1978. Mineralogical constraints on Curie isotherms in deep crustal magnetic anomalies. Geophys. Res. Lett., 5: 105–108.

Haggerty, S.E., 1979. The aeromagnetic mineralogy of igneous rocks. Can. J. Earth Sci., 16: 1281–1293.

Haggerty, S.E. and Toft, P.B., 1985. Native iron in the continental lower crust: Petrological and geophysical implications. Science, 229: 647–649.

Hahn, A., Kind, E.G. and Mishra, D.C., 1976. Depth estimation of magnetic sources by means of Fourier amplitude spectra. Geophys. Prospect., 24: 287–308.

Hall, D.H., 1968. Regional magnetic anomalies, magnetic units, and crustal structure in the Kenora district of Ontario. Can. J. Earth Sci., 5: 1277-1296.

Hall, D.H., 1974. Long-wavelength aeromagnetic anomalies and deep crustal magnetization in Manitoba and northwestern Ontario, Canada. J. Geophys., 40: 403–430.

Hall, D.G., Coles, R.L. and Hall, J.M., 1979. The distribution of surface magnetization in the English River and Kenora subprovince of the Archean shield in Manitoba and Ontario. Can. J. Earth Sci., 16: 1764–1777.

Hall, D.H., Noble, I.A. and Millar, T.W., 1985. Crustal structure of the Churchill-Superior boundary zone between 80° and 98°W longitude from Magsat anomaly maps and stacked passes. J. Geophys. Res., 90: 2621–2630.

Han, T.-M., 1982. Iron-formations of Precambrian age: Hematite-magnetite relationships in some Proterozoic iron deposits - A microscopic observation. In: G.C. Amstutz, A. El Goresy, G. Grenzel, C. Kluth and G. Moh (Editors), Ore Genesis - The State of the Art. Springer-Verlag, Heidelberg, pp. 451–459.

Hargraves, R.B., 1959. Magnetic anisotropy and remanent magnetism in hemo-ilmenite from ore deposits at Allard Lake, Quebec. J. Geophys. Res., 64: 1565–1578.

Harrison, C.G.A., Carle, H.M. and Hayling, K.L., 1986. Interpretation of satellite elevation magnetic anomalies. J. Geophys. Res., 91: 3633–3650.

Hildenbrand, T.G., 1985. Magnetic terrains in the central United States determined from the interpretation of digital data. In: W.J. Hinze (Editor), The Utility of Regional Gravity and Magnetic Anomaly Maps. Soc. Explor. Geophys., Tulsa, OK, pp. 248–266.

Himmelberg, G.R. and Ford, A.B., 1977. Iron-titanium oxides of the Dufek Intrusion, Antarctica. Am. Mineral., 62: 623–633.

Hoffman, P.F., 1988. United plates of America, the birth of a craton. Annu. Rev. Earth Planet. Sci., 16: 543-603.

Irvine, T.N., 1965. Sedimentary structures in igneous intrusions with particular reference to the Duke Island ultramafic complex. Soc. Econ. Paleontol. Miner., Spec. Publ., 12: 220-232.

Irvine, T.N., 1974. Petrology of the Duke Island ultramafic complex, southeastern Alaska. Geol. Soc. Am. Mem., 138, 240 pp.

Irvine, T.N. and Smith, C.H., 1969, Primary oxide minerals in the layered series of the Muskox Intrusion. Econ. Geol. Monogr., 4: 76–94.

Ishihara, S., 1979. Lateral variation of magnetic susceptibility of the Japanese granitoids. J. Geol. Soc. Jpn., 85: 509–523.

Ishihara, S., 1981. The granitoid series and mineralization, Econ. Geol. 75th Anniv. Vol., 458–484.

James, H.L., 1954. Sedimentary facies of iron-formation. Econ. Geol., 49: 235–291.

James, H.L., 1966. Chemistry of iron-rich sedimentary rocks. U.S. Geol. Surv., Prof. Pap. 440-W, 61 pp.

Jensen, H., 1961. The airborne magnetometer. Sci. Am., 204: 151–162.

Johnson, B.D., 1985. Viscous remanent magnetization model for the Broken Ridge satellite magnetic anomaly. J. Geophys. Res., 90: 2640–2646.

Jordan, T.H. and Frazer, L.N., 1975. Crustal and upper mantle structure from S_p phases. J. Geophys. Res., 80: 1504-1508.

King, E.R. and Zietz, I., 1971. Aeromagnetic study of the midcontinent gravity high of central United States. Geol. Soc. Am. Bull., 82: 2187-2208.

Kling, J.F., 1986. Petrology of the Strong Creek Prospect, Albany Co., Wyoming and in the system: Fe-Ti-Si-O-S. M.S. Thesis, University of Wyoming, Laramie, 109 pp.

Krutikhovskaya, Z.A. and Pashkevich, I.K., 1977. Magnetic model for the Earth's crust under the Ukrainian shield. Can. J. Earth Sci., 14: 2718-2728.

Krutikhovskaya, Z.A. and Pashkevich, I.K., 1979. Long-wavelength magnetic anomalies as a source of information about deep crustal structure. J. Geophys., 46: 301–317.

Krutikhovskaya, Z.A., Silina, I.M., Bondareva, N.M. and Podolyanko, S.M., 1979. Relation of magnetic properties of the rocks of the Ukranian shield to their composition and metamorphism. Can. J. Earth Sci., 16: 984–991.

LaBrecque, J.L., Cande, S.C. and Jarrard, R.D., 1985. Intermediate-wavelength magnetic anomaly field of the North Pacific and possible source distributions. J. Geophys. Res., 90: 2549–2564.

Langel, R.A. and Estes, R.H., 1982. A geomagnetic field spectrum. Geophys. Res. Lett., 9: 250–253.

Langel, R.A. and Thorning, L., 1982. A satellite magnetic anomaly map of Greenland. Geophys. J. R. Astron. Soc., 77: 599–612.

Langel, R.A., Phillips, J.D. and Horner, R.J., 1982. Initial scalar magnetic anomaly map from Magsat. Geophys. Res. Lett., 9: 269–272.

Larson, W., Ozima, M., Nagata, T. and Strangway, D., 1969. Stability of remanent magnetization of igneous rocks. Geophys. J. R. Astron. Soc., 17: 263–292.

Leinert B.R. and Wasilewski, P.J., 1979. A magnetic study of the serpentinization process at Burro Mountain, California. Earth Planet. Sci. Lett., 43: 406–416.

Lidiak, E.G., 1974. Magnetic characteristics of some Precambrian basement rocks. J. Geophys., 40: 549–654.

Lowes, F.J., 1966. Mean square values on sphere of spherical harmonic vector fields. J. Geophys. Res., 71: 2179.

Mathison, C.I., 1975. Magnetite and ilmenite in the Somerset Dam layered basic intrusion, southeastern Queensland. Lithos, 8: 93–111.

Mayhew, M.A., 1985. Curie isotherm surfaces inferred from high-altitude magnetic anomaly data. J. Geophys. Res., 90: 2647–2654.

Mayhew, M.A. and Galliher, S.G., 1982. An equivalent layer magnetization model for the United States derived from Magsat data. Geophys. Res. Lett., 9: 311–313.

Mayhew, M.A., Johnson, B.D. and Wasilewski, P.J., 1985a. A review of problems and progress in studies of satellite magnetic anomalies. J. Geophys. Res., 90: 2511–2522.

Mayhew, M.A., Estes, R.H. and Myers, D.M., 1985b. Magnetization models for the source of the "Kentucky anomaly" observed by Magsat. Earth Planet. Sci. Lett., 74: 117–129.

Mayhew, M.A., Thomas, H.H. and Wasilewski, P.J., 1982. Satellite and surface geophysical expression of anomalous crustal structure in Kentucky and Tennessee. Earth Planet. Sci. Lett., 58: 395–405.

McSwiggin, P.L., Morey, G.B. and Chandler, V.W., 1987. New model of the midcontinent rift in easter Minnesota and western Wisconsin. Tectonics, 6: 677-685.

Mørk, H.B.E., 1985. A gabbro to eclogite transition in Flemsøy, Sunnmøre, Western Norway. Chem. Geol., 50: 283–310.

Morley, L.W., MacLaren, A.S. and Charbonneau, B.W., 1967. Magnetic anomaly map of Canada. Geol. Soc. Can., Map 1255A.

Morley, L.W., MacLaren, A.S. and Charbonneau, B.W., ed. first edition, second edition anonymous, McGrath, P.H., Hood, P.J., and Darnley, A.G., ed. third edition, Dods, S.D., Hood, P.J., Teskey, D.J.,and McGrath, P.H., ed. fourth edition, Magnetic anomaly map of Canada. Geol. Surv. Can. Map 1255A, scale 1:5,000,000, 1984.

Mutton, A.J. and Shaw, R.D., 1979. Physical property measurements as an aid to magnetic interpretation in basement terrains. Bull. Aust. Soc. Explor. Geophys., 10: 79–91.

Nagata, T., 1961. Rock Magnetism. Maruzen Co. Ltd., Tokyo, 355 pp.

Nesbitt, B.E., 1986. Oxide-sulfide-silicate equilibria associated with metamorphosed ore deposits. Part II. Pelitic and felsic volcanic terrains. Econ. Geol., 81: 841–856.

Ohmoto, H. and Kerrick, D., 1977. Devolatilization equilibria in graphitic systems. Am. J. Sci. 277: 1013–1044.

Okubo, Y., Graf, R.J., Hansen, R.O., Ogawa, K. and Tsu, H., 1985. Curie point depths of the island of Kyushu and surrounding areas, Japan. Geophysics, 53: 481–494.

Patchen, A.D., 1987. Petrology and stratigraphy of the Lake Owen layered mafic complex, S.E. Wyoming. M.S. thesis, University of Wyoming, 163 pp.

Pauthenet, R. and Bochirol, L., 1951. Spontaneous magnetization of ferrites. J. Physique Radium, 12: 249–251.

Pinet, M. and Smith, D.C., 1985. Petrochemistry of opaque minerals in eclogites from Western Gneiss region, Norway. 1. Petrology of the oxide microassemblages. Chem. Geol., 50: 225–249.

Piper, J.D.A., 1979. The paleomagnetism of the central zone of the Lewisian foreland, north-west Scotland. Geophys. J. R. Astron. Soc., 59: 101–122.

Plafker, G., Jones, D.L. and Pessagno, E.A., 1977. A Cretaceous accretionary flysch and melange terrane along the Gulf of Alaska margin. U.S. Geol. Surv., Circ., 751-B: B41-B43.

Pognante, U., 1985. Coronitic reactions and ductile shear zones in eclogitized ophiolite metagabbro, Western Alps, North Italy. Chem. Geol., 50: 99–109.

Powell, D.W., 1970. Magnetized rocks within the Lewisian of western Scotland and under the Southern Uplands. Scot. J. Geol., 6: 353–369.

Pullaiah, G., Irving, E., Buchan, K.L. and Dunlop, D.J., 1975. Magnetization changes caused by burial and uplift. Earth Planet. Sci. Lett., 28: 133-143.

Ramaswamy, K. and Ramanathan, P., 1982. Paleomagnetic study of some charnockites of the Palni Hills. Phys. Earth Planet. Inter., 28: 269–272.

Raymond, C.A. and LaBrecque, J.L., 1987. Magnetization of oceanic crust: thermoremanent magnetization or chemical remanent magnetization? J. Geophys. Res., 92: 8077–8088.

Regan, R.D., Cain, J.C. and Davis, W.M., 1975. A global magnetic anomaly map, J. Geophys. Res., 80: 794–802.

Regan, R.D. and Marsh, B.D., 1982. The Bangui magnetic anomaly: Its geological origin. J. Geophys. Res., 87: 1107–1120.

Riddihough, R.P., 1972. Regional magnetic anomalies and geology in Fennoscandia: a discussion. Can. J. Earth Sci., 9: 219–232.

Roeske, S., 1986. Field relations and metamorphism of the Raspberry Schist, Kodiak Islands, Alaska. Geol. Soc. Am. Mem., 164: 169–184.

Ruder, M.E. and Alexander, S.S., 1986. Magsat equivalent source anomalies over the southeastern United States: implications for crustal magnetization. Earth Planet. Sci. Lett., 78: 33–43.

Saad, A.H., 1969. Magnetic properties of ultramafic rocks from Red Mountain, California. Geophysics, 34: 974–987.

Sautter, V., 1985. An eclogite paragenesis from Aleksod basement, Central Hoggar, South Algeria. Chem. Geol., 50: 331–347.

Schlinger, C.M., 1985. Magnetization of the lower crust and interpretation of regional magnetic anomalies: example from Lofoten and Vesterålen, Norway. J. Geophys. Res., 90: 11,484–11,504.

Schlinger, C.M. and Veblen, D.R., 1989. Magnetism and transmission electron microscopy of Fe-Ti oxides and pyroxenes in a granulite from Lofoten, Norway. J. Geophys. Res., 94: 14,009–14,026.

Schlinger, C.M., Kahn, M.J. and Wasilewski, P., 1992. Rock magnetism of the Kohistan Complex, Pakistan. Geol. Bull. Univ. Peshwar, in press.

Schnetzler, C.C., 1985. An estimation of continental crust magnetization and susceptibility from Magsat data for the conterminous United States. J. Geophys. Res., 90: 2617–2620.

Schnetzler, C.C., Taylor, P.T., Langel, R.A., Hinze, W.J. and Phillips, J.D., 1985. Comparison between the recent U.S. composite magnetic anomaly map and Magsat anomaly data. J. Geophys. Res., 90: 2543–2548.

Schouten, J.A., 1971. A fundamental analysis of magnetic anomalies over oceanic ridges. Mar. Geophys. Res., 1: 111–114.

Schult, A., 1970. Effects of pressure upon the Curie temperature of titanomagnetites. Earth Planet. Sci. Lett., 10: 81–86.

Segawa, J. and Oshima, S., 1975. Buried Mesozoic volcanic-plutonic fronts of the northwestern Pacific island arcs and their implications. Nature, 256: 15–19.

Shimizu, Y., 1960. Magnetic viscosity of magnetite. J. Geomag. Geoelec., 11: 125–133.

Shive, P.N., 1989a. Can remanent magnetization in the deep crust contribute to long wavelength magnetic anomalies?. Geophys. Res. Lett., 16: 89–92.

Shive, P.N., 1989b. Reply. Geophys. Res. Lett., 16: 599-600.

Shive, P.N. and Morel, J.A., 1980. Paleomagnetism of the Precambrian Lake Owens Complex, Medicine Bow Mountains, Wyoming. Contrib. Geol., 19: 1–8.

Shive, P.N. and Fountain, D.M., 1988. Magnetic mineralogy in an Archean crustal cross section: implications for crustal magnetization. J. Geophys. Res., 93: 12,177–12,186.

Shive, P.N., Frost, B.R. and Peretti, A., 1988. The magnetic properties of metaperidotitic rocks as a function of metamorphic grade: implications for crustal magnetic anomalies. J. Geophys. Res., 93: 12,187–12,195.

Shive, P.N., Houston, R.S. and Blakely, R.J., 1990. Modeling of aeromagnetic data from the Precambrian Lake Owens Mafic Complex, Wyoming. Geol. Soc. Am., Bull., 102: 1317-1322.

Shuey, R.T., Schellinger, D.K., Johnson, E.H. and Alley, L.B., 1973. Aeromagnetics and the transition between the Colorado Plateau and Basin and Range provinces. Geology, 1: 107–110.

Shuey, R.T., Schellinger, D.K., Tripp, A.C. and Alley, L.B., 1977. Curie depth determination from aeromagnetic spectra. Geophys. J. R. Astron. Soc., 50: 75–101.

Smith, G.M., 1984. A new approach to the theory of single domain TRM. Geophys. Res. Lett., 11: 201–204.

Smulikowski, K. and Smulikowski, W., 1985. On the porphyroblastic eclogites of the Snieznik Mountains in the Polish Sudetes. Chem. Geol., 50: 201–222.

Snoke, A.W., Quick, J.E. and Bowman, H.R., 1981. Bear Mountain igneous complex, Klamath Mountains, California: an ultrabasic to silicic calc-alkaline suite. J. Petrol., 22: 501–552.

Speer, J.A., 1981. The nature and magnetic expression of isograds in the contact aureole of the Liberty Hill pluton, South Carolina: Summary. Geol. Soc. Am. Bull., Part I, 92: 603–609.

Stacey, F.D. and Banerjee, S.K., 1974. The Physical Principles of Rock Magnetism. Elsevier, Amsterdam, 195 pp.

Subrahmanyan, C. and Verma, R.K., 1981. Densities and magnetic susceptibilities of Precambrian rocks of different metamorphic grade (southern Indian Shield). J. Geophys., 49: 101-107.

Takahashi, M., Aramake, S. and Ishihara, S., 1980. Magnetite- series/illmenite-series vs. I-type/S-type granitoids. Min. Geol. Spec. Issue, 8: 13–28.

Taylor, P.T., 1983. Magnetic data over the Arctic from aircraft and satellites. Cold Regions. Sci. Technol., 7: 35–40.

Thiel, E., 1956. Correlation of gravity anomalies with the Keweenawan geology of Wisconsin and Minnesota. Geol. Soc. Am. Bull., 95: 1150-1157.

Thompson, G.A. and Talwani, M., 1964. Crustal structure of the Pacific Basin to central Nevada. J. Geophys. Res., 69: 4813-4837.

Toft, P.B. and Haggerty, S.E., 1988. Limiting depth of magnetization in cratonic lithosphere. Geophys. Res. Lett., 15: 530–533.

Treloar N.A., Shive, P.N. and Fountain, D.M., 1986. Viscous remanence acquisition in deep crustal rocks (abstract). EOS, Trans. Am. Geophys. Union, 67: 266.

U.S. Geological Survey, and Society of Exploration Geophysicists, 1982. Composite magnetic anomaly map of the United States, Part A: Conterminous United States. U.S. Geol. Surv., Map GP 954-A, scale 1:1,250,000, 2 sheets.

Vasicek, J.M., Frey, H.V. and Thomas, H.H., 1988. Satellite magnetic anomalies and the Middle American Trench. Tectonophysics, 154: 19–24.

van Schmus, W.R. and Hinze, W.J., 1985. The midcontinent rift system. Ann. Rev. Earth Planet. Sci., 13: 345-383.

Vincent, E.A. and Phillips, R., 1954. Iron-titanium oxide minerals in layered gabbros of the Skaergaard intrusion, East Greenland. Geochim. Cosmochim. Acta, 6: 1–26.

Vine, F.J., Poster, C.K. and Gass, I.G., 1973. Aeromagnetic survey of the Troodos Igneous Massif, Cyprus. Nature, Phys. Sci., 244: 34–38.

Wager, L.R. and Brown, G.M. 1968. Layered Igneous Rocks. W.H. Freeman and Co, San Francisco, CA, 588 pp.

Wagner, J.J., 1984. Petrophysical properties of the Ivrea Zone and adjacent areas. In: J.J. Wagner and St. Mueller (Editors), Geomagnetic and Gravimetric Studies of the Ivrea Zone. Beitr. Geol. Schweiz, Geophys., 21: 31–38.

Wasilewski, P.J. and Fountain, D.M., 1982. The Ivrea Zone as a model for the distribution of magnetization in the continental crust. Geophys. Res. Lett., 9: 333–336.

Wasilewski, P.J. and Mayhew, M.A., 1982. Crustal xenoliths magnetic properties and long wavelength anomaly source requirements. Geophys. Res. Lett., 9: 329–332.

Wasilewski, P. and Warner, R.D., 1988. Magnetic petrology of deep crustal rocks - Ivrea Zone, Italy. Earth Planet. Sci. Lett., 87: 347–361.

Wasilewski, P.J., Thomas, H.H. and Mayhew, M.A., 1979. The Moho as a magnetic boundary. Geophys. Res. Lett., 6: 541–544.

Willemse, J., 1969. The vanadiferous magnetic iron ore of the Bushveld Complex. Econ. Geol. Monogr., 4: 187–208.

Williams, M.C., Shive, P.N., Fountain, D.M. and Frost, B.R., 1985/86. Magnetic properties of exposed deep crustal rocks from the Superior Province of Manitoba. Earth Planet. Sci. Lett., 76: 176–184.

Wörm, H.-U., 1989. Comment on "Can remanent magnetization in the deep crust contribute to long wavelength magnetic anomalies," by Peter N. Shive. Geophys. Res. Lett., 16: 595–597.

Zietz, I., 1982. Composite magnetic anomaly map of the United States, part A, Conterminous United States. U.S. Geol. Surv., Map GP-954A.

Chapter 5

Thermal state of the continental lower crust

DAVID S. CHAPMAN and KEVIN P. FURLONG

1. Introduction

Temperature distributions within the lithosphere affect a wide variety of physical properties and processes. Rock density, electrical and magnetic properties and many seismic properties are temperature dependent. So are mineral phase boundaries, the rates of chemical reactions and many processes of rock deformation. For these reasons it is important to obtain as reliable an estimate of temperature within the crust as possible.

For the uppermost few kilometers of the crust, temperatures can be either measured directly in drill holes, or extrapolated confidently from measurement of surface heat flow and a knowledge of the behavior of thermal properties with depth. For depths greater than about 10 km the actual temperature estimate is less certain and depends both on the model of heat transfer adopted and on accuracy of the values of thermal conductivity and heat production estimated for the lower crust and uppermost mantle. The uncertainty of temperature estimates becomes progressively greater with greater depth.

Commonly cited temperature profiles for the continents, especially in the petrologic literature, are those published more than 25 years ago by Clark and Ringwood (1964). The long standing popularity of these geotherms is all the more remarkable because their paper predated the general acceptance of plate tectonics and sea floor spreading in the late 1960s, predated the discovery of the linear heat flow–heat production relationship in 1968, a cornerstone of crustal geotherm calculations, and predated several important high temperature experiments concerning the thermal conductivity of mantle material. Furthermore, at that time there were only 73 continental heat flow measurements to serve as model constraints and to provide information on the thermal state of the continental lithosphere; the corresponding number at the end of 1990 exceeded 10,000.

Although the single steady-state oceanic geotherm of Clark and Ringwood (1964) has been abandoned in favor of a family of transient geotherms (cf. Sclater and Francheteau, 1970) related to the age of ocean floor and consistent with the model of sea floor spreading, three continental geotherms of Clark and Ringwood (1964, their fig. 2) are still in use. However, some parameter values used by Clark and Ringwood, in particular high-temperature thermal conductivity, have been shown by subsequent experiments (Schatz and Simmons, 1972) to be more than 100% in error. Fortunately, geotherm calculations are not totally dependent on a single parameter, and thus through an acknowledged use of "constraints imposed on the temperatures by considerations not of a purely thermal nature" Clark and Ringwood (1964,

p. 53), arrived at a basic thermal state for shields and platforms which is not in great dispute.

The purpose of this paper is to provide an updated rationale for the computation of steady-state, static, continental geotherms and then to describe geological processes that modify these framework geotherms. In doing so we use refinements in modeling lithospheric geotherms made by others over the last two decades (Blackwell, 1971; Roy et al., 1972; Lachenbruch and Sass, 1977; Pollack and Chapman, 1977). Of particular importance are comprehensive compilations of thermophysical properties (Roy et al., 1981; Rybach and Cermak, 1982; Cermak and Rybach, 1982). A preferred family of geotherms is derived and presented; a detailed parameter sensitivity analysis and a discussion of applicability of the framework geotherms is given in Chapman (1986).

2. Geotherm models

Although both conductive and convective heat transfer may be important in determining temperature distributions in the lithosphere, especially in active orogenic regions, conductive processes mainly dominate in stable areas. Once an appropriate model of heat transfer is established, and boundary conditions are imposed, the calculation of temperature distributions is straightforward. The governing differential equation of time-dependent conductive heat transfer is:

$$-\text{div}\,(-k\ \text{grad}\ T) + A = \rho c \frac{\partial T}{\partial t} \tag{1}$$

where T is temperature, t is time, A is volumetric heat production, k is thermal conductivity, ρ is density and c is specific heat.

For one-dimensional heat transfer in a homogeneous and isotropic medium, equation (1) can be written as:

$$\frac{\partial^2 T}{\partial z^2} + \frac{A}{k} = \frac{1}{\alpha}\frac{\partial T}{\partial t} \tag{2}$$

where $\alpha = k/\rho c$ is thermal diffusivity.

The most common simplification of (2), and one which is employed here involves steady-state heat transfer in a medium with arbitrary heat production. This leads to Poisson's equation:

$$\frac{\partial^2}{\partial z^2} = -\frac{A}{k} \tag{3}$$

Throughout the lithosphere, heat production, A, varies with depth following the distribution of radiogenic isotopes. Thermal conductivity k varies with composition and with both pressure and temperature. Analytic solutions for $T(z)$ can be found when the heat production $A(z)$ follows some simple analytical forms and when thermal conductivity is constant or temperature dependent (cf. Rybach, 1981), however, these requirements are restrictive. The algorithm incorporated here uses the solution of (3) in a layer of constant heat generation and constant thermal conductivity:

$$T(z) = T_{\mathrm{T}} + \frac{q_{\mathrm{T}}}{k}z - \frac{Az^2}{2k} \tag{4}$$

where T_T and q_T are the temperature and heat flow respectively at the top of the layer where $z = 0$. If the layer has thickness Δz then the temperature at, and heat flow through, the bottom of the layer (T_B, q_B) can be expressed in terms of the temperature and heat flow at the top of the layer (T_T, q_T) and properties (A, k) of the layer:

$$T_B = T_T + \frac{q_T}{k}\Delta z - \frac{A\,\Delta z^2}{2k} \tag{5}$$

$$q_B = q_T A\,\Delta z \tag{6}$$

Equations (5) and (6) are applied to successive layers, resetting T_T and q_T at the top of each new layer with the values T_B and q_B solved for the bottom of the previous layer. The layer thickness can be made small enough so that complicated and discontinuous distributions, $A(z)$, are adequately approximated. Thermal conductivity effects, $k(z, T)$, are incorporated and updated at each step in an iterative loop. In practice, heat production and thermal conductivity are described by piecewise continuous functions and the computations are carried out with a 0.1-km depth increment.

3. Thermophysical parameters

The gross pattern of thermal conductivity in the lithosphere is controlled by composition, to a lesser extent by temperature, and finally by pressure effects. Although the structural and compositional complexity of both the upper and lower crust (cf. Kay and Kay, 1981) indicates a concomitant complexity in physical and thermal properties, it is necessary to generalize a structure for the purpose of computing geotherms. The generalized lithosphere model used here consists of an upper crust of granite to andesitic composition, a lower crust consisting of gabbro or granulite facies metamorphic rocks and an ultramafic upper mantle. Although one could design a variety of generalized composiitonal models, we have evaluated this simple layered model which would be in general agreement with average seismic velocity characteristics of the crust.

Recent compilations of thermal conductivity determinations (Roy et al., 1981; Cermak and Rybach, 1982) on several thousand rock samples provide a convenient summary from which an appropriate crustal thermal conductivity profile may be assembled. A value of 3.0 W m^{-1} K^{-1} is an appropriate value to use for a granitic upper crust at conditions replicating the laboratory conditions (room temperature and atmospheric pressure) under which the conductivities were determined. The thermal conductivity of possible lower crustal rocks is constrained to a narrow range by the average values given by Cermak and Rybach (1982) for diorite (2.91 W m^{-1} K^{-1}), granodiorite (2.65 W m^{-1} K^{-1}), gabbro (2.63 W m^{-1} K^{-1}), amphibolite (2.46 W m^{-1} K^{-1}) and gneisses (2.44 W m^{-1} K^{-1}). Thermal properties of "granulites" rarely appear in standard thermal properties tabulations, but in specific studies (e.g., Fountain et al., 1987) both thermal conductivity and heat production of granulitic rocks have been measured. The thermal conductivity of those samples and of rocks compositionally equivalent to granulites generally lies between 2.4 and 2.9 W m^{-1} K^{-1}. A value of 2.6 W m^{-1} K^{-1} is therefore reasonable for our generalized lower crustal rocks, but again only at laboratory measurement conditions.

Although in the past it has been common to consider thermal conductivity as being constant throughout the crust, it seems more appropriate now to use the growing experimental data base and to calculate thermal conductivity as a function of pressure and temperature directly in the geotherm computation. Thermal conductivity of most rocks at crustal conditions varies inversely with temperature and directly with pressure (or depth) according to the relation:

$$k(T,z) = k_0\,(1 + cz)/(1 + b\,\Delta T) \tag{7}$$

where T is temperature in degrees Celsius, and b and c are constants, and k_0 is a conductivity measured at 0°C and one atmosphere pressure. The constants b and c are determined for specific lithologies by a simple least-squares fit of the function in (7) to laboratory data.

Details of the thermal conductivity profile for this average lithosphere model are as follows. For the upper crust, a temperature coefficient b of 1.5×10^{-3} K^{-1} is selected (a value intermediate between experimentally determined values for granite and granodiorite). For the lower crust a temperature coefficient b of 1.0×10^{-4} K^{-1} is used. A pressure coefficient c of 1.5×10^{-3} km^{-1} is used throughout the crust. Zero depth and zero temperature values k_0 are 3.0 and 2.6 W m^{-1} K^{-1} for the upper and lower crust respectively. The relative effects of temperature and pressure depend on the particular geotherm and on depth in the crust. For the coolest thermal regime corresponding to a surface heat flow of 40 mW m^{-2}, the base of the upper crust attains a temperature of 199°C which lowers the surface thermal conductivity of 3.0 W m^{-1} K^{-1} to 2.37 W m^{-1} K^{-1}. The corresponding decrease in the upper crust associated with a 90-mW m^{-2} geotherm is from 3.0 to 1.74 W m^{-1} K^{-1}. In the lower crust, the temperature coefficient is sufficiently small that the temperature and pressure effects nearly cancel. For the 40-mW m^{-2} geotherm, pressure effects are greater and conductivity increases slowly from 2.61 to 2.64 W m^{-1} K^{-1} in lower crust. For the 90-mW m^{-2} geotherm, temperature effects dominate, and conductivity decreases from 2.53 to 2.50 W m^{-1} K^{-1} between 16 and 35 km. For the mantle portion of the lithosphere, the hypothetical mantle model of Schatz and Simmons (1972) is followed.

Although the structural and geochemical complexity inherent in the formation and consolidation of the continental lithosphere may well produce a complex vertical distribution of radioactive heat production, it is again necessary to derive a generalized model for the calculation of geotherms. However, any general model should adhere to the following constraints: (1) average levels of heat generation measured on surface felsic rocks cannot be sustained throughout the crust because the heat flow produced by such a vertical column would exceed the surface heat flow of 40 mW m^{-2} characteristic of shields; (2) a vertical distribution of heat generation in the crust should satisfy the linear heat flow–heat production relationship (although the vertical distribution may not be the sole cause of that relationship).

For the upper crust we use an exponentially decreasing heat production, $A(z) = A_0 \exp(-z/D)$, which is but one (but the most widely applied) of various heat production distributions which satisfies the linear heat flow–heat production relationship:

$$q_o = q_r + DA_0 \tag{8}$$

were q_o is surface heat flow, q_r is reduced heat flow and A_0 is surface heat generation. D is a parameter with dimensions of length and characterizes the depth distribution $A(z)$. In order to parameterize the geotherm calculation in terms of surface heat flow, we further use an empirical relationship that partitions the observed surface heat flow, with 40% being attributed to upper crustal radiogenic sources and 60% to deeper sources (Pollack and Chapman, 1977; Vitorello and Pollack, 1980). The crustal mechanism which causes this link between surface heat flow and deeper crustal heat flow is at best tenuous. The application of this partitioning serves to increase lower crust and upper mantle heat flow in regions of high surface heat flow and allows an easier generalization of the thermal parameters in our standard crustal section. Thus surface radioactivity A_0 can be related simply to surface heat flow using $A_0 = 0.4q_0/D$. The exponential distribution is continued downwards until the heat production falls to the lower crust heat production A_L at $z = D + \ln(A_0/A_L)$, or until the Conrad discontinuity (assumed at 16 km in the general model) is reached.

Specifying heat production in the lower crust is difficult. The actual value is best determined by comparison with probable lower crustal rock types and with measurements on lower crustal xenoliths. There appears to be substantial variability in lower crustal samples and in samples which may represent the lower crust: from 0.02 μW m^{-3} for metapelites found as xenoliths in the Massif Central, France (Dupuy et al., 1979) to values approaching 1 μW m^{-3} (e.g., Fountain et al., 1987). However, these extreme values are not likely to represent the entire lower crust. When rock types are weighted as to their probable abundance in the lower crust, then the range of average heat production values is narrowed to about 0.3–0.7 μW m^{-3}. This range is similar to the estimate of Nicolaysen et al. (1981) who calculated a lower crust heat production of 0.56 μW m^{-3} for the Vredefort area of South Africa based on an estimated lower crust composition of 50% leucogranofels and 50% mafic granulite. In this study the general model lower crust heat production is assumed to be 0.45 μW m^{-3}, similar to that assumed by Fountain et al. (1987) for the Pikwitonei–Sachigo crustal section. Finally, for the upper mantle section of the lithosphere a uniform heat production of 0.02 μW m^{-3} is assumed for the depleted upper mantle down to 200 km and 0.045 μW m^{-3} from 200 to 250 km, the depth limit for the calculation.

To determine specific geotherms for the continental crust, a crustal structure must be specified. In these generalized geotherms we have maintained a constant crustal thickness and constant layer thickness within the crust. The standard crust is 35 km thick, with the transition from upper crust to lower crust occurring at 16 km.

4. Framework geotherms

A family of steady-state, static geotherms for stable continental crust, parameterized as described above, is shown in Figure 5-1. The most important characteristic of this family of geotherms is the divergence from a common temperature at the surface. The geotherms will converge again when they are truncated by the mantle 1300°C adiabat, the deepest convergence to the adiabat being at about 250 km for the 40-mW m^{-2} geotherm (Chapman, 1986). Between these depths the limiting geotherms of 90 and 40 mW m^{-2} bracket a broad temperature–depth (equivalently *P-T*) space. Temperature differences at a common depth between these bracketing geotherms can be large: 500°C at 30 km depth and a maximum of 800°C at 60 km depth. Thus it is clear that a single continental geotherm is no more appropriate than a single oceanic geotherm has been. Instead, one must recognize the possible diversity of thermal states available for the continental crust and try to find the appropriate geotherm for a particular region of interest.

The geotherms in Figure 5-1 are furthermore characterized by roughly constant gradients throughout the upper crust, lower but changing gradients throughout the lower crust and slowly decreasing gradients throughout the mantle portion of the lithosphere. The upper crustal gradients remain almost constant because the effect of decreasing heat flow with depth (i.e. within a heat-producing medium) acting to decrease the thermal gradient is almost exactly counteracted by a decrease in thermal conductivity caused by temperature increase

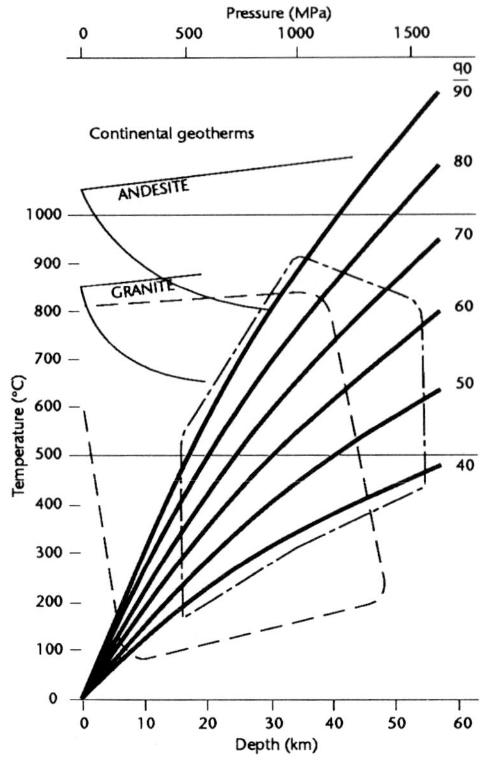

Fig. 5-1. Framework geotherms for the continental crust. Geotherms were computed assuming steady-state, static conditions, and conductive heat transfer. Geotherm parameter is characteristic surface heat flow q_0 in mW m^{-2}. Solid lines marked by labels *granite* and *andesite* show possible solidii brackets for wet (minimum) and dry (maximum) melting in the crust. Dashed line encloses *P-T* field appropriate for metamorphic rocks; dash-dot line polygon encloses steady-state *P-T* field of lower crust in stable regions. Lack of complete overlap in temperature space can be explained by dynamic tectonic processes. (Modified from Chapman, 1986)

that acts in the opposite sense to increase the thermal gradient. For high heat flow regions the thermal conductivity effect is pronounced and can lead to a maximum in the thermal gradient at the base of the upper crust.

Also shown in Figure 5-1 are representative solidii curves for the crust (cf. Mysen, 1981). The hottest thermal states characterized by 80 and 90 mW m^{-2} geotherms could provide for deep crustal melting under hydrous conditions. Dry melting in the crust seems to be precluded in regions where these steady-state static geotherms apply. Below the crust, the geotherms consistent with surface heat flow between 90 and 50 mW m^{-2} all exceed temperatures defined by a mixed volatile mantle solidus but the 45 and 40 mW m^{-2} geotherms do not. This is consistent with the observation that low heat flow regions such as shields do not have pronounced seismic low-velocity zones, which are thought to be caused by partial melting.

Finally, one should recall the assumption of partitioning the source of continental heat flow with 40% being ascribed to upper crustal radioactive heat production and 60% to deeper sources, be they radioactive, advective or heat liberated from cooling the interior of the Earth. Thus, the reduced heat flow values implicit in the Figure 5-1 geotherms are simply 0.6 q_0 and vary from 24 to 54 mW m^{-2} for the range of geotherms shown. Mantle heat flow is generally less than reduced heat flow (e.g., Furlong and Chapman, 1987); values of heat flow at a depth of 55 km for the Figure 5-1 geotherms vary from 17 to 50 mW m^{-2}.

Average crustal heat production values implicitly assumed for these geotherms are very nearly the difference between q_0 and q_{55} divided by the crustal thickness of 35 km used for computations. For most continental regions characterized by a heat flow of 50–70 mW m^{-2} the average crustal heat production assumed is 0.74–0.94 μW m^{-3}. These values fall within the range suggested independently for geochemical models of the continental crust (Smithson and Decker, 1974; Haack, 1983; Weaver and Tarney, 1984).

Alternative assumptions about reduced heat flow and radioactivity distributions can be made and these assumptions may lead to somewhat different geotherms. To cover all possible cases one would need many suites of geotherms for different combinations of reduced and surface heat flow. However, as our limited and imprecise knowledge of reduced heat flow precludes detailed knowledge, we feel the averaging assumptions which result in a single set of geotherms are justified. The sensitivity of calculated temperatures to assumed values of model parameters is given by Chapman (1986).

5. Processes and conditions that affect or modify the thermal state of the continental lower crust

The framework geotherms described in the previous section adequately describe the thermal state of most continental areas through much of their geologic history, but significant modifications of the steady-state thermal regime depicted by our framework geotherms must be made for active orogens. Several tectonic processes affect the thermal state of the crust; some processes create a transient thermal perturbation that decays once the orogenic process ceases, while other processes permanently modify the structure and composition of the crust, thereby altering its long-term thermal state.

The following sections explore the thermal effects of (1) burial by sedimentation, (2) exhumation by erosion, (3) burial by thrusting, (4) extensional unroofing, and (5) magmatic underplating. The thermal consequences of these processes for the lower crust are explored. The initial crustal structure for each of these models is the same as that used to produce the family of static geotherms.

The tectonic events which cause burial, unroofing, and crustal modification potentially have related deep seated thermal processes. We have not evaluated those in this analysis, but rather have focused on the consequences of the crustal modification itself.

5.1. Burial by sedimentation

Thick sequences of sedimentary rock formed in continental interiors or on continental shelf environments are evidence that significant areas of continental crust have experienced as much as 10 or 15 km burial. The thermal consequence for the underlying crust of such burial depends on the amount of burial, thermal conductivity and radiogenic heat production of the sedimentary fill, and the rates of burial. These effects are shown schematically in Figure 5-2. The framework geotherm (Fig. 5-2a) is selected from the family of geotherms shown in the previous section, appropriate for the background heat flow and thermal conditions existing at the onset of sedimentation. The curvature in the temperature–depth profile is caused primarily by the heat production distribution and thermal conductivity distribution as described in the previous sections.

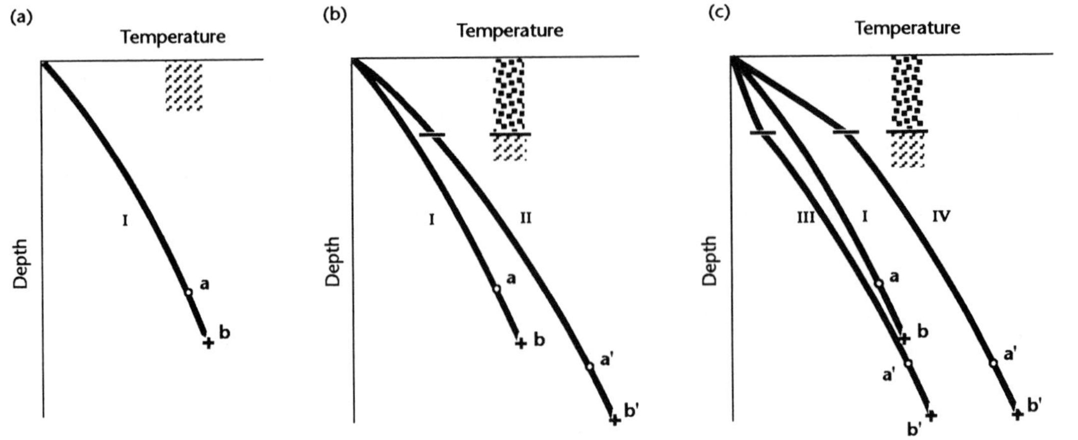

Fig. 5-2. Long-term thermal effects of burial by sedimentation. (a) Sample framework geotherm I. Characteristic temperature–depth profile is determined by surface heat flow and by thermal conductivity and heat production distributions. Points *a* and *b* depict *P-T* conditions at two depths in the lower crust.
(b) Effect of heat production. Sediment and basement have similar thermal conductivity and heat production. Increased total heat production after sedimentation creates a new and warmer geotherm II. Points *a* and *b* move to *a'* and *b'* in *P-T* space after burial and when transient effects of sedimentation have decayed.
(c) Effect of thermal conductivity. Sediments have no heat production, and sediments and basement have different thermal conductivity. High-conductivity sediments create a cooler new geotherm III, whereas low-thermal-conductivity sediments create a new, hotter geotherm IV after burial and when transient effects of sedimentation have decayed. Shaded bars indicate the extent of sediment burial and the location of sediment–basement interface.

Figures 5-2b and 2c show the long-term effects of sediment heat production and thermal conductivity, respectively. Heat production in the sediments increases total crustal heat production and thus produces a new and warmer geotherm. Points a and b (Fig. 5-2b) indicating initial *P-T* conditions within the lower crust prior to sedimentation are shifted to a' and b' respectively after transient thermal effects of sedimentation have decayed. It is important to note that in *P-T* space, the points a and b do not simply move down the geotherm, increasing in depth by the thickness of sediments, even when transient thermal effects and thermal conductivity contrasts are ignored. The thermal effect of different sediment thermal conductivity is illustrated in Figure 5-2c. For simplicity, a case is shown for negligible sediment heat production. Geotherm I is the reference framework geotherm appropriate to the prevailing heat flow prior to the onset of sedimentation. Geotherm III results when the sediment has a high thermal conductivity; the illustration shows sediment conductivity twice that of the basement as might exist, for example, in a clean, quartz-rich sandstone. High conductivity produces a low thermal gradient and a discontinuity in gradient at the sediment–basement contact to conserve vertical heat flow. In contrast, sediment with a low thermal conductivity produces a high thermal gradient through the sedimentary section as shown by geotherm IV. Temperature increases in the lower crust in response to low thermal conductivity cover can be significant. For example, a 5-km shaley section of basin fill with average thermal conductivity of 2 W m^{-1} K^{-1} deposited on basement with typical heat flow of 60 mW m^{-2} produces an increase of 150°C throughout the crystalline crust. The same burial event with a clean sandstone having a conductivity of 4 W m^{-1} K^{-1} produces an increase of only about 75°C. These are the qualitative and isolated end-effects of accumulating a sedimentary layer on top of a crystalline continental crust.

The complete effects of burial by sedimentation involve important transient effects that are significant not only during burial but also while the lithosphere recovers to a thermal steady-state condition. Sedimentation constitutes a downward advection of cold material from the surface. For sedimentation rates greater than ~0.25 mm y^{-1} (numerically equal to km Ma^{-1}) thermal diffusion cannot keep pace with the material advection, resulting in a depression of isotherms. As the entire lithosphere is in thermal imbalance after rapid sedimentation, the recovery period is governed by the time constant of the lithosphere (computed as $L^2/4\alpha$, where L is lithosphere thickness and α is thermal diffusivity; e.g., for a 80-km lithosphere the time constant is 50 Ma [$(80\ \mathrm{km})^2/(4 \times 32\ \mathrm{km}^2/\mathrm{Ma})$] rather than a much shorter time constant for a 5–10-km sediment layer alone.

Although analytic solutions for the transient thermal state exist for constant sedimentation rates and homogeneous material (e.g., Kappelmeyer and Haenel, 1974), more realistic basin evolution models simulating deposit of several sedimentary layers each with their own sedimentation rate and thermal properties must be solved by numerical methods. Simulation of burial by sedimentation, illustrating the full transient response of the lithosphere is shown in Figure 5-3. The simplest case of simple burial and no erosion is treated. Total burial of 10 km occurs at a constant rate of 0.5 mm y^{-1} over 20 Ma. The thermal conductivity of sediments is 2.6 W m^{-1} K^{-1}, equal to that of the basement. Sediment heat production is 1.0 μW m^{-3}, adding 10 mW m^{-2} to the equilibrium surface heat flow after burial. The

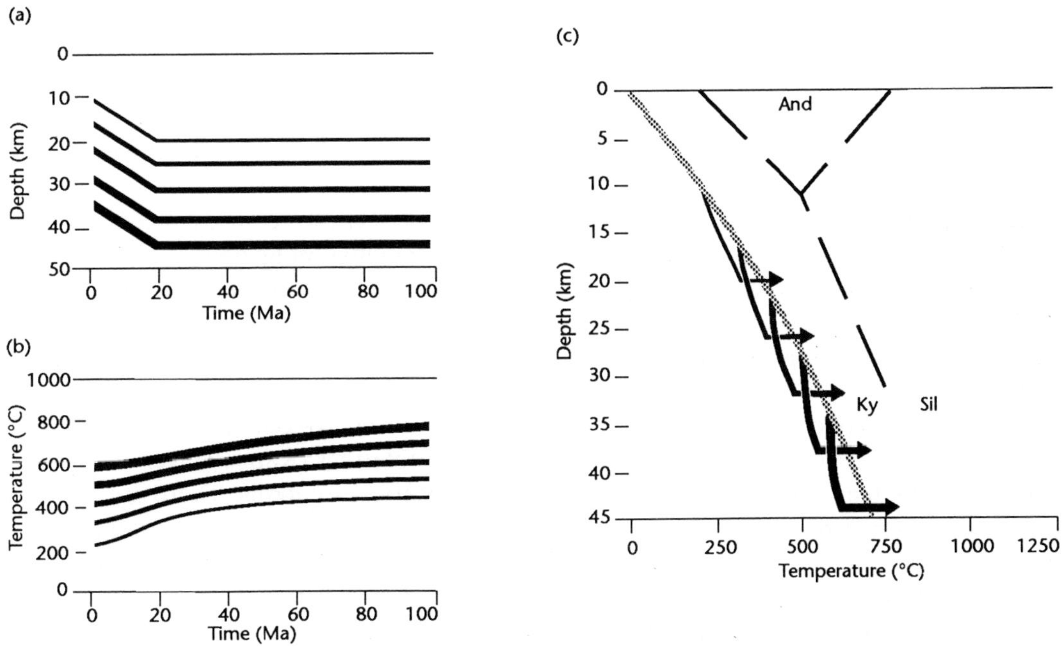

Fig. 5-3. Thermal effects of burial by sedimentation.
(a) Depth–time plot showing 10 km subsidence over a time span of 0–20 Ma; thermal computations are continued for an additional 80 Ma. Particle paths are traced for rock units at original depths of 10, 16, 22, 28, and 34 km.
(b) Temperature–time paths for the same rock units.
(c) Depth–temperature or P-T paths for the same rock units. The initial geotherm (dashed line) corresponds to a surface heat flow of 65 mW m^{-2}, sediment heat production is 1.0 μW m^{-3}, thermal conductivity of sediments and basement is 2.6 W m^{-1} K^{-1}, surface temperature is held at 10°C. Line weights indicate equivalent curves in each part of this figure.

transient thermal effect is shown both in Figure 5-3b as temperature–time plots for crustal positions initially at depths of 10, 16, 22, 28, and 34 km, and in Figure 5-3c for the same crustal positions in depth–temperature (or *P-T*) space. The transient thermal effect in this simulation is considerable. At the end of burial the entire lower crust is 150–160°C cooler than its respective equilibrium temperature. Note also that equilibrium temperatures in the lower crust are about 80°C higher than the initial geotherm due to the additional radioactive heat of the sediments that requires a new equilibrium geotherm. Response time for thermal recovery is more than 50 Ma because the entire lithosphere must readjust to a new equilibrium thermal state.

Further complexity in the transient thermal response of the crust to burial by sedimentation is caused by different sediment thermal conductivity. Low thermal conductivity sediments create high thermal gradients in the sedimentary section and result in final temperatures up to 140°C warmer than the initial geotherm at mid crustal depths. High thermal conductivity sediments in contrast, result in low thermal gradients in the sedimentary section and equilibriumn temperatures that are about 30°C less than the initial framework geotherm for their final depths in spite of the incremented crustal heat production.

In summary, burial by sedimentation produces characteristic *P-T-t* paths for lower crustal rocks. Depending on the thermal conductivity and heat production of the sediments, the final *P-T* conditions may either exceed or fall below the initial geotherm. Very slow sedimentation produces a pressure and temperature increase subparallel to the initial geotherm; rapid sedimentation forces the *P-T-t* path towards a two-stage process, a relatively isothermal pressure increase followed by isobaric heating. The transient thermal recovery from rapid sedimentation events requires several lithosphere time constants, on the order of 100 Ma for thick continental lithosphere.

5.2. Exhumation

Existence of vast tracts of rocks now at the surface that record pressures of 600–800 MPa (corresponding to depths of 22–30 km) is evidence that exhumation by uplift and erosion is a common geologic process (Bohlen and Mezger, 1989). As with the opposite process of burial, the thermal consequence of exhumation for the crust depends on the amount and rate of uplift and erosion, and on the thermal conductivity and heat production of the crustal section that is removed.

For simple cases of constant thermal properties and uniform erosion, the resulting transient thermal regime of the crust may be found from analytical solutions (Kappelmeyer and Haenel, 1974); more complex situations are simulated using numerical solutions. There are two end-member cases for exhumation: (1) instantaneous or very rapid uplift and erosion, and (2) very slow erosion. For instantaneous uplift and erosion the thermal event recorded in the crust would be one of nearly isothermal pressure decrease because the vertical advection of material exceeds cooling by heat conduction. Rapid cooling is experienced by rocks only just preceding removal by erosion as the cold surface temperature condition enhances heat loss through the near surface rocks. For very slow erosion rates (less than 0.1 mm y^{-1}) heat conduction keeps pace with heat advection and rocks experience a *P-T* path sub-parallel to the original geotherm. The final equilibrium geotherm following erosion is cooler than the original geotherm because a fraction of the crustal heat production will have been stripped.

An example of the thermal effects of exhumation intermediate between the two end-members described above is shown in Figure 5-4. In this case 10 km of crust is eroded over

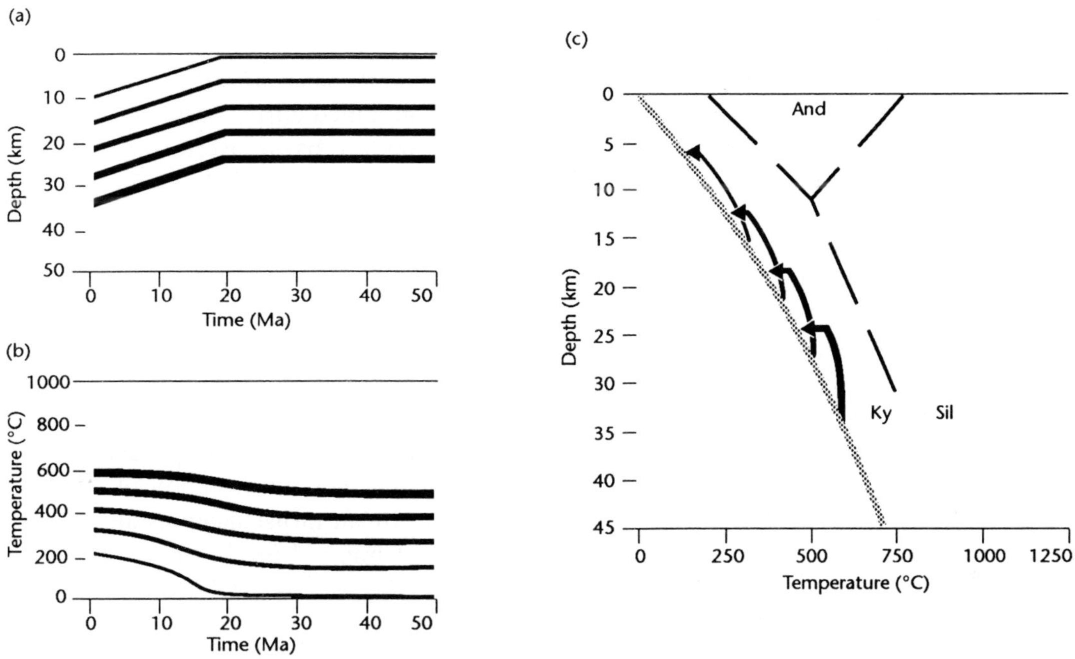

Fig. 5-4. Thermal effects of crustal exhumation by erosion.
(a) Depth–time plot showing removal of 10 km of material over a time span of 0–20 Ma; thermal computations are continued for an additional 30 Ma. Particle paths are traced for rock units at original depths of 10, 16, 22, 28, and 34 km.
(b) Temperature–time paths for the same rock units.
(c) Depth–temperature or *P-T* paths for the same rock units. The initial geotherm (shaded line) corresponds to a surface heat flow of 65 mW m^{-2}, surface heat production is 3 μW m^{-3}, thermal conductivity is 2.6 W m^{-1} K^{-1}, surface temperature is 10°C. Line wights indicate equivalent curves.

20 Ma at a rate of 0.5 mm y^{-1}. The paths of rocks initially at depths between 16 and 34 km are shown in depth–time space (Fig. 5-4a), in temperature–time space (Fig. 5-4b), and in depth–temperature or *P-T* space (Fig. 5-4c). The *P-T* path followed by a lower crustal rock unit is initially isothermal. At shallow depths, where the cool surface boundary is sensed, the *P-T* path assumes a form sub-parallel to the initial geotherm, although exceeding it by about 50°C; at lower crustal depths of 30–40 km the surface cooling effect is progressively lessened and the *P-T* paths reflect more closely an isothermal pressure release. For the simulation shown in Figure 5-4c the temperature of a sample at an initial depth of 34 km exceeds the initial geotherm by 120°C when erosion is stopped. At the cessation of erosion the lithosphere restores its thermal equilibrium slowly, the new geotherm being governed by sub-lithospheric heat flow and the new vertical distribution of thermal conductivity and heat production after erosional stripping. If enough time passes for thermal equilibrium to be reached, the curves shown in Figure 5-4c would cross the initial geotherm and define a new cooler geotherm. As shown in Figure 5-4, this effect requires more than 50 Ma.

5.3. Burial by thrusting

A second geological process that produces pressure and temperature increases in lower crustal rocks is burial by thrusting. Overthrusting produces similar thermal effects to those of rapid sedimentation, a rapid pressure increase followed by near isobaric warming. The heating phase, however, is commonly interrupted by the process of erosion as the crust is exhumed. Erosion is more common after thrusting than with basin development. Whereas the lithospheric structure and composition are commonly modified during basin development, creating a new isostatic equilibrium accommodating the mature basin, overthrust sheets are relatively thin-skinned phenomena that create a temporary isostatic imbalance before they are partially eroded.

Two cases of overthrusting are simulated. In both cases the overthrust is a repetition of the crustal structure and properties (k, A) of the initial crust. Figure 5-5a shows the thermal consequences for the lower crust of emplacing a 10-km thrust sheet instantaneously. An isothermal pressure increase of 260 MPa (equivalent to the load of a 10-km thrust sheet) is experienced, followed by isobaric warming (in the absence of erosion) to the new equilibrium geotherm. Note that the new geotherm is warmer than the initial geotherm because of radioactive heat production in the thrust sheet increasing the total heat production and thus heat flow. Rapid erosion of the thrust sheet alters the thermal evolution of the crust considerably. Figure 5-5a also portrays the case where the 10 km thrust sheet instantaneously emplaced is followed 1 Ma later by erosion at a rate of 0.5 mm y^{-1} for 14 Ma until 7 km of the thrust sheet has been eroded. Very little isobaric heating is now exhibited by the lower crust because of the rapid onset of erosion following thrusting. The little isobaric heating that is evident is most pronounced in the shallow part of the crust whose thermal recovery from overthrusting outpaces the deeper crust. At a depth of 25–40 km after thrust emplacement

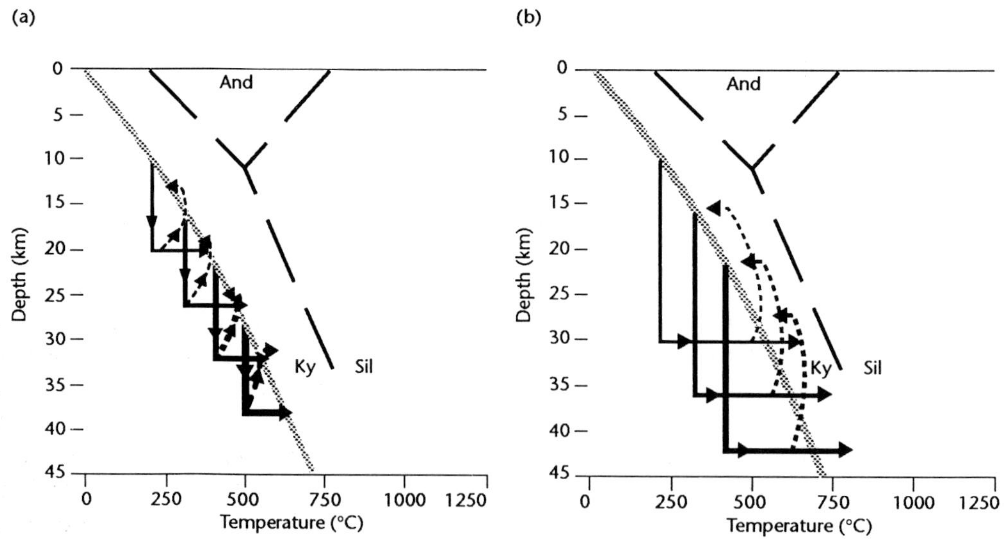

Fig. 5-5. Thermal effects of thrusting.
(a) Depth–temperature or P-T paths for the case of a 10-km thrust sheet emplaced instantaneously. Two post-thrust paths are shown for a 50-Ma period following thrusting: simple thermal relaxation without erosion, and thermal response if 7 km of the thrust sheet is eroded at a rate of 0.5 mm y^{-1} starting 1 Ma after thrusting (dashed line).
(b) A similar plot for the case of a 20-km thrust sheet emplaced instantaneously. Two post-thrust paths are shown for a 100-Ma period following thrusting: simple thermal relaxation without erosion, and thermal response if 15 km of the thrust sheet is eroded at a rate of 0.5 mm y^{-1} starting 30 Ma after thrusting (dashed line). The initial geotherm (dashed line) corresponds to a surface heat flow of 65 mW m^{-2}, surface heat production is 3 μW m^{-3}, thermal conductivity is 2.6 W m^{-1} K^{-1}, surface temperature is 10°C.

no isobaric heating is evident because that section of the crust has not significantly recovered thermally from overthrusting before erosional effects are seen.

The thermal consequences of a thicker overthrust package are shown in Figure 5-5b. Instantaneous emplacement of a 20-km thrust sheet with typical upper crustal material properties results in an isothermal pressure increase of 800 MPa followed by isobaric warming. In the absence of erosion, new equilibrium temperatures, consistent with the new and modified heat production distribution would be 100–150°C higher than the initial geotherm.

Inception of thrust sheet erosion at any time causes an abrupt departure from the isobaric warming with all rock units following a characteristic *P-T-t* path whose details reflect the initial depth, erosion rate, and delay between thrusting and onset of erosion. Figure 5-5b shows the case of a thicker thrust sheet and a longer interval between thrusting and onset of erosion. A 20-km thrust sheet is instantaneously emplaced and after 30 Ma of thermal relaxation, 15 km of the thrust sheet is eroded at 0.5 mm y^{-1}. The *P-T-t* paths in Figure 5-5b show a characteristic cycle of an isothermal pressure increase, followed first by isobaric warming, and finally by near-isothermal pressure decrease. If erosion is avoided, the lower crust warms in 100 Ma to temperatures 120°C warmer than the initial geotherm because new equilibrium conditions accommodate the additional heat production in the thrust sheet. If the post-thrust thermal relaxation is interrupted by uplift and erosion prior to temperatures regaining the original geotherm (Fig. 5-5b), the vertical advection of heat accompanying uplift causes the *P-T-t* paths to cross and exceed the original geotherm by 120°C also. These cases of overthrusting illustrate the point (e.g., Fowler and Nisbet, 1982; Hodges and Royden, 1984; Thompson and England, 1984) that lower crustal rocks may follow a wide variety of *P-T-t* paths. Pressure–temperature conditions recorded in rocks may result from transient excursions in *P-T* space and not reflect any of the framework geotherms directly. It is also important to note that it is difficult even in cases of relatively thick overthrusts to place lower crustal rocks into the sillimanite field. Pre-thrusting heat flow within the crust must significantly exceed 65 mW m^{-2} and there must be a substantial lag between thrusting and significant erosion in order for the *P-T-t* path to cross into the silliminite field.

5.4. Extensional unroofing

Erosion is just one of several processes that bring deeply buried rocks to shallow levels. In extension or rift zones, crustal rocks are brought to shallow levels by the extension itself. The extensional mechanism may vary from one extensional environment to another. For example, in the Basin and Range Province of Utah, Arizona, and Nevada, a model has been proposed for an extensional mechanism which is almost the inverse of thrusting. In this proposed mechanism (Wernicke, 1985) lower crustal rocks have been brought to the surface along low-angle detachments that may cut through the crust and perhaps the lithosphere (Fig. 5-6a). Rocks now exposed in the extensional allochthons were once buried at depths as much as 20 km. The unroofing of these rocks along low-angle normal faults records a thermal history that differs from that experienced during simple erosion or the thrust cycle. It is important to be able to make this distinction because a major point of contention regarding the evolution of these metamorphic core complexes (the extensional allochthons of Fig. 5-6a) is whether the metamorphism recorded in the rocks occurred during the extensional phase, or was inherited from a previous thrust cycle. In addition, if the thermal fingerprint did develop during the

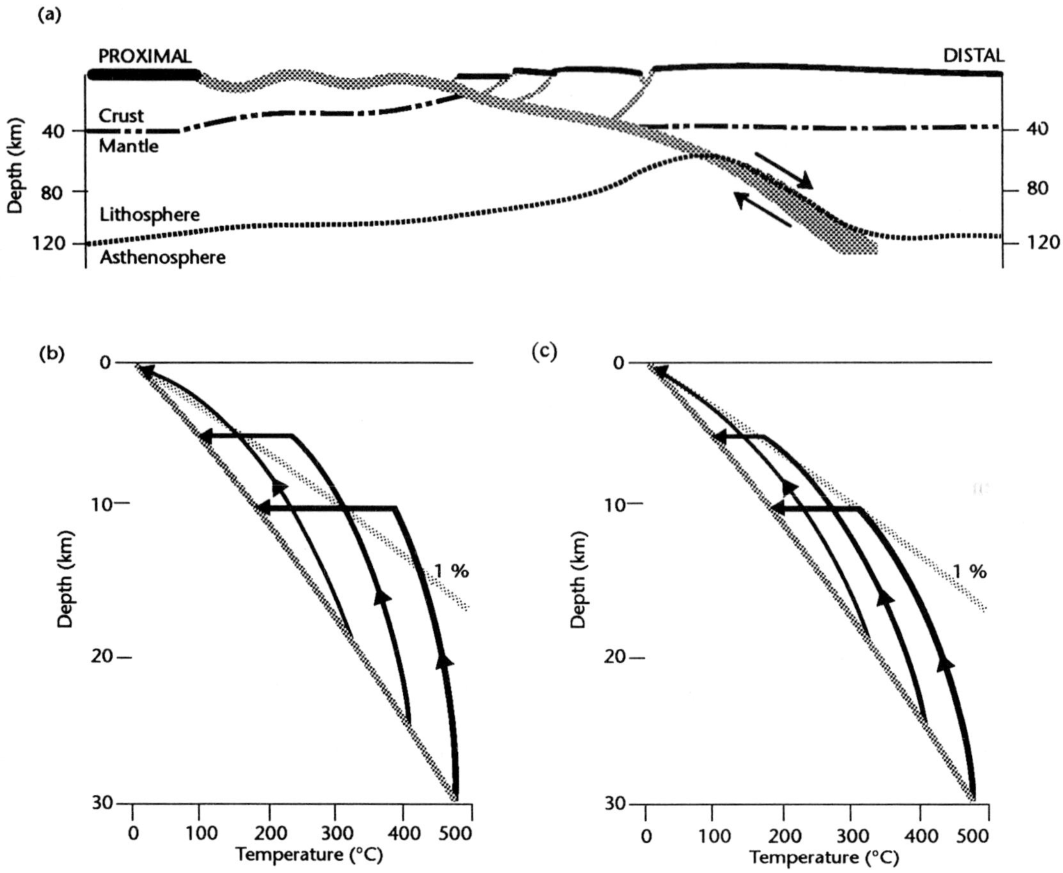

Fig. 5-6. Thermal effects of extensional unroofing.
(a) Schematic cross-section for proposed extensional unroofing mechanism of Wernicke (1985). Rocks exposed may contain the thermal fingerprint of the extensional event.
(b) Temperature–depth paths for rocks at different crustal levels during extensional unroofing event. Unlabeled curve is framework geotherm; curve labeled 1% is the modified geotherm if the region is undergoing steady-state extension at a rate of 1% per million years. Duration of extension event is 5 Ma.
(c) Same as (b) but duration of extensional event is 15 Ma.

extensional phase, it may contain information on the rates and duration of the unroofing event.

Figure 5-6b shows the thermal signature, or *P-T* paths followed by crustal rocks brought to shallower levels in the crust by this extensional mechanism. Although the rocks overshoot the equilibrium geotherm, as in the case of the thrust cycle, the two types of fingerprints are distinct. In this extensional event, the rocks continually cool during unroofing, while in the thrust cycle there was initial heating. The amount by which the temperatures overshoot the geotherm is a function of the uplift rate. Thus if we have knowledge of the equilibrium geotherm, the thermal fingerprint can place limits on the extensional rates. The metamorphic data from these "core complexes" of the western U.S. needed to determine this uplift rate (that is to distinguish between the thermal histories in Figs. 5-6b and c) is not yet available, but continued improvements in the geobarometers and geothermometers used may soon provide the needed resolution.

Other thermal models have been proposed for extensional regimes (e.g., Lachenbruch et al., 1985) which also provide testable *P-T-t* histories to be validated by observations of crustal structure and petrology.

5.5. Magmatic underplating

Mafic magma emplaced at the base of the continental crust has long been considered an important process in crustal formation, growth, and modification [see discussions in Furlong and Fountain (1986) and McKenzie and Bickle (1988)]. As far as the thermal effects on crustal rocks are concerned, magmatic underplating can be simulated by a thermal model shown schematically in Figure 5-7a. Mantle material rising along an asthenosphere adiabat moves into the supersolidus field, generating melt. This melt is assumed to collect ultimately at the base of the crust; the quantity of melt produced along the upwelling path and the resulting underplate thickness are given in figures 4 and 5 of Furlong and Fountain (1986). After emplacement the crustal underplate cools, giving up both its latent heat and sensible heat (product of heat capacity and its excess temperature above surroundings) to the overlying crust.

Thermal modification of the crust in response to magmatic underplating depends primarily on the position in the crust and the depth of magmatic accretion. Simulations of magmatic underplating at a depth of 35 km are shown in Figures 5-7b–d. The accumulated melt thickness was 12.5 km, near the maximum melt that can be generated by mantle upwelling from an initial depth of 125 km. The latent heat is liberated over 5 Ma.

We first demonstrate the thermal effects of pure underplating and then combine underplating with subsidence or uplift and erosion of the crust depending on the density changes that the uderplating may have caused. In the pure underplating case (Fig. 5-7b) the lowermost crust experiences isobaric heating of 250°C; the magnitude of temperature change decreases monotonically towards the surface. Rocks below 25 km may be sufficiently heated to reach the sillimanite stability field. Figure 5-7c illustrates the thermal effects of combining underplating with subsequent burial. In this case the lowermost crust, adjacent to the magmatic intrusion, senses the heating event long before the thermal effects of burial have propagated from the surface. In contrast, at shallow crustal levels the burial thermal effect preceeds the underplate heating and the *P-T-t* path subsequently never exceeds the new equilibrium geotherm. If underplating is accompanied by uplift and erosion, the *P-T-t* path exhibited by lower crustal rocks is markedly changed (Fig. 5-7d). Both processes produce a higher-temperature transient geotherm. Rocks at intermediate depths of 15–25 km may be heated by as much as 200–300°C above the original geotherm before cooling as they are brought towards the cool surface.

Additionally, as shown in Figure 5-7d, with crustal underplating and unroofing some sections of the lower crust cross into the sillimanite field. The production of middle and lower crustal granulite facies rocks will be aided by underplating. Additionally, the greater the heat flow prior to the underplating event or the shallower the emplacement of the underplate, the more effective this mechanism will be in generating granulites.

A process related to magmatic underplating which also causes a perturbation of the thermal state of the crust is the emplacement of sills within the crust, a process referred to as crustal inflation. Compared to magmatic underplating of an equal volume of melt, crustal inflation produces higher temperature local heating at the sill margins but shorter lived thermal events (Armstrong and Chapman, 1990). Figure 5-8a shows the thermal effects of emplacing a 7.5-km sill at a depth of 15 km. The sill loses heat (latent heat and sensible heat) in both vertical

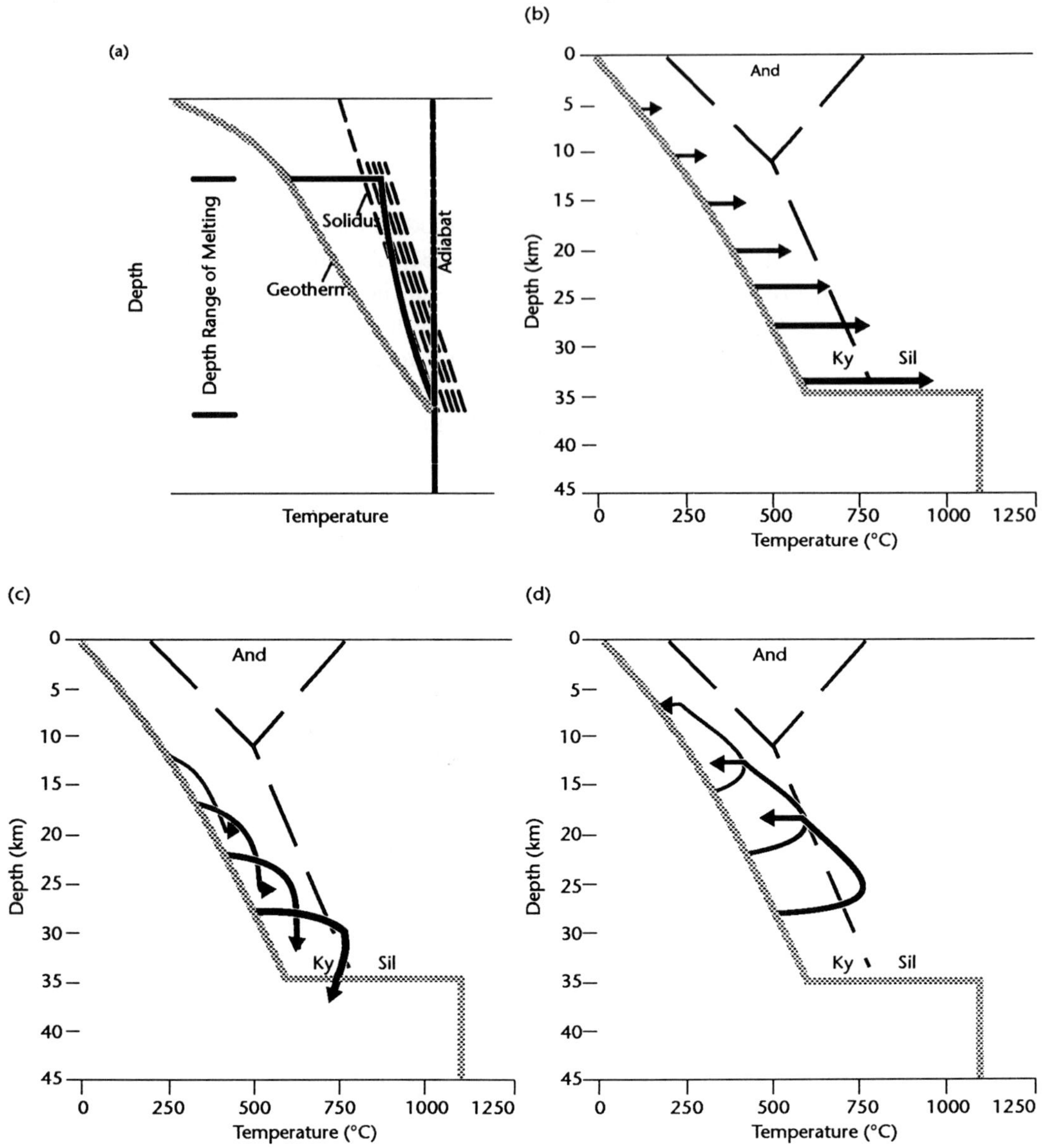

Fig. 5-7. Thermal effects of magmatic underplating.

(a) Conditions governing the melt fraction and temperature for crustal magmatic underplating. Bold line shows schematic temperature–depth profile for mantle material rising adiabatically. Adiabat, solidus, and preemplacement geotherm are labeled; unlabeled dashed lines represent contours of constant melt fraction in temperature–depth space. (After Furlong and Fountain, 1986).

(b) Depth–temperature plot showing the thermal effects of sub-crustal magmatic undeplating on the crustal thermal regime. The inital geotherm (dashed line) corresponds to a surface heat flow of 65 mW m^{-2} and includes the step temperature increase at the underplated layer. A partially molten layer 12.5 km thick is emplaced at the base of a 35-km-thick crust and is allowed to give up its latent heat over 5 Ma. The simulation is continued for 50 Ma.

(c) Underplating followed immediately by 10 km burial in 20 Ma, other parameters unchanged.

(d) Underplating followed immediately by 10 km erosion in 20 Ma, other parameters unchanged.

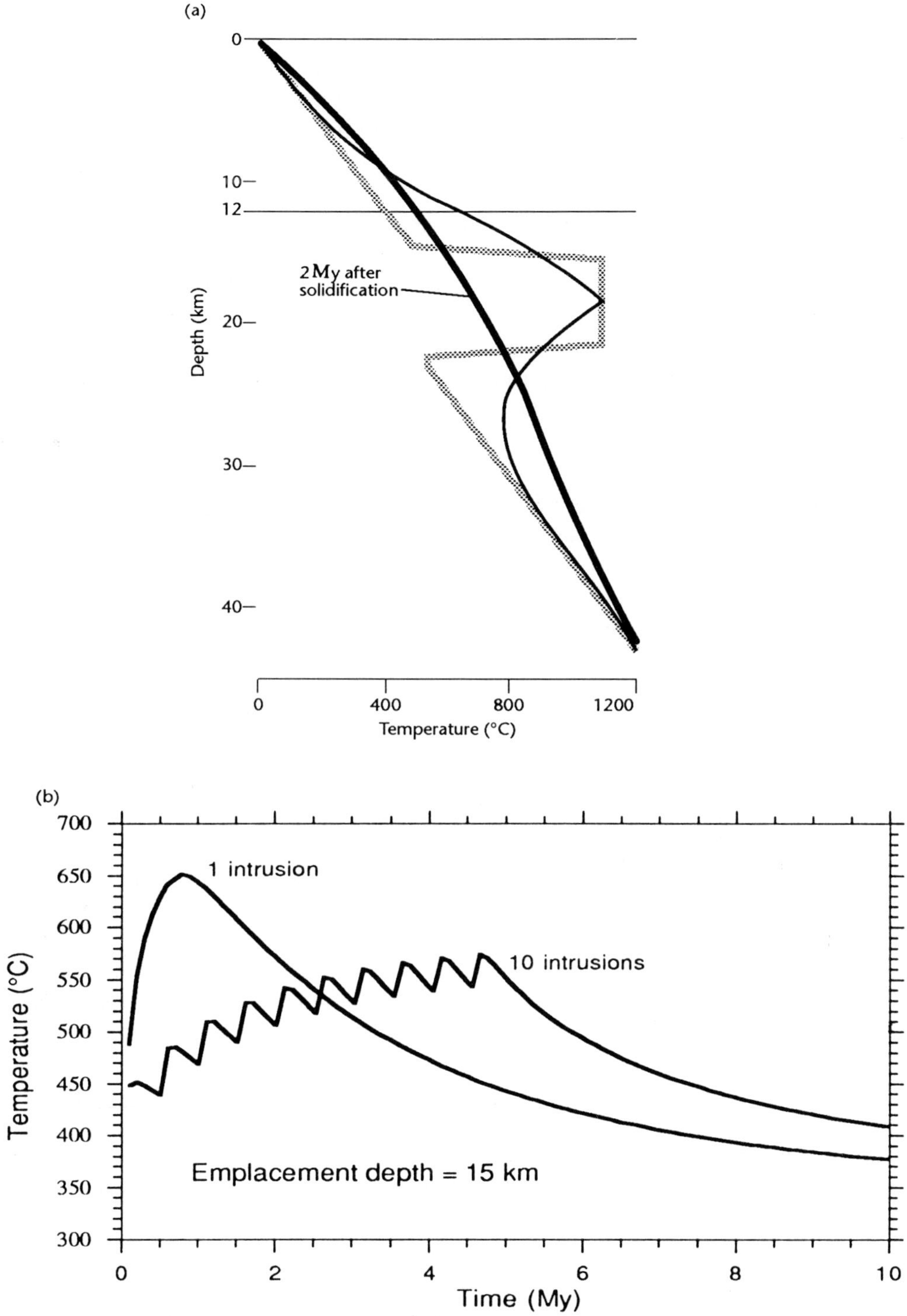

Fig. 5-8. Thermal effects of magmatic inflation of the crust.

(a) Temperature–depth plot for a 7.5-km-thick sill emplaced 15 km below the surface. Three curves show the temperature field at emplacement, when the sill has just solidified, and 2 Ma after solidification.

(b) Temperature–time plots for crust at 12 km depth showing the effect of multiple intrusions forming a 10 km sill.

directions, heating the country rock at the sill margins by about 400°C. The heating effect is diminished with distance from the sill and, for this simulation, is negligible after a few million years. Figure 5-8b indicates that heating can be prolonged by emplacing the sill gradually with repeated intrusions, but the magnitude of crustal heating is lessened at any position because the early pulses of magma have cooled significantly before the last pulses are emplaced.

6. Crustal heterogeneities

Dynamic processes such as crustal extension, compression and volcanism produce important transient thermal perturbations to the steady-state thermal regime of the continental crust. But these processes also modify the crust, structurally and compositionally. If the structural modification results in changed thermal conductivity and/or changed heat production distributions, the long-term or quasi steady-state thermal regime may also be modified.

Temperatures in the lower crust are very sensitive to heat flow because temperature gradients and temperatures must adjust to conduct all of the sub-crustal heat flow as well as heat generated within the crust to the Earth's surface. The variation in temperature at 40 km depth between continental rift zones (characterized by heat flow of 90 mW m^{-2}) and cratons (characterized by heat flow of 40 mW m^{-2}) is 600°C (Fig. 5-1). If an upper crust whose radiogenic heat production integrates to a heat flow of 30 mW m^{-2} is thickened (e.g., doubled) by a sequence of thrusting, then the thrust terrain could have its heat flow altered from 60 to 90 mW m^{-2}. Such an enhancement of heat flow would produce a 300°C increase throughout the lower crust (Fig. 5-1).

Two- and three-dimensional compositional heterogeneities may also produce strong lateral temperature variations within the continental crust. Figure 5-9 shows a generic model of a cratonic region and the thermal consequences of heterogeneities in thermal properties. The particular section shown in Figure 5-9a was adapted from published crustal sections by Smithson and Brown (1977). Thermal properties were assumed for representative rock units (see table 1 of Furlong and Chapman, 1987) with the geometry shown in Figure 5-9b. The steady-state and temperature fields were found by numerical methods. Figure 5-9c indicates that temperatures in a compositionally heterogeneous crust may vary by as much as 100°C at mid-crustal levels over a horizontal distance as small as 50–75 km, even where background heat flux is constant. Furlong and Chapman (1987) and Fountain et al. (1987) have further shown that crustal heterogeneity such as shown in Figure 5-9 can lead to an apparent linear heat flow–heat production relationship without a regular step-wise, linear, or exponential decrease of heat production with depth as has often been supposed (Roy et al., 1968; Lachenbruch, 1970). Aquisition of detailed heat production and thermal conductivity profiles through deeply exposed sections of continental crust such as accomplished by Fountain et al. (1987) provide an important contribution in resolving the extent and thermal consequenses of crustal heterogeneities.

7. Conclusion

The thermal state of the lower crust is controlled by a variety of factors. The background conditions are dependent on crustal composition and resulting thermal parameters (conductivity, heat production) and also the heat flux from the mantle. We have shown a family of

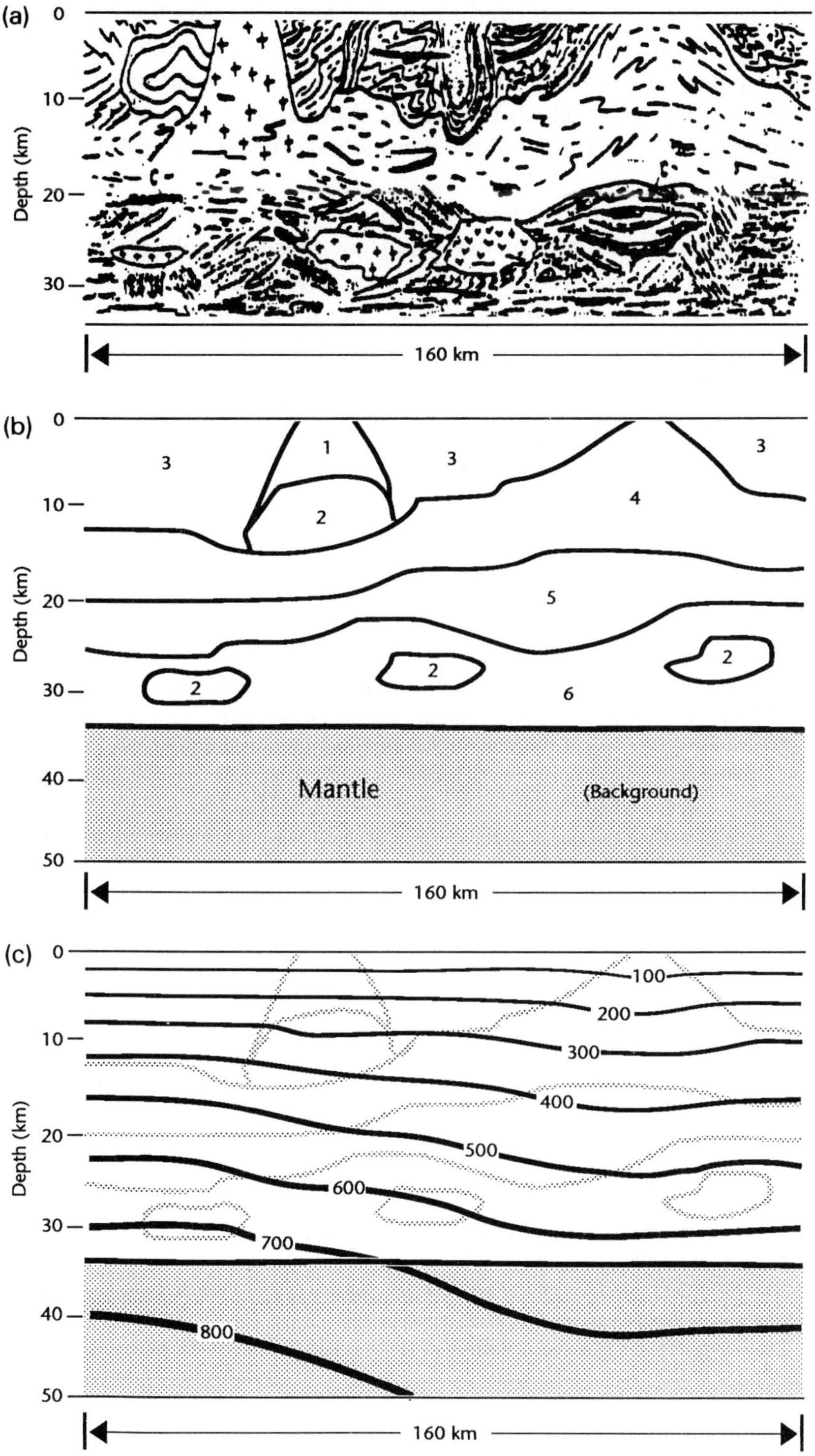

Fig. 5-9. Thermal effects of crustal heterogeneities.
(a) Imaginative cross section of continental crust (after Smithson and Brown, 1977).
(b) Crustal section used in thermal model. Units are: *1* = enriched granite; *2* = depleted granite; *3* = metasediment; *4* = enriched migmatite; *5* = depleted migmatite; *6* = mafic granulite. Heat production and thermal conductivity details given in Furlong and Chapman (1987).
(c) Thermal cross section. Dark lines are isotherms labeled in °C; lighter lines separate lithologies.

static continental geotherms parameterized by surface heat flow which can serve as a framework for evaluations of crustal thermal regimes. Of critical importance in crustal regimes is the transient response of the crust to tectonic events. The relaxation time following the cessation of most tectonic events which modify the thermal structure of the crust is 100 Ma or longer. As a result, very few regions of the continental crust are in truly static steady-state thermal conditions.

The effects of tectonic events on the thermal structure have been evaluated using simplified models of the tectonism. Important in the results of these modeling studies is the large deviations from equilibrium geotherms which lower crustal rocks experience during even relatively simple tectonic events. Additionally, these tectonic events leave a permanent record in that rarely is the final equilibrium geotherm identical to the starting geotherm.

The modeling results also indicate that there are only a few conditions of lower crustal thermal structure and tectonism which can potentially lead to the development of granulite terrains. The widespread occurrence of such regions requires further evaluation of the conditions under which they develop and the role of various tectonic events in the metamorphism.

References

Armstrong, P.A. and Chapman, D.S., 1990. Can counterclockwise *P-T-t* paths in regional granulite terranes be produced by sill intrusion and underplating? EOS, Trans. AGU, 71: 1618.

Blackwell, D.D., 1971. The thermal structure of the continental crust. In: The Structure and Physical Properties of the Earth's Crust. Geophys. Monogr. Ser., Am. Geophys. Union, 14: 169–184.

Bohlen, S.R. and Mezger, K., 1989. Origin of granulite terranes and the formation of the lowermost continental crust. Science, 244: 326–329.

Cermak, V. and Rybach, L., 1982. Thermal conductivity and specific heat of minerals and rock. In: G. Angenheister (Editor), Landolt-Bornstein Numerical Data and Functional Relationships in Science and Technology. New Series, Group V, vol. 16, Springer–Verlag, Berlin.

Chapman, D.S., 1986. Thermal gradients in the continental crust. In: J.B. Dawson et al. (Editors), The Nature of the Lower Continental Crust. Geol. Soc. Spec. Publ., 24: 63–70.

Clark, S.P., Jr. and Ringwood, A.E., 1964. Density distribution and constitution of the mantle. Rev. Geophys., 2: 35–88.

Dupuy, C., Leyreloup, A. and Vernier, J., 1979. The lower continental crust of the Massif Central (Bournac, France) with special references to REE, U and Th composition, evolution, heat flow production. Phys. Chem. Earth, 11: 401–415.

Fountain, D.M., Salisbury, M.H. and Furlong, K.P., 1987. A heat production model for the Superior Province crust based on the Pikwitonei–Sachigo continental cross section, central Manitoba: implications for the thermal structure of Archean crust. Can. J. Earth Sci., 24: 1583–1594.

Fowler, C.M.R. and Nisbet, E.G., 1982. The thermal background to metamorphism II. Simple two-dimensional conductive models. Geosci. Can., 9: 208–214.

Furlong, K.P. and Chapman, D.S., 1987. Crustal heterogeneities and the thermal structure of the continental crust. Geophys. Res. Lett., 14: 314–317.

Furlong, K.P. and Fountain, D.M., 1986. Continental crustal underplating: thermal considerations and seismic-petrologic consequences. J. Geophys. Res., 91: 8285–8294.

Haack, U., 1983. On the content and vertical distribution of K, Th, and U in the continental crust. Earth Planet. Sci. Lett., 62: 360–366.

Hodges, K.V. and Royden, L., 1984. Geologic thermobarometry of retrograded metamorphic rocks: an indication of the uplift trajectory of a portion of the Northern Scandinavian Caledonides. J. Geophys. Res., 89: 7077–7090.

Kappelmeyer, O. and Haenel, R., 1974. Geothermics with Special Reference to Application. Geoexploration Monograph 1, Gebruder Borntraeger, 238 pp.

Kay, R.W. and Kay, S.M., 1981. The nature of the lower continental crust: inferences from geophysics, surface geology, and crustal xenoliths. Rev. Geophys. Space Phys., 19: 271–297.

Lachenbruch, A.H., 1970. Crustal temperature and heat production: Implications of the linear heat flow relation. J. Geophys. Res., 75: 3291–3300.

Lachenbruch, A.H. and Sass, J.H., 1977. Heat flow in the United States and the thermal regime of the crust. In: J.G. Heacock (Editor), The Earths Crust –Its Nature and Physical Properties. Geophys. Monogr., Am. Geophys. Union, 20: 503–548.

Lachenbruch, A.H., Sass, J.H. and Galanis, S.P., 1985. Heat flow in southernmost California and the origin of the Salton Trough. J. Geophys. Res., 90: 6709–6736.

McKenzie, D.P. and Bickle, M.J., 1988. The volume and composition of melt generated by extension of the lithosphere. J. Petrol., 29: 625–679.

Mysen, B., 1981. Melting curves of rocks and viscosity of rock-forming melts. In: Y.S. Touloukian, W.R. Judd and R.F. Roy (Editors), Physical Properties of Rocks and Minerals. Vol. II, McGraw-Hill/CINDAS Data Series on Material Properties. McGraw-Hill, New York, NY, pp. 361–407.

Nicolaysen, L.O., Hart, R.J. and Gale, N.H., 1981. The Vredefort radioelement profile extended to supracrustal strata at Carletonville, with implications for continental heat flow. J. Geophys. Res., 86: 10,653–10,661.

Pollack, H.N. and Chapman, D.S., 1977. On the regional variation of heat flow, geotherms and lithosphere thickness. Tectonophysics, 38: 279–396.

Roy, R.F., Blackwell, D.D. and Birch, F., 1968. Heat generation of plutonic rocks and continental heat flow provinces. Earth Planet. Sci. Lett., 5: 1–12.

Roy, R.F., Blackwell, D.D. and Decker, E.R., 1972. Continental heat flow. In: E.C. Robertson (Editor), The Nature of the Solid Earth. McGraw-Hill, New York, NY, pp. 506–543.

Roy, R.F., Beck, A.E. and Touloukian, Y.S., 1981. Thermophysical properties of rocks. In: Y.S. Touloukian, W.R. Judd and R.F. Roy (Editors), Physical Properties of Rocks and Minerals. Hemisphere Publishing, New York, NY, pp. 409–502.

Rybach, L., 1981. Geothermal systems, conductive heat flow, geothermal anomalies. In: L. Rybach and L.J.P. Muffler (Editors), Geothermal Systems: Principles and Case Histories. Wiley, Chichester, pp. 3–36.

Rybach, L. and Cermak, V., 1982. Radioactive heat generation in rocks. In: G. Angenheister (Editor), Landolt-Bornstein Numerical Data and Functional Relationships in Science and Technology. Group V, Vol. 1a, Springer-Verlag, Berlin, pp. 353–371.

Schatz, J.P. and Simmons, G., 1972. Thermal conductivity of Earth materials at high temperatures. J. Geophys. Res., 77: 6922–2983.

Sclater, J.C. and Francheteau, J., 1970. The implication of terrestrial heat flow observations on current tectonic and geochemical models of the crust and upper mantle of the Earth. Geophys. J. R. Astron. Soc., 20: 609–542.

Smithson, S.B. and Decker, E.R., 1974. A continental crustal model and its geothermal implications. Earth Planet. Sci. Lett., 22: 215–225.

Smithson, W.B. and Brown, S.K., 1977. A model for the lower continental crust. Earth Planet. Sci. Lett., 35: 134–144.

Thompson, A.B. and England, P.C., 1984. Pressure-temperature-time paths of regional metamorphism II. Their inference and interpretation using mineral assemblages in metamorphic rocks. J. Petrol., 25: 929–955.

Vitorello, I. and Pollack, H.N., 1980. On the variation of continental heat flow with age and the thermal evolution of the continents. J. Geophys. Res., 85: 983–995.

Weaver, B.L. and Tarney, J., 1984. Empirical approach to estimating the composition of the continental crust. Nature, 310: 575–577.

Wernicke, B., 1985. Uniform-sense simple shear of the continental lithosphere. Can. J. Earth Sci., 22: 108–125.

Chapter 6

Rheology of the lower crust

E.H. RUTTER and K.H. BRODIE

1. Introduction

The rocks of the lower continental crust experience a wide range of physical environments, principally dependent upon the thickness of the crustal section and the heat-flow regime, and encompassing some of the most extreme pressure/temperature conditions to which quartzo-feldspathic and some metabasic rocks are subjected. Additionally, the rock types span a wider range of mineralogical compositions than those of the oceanic crust and upper mantle. Deformation may occur under dry conditions with a virtual absence of pore fluid, or it may be accompanied or perhaps facilitated by coeval prograde or retrograde metamorphic transformations. During devolatilization reactions, pore-fluid pressures may approach lithostatic pressures. Given the foregoing, a wide range of rheologic behaviour is to be expected, hence it may be unwise to try to generalize too far about 'the rheology of the lower continental crust'. It is important to remember that in any large-scale tectonic event involving the continental crust, whether it be collisional (possibly requiring stacking of crustal slices), extensional (leading to the formation of sedimentary basins or even oceanic crust), or strike-slip (leading to the assembly of tectonic 'terranes'), the entire thickness of the lithosphere must fail. An understanding of lower crustal rheology is therefore essential to the understanding of the dynamics of large-scale tectonic processes.

Although the nature of the rock-types themselves and the physical environment of deformation exercise important controls on the mechanical response of the lower crust, the mode of rock failure (whether it is localized into bands or uniformly distributed) can depend strongly on strain-dependent changes in rock microstructure. One should not presume that the concept of 'steady-state' flow, a term used to describe the behaviour of rocks in laboratory experiments, is applicable to large deformations in nature.

In this chapter we review the information pertinent to lower crustal rheology that has been gathered from experimental rock mechanics. We also consider what can be inferred from the structural and microstructural study of rocks naturally deformed under lower crustal conditions, and how inferences can be made about aspects of lower crustal rheology from the observation of the response of the continental crust to loads (topographic or buried) placed upon it. Finally, we examine attempts to use rheological data to erect theoretical and analog models of tectonic processes and discuss how a consideration of rheological properties of the lower crust influences the interpretation of deep crustal soundings by geophysical methods.

2. Deformation mechanisms

Three main categories of rock-deformation mechanism may be recognized, although it must be noted that there are some variants which can blur the distinctions between them:

(a) Cataclastic deformation, which involves some combination of brittle fracture and frictional sliding of grains or rock fragments relative to one another. There is no internal distortion of the crystal structure of minerals, hence void-space creation must accompany cataclasis and rock strength must increase with effective mean pressure. Effective mean pressure is the average value of the normal stress components on the rock minus the pore-fluid pressure. Upper crustal faulting is the most important geological expression of cataclastic deformation, but distributed strains of a few percent often precede fault localization.

(b) Intracrystalline plasticity, in which distortion of crystal structure occurs through the glide motion of crystal dislocations. If sufficient slip systems are available to satisfy strain-compatibility conditions, a polycrystal can deform at constant volume, hence the process is largely pressure-insensitive. The flow stress is, however, reduced by increasing temperature or decreasing strain rate. Finer-grained rocks are usually slightly stronger than their coarser equivalents. Plastic strains of a few tens of percent result in the formation of crystallographic preferred orientation. At elevated temperatures (above about 0.3 of the absolute melting temperature at geological strain rates) large plastic strains are accompanied by dynamic recrystallization, in which the new grains are commonly smaller than the parent grains. In the case of the latter type of flow, or when the resistance to plastic flow is sensitive to pore-fluid chemistry, finer-grained rocks may be markedly weaker than their coarser equivalents.

(c) Flow by diffusive mass transfer, which proceeds by stress-induced diffusion of matter from grain-boundaries sustaining a high normal stress to boundaries subjected to low normal stress. A grain-shape fabric may be produced but crystallographic preferred orientation generally does not result. The process is favoured by high temperatures (because diffusion is temperature-sensitive) and small grain diameters (the distance between source and sink for diffused matter). The process ideally takes place at constant volume and hence is pressure-insensitive. Sliding displacements across grain boundaries must accompany diffusive mass transfer in order to maintain strain compatibility. At high temperatures and small grain sizes, surface tension forces maintain equant grain shapes and a stable flow at constant microstructure can occur. This flow has been likened to the granular flow of an oily sand, but large strains are achieved by grain-boundary sliding accommodated by some combination of diffusive mass transfer and intracrystalline plasticity in the grain-boundary regions.

Each of the above-mentioned deformation mechanisms is characterized by a diagnostic set of microstructures and features of mechanical behaviour. In naturally deformed rocks, however, changes in environmental conditions, deformation rate or strain-induced microstructural changes can result in the microstructural imprint of more than one type of deformation mechanism being preserved. On the other hand, accompanying effects of metamorphism involving microstructural and/or mineralogical changes may partially or wholly obscure microstructural evidence for the dominant deformation mechanism.

In what follows we will frequently refer to 'grain size-sensitive flow', a term used to describe flow when the strength decreases with decreasing grain size, but without necessarily implying the mechanism by which it occurs.

3. Data from experimental rock mechanics

3.1. Limitations of experimental data

Rocks of the lower continental crust experience mean pressures in the range 500–1500 MPa and temperatures in the range 200 to perhaps more than 900°C, depending on the crustal thickness and geothermal gradient. In experimental studies it has been common to operate at the higher end of the temperature range, in order to enhance deformation kinetics over the experimentally accessible time scale. The utility of this device is limited, however, according to whether solid-state mineral reactions, phase changes or melting occur in the sample under the higher-temperature conditions.

Experimentation aimed at determining constitutive flow laws for rocks in the temperature range 600 to 1000°C at pressures of 500 MPa and above has usually been carried out using the solid-confining medium technique (see review of techniques by Tullis and Tullis, 1986). With this technique, lateral expansion of the specimen during deformation is inhibited by the confining medium. This factor, coupled with high piston friction and the need to measure the differential force on the sample from a load cell outside the pressure vessel, means that the technique suffers from poor differential stress resolution at low stress levels. This is further compounded by the development of relatively large thermal gradients across the test specimen. Use of an internally heated gas-medium testing machine (e.g., Paterson, 1970; Heard, 1985) in the above temperature range, or even at higher temperatures, eliminates these problems, but it is far more difficult to operate such a machine routinely at pressures above 500 MPa. Because mean pressure does not strongly influence the resistance to plastic flow of most materials, provided one still operates within the stability field of the phase or phases of interest, it is possible to obtain meaningful results at these lower pressures. Extra care has to be taken to ascertain whether significant cataclastic deformation (deformation involving fractures) occurs in addition to plastic flow. Cataclastic deformation can be suppressed by the application of sufficiently high hydrostatic pressure.

A major deficiency of all experimental studies is that meaningful mechanical data can only be obtained over relatively small strains, typically up to 20 or 30%. As the (typically cylindrical) specimen deforms, the constraint of the loading pistons causes most strain to accumulate in the centre of the test piece, so that the specimen becomes barrel-shaped. At strains above about 25% the central part of the specimen begins to extrude around the loading pistons, so that strain becomes strongly heterogeneous. In data-processing it is usual to assume homogeneous deformation at constant volume in order to calculate the differential stress on the sample and the strain rate. Thus, with increasing strain the shape of the stress-strain curve and the calculated strain rate become progressively less satisfactory representations of the real behaviour of the material.

One of the principal aims of experimental rock-deformation studies is to obtain constitutive flow laws for particular deformation mechanisms, from which extrapolations to geological temperature, pressure and strain-rate conditions can be made. Experimental data from a wide range of materials at high temperature (above about half of the absolute melting temperature) and at strain rates where thermally activated recovery processes are significant, are usually interpreted in terms of power-law creep of the form:

$$\dot{e} = A \exp(-H/RT)(\sigma^n/d^m) \tag{1}$$

in which $\dot{e}$ is strain rate, H is the activation enthalpy for flow, R is the gas constant, T is absolute temperature, σ is differential stress for 'steady-state' flow, d is grain size and n, m and A are material constants. As m varies from zero the flow becomes progressively more grain-size-sensitive. Flow by diffusive mass transfer processes is grain-size-sensitive with m about 2 or 3. For flow by intracrystalline plasticity m is expected to be about -0.5, but little is known of circumstances under which m may take a larger value.

For power-law creep, when the deformation rate is controlled by the kinetics of dynamic recovery processes such as dislocation climb or dynamic recrystallization, the stress exponent, n, is often about 3 or 4. A value of this order is expected if dislocation climb velocity is linearly proportional to stress, and free dislocation density (and hence strain rate) varies as the square of stress. Higher values of n (up to 7 or 8) may arise if geometric effects lead to a concentration of the local stress value at the point in the substructure at which creep rate is determined (relative to the externally measured stress on the sample) (Weertman, 1968). Smaller values of n (down to 1) are associated with creep controlled by processes at grain boundaries, such as flow by diffusive mass transfer.

At high stress levels, power-law creep breaks down and mechanical data are usually better described by a relationship of the form:

$$\dot{e} = B\exp(-H/RT)\exp(D\sigma)/d^m \tag{2}$$

in which B and D are empirical constants. This is similar to power-law creep in which n increases with increasing stress, but in this case strain rate increases more rapidly with stress than would be characteristic of power-law creep. It is commonly inferred that in this regime the glide motion of dislocations in their slip planes controls the overall creep rate. Tsenn and Carter (1987) have reviewed power-law breakdown characteristics for geological materials, and conclude that for carbonates it occurs at a differential stress of about 10^{-3} G (on the order of 200 MPa), and for many silicates at about 10^{-2} G (on the order of 1000 MPa), where G is the shear modulus. Under conditions such that cataclastic deformation is easier than plastic flow (e.g., up to a few hundreds of MPa and ca. 500°C at laboratory strain rates) the strengths of silicate rocks are very sensitive to effective mean pressure. These conditions arise in the upper crust, or under any conditions where pore-fluid pressure approaches the overburden (lithostatic) pressure. Mean pressure also affects the resistance to plastic flow (to a much lesser extent) as a result of small volume changes accompanying the formation and motion of crystal defects (e.g., Poirier, 1985). The effect for dry rocks is commonly small over the pressure ranges of crustal deformation, but a complete constitutive flow law properly should incorporate mean pressure as a variable. Borch and Green (1987) argued that pressure affects creep rate of minerals according to its effect on the solidus temperature of the phase concerned.

Although flow laws of the above types have some physical basis, and over the limited ranges of experimental investigations reasonably satisfactory fits of these types of law can usually be obtained, extrapolations outside the experimental range should always be regarded with due scepticism. It should be remembered that other mathematical descriptions of the data, which may give radically different extrapolations, can generally be equally well fitted. A more complete description of the flow behaviour would include a function describing the structural state of the material in addition to grain size (e.g., grain shape, phase composition, dislocation density and geometric configuration, crystallite orientation distribution, etc.). These parameters generally vary, some slowly, some rapidly, with strain. The description of transient creep requires the incorporation of a 'structure factor' into the flow stress/strain rate

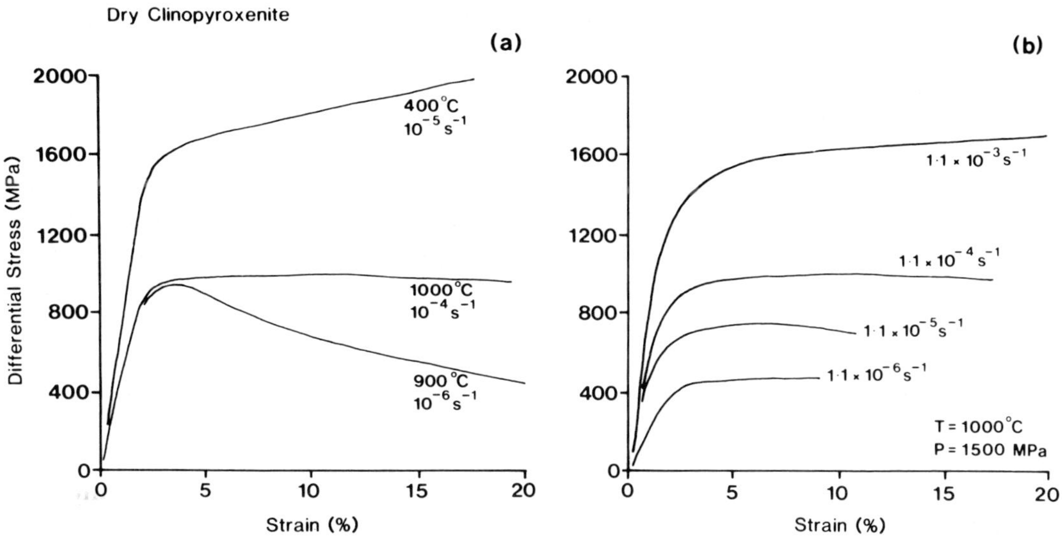

Fig. 6-1. Examples of stress/strain curves from experimental rock deformation studies on dry clinopyroxenite at 1500 MPa confining pressure. Strain rate is shown beside each curve. Data obtained using the solid confining medium technique by Kirby and Kronenberg (1984).

(a) Curves showing a range in behaviour from low-temperature (400°C), strain-hardening plastic flow, through 'steady-state' flow (flat-topped curve at 1000°C) to strain-softening behaviour (900°C). In the latter case, a shear zone formed, in which the grain-size had been dynamically reduced.

(b) Example of the effect of reducing the strain-rate at constant temperature (1000°C) and confining pressure. Data like these are used to obtain constitutive flow laws, which can be extrapolated to 'natural' strain-rates, as illustrated in Fig. 6-2.

relationship. The omission of a structure factor from the flow law implies truly steady-state flow that, although it might be appropriate for the small strains induced in experiments, is rather more questionable for the large strain ranges of natural deformations.

In natural deformation by intracrystalline plasticity, significant microstructural changes occur with large strains. These include development of a grain-shape fabric with attendant crystallographic preferred orientation, and tectonic grain-size reduction by dynamic recrystallization. True 'steady-state' flow requires constant strain rate, constant stress, constant temperature and constant microstructure. During plastic deformation microstructure commonly changes only slowly with strain, so that 10 or 20% strain in an experiment may not be accompanied by a large microstructural adjustment. The stress-strain curve can therefore be flat-topped, a characteristic of 'steady-state' flow, over the experimental strain range (Fig. 6-1). Lack of stability of the microstructure with respect to a large strain is not always apparent from experimental studies, but may be of great importance for extrapolation to large natural deformations (shear strains from 5 to 5000+). Although constitutive flow laws for a range of rock types undergoing 'steady-state' flow have been reported (see summaries by Kirby and Kronenberg, 1987; Carter and Tsenn, 1987) and even more widely applied in the construction of mathematical models of the deformation of the lithosphere (e.g., Meissner and Strehlau, 1982; Houseman and England, 1986; White and Bretan, 1985; White et al., 1986; Strehlau and Meissner, 1987; Kusznir and Park, 1986), their applicability to nature is subject to limitations dependent upon the degree of microstructural stability, quite apart from the errors of measurement intrinsic to the testing technique used. The problem of the effects of strain-dependent microstructural evolution on flow stress must be addressed in future mechanical studies.

In the summary given below of experimental data on rocks and minerals likely to be of importance in the lower continental crust, the above limitations should be borne in mind. We summarize data from which extrapolations to geological conditions can be made, on quartz, feldspars, ortho- and clinopyroxenes, and olivine. These minerals occur either as substantial monomineralic rock bodies or in sufficient modal proportions in polymineralic rocks that they are presumed to constitute the principal load-supporting or most easily deformable phase.

In Figure 6-2, flow stresses from experimentally obtained constitutive flow-law data have been extrapolated to a constant strain rate of 3×10^{-14} s^{-1} and plotted against temperature, so that the relative strengths of different rock types can be compared at a geologically realistic rate. The shapes of these curves depend both on H and on n. The extrapolations have been carried out without attempting to normalize to constant pressure or to a particular geobaric gradient. The extrapolation is several orders of magnitude in strain rate beyond the range of conditions over which experimentation was carried out. However, the deformation microstructures of the experimental materials generally correspond to those observed in naturally deformed rocks, and the relative strengths of different phases in naturally deformed polyphase rocks inferred from microstructural studies usually conform to relative strengths inferred from flow-law extrapolation. These facts provide some degree of confidence in the qualitative validity of the extrapolations.

3.2. Quartz

Quartz is of special importance in crustal deformation by virtue of its common occurrence, frequently comprising the load-supporting framework of grains in the rock, and because many microstructural studies of naturally deformed rocks show it to be the most deformable phase. As a result, a large amount of effort has been expended in studies of quartz plasticity. Early attempts to deform quartz by intracrystalline plasticity in mechanical experiments were unsuccessful, until the adaptation of the solid-confining medium experimental technique for deformation studies (Griggs and Blacic, 1964, 1965; Griggs, 1967). Griggs and Blacic (1965) showed that synthetic quartz crystals, containing structure-bound hydroxyl groups at concentration levels of several 1000 $H/10^6$ Si deformed by plastic flow at differential stresses below 1000 MPa at temperatures above about 400°C. Through this and later studies it was established that the greater the OH content the greater the weakening, and that natural quartzites could be rendered weak by deformation in the presence of pore water, but *only* at high (>1000 MPa) confining pressures (Griggs and Blacic, 1965; Hobbs et al., 1972; Tullis et al., 1973, 1979; Kekulawala et al., 1981). Such 'water' or 'hydrolytic' weakening was assumed to be important in the natural deformation of quartz, because water is assumed generally to be available during natural rock deformation under a wide range of crustal conditions. The apparent pressure-sensitivity of the water-weakening effect is far greater than the amount of pressure-sensitivity usually associated with deformation by intracrystalline plasticity. It is not clear to what extent the sensitivity owes to the effect of pressure on the properties of environmental water or to volume changes associated with the formation of intracrystalline defects.

Continued efforts to induce water-weakening at lower (300 MPa) confining pressures in gas-medium testing machines, have not been generally successful (Kekulawala et al., 1978, 1981; Paterson, 1989), although the reverse effect, of causing hardening through precipitation

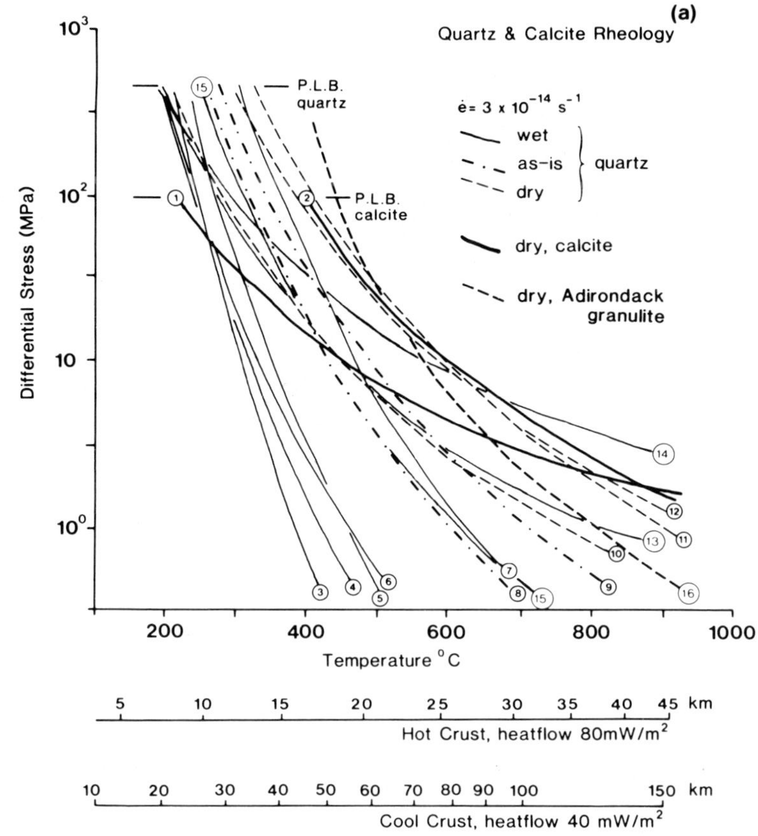

Fig. 6-2. Assembly of experimentally obtained constitutive flow law data, extrapolated to a geologically representative strain-rate of 3×10^{-14} s^{-1}. Flow stress is plotted against temperature. Extrapolations are made for the confining pressure at which the experiments were carried out. The effect of pressure variations is not likely to exceed the variability between experimental studies on similar rock types. Scales are also shown for the temperature/depth relations expected for heat-flows of 40 (cool lithosphere) and 80 mW m^{-2} (hot lithosphere). For the former the lower crust is expected to be strong and elastic, and for the latter it is expected to be weak and plastic. *P.L.B.* = Power Law Breakdown stress (after Tsenn and Carter, 1987).
(a) Data for quartz rocks, granitic rocks and diorite (in which quartz is likely to have been the weakest phase). Data for calcite rocks are shown for comparison.

Data sources:

1	Yule Marble	Heard and Raleigh, 1972	dry	9	Heavitree quartzite	Jaoul et al., 1984	as-is
2	Carrara Marble	Schmid et al., 1980	dry	10	Simpson quartzite	Koch, 1983	dry
3	Heavitree quartzite	Jaoul et al., 1984	wet	11	Heavitree quartzite	Jaoul et al., 1984	dry
4	Heavitree quartzite	Jaoul et al., 1984	wet	12	Westerly granite	Hansen and Carter, 1983	dry
5	Quadrant quartzite	Hansen, 1982	wet	13	Quartzite (average)	Paterson and Luan, 1990	wet
6	Westerly granite	Hansen and Carter, 1983	wet	14	Synthetic quartzite	Paterson and Luan, 1990	wet
7	Quartz diorite	Hansen, 1982	wet	15	Simpson quartzite	Koch et al., 1989	wet
8	Heavitree quartzite	Jaoul et al., 1984	as-is	16	Adirondack granulite	Wilks and Carter, 1989	dry

of molecular water by heat treatment, is easily produced (Kirby and McCormick, 1979). The experimental data gathered indicate the complexity of the phenomenon of the incorporation of water and related point defects in quartz, and there remains no general model for the solubility in quartz of such defects and of the way that this depends on temperature and

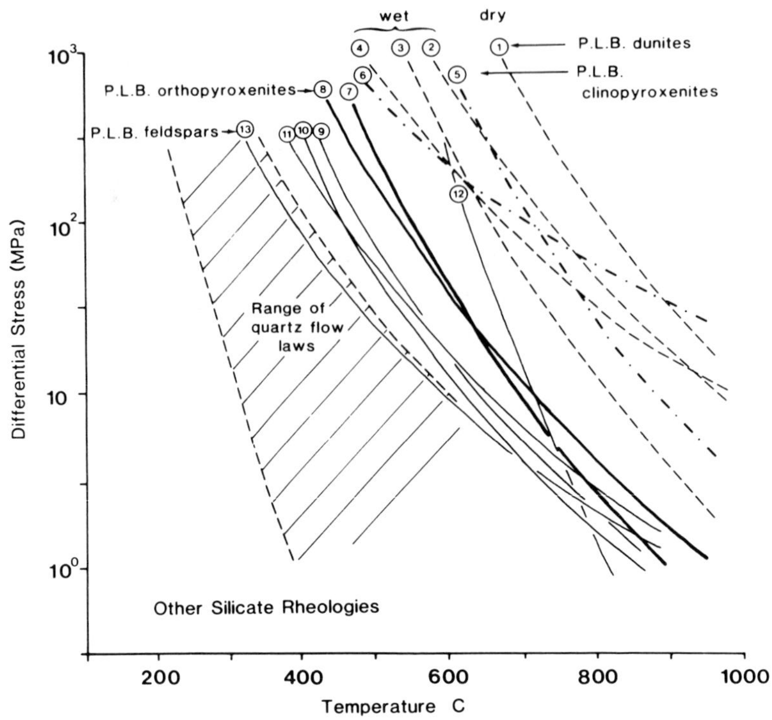

Fig. 6-2 (continued).
(b) Data for feldspar rocks, orthopyroxenite, clinopyroxenite and dunite. The field of quartz behaviour is shown for comparison.

Data sources:

1	Aheim and Anita Bay dunite	Chopra and Paterson, 1981	dry	8	Orthopyroxenite	Ross and Nielsen, 1978	wet
2	Aheim dunite	Chopra and Paterson, 1981	wet	9	Maryland diabase	Caristan, 1982	dry
3	Anita Bay dunite	Chopra and Paterson, 1981	wet	10	Hale albite	Shelton and Tullis, 1981	dry
4	Mt Burnet dunite	Post, 1977	wet	11	Anorthosite	Shelton and Tullis, 1981	dry
5	Clinopyroxenite	Boland and Tullis, 1986	wet	12	Albite (uniaxial)	Ji and Mainprice, 1986	dry
6	Clinopyroxenite	Kirby and Kronenberg, 1984	dry	13	Enfield aplite	Shelton and Tullis, 1981	dry
7	Orthopyroxenite	Raleigh et al., 1971	dry				

pressure. On a laboratory time scale, the diffusion of water into quartz appears to be too slow for equilibrium to be attained except in very fine-grained materials (Paterson, 1989; Gerretsen et al., 1989; Paterson and Luan, 1990), and incorporation of water during some experiments depends on infiltration along microfractures or incorporation during grain-boundary migration. There now appears to be a consensus that the water-weakening effect correlates with the presence of small clusters of molecular water within the quartz structure (Aines and Rossman, 1984; Aines et al., 1984). The theories of the mechanism(s) of hydrolytic weakening fall into two groups:

(a) Water hydrolyses silicon-oxygen bridges in dislocation cores and hence facilitates dislocation glide, as originally envisaged by Griggs and Blacic (1965). Dislocation velocity is therefore controlled by the rate of diffusion of water-related defects in quartz.

(b) Water-related defects act as dopants that, by accepting or donating electrons, influence the concentration of charged defects, such as kinks on dislocation lines, which are required for plastic deformation (Hirsch, 1981; Hobbs, 1981; Ord and Hobbs, 1986).

Experimental evidence has been adduced in support of both types of interpretation. Linker and Kirby (1981) and Linker et al. (1984) demonstrated slip anisotropy in single crystals of synthetic quartz, with easiest slip being in the <*c*> direction in a prism plane. These results are consistent with observations made in an earlier study by Hobbs et al. (1972). Linker et al. (1984) pointed out that oxygen diffusion shows a similar anisotropy (Giletti and Yund, 1984; Dennis, 1984), and suggested that plasticity depends on water-diffusion-controlled dislocation glide. By means of solid-medium testing machine experiments in which oxygen activity was controlled in sealed capsules using solid-phase oxygen buffers, Ord and Hobbs (1986) showed that the resistance to plastic flow was significantly decreased as either water or oxygen activity in the environmental fluid was increased. Constitutive flow laws for quartz are summarized by Kirby and Kronenberg (1987), Carter and Tsenn (1987) and Paterson and Luan (1990), and these are represented graphically in Figure 6-2 as a plot of log flow stress vs. temperature. Jaoul et al. (1984) presented experimental data which showed that in progressing from vacuum dried, through increasing amounts of water encapsulated with experimental runs, the general level of resistance to plastic flow, the activation enthalpy and the stress exponent value decreased.

Water-weakening of initially dry quartz appears to be an exclusively high-pressure phenomenon (at least over laboratory-accessible strain rates) and plasticity of dry quartz can only be studied without the complication of cataclastic deformation at very high pressures. As a result, the available experimentally determined constitutive flow laws for quartz and quartzites (excluding synthetic quartz and quartzites) have been obtained from solid-medium machines. Confining pressures of ca. 1500 MPa which have typically been used for the above studies correspond to burial depths of ca. 50 km, yet from microstructural studies on naturally deformed rocks quartz is clearly very plastic even at confining pressures on the order of 400–500 Mpa (e.g., White, 1976). Therefore, is quartz hydrolytically weakened during deformation under mid-crustal conditions? For the equilibration of the particular water-related defects in the quartz structure, is there a trade-off between water pressure and time, so that in lower-pressure experiments to date such equilibration has not been achieved (Kekulawala et al., 1981)?

In the geological literature it has certainly become fashionable to presume that during natural rock deformation quartz is hydrolytically weakened, but clear evidence is usually lacking (but see Carter et al., 1990). Linker et al. (1984) showed that in hydrolytically weakened synthetic quartz $(2\bar{1}\bar{1}0)$<0001> was the easiest slip system. In the natural deformation of quartz, however, it is almost universally reported that the preferred orientation patterns produced are controlled by easy <*a*> slip on a variety of slip planes, including basal, first-order prism and rhombs (Bouchez, 1978; Schmid and Casey, 1986) (Fig. 6-3). Does this mean that in the continental crust hydrolytic weakening is unimportant, or is the effect of water on <*a*> slip rate-dependent? Mainprice et al. (1986) comment on the rarity of preferred orientations produced by dominant <*c*> slip, and describe two natural occurrences, both from apparently high-temperature (>650°C) and possibly hydrous environments. Based on an extrapolation of existing diffusion data, Paterson and Luan (1990) argue that equilibration of quartz in medium-grained rocks with respect to intergranular water probably does occur over geological time scales at temperatures above about 300°C, but the extent to which it happens in particular instances (and hence the flow behaviour) may depend upon the equilibration mechanism.

As shown on Figure 6-2, extrapolation of the available flow laws (at a constant pressure of 1500 Mpa) would leave wet quartz extremely weak over the temperature range expected

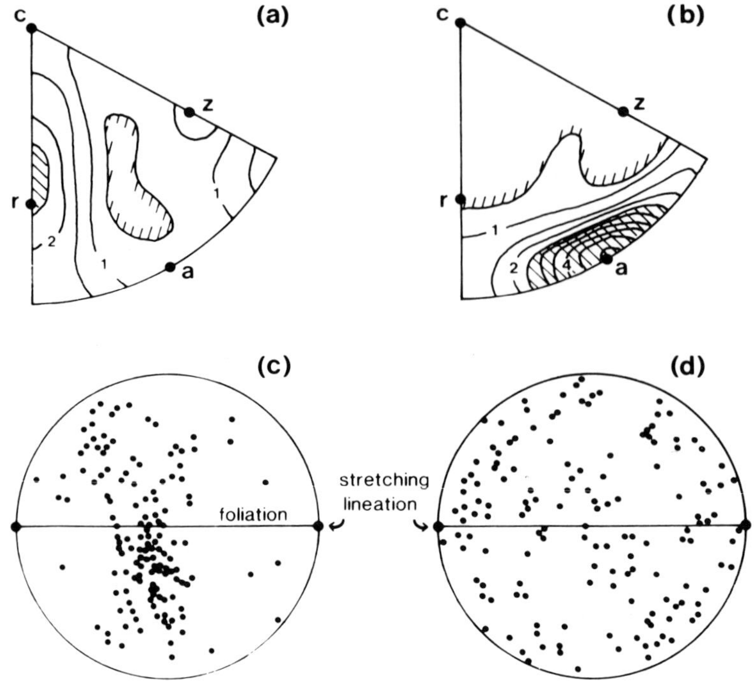

Fig. 6-3. Aspects of preferred crystallographic orientation development in deformed rocks.
(a) and (b) show inverse pole figures for the direction normal to a shear zone and parallel to the shear direction in a quartz mylonite from a shear zone at Cap de Creuz, northwestern Spain (after Schmid and Casey, 1986). Contours are at intervals of ×0.5 uniform concentration. The pole to r tends to lie parallel to the shear zone normal and the intracrystalline slip vector a parallel to the shearing direction. The alignment of the easiest slip vector parallel to the shear direction may have caused fabric-softening of the rock within the shear zone.
(c) and (d) show the patterns of *c* axis preferred orientation of quartz in a sheared rock. (c) shows the strong fabric of the large quartz porphyroclasts, and (d) the lack of preferred orientation in the fine-grained, recrystallized matrix (after Behrmann and Mainprice, 1987). The authors interpreted these patterns to mean that whilst the coarse grains were deforming by intracrystalline plasticity, the finer-grained matrix was flowing by diffusion-accommodated grain-boundary sliding, leading to weakening and hence localization of the flow into the shear zone.

of lower crustal depths. It would be weaker than calcite marble deforming by dislocation creep! The extrapolation of the marble data shown in Figure 6-1 is probably the most reliable for any rock type, and would be expected to apply to a coarse-grained marble, so that grain-size-sensitive flow processes do not supervene. Environmental water does not appear to affect the high-temperature rheology of calcite rocks (Rutter, 1974). Structural relations (e.g., folds, boudins) between marble and quartz-rich rocks can often be observed in high-grade metamorphic rocks and marbles generally appear weaker than the quartz-rich rocks.

Paterson and Luan (1990) re-examine critically the scatter of existing experimental rheological data for dry and wet quartz and conclude that they are indistinguishable statistically with respect to values of stress exponent and activation enthalpy. They attribute differences in behaviour to varying degrees of equilibration with respect to intergranular water, according to the experimental methods used, and infer that these differences affect mainly the value of the pre-exponential term A. They draw attention to the possible effects of small amounts of grain-boundary melting in some studies, tending to reduce the apparent n value, and hence the strength when extrapolated to geological strain rates. They present two new flow laws which they consider could represent the behaviour of quartz equilibrated with respect to water:

$$\log \dot{e} = -7.12 - (135\ \text{kJ}/RT) + 3.1 \log \sigma \tag{3}$$

based on pre-existing data, and

$$\log \dot{e} = -9.39 - (135\ \text{kJ}/RT) + 4.0 \log \sigma \tag{4}$$

from the behaviour of polycrystals fabricated from silicic acid. These are plotted on Figure 6-2. They show a more satisfactory behaviour compared to calcite rocks when extrapolated to geological strain rates, and may represent a lower bound on the behaviour of quartzites in nature.

3.3. Olivine

Olivine rheology is expected largely to determine the mechanical behaviour of the upper mantle, and probably also that of the olivine-rich ultramafic bodies of various sizes and shapes commonly found in exposed lower crustal sections. However, it is the comparative behaviour of olivine-rich rocks with that of more typical lower crustal rocks that is of particular significance here. Owing to its importance as a major constituent of the upper mantle a great deal of effort has been expended during the last 20 years to determine the mechanical properties of olivine.

The high solidus temperature of dry olivine means that it is one of the strongest of the common rock-forming silicate minerals at a given temperature. Like quartz, its resistance to plastic deformation is affected by the presence of environmental water, but this is not expected to be of great importance in many natural environments because fluid inclusions in olivine rocks are usually dominated by CO_2 (e.g., Roedder, 1965; Green and Gueguen, 1983). The flow of iron-bearing olivines is enhanced by an increase in oxygen partial pressure (e.g., Kohlstedt and Hornack, 1981). It has been suggested that the water-weakening effect arises from an increase in oxygen activity resulting from the dissociation of environmental water.

Experimentally determined constitutive laws for flow of olivine by intracrystalline plasticity show a high degree of consistency in the values of the flow-law parameters (Chopra and Paterson, 1981, 1984; Karato et al., 1982, 1986; Zeuch, 1982). In particular, the problems of a markedly pressure-sensitive water-weakening effect of the type observed in quartz are absent, and high-quality experimental data have been obtained over a wide range of pressures. Water causes an apparent lowering of the activation enthalpy and stress exponent for flow. Borch and Green (1987) showed that the effect of pressure and volatiles on the flow behaviour of olivine scale with the effect of these variables on the melting temperature of the material. This considerably aids the extrapolation of laboratory data to nature. Flow laws for intracrystalline plasticity extrapolated to a low strain rate are illustrated on the log stress vs. temperature plot presented in Figure 6-2. The high strength of olivine relative to all other minerals shown should be noted.

Chopra (1986) and Karato et al. (1986) have successfully prepared high-density hot-pressed aggregates of fine-grained olivine, from which it has been possible to demonstrate the transition from grain-size-independent flow by intracrystalline plasticity to grain-size-sensitive flow (Fig. 6-4), characterized by nearly linear viscosity. The latter type of flow was interpreted in terms of rate control by diffusive mass transfer around grain boundaries or

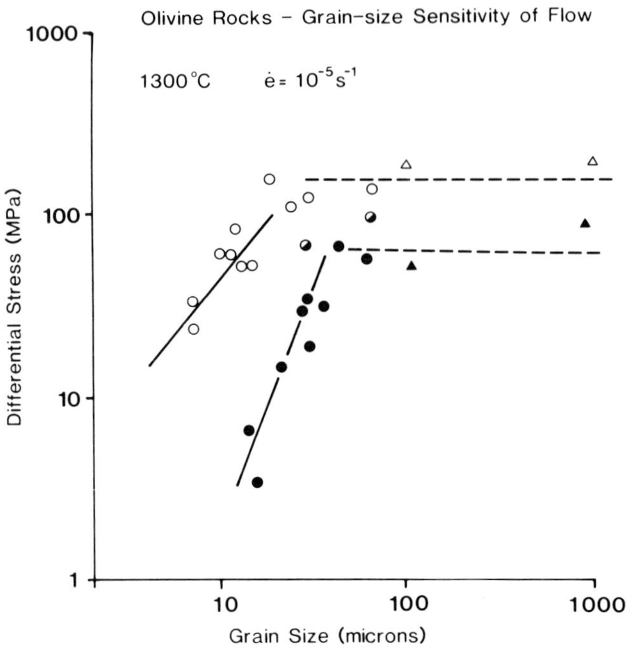

Fig. 6-4. Experimentally observed transition from grain-size insensitive plastic flow of olivine at large grain-sizes, to grain-size sensitive flow at small grain-sizes; 1300°C, strain-rate = 10^{-5} s^{-1}. Closed symbols: wet tests. Open symbols: dry tests. Half-closed symbols: as-is specimens containing a small amount of water. Continuous lines show best-fits to the data in the grain-size sensitive regime. Dashed lines show trends in the grain-size insensitive regime. Open triangles: dry dunite data of Chopra and Paterson (1984); closed triangles: wet dunite data of Chopra and Paterson (1981); other data of Karato et al. (1986).

through grains. Diffusion-accommodated, grain-size- sensitive flow, facilitated by the presence of intergranular basaltic melt, has also been reported by Cooper and Kohlstedt (1984).

3.4. Feldspars

Feldspar rheology is likely to be of particular importance in the lower continental crust owing to its volumetric importance in metabasic and dioritic rocks, in which quartz does not form the principal load-bearing framework. Such rocks may form a substantial proportion of the lower continental crust. However, feldspars have received far less attention from experimental rock mechanics than that devoted to quartz and olivine, arising perhaps largely from their evident structural and chemical complexity.

The mechanical behaviour of feldspars has been reviewed by Tullis (1983) and Gandais and Willaime (1984). To a first approximation, the mechanical behaviour observed in experiments is not strongly sensitive to feldspar composition. At low temperatures, all feldspars fail by cleavage fracture, usually at stress levels lower than those required to cause flow or fracture in quartz. All feldspars begin to show evidence for intracrystalline plasticity by dislocation glide at about 700°C in short-duration laboratory experiments carried out at high confining pressures to inhibit fracture. Mechanical twinning in plagioclase is also produced only at high temperature. Plastic deformation generally results in the formation of planar faults, which trail behind moving dislocations (Marshall and McLaren, 1977), and which are thought to result from loss of short range order. Small fault vectors both normal and parallel to the slip plane have been reported. Dissociation of dislocations has been observed in experimentally deformed sanidine (Kovacs and Gandais, 1980) and this, together with large Burgers vectors for perfect dislocations and the tendency to form superdislocations in the low-temperature ordered structures, is expected to inhibit recovery by dislocation climb. Ji and Mainprice (1987) report that the brittle–ductile transition in experimentally deformed sintered albite coincides with the order–disorder transition. Evidence for recovery by dislocation climb is seen at 800–900°C and above, and at high strains recovery by dynamic recrystallization has been produced (Marshall et al., 1976; Tullis and Yund, 1977, 1985) (Plate 6-1). As with quartz, the plasticity of feldspars appears to be enhanced substantially by deformation in the presence of water, especially at high pressure (Tullis et al., 1979; Tullis and Yund, 1980).

No generalized flow laws for high-temperature intracrystalline plasticity of feldspars are available, but Shelton and Tullis (1981) and Tullis (1983) presented preliminary flow-law parameters for creep of Hale albite rock and an anorthosite, obtained using the solid-medium testing technique. Ji and Mainprice (1986) obtained a flow law for plastic flow of dry albite rock ($n = 2.5$) under unconfined conditions at 1000°C, and demonstrated a transition to linear viscous flow as partial melt began to appear above 1060°C. Under various different conditions, the stress exponent may vary between 1 and about 5, and the activation enthalpy varies from 524 kJ mol^{-1} (dry, unconfined) through 430 kJ mol^{-1} (dry, confined) to 230 kJ mol^{-1} (air-dry, confined), presumably varying according to water content. Extrapolation of these data to a strain rate of 3×10^{-14} s^{-1} is shown in Figure 6-2. For plastic deformation the feldspars appear to be stronger than even dry quartz but weaker than mafic minerals (amphiboles, pyroxenes and garnets), and this is what would be expected from natural observations of the behaviour of these minerals when deformed together in polyphase rocks or in adjacent monomineralic rock masses.

3.5. Amphiboles

Little is known of the plasticity of amphiboles, with hornblendic amphibole being the only one that has been experimentally investigated (Rooney and Riecker, 1973; Rooney et al., 1974). Morrison-Smith (1976) made transmission electron microscope observations on these samples, and reported that the dominant slip system was (100)<001>. It is one of the strongest silicate minerals, behaving passively both in experiments and in nature unless it is present in sufficient quantity to form the load-supporting framework.

Mechanical twinning on (101) was the dominant mode of deformation in the experimental studies of Rooney et al. (1974). From experimental studies on a hornblendite, Dollinger and Blacic (1975) observed slip on (100)<001> at 600–750°C to produce kink bands, and at 800°C observed (100) slip in various directions. They did not observe (101) twinning nor evidence of recovery. No constitutive flow law data are available.

3.6. Pyroxenes

Pyroxenes form important components of both the lower crust and upper mantle, yet these minerals have not been studied as exhaustively as olivine. Twinning and high-temperature plastic flow of single crystals of clinopyroxenes were studied by Kollé and Blacic (1982,1983). High-pressure/temperature studies of clinopyroxenites from which flow-law data were obtained have been made by Kirby and Kronenberg (1984) using a solid-medium testing machine (400–1100°C, 170–2000 MPa confining pressure) and Boland and Tullis (1986) using

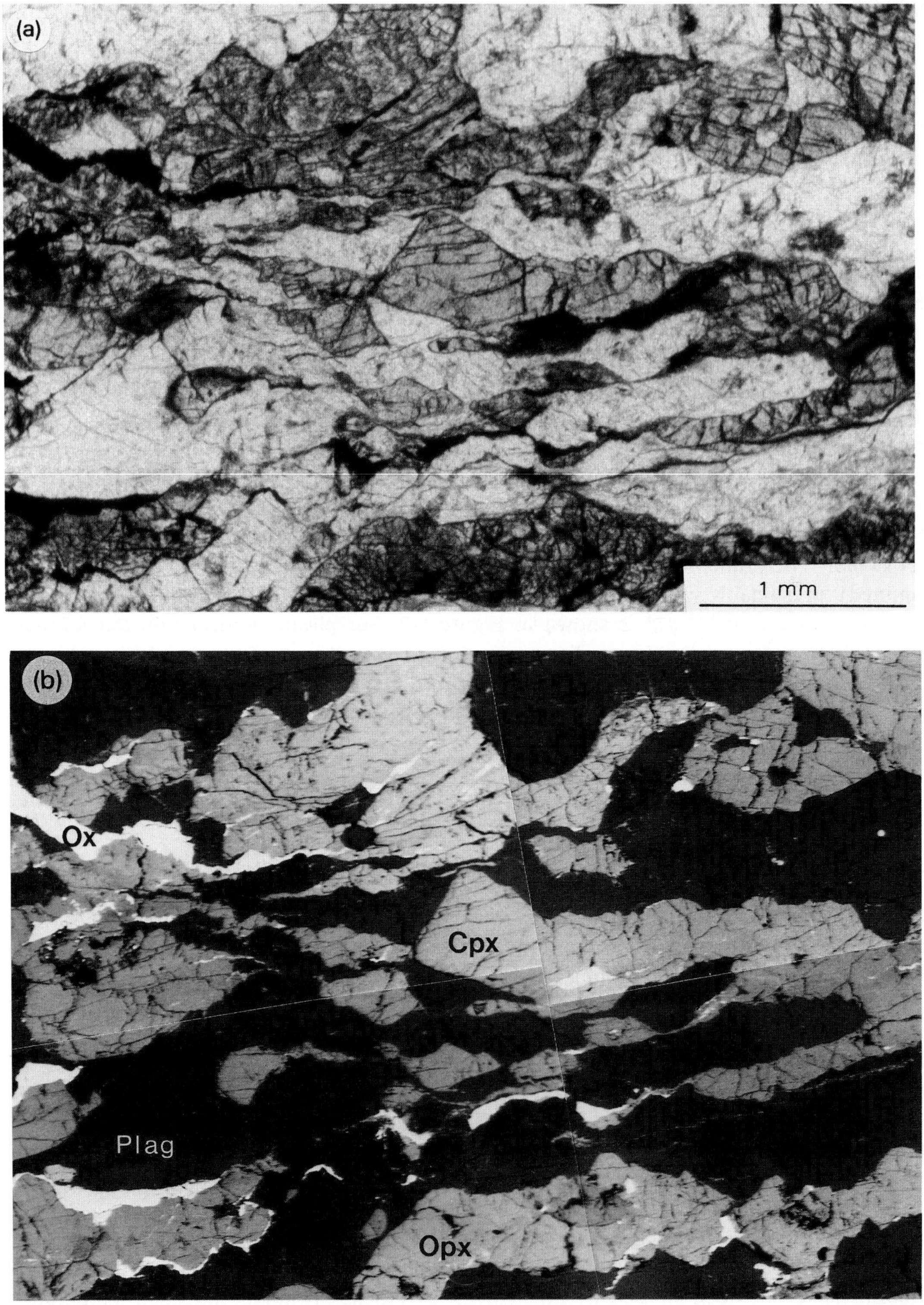

PLATE 6-1

Experimentally deformed clinopyroxene-orthopyroxene-plagioclase (An_{60}) granulite facies metabasic rock (original rock from the Ivrea–Verbano Zone, northern Italy). Deformation conditions: 800°C, 700 MPa confining pressure, $\dot{e} = 10^{-4}\ s^{-1}$, dry, 27% shortening.

(a) Optical micrograph (ppl) and (b) backscattered electron image of the same area and to a similar scale. The initial rock was relatively equigranular and a strong foliation has developed in a zone in the centre of the sample (illustrated) due to plastic deformation and grain refinement of the plagioclase and extension of the Fe-oxide grains. Deformation of the pyroxene appears to be almost totally cataclastic.

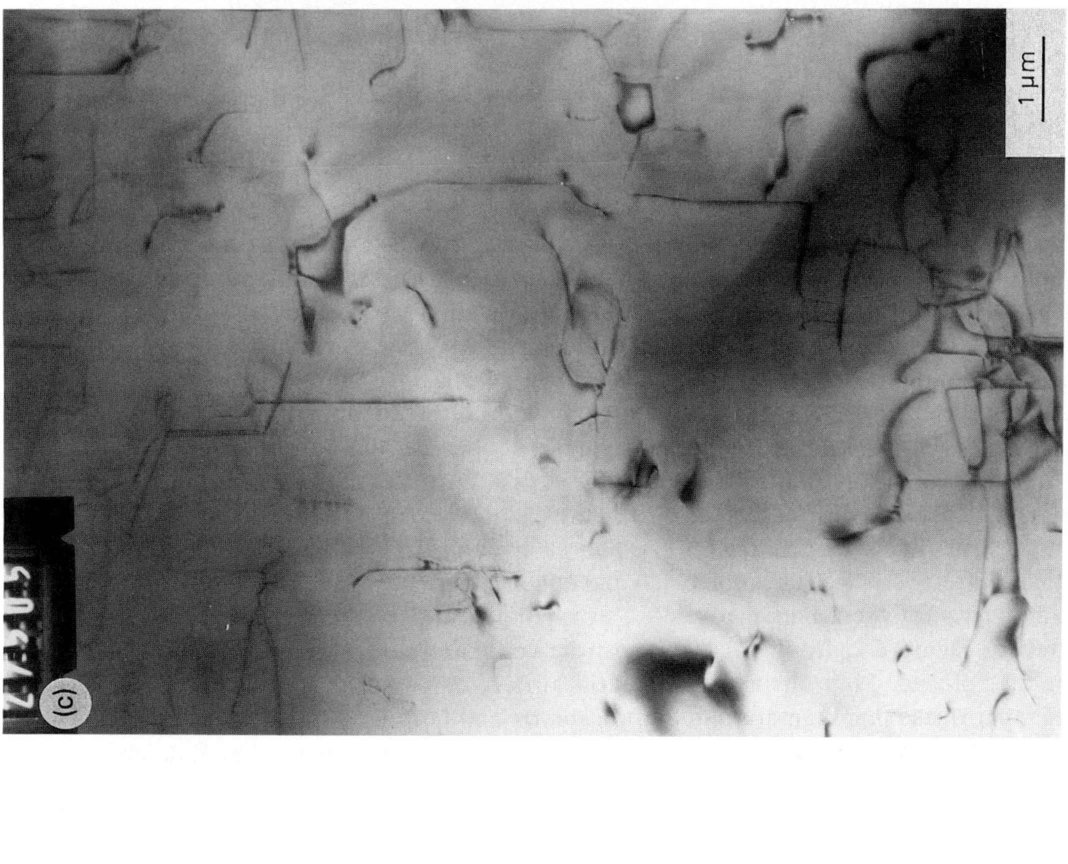

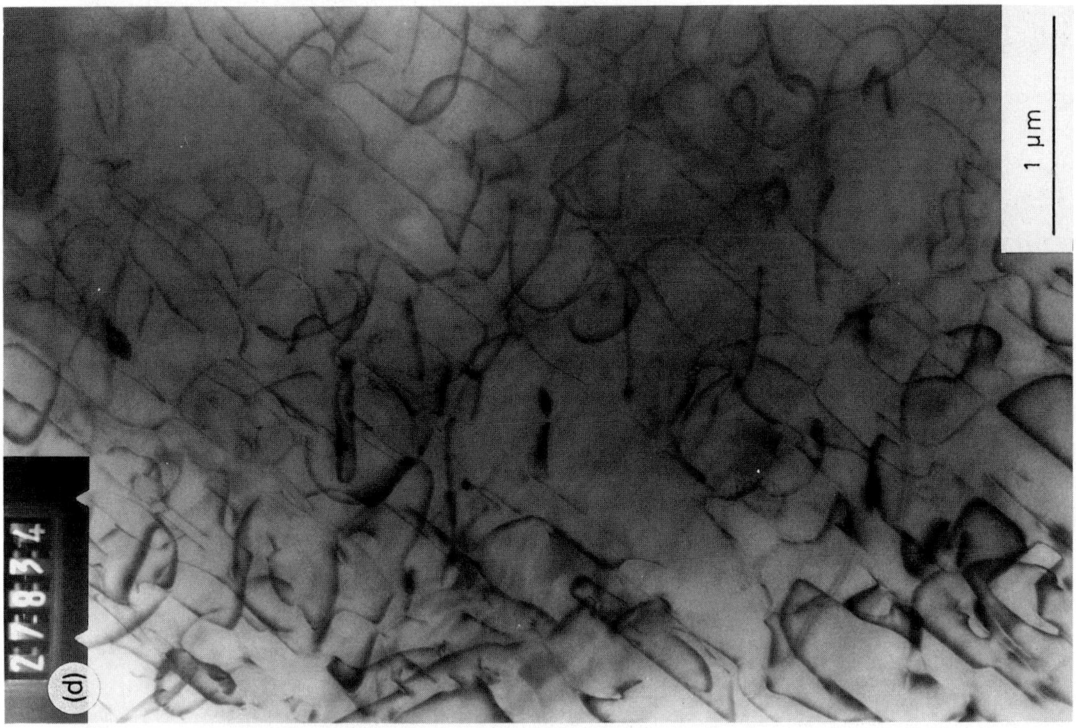

PLATE 6-1 (continued)

(c) TEM micrograph of plagioclase prior to the experimental deformation showing the relatively low density of free dislocations.
(d) TEM micrograph of the plagioclase after deformation showing the increase in dislocation density.

PLATE 6-1 (continued)

(e) TEM micrograph of relict and recrystallized plagioclase in the deformed sample. Note the abundance of deformation twins in the relict plagioclase, and the development of recrystallized grains with low dislocation densities.

a gas-medium apparatus (990–1270°C, 300 MPa confining pressure). In both cases the deformation conditions spanned the broad transition from brittle to plastic behaviour. Kirby and Kronenberg (1984) also demonstrated the transition from power-law creep at low differential stress (below about 1200 MPa) to exponential-law creep at higher stress, at 1500 MPa confining pressure. Additionally, Boland and Tullis (1986) investigated the effect of encapsulating 1% water with the sample (water pressure unknown). Although they found that the flow behaviour of the dry material tested by Kirby and Kronenberg (1984) was consistent with their results when extrapolated, addition of water caused a 60% weakening. For the wet material, the activation enthalpy was larger and the stress exponent smaller than found for the dry pyroxenite of Kirby and Kronenberg (1984). The water-weakening was shown to be associated with enhanced recovery by dislocation climb and enhanced grain-boundary mobility.

The flow laws for these clinopyroxenites, extrapolated to a strain rate of 3×10^{-14} s^{-1}, are shown in Figure 6-2, from which they can be compared with the behaviour of quartz, feldspar and olivine. Like olivine, significantly higher temperatures are required for the onset of rapid weakening than are required for quartz and feldspar. Experimental studies on orthopyroxenites from which flow-law data were obtained have been reported by Raleigh et al.

(1971, gas and solid medium, 840–1100°C, 500–2000 MPa confining pressure, dry), Ross and Nielsen (1978, solid medium, 950–1300°C, 1000 MPa confining pressure, inferred to be wet through dehydration of talc-confining medium) and Avé Lallemant (1978, websterite, 68% cpx, 32% opx, solid medium, 1000–1200°C, 1000 MPa confining pressure). The parameters of the flow laws reported for the two orthopyroxenites are remarkably similar, implying that there is no water weakening effect. These data, extrapolated to a strain rate of 3×10^{-14} s^{-1}, are also shown in Figure 6-2. The behaviour of the websterite appears to be dominated by the clinopyroxene, but the orthopyroxenites are distinctly weaker than the olivine and clinopyroxene rocks, an observation which is not always consistent with observations on naturally deformed rocks.

3.7. Other minerals and polymineralic rocks

The mechanical properties of other minerals which are important in lower crustal rocks, such as garnets, phyllosilicates and aluminium silicates are not well known from experimental studies. Commonly they do not form the load-supporting framework of rocks. In the section dealing with natural rock microstructures we will consider their behaviour by microstructural comparison with the behaviour of the volumetrically more important minerals discussed above.

Carbonates are not likely to be volumetrically important constituents of the lower crust, but where they occur they are striking owing to their apparent weakness compared with their silicate host-rocks. Experimentally determined rheological characteristics of calcite rocks are amongst the best available, because they flow easily at moderate temperatures and pressures.

Most work has been carried out on Yule Marble (Heard and Raleigh, 1972), Solnhofen Limestone (Heard, 1960; Rutter, 1972, 1974; Schmid et al., 1977) and Carrara Marble (Rutter, 1972, 1974; Schmid et al., 1980). Intracrystalline plastic flow of relatively coarse-grained marble by dislocation creep is characterized by an unusually high stress exponent (4.6–8), so that the flow stress falls only slowly with reduction in strain rate. Extrapolation of flow law data for Carrara Marble to a low strain rate is shown in Figure 6-2a.

The foregoing discussions are based around the behaviour of monomineralic polycrystals, and the presumption that lower crustal rocks are either monomineralic or that their mechanical behaviour is dominated by the properties of the most deformable phase. Such presumptions can only lead to approximate descriptions of lower crustal rheology. Systematic experimental studies of the flow of polymineralic rocks at high temperatures are in their infancy (but see, for example, Tullis and Yund, 1977; Tullis et al., 1991; and Plate 6-1). Handy (1990) considers the question of the inference of the flow behaviour of polymineralic rocks from the behaviour of their constituent minerals, with particular reference to deformation under mid- and lower crustal conditions. The strengths of two-phase rocks lie between the strengths of monomineralic rocks of the component phases (Jordan, 1987; Handy, 1990). Strength depends upon how the component phases are arranged in the aggregate or on how this is modified during strain (for example, as a result of foliation development). Often the component minerals fail in strikingly different ways. For example, brittle feldspar grains may be carried about in a matrix of flowing quartz, or brittle garnet clasts may be carried about in a matrix of ductile, recrystallized plagioclase (e.g., Brodie and Rutter, 1985) (Plates 6-2 and 6-3).

4. Localization of strain into shear zones

Constitutive flow laws obtained from high-temperature mechanical experiments are meant to describe steady-state flow (flow at constant stress for the duration of the experiment). Although shear zones can be nucleated in materials characterized by strain-independent microstructure (Hobbs et al., 1990), the maintenance of a concentration of flow in a shear zone probably requires the operation of a strain-dependent weakening mechanism, so that, once a shear zone has been nucleated, it is easier for deformation to continue within it than to spread throughout the rock volume. There is ample field evidence to indicate that localization of flow into shear zones plays an important role in the mechanical behaviour of the lower crust (e.g., Coward, 1984).

The production of a shear zone in naturally deformed rocks appears always to be accompanied by a significant microstructural change. The amount of strain required to modify the microstructure is strongly sensitive to deformation mechanism and the nature of the deforming material. White et al. (1980) and Kirby (1985) review many of the processes of microstructural evolution which lead to strain localization or are associated with it.

Once established, a shear zone may accommodate a very large displacement, and an important task of experimental rock mechanics is to establish flow laws for the steady-state flow of the microstructurally modified material inside shear zones. However, the microstructural breakdown process associated with localization deserves special attention. The tip of a spreading shear zone can be modelled as a shear-mode crack tip. A state of intensified stress exists around it, fed by the displacements that accumulate across the shear zone. The stress state can be analyzed by the methods of fracture mechanics (e.g., Rutter and Brodie, 1985; Blenkinsop and Drury, 1988). It is in this region that the microstructural

PLATE 6-2

Optical micrograph (crossed polars) of quartzo-feldspathic mylonite showing brittle deformation of plagioclase in a fine-grained dynamically recrystallized quartz matrix (mylonite in Lewisian gneiss above Arnaboll Thrust, NW Scotland).

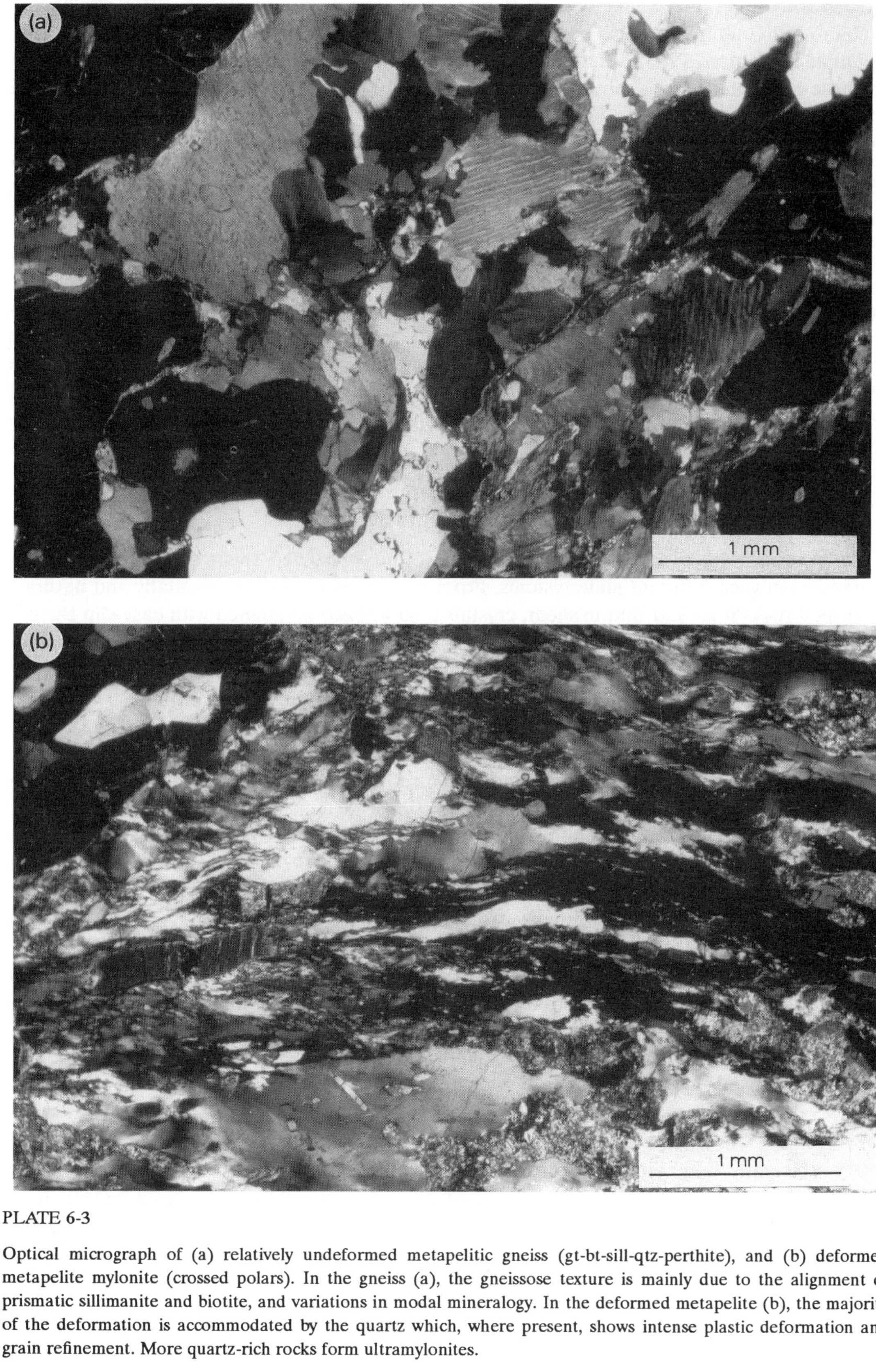

PLATE 6-3

Optical micrograph of (a) relatively undeformed metapelitic gneiss (gt-bt-sill-qtz-perthite), and (b) deformed metapelite mylonite (crossed polars). In the gneiss (a), the gneissose texture is mainly due to the alignment of prismatic sillimanite and biotite, and variations in modal mineralogy. In the deformed metapelite (b), the majority of the deformation is accommodated by the quartz which, where present, shows intense plastic deformation and grain refinement. More quartz-rich rocks form ultramylonites.

'breakdown' takes place. Very large stresses may locally be produced, stresses that are probably high enough to cause power-law breakdown and in some cases brittle fracture (e.g., Segall and Simpson, 1986). Microstructures used to infer palaeostress magnitude may be set by these high stress levels, and may not be representative of the remotely applied tectonic stress field.

A number of factors may be important in favouring localization of deformation:

4.1. Cataclastic deformation

The concentration of cataclastic deformation into a brittle fault provides a dramatic example of shear localization. The microstructural change is the mechanical granulation of the host-rock. Because this is more typically characteristic of upper crustal deformation we will not consider it in any more detail here, although there is no reason why brittle faulting should not occur at any temperature and total pressure, provided pore-fluid pressure is high. Microstructural evidence for cataclastic deformation under high-temperature conditions is unlikely to be preserved owing to subsequent microstructural adjustments (sintering, grain growth, metamorphic reactions, etc.).

4.2. Fabric (geometric) softening

In low-symmetry rock-forming minerals, the critical resolved shear stress for slip varies markedly between different glide systems. From studies on both experimentally and naturally deformed rocks it is clear that in shear, crystals tend to become aligned with easy-slip Burgers vectors parallel to the shear direction (e.g., Schmid et al., 1987a) (Fig. 6-3). The resultant weakening of the aggregate of grains implies a propensity for shear localization.

4.3. Reaction softening

Natural shear zones frequently are mineralogically distinct from their host rocks because they contain the products of metamorphic reactions. 'Softer' reaction products (e.g., talc, serpentine, chlorite, micas, actinolite) favour shear localization, and are more characteristic of retrograde than prograde reactions. White and Knipe (1978) coined the term 'reaction softening' for this type of deformation/metamorphism interrelationship.

Reactions and phase transformations of various kinds are commonly localized within shear zones. Often it is not clear whether the transformation was causally linked to shear localization, or whether the shear zone merely provided access for one or more of the reactants involved in the transformation (e.g., water, dissolved ions, CO_2). The general question of deformation/metamorphism interrelationships is considered by Brodie and Rutter (1985) and Rutter and Brodie (1988b).

4.4. Potential effects of grain-size reduction

At sufficiently large strains (typically 30% to more than 90% shortening), a substantial volume of the original rock may be replaced by a dynamically recrystallized aggregate of much finer grains of the same phase (e.g., Post, 1977; Kirby and Kronenberg, 1984; Schmid et al., 1987a). This is provided the temperature is not so high that recrystallized grain size is commensurate with initial grain size. Alternatively, if in a polymineralic rock a metamorphic reaction accompanies deformation, one or more product phases may be fine-grained (e.g., Rubie, 1983; Brodie and Rutter, 1987a). If pore-fluid pressure is high, grain-size reduction may also occur through cataclasis. Weakening associated with a sufficient degree of grain-size reduction, so that a grain-size-sensitive flow mechanism becomes activated, is thought to be one of the most important ways in which flow can become localized into shear zones (Figs. 6-5 and 6-6). Such weakening can be brought about by a number of mechanisms (Kirby, 1985). Flow involving solid-state diffusion between sources and sinks for the diffusing species on grain boundaries is grain-size-sensitive, and appears to be characterized by a low value of the stress exponent in the constitutive flow law (between 1 and 2). For creep controlled by through-the-grain diffusion (Nabarro-Herring creep), $\dot{e} \propto 1/d^2$, where $\dot{e}$ is strain rate and d is grain size. When grain-boundary diffusion is rate-controlling (Coble creep), $\dot{e} \propto 1/d^3$. These two processes lead to development of a grain-shape fabric. Coble creep was invoked by Schwenn and Goetze (1978) to interpret grain-size-sensitive flow of olivine aggregates during hot-pressing. Karato et al. (1986) (Figs. 6-4 and 6-5) and Chopra (1986), using synthetic, hot-pressed aggregates of olivine, demonstrated experimentally a transition from grain-size-independent dislocation creep to grain-size-sensitive flow. Grain boundary diffusion may be enhanced if intergranular water or melt exists (e.g., Cooper and Kohlstedt, 1984).

Flow by diffusion- and/or plasticity-accommodated grain-boundary sliding also shows grain-size sensitivity of the above types, according to which diffusion path is dominant. This mechanism does not result in development of a shape fabric, and it is widely held that any pre-existing crystallographic preferred orientation tends to be destroyed as a result of the random grain rotations which can take place. Grains remain equidimensional, hence truly steady-state flow can occur at a constant microstructure (Fig. 6-3). This mechanism has been demonstrated

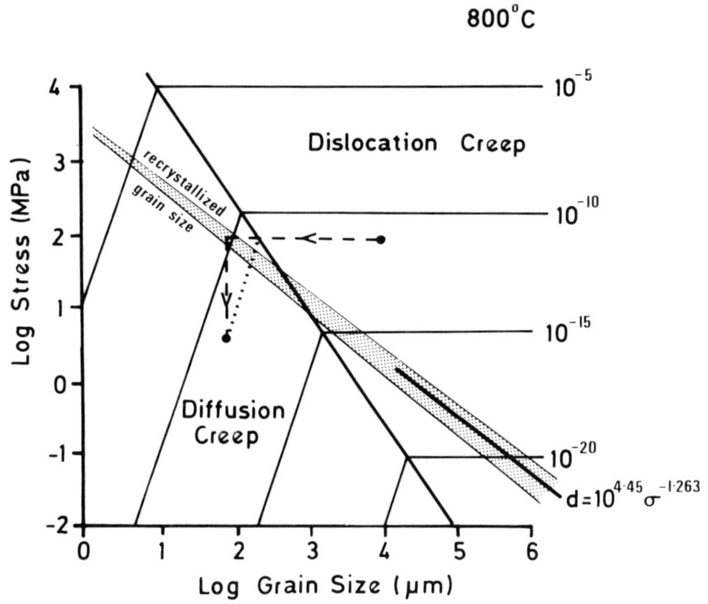

Fig. 6-5. Deformation mechanism map for wet olivine at 800°C, extrapolated from the experimental data of Karato et al. (1986) and Rutter and Brodie (1988b) to show how dynamic recrystallization might cause weakening through a shift at constant strain-rate from dislocation creep (flow controlled by dislocation climb) into the grain-size sensitive (diffusion creep) flow field (after Rutter and Brodie, 1988a)

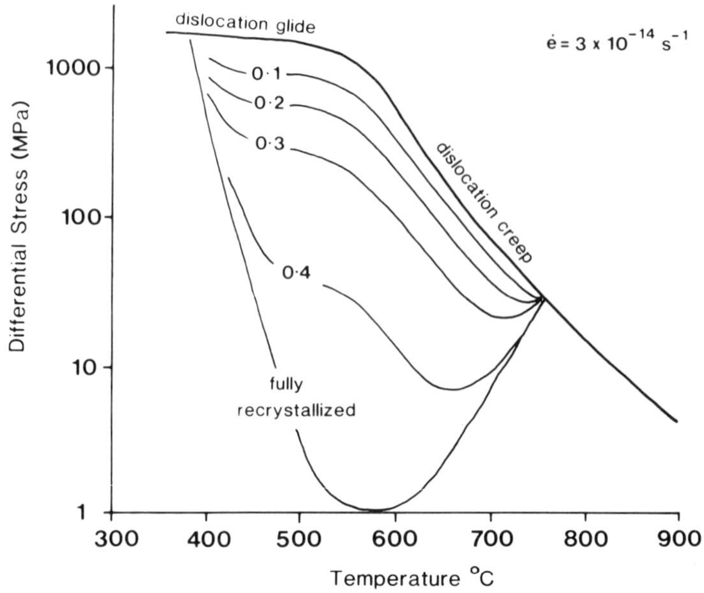

Fig. 6-6. Profile of strength vs. temperature for wet olivine, showing the weakening effect expected for varying fractions (indicated on curves) of dynamic recrystallization at a constant strain-rate of $3 \times 10^{-14}\ s^{-1}$. Within the interval 400 to 700°C, dynamic recrystallization would be expected to favour flow localization into shear zones; outside that range it would not. Based on Rutter and Brodie (1988b).

to operate in the flow of a natural fine-grained limestone (Schmid et al., 1977) and the explicit form of the grain-size sensitivity for calcite has been determined by Walker et al. (1990) (Fig. 6-7) using synthetic hot-pressed aggregates of controlled grain size. Rutter and Brodie (1988b) interpreted weakening of serpentinite during experimental dehydration to olivine to flow of the transiently ultrafine-grained olivine by diffusion-accommodated grain-boundary sliding.

Recovery from work hardening by dynamic recrystallization may produce weakening because the new grains are strain-free and hence able to flow transiently at a lower stress. Cyclic recrystallization continually 'refreshes' the microstructure and allows large strains to accumulate by intracrystalline plasticity. Dynamic recrystallization commonly proceeds either by grain-boundary migration, driven by differences in dislocation density across the boundary, or by progressive subgrain misorientation (see review by Urai et al., 1986). From studies on metals (Sellars, 1978) and halite (Guillopé and Poirier, 1979) the former mechanism is favoured by high temperatures and stresses and the latter by lower temperatures and stresses. The possibility of a catastrophic switch between recrystallization mechanisms is possible if initially (impurity) pinned boundaries can break free (Guillopé and Poirier, 1979; Urai et al., 1986). Tullis and Yund (1985) appealed to dynamic recrystallization to explain strain-dependent weakening of synthetic albite aggregates, and Zeuch (1982) invoked it to explain the results of Post (1977) for olivine, in which 'ductile faulting' appeared to be associated with the grain-size reduction. Kirby and Kronenberg (1984) reported strain-dependent weakening in the experimental deformation of clinopyroxenite (Fig. 6-1). This may have been related to the development of shear bands of very fine-grained, dynamically recrystallized pyroxene, which exhibited a strong preferred crystallographic orientation.

It should be noted, however, that dynamic recrystallization may not always produce weak-

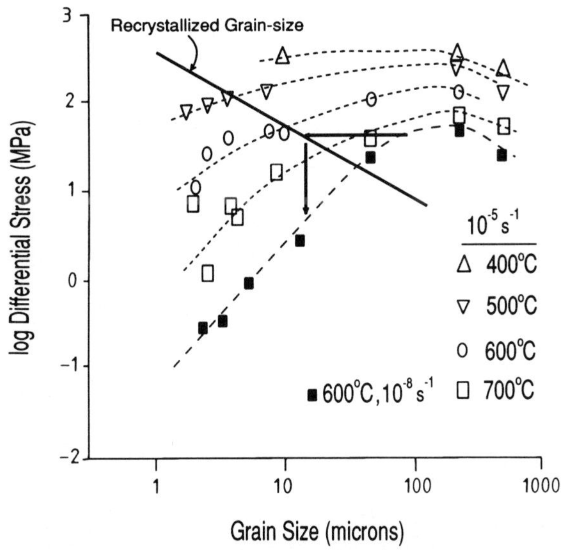

Fig. 6-7. Experimental data for calcite rocks (after Walker et al., 1990) showing the transition from intracrystalline plastic flow at coarse grain-size to grain-size sensitive flow at small grain-sizes. Open symbols are for a strain-rate of 10^{-5} s^{-1}, whereas the 600°C data in closed symbols are at a strain-rate of 10^{-8} s^{-1}. Data at the two largest grain-sizes are from experiments on Carrara marble and Taiwan marble, respectively. The decrease in strength with increasing grain-size here is due partly to the growing importance with increasing grain-size of twinning as a deformation mechanism. All other points refer to synthetic, hot-pressed calcite aggregates of controlled grain-size. For the coarse-grained rocks, the stress exponent of the flow law is approx. 4.5. In the grain-size sensitive regime at fine grain-sizes the flow approaches linear viscous. The relationship between recrystallized grain-size and stress (from Schmid et al., 1980) is shown as a continuous straight line. The two straight arrows show how a rock recrystallizing from about 100 μm grain-size to 15 μm (if the strain were sufficiently high) might be weakened by a factor of almost ten times.

ening. Schmid et al. (1980) observed that grain-size reduction by dynamic recrystallization in marble did not affect flow stress up to 30% strain. We have made similar observations for marble deformed up to 65% shortening. Schmid et al. (1987a) reported minor strain softening in direct shear experiments to high strains, however. In these instances the grain size was not reduced sufficiently to activate grain-size-sensitive creep. The net lack of any weakening, however, may be accentuated because grain-size reduction in the intracrystalline plastic flow regime causes hardening through the Hall-Petch effect (e.g., Poirier, 1985), which may counteract the softening associated with strain-free grain formation.

From experiments on quartzites, Kronenberg and Tullis (1984) suggested that in the presence of intergranular water, grain refinement facilitated the diffusive ingress of water, leading to hydrolytic weakening. In this case, and for recovery by dynamic recrystallization, it is expected that the stress exponent in the flow law will be that characteristic of dislocation creep. However, it is not yet clear what should be the nature of the grain-size sensitivity of the flow stress. Because plastic flow causes development of preferred crystallographic orientation (fabric), the commonly observed existence of a strong fabric in natural shear zones characterized by grain refinement may point to an important or dominant role of intracrystalline plastic flow in such zones (e.g., Plates 6-3d and 6-4c). More comprehensive experimental studies will be needed before it will be possible objectively to interpret deformation mechanisms in natural shear zones characterized by grain-size reduction.

PLATE 6-4

Sequence of optical micrographs (crossed polars) showing progressive grain refinement and the development of a mylonite in a pyroxene-plagioclase-garnet granulite facies metabasic rock (Premosello, Ivrea–Verbano Zone, northern Italy).

(a) Relatively undeformed granulite facies granofels showing equigranular, near-equilibrium microstructure. Low intensity optical strain features are present in the plagioclase grains.

(b) Protomylonite, showing initial recrystallization at the margins of the plagioclase grains and internal strain in the larger plagioclase grains. Pyroxene and garnet remain undeformed.

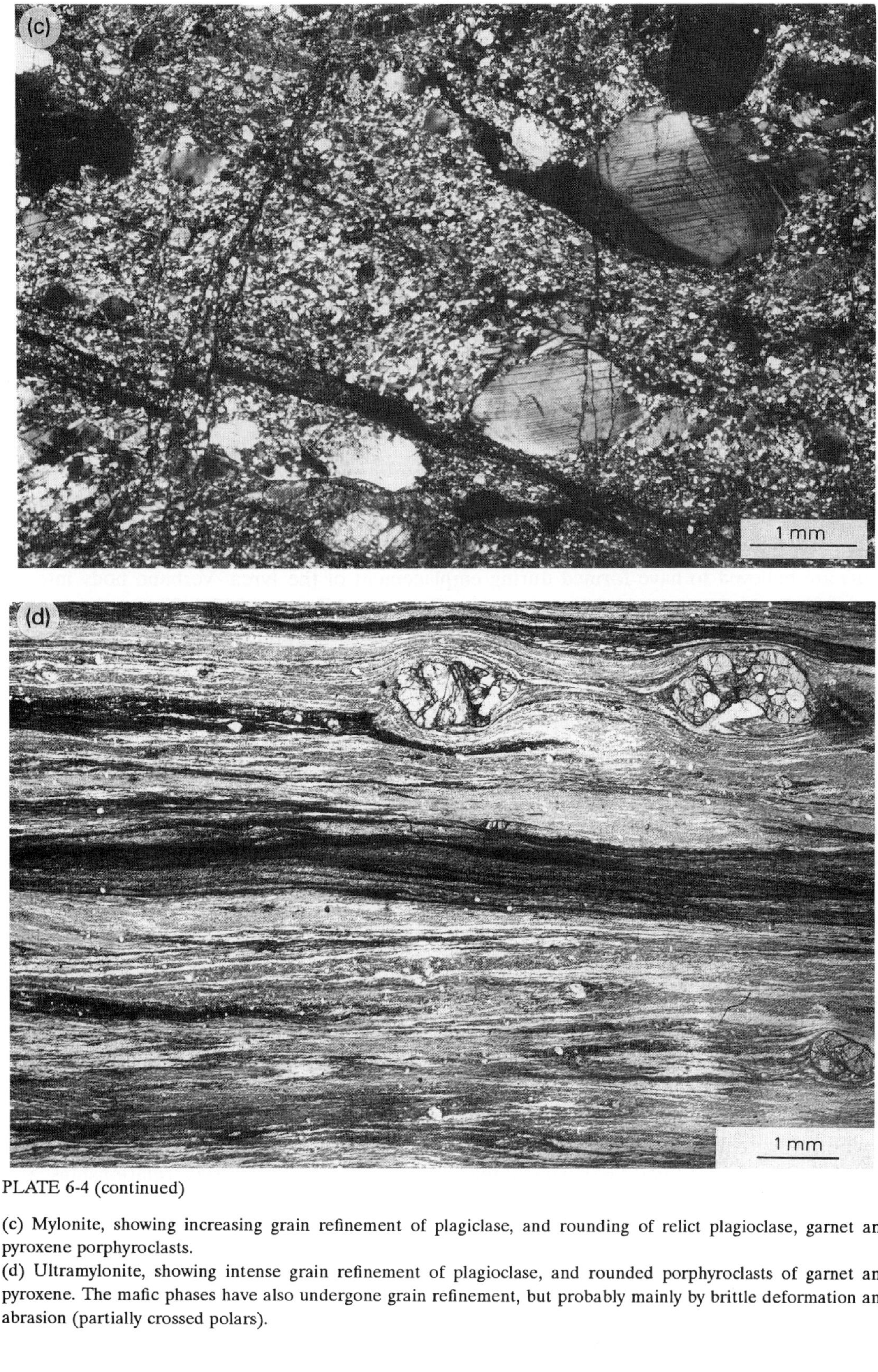

PLATE 6-4 (continued)

(c) Mylonite, showing increasing grain refinement of plagiclase, and rounding of relict plagioclase, garnet and pyroxene porphyroclasts.

(d) Ultramylonite, showing intense grain refinement of plagioclase, and rounded porphyroclasts of garnet and pyroxene. The mafic phases have also undergone grain refinement, but probably mainly by brittle deformation and abrasion (partially crossed polars).

4.5. Shear heating

Although it is unlikely that shear zone localization can be caused by a thermal-softening 'fluctuation', under some circumstances shear heating may be important in stabilizing the propagation and continued development of a shear zone. Significant thermal-softening is favoured by:

(a) a large activation enthalpy for flow, which maximizes the strain rate increase for a given temperature change;

(b) a wide shear zone (for a given shear strain rate), so that heat loss occurs more slowly;

(c) a high shear stress and/or shear strain rate, because rate of heat generation is proportional to the product of shear stress with shear strain rate. As pointed out by Brun and Cobbold (1980), the possible effects of shear heating cannot be ignored in the construction of mathematical models to describe the behaviour of the crust when localized deformation zones can develop.

Under high shear stresses and strain rates, seismogenic faulting can produce frictional melting. Narrow, brittle fault zones ramified with frictional melt (pseudotachylite) appear to be not uncommon in terrains of high-grade metamorphic rocks (e.g., Sibson, 1977). The mineralogy and low porosity of such rocks mean than under even moderate confining pressures at low temperatures (probably less than 300°C) they will not fail at differential stresses less than several hundreds of MPa. In the Ivrea–Verbano zone of northern Italy, spectacular developments of pseudotachylite-bearing faults in otherwise pristine high-grade gneiss are believed to have formed during emplacement of the Ivrea–Verbano body into its present shallow level in the Alpine orogen (e.g., Brodie and Rutter, 1987b). The cool, resistant slab may even have behaved as an indenter around which the Alpine arc is curved (e.g., Platt et al., 1989), and large differential stress levels may have been concentrated into it by the stress-amplification mechanism of Kusznir and Park (1986).

Hobbs et al. (1986) argued that pseudotachylites associated genetically with ultramylonites could be formed as a result of plastic instabilities developing through heating during otherwise steady plastic flow. Their development is more likely in the stronger, metabasic rocks than in felsic rock types. Such associations are not uncommon in nature, and several examples are documented by Hobbs et al. (1986). Many examples can be found in the Ivrea–Verbano zone, particularly within metabasic mylonite sheets. In the latter case, however, they may equally have formed thorough the reactivation of old mylonite zones as brittle faults during Alpine deformation.

5. The size of recrystallized grains

Following the experience of materials scientists, it has been found that for several rock-forming minerals the size of dynamically recrystallized grains varies inversely with stress. After experimental calibration therefore, and subject to various provisos, recrystallized grain size can be used to estimate palaeostress values (e.g., Mercier et al., 1977; Twiss, 1977; Etheridge and Wilkie, 1978; White, 1979; Kohlstedt and Weathers, 1980; Christie and Ord, 1980; Hacker et al., 1990). Experimentally determined stress/grain-size relationships are summarized in Figure 6-8. Other microstructural features may carry information about palaeostress magnitudes (e.g., free dislocation density, subgrain size), but recrystallized grain size is likely to be more resistant to resetting by later, transient tectonic events or through stresses induced during cooling and depressurization (White, 1979). The resistance of

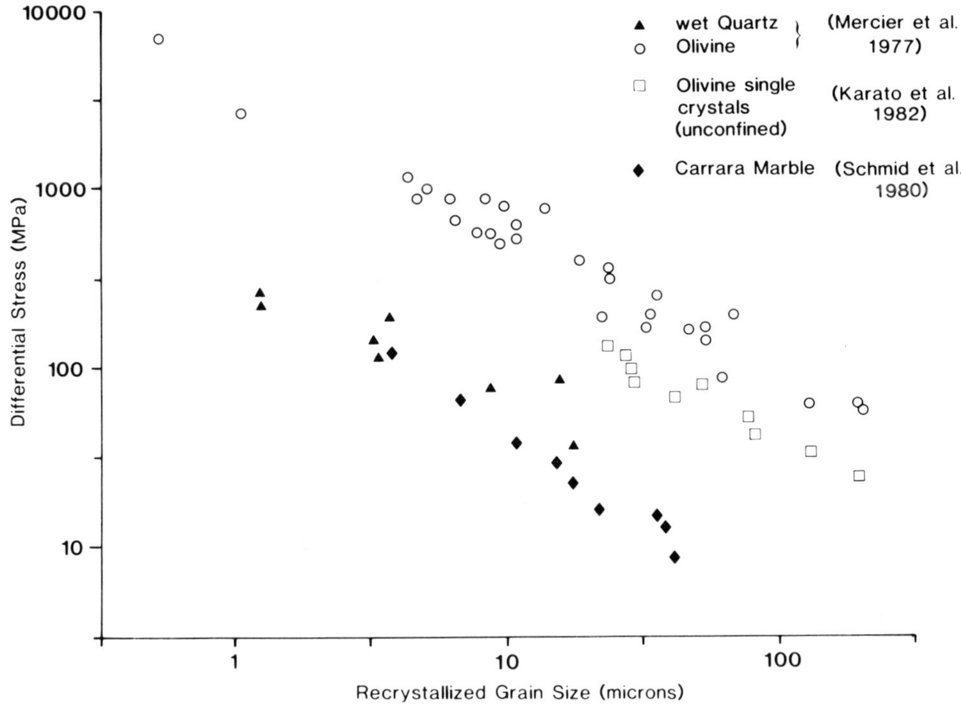

Fig. 6-8. Experimental data showing how the size of dynamically recrystallized grains (for olivine, quartz and calcite) decreases at higher flow stresses, which are in turn a consequence of lower temperatures and/or higher strain-rates.

dynamically recrystallized grains to static grain-growth is enhanced by the presence of a small amount of an intergranular second phase.

Numerous attempts have been made to estimate palaeostress values using quartzites, particularly from mylonites produced during overthrusting. Differential stresses in the range 10 to 150 MPa have commonly been recorded (e.g., Christie and Ord, 1980; Kohlstedt and Weathers, 1980; Hacker et al., 1990), but no consistent picture has yet emerged relating flow stress magnitude to tectonic environment for rocks of mid- to lower crustal provenance. On the other hand, a tentative set of stress vs. depth profiles for subcontinental upper mantle from intracratonic, continental extension and rift environments has been presented by Mercier (1980). Recrystallized olivine grainsizes were measured in peridotite xenoliths brought to the surface in basaltic lavas, their depths of origin being estimated by pyroxene geobarometry. In each case stress falls to a low level of about 5 MPa with increasing depth. The rift environment records the most rapid weakening with depth and the intracratonic profile the least rapid weakening.

6. Structures and microstructures of naturally deformed rocks

A number of regions of the Earth are considered to represent sections through lower continental crust which have been brought to the surface, commonly through displacement on a shear zone which transects the entire crust [e.g., the Ivrea–Verbano zone (Zingg, 1983; Schmid et al., 1987b), the Arunta zone, Australia (Goleby et al., 1989), Fiordland, South Island, New Zealand (Oliver, 1980; Gibson et al., 1988), southern Calabria, Italy (Schenk,

1984), the Kapuskasing zone (Percival and Card, 1983; Percival and McGrath, 1986; Percival, 1989)] or through erosion [e.g., the Lewisian Complex, Scotland (Beach, 1980; Coward, 1984)].

Volumetrically the most important rock types in all instances tend to be quartz-feldspar-mica±garnet ortho- and paragneisses and schists, plagioclase-hornblende amphibolites, plagioclase-pyroxene±quartz±garnet granulite facies granofels and gneisses, pyroxenites, peridotites, anorthosites and, in some circumstances, eclogites. The mechanical properties of the lower crust are determined by the properties of these rock types.

The rocks generally show evidence of having arrived at their microstructural states after having suffered at least one cycle of high-grade regional metamorphism and deformation, with the final texture being developed under conditions of low to zero deviatoric stress (Myers, 1984; Brodie and Rutter, 1987b; Rutter and Brodie, 1990) (Plates 6-3a and 6-4a). This can be inferred from the occurrence of equilibrium mineral assemblages and coarse, isotropic equilibrium textures, with straight or gently curving grain boundaries and a lack of optical strain features within grains. On a larger-than-grain-scale, deformation-induced features are generally present. These include folds, boudinage and foliations defined by mineral banding and/or preferred shape orientation (schistosity) of the phases which tend to crystallize with an inequidimensional habit (e.g., micas, sillimanite). When the phases defining the schistosity exhibit equilibrium textural features on the grain scale, it is possible that the schistosity is simply a mimetic feature, guided by a pre-existing deformation-induced planar feature such as the shape of flattened grains or a pre-existing cleavage. Linear features in high-grade rocks are frequently defined by elongate clusters of equidimensional grains replacing features with a pre-existing elongate morphology, or defined by the intersection of two planar foliations. Microstructural evidence for the deformation mechanism(s) giving rise to perhaps most of the strain is often not preserved.

A caveat is required at this point, however. During deformation at high homologous temperatures and low stresses, the effects of interfacial forces on grain-to-grain textural relations can become comparable to the effects of shear forces. Textural coarsening may occur even in a material which is being slowly strained. Hence, dynamic textures may become difficult to distinguish from the effects of static recovery and grain growth (Plate 6-5). Rock masses which have clearly undergone high strains (on the basis of their strong foliation, mesoscale folds, deformed xenoliths, etc.) often show little evidence for this on the scale of a thin-section (e.g., Plate 6-5a,b). It is, therefore, important to integrate both field and microstructural observations in order correctly to deduce the deformation history of a region.

Commonly, the temperature peak of a regional metamorphic event tends to develop after the main deformation events which formed the thermal anomaly. It is widely believed that the strength of rocks during prograde metamorphism is lower than the resistance to intracrystalline plastic flow in the absence of metamorphic transformations (see below). The metamorphism arises through the relaxation of the distorted pattern of crustal isotherms back to an equilibrium configuration, through conductive and advective heat transfer, coupled with radiogenic heating. In regions of collisional tectonics, this principle is easily perceived from the observation that large-scale, complex, early formed structures are often overprinted by a simple structure of metamorphic isograd surfaces (e.g., the Lepontine metamorphism of the Western Alps; the Barrovian metamorphism of the Scottish Dalradian). In Archaean high-grade metamorphic terrains, the frequently observed synmetamorphic formation of structural gneiss domes may represent a mechanism of synmetamorphic advective heat transfer, and hence may testify to the weakness of high-grade terrains dominated by sialic rocks during

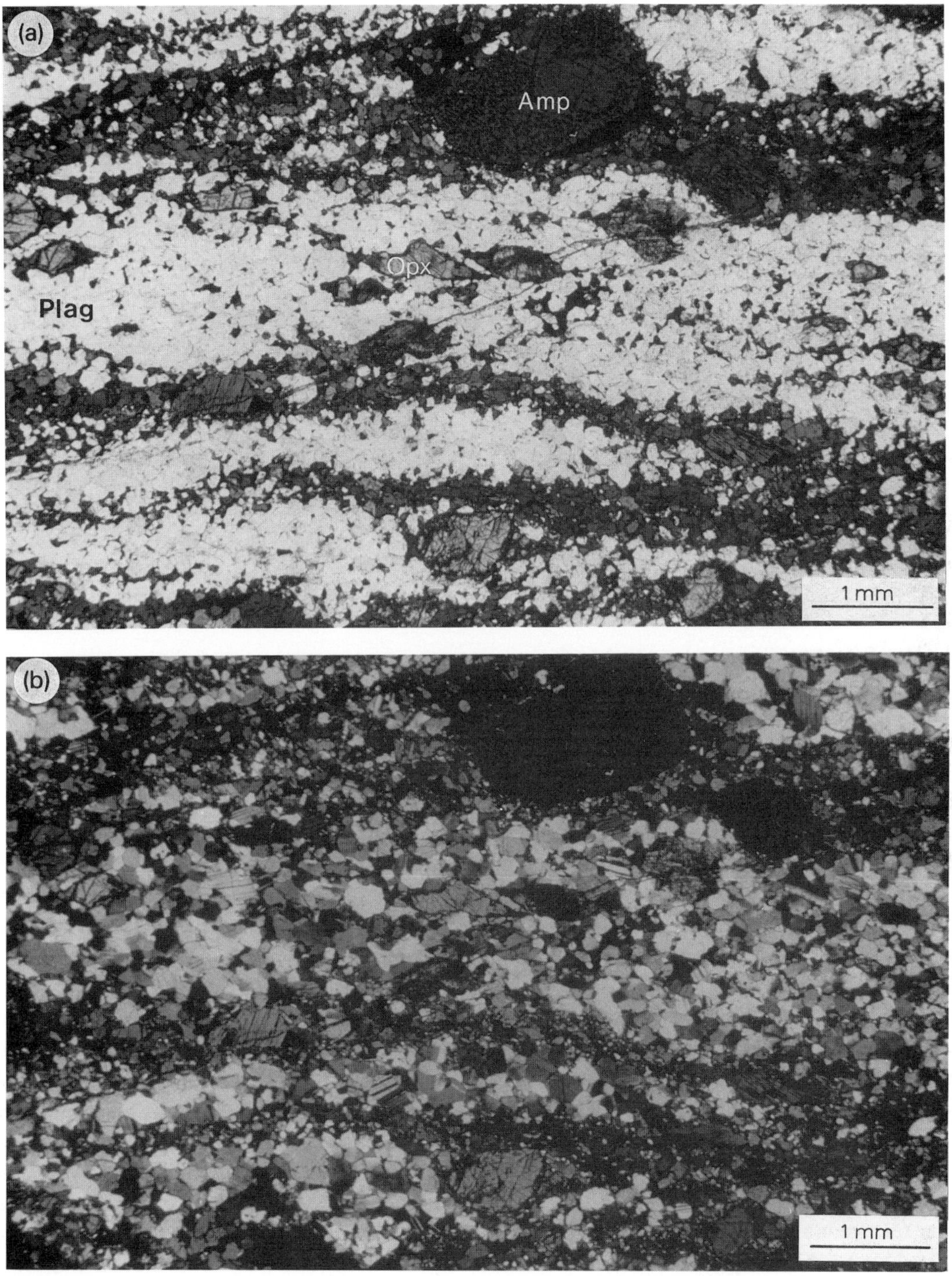

PLATE 6-5

(a,b) Optical micrograph of annealed metabasic mylonite showing porphyroclasts of amphibole and orthopyroxene (En_{65}) in an equigranular matrix of alternating bands of plagioclase (An_{55-60}), showing equilibrium grain shapes and pargasitic to ferroan-pargasitic hornblende. Fe-Ti oxides and minor quartz are also present (Kamila shear zone, Kohistan, Himalayas). (a) plane-polarized light, (b) crossed polars.

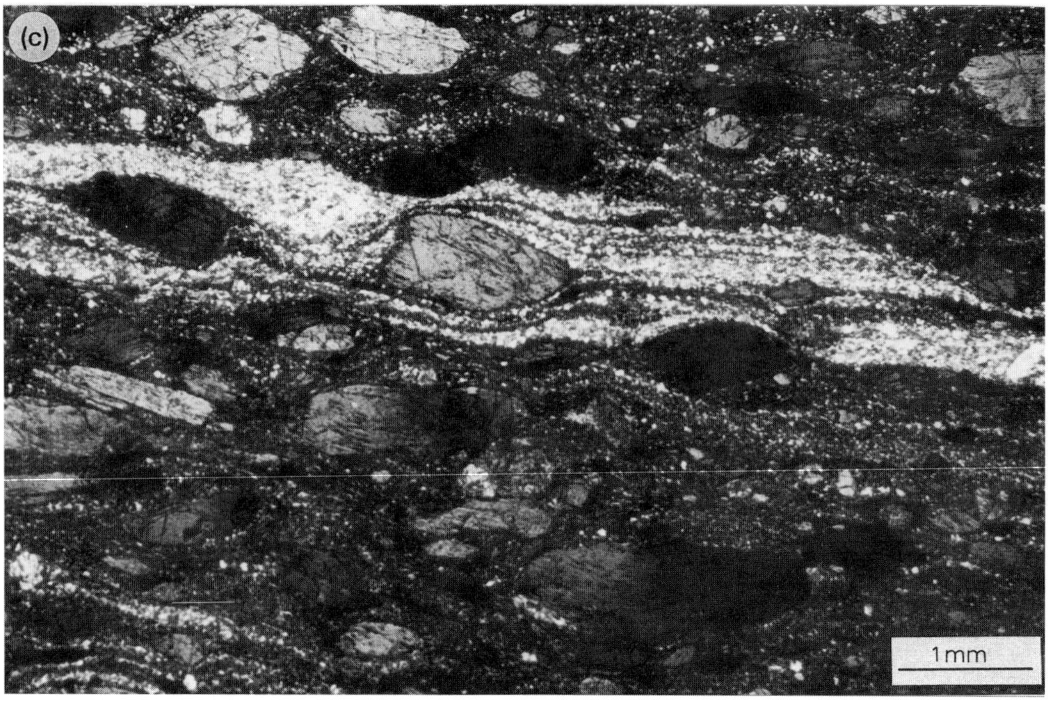

PLATE 6-5 (continued)

(c) Optical micrograph (crossed polars) of amphibole-plagioclase amphibolite facies metabasic mylonite showing development of a strong shape and crystallographic preferred orientation in both the plagioclase matrix (An_{30}) and the amphibole porphyroclasts and matrix (Ivrea–Verbano Zone, northern Italy).

metamorphism (e.g., Kröner, 1984; Hickman, 1984). Superimposed upon early formed pervasive deformation structures, evidence of subsequent deformation is commonly found in the form of more-or-less localized zones of high shear strain (e.g., Fig. 6-9) often associated with the uplift and exhumation of the high-grade rocks [e.g., Ivrea–Verbano zone, Italy (Hodges and Fountain, 1984; Handy, 1987; Brodie and Rutter, 1987b); Grenville Province, Canada (Davidson, 1984); Fiordland, New Zealand (Gibson et al., 1988); Kapuskasing uplift, Canada (Boland et al., 1988); Arunta Block, Australia (Goleby et al., 1989)]. The width of such localized zones may range from centimetric (e.g., Plate 6-6) to kilometric (e.g., the Nordre Stromfjord shear zone, Greenland; Sorensen, 1983). They may be accompanied by local effects of metamorphic retrogression, often involving metasomatic introduction of water and gain or loss of other chemical components. Absence of retrogressive effects may mean that the superimposed deformation occurred at high temperatures so that the high-grade mineral assemblage remained stable, or that the rock mass remained chemically closed owing to its low permeability. The observation of intracrystalline plastic flow of particular minerals may often be the only way to estimate a minimum temperature for the deformation. Non-retrogressive zones of superimposed deformation are easily recognized in the field through the development of a characteristic shape fabric of strained grains and a distortion or offset of pre-existing structural features. Microstructurally, they are recognized through the formation of optical strain features and, commonly, intense grain-size reduction through dynamic

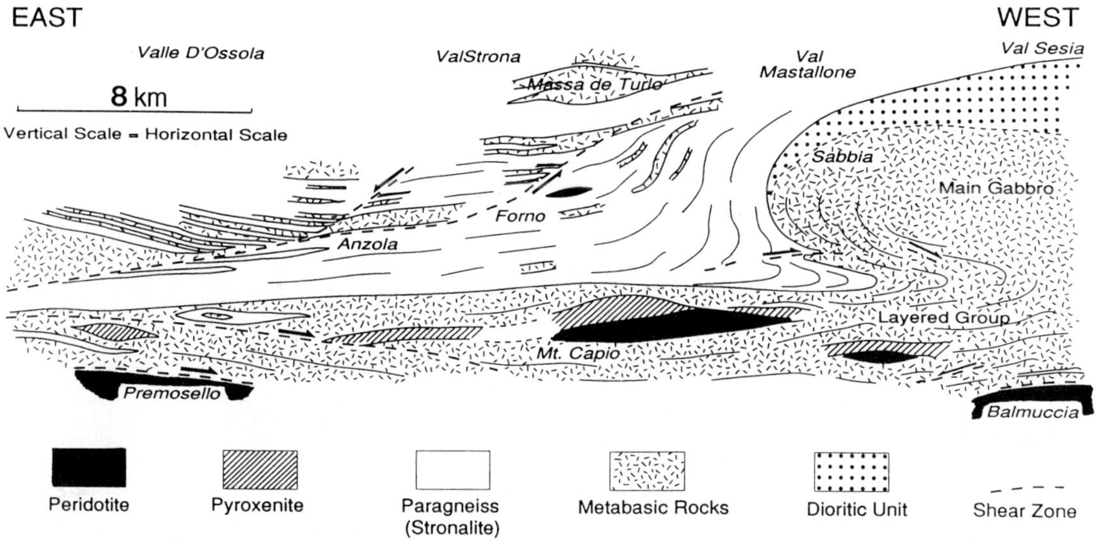

Fig. 6-9. Approximate restored geological cross-section, after removing the effects of post-Triassic folding and faulting, showing the principal features of the structurally lower part of the northern Ivrea–Verbano zone, between Valle d'Ossola and Val Sesia, as it would have appeared at the end of its period of crustal extension (probably during Triassic time). The section is drawn parallel to the slip direction of the high-temperature shear zones (shown as dashed lines), and represents about half of the presently preserved total thickness of the Ivrea–Verbano zone (after Brodie and Rutter, 1987; Brodie et al., 1992).

recrystallization (e.g., Plates 6-3, 6-6 and 6-7). Mineral assemblages remain unchanged by the deformation, except for mineral chemistry adjustments in response to small changes in the pressure/temperature conditions. If high temperatures persist after deformation, annealing (static) recrystallization of the microstructure may occur, removing intracrystalline strain features. Retrogressive shear zones are recognizable through the formation of localized zones of the new mineral assemblage together with a new grain-shape fabric. The relative timing of the retrogression with respect to the deformation may vary. In the case of either retrogressive or non-retrogressive shear zones, the localized nature of the deformation means that both the deformed rock and undeformed host rock are available for comparison. The stable localization of strain probably also means that mechanical softening processes operated.

The rheology and other properties of the lower continental crust are likely to have been very different during the prograde part of a metamorphic cycle and the subsequent history following the metamorphic peak. In the remainder of this section we attempt to compare these stages.

6.1. Crustal rheology during prograde metamorphism

6.1.1. Effects of metamorphic reactions on deformability

Prograde metamorphic cycles are generally short-lived on a geological time scale (ca. 10 Ma, from stratigraphic considerations). Existing pieces of lower continental crust are likely to experience further prograde effects only in the central parts of continental collision zones, or

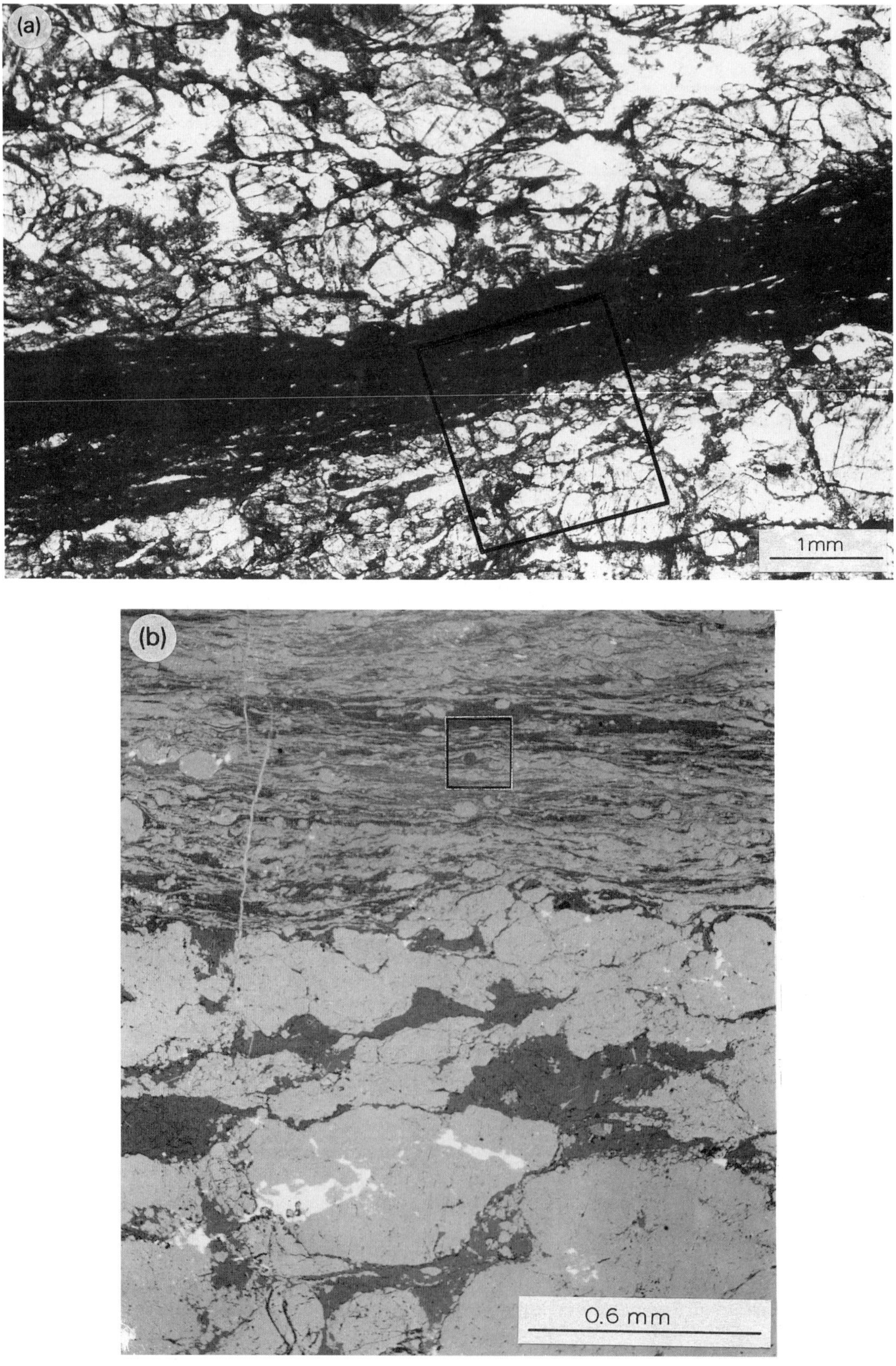
(a)
1mm
(b)
0.6 mm

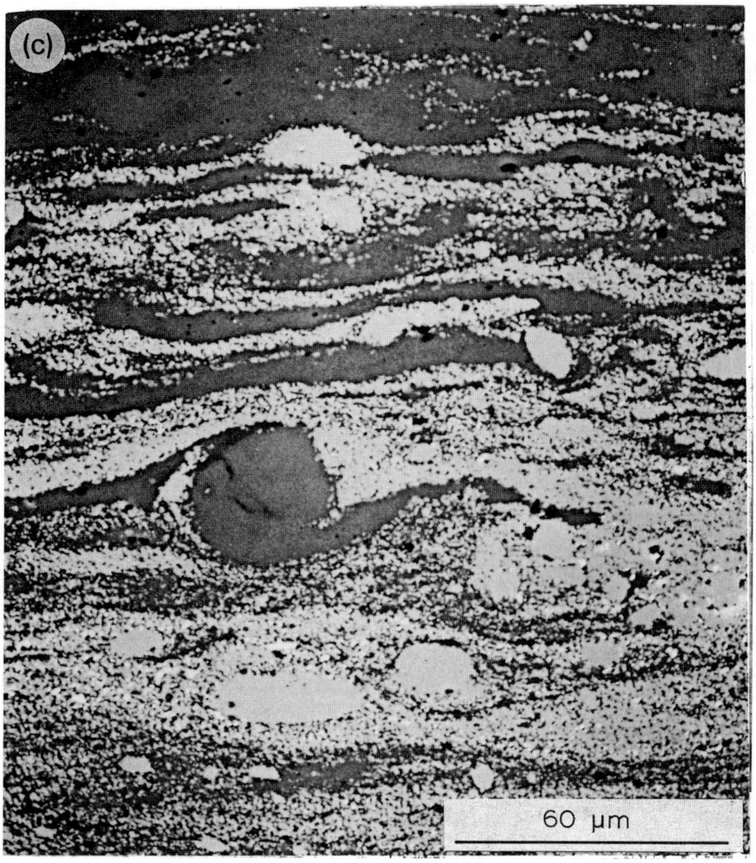

PLATE 6-6

Naturally deformed clinopyroxenite containing a narrow ultramylonite zone (Alpe Lut, Valle d'Ossola, Ivrea–Verbano Zone, northern Italy).
(a) Optical micrograph (crossed polars) showing shear strain of ca 1 in the majority of the rock (indicated by the distortion of the plagioclase grains (see also b) and more resistant pyroxene. A narrow ultramylonite band runs through the centre of the field of view, the dark colour being due to the very fine grain-size.
(b) Backscattered electron micrograph of the area outlined in a) at the host-rock–mylonite contact. Plagioclase (An_{80-82}, dark contrast), diopside-rich clinopyroxene (intermediate contrast), Fe-oxides (bright contrast).
(c) Higher-magnification backscattered electron micrograph of ultramylonite zone (area indicated in b). The matrix is composed of intergrown plagioclase (An 80) and clinopyroxene. The grain size is on the order of 1μm. A rotated relict porphyroclast of plagioclase (centre) of similar composition to the matrix plagioclase can be used to infer the sense of shear.

through an unusual accession of heat high into the continental lithosphere, for example in association with crustal stretching or rifting.

It can often be argued that the strength of the rocks during the observed pervasive deformation of the rocks was related to the progression of metamorphic reactions, particularly dehydration reactions (Fyfe, 1976; Fyfe et al., 1978; Brodie and Rutter, 1985; Murrell, 1985; Rutter and Brodie, 1990). Some metamorphic transformations may favour fault localization and even seismicity at high pressures (Burnley and Kirby, 1982; Kirby, 1987; Rutter and Brodie, 1988b; Dell'Angelo and Kirby, 1988; Green and Burnley, 1990). Deformability may be influenced in several ways. Within actively recrystallizing rocks the transient existence

PLATE 6-7

Metabasic mylonite showing grain refinement of plagioclase (An 80–85), elongation of the orthopyroxene (En 70) porphyroclasts oriented for easy <001> slip (arrowed), and rounding of garnet and other orthopyroxene porphyroclasts. Fine-grained orthopyroxene is present in the trails of the pyroxene porphyroclasts (crossed polars, Colloro, Ivrea–Verbano Zone, northern Italy).

of fine-grained reaction products may permit grain-size-sensitive flow processes until grain growth supervenes. Solid-phase volume changes may aid intracrystalline plasticity (Poirier, 1982; Paterson, 1983), and there is growing evidence that shear stress may influence both the mechanisms and kinetics of some metamorphic transformations (Burnley and Green, 1989). It is possible that the elevation of the activity of chemical components of the vapour phase during dehydration may facilitate the plasticity of silicates by 'hydrolytic weakening', particularly of the fine-grained phases (but see earlier). Finally, the elevation of pore-fluid pressure and the concomitant lowering of effective pressure between grains may facilitate cataclasis. Fluid permeation into adjacent rock masses not actively involved in reactions will tend to spread the enhanced deformability throughout the rock volume.

Whereas constitutive flow laws for intracrystalline plasticity have been obtained for a range of rock types, and these can be extrapolated to temperatures and strain rates of geological interest, it is not so easy to obtain constitutive flow laws for rocks undergoing metamorphic transformations. The use of suitable rock mechanics data to help understand synmetamorphic deformation in nature would also require it to be incorporated into a complete mathematical model of a section of crust, with prescribed thermal, displacement and fluid-flow boundary conditions.

At present there is only a very limited understanding of some of the above types of mechanical interactions between deformation and metamorphism. The facilitation of deformability through the formation of transiently fine-grained reaction products has been demonstrated during the olivine-spinel transformation using a magnesium germanate analog (Vaughan and Coe, 1981), and in the experimental dehydration of serpentinite under controlled pore pressure (Rutter and Brodie, 1988b). Microstructural evidence for such effects in nature is widespread (e.g., Rubie, 1983, 1990; Brodie and Rutter, 1987a). Simple measurements have been made of weakening and embrittlement during dehydration reactions under undrained conditions, in which evolved pore fluid is not allowed to escape from the sample volume [e.g., gypsum (Heard and Rubey, 1966; Murrell and Ismail, 1976); serpentinite (Raleigh and Paterson, 1965; Murrell and Ismail, 1976); chloritite (Murrell and Ismail, 1976)]. Detailed studies of the magnitude of the evolved pore pressure and of the mode of failure and microstructural evolution when the reaction progress is controlled by the heat input (as it would be in nature) have yet to be made. From observations on natural rocks, localization of flow and hence weakening associated with retrograde mineral transformations can often be inferred, but it can be difficult to deduce the mechanisms involved. Figure 6-10 shows theoretical speculations on how deformability, permeability and elastic wave velocity might vary during the progression of the dehydration of muscovite + quartz to orthoclase + sillimanite, an important reaction in pelitic rocks under lower crustal conditions (after Rutter and Brodie, 1990). The maximum pore pressure attained depends upon the rate of heat input and how quickly excess pore pressure is dissipated by permeation, and determines the precise amount by which the strength is reduced and the mode of failure (whether the deformation is distributed or localized into a fault). Whilst the pore pressure is close to the lithostatic pressure the strength is expected to be vanishingly small, P and S wave velocities are likely to be lowered by ca. 1 km s^{-1} and electrical conductivity should be anomalously high.

The low-velocity channel in the lower crust beneath the Pennine Alps (Mueller et al., 1976; Miller et al., 1982) may represent a zone in which dehydration reactions are actively occurring. Hyndman (1988) interprets gently dipping seismic reflectors observed in lower continental crust above the northern Cascadia subduction zone to be layers containing trapped pore water produced from dehydration reactions in the subjacent subducting slab. Such conditions are only likely to be seen in tectonically active regions and may be short-lived on a geological time scale, although Hyndman (1988) discusses circumstances in which they may be long-lived. However, the common preservation of high-grade, relatively anhydrous mineral assemblages in exposed geological sections of lower crustal provenance must mean that once-high pore pressures were dissipated to sufficiently low values such that retrogressive (re-hydration) reactions were precluded during uplift and cooling (e.g., Yardley, 1986). The idea that the 'normal' state of the lower crust is one of low pore pressure is supported by the fact that lower crustal seismic velocities typically correspond closely to high-pressure velocity measurements on dry crystalline rocks (Fountain, 1976; Kern, 1978; Meissner, 1986; Rutter and Brodie, 1989; Fountain et al., 1990).

6.1.2. Role of partial melting

From field studies (e.g., Paquet and Francois, 1980, Paquet et al., 1981; Myers, 1984; Hollister and Crawford, 1986; Davidson et al., 1992) it is clear that partial melts frequently play a role in the deformation of rocks undergoing high-grade metamorphism. The release of water during dehydration reactions may promote the formation of partial melts of granitic composition, which in turn dissolve the pore water. Extraction of melt, therefore, extracts water from the refractory residue. The injection of mantle-derived, hotter basic magma may

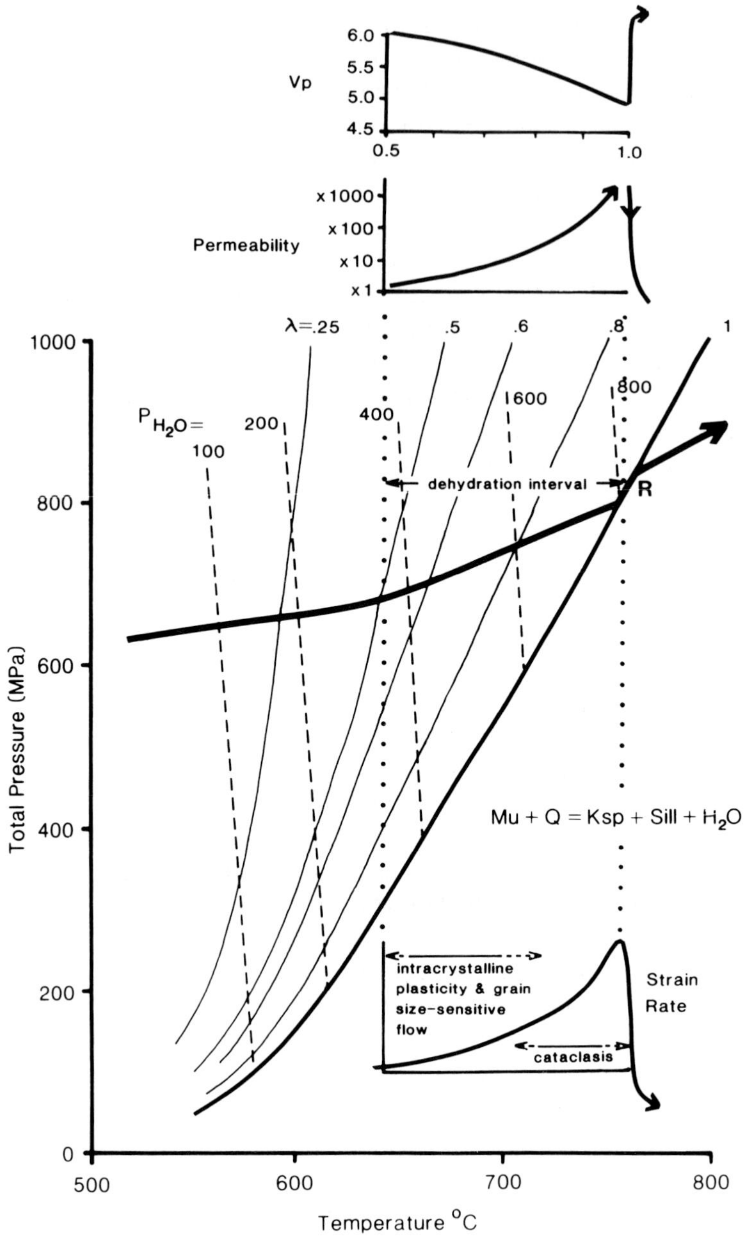

Fig. 6-10. Graph of P_{total} vs. temperature for the reaction muscovite + quartz = K-feldspar + sillimanite + water (after Rutter and Brodie, 1990), subject to the assumption that water activity corresponding to the pore pressure applies to stressed grain boundaries, and nelecting the effects of melting which may occur during the dehydration interval. Equilibrium boundaries for constant values of pore pressure, P_{H_2O} (and $\lambda = P_{H_2O}/P_{total}$) are shown. The approximate form of a possible pressure-temperature path through the dehydration interval is also shown by a bold line (for a constant specific volume of water = 0.0014 m^3 kg^{-1}, a constant heat flux (dQ/dt) and a constant range of total pressure (dP_{total}/dt). When dehydration begins (at λ = 0.5 in this case), the *P-T* path is deflected upwards because part of the heat input is now required to drive the reaction. When the path reaches the λ = 1 curve, it follows this curve (path segment R) until the reaction is complete and excess (relative to available void space) fluid has been dissipated through permeation.

Inset curves show how P-wave velocity (V_P) and (qualitatively) strain-rate are likely to vary over the dehydration interval.

be important in the supply of heat for the partial melting of crustal rocks, and it may of itself enhance the deformability of the rock pile.

In many respects, the role of melt during deformation is expected to be similar to the role of pore water released in dehydration reactions: it acts as a pressurized pore fluid and facilitates cataclastic deformation at low stresses. Whether melting occurs under pore-fluid-present or pore-fluid-absent conditions may influence the mechanical behaviour. For fluid-absent melting the volume change is positive, hence conditions favouring hydraulic fracturing are attained more rapidly than in the case of fluid-present melting (Clemens, 1984). Unfortunately, far less is known about the rheology of partially melted rocks than the effects of pore pressure on rock mechanics. Experimental studies have demonstrated two types of effect of partial melting:

(a) The melt acts like a pore fluid at a pressure equal to the least principal stress (Van der Molen and Paterson, 1979; Paquet and Francois, 1980, 1981; Dell'Angelo and Tullis, 1988; Wilshire and Kirby, 1986; Wilshire, 1990). Under non-hydrostatic stress this favours brittle fracture, with the opening of liquid-filled cracks parallel to the maximum stress. The strength corresponds to the uniaxial strength of the rock. Up to melt fractions of about 20–30%, the solid fragments form an interlocking network, so that whilst it is possible to form a cataclastic fault 'lubricated' by pressurized melt, particulate flow of the fragments is not possible. In natural migmatite terrains, planar, melt-lubricated fault zones containing entrained fragments are frequently observed. Van der Molen and Paterson (1979) observed from experiments that rock strength was not sensitive to melt fraction nor to strain-rate variations when the melt fraction was less than about 20%. At higher melt fractions the interlocking framework of fragments collapses, fragments are able to roll past one another, and the strength falls rapidly to a level comparable with the flow resistance of the liquid. Under this condition, melt would be easily squeezed through the rock mass under very small deviatoric stresses (< 1 MPa), hydraulically fracturing the surrounding rocks as it moves.

(b) Diffusion creep may be kinetically accelerated owing to the high-diffusivity pathway provided by melt which wets the grain boundaries. Cooper and Kohlstedt (1984) and Ji and Mainprice (1986) reported Newtonian-viscous flow of fine-grained olivine and albite rocks respectively when grain boundaries contained melt. Dell'Angelo and Tullis (1988) also inferred this process to be active in the experimental deformation of partially molten, fine-grained granitic rock. Little attention has been given to the question of the recognition of this effect in naturally deformed rocks.

6.2. Crustal rheology of high-grade rocks after peak metamorphism

6.2.1. Isochemical high-temperature deformation

It is to the post-peak-metamorphic history of crystalline rocks that flow laws obtained from the experimental deformation of rocks and minerals by intracrystalline plasticity are most relevant. Good correspondence has been found between the microstructures of naturally and experimentally deformed rocks (e.g., Plates 6-1 and 6-7) when the deformation is not accompanied by chemical or mineralogical changes. Plates 6-3 and 6-6 to 6-8 show typical microstructures of metapelitic, metabasic and ultrabasic rocks from the Ivrea–Verbano zone, which were deformed under lower crustal conditions.

From microstructural studies, the relative ease of plastic deformation of different mineral phases in the same rocks can generally be inferred. Deformation of metapelites and metagranitoids is controlled by the behaviour of quartz (e.g., Hanmer, 1982; White et al., 1980) and

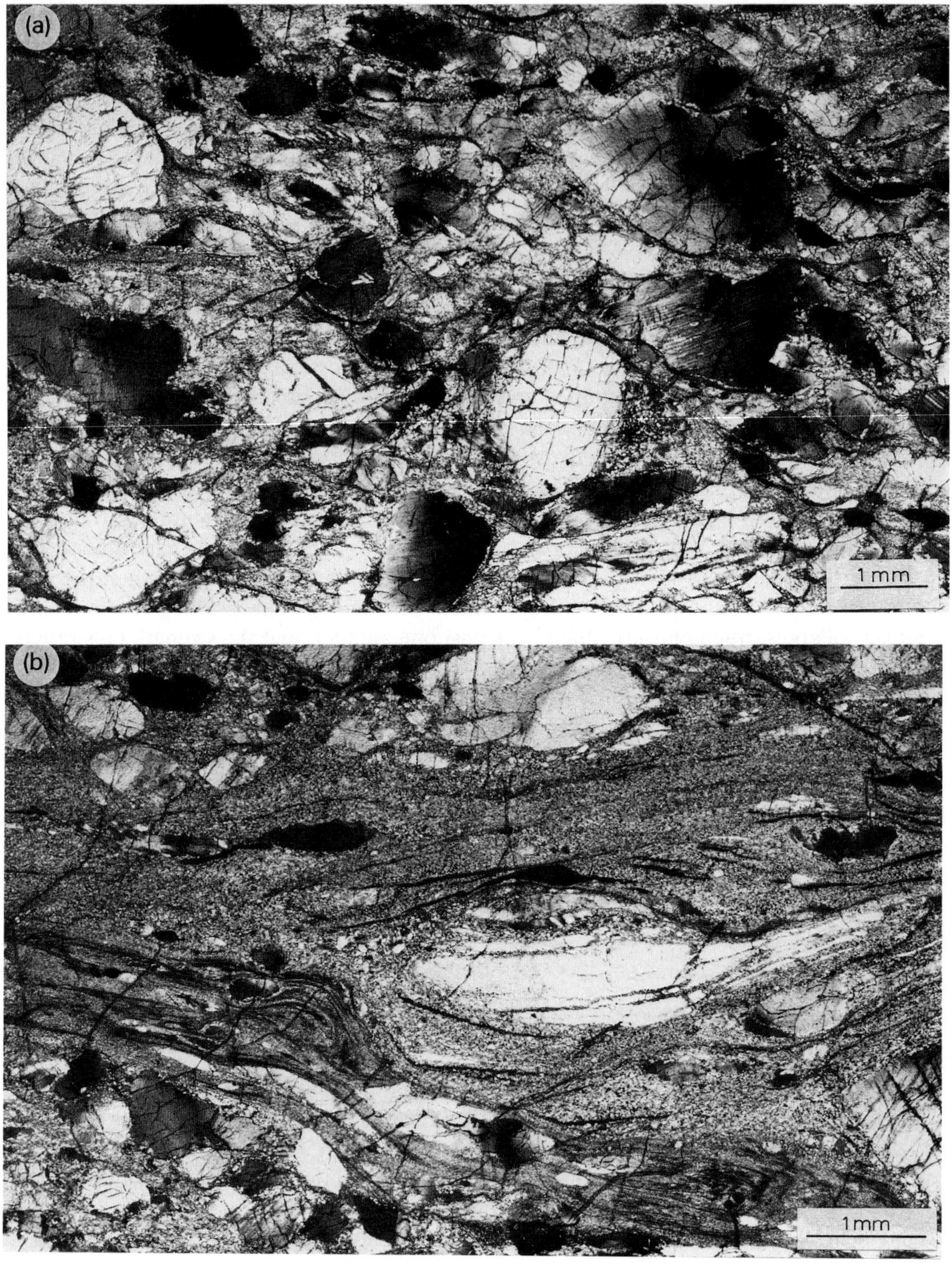

PLATE 6-8

Deformed olivine-clinopyroxene-spinel dunite (Premosello, Ivrea–Verbano Zone, northern Italy).

(a) Optical micrograph of protomylonitic dunite showing initial stages of recrystallization of the olivine. Optical strain features are visible in the porphyroclasts (crossed polars).

(b) Mylonitic dunite showing the development of fine- and ultrafine-grained zones of olivine and clinopyroxene surrounding the relict porphyroclasts (crossed polars).

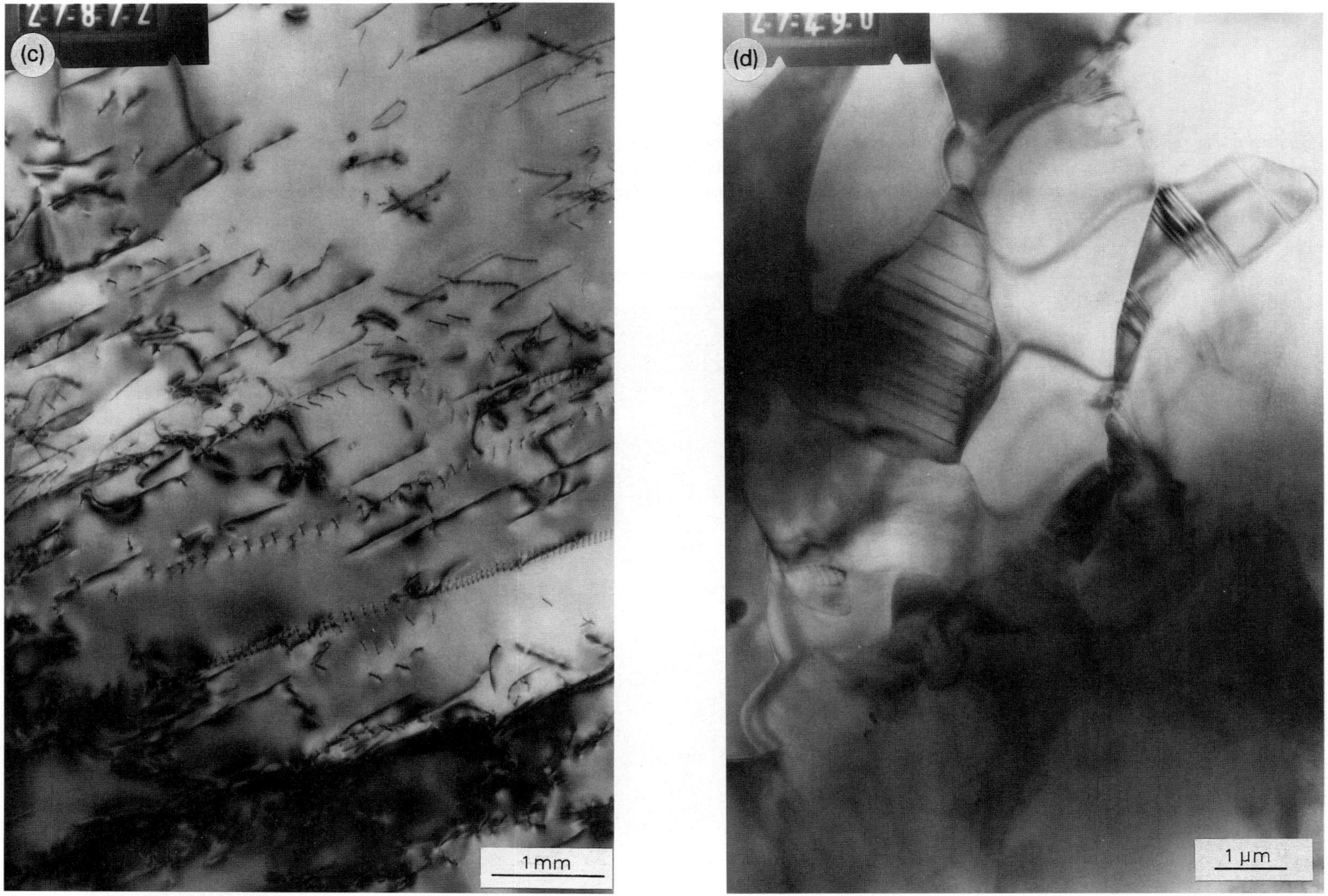

PLATE 6-8 (continued)

(c) TEM micrograph showing the dislocation structure of an olivine porphyroclast (indicative of high-temperature plastic deformation) and development of bound dislocations in subgrain walls. Free dislocations are also relatively abundant.

(d) TEM micrograph of the ultrafine-grained matrix (see (b)) showing intergrown clinopyroxene (containing planar stacking faults) and olivine, of grain size 1–2 μm.

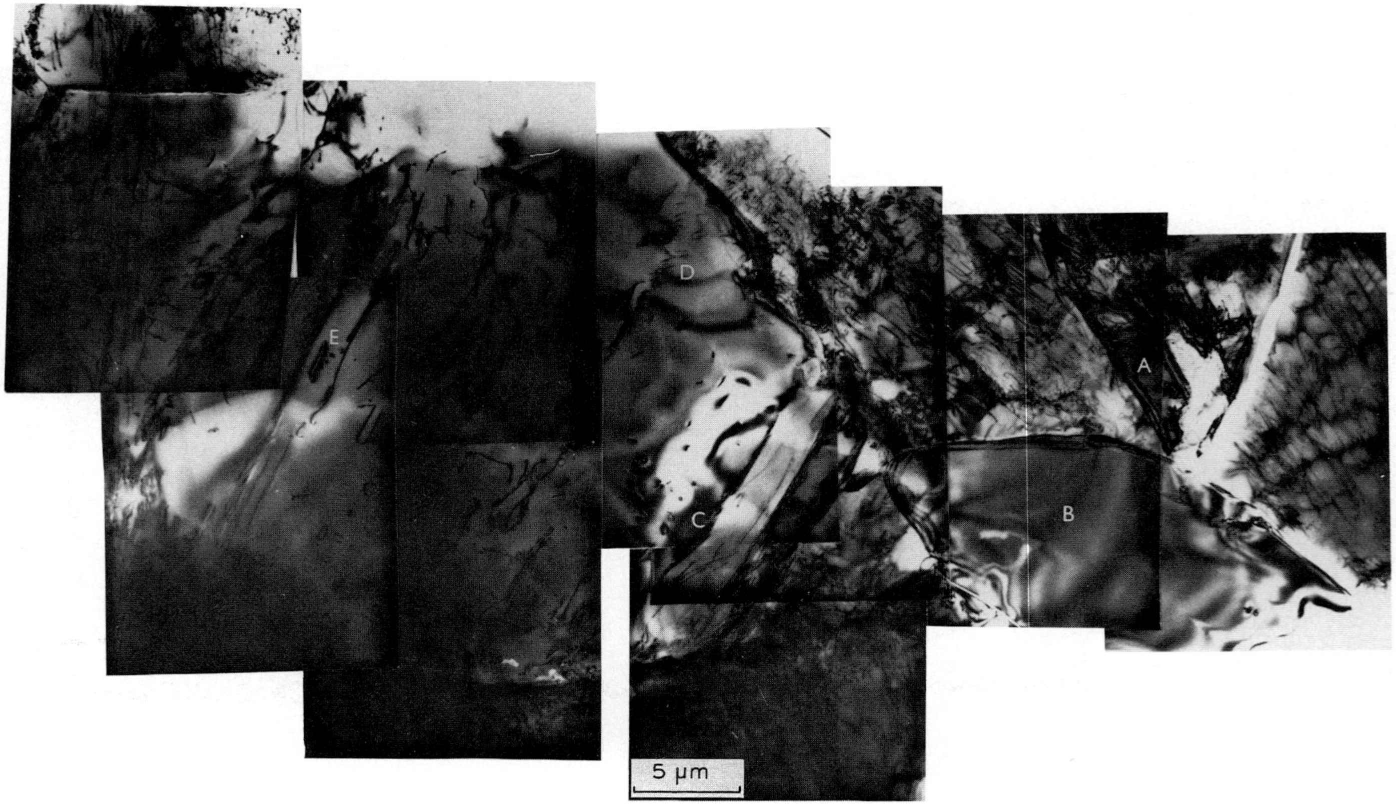

PLATE 6-9

High-voltage TEM image of a boundary region between a relict olivine porphyroclast, with a high dislocation density and (100) subgrain boundaries (region *A*), and recrystallized matrix olivine (regions *B*, *C*, *D*). Small, newly recrystallized grains contain low dislocation densities (*B*), but earlier-formed recrystallized grains show evidence of high-temperature plasticity (subgrain walls (*D*), curved dislocation lines and loops (*E*)). In due course such grains would recrystallize again. This is the type of microstructure expected of a material deforming with cyclic dynamic recrystallization. Olivine mylonite, Anita Bay, New Zealand.

perhaps also by the plasticity of micas (if present). Over a wide range of metamorphic grades, when quartz and feldspar occur together the quartz is the more ductile and the feldspar is either resistant to flow or deforms by brittle fracture (Plates 6-2 and 6-3). With increasing grade the proportions of quartz and mica generally decrease, and feldspar may become the principal load-supporting phase. From a transmission electron microscopy study of naturally deformed calcic plagioclase, Montardi and Mainprice (1987) argued that pressure-sensitive cross slip of dissociated <001> dislocations may be the most important rate-controlling process in lower crustal plagioclase plasticity. Olsen and Kohlstedt (1985) provide a comprehensive account of intracrystalline plasticity and progressive dynamic recrystallizaion in plagioclase from the Nordre Stromfjord shear zone, west Greenland. White and Mawer (1986, 1988) have described perthitic feldspar plasticity in mylonites from Parry Sound, Ontario (Grenville Province). Natural rock microstructures confirm the general pattern of behaviour observed in experimental studies, that the more mafic minerals are the most resistant to intracrystalline plastic deformation (review by Brodie and Rutter, 1985). Although in naturally deformed (hornblendic) amphibolites evidence is commonly seen for limited plasticity and dynamic recrystallization (e.g., Dollinger and Blacic, 1975; Brodie, 1981; Biermann and van Roermund, 1983), the deformability of associated plagioclase is generally orders of magnitude greater. Both clinopyroxene and orthopyroxene are also generally more resistant than plagioclase. Garnets almost always behave as resistant inclusions in a flowing matrix of other minerals, or they may fail by brittle fracture. Garnet-pyroxene eclogite units are generally found to be extremely resistant to deformation, except where accompanied by metamorphic retrogression.

Many descriptions of the natural plastic deformation of olivine-rich rocks have been reported. Ultramafic xenoliths from basalt flows and kimberlite pipes, coupled with samples from ultramafic massifs and pods in continental crustal rocks, have allowed plastic flow and recrystallization (Plate 6-9) over an extremely wide range of pressure/temperature environments to be studied (e.g., Green and Radcliffe, 1972; Boland and Buiskool-Toxopeus, 1976; Mercier et al., 1977; George, 1977; Kirby and Green, 1980). In peridotites, olivine generally flows more easily than the pyroxenes, but the pyroxenes generally recrystallize to finer grain sizes after smaller amounts of strain (e.g., Boullier and Gueguen, 1975).

The microstructural changes which typically accompany intracrystalline plastic deformation, viz. dynamic recrystallization and/or development of an intense crystallographic preferred orientation, commonly lead to localization of the flow into shear zones. In the structurally lowermost part of the Ivrea–Verbano zone crustal section for example, high-temperature deformation after the peak of regional amphibolite and granulite facies metamorphism (dated ca. 300 Ma) was localized into conjugate arrays of plastic shear zones, which accommodated pure-shear extension of the layered sequence of amphibolites, metapelites, basic granulite facies rocks and ultrabasic (pyroxenite and peridotite) lenses (Brodie and Rutter, 1987b). Activity of the shear zones was associated with cooling of the rocks, as documented by radiometric dating (Zingg, 1980, 1983; Handy, 1987; Brodie et al., 1989). Deformation was isochemical (Brodie, 1981) and usually isomineralic, with only small changes in mineral chemistry of individual phases being recorded between host and sheared rocks. The central parts of the shear zones, which range up to ca. 30 m in thickness, are characterized by intense grain-size reduction of feldspar, olivine or quartz (according to which is the principal load-supporting phase) by dynamic recrystallization (Plates 6-3, 6-4, 6-5c, 6-6, 6-8). The shear offset on one zone was estimated to be about 5 km. The shear zones pass laterally into their host-rocks via zones (up to several hundreds of metres wide) of decreasing intensity of foliation defined by the shape of plastically deformed grains.

In metabasic rocks of the Ivrea–Verbano zone, plagioclase is the softest phase, but dynamic recrystallization becomes effective after strains of only a few tens of percent. Mafic phases (amphiboles and pyroxenes) in metabasic rocks show more limited plasticity, but both can become deformed sufficiently to drive dynamic recrystallization (Plates 6-6 and 6-7). The rims of porphyroclasts of hornblende and the entire volume of recrystallized grains can be compositionally different from the parent grains, reflecting the (small) difference in pressure/temperature conditions of original crystallization and subsequent deformation (Brodie, 1981). From the constraints of the original mineral assemblages in the Ivrea–Verbano zone and the radiometrically determined cooling curve for the region, this deformation can be inferred to have occurred in the temperature interval between 750 and about 450°C. In all rock types, garnet is extremely resistant to plastic deformation. It is often carried about as passive, equidimensional grains in a flowing matrix of other phases (Plate 6-4d). Some granulation of garnet around porphyroclasts occurs, leaving them highly rounded (Plates 6-4d and 6-7). From the microstructure and the observation that the fine grains are generally compositionally identical to the host grain, it has been inferred that the granulation may be by brittle fracturing (Brodie and Rutter, 1985). Brittle abrasion may also be responsible for rounding grains of other mafic phases when carried about in a matrix of flowing plagioclase. From the general correlation between feldspar grain refinement and shear localization seen in the Ivrea–Verbano zone rocks, it is tempting to infer that the rock within the shear zone has been weakened by grain-size-sensitive processes of the types discussed in the previous section. Sometimes, the recrystallized feldspars in metabasic shear zones have a strong preferred crystallographic orientation (Plate 6-5c), which suggests that intracrystalline plastic processes have been involved at least late in the evolution of the zone. Equally, randomization of fabric in fine-grained, recrystallized matrix feldspars of a high-grade mylonite from Cucamonga Canyon, California, has also been reported (Behrmann and Mainprice, 1987) and attributed to the effects of flow by grain-boundary sliding. Sodre-Borges and White (1980) described marked grain-size reduction in mylonitic shear zones in an anorthosite in the Lewisian complex of South Harris. Scotland. They demonstrated that the grain-size reduction had been through dynamic recrystallization. From the observation that recrystallized grains from pre-existing bands of different composition had become mixed in the shear zone, they inferred that grain-boundary sliding had become important in the flow after the grain-size reduction had occurred.

Shear zones with the above characteristics are very common in geological sections of lower crustal provenance (e.g., Etheridge, 1975; Coward, 1984; Daly, 1986; White and Mawer, 1986, 1988; Behrmann and Mainprice, 1987). Localized zones have been described ranging in width from centimetric to several kilometres. The reactivation of lower continental crust, which has been 'hardened' in a regional metamorphic event, by new tectonothermal activity probably typically proceeds by the formation of localized shear zones. Hence modelling of the behaviour of the crust at such times requires (a) information on the geometry of shear zone arrays, (b) mechanical data on the change of rheological properties as flow localization develops and, (c) constitutive flow laws for the fully evolved rocks within shear zones.

6.2.2. Reactivation accompanied by metamorphic retrogression

The above examples were of near-isochemical, post-peak-metamorphism deformation of crystalline rocks. The tectonic reactivation of mid- and lower-crustal rocks often involves the formation of shear zones which act as loci for the introduction of fluids (e.g., Beach, 1980;

Kerrich et al., 1980; Hanmer, 1982; LaTour and Kerrich, 1982; Brodie and Rutter, 1985; Krishna-Sinha et al., 1986). The sheared rocks frequently appear as highly foliated products of the retrogressive metamorphism of the parent rocks. Static hydration of the host-rocks is also common on either side of the shear zone, the host-rock texture often being pseudomorphed by mineral products of the retrograde reaction (e.g., Beach, 1980; Rutter and Brodie, 1985). Because such zones soak up water it is likely that high pore pressures do not play an important role in their progressive development (although they may be required for their initiation). Most of the possible mechanistic interrelationships mentioned above between deformation and metamorphism may be involved in the behaviour of retrogressive shear zones (Plate 6-10).

Often it is not clear to what extent the metamorphism has facilitated the weakening of the rock in the shear zone or vice versa. Segall and Simpson (1986) have noted field evidence that brittle fracture may provide the dilatancy necessary to allow the initial influx of water into the shear zone, an inference also made by Beach (1980) and McCaig (1989). However, Fischer and Paterson (1989) have shown that, especially under low (but non-zero) effective pressures substantial increases in porosity (and presumably also permeability) may accompany deformation even in marble at temperatures as high as 600°C. This may be attributable to microcracking, but it should also be considered that dynamic creation and maintenance of significant intergranular porosity (especially in silicate rocks) may be a strain accommodation process if insufficient slip systems are available within grains to allow polycrystal flow at constant volume. Additionally, it is possible that in flow by high-temperature grain-boundary sliding, the formation of a small but significant, dynamically maintained porosity is common (Behrmann and Mainprice, 1987; Walker et al., 1990; Carter et al., 1990). Such porosity in shear zones in the middle and deep crust may permit the transport of the fluids which cause metamorphic retrogression without the need always to appeal to hydraulic fracture.

Rutter et al. (1985) reported experiments on the 'syntectonic' hydration of basalt at 600°C under controlled pore-water pressure, wherein the original plagioclase+pyroxene assemblage was replaced by plagioclase+amphibole. Although oriented growths of amphibole were produced, corresponding to naturally observed microstructures, only a slight mechanical weakening of the rock resulted. At present, therefore, there is little experimental rock mechanics basis for modelling the behaviour of crystalline rocks reactivated by shear zones in which localized retrogressive metamorphic reactions occur. Yet the progressive reactivation of pre-existing basement rocks is an important process in orogenic cycles. The Western Alps provides an excellent example of this point. The Hercynian basement massifs (e.g., Aar, Aiguilles Rouges, Belledonne) of the external Helvetic zone show little effect of Alpine metamorphism and deformation, behaving instead as rigid masses around and over which the more ductile Mesozoic and Tertiary cover rocks have been deformed (e.g., Ramsay, 1963). Passing up the steep thermal gradient of Alpine metamorphism (through the Gotthard massif), retrogressive shear zones begin to affect the basement rocks, increasing in frequency until within the Pennine zone the basement rocks deform homogeneously with the cover sediments, the retrogressive metamorphism (and the necessary pore water) having fully penetrated the basement.

In the foregoing, we have tried to illustrate the range of ways in which rocks of the lower crust can become deformed, and to indicate that what we know of rock-deformation mechanisms and constitutive flow laws from laboratory experiments at present only describes flow by intracrystalline plasticity at constant chemistry and over small strains. In nature, a wider range of deformation mechanisms is possible, and significant microstructural, mineralogical and chemical changes may take place throughout a given tectonothermal cycle.

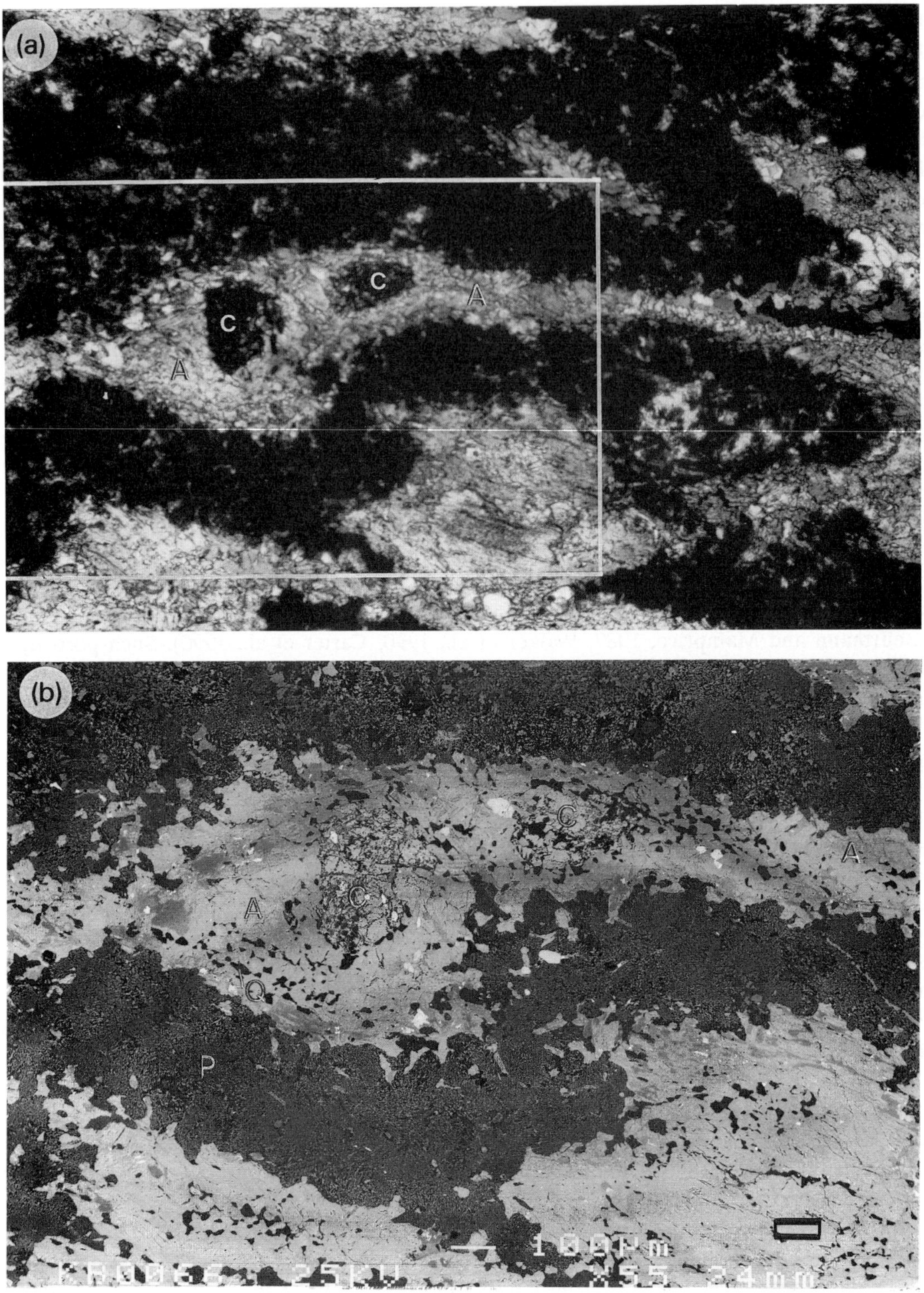

PLATE 6-10

Optical (a) and back-scattered electron (b) micrographs of a naturally deformed metagabbro (Cabo Ortegal Complex, northwest Spain), showing deformation apparently related to a retrogressive metamorphic reaction. The area of (b) is outlined in (a). The scale bar in (b) is 100 μm. The clinopyroxene (*C*), originally rimmed by granulite facies garnet, has partially reacted to a fine-grained intergrowth of amphibole and quartz (*A* and *Q*), which has been drawn out during the deformation. Plagioclase (*P*) has been altered at lower temperatures to a fine-grained symplectite of epidote and sodic plagioclase.

7. Flexure of the continental lithosphere

The mechanical response of the lithosphere to the imposition of localized loads depends on its rheology through a wide depth interval. Therefore, from the form of topographic profiles or the geometry of sedimentary basins it may be possible to constrain independently aspects of crust/mantle rheology, which can then be compared with theoretical modelling based on experimentally obtained flow laws.

Owing to the relatively simple structure of the oceanic lithosphere, the mechanical analysis of its flexure due to seamounts and subduction has provided good agreement with geophysical constraints such as gravity anomalies, topographic profiles and localization of seismicity. The problem has usually been approached by assuming the lithosphere to behave as an elastic (e.g., Walcott, 1970; Hanks, 1971; McNutt and Menard, 1978), elastic-plastic (e.g., Bodine and Watts, 1979) or viscoelastic (e.g., De Rito et al., 1986) slab overlying an inviscid asthenosphere. More realistic layered rheologic models have also been explored (e.g., Goetze and Evans, 1979; Kirby, 1983). The weight of the localized load is supported by the elasticity of the slab together with buoyancy forces arising from the lateral displacement of the upper mantle asthenosphere.

Similar approaches have also been made to the analysis of vertical displacements of the continental lithosphere, such as are associated with intracratonic sedimentary basins (Watts et al., 1982; Bills, 1983; Lambeck, 1983; Nunn and Sleep, 1984; Garner and Turcotte, 1984), subsidence of continental margins during continental extension and rifting (Park and Westbrook, 1983; Alvarez et al., 1984), and the isostatic compensation of fold-and-thrust belts formed during continental collision (Lyon-Caen and Molnar, 1983; Karner and Watts, 1983; Hall and Chase, 1989). The problem is generally more complicated than for oceanic lithosphere flexure, owing to the existence of more significant vertical and horizontal lithologic heterogeneities, a more complex thermal history and the requirement to take into account loading arising from sediment deposition.

Elastic flexure theory can be used to define the effective thickness of the 'mechanical' lithosphere. Flexural rigidity, which measures the resistance to bending of an elastic plate, is proportional to the third power of the plate thickness. The effective thickness is the zone within which bending stresses may be stored for long periods of time and below which deviatoric stresses drop virtually to zero. The apparent, or effective thickness is measured by adjusting the thickness of a model elastic plate until its profile gives the best fit with the observed topographic profile.

Departures from a perfect fit arise because the lithosphere is not simply a perfectly elastic plate. For simple elastic flexure, the fibre stresses increase linearly with distance from the neutral surface, reaching large tensile values on the outer arc and large compressive values on the inner arc. In reality, near the free surface the tensile stress profile will initially increase with depth according to the rate of increase of brittle strength with depth, and around the inner arc of the flexure the compressive stresses will be relaxed by time-dependent creep because temperature increases with depth. The time-dependent relaxation of bending stresses at depth reduces the apparent mechanical thickness, so that the compensation for the applied load is spread over a narrower zone than would be required if the plate remained perfectly elastic. Thus, the seaward walls of oceanic trenches are steeper than predicted from elastic models (e.g., McAdoo et al., 1978; Bodine and Watts, 1979) and sedimentary basins may become narrowed (Fig. 6-11), with uplift of the margins as peripheral bulges (Nunn and Sleep, 1984; Willett et al., 1985; De Rito et al., 1986). The effective mechanical thickness

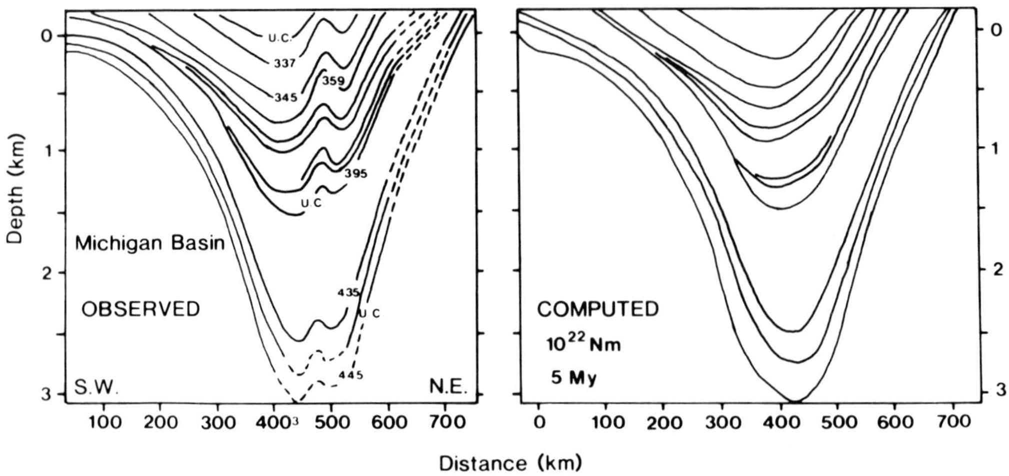

Fig. 6-11. Production of an intracratonic sedimentary basin through lithospheric flexure. The left figure shows a cross-section through the Palaeozoic Michigan basin. The numbers indicate the age B.P. of particular stratigraphic levels in millions of years; *u.c.* = unconformity. The right figure shows the computed profile assuming a viscoelastic lithosphere of flexural rigidity 10^{22} Nm and a viscoelastic decay time of 5 Ma. The cross-section was computed by filling the computed deflection of the lithosphere with the restored thickness of each sedimentary layer (after Nunn and Sleep, 1984). The narrowing of the basin with time as a result of viscous relaxation of the lower lithosphere is apparent through the tendency to expose lower stratigraphic levels with increasing distance from the basin centre.

of the lithosphere is commonly much less than the seismic or thermal thicknesses (ca. 100 km), owing to the effects of anelastic relaxation processes within the plate. Because creep is temperature-sensitive, the mechanical thickness might be expected to vary inversely as the thermal gradient so that the base of the mechanical lithosphere would correspond to a particular geotherm. Although this is true of the oceanic lithosphere [the base of the mechanical lithosphere being inferred to correspond to the 750°C isotherm (Watts et al., 1980)], the lithologic heterogeneity and potentially complex thermal history of continental lithosphere might be expected to prevent the principle being applicable to the latter. Willett et al. (1985) show that there is no strong correlation between heat flow and apparent mechanical thickness in the continental lithosphere, so that for a given present-day heat flow the base of the mechanical lithosphere may correspond to temperatures lying in the wide range between 300 and 900°C (Fig. 6-12). They describe the results of numerical models based on rock mechanics data for the mechanical properties of the continental crust and upper mantle. These very effectively demonstrate the temporal evolution of flexural topographic profiles and stress profiles with depth and effective mechanical lithosphere thickness for two constant values of surface heat flow (40 and 90 mW m^{-2}) (Figs. 6-13 and 6-14). The effects of stress relaxation are virtually complete after about 1–10 Ma.

Tending to counteract the effects of effective lithospheric thinning caused by stress relaxation, the general cooling of the lithosphere causes effective lithospheric thickening, but over a much longer time scale (hundreds of millions of years). As a result, effective elastic thicknesses can exceed 100 km in old continental platforms and shields (Lyon-Caen and Molnar, 1983; Karner et al., 1983; Peltier, 1984; McNutt, 1987; McNutt and Kogan, 1987; Grotzinger and Royden, 1990) and in old, eroded mountain belts (Karner and Watts, 1983) (Fig. 6-15). Garner and Turcotte (1984) argue that cooling during the evolution of the Anadarko basin (Oklahoma) through the Carboniferous period was associated with effective

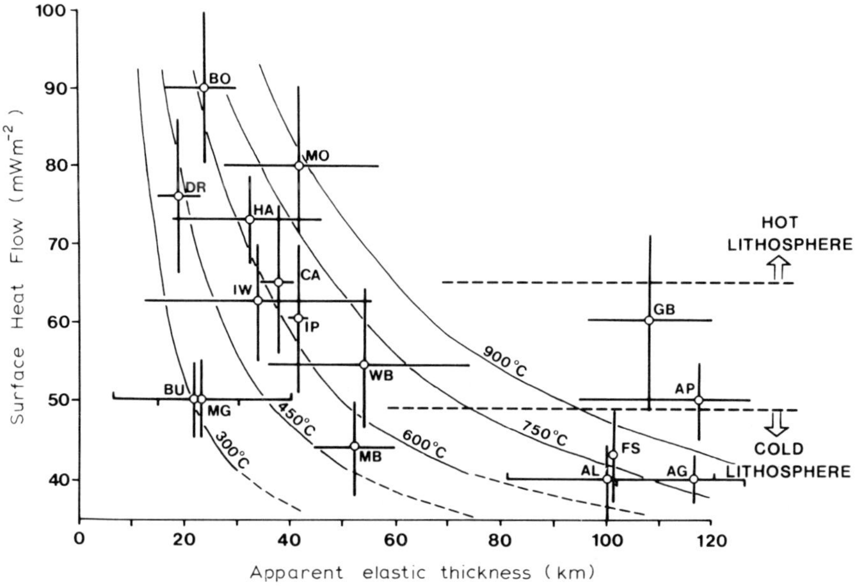

Fig. 6-12. Relationship between present-day surface heat flow and apparent elastic thickness at selected flexure sites. The solid lines indicate depths to isotherms at a given heat flow for a steady-state thermal model. These data show that the thermal structure of the lithosphere is not a simple function of its age, and that apparent elastic thicknesses for continental lithosphere do not correspond to a single isotherm, as they do for oceanic lithosphere. *AG* = Lake Agassis; *AL* = Lake Algonquin; *FS* = Fennoscandia; *CA* = Caribou Mtns; *IP* = Interior plains; *GB* = Ganges basin; *IW* = Idaho–Wyoming thrust; *MB* = Michigan basin; *BU* = Boothia uplift; *MG* = mid-continent gravity high; *HA* = Lake Hamilton; *BO* = Lake Bonneville; *DR* = North Great Dividing range; *AP* = Appalachian foreland basin; *MO* = Molasse basin; *WB* = Williston basin. After Willett et al. (1985).

lithospheric thickening from ca. 15 to 70 km. Karner et al. (1983) suggested that the long-term tendency for the continental lithosphere to thicken with age follows the same trend as the thickening of the oceanic lithosphere as it cools. In the Alps, Appalachians and Himalayas, flexural foreland basins are preserved to some extent by the weight of surface topography produced through crustal thickening and, in the case of the Alps and Appalachians, by buried loads formed from intracrustal wedges of upper mantle rocks. From an analysis of gravity and topographic data, Karner and Watts (1983) showed that the effective elastic thickness of the lithosphere under the Himalayan and Appalachian foreland basins was ca. 100 km, which they argued is consistent with the basins being developed on old, cold (1000 Ma) continental crust. In contrast, effective lithospheric thicknesses ranging from 25 to 50 km beneath the Alpine foreland basin was interpreted in terms of the basin being developed on young (300 Ma), warm continental basement which had suffered high-grade metamorphism during the Hercynian orogeny (Fig. 6-16). The correlation between apparent elastic thickness of continental lithosphere and present-day surface heat flow is pointed out by Willett et al. (1985) (Fig. 6-12).

The above arguments suggest that old, cold [present-day heat flow 30–50 mW m^{-1} (Meissner, 1986)] continental lithosphere is strong and elastic throughout its entire thickness, whereas in areas of higher heat flow (66–90 mW m^{-1}) the elastic thickness has been reset to a thinner value as a result of a thermal event associated, for example, with rifting.

Several lines of evidence are consistent with the idea that in high heat-flow regions the lower continental crust behaves as a weak zone within the lithosphere. The results of

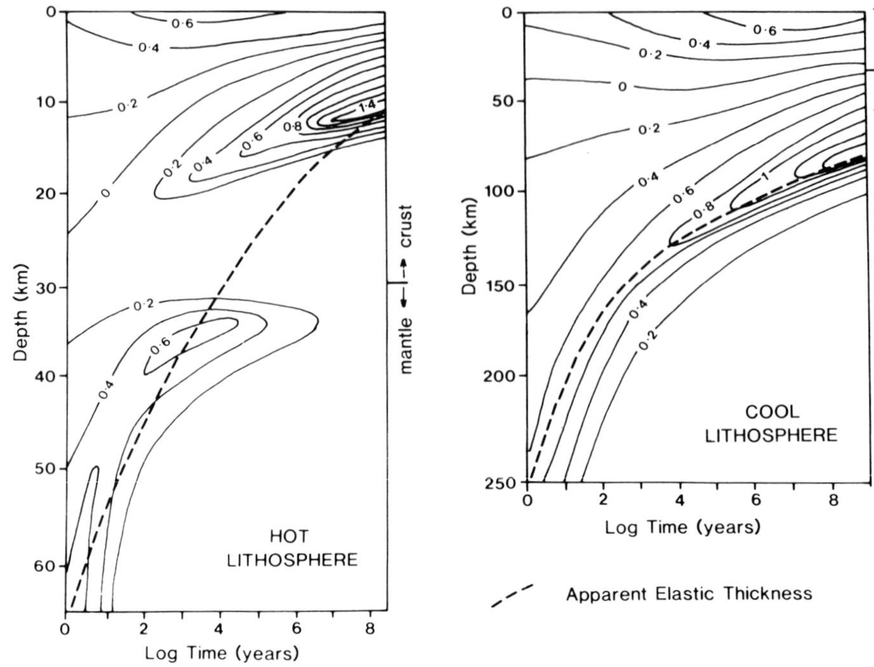

Fig. 6-13. Computed temporal evolution of a hot (surface heat flow = 90 mW m^{-2}) and a cool (surface heat flow = 40 mW m^{-2}) viscoelastic lithosphere during flexure, following instantaneous application of load. The contours show values of deviatoric stress (x100 MPa) on a plane directly below the centre of the applied load. The neutral surface is indicated by the 0 contour. The dashed line shows the time-variation of the apparent elastic thickness of the lithosphere. For both crust and mantle $n = 3$, but the crust is characterized by a lower activation enthalpy for flow than the mantle. For the cool lithosphere, the apparent elastic thickness decreases, owing to the viscous relaxation of the stresses in the lower part of the lithospheric mantle only. For the hotter lithosphere, stresses are relaxed quickly in the lower crust as well as the lower lithosphere, which forces the upper crust to carry the entire fibre load, both extensional above and compressional below the neutral surface. After Willett et al. (1985).

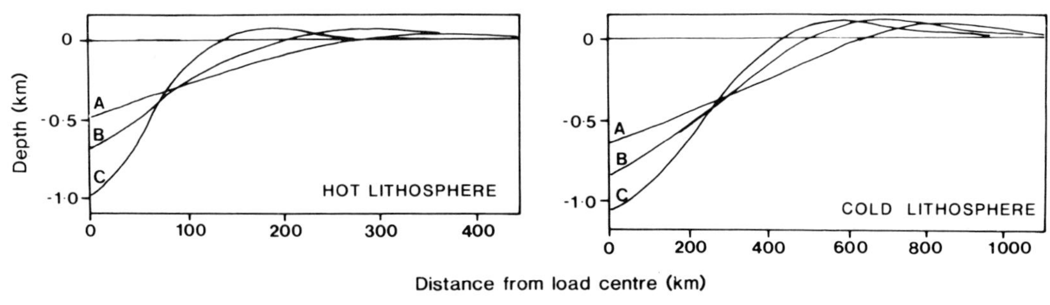

Fig. 6-14. Profiles of the vertical displacements of the free surface whose flexure behaviour is described in Fig. 6-13. Profiles are shown for (A) 1 year, (B) 10^4 years and (C) 10^7 years. In both cases the initial broad, elastic depression narrows with time and a peripheral bulge develops as asthenospheric relaxation transfers mass away from beneath the centre point of the applied load. After Willett et al. (1985).

rock deformation studies described earlier predict that the lower crust should behave as an asthenosphere relative to the cooler upper crust and a stronger upper mantle (e.g., Meissner and Strehlau, 1982; Willet et al., 1985; see Fig. 6-13). Vink et al. (1984) infer that the continental lithosphere is intrinsically weaker than the oceanic one, because it is a preferred

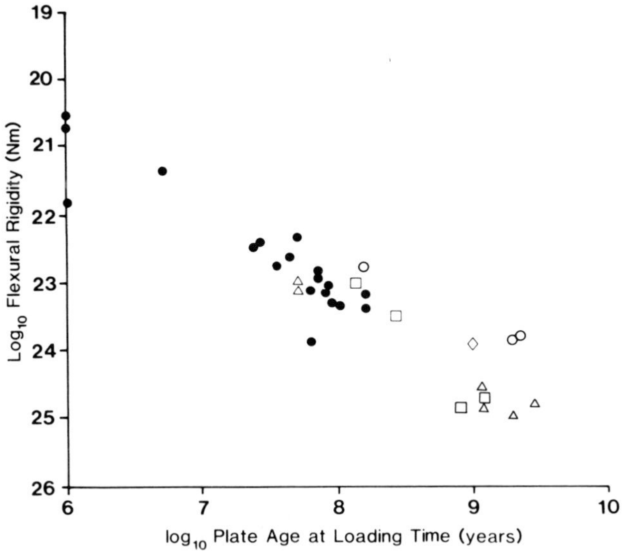

Fig. 6-15. Log-log plot of plate age at time of loading vs. effective flexural rigidity of the lithosphere. Closed symbols: oceanic lithosphere. Open symbols: continental lithosphere. diamond, Michigan basin; squares: foreland basins; triangles: ice loads; circles: minimization of gravity anomalies. That both groups of data lie approximately along the same trend suggests that the mechanical properties of both lithospheres are similar and that the implied effective lithospheric thickening with time is a consequence of cooling since the last 'thermal resetting' event. Although this tendency is the opposite to that predicted by visco-elastic models, the time-scale for continental flexure considered here is one to two orders of magnitude greater than that over which most of the visco-elastic relaxation of the lithosphere occurs following loading (see Fig. 6-13). Uncertainty bars on the data have been omitted. After Karner et al. (1983).

site for rifting once it has been heated. The post-glacial flexural uplift of Pleistocene Lake Bonneville has been interpreted in terms of a lithosphere restricted to the upper crust, with the lower crust behaving as a weak layer (Crittenden, 1963; Walcott, 1970), although other studies have concluded that the elastic layer extends through the entire thickness of the crust (Passey, 1981; Nakiboglu and Lambeck, 1983). There is a general lack of seismicity in the lower continental crust (Meissner and Strehlau, 1982; Sibson, 1982), yet in places upper mantle earthquakes occur (Chen and Molnar, 1983), which may indicate that the upper mantle is locally stronger than the lower crust. Fletcher and Hallet (1983) and Ricard and Froidevaux (1986) modelled periodic necking instabilities in models of a stiff upper layer underlain by plastic and power-law viscous substrates. Fletcher and Hallet (1983) argued that the periodicity of horst-and-graben features in the Basin and Range province is consistent with an upper crust up to 15 km thick above a weaker lower crust (see also Zuber et al., 1986). A weak, lower crustal layer reduces the total bending moment in flexure (De Rito et al., 1986), and it may be reduced still further if preferential shear deformation occurs within the weak layer parallel to its boundaries. The effect on the effective elastic thickness, however, depends on where the weak layer is situated within the lithosphere. When there is no slip within the weak layer it is most effective in reducing the effective thickness when it lies near the base of the layer. At mid-lithospheric depths it has almost no effect at all. When slip can occur, it is most effective at the middle of the section (where layer-parallel shear stresses are highest). De Rito et al. (1986) calculate that for the intracratonic Michigan basin, with a crustal thickness of 35 km, a weak lower crust (if it were present) would be expected to reduce the effective

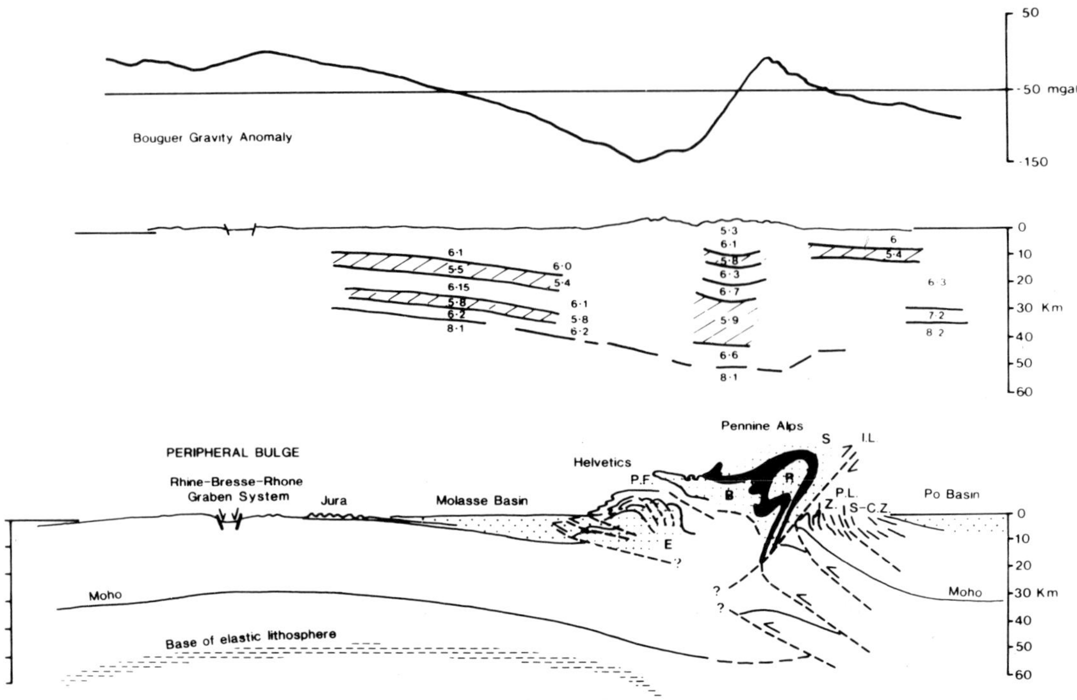

Fig. 6-16. Illustration of lithospheric flexure in connection with continental collision by means of a cross-section through the Western Alps. The loading is part topographic and partly arising from buried 'flakes' of upper mantle material which correspond to the geophysical 'Ivrea Body'. Flexure causes the development of the Molasse basin and a flexural bulge represented by the Rhone–Bresse–Rhine graben system. The Bouguer anomaly pattern is characteristic of many regions in which lower crust has been upthrust to intersect the erosion level (Fountain and Salisbury, 1981). The seismic P-velocity data (Mueller et al., 1976, section line further east than the geologic section) demonstrates the existence of several low-velocity areas, some of which may correspond to regions of active metamorphism involving dehydration reactions. The section through the Helvetic and Pennine Alps is based on Schmid et al. (1987b), and shows the change in mechanical response of the Hercynian basement (+symbol) from brittle faulting in the Helvetics to pervasive flow in the cores of nappes in the (hotter) Pennine zone. The lower crust is remobilized only in the core of the collision zone. The multiple flake structure for the upthrust mantle forming the buried 'Ivrea body' is based on gravity modelling by Menard and Thouvenot (1984). *E* = external massifs; *P.F.* = Pennine front; *B* = Gt. St. Bernard nappe; *R* = Monte Rosa nappe; *S* = Sesia–Dente Blanche nappe; *I.Z.* = Ivrea–Verbano zone; *I.L.* = Insubric line; *P.L.* = Pogallo Line; *S–C.Z.* = Strona–Ceneri zone. Figure based on Karner and Watts (1983).

elastic thickness from the observed 55 km based on a single-layer visco-elastic model to about 28 km. A value of ca. 45–60 km had previously been obtained by Haxby et al. (1976) using the elastic curve-fitting technique.

Although the geometry of lithospheric flexure provides an independent way to approach lithospheric rheology, its interpretation depends upon constructing models that involve assumptions about lithospheric structure and the type of mechanical response it makes to loading. It is not possible to make generalizations about whether the lower crust is always weak, or merely sometimes. It seems likely that the lower crust is weak in continental lithosphere which has 'recently' suffered a thermal event such as is associated with rifting or continental collision, but strong enough in cratonic regions to sustain buried loads even over hundreds of millions of years.

A fundamental distinction must be made between the rheology appropriate to lithospheric flexure, in which fibre strains are small (on the order of 1–5%, Kirby, 1983) and the rheology appropriate to tectonic processes at plate margins that are characterized by enormous strains, sometimes measured in thousands. Fibre strains on the order of 1% can probably be accommodated by transient creep at differential stress levels considerably below those required for 'steady-state' creep. Hence the use of 'steady-state' rheology to estimate flexural behaviour may overestimate the flexural rigidity (Kirby, 1983). Rheology based on experimentally observed 'steady-state' creep behaviour, however, is probably only applicable to processes producing permanent shear strains on the order of 1–10, such as occur at the initiation of large-scale tectonic events. As argued in previous sections, the rheology appropriate to lithospheric rifting, large-scale strike-slip displacements and continental collision is likely to be that of the microstructurally distinctive rocks which occur in shear zones.

8. Discussion

We have examined the experimental data from rock mechanics pertinent to the rheology of the lower continental crust, the inferences that can be drawn from the structural and microstructural study of rocks of lower crustal provenance, and the implications of studies of lithospheric flexure. In this section we consider the implications of these data for the state of stress, the interpretation of geophysical remote sensing data and the construction of theoretical models.

8.1. The state of stress and rheological stratification

Assuming that upper and lower bounds on geological strain rates can be estimated from stratigraphic constraints and the kinematic framework of plate tectonics (e.g., Fyfe et al., 1978), rock mechanics data for flow by intracrystalline plasticity constrain the maximum stress levels to be expected in the lower crust where temperatures are sufficiently high for flow to occur (Brace and Kohlstedt, 1980). These are likely to be of the order of those shown in Figures 6-2 and 6-8. This range of values is consistent with figures inferred from recrystallized grain sizes of naturally deformed rocks. The phenomenon of lithospheric flexure tells us that when the thermal gradient is too low, maximum differential stress levels are limited by the elastic stress distribution required by the flexural geometry. These can be several 100 MPa in parts of the depth profile. The point of difference between 'high' and 'low' thermal gradients, with respect to the difference between plastic or elastic behaviour in 35-km-thick crust, probably corresponds to a surface heat flow of about 50 $mW\ m^{-1}$ (Willett et al., 1985).

Because it is widely believed that the continental lithosphere is broadly stratified (with increasing depth) into sedimentary rocks, quartzo-feldspathic basement rocks, metabasic$\pm$ quartzo-feldspathic rocks, and ultramafic rocks, it has been argued that the lithosphere will also be rheologically stratified (Meissner and Strehlau, 1982; Ranalli and Murphy, 1987; Carter and Tsenn, 1987). Rheological discontinuities would then develop at major petrologic interfaces, such as the Moho, with the rocks above the discontinuity being weaker than those below it. The effect can be readily appreciated from Figures 6-2 and 6-8, and such data have been combined to draw generalized strength versus depth profiles such as that illustrated schematically in Figure 6-17.

At shallow depths, the resistance of rocks to cataclastic deformation and frictional sliding on faults is less than that for intracrystalline plastic flow, hence the strength of the upper

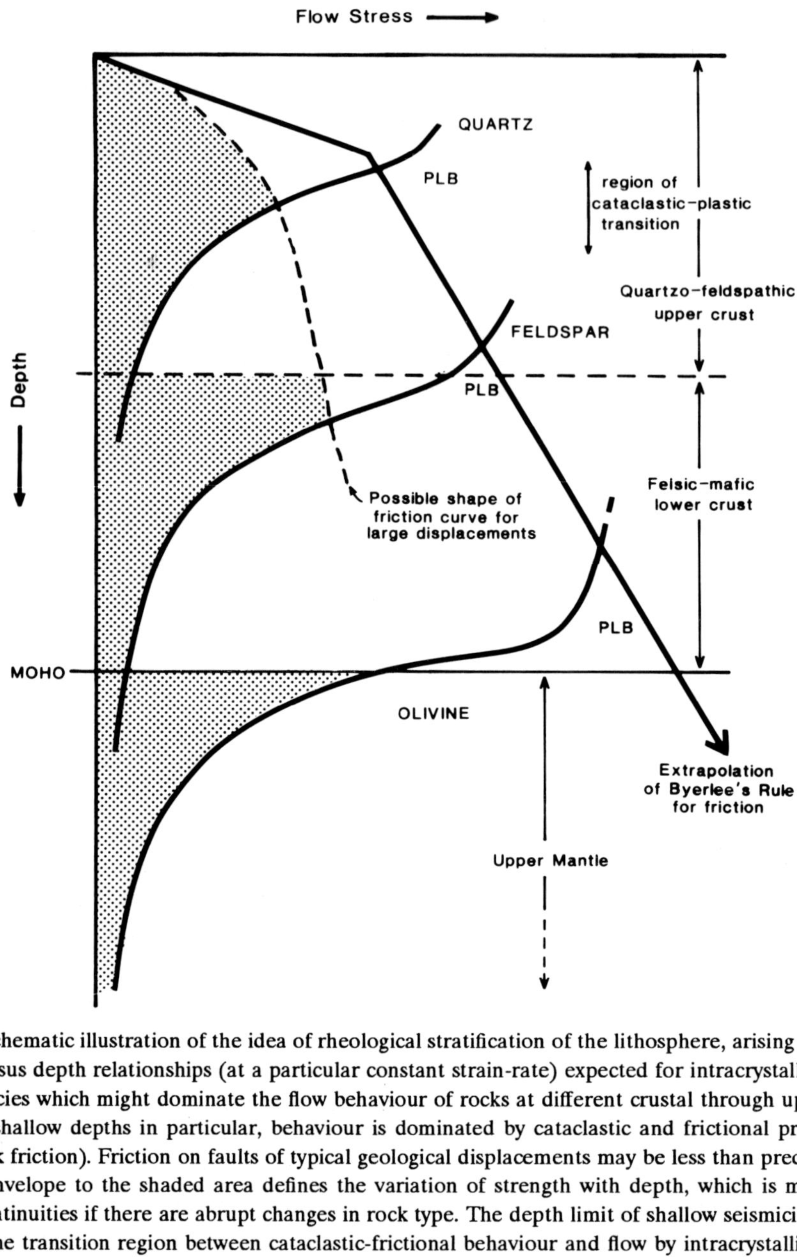

Fig. 6-17. Schematic illustration of the idea of rheological stratification of the lithosphere, arising from the different strength versus depth relationships (at a particular constant strain-rate) expected for intracrystalline plastic flow of mineral species which might dominate the flow behaviour of rocks at different crustal through upper mantle depth ranges. At shallow depths in particular, behaviour is dominated by cataclastic and frictional processes (Byerlee's rule for rock friction). Friction on faults of typical geological displacements may be less than predicted by Byerlee's rule. The envelope to the shaded area defines the variation of strength with depth, which is marked by possibly sharp discontinuities if there are abrupt changes in rock type. The depth limit of shallow seismicity is thought to be related to the transition region between cataclastic-frictional behaviour and flow by intracrystalline plasticity. *PLB* = region of power-law breakdown.

crust is inferred to be pressure-sensitive and, therefore, to increase with depth. This trend is eventually reversed when the temperature becomes high enough to allow temperature and rate-sensitive intracrystalline plastic flow, at approximately mid-crustal depths.

According to a useful generalization proposed by Byerlee (1968, 1978), rock friction is approximately independent of rock type. It is also relatively insensitive to temperature and strain rate. Hence, the behaviour of rocks deforming by frictional sliding can be approximated by a single curve of strength increasing with depth (Fig. 6-17). The slope of the curve is greater in a crustal section under tangential compression than under extension. On the assumption that earthquakes will only occur in the brittle, frictional regime, it has been proposed that the depth of transition from deformation dominated by frictional sliding to plastic flow can be correlated with the typically mid-crustal depth limit of shallow earthquakes (Sibson, 1982; Meissner and Strehlau, 1982) in regions of moderate heat flow. This model predicts that in regions of low heat flow, seismicity through brittle failure may extend to the lower crust or even upper mantle, because feldspathic-mafic and ultramafic rocks may be so resistant to plastic failure that brittle faulting is the only accessible deformation mechanism. In broad terms this is consistent with observations (e.g., Chen and Molnar, 1983) but anomalies do exist. Shudovsky et al. (1987) discuss the origins of deep crustal (to 30 km) earthquakes in the East African Rift. They conclude that the high strength of the lower crust implied by application of the above model arises from the combined effects of a very mafic, lower crustal petrography and an 'anomalous' heat-flow regime (a warm crust but a hot upper mantle). Through its obvious attractiveness and broad consistency with observations, this simple concept has acquired a degree of respectability through force of repetition. However, the extent of this respectability may not be wholly warranted by the factual data upon which the concept is based, nor is it consistently supported by more recent work. First, Byerlee's rule is based on rock friction experiments involving fault displacements of only a few mm. A recent study of rock friction to much larger displacements (hundreds of mm) has shown a marked decrease in friction with displacement (Blanpied et al., 1988). These results, should they prove general, would place serious doubts on the utility of Byerlee's generalization. Secondly, aseismic (velocity-strengthening) brittle faulting (perhaps modified by processes involving interactions with pore fluids, such as hydrothermal cementation and pressure-solution creep) as a principal mode of rock failure may extend far below the depth at which velocity-weakening (seismogenic) behaviour holds (Scholz, 1988). Thirdly, it has been shown (Hobbs et al., 1986; Ord and Hobbs, 1989) that earthquakes may nucleate from instabilities in the plastic flow of rocks, at depths greater than the so-called brittle-plastic transition.

The brittle-plastic 'transition' remains a useful concept, but the transition is likely to be very broad, perhaps extending typically over as much as 5–10 km of depth. Within it, rock strengths are likely to be lower than expected from extrapolation of Byerlee's rule (Fig. 6-17), and within or near it the depth limit of seismicity may also lie. Below the depth at which intracrystalline plastic flow becomes the dominant deformation mechanism (in coarse-grained rocks and when pore pressure is significantly less than the least principal stress), lithological discontinuities are also likely to be rheological discontinuities. The most important of these is likely to be the Moho, but substantial effects may arise if there is a significant change in, for example, the quartz/feldspar modal ratio across a lithological contact (e.g., Fig. 6-17). These strength contrasts across lithological boundaries become more marked as the geothermal gradient is lowered. It has been argued that the weak zones on one side of strength discontinuities will be preferred sites for the development of major, sub-horizontal detachment faults (e.g., White and Bretan, 1985; Ord and Hobbs, 1989).

In order to accommodate large displacements at plate boundaries, however, deformation involving large local strains must develop across the whole thickness of the lithosphere (Kusznir and Park, 1986), so that failure must spread across all of the yield profile (shaded in Fig. 6-17). It is reasonable to suppose that initial yielding of the lithosphere will be concentrated in the the weaker intervals, but this causes stress amplification in the stronger intervals (Kusznir and Park, 1986). As the stronger intervals fail plastically, these will be expected to produce the smallest recrystallized grain sizes. Hence these, rather than the weaker intervals, might eventually be expected to be preferred sites for shear zone localization (Rutter and Brodie, 1988a) (see Fig. 6-6). In contrast, the weaker intervals are likely to flow relatively homogeneously.

Supporting this argument is the commonly cited tendency for plastic detachment faults to develop in the middle crust (e.g., Miller et al., 1983; Lister et al., 1986; Davis et al., 1986; Gibson et al., 1988) and the occurrence of gently dipping seismic reflectors within the sub-crustal upper mantle (McGeary and Warner, 1985). In the upper mantle, rocks are expected to be strong. However, Boillot et al. (1987) have described intense shearing and grain refinement in peridotites believed to be of sub-continental upper mantle provenance, and Rutter and Brodie (1988a) have described preferential shearing and mylonite formation on the peridotite side of a peridotite-metabasite interface at the structural base of the Ivrea–Verbano zone. Therefore, we conclude that whilst the simple argument for rheological stratification makes sense for limited amounts of permanent deformation, it is by no means clear that 'weak' zones identified on the basis of increasing temperature with depth coupled with lithological stratification will remain the weakest intervals during large finite deformations.

8.2. Deep seismic soundings

A consensus view that appears to be emerging from the growing number of deep seismic reflection profiles of the continental lithosphere is that bright, reflective lamellar structures are characteristic of lower continental crust that has suffered tectonothermal activity during the Phanerozoic era. This appears to be true of much of continental Europe. In contrast, the lower continental crust of old, cold cratons is less reflective and often characterized by intersecting, dipping reflectors which bound wedge-shaped blocks (see review by Meissner, 1989). Cratonic massifs within Europe that were not reactivated during either the Caledonian, Variscan or Alpine orogenies (Brabant, Rhenish and Bohemian massifs) do not show lower crustal lamellae (Bois et al., 1987; Meissner, 1989).

The principal theories of the origin of lower crustal lamellae are that: (a) they represent interlayering (either tectonic or intrusive) of rocks such as metapelites, metabasites and ultramafics; (b) they represent flow-induced features such as an intense crystallographic fabric or anastomosing, conjugate arrays of low-angle shear zones; or (c) they represent sub-horizontal zones containing trapped pore fluids. Hypotheses (a) and (b) are attractive and it is likely that both are correct to different degrees in different places. If compositional layering arises from the emplacement of mafic sills into the lower crust (e.g., Meissner, 1967; Klemperer et al., 1986; Allmendinger et al., 1987; Warner, 1987; Rutter et al., 1992), then a mechanism is also provided for the unusual accession of heat into high levels within the continental lithosphere, driving the tectonics, high-grade regional metamorphism and perhaps crustal melting, which are otherwise difficult to explain. Herzberg et al. (1983) have shown that the density of basic magma at 1000 MPa pressure is greater than that expected of lower crustal rocks, hence pooling of magma as sills in the lower crust is to be expected. The Ivrea–Verbano zone provides an excellent example of a petrographically well layered lower crustal section (Fig. 6-9) that would display a lamellar structure on a seismic reflection profile. In the Ivrea–Verbano zone low-angle shear zones also occur, but it is not clear whether they would give rise to lamellae. It has been suggested that the emplacement of a large amount of basic magma (the mafic formation) into the section during the Variscan orogeny may have been largely responsible for the attainment of granulite facies conditions and the 'degranitization' of the pelitic rocks (e.g., Schmid and Wood, 1976; Schmid, 1978). However, much of the observed interlayering of mafic and pelitic rocks in the section had developed earlier in the history of the complex.

Hypothesis (c), that lamellae might represent planar zones containing free pore water (e.g., Matthews, 1986), has also been invoked to explain the anomalously high electrical conductivity frequently reported for the lower crust (e.g., Shankland and Anders, 1983; Bailey et al., 1989). Unless a mechanism were available for removal of such water prior to uplift and erosion, it is difficult to reconcile such an hypothesis with the preservation of extensive tracts of high-grade metamorphic rocks (e.g., Yardley, 1986). An attractive alternative explanation for high electical conductivity of the lower crust has been proposed by Frost et al. (1989). They demonstrated continuous grain-boundary films of graphite in rocks from the Laramie Anorthosite Complex, the origin of which they ascribed to the breakdown of a carbon dioxide fluid phase by reaction with titanomagnetite during cooling. In addition, experimental measurements on rock conductivity under pressure and at temperatures up to 900°C (Glover, 1989) have shown that, even in the absence of pore water, metabasic rocks are sufficiently conductive to account for the remote-sensing observations on the lower crust.

It has been noted (e.g., Meissner, 1989) that lamellar structures occur where the Moho can be inferred to be relatively flat over large regions, where the thickened roots of mountain chains have been removed and/or where the lithosphere has recently been, or is presently being, stretched (e.g., Basin and Range Province, U.S.A., and the Mesozoic sedimentary basins of NW Europe). The processes of continental delamination as described by Turcotte (1989) and the injection of magmatic bodies as sills (e.g., Fountain, 1989; Wilshire, 1990) may be involved in the development of these features. It is also tempting to believe that in such regions the lower crust is so weak that it flows laterally in response to shear forces induced through Moho or upper crust topography, tending to remove such topography and perhaps forming deformation-induced layering features. A Moho topography of 10 km would be expected to induce stress differentials on the order of 50 MPa as a result of lateral density contrasts. For a strain rate of ca. 10^{-14} s^{-1} and a heat flow above about 60 mW m^{-1} this is commensurate with the flow strength expected of quartzo-feldspathic rocks at ca. 30 km depth (Fig. 6-2).

8.3. Modelling tectonic processes

Analytic and numerical modelling of the development of extensional, transcurrent, compressional and flexural deformations is becoming progressively more representative of real geometries (e.g., Bird and Piper, 1980; Fletcher and Hallet, 1983; England, 1983; England et al., 1985; Ricard and Froidevaux, 1986; Houseman and England, 1986; Vilotte et al., 1986; Kusznir and Park, 1986; De Rito et al., 1986). For analytic and computational simplicity, however, it has been usual to employ a vertically averaged rheology. This involves integrating the deviatoric stress with respect to depth across the layer or layers which effectively determine the overall strength of the lithosphere. A 'representative' power-law rheology of the form of eq. 1 is chosen, and the force per unit depth, F, required to deform the lithosphere is calculated from:

$$F = B\dot{e}nRT_o^2 \exp[(H/nRT_o)/H]\ (\mathrm{d}T/\mathrm{d}z) \qquad (5)$$

where $B = A^{-1/n}\,2^{(1-n)/2n}$ and T_o is the temperature at the top of the strength-controlling layer (England, 1983). Houseman and England (1986) argue that olivine rheology with $n \sim 3$ is the most important control on lithosphere strength. The effects of other, more non-linear contributions to the overall behaviour (such as upper crustal faulting or elastic-plastic failure) can be satisfactorily modelled assuming an 'effective' $n > 3$. Houseman and England (1986) explore the effects of $1 < n < 10$ on the behaviour of continental convergence zones. It is clear that this simplifying concept allows a considerable development of understanding of continental lithosphere deformation to be made without the need to consider local heterogeneities of rock type and failure modes.

Scaled analog modelling of the development of large-scale tectonic features in a rheologically stratified lithosphere is still in its infancy, but it is a technique that shows great promise (e.g., Allemand and Brun, 1991; Davy and Cobbold, 1991). Davy and Cobbold (1991) constructed a 4-layer 'lithosphere', scaled for the effects of gravity, in which the upper crust was modelled by loose sand and ductile lower crust, upper mantle and lower mantle were modelled by Newtonian-viscous silicones of different viscosities. Deformed in compression and in extension, or by rigid or deformable indenters, it was possible to reproduce geometric features of natural tectonic processes that could be analyzed in three dimensions.

9. Conclusions

The rheology of the lower continental crust is determined principally by (a) its mineralogical composition and whether metamorphic reactions occur, (b) the thermal regime and (c) the applied displacement field.

Plastic flow is facilitated when quartz and/or feldspar form the principal load-supporting phases. Mafic minerals are considerably more resistant. It is believed that syntectonic metamorphism and partial melting causes weakening relative to the plastic flow strength, but there is presently no basis for quantitative modelling of deformation accompanied by metamorphism.

From lithospheric flexure observations, the distribution of seismicity with depth, extrapolation of flow laws for plastic deformation of rocks and interpretation of deep seismic reflection profiles, it is clear that the question of whether or not the lower continental crust behaves as a weak zone depends sensitively on the thermal regime. Old, cold cratonic lithosphere, characterized by low present-day heat flow (less than 50 mW m^{-1}) behaves elastically over long time periods (hundreds of millions of years). Lower continental crust that has been reactivated during a Phanerozoic tectonothermal event and is characterized by a higher than average present-day heat flow is likely to be weaker than the underlying upper mantle, weaker than overlying middle crust, and be weak enough to flow under stresses arising from lateral density contrasts.

For processes involving moderate strains (5–50%), such as the initiation of rifting or intracratonic extensional basin formation, and provided no metamorphism or melting is involved, extrapolation of suitable experimentally determined constitutive laws for 'steady-state flow' to geological temperatures and strain rates is probably adequate for modelling purposes. It is not certain whether 'wet' or 'dry' rheology is most applicable. For processes involving smaller strains, such as lithospheric flexure, transient creep laws may be more appropriate, and rocks may behave as if significantly weaker than steady-state creep laws would predict. For large displacement processes at plate margins, where the entire thickness of the lithosphere must rupture, steady-state creep laws for intracrystalline plasticity of coarse-grained rocks are probably of limited applicability. The effects of large strains on the microstructure and rheology of rocks, leading to a tendency to localize flow into plastic shear zones remains an outstanding problem for experimental investigations and a target for future incorporation into theoretical models of crustal deformation.

Acknowledgements

This chapter was written whilst we were in transition from Imperial College London to Manchester University. Most of our microstructural work described here was carried out at Imperial College, and we are grateful to technical staff in the Geology and Materials Science (HVTEM) departments for their help. Our work was supported by U.K. Natural Environment Research Council grants GR3/5636, GR3/6174 and GR9/18. Stephen Covey-Crump provided a very helpful and detailed critique of the first draft of the text and Stephen Kirby a helpful and constructive review.

References

Aines, R.D., Kirby, S.H. and Rossman, G.R., 1984. Hydrogen speciation in synthetic quartz. Phys. Chem. Miner., 11: 204–212.

Aines, R.D. and Rossman, G.R., 1984. Water in minerals? A peak in the infrared. J. Geophys. Res., 89: 4059–4071.

Allemand, P. and Brun, J.P., 1991. Continental rifts in nature and experiment. In: P.R. Cobbold (Editor), Experimental and Numerical Modelling of Continental Deformation. Tectonophysics, 188: 63–69.

Allmendinger, R.W., Nelson, K.D., Potter, C.J., Barazangi, M., Brown, L.D. and Oliver, J.E., 1987. Deep reflection characteristics of the continental crust. Geology, 15: 304–310.

Alvarez, F., Virieaux, F. and Le Pichon, X., 1984. Thermal consequences of lithosphere extension over continental margins: the initial stretching phase. Geophys. J. R. Astron. Soc., 78: 389–411.

Avé Lallemant, H.G., 1978. Experimental deformation of diopside and websterite. Tectonophysics, 48: 1–27.

Bailey, R.C., Craven, J.A., Macnae, J.C. and Polzer, B.D., 1989. Imaging of deep fluids in the Archaean crust. Nature, 340: 136–138.

Beach, A., 1980. Retrogressive metamorphic processes in shear zones with special reference to the Lewisian complex. J. Struct. Geol., 2: 257–263.

Behrmann, J.H. and Mainprice, D.H., 1987. Deformation mechanisms in a high temperature quartz-feldspar mylonite: evidence for superplastic flow in the lower continental crust. Tectonophysics, 140: 297–305.

Biermann, C. and van Roermund, H.L.M., 1983. Defect structures in naturally deformed clinoamphiboles — a TEM study. Tectonophysics, 95: 267–278.

Bills, B.G., 1983. Thermoelastic bending of the lithosphere: implications for basin subsidence. Geophys. J. R. Astron. Soc., 75: 169–200.

Bird, P. and Piper, K., 1980. Plane stress finite element models of tectonic flow in southern California. Phys. Earth Planet. Inter., 21: 158–175.

Blanpied, M.L., Tullis, T.E. and Weeks, J.D., 1988. Textural and mechanical evolution of granite gouge in high displacement sliding experiments (abstract). EOS, Trans. Am. Geophys. Union, 69: 1463.

Blenkinsop, T.G. and Drury, M.R., 1988. Stress estimates and fault history from quartz microstructures. J. Struct. Geol., 1:, 673–684.

Bodine, J.H. and Watts, A.B., 1979. Lithospheric flexure seaward of the Bonin and Mariana trenches. Earth Planet. Sci. Lett., 43: 132–148.

Boillot, G., Recq, M., Winterer, E.L., Meyer, A.W., Applegate, J., Baltuck, M., Bergen, J.A., Comas, M.C., Davies, T.A., Dunham, K., Evans, C.A., Girardeau, J., Goldberg, G., Haggerty, J., Jansa, L.F., Johnson, J.A., Kasahara, J., Loreau, J.P., Luna-Sierra, E., Moullade, M., Ogg, J., Sarti, M., Thurow, J. and Williamson, M., 1987. Tectonic denudation of the upper mantle along passive margins: a model based on drilling results (ODP leg 103, western Galicia margin, Spain). Tectonophysics, 132: 335–342.

Bois, C., Cazes, M., Hirn, A., Matte, P., Mascle, A. Montadert, L., and Pinet, B., 1987. Crustal laminations on deep seismic reflection profiles in France and neighbouring areas. Geophys. J. R. Astron. Soc., 89: 297–304.

Boland, A.V., Ellis, R.M., Northey, D.J., West, G.F., Green, A.G., Forsyth, D.A., Mereu, R.F., Meyer, R.P., Morel-a-l'Huissier, P., Buchbinder, G.G.R., Asudeh, I. and Haddon, R.A.W., 1988. Seismic delineation of upthrust Archaean crust in Kapuskasing, Northern Ontario. Nature, 335: 711–713.

Boland, J.N. and Buiskool-Toxopeus, J.M.A., 1976. Petrofabrics, microtextures and dislocation substructures of olivine in a peridotite mylonite (Alpe Arami, Switzerland). Leidse Geol. Meded., 51: 1–36.

Boland, J.N. and Tullis, T.E., 1986. Deformation behaviour of wet and dry clinopyroxenite in the brittle to ductile transition region. In: B.E. Hobbs and H.C. Heard (Editors), Mineral and Rock Deformation: Laboratory studies. The Paterson Volume. Am. Geophys. Union, Monogr., 36: 35–50.

Borch, R.S. and Green, H.W., 1987. Dependence of creep in olivine on homologous temperature and its implications for flow in the mantle. Nature, 330: 345–348.

Bouchez, J.L., 1978. Preferred orientations of quartz a axes in some tectonites. Tectonophysics, 49: T25–T30.

Boullier, A.M. and Gueguen, Y., 1975. SP-mylonites: origin of some mylonites by superplastic flow. Contrib. Mineral. Petrol., 50: 79–92.

Brace, W.F. and Kohlstedt, D.L., 1980. Limits on lithospheric stress imposed by laboratory experiments. J. Geophys. Res., 85: 6248–6252.

Brodie, K.H., 1981. Variation in amphibole and plagioclase composition with deformation. Tectonophysics, 78: 385–402.

Brodie, K.H. and Rutter, E.H., 1985. On the relationship between deformation and metamorphism, with special reference to the behaviour of basic rocks. In: A.B. Thompson and D.C. Rubie (Editors), Metamorphic Reactions, Kinetics, Textures and Deformation. Adv. Phys. Geochem., 4, Springer, Berlin, pp. 138–179.

Brodie, K.H. and Rutter, E.H., 1987a. The role of transiently fine-grained reaction products in syntectonic metamorphism: natural and experimental examples. Can. J. Earth Sci., 24: 554–564.

Brodie, K.H. and Rutter, E.H., 1987b. Deep crustal extensional faulting in the Ivrea zone of northern Italy. Tectonophysics, 140: 193–212.

Brodie, K.H., Rex, D. and Rutter, E.H., 1989. On the age of deep crustal extensional faulting in the Ivrea zone, northern Italy. In: D. Dieterich and M.P. Coward (Editors), Alpine Tectonics. Geol. Soc. London, Spec. Publ., 45: 203–210.

Brodie, K.H., Rutter, E.H. and Evans P.E., 1992. On the structure of the Ivrea–Verbano zone (N. Italy) and its implications for present day lower crust geometry. Terra Nova, 4: 34–40.

Brun, J.P. and Cobbold, P.R., 1980. Strain heating and thermal softening in continental shear zones. J. Struct. Geol., 2: 149–158.

Burnley, P.C. and Green, H.W., 1989. Stress dependence of the olivine-spinel transformation. Nature, 338: 753–756.

Burnley, P.C. and Kirby, S.H., 1982. Pressure-induced embrittlement of polycrystalline tremolite (abstract). EOS, Trans. Am. Geophys. Union, 6: 1095.

Byerlee, J.D., 1968. The brittle-ductile transition in rocks. J. Geophys. Res., 73: 4741–4750.

Byerlee, J.D., 1978. Friction of rocks. Pure Appl. Geophys., 116: 615–626.

Caristan, Y., 1982. The transition from high-temperature creep to fracture in Maryland diabase. J. Geophys. Res., 87: 6781–6790.

Carter, N.L. and Tsenn, M.C., 1987. Flow properties of the continental lithosphere. Tectonophysics, 136: 27–63.

Carter, N.L., Kronenberg, A.K., Ross, J.V. and Wiltschko., 1990. Control of fluids on deformation of rocks. In: R.J. Knipe and E.H. Rutter (Editors), Deformation Mechanisms, Rheology and Tectonics. Geol. Soc. London, Spec. Publ., 54: 1–14.

Chen, W.P. and Molnar, P., 1983. Focal depths of intracontinental and intraplate earthquakes and their implications for the thermal and mechanical properties of the lithosphere. J. Geophys. Res., 88: 4183–4214.

Chopra, P.N., 1986. The plasticity of some fine-grained aggregates of olivine at high pressure and temperature. In: B.E. Hobbs and H.C. Heard (Editors), Mineral and Rock Deformation: Laboratory Studies. The Paterson Volume. Am. Geophys. Union, Monogr., 36: 25–34.

Chopra, P.N. and Paterson, M.S., 1981. The experimental deformation of dunite. Tectonophysics, 78: 453–473.

Chopra, P.N. and Paterson, M.S., 1984. The role of water in the deformation of dunite. J. Geophys. Res., 89: 7861–7876.

Christie, J.M. and Ord, A., 1980. Flow stress from microstructure of mylonites: examples and current assessment. J. Geophys. Res., 85: 6253–6262.

Clemens, J.D., 1984. Water contents of silicic to intermediate magmas. Lithos, 17: 273–287.

Cooper, R.F. and Kohlstedt, D.L., 1984. Rheology and structure of olivine basalt partial melts. J. Geophys. Res., 91: 9315–9323.

Coward, M.P., 1984. Major shear zones in the Precambrian crust; examples from NW Scotland and southern Africa and their significance. In: A. Kröner and R. Greiling (Editors), Precambrian Tectonics Illustrated. E.Schweitzerbart'sche Verlagsbuchhandlung (Nagele u. Obermiller), Stuttgart, pp. 207–236.

Crittenden, M.D., 1963. Effective viscosity of the Earth derived from isostatic loading of Pleistocene Lake Bonneville. J. Geophys. Res., 68: 5517–5530.

Daly, M.C., 1986. Crustal shear zones and thrust belts: their geometry and continuity in Central Africa. Philos. Trans. R. Soc. London, Ser. A, 317: 111–128.

Davidson, A., 1984. Identification of ductile shear zones in the southwestern Grenville province of the Canadian Shield. In: A. Kröner and R. Greiling (Editors), Precambrian Tectonics Illustrated. E.Schweitzerbart'sche Verlagsbuchhandlung (Nagele u. Obermiller), Stuttgart, pp. 263–279.

Davidson, C., Hollister, L.S. and Schmid, S.M., 1992. Role of melt in the formation of a deep-crustal compressive shear zone: the Maclaren Glacier metamorphic belt, south central Alaska. Tectonics 11: 348–359.

Davis, G.H., Lister, G.S. and Reynolds, S.J., 1986. Structural evolution of the Whipple and South Mountains shear zones, southwestern United States. Geology, 14: 7–10.

Davy, P. and Cobbold, P.R., 1990. Some experiments on shortening of a 4-layer model of the continental lithosphere. In: P.R. Cobbold (Editor), Experimental and Numerical Modelling of Continental Deformation. Tectonophysics, 188: 1–26.

Dell'Angelo, L.N. and Kirby, S.H., 1988. Anomalous sliding behavior of aplite under conditions that favor transformation of albite to jadeite + quartz (abstract). EOS, Trans. Am. Geophys. Union, 69: 1463.

Dell'Angelo, L.N. and Tullis, J., 1988. Experimental deformation of partially melted granitic aggregates. J. Met. Geol., 6, 495–516.

Dennis, P.F., 1984. Oxygen self-diffusion in quartz under hydrothermal conditions. J. Geophys. Res., 89: 4047–4058.

DeRito, R.F., Cozzarelli, F.A, and Hodge, D.S., 1986. A forward approach to the problem of non-linear viscoelasticity and the thickness of the mechanical lithosphere. J. Geophys. Res., 91: 8295–8313.

Dollinger, G.S. and Blacic, J.D., 1975. Deformation mechanisms in naturally and experimentally deformed amphiboles. Earth Planet. Sci. Lett., 26: 409–416.

England, P., 1983. Constraints on extension of continental lithosphere. J. Geophys Res., 88: 1145–1152.

England, P., Houseman, G. and Sonder, L., 1985. Length scales for continental deformation in convergent, divergent and strike-slip environments: analytical and approximate solutions for a thin viscous sheet. J. Geophys. Res., 90: 3551–3558.

Etheridge, M.A., 1975. Deformation and recrystallization of orthopyroxene from the Giles complex, Central Australia. Tectonophysics, 25: 87–114.

Etheridge, M.A. and Wilkie, J.C., 1978. Grainsize reduction, grain boundary sliding and the flow strength of mylonites. Tectonophysics, 58: 159–178.

Fischer, G.J. and Paterson, M.S., 1989. Dilatancy during the deformation of rocks at high temperatures and pressures. J. Geophys. Res., 94: 17,607–17,617.

Fletcher, R.C. and Hallet, B., 1983. Unstable extension of the lithosphere: a mechanical model for basin-and-range structure. J. Geophys. Res., 88: 7457–7466.

Fountain, D., 1976. The Ivrea–Verbano and Strona–Ceneri zones, northern Italy: a cross-section of the continental crust — new evidence from seismic velocities of rock samples. Tectonophysics, 33: 145–165.

Fountain, D.M., 1989. Growth and modification of lower continental crust in extended terrains: The role of extension and magmatic underplating. In: R.F. Mereu, S. Mueller and D.M. Fountain (Editors), Properties and Processes of Earth's Lower Crust. Geophys. Monogr., 51 (IUGG Volume 6), pp. 287–300.

Fountain, D.M. and Salisbury, M.H., 1981. Exposed cross-sections through the continental crust: implications for crustal structure, petrology and evolution. Earth Planet. Sci. Lett., 56: 263–277.

Fountain, D.M., Salisbury, M.H. and Percival, J., 1990. Seismic structure of the continental crust based on rock velocity measurments from the Kapuskasing Uplift. J. Geophys. Res., 95: 1167–1186.

Frost, B.R., Fyfe, W.S., Tazaki, K. and Chan, T., 1989. Grain-boundary graphite in rocks and implications for high electical conductivity in the lower crust. Nature, 340: 134–136.

Fyfe, W.S., 1976. Chemical aspects of rock deformation. Philos. Trans. R. Soc. London, Ser. A, 283: 221–228.

Fyfe, W.S., Price, N.J. and Thompson, A., 1978. Fluids in the Earth's Crust. Elsevier, Amsterdam.

Gandais, M. and Willaime, C., 1984. Mechanical properties of feldspars. In: W.L. Brown (Editor), Feldspars and Feldspathoids. NATO ASI Series C.137, D. Reidel, Dordrecht, pp. 207–280.

Garner, D.L. and Turcotte, D.L., 1984. The thermal and mechanical evolution of the Anadarko Basin. Tectonophysics, 107: 1–24.

George, R.P., 1977. Subsolidus deformation mechanisms and hypersolidus creep of peridotite. In: H.J.B. Dick (Editor), Magma Genesis. Proc. Chapman Conference on Partial Melting in the Upper Mantle. Oreg. Dep. Geol. Miner. Ind., Bull., 96: 233–262.

Gerretsen, J., Paterson, M.S. and McLaren, A.C., 1989. The uptake and solubility of water in quartz at elevated pressure and temperature. Phys. Chem. Miner., 16: 334–342.

Gibson, G.M., McDougall, I. and Ireland, T.R., 1988. Age constraints on metamorphism and the development of a metamorphic core complex in Fiordland, southern New Zealand. Geology, 16: 405–408.

Gilletti, B.J. and Yund, R.A., 1984. Oxygen diffusion in quartz. J. Geophys. Res., 89: 4039–4046.

Goleby, B.R., Shaw, R.D., Wright, C., Kennett, B.L. and Lambeck, K., 1989. Geophysical evidence for "thick-skinned" crustal deformation in central Australia. Nature, 337: 325–330.

Goetze, C. and Evans, B., 1979. Stress and temperature in the bending lithosphere as constrained by experimental rock mechanics. Geophys. J. R. Astron. Soc., 59: 463–478.

Glover, P., 1989. Electrical conductivity of rock samples subjected to high temperatures aind pressures. PhD Thesis, University of East Anglia.

Green, H.W. and Gueguen, Y., 1983. Deformation of peridotite in the mantle and extraction by kimberlite: a case history documented by solid and fluid inclusions in olivine. Tectonophysics, 92: 71–92.

Green, H.W. and Radcliffe, S.V., 1972. Deformation processes in the upper mantle. In: H.C. Heard, I.Y. Borg, N.L. Carter and C.B. Raleigh (Editors), Flow and Fracture of Rocks. The Griggs Volume. Geophys. Monogr., 16: 139–156.

Green, H.W. and Burnley, P.C., 1990. The failure mechanism for deep-focus earthquakes. In: R.J. Knipe and E.H. Rutter (Editors), Deformation Mechanisms, Rheology and Tectonics. Geol. Soc. London, Spec. Publ., 54: 133–142.

Griggs, D.T., 1967. Hydrolytic weakening of quartz and other silicates. Geophys. J. R. Astron. Soc., 14: 19–32.

Griggs, D.T. and Blacic, J.D., 1964. The strength of quartz in the ductile regime. Trans. Am. Geophys. Union, 45: 102–103.

Griggs, D.T. and Blacic, J.D., 1965. Quartz: anomalous weakness of synthetic crystals. Science, 147: 292–295.

Grotzinger, J. and Royden, L., 1990. Elastic strength of the Slave craton at 1.9 Gyr and implications for the thermal evolution of continents. Nature, 347: 64–66.

Guillopé, M. and Poirier, J.P., 1979. Dynamic recrystallization during creep of single-crystalline halite: an experimental study. J. Geophys. Res., 84: 5557–5567.

Hacker, B.R., Yin, A., Christie, J.M. and Snoke, A.W., 1990. Differential stress, strain rate, and temperatures of mylonitization in the Ruby mountains, Nevada: implications for the rate and duration of uplift. J. Geophys. Res., 95: 8569–8580.

Hall, M.K. and Chase, C.G., 1989. Uplift, unbuckling and collapse: flexural history and isostasy of the Wind River range and Granite Mountains, Wyoming. J. Geophys. Res., 94: 17,581–17,593.

Handy, M.R., 1987. The structure, age and kinematics of the Pogallo fault zone; southern Alps, northwestern Italy. Eclogae Geol. Helv., 80: 593–632.

Handy, M.R., 1990. The solid-state flow of polymineralic rocks. J. Geophys. Res., 95: 8647–8661.

Hanks, T.C., 1971. The Kuril trench–Hokkaido rise system: large shallow earthquakes and simple models of deformation. Geophys. J. R. Astron. Soc., 23: 173–189.

Hanmer, S.K., 1982. Microstructures and geochemistry of plagioclase and microcline in naturally deformed granite. J. Struct. Geol., 4: 197–213.

Hansen, F.D., 1982. Semibrittle creep of selected crustal rocks at 1000 MPa. PhD Thesis, Texas A&M University, 224 pp.

Hansen, F.D and Carter, N.L., 1983. Semibrittle creep of dry and wet Westerly granite at 1000 MPa. Proc. 24th U.S. Symposium on Rock Mechanics, Texas A&M University, pp. 429–447.

Haxby, W.F., Turcotte, D.L. and Bird, J.M., 1976. Thermal and mechanical evolution of the Michigan basin. Tectonophysics, 36: 57–75.

Heard, H.C., 1960. Transition from brittle to ductile behaviour in Solenhofen limestone as a function of temperature, confining pressure and interstitial fluid pressure. Geol. Soc. Am. Mem., 79: 193–226.

Heard, H.C., 1985. Experimental determination of mechanical properties. In: H.R. Wenk (Editor), Preferred Orientation in Deformed Materials. Academic Press, Orlando, FL, pp. 485–506.

Heard, H.C. and Rubey, W.W., 1966. Tectonic implications of gypsum dehydration. Geol. Soc. Am. Bull., 77: 741–760.

Heard, H.C. and Raleigh, C.B., 1972. Steady-state flow of marble at 500 to 800°C. Bull. Geol. Soc. Am., 83: 935–956.

Herzberg, C.T., Fyfe, W.S. and Carr, M.J., 1983. Density constraints on the formation of the continental Moho and crust. Contrib. Mineral. Petrol., 84: 1–5.

Hickman, A.H., 1984. Archaean diapirism in the Pilbara block, Western Australia. In: A. Kröner and R. Greiling (Editors), Precambrian Tectonics Illustrated. E.Schweitzerbart'sche Verlagsbuchhandlung (Nagele u. Obermiller), Stuttgart, pp. 113–128.

Hirsch, P.B., 1981. Plastic deformation and electronic mechanisms in semiconductors and insulators. J. Phys. Colloq., 42: C3-149–C3-159.

Hobbs, B.E., 1981. The influence of metamorphic environment on the deformation of minerals. Tectonophysics, 78: 335–383.

Hobbs, B.E., McLaren, A.C. and Paterson, M.S., 1972. Plasticity of single crystals of synthetic quartz. In: H.C. Heard, I.Y. Borg, N.L. Carter and C.B. Raleigh (Editors), Flow and Fracture of Rocks. The Griggs Volume. Geophys. Monogr., 16: 29–53.

Hobbs, B.E, Ord, A. and Teyssier, C., 1986. Earthquakes in the ductile regime? Pure Appl. Geophys., 124: 309–336.

Hobbs, B.E., Muhlhaus, H.-B., and Ord, A., 1990. Instability, softening and localization of deformation. In: R.J. Knipe and E.H. Rutter (Editors), Deformation Mechanisms, Rheology and Tectonics. Geol. Soc. London, Spec. Publ., 54: 143–166.

Hodges, K.V. and Fountain, D.M., 1984. Pogallo Line, south Alps, Northern Italy: an intermediate crustal level, low-angle normal fault? Geology, 12: 151–155.

Hollister, L.S. and Crawford, M.L., 1986. Melt-enhanced deformation: a major tectonic process. Geology, 14: 558–561.

Houseman, G. and England, P., 1986. Finite strain calculations of continental deformation, 1. Method and general results for convergent zones. J. Geophys. Res., 91: 3651–3663.

Hyndman, R.D., 1988. Dipping seismic reflectors, electrically conductive zones, and trapped water in the crust over a subducting plate. J. Geophys. Res., 93: 13,391–13,405.

Jaoul, O., Tullis, J. and Kronenberg, A.K., 1984. The effect of varying water contents on the creep behaviour of Heavitree quartzite. J. Geophys. Res., 89: 4298–4312.

Ji, S. and Mainprice, D.H., 1986. Transition from power law to Newtonian creep in experimentally deformed dry albite rock (abstract). EOS, Trans. Am. Geophys. Union, 67: 1235.

Ji, S. and Mainprice, D.H., 1987. Experimental deformation of sintered albite above and below the order–disorder transition. Geodyn. Acta, 1: 113–124.

Jordan, P.G., 1987. The deformational behaviour of bimineralic limestone-halite aggregates. Tectonophysics, 135: 185–197.

Karato, S., Toriumi, M. and Fujii, T., 1982. Dynamic recrystallization of olivine single crystals during high temperature creep. Earth Planet. Sci. Lett., 7: 649–652.

Karato, S., Paterson, M.S. and Fitzgerald, J.D., 1986. Rheology of synthetic olivine aggregates: influence of grain size and water. J. Geophys. Res., 91: 8151–8176.

Karner, G.D. and Watts, A.B., 1983. Gravity anomalies and flexure of the lithosphere at mountain ranges. J. Geophys. Res., 88: 10,449–10,477.

Karner, G.D., Steckler, M.S. and Thorne, J.A., 1983. Long-term thermo-mechanical properties of the continental lithosphere. Nature, 304: 250–253.

Kekulawala, K.R.S.S., Paterson, M.S. and Boland, J.N., 1978. Hydrolytic weakening in quartz. Tectonophysics, 46: T1–T6.

Kekulawala, K.R.S.S., Paterson, M.S. and Boland, J.N., 1981. An experimental study of the role of water in quartz deformation. In: N.L. Carter, M. Friedman, J.M. Logan and D.W. Stearns (Editors), Mechanical Behaviour of Crustal Rocks. The Handin Volume. Geophys. Monogr., 24: 49–60.

Kern, H., 1978. The effect of high temperature and high confining pressure on compressional wave velocities in quartz-bearing and quartz-free igneous and metamorphic rocks. Tectonophysics, 44: 185–203.

Kerrich, R., Allison, I., Bennett, R.L., Moss, S. and Starkey, J., 1980. Microstructural and chemical transformation accompanying deformation of granite in a shear zone at Mieville, Switzerland: with implications for stress-corrosion cracking and superplastic flow. Contrib. Mineral. Petrol., 73: 221–242.

Kirby, S.H., 1983. Rheology of the lithosphere. Rev. Geophys. Space Phys., 21: 1458–1487.

Kirby, S.H., 1985. Rock mechanics observations pertinent to the rheology of the continental lithosphere and the localization of strain along shear zones. Tectonophysics, 119: 1–27.

Kirby, S.H., 1987. Localized polymorphic phase transformations in high pressure faults and applications to the physical mechanism of deep earthquakes. J. Geophys. Res., 92: 13,789–13,800.

Kirby, S.H. and Green, H.W., 1980. Dunite xenoliths from Hualalai volcano: evidence for mantle diapiric flow beneath the island of Hawaii. Am. J. Sci., 280A: 550–575.

Kirby, S.H. and Kronenberg, A.K., 1984. Deformation of clinopyroxenite: evidence for a transition in flow mechanism and semibrittle behaviour. J. Geophys Res., 89: 3177–3192.

Kirby, S.H. and Kronenberg, A.K., 1987. Rheology of the lithosphere: selected topics. Rev. Geophys., 25: 1219–1244.

Kirby, S.H. and McCormick, J.W., 1979. Creep of hydrolytically weakened synthetic quartz crystals oriented to promote (2$\bar{1}\bar{1}$0)<001>slip: a brief summary of work to date. Bull. Minéral., 102: 124–137.

Klemperer, S.L., Hauge, T.A., Hauser, E.C., Oliver, J.E. and Potter, C.J., 1986. The Moho in the northern Basin and Range province, Nevada, along the COCORP 40°N seismic reflection transect. Bull. Geol. Soc. Am., 97: 603–618.

Koch, P.S., 1983. Rheology and microstructure of experimentally deformed quartz aggregates. PhD Thesis, Univ. Calif., Los Angeles, CA, 464 pp.

Koch, P.S., Christie, J.M., Ord, A. and George, R.P. Jr., 1989. Effects of water on the rheology of experimentally deformed quartzite. J. Geophys. Res., 94: 13,975–13,996.

Kohlstedt, D.L. and Hornack, P., 1981. Effect of partial pressure on the creep of olivine. In: F.D. Stacey, M.S. Paterson and A. Nicholas (Editors), Anelasticity of the Earth. Am. Geophys. Union, Geodyn. Ser., 4: 101–107.

Kohlstedt, D.L. and Weathers, M.S., 1980. Deformation induced microstructures, palaeopiezometers, and differential stresses in deeply eroded fault zones. J. Geophys. Res., 85: 6269–6285.

Kollé, J.J and Blacic, J.D., 1982. Deformation of single-crystal clinopyroxenes, 1. Mechanical twinning in diopside and hedenbergite. J. Geophys. Res., 87: 4019–4034.

Kollé, J.J. and Blacic, J.D., 1983. Deformation of single-crystal clinopyroxenes, 2. Dislocation controlled flow processes in hedenbergite. J. Geophys. Res., 88: 2381–2394.

Kovacs, M.P. and Gandais, M., 1980. Transmission electron microscope study of experimentally deformed K-feldspar crystals. Phys. Chem. Miner., 6: 61–76.

Krishna-Sinha, A., Hewitt, D.A. and Rimstidt, J.D., 1986. Fluid infiltration and element mobility in the development of ultramylonites. Geology, 14: 833–866.

Kronenberg, A.K. and Tullis, J., 1984. Flow strengths of quartz aggregates: grain-size and pressure effects due to hydrolytic weakening. J. Geophys. Res., 89: 4281–4297.

Kröner, A., 1984. Dome structures and basement reactivation in the Pan-African Damara belt of Namibia. In: A. Kröner and R. Greiling (Editors), Precambrian Tectonics Illustrated. E.Schweitzerbart'sche Verlagsbuchhandlung (Nagele u. Obermiller), Stuttgart, pp. 191–206.

Kusznir, N.J. and Park, R.G., 1986. Continental lithosphere strength: the critical role of lower crustal deformation. In: J.B. Dawson, D.A. Carswell, J. Hall and K.H. Wedepohl (Editors), The Nature of the Lower Continental Crust. Geol. Soc., Spec. Publ., 24: 79–93.

Lambeck, K., 1983. The role of compressive forces in intracratonic basin formation and mid-plate orogenies. Geophys. Res. Lett., 10: 845–848.

LaTour, T.E. and Kerrich, R., 1982. Microstructures, mineral chemistry and oxygen isotopes of two adjacent mylonite zones: a comparative study. In: F.E. Heuze and R.E. Goodman (Editors), Issues in Rock Mechanics. Proc. 23rd U.S. Symposium on Rock Mechanics, Berkeley, CA, pp. 389–396.

Linker, M.F. and Kirby, S.H., 1981. Anisotropy in the rheology of hydrolytically weakened synthetic quartz crystals. Am. Geophys. Union, Monogr., 24: 29–48.

Linker, M.F., Kirby, S.H., Ord, A. and Christie, J.M., 1984. Effects of the compression direction on the plasticity and rheology of hydrolytically weakened synthetic quartz crystals at atmospheric pressure. J. Geophys. Res., 89: 4241–4256.

Lister, G.S., Etheridge, M.A. and Symonds, P.A., 1986. Detachment faulting and the evolution of passive continental margins. Geology, 14: 246–250.

Lyon-Caen, H. and Molnar, P., 1983. Constraints on the structure of the Himalaya from an analysis of gravity and flexural model of the lithosphere. J. Geophys. Res., 88: 8171–8192.

Mainprice, D.H., Bouchez, J.L., Blumenfeld, P. and Tubia, J.M., 1986. Dominant c slip in naturally deformed quartz: implications for dramatic plastic softening at high temperature. Geology, 14: 819–822.

Marshall, D.B., Vernon, R.H. and Hobbs, B.E., 1976. Experimental deformation and recrystallization of a peristerite. Contrib. Mineral. Petrol., 57: 49–54.

Marshall, D.B. and McLaren, A.C., 1977. Deformation mechanisms in experimentally deformed plagioclase feldspars. Phys. Chem. Miner., 1: 315–370.

Matthews, D.H., 1986. Seismic reflections from the lower crust around Britain. In: J.B. Dawson, D.A. Carswell, J. Hall and K.H. Wedepohl (Editors), The Nature of the Lower Continental Crust. Geol. Soc. London, Spec. Publ., 24: 11–21.

McAdoo, D.C., Caldwell, J.G. and Turcotte, D.L., 1978. On the elastic perfectly-plastic bending of the lithosphere under generalized loading with application to the Kurile trench. Geophys. J. R. Astron. Soc., 54: 11–26.

McCaig, A.M., 1989. Fluid flow through fault zones. Nature, 340: 600.

McGeary, S. and Warner, M.R., 1985. Seismic profiling in the continental lithosphere. Nature, 317: 795–797.

McNutt, M. and Menard, H.W., 1978. Lithospheric flexure and uplifted atholls. J. Geophys. Res., 83: 1206–1212.

McNutt, M., 1987. Lithospheric stress and deformation. Rev. Geophys., 25: 1245–1253.

McNutt, M. and Kogan, M.G., 1987. Isostasy in the USSR II: interpretation of admittance data. In: K.A.F. Fuchs (Editor), The Composition, Structure and Dynamics of the Lithosphere. Geodyn. Ser., Am. Geophys. Union, pp. 309–327.

Meissner, R., 1967. Zum aufbau der Erdekruste. Ergbnisse der Weitwinkelmessungen im bayerischen Molasse-becken. Gerl. Beitr. Geophys., 76: 241–254, 295–314.

Meissner, R., 1986. The Continental Crust — A Geophysical Approach. Academic Press, Orlando, FL, 426 pp.

Meissner, R., 1989. Rupture, creep, lamellae and crocodiles: happenings in the continental crust. Terra Nova, 1: 17–28.

Meissner, R. and Strehlau, J., 1982. Limits of stresses in the continental crust and their relation to depth-frequency distribution of shallow earthquakes. Tectonics, 1: 73–89.

Menard, G. and Thouvenot, F., 1984. Ecaillage de la lithosphere Européene sous les Alpes Occidentales: arguments gravimetriques et sismiques liés a l'anomalie d'Ivrea–Verbano. Bull. Geol. Soc. Fr., 7, Tome 26: 875–884.

Mercier, J.-C., 1980. Magnitude of continental lithospheric stresses inferred from rheomorphic petrology. J. Geophys. Res., 85: 6293–6303.

Mercier, J.-C., Anderson, D.A. and Carter, N.L., 1977. Stress in the lithosphere: inferences from steady flow of rocks. Pure Appl. Geophys., 115: 199–226

Miller, E.L., Gans, P.B. and Garing, J., 1983. The Snake Range décollement: an exhumed mid-Tertiary brittle–ductile transition. Tectonics, 2: 239–263.

Miller, H., Mueller, S. and Perrier, G., 1982. Structure and dynamics of the Alps: a geophysical inventory. In: H. Berckhemer and K. Hsü (Editors), Alpine-Mediterranean Geodynamics. Am. Geophys. Union, Geodyn. Ser., 7: 175–204.

Montardi, Y. and Mainprice, D., 1987. A transmission electron microscopic study of the natural plastic deformation of calcic plagioclases (An 68–70). Bull. Minéral., 110: 1–17.

Morrison-Smith, D.J., 1976. Transmission electron microscopy of experimentally deformed hornblende. Am. Mineral., 61: 272–280.

Mueller, S., Egloff, R. and Ansorge, J., 1976. Struktur des tieferen untergrundes entlang der Schweizer Geotraverse. Schweiz. Mineral. Petrogr. Mitt., 56: 685–692.

Murrell, S.A.F., 1985. Aspects of relationships between deformation and metamorphism which causes evolution of water. In: A.B. Thompson and D.C. Rubie (Editors), Metamorphic Reactions, Kinetics, Textures and Deformation. Adv. Phys. Geochem. 4, Springer, Berlin, pp. 211–241.

Murrell, S.A.F. and Ismail, I.A.H., 1976. The effect of the decomposition of hydrous minerals on the mechanical properties of rocks. Tectonophysics, 31: 207–258.

Myers, J.S., 1984. Archaean tectonics in the Fiskenaesset region of southwest Greenland. In: A. Kröner and R. Greiling (Editors), Precambrian Tectonics Illustrated. E.Schweitzerbart'sche Verlagsbuchhandlung (Nagele u. Obermiller), Stuttgart, pp. 95–112.

Nakiboglu, S.M. and Lambeck, K., 1983. A reevaluation of the isostatic rebound of Lake Bonneville. J. Geophys. Res., 88: 10,439–10,448.

Nunn, J.A. and Sleep, N.H., 1984. Thermal contraction and flexure of intracratonal basins: a three-dimensional study of the Michigan Basin. Geophys. J. R. Astron. Soc., 76: 587–635.

Oliver, G.J.H., 1980. Geology of the granulite and amphibolite facies gneisses of Doubtful Sound, Fiordland, New Zealand. N.Z. J. Geol. Geophys., 23: 27–41.

Olsen, T.S. and Kohlstedt, D.L., 1985. Natural deformation and recrystallization of some intermediate plagioclase feldspars. Tectonophysics, 111: 107–131.

Ord, A. and Hobbs, B.E., 1986. Experimental control of the water weakening effect in quartz. Am. Geophys. Union, Monogr., 36: 51–72.

Ord, A. and Hobbs, B.E., 1989. The strength of the continental crust, detachment zones and the development of plastic instabilities. Tectonophysics, 158: 269–289.

Paquet, J. and François, P., 1980. Experimental deformation of partially melted granitic rocks at 600–900°C and 250 MPa confining pressure. Tectonophysics, 68: 131–168.

Paquet, J., François, P. and Nedelec, A., 1981. Effect of partial melting on rock deformation: experimental and natural evidence on rocks of granitic compositions. Tectonophysics, 78: 545–565.

Park, M.J.M. and Westbrook, G.K., 1983. Visco-elastic modelling of lithospheric stresses arising from density contrasts. Geophys. J. R. Astron. Soc., 74: 905–914.

Passey, Q.R., 1981. Upper mantle viscosity derived from the difference in rebound of the Provo and Bonneville shorelines: Lake Bonneville Basin, Utah, J. Geophys. Res., 86: 11,701–11,708.

Paterson, M.S., 1970. A high pressure, high temperature apparatus for rock deformation. Int. J. Rock Mech. Min. Sci., 7: 517–526.

Paterson, M.S., 1983. Creep in transforming polycrystalline materials. Mech. Mater., 2: 103–109.

Paterson, M.S., 1989. The interaction of water with quartz and its influence on dislocation flow — an overview. In: S. Karato and M. Toriumi (Editors), Rheology of Solids and of the Earth. Oxford Sci. Publ., Oxford, pp. 107–142.

Paterson, M.S. and Luan, F-C., 1990. Quartz rheology under geological conditions. In: R.J. Knipe and E.H. Rutter (Editors), Deformation Mechanisms, Rheology and Tectonics. Geol. Soc. London, Spec. Publ., 54: 299–300 .

Peltier, W.R., 1984. The thickness of the continental lithosphere. J. Geophys. Res., 89: 11,303–11,316.

Percival, J.A., 1989. Granulite terranes and the lower crust of the Superior Province. In: R.F. Mereu, S. Mueller and D.M. Fountain (Editors), Properties and Processes of Earth's Lower Crust. Geophys. Monogr. 51 (IUGG Volume 6), pp. 301–310.

Percival, J.A. and Card, K.D., 1983. Archaean crust as revealed in the Kapuskasing Uplift, Superior Province, Canada. Geology, 11: 323–326.

Percival, J.A. and McGrath, P.H., 1986. Deep crustal structure and tectonic history of the northern Kapuskasing uplift of Ontario: and integrated petrological-geophysical study. Tectonics, 5: 553–572.

Platt, J.P., Behrmann, J.H., Cunningham, P.C., Dewey, J.F., Helman, M., Parish, M., Shepley, M.G., Wallis, S. and Weston, P.J., 1989. Kinematics of the Alpine arc and the motion history of Adria. Nature, 337: 158–161.

Poirier, J.P., 1982. On transformation plasticity. J. Geophys. Res., 87: 6971–6797.

Poirier, J.P., 1985. Creep of Crystals. Cambridge University Press, Cambridge, 260 pp.

Post, R.L., 1977. High temperature creep of Mt. Burnet dunite. Tectonophysics, 42: 75–110.

Raleigh, C.B. and Paterson, M.S., 1965. Experimental deformation of serpentinite and its tectonic implications. J. Geophys. Res., 70: 3965–3985.

Raleigh, C.B., Kirby, S.H., Carter, N.L. and Avé Lallemant., 1971. Slip and the clinoenstatite transformation as competing rate processes in enstatite. J. Geophys. Res., 76: 4011–4022.

Ramsay, J.G., 1963. Stratigraphy, structure and metamorphism of the Western Alps. Proc. Geol. Assoc., 74: 357–392.

Ranalli, G. and Murphy, D.C., 1987. Rheological stratification of the lithosphere, Tectonophysics, 132: 281–295.

Ricard, Y. and Froidevaux, C., 1986. Stretching instabilities and lithospheric boudinage. J. Geophys. Res., 91: 8314–8324.

Roedder, E., 1965. Liquid CO_2 inclusions in olivine-bearing nodules and phenocrysts from basalts. Am. Mineral., 50: 1746–1782.

Rooney, T.P. and Riecker, R.E., 1973. Constant strain rate deformation of amphibole minerals. Air Force Cambridge Res. Lab., Environ. Res. Pap. 430, 33 pp.

Rooney, T.P, Gavasci, A.T. and Riecker, R.E., 1974. Mechanical twinning in experimentally and naturally deformed hornblende. Air Force Cambridge Res. Lab., Environ. Res. Pap. 484, 21 pp.

Ross, J.V. and Nielsen, K.C., 1978. High temperature flow of wet polycrystalline enstatite. Tectonophysics, 44: 233–261.

Rubie, D.C., 1983. Reaction enhanced ductility: the role of solid–solid univariant reactions in the deformation of the crust and mantle. Tectonophysics, 96: 331–352.

Rubie, D.C., 1990. Mechanisms of reaction-enhanced deformability in minerals and rocks. In: D.J. Barber and P.G. Meredith (Editors), Deformation Processes in Minerals, Ceramics and Rocks. Mineral. Soc. Ser., 1: 262–295.

Rutter, E.H., 1972. The influence of interstitial water on the rheological behaviour of calcite rocks. Tectonophysics, 14: 13–33.

Rutter, E.H., 1974. The influence of temperature, strain rate and interstitial water in the experimental deformation of calcite rocks. Tectonophysics, 22: 311–334.

Rutter, E.H. and Brodie, K.H., 1985. The permeation of water into hydrating shear zones. In: A.B. Thompson and D.C. Rubie (Editors), Metamorphic Reactions, Kinetics, Textures and Deformation. Adv. Phys. Geochem. 4, Springer, Berlin, pp. 242–249.

Rutter, E.H. and Brodie, K.H., 1988a. The role of tectonic grain-size reduction in the rheological stratification of the lithosphere. Geol. Rundsch., 77: 295–308.

Rutter, E.H. and Brodie, K.H., 1988b. Experimental syntectonic dehydration of serpentinite under controlled pore water presssure. J. Geophys. Res., 93: 4907–4932.

Rutter, E.H. and Brodie, K.H., 1990. Some geophysical implications of the deformation and metamorphism of the Ivrea–Verbano zone, northern Italy. Tectonophysics, 182: 147–160.

Rutter, E.H., Brodie, K.H. and Evans, P.E., 1992. Structural geometry, lower crustal magmatic underplating and lithospheric stretching in the Ivrea–Verbano zone, northern Italy. J. Struct. Geol., in press.

Rutter, E.H., Peach, C.J., White, S.H. and Johnson, D., 1985. Experimental syntectonic hydration of basalt. J. Struct. Geol., 7: 251–266.

Schenk, V., 1984. Petrology of felsic granulites, metapelites, metabasics, ultramafics and metacarbonates from southern Calabria, Italy: prograde metamorphism, uplift and cooling of a former lower crust. J. Petrol., 25: 255–298.

Schmid, R., 1978. Are the metapelites of the Ivrea–Verbano zone restites? Mem. Sci. Geol. Univ. Padova, 33: 67–69.

Schmid, R. and Wood, B.J., 1976. Phase relationships in granulitic metapelites from the Ivrea–Verbano zone (northern Italy). Contrib. Mineral. Petrol., 54: 255–279.

Schmid, S.M., Boland, J.N. and Paterson, M.S., 1977. Superplastic flow in fine-grained limestone. Tectonophysics, 43: 257–291.

Schmid, S.M., Paterson, M.S. and Boland, J.N., 1980. High temperture flow and dynamic recrystallization in Carrara marble. Tectonophysics, 65: 245–280.

Schmid, S.M. and Casey, M., 1986. Complete fabric analysis of some commonly observed quartz c-axis patterns. Am. Geophys. Union, Monogr., 36: 263–268.

Schmid, S.M., Panozzo, R. and Bauer, S., 1987a. Simple shear experiments on calcite rocks: rheology and microfabric. J. Struct. Geol., 9: 747–778.

Schmid, S.M., Zingg, A. and Handy, M.R., 1987b. The kinematics of movements along the Insubric line and the emplacement of the Ivrea–Verbano zone. Tectonophysics, 135: 47–66.

Schwenn, M.B. and Goetze, C., 1978. Creep of olivine during hot pressing. Tectonophysics, 48: 41–60.

Scholz, C.H., 1988. The brittle–plastic transition and the depth of seismic faulting. Geol. Rundsch., 77: 319–328.

Segall, P. and Simpson, C., 1986. Nucleation of ductile shear zones on dilatant fractures. Geology, 14: 56–59.

Sellars, C.M., 1978. Recrystallization of metals during hot deformation. Philos. Trans. R. Soc. London, Ser A, 228: 147–158.

Shankland, T.J. and Anders, M.E., 1983. Electrical conductivity, temperature and fluids in the lower crust. J. Geophys. Res., 88: 9475–9484.

Shelton, G.L. and Tullis, J., 1981. Experimental flow laws for crustal rocks. EOS, Trans. Am. Geophys. Union, 62: 396.

Shudovsky, G.N., Cloetingh, S., Stein, S. and Wortel, R., 1987. Unusually deep earthquakes in east Africa: constraints on the thermo-mechanical structure of a continental rift system. Geophys. Res. Lett., 14: 741–744.

Sibson, R.H., 1977. Fault rocks and fault mechanisms. J. Geol. Soc. London, 133: 191–214.

Sibson, R.H., 1982. Fault zone models, heat flow and the depth distribution of earthquakes in the continental crust. Bull. Seismol. Soc. Am., 72: 151–163.

Sodre-Borges, F. and White, S.H., 1980. Microstructural and chemical studies of sheared anorthosites, Roneval, South Harris. J. Struct. Geol., 2: 273–280.

Sorensen, K., 1983. Growth and dynamics of the Nordre Stromfjord shear zone. J. Geophys. Res., 88: 3419–3437.

Strehlau, J. and Meissner, R., 1987. Estimation of crustal viscosities and shear stresses from an extrapolation of experimental steady state flow data. In: K. Fuchs and C. Froidevaux (Editors), Composition, Structure and Dynamics of the Lithosphere–Asthenosphere System. Geodyn. Ser., 16: 69–87.

Tsenn, M.C. and Carter, N.L., 1987. Upper limits of power law creep of rocks. Tectonophysics, 136: 1–26.

Tullis, J., 1983. Deformation of Feldspars. In: P.H.Ribbe (Editor), Feldspar Mineralogy, Vol. 2, 2nd ed. Mineral. Soc. Am., pp. 297–323.

Tullis, J. and Yund, R.A., 1977. Experimental deformation of dry Westerly granite. J. Geophys. Res., 82: 5705–5718.

Tullis, J. and Yund, R.A., 1980. Hydrolytic weakening of experimentally deformed Westerly granite and Hale albite rock. J. Struct. Geol., 2: 439–451.

Tullis, J. and Yund, R.A., 1985. Dynamic recrystallization of feldspar: a mechanism for ductile shear zone formation. Geology, 13: 238–241.

Tullis, J., Christie, J.M. and Griggs, D.T., 1973. Microstructures and preferred orientations of experimentally deformed quartzites. Bull. Geol. Soc. Am., 84: 397–314.

Tullis, J., Shelton, G.L. and Yund, R.A., 1979. Pressure dependence of rock strength: implications for hydrolytic weakening. Bull. Minéral., 102: 110–114.

Tullis, T.E. and Tullis, J., 1986. Experimental rock deformation techniques. Am. Geophys. Union, Monogr., 36: 297–234.

Tullis, T.E., Horowitz, F.G. and Tullis, J., 1991. Flow laws of polyphase aggregates from end–member flow laws. J. Geophys. Res., 96: 8081–8096.

Turcotte, D.L., 1989. Geophysical processes influencing the lower continental crust. In: R.F. Mereu, S. Mueller and D.M. Fountain (Editors), Properties and Processes of Earth's Lower Crust. Geophys. Monogr. 51 (IUGG Volume 6), pp. 321–330.

Twiss, R.J., 1977. Theory and applicability of a recrystallized palaeopiezometer. Pure Appl. Geophys., 115: 227–244.

Urai, J.L., Means, W.D. and Lister, G.S., 1986. Dynamic recrystallization of minerals. Am. Geophys. Union, Monogr., 36: 161–199

Vaughan, P.J. and Coe, R.S., 1981. Creep mechanism in Mg_2GeO_4: effects of a phase transition. J. Geophys. Res., 86: 389–404.

Van der Molen, I. and Paterson, M.S., 1979. Experimental deformation of partially melted granite. Contrib. Mineral. Petrol., 70: 299–318.

Vilotte, J.P., Daignieres, M., Madariaga, R. and Zienkewicz, O., 1986. The role of a heterogeneous inclusion during continental collision. Phys. Earth Planet. Inter., 36: 236–259.

Vink, G.E., Morgan, W.J. and Zhao, W.L., 1984. Preferential rifting of continents, a source of displaced terranes. J. Geophys. Res., 89: 10,072–10,076.

Walcott, R.I., 1970. Flexural rigidity, thickness and viscosity of the lithosphere. J. Geophys. Res., 75: 3941–3954.

Walker, A., Rutter, E.H. and Brodie, K.H., 1990. Experimental study of grain-size sensitive flow of synthetic, hot-pressed calcite rocks. In: R.J. Knipe and E.H. Rutter (Editors), Deformation Mechanisms, Rheology and Tectonics. Geol. Soc. London, Spec. Publ., 54: 259–282.

Warner, M., 1987. Basalts in the lower continental crust — evidence from deep seismic profiles. Terra Cognita, 7: 158.

Watts, A.B., Bodine, J.H. and Steckler, M.S., 1980. Observations of flexure and the state of stress in the oceanic lithosphere. J. Geophys. Res., 85: 6369–6376.

Watts, A.B., Karner, G.D. and Steckler, M.S., 1982. Lithospheric flexure and the evolution of sedimentary basins. Philos. Trans. R. Soc. London, Ser. A, 305: 245–281.

Weertman, J., 1968. Dislocation climb theory of steady state creep. Trans. Am. Soc. Metals, 61: 681–694.

White, J.C. and Mawer, C.K., 1986. Extreme ductility of feldspars from a mylonite, Parry sound, Canada. J. Struct. Geol., 8: 133–143.

White, J.C. and Mawer, C.K., 1988. Dynamic recrystallization and associated exsolution in perthites: evidence of deep crustal thrusting. J. Geophys. Res., 93: 325–337.

White, S.H., 1976. The effects of strain on the microstructures, fabrics and deformation mechanisms in quartzites. Philos. Trans. R. Soc. London, Ser. A, 283: 69–86.

White, S.H., 1979. Difficulties associated with palaeostress estimates. Bull. Minéral., 102: 210–215.

White, S.H. and Bretan, P.G., 1985. Rheological controls on the geometry of deep faults and the tectonic delamination of the continental crust. Tectonics, 4: 303–309.

White, S.H. and Knipe, R.J., 1978. Transformation and reaction enhanced ductility in rocks. J. Geol. Soc. London, 135: 513–516.

White, S.H., Bretan, P.G. and Rutter, E.H., 1986. Fault zone reactivation: kinematics and mechanisms. Philos. Trans. R. Soc. London, Ser. A, 317: 81–97.

White, S.H., Burrows, S.E., Carreras, J., Shaw, N.D. and Humphreys, F.J., 1980. On mylonites in ductile shear zones. J. Struct. Geol., 2: 175–187.

Wilks, K.R. and Carter, N.L., 1989. Mechanical behavior of the Adirondack and Pikwitonei granulites. EOS, Trans. Am. Geophys. Union, 70: 477.

Willett, S.D., Chapman, D.S. and Neugebauer, H.J., 1985. A thermo-mechanical model of the continental lithosphere. Nature, 314: 520–523.

Wilshire, H.G., 1990. Lithology and evolution of the crust–mantle boundary region in the southwestern Basin and Range Province. J. Geophys. Res., 95: 649–665.

Wilshire, H.G. and Kirby, S.H., 1986. Dikes, joints and faults in the upper mantle. Tectonophysics, 121: 23–31.

Yardley, B.W.D., 1986. Is there water in the deep continental crust? Nature, 323: 111.

Zeuch, D.H., 1982. Ductile faulting, dynamic recrystallization and grain-size sensitive flow of olivine. Tectonophysics, 83: 293–308.

Zingg, A., 1980. Regional metamorphism in the Ivrea–Verbano zone (southern Alps, northern Italy): field and microscopic investigations. Schweiz. Mineral. Petrogr. Mitt., 60: 153–179.

Zingg, A., 1983. The Ivrea–Verbano and Strona–Ceneri zones (southern Alps, Ticino and N. Italy) — a review. Schweiz. Mineral. Petrogr. Mitt., 63: 361–392.

Zuber, M.T., Parmentier, E.M. and Fletcher, R.C., 1986. Extension of continental lithosphere: a model for two scales of basin and range deformation. J. Geophys. Res., 91: 4826–4838.

Chapter 7

Xenoliths — Samples of the lower continental crust

ROBERTA L. RUDNICK

1. Introduction

In order to understand the physical properties, composition, age and growth processes of the lower continental crust, it is necessary to obtain samples that have resided at deep crustal levels for substantial periods of time. Rocks that equilibrated at lower crustal temperatures and pressures occur either in large tracts of surface outcrop (high-pressure granulite facies terrains, see Mezger, this volume; Percival et al., this volume) or occur as accidental fragments (xenoliths) that are carried rapidly to the Earth's surface by fast-erupting volcanics.

Major differences exist between these two types of lower crustal rocks. First, if one considers the age of crust in which the two types of lower crustal samples occur (Fig. 7-1), it is apparent that many granulite facies terrains are Archean whereas very few granulite xenolith localities are situated in Archean crust. Of the high-pressure terrains shown in Figure 7-1a, 51% are Archean, 32% are Proterozoic, and the remaining 17% are Phanerozoic. In contrast, ages of crust through which xenoliths are erupted are much younger. Of the lower crustal xenolith localities shown in Figure 7-1b, 34% are situated in Phanerozoic crustal blocks, whereas 52% are erupted within Proterozoic crustal blocks and only 14% erupt through Archean crust.

A second difference between granulites that occur as terrains and those that occur as xenoliths is their composition. Figure 7-2 shows SiO_2 content plotted against Mg number, taken from a compilation of granulite data in the literature (Rudnick and Presper, 1990). If the literature data are representative, it shows that Archean granulite terrains tend to be dominated by evolved compositions, and post-Archean granulite terrains span a large range in compositions, with many falling at intermediate values, whereas granulite xenoliths are clearly dominated by mafic lithologies (Fig. 7-2).

What is the significance of these differences? It is possible to explain them in a number of ways.

First, one could argue that granulite terrains are not representative of the lower continental crust. Because all high-pressure granulite facies terrains contain some amount of metasedimentary rocks, which originally formed at the Earth's surface, and because these terrains currently sit at the Earth's surface above a normal thickness crust, it is possible that granulite terrains form primarily through continent–continent collision. This argument has previously been put forth on the basis that maximum pressures recorded in granulite terrains are nearly constant at 0.89 ± 0.15 GPa, compatible with equilibration at the base of the upper

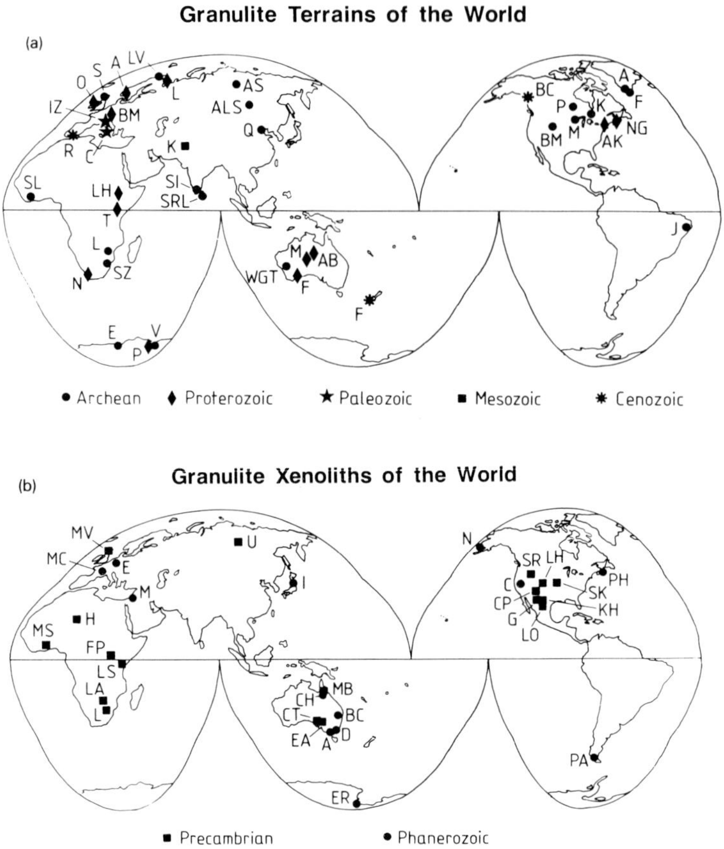

Fig. 7-1. Locations of world-wide granulite terrains (A) and granulite xenoliths (B).
(A) Key to terrain localities:
Archean: A = Amîtsoq gneisses, ALS = Aldan Shield, AS = Anabar Shield, BM = Beartooth Mountains, E = Enderby Land, F = Fiskenaesset, J = Jequie Complex, K = Kapuskasing structural zone, L = Limpopo, LV = Lofoten–Vesterålen, M = Minnesota River valley, P = Pikwitonei, Q = Qianxi, S = Scourian, SI = southern India, SL = Sierra Leone, SRL = Sri Lanka, SZ = Ancient Gneiss Terrain, Swaziland, V = Vestfold Hills, WGT = western gneiss terrain.
Proterozoic: A = Arendal, AB = Arunta Block, AK = Adirondacks, F = Fraser Ranges, L = Lapland, LH = Labwor Hills, M = Musgrave Ranges, N = Namaqualand, NG = north Grenville, O = Ox Mt. granulites, Ireland, P = Prydz Bay, T = Furua Complex, Tanzania.
Paleozoic: C = Calabria, IZ = Ivrea Zone, F = Fiordland.
Mesozoic: K = Kohistan.
Cenozoic: BC = Coast Ranges, British Columbia, R = Ronda Massif.
(B) Key to xenolith localities:
A = Anakies, Victoria, BC = Boomi Creek, N.S.W., C = Chinese Peaks, California, CC = Camp Creek, Arizona, CH = Chudleigh Province, Queensland, CP = Colorado Plateau, CT = Calcutteroo, South Australia, D = Delegate, N.S.W., E = Eifel, West Germany, EA = El Alamein, South Australia, ER = Mt. Erebus volcanic field, Antarctica, G = Geronimo Volcanic Field, Arizona, H = Hoggar, Algeria, I = Ichinomegata, Japan, KH = Kilbourne Hole, New Mexico, L = Lesotho, LA = Lace kimberlite, South Africa, LH = Leucite Hills, Wyoming, LO = La Olivina, Mexico, LS = Lashaine, Tanzania, M = Mt. Carmel, Israel, MB = McBride Province, Queensland, MC = Massif Central, France, MS = Man Shield, Liberia, MV = Midland Valley, Scotland, N = Nunivak Island, Alaska, PA = Pali Aike, Chile, PH = Popes Harbour, Nova Scotia, RC = Riley County, Kansas, S = Sardinia, Italy, SR = Snake River Plains, Idaho, U = Udachnaya, Siberia.

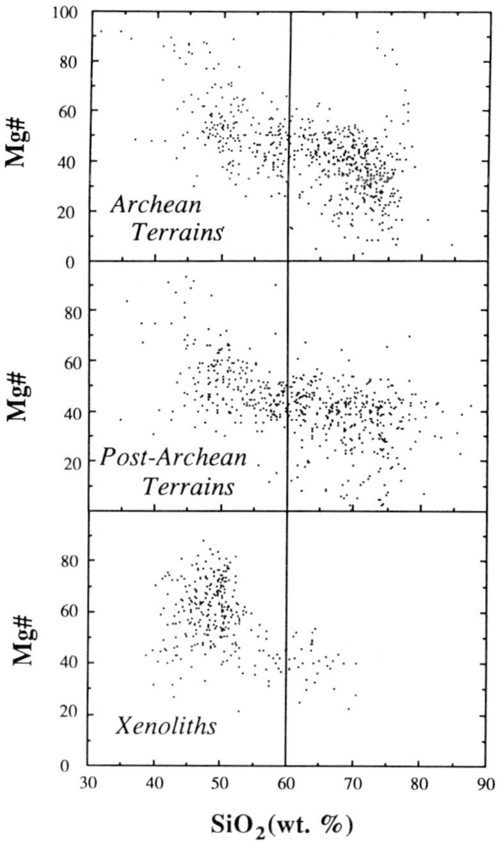

Fig. 7-2. SiO_2 versus Mg# for granulites where Mg# = 100 Mg/(Mg + Fe) (from Rudnick and Presper, 1990).

plate in a continental collision zone (Newton and Perkins, 1982). If this is true, the high pressures and temperatures recorded by the granulites are only a transient feature, recorded at the maximum temperature the rocks reached along a clockwise pressure-temperature path (England and Thompson, 1984). Granulites formed in such a setting, therefore, cannot be considered representative of the lower crust. However, recent studies of individual granulite terrains, in particular delineation of *P-T* histories, show that many granulite terrains have experienced isobaric cooling paths (Bohlen, 1987). In cases where time constraints can be made [e.g., Enderby Land, Antarctica (Harley and Black, 1987), Arunta Block, Central Australia (Windrim and McCulloch, 1986), and Pikwitonei, Canada (Mezger et al., 1988)], it is clear that the granulites resided in the lower crust for considerable periods of time.

Second, one could argue that granulite xenoliths are not representative of the lower continental crust. It is possible that the xenoliths carried by alkali basalts are dominated by mafic compositions because they represent an earlier phase of basaltic intrusion into the lower crust that culminated in the eruption of the xenolith's host basalt. That is, the composition of granulite xenoliths may be biased toward a mafic end-member simply because they occur in areas of prolonged, recent basaltic activity. In order to evaluate this hypothesis it is necessary to examine the chemical compositions of xenoliths and, ideally, determine the age of granulite xenoliths. This topic will be considered in more detail later in the chapter.

Third, the compositional differences observed between xenoliths and terrains may be due to differences in the crustal levels from which the granulites are derived (see Mezger, this volume; Percival et al., this volume). Bohlen (1987) suggested that equilibration pressures for granulite xenoliths are, in general, higher than those of granulite terrains. He envisions a chemically stratified crust composed of mafic, magmatically underplated material at the deepest levels underlying diverse lithologies, more typical of granulite facies terrains, at higher levels. One piece of evidence in support of this hypothesis is that seismic refraction profiles commonly exhibit velocities greater than 7 km s^{-1} (indicative of denser, mafic material) at depths of 30 km or more (Wever and Sadowiak, 1989; Jackson et al., 1990). However, Harley (1989) has recently shown that equilibration pressures for granulite terrains are highly variable (from >1.2 GPa to 0.4 GPa), as well as their *P-T* paths; and although it is true that many granulite xenoliths exhibit high equilibration pressures, a significant number also contain simply two pyroxenes without garnet, making it impossible to evaluate their equilibration depths. Therefore, it is likely that granulite xenoliths and granulite terrains are both derived from a wide range of crustal depths.

Finally, the age differences between granulite terrains and xenoliths may reflect a change in lower crustal composition with time, which may in turn reflect changing crustal growth processes. In this respect it is noteworthy that felsic and metasedimentary granulite xenolith types occur mainly in vents that transect Proterozoic crustal blocks [e.g., Kilbourne Hole, N.M. (Padovani and Carter, 1977a); various Mexican localities (Nimz et al., 1986; Ruiz et al., 1988a,b); Hoggar, Algeria (Leyreloup et al., 1982); McBride Province, Australia (Rudnick and Taylor, 1987); Calcutteroo, South Australia (Arculus et al., 1988)]. In addition, Phanerozoic granulite facies terrains often contain significant volumes of mafic lithologies; the Ivrea zone (Mehnert, 1975; Mazzucchelli, 1983) and Fiordland, New Zealand (Oliver, 1977, 1980), are two notable examples where mafic rocks are volumetrically important. In order to test this hypothesis it will be important to obtain geochemical and geochronological information for the few xenoliths that are erupted within Archean crustal blocks.

This chapter reviews information about the lower crust obtained from studies of granulite facies xenoliths. After a review of xenolith occurrences and general mineralogical, physical and chemical features, several case studies will be summarized. Emphasis will be placed on the geochemical nature of lower crustal xenoliths, in particular their chemical and isotopic characteristics and how these data allow development of petrogenetic models of lower crustal growth and modification. Finally, the question of to what degree granulite xenoliths represent the lower crust will be addressed, with presentation of some unresolved questions that await future research.

2. The samples

2.1. Definitions

Before discussing the results of various xenolith studies it is necessary to define what a lower crustal xenolith is. In this paper I define the lower crust as the lower half of the crust: its lower boundary is the Moho [based on the seismic definition, see White (1988)]* and

*The recognition that the crust–mantle boundary may be transitional, with interleaved mafic granulite and lherxzolite, has led to the definition of a "petrologic Moho" (Griffin and O'Reilly, 1978a), i.e.: the depth at which lherzolites become the "dominant" lithology. However, this depth is often difficult to determine unequivocally and has only been attempted for eastern Australia. For this reason, I have chosen the seismic Moho as the crust–mantle boundary, which generally corresponds to depths ≤45 km ($P \leq 1.2$ GPa), including eastern Australia.

its upper boundary is the mid-point of the continental crust. Lower crustal xenoliths are, therefore, fragments of rock that can be shown to have been entrained from these depths. Generally, lower crustal xenoliths have mineral assemblages characteristic of the granulite facies. However, in several cases amphibolite facies rocks may constitute the lower crust, especially where water content is high, as above subduction zones [e.g., the Ichinomegata locality in Japan (Kuno, 1967)].

Several lines of evidence can be used to judge whether xenoliths are derived from the lower crust. The first, and most obvious, is that xenolith mineralogy and estimated *P-T* equilibration conditions should correspond to appropriate lower crustal depths (0.6 to $\leq$1.4 GPa)(Griffin and O'Reilly, 1987). Secondly, the primitive composition of the host and the presence of lherzolite xenoliths in the same vent, while not definitive, do indicate that the host magma erupted quickly (in a matter of hours) from great depths. If coexisting lherzolites are not observed and the host magma is evolved, as is the case for the Snake River Plains xenolith suite (Leeman et al., 1985) and the east Eifel xenolith suite (Wörner et al., 1982; Grapes, 1986), it may indicate that the magma ponded and fractionated at a mid- to upper crustal level, where it entrained the crustal xenoliths. Thirdly, decompression features, such as preferential melting of mafic phases (Padovani and Carter, 1977b) and kelyphite formation on rims and in cracks in garnets (Garvie and Robinson, 1984) are interpreted to have resulted from rapid ascent of the xenolith from deep crustal levels. Finally, perhaps the most definitive evidence for derivation of a xenolith from the lower crust is isotopic homogenization of minerals, yielding ages equal to the eruption age of the host basalt (van Breeman and Hawkesworth, 1980; Padovani and Hart, 1981; McDonough et al., 1990). However, it is possible that isotopic homogenization on a mineral scale was not attained in the lower crust if the geothermal gradient attending the basaltic or kimberlitic eruption was not high (Harte et al., 1981). Therefore, when metamorphic mineral ages equal eruption ages it is safe to assume the xenolith was derived from the lower crust at the time of eruption. When mineral ages are older than eruption ages other information is required to ascertain whether the xenolith is truly lower crustal. Metamorphic mineral ages or zircon ages from xenoliths that are older than eruption age may also provide evidence of lower crustal derivation if it can be shown that little uplift and erosion has occurred in the region since the growth of granulite facies minerals (e.g., Rudnick and Williams, 1987). Determining whether or not some granulite xenolith suites actually come from near-surface granulite facies sequences [e.g., Massif Central suite and Hoggar suite, (Griffin and O'Reilly, 1987)], must await accumulation of geochronologic data for these samples.

2.2. Occurrence

Lower crustal xenoliths have been reported from every continent, however the number of integrated petrographic and geochemical studies on these xenolith suites is limited when compared with those of granulite terrains. Table 7-1 summarizes the information available on xenolith suites for which chemical and/or mineral chemical data have been published and Figure 7-1b shows lower crustal xenolith localities of the world.

Lower crustal xenolith-bearing alkali basalts tend to be located in non-cratonic areas, with many erupting within Phanerozoic fold belts or rifts. The specific tectonic settings of the basaltic magmatism have been summarized in Nixon (1987b). These settings include continental rifts (for example the Rio Grande rift, the Rhinegraben; and the east African rift) back-arc settings (such as Nunivak island, Alaska, Ichinomegata, Japan, Pali Aike, Chile)

TABLE 7-1

Geochemical and mineral chemical studies of lower crustal xenoliths

Locality	Host	Xenolith types	Types of analyses	Pipe age	Crust age	Protolith age	Reference[1]
NORTH AMERICA							
Nunivak Island, Alaska	AB	MG	ME, Min.	<5 Ma	Phanerozoic		Francis (1976)
Popes Harbour *, Nova Scotia	K	MG, PG, FG	ME, TE, Min., Sr, Nd	<400 Ma	Phanerozoic		Owen et al. (1988), Eberz et al. (1991)
Snake River Plains *, Idaho	Evol. B	PG	Sr, Nd, Pb	<2 Ma	Archean	~2.8 Ga	Leeman et al. (1985)
Bearpaw Mts., Montana	K	MGG, Mg, FG	ME, TE, O	45 Ma	Archean		Collerson et al. (1992), Kempton and Harmon (1992)
Chinese Peak, Sierra Nevada	AB	MGG	ME, TE, Min, Sr, O, Zir	10 Ma	Phanerozoic	180 Ma	Dodge et al. (1986)
Big Creek, Sierra Nevada	AB	MGG, Ec, FG	ME, TE, Min	8 Ma	Phanerozoic		Dodge et al. (1988), Domenick et al. (1983)
Colorado/Wyoming	K	MGG, MG	ME, Min.		Proterozoic		Bradley and McCallum (1984)
Navajo Volcanic Field, Colorado Plateau	K	MGG, FG, Ec	ME, Min.	25–30 Ma	Proterozoic		Ehrenberg and Griffin (1979), Broadhurst (1986)
Camp Creek, Arizona	L	MGG, Am, Ec	ME, TE, Min., Sr, Nd, Pb	23–27 Ma	Proterozoic	1.2–1.9 Ga	Esperança et al. (1988)
Chino Valley, Arizona	L	MGG, Am	ME, TE, Min.	25 Ma	Proterozoic		Arculus and Smith (1979), Schulze and Helmstadt (1979), Arculus et al. (1988)
Geronimo Volcanic Field New Mexico	AB	MG, Int.G.	ME, TE, Min., Sr, Nd, Pb, O	<3 Ma	Proterozoic	1.1–1.4 Ga	Kempton et al. (1990) Kempton and Harmon (1992)
Kilbourne Hole, New Mexico	AB	MG, An, PG, FG	Min., ME, Sr, Nd, Pb, O Zir	<1 Ma	Proterozoic	1.5 Ga	Padovani and Carter (1977), Davis and Grew (1977), Padovani et al. (1981), James et al. (1980), Reid et al. (1989)
Elephant Butte, New Mexico	AB	MG	Min., ME	<3 Ma	Proterozoic		Baldridge (1979)
Engle Basin, New Mexico	AB	MG	Min.	<3 Ma	Proterozoic		Warren et al. (1979)
Northern Mexico	AB	MG, PG, FG	ME, TE, Min., Sr, Nd, Zir	<25 Ma	Proterozoic	from 1 to 1400 Ma	Nimz et al. (1986), Ruiz et al. (1988a,b), Roberts et al. (1989), Hayob et al. (1989), Rudnick and Cameron (1991)
Riley County, Kansas	K	MGG, Ec	Min.	<230 Ma	Proterozoic		Meyer and Brookins (1976)

SOUTH AMERICA							
Pali Aike	AB	MG	ME, Min.	<3 Ma	Phanerozoic		Selverstone and Stern (1983)
EUROPE							
Scotland	AB	MG, An, PG, MGG	ME, Min., Nd, Sr, Zir	~350 Ma	Archean/ Proterozoic	360 Ma	van Breeman and Hawkesworth (1980), Halliday et al. (1984, 1985), Hunter et al. (1984), Upton et al. (1983)
Kampernich, E. Eifel, Germany	AB	MGG, Am	ME, TE, Min., Sr, Nd, Pb, O	<1 Ma	Phanerozoic	1.5 Ga or ~450 Ma?	Okrusch et al. (1979), Stosch and Lugmair (1984), Stosch et al. (1986), Rudnick and Goldstein (1990), Loock et al. (1990) Kempton and Harmon (1992)
Wehr Volcano *, E. Eifel, Germany	AB	Am	ME, TE, Min.	<1 Ma	Phanerozoic		Wörner et al. (1982), Grapes (1986)
N. Hessian Depression, Germany	AB	MGG, MG, PG, FG	ME, TE, Min., O	<50 Ma	Phanerozoic		Mengel and Wedepohl (1983) Mengel (1990)
Massif Central, France	AB	MGG, MG, PG, FG	ME, TE, Nd, Sr, Pb, O	<5 Ma	Phanerozoic	~350 Ma	Leyreloup et al. (1977), Dostal et al. (1980), Vidal and Postaire (1985), Downes and Leyreloup (1986), Downes et al. (1990) Kempton and Harmon (1992)
Tallante, Spain	AB	PG	Min.	<20 Ma	Phanerozoic		Vielzeuf (1983)
Sardinia, Italy	AB	MG	Min.	~3 Ma	Phanerozoic		Rutter (1987)
AFRICA							
Hoggar, Algeria	AB	MG, An, PG	ME, TE	<20 Ma	Proterozoic		Leyreloup et al. (1982)
Man Shield, Liberia	K	MGG, An, Ec	ME, Min.	90–120 Ma	Archean		Toft et al. (1989)
Lashaine, Tanzania	AB	MGG, Ec, An	ME, TE, Min., Sr, Nd, Pb	<20 Ma	Proterozoic		Dawson (1977), Jones et al. (1983), Cohen et al. (1984)
Fort Portal, Uganda	AB	MGG, MG	Min., ME, TE	<3 Ma	Proterozoic		Thomas and Nixon (1987)
Lace Kimberlite, S. Africa	K	PG	Min.	90–140 Ma	Archean		Dawson and Smith (1987)

TABLE 7-1 (continued)

Locality	Host	Xenolith types	Types of analyses	Pipe age	Crust age	Protolith age	Reference[1]
AFRICA (continued)							
Lesotho, S. Africa	K	MGG, MG, FG, Ec	ME, TE, Min., Sr, Nd, Pb, Zir	90–140 Ma	Proterozoic	1.4 Ga	Davis (1976), Rogers (1977), Griffin et al. (1979), Harte et al. (1981), Rogers and Hawkesworth (1982), van Calsteren et al. (1986)
Markt, S. Africa	K	MGG	Min.	90–140 Ma	Archean		van Calsteren et al. (1986)
MIDDLE EAST–ASIA							
Mt. Carmel, Israel	AB	MGG	Min.	~100 Ma	Phanerozoic		Esperança and Garfunkel (1986), Mittlefehldt (1986)
Birket Ram, Israel	AB	MGG, Am	TE	10,000 y	Phanerozoic		Mittlefehldt (1984)
Udachnaya, Siberia, Russia	K	MGG	Min.		Phanerozoic		Shatsky et al. (1983, 1990)
Ichinomegata, Japan	And.	Am, Gab	ME, TE, Min., Sr	10,000 y	Phanerozoic		Kuno (1967), Aoki (1971), Zashu et al. (1980), Tanaka and Aoki (1981)
AUSTRALIA–ANTARCTICA							
McBride Prov., N. Queensland	AB	MGG, MG, PG, FG	ME, TE, Min., Sr, Nd, Pb, O Zir	<3 Ma	Proterozoic	300 Ma and ~1.6 Ga	Kay and Kay (1983), Rudnick and Taylor (1987), Rudnick and Williams (1987), Stolz and Davies (1989), Stolz (1987), Rudnick (1990), Rudnick and Goldstein (1990) Kempton and Harmon (1992)
Chudleigh Prov., N. Queensland	AB	MGG, MG	ME, TE, Min., Sr, Nd, Pb, O	<1 Ma	Phanerozoic	<100 Ma	Kay and Kay (1983), Rudnick et al. (1986), Rudnick and Taylor (1990), O'Reilly et al. (1988), Rudnick and Goldstein (1990) Kempton and Harmon (1992)

Central Queensland	AB	MG, MGG	ME, TE, Min., Sr, Nd	$\leq$50 Ma	Phanerozoic	Griffin et al. (1987), O'Reilly et al. (1988)
Gloucester, N.S.W.	AB	MGG, MG	ME, TE, Sr, Nd	$\leq$50 Ma	Phanerozoic	Griffin et al. (1986), O'Reilly et al. (1988)
Sydney Basin	AB	MG	ME, TE, Sr, Nd	$\leq$50 Ma	Phanerozoic	Griffin et al. (1986), O'Reilly et al. (1988)
Boomi Creek, N.S.W.	AB	MG	ME, TE, Min.	$\leq$50 Ma	Phanerozoic	Wilkinson (1975), Wilkinson and Taylor (1980)
Delegate, N.S.W.	AB	MG, MGG, FG, Ec	ME, TE, Min., Sr, Nd	~140 Ma	Phanerozoic	Lovering and White (1964, 1969), Griffin and O'Reilly (1986) O'Reilly et al. (1988), Arculus et al. (1988)
Jugiong, N.S.W.	K	MG, MGG	Min.	<17 Ma	Phanerozoic	Arculus et al. (1988)
White Cliffs, N.S.W.	K	MGG	Min.	~260 Ma	Proterozoic	Arculus et al. (1988)
Anakies, Victoria	AB	MG, MGG	ME, TE, Min., Sr, Nd	<2 Ma	Phanerozoic	Sutherland and Hollis (1982), Wass and Hollis (1983), Reilly et al. (1988)
El Alamein, S. Australia	K	MGG, Ec	ME, TE, Min.	~170 Ma	Proterozoic	Edwards et al. (1979), Arculus et al. (1988)
Calcutteroo, S. Australia	K	MGG, FG, Ec	ME, TE, Min., Sr, Nd	~170 Ma	Proterozoic	McCulloch et al. (1982), Arculus et al. (1988)
Mt. Erebus Volcanic Field, Antarctica	AB	MG, MGG	ME, TE, Min., Sr, O	<5 Ma	Phan./Prot.	Kyle et al. (1987), Kalamarides et al. (1987), Berg et al. (1989)

[1] Only papers in which data are reported are listed here.

* Xenoliths from these localities are probably derived from mid-crustal levels based on either: equilibration pressures, lack of mantle derived xenoliths in the same hosts, or chemically evolved character of the host.

Abbreviations

Host types: AB = alkali basaltic association; K = kimberlitic association (including lamproites, minettes, kimberlites); Evol. B. = evolved basalt; L = latite.
Xenolith types: Am = amphibolite; An = anorthosite; Ec = eclogite; FG = felsic granulite; MG = mafic granulite; MGG = mafic garnet granulite; PG = paragneiss.
Types of analyses: ME = major-element analyses; Min = mineral analyses; TE = trace-element analyses; Sr = Sr isotope analyses; Nd = Nd isotope analyses; Pb = Pb isotopic analyses; O = oxygen isotopic analyses; Zir = U-Pb zircon analyses.

or hot-spot traces as found in eastern Australia. In other areas it is difficult to ascribe the volcanism to any given tectonic setting. For example, the numerous xenolith occurrences in the Colorado Plateau do not fall into a simple tectonic setting although these extension-related volcanics may be associated with mantle plumes or back-arc spreading (Best and Brimhall, 1974; Omerod et al., 1988).

In contrast to alkali basalts, kimberlitic hosts (e.g., kimberlites, lamprophyres and minettes) generally erupt through stable continental regions, although eruption of the kimberlite may be associated with super-continent break up (Nixon, 1987a). It is the kimberlitic hosts that provide samples of lower crust from Archean regions, for example, the Man Shield (Toft et al., 1989); the Lace Kimberlite, South Africa (Dawson and Smith, 1987); the Bearpaw Mountains, Montana (Collerson et al., 1988) and the Siberian platform (Shatsky et al., 1990).

3. Effects of transport

The processes acting on xenoliths from the time of entrainment by the host magma to the time of sample collection may cause chemical and physical modification of the rock. These processes include: heating by the host, infiltration by the host, hydrothermal alteration by fluids derived from the host magma, decompression-induced melting or mineral breakdown and weathering and alteration at the Earth's surface. Any study of the chemical or physical properties of lower crustal xenoliths should first address the possibility of post-entrainment modifications.

Post-entrainment physical changes to xenoliths are caused by rapid decompression that the rock undergoes on its way to the Earth's surface [estimated travel time of xenoliths in alkali basalts are on the order of hours (Spera, 1980)]. Decompression induced changes include development of microcracks, partial melting along grain boundaries and of hydrous phases (this is particularly important in intermediate to felsic lithologies) and formation of kelyphite rims on garnets. Padovani and Carter (1977b) presented a detailed microprobe study of decompression features in paragneiss and anorthosite xenoliths from Kilbourne Hole, New Mexico. They found the chemical composition of melts along grain boundaries bore no relation to the host basalt but instead resembled nearby mineral phases. Heterogeneities in the glasses suggested little mixing of melts occurred and elongated crystals within the glass were interpreted as quench crystals. Garnet is particularly susceptible to decompression effects. For example, garnets from many granulite xenoliths have a dark, kelyphite rim of variable thickness. This rim is often extremely fine-grained, with crystals less than 1 μm across (Fig. 7-3). Such kelyphitic rims are generally interpreted as being due to decompression melting with subsequent quench crystallization at the Earth's surface (Padovani and Carter, 1977b; Jones et al., 1983; Garvie and Robinson, 1984) An alternative explanation was given by Griffin et al. (1987), who interpreted these rims as being due to localized heating in the lower crust. However, the observation that kelyphite forms along cracks in garnets (Fig. 7-3) suggests that it formed after xenolith entrainment.

The degree of post-entrainment chemical modification of xenoliths appears to depend on the nature of the host. Kimberlite-hosted xenoliths often show clear evidence of hydrothermal alteration in thin-section, with the development of greenschist facies mineral assemblages in minerals and on grain boundaries (Griffin et al., 1979; Padovani et al., 1982). This is generally attributed to circulation of host-derived hydrothermal fluids. Such hydrothermal alteration is probably associated with chemical changes. For example, Ba and Sr contents are very high in

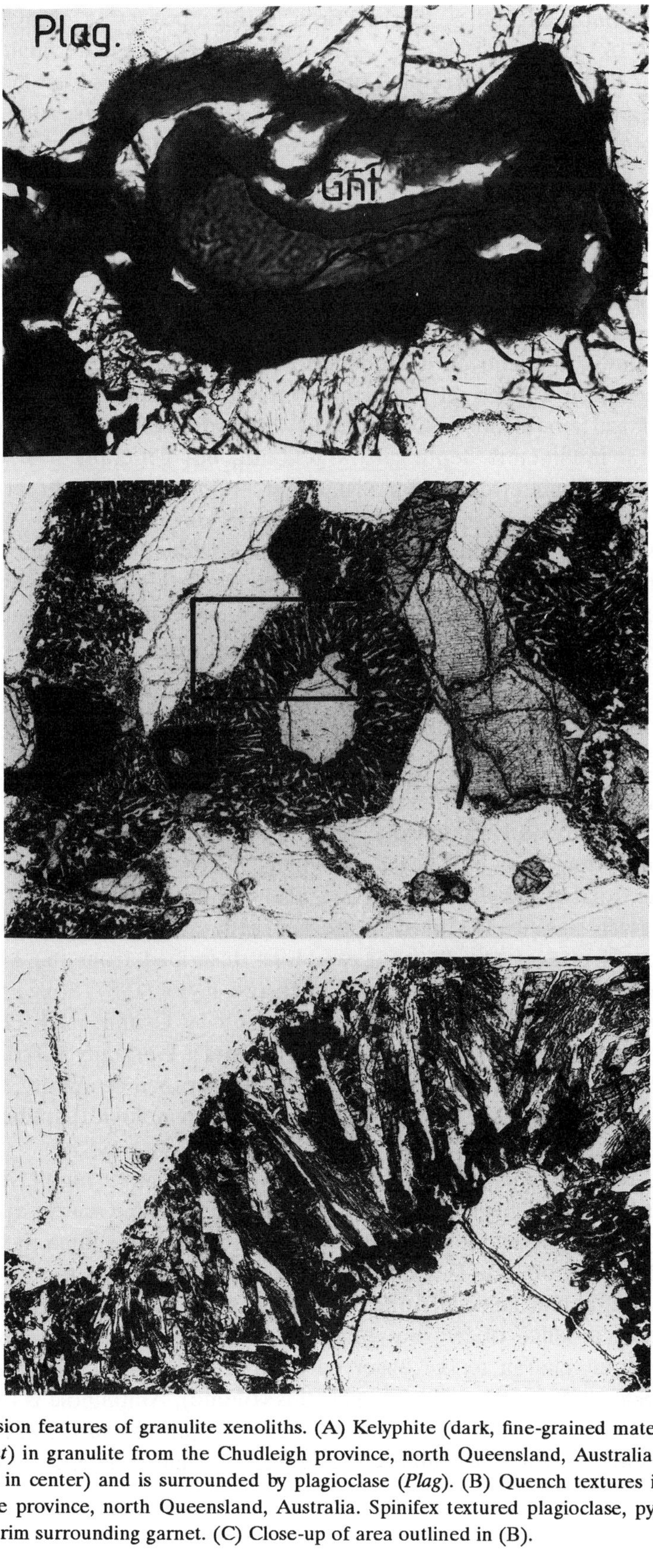

Fig. 7-3. Decompression features of granulite xenoliths. (A) Kelyphite (dark, fine-grained material) on rim and in cracks in garnet (*Gnt*) in granulite from the Chudleigh province, north Queensland, Australia. Garnet surrounds spinel (dark mineral in center) and is surrounded by plagioclase (*Plag*). (B) Quench textures in kelyphite rim on garnet from McBride province, north Queensland, Australia. Spinifex textured plagioclase, pyroxenes and spinel form a still euhedral rim surrounding garnet. (C) Close-up of area outlined in (B).

some granulite xenoliths carried by kimberlitic or latitic hosts [#217, Rogers (1977); #156, Arculus (1979); #15, Arculus (1988); see section 7, below]. These anomalously high values are interpreted as being due to precipitation of secondary phases on grain boundaries after eruption (Rudnick and Taylor, 1987; Emery, 1985; R.J. Arculus, pers. commun., 1987), with the Ba and Sr presumably derived from the host. Beyond alkali earth elements, it is not clear which other elements may be affected by post-entrainment alteration, however, Rogers and Hawkesworth (1982) suggested that the REE elements in some granulite xenoliths from Lesotho may be affected by contamination from kimberlite. In addition, the high K_2O and Rb content of kimberlite-hosted xenoliths may also be an artifact (see sections 6.3 and 7 below). Because of the possibility of multiple processes acting on kimberlite-hosted xenoliths (e.g., decompression melting of certain minerals overprinted by hydrothermal exchange with the host) it may be difficult to effectively document post-entrainment chemical changes.

In contrast, xenoliths carried by alkali basalts generally show little evidence for post-entrainment chemical modification. Occasionally, thin veins of host basalt can be observed penetrating several millimeters depth into a xenolith, but generally host-xenolith contacts are sharp and there is no evidence for significant grain-boundary alteration by the host. Weathering effects, such as formation of iddingsite in olivines, appear to be isochemical. Additionally, development of caliche, which has been documented in some mantle xenoliths (Jagoutz et al., 1980), may pose a problem in the interpretation of the whole-rock chemistry and isotope systematics of basalt-hosted granulite xenoliths coming from arid regions.

4. Mineralogy

The mineralogy of lower crustal xenoliths is a function of bulk composition, pressure and temperature of equilibration and fluid activity. As most granulite xenoliths are mafic, their mineralogies are dominated by pyroxene, plagioclase and garnet, with the proportions of these minerals depending upon bulk composition and *P-T* history. Olivine, spinel, amphibole, biotite/phlogopite, quartz and Fe-Ti-oxides are also commonly reported. Scapolite and sapphirine, though given significant attention, are generally rare in mafic granulite suites [e.g., scapolite is reported in 6 of the 80 mafic xenoliths described from the McBride province Northern Queensland (Kay and Kay, 1983; Rudnick and Taylor, 1987; Stolz and Davies, 1989), 6 of the 33 granulite xenoliths from Lesotho described by Griffin et al. (1979) and occurs in 5 of the 70 xenoliths described from the Eifel, western Germany (Okrusch et al., 1979; Loock, 1988; Rudnick, unpubl.)]. Common accessory phases include apatite, rutile and zircon, the latter being easily overlooked but important for geochronological studies. Less common accessories are sulfides, sphene and native Fe (Haggerty and Toft, 1985).

Mineralogy of intermediate to felsic granulite xenoliths is more highly variable, as might be expected considering the likely diverse origins of these rocks. Quartz, alkali feldspar and plagioclase are generally the major phases, with pyroxenes, garnet, sillimanite, kyanite, biotite, amphibole, rutile, graphite, Fe-Ti oxides, apatite, monazite, zircon, sulfides, corundum, scapolite, spinel and sapphirine occurring as minor or accessory phases.

The presence or absence of hydrous phases in the lower crust has received attention recently, especially in relation to the observed deep-seated low-resistivity zones present in some areas (Gough, 1986; Yardley, 1986; Jones, this volume). Amphibole is the most common hydrous phase in lower crustal xenoliths and its modal proportion varies widely from suite to suite. Xenolith suites in which amphibole is abundant include The Eifel, western Germany

(Okrusch et al., 1979; Stosch et al., 1986), The Anakies, southeastern Australia (Wass and Hollis, 1983), Hill 32, northern Queensland, Australia (Kay and Kay, 1983; Stolz and Davies, 1989), Camp Creek, Arizona (Esperança et al., 1988) and Ichinomegata, Japan (Aoki, 1971; Kuno, 1967). In all but the latter two suites, amphibole generally makes up less than 50 modal percent of the rock. Compositionally, the amphiboles are pargasites with variable but high Ti contents. No fluorine or OH^- analyses have been reported, but Wass and Hollis (1983) found that whereas coexisting apatites contain up to 4% F, no fluorine was detectable in amphiboles in granulite xenoliths from The Anakies. Amphibole commonly occurs in several morphologies within a given specimen, yet generally each of these texturally distinct amphiboles have similar compositions (Okrusch et al., 1979; Wass and Hollis, 1983; Stosch et al., 1986; Stolz and Davies, 1989).

Multiple origins have been ascribed to explain the occurrence of amphibole in granulite facies xenoliths. In some cases amphibole is demonstrably secondary and forms rims on clinopyroxenes (Okrusch et al., 1979). Wass and Hollis (1983) suggested that amphibole in the Anakies xenoliths was relict igneous amphibole which survived dehydration caused by influx of mantle-derived CO_2. On the other hand, Stolz (1987) suggested that whereas some interstitial amphibole may be primary, amphibole occurring as intergrowths with pyroxene and as lamellae within pyroxene probably formed by reaction of pre-existing clinopyroxene with a mantle-derived hydrous fluid phase. Support for this model is found in the observation that amphibole from an Eifel xenolith measured by Stosch and Lugmair (1984) had the lowest $^{87}Sr/^{86}Sr$ ratio yet the highest Rb/Sr ratio of all minerals in the rock, requiring amphibole growth to be a recent event.

5. *P-T* estimates

Reviews of geothermobarometry applied to granulites are found in Kay and Kay (1981), Essene (1982), Newton (1983) and Frost and Chacko (1989). Equilibration conditions may be estimated for a granulite xenolith provided the appropriate mineral assemblages are present and these minerals are in equilibrium. Establishing the existence of mineral equilibria is difficult. Lack of compositional zoning is commonly cited as evidence for mineral equilibration, although it is conceivable that if a rock cools quickly, it may pass below the blocking temperature before significant diffusion occurs. However, in the lower crustal environment it is unlikely that such rapid cooling would occur. Isotopic analyses of mineral phases may also be used to establish whether equilibria is attained. Xenoliths erupted in areas of high heat flow commonly show mineral equilibration [e.g., Kilbourne Hole, New Mexico (Reid et al., 1989), southeastern Australia (McDonough, 1990)], whereas those erupted in cooler regions may not (Harte et al., 1981). In samples containing zoned minerals, P and T are often calculated assuming equilibration between adjacent mineral rims (e.g., Rudnick et al., 1986; Griffin et al., 1987), although if mineral dissolution has occurred (Duebendorfer and Frost, 1988), this may yield spurious results. Duebendorfer and Frost (1988) recommend making microprobe maps in order to evaluate rim-rim equilibration.

Coronal textures are clear indications that equilibration has not been attained in a given specimen, nevertheless important information on P-T paths may be obtained from a rock which displays such textures. Corona-bearing granulite xenoliths have been described from Nunivak Island, Alaska (Francis, 1976), the Eifel, western Germany (Okrusch et al., 1979) and from the Chudleigh province, Queensland, Australia (Kay and Kay, 1983; Rudnick et al., 1986). The Nunivak Island coronas are set in an olivine matrix and are composed of, from rim to core: orthopyroxene, spinel and spinel-clinopyroxene symplectite. Francis (1976) interpreted these coronas as due to the reaction of olivine, spinel and Ca-plagioclase to form Na-plagioclase, clinopyroxene, orthopyroxene and spinel during either isobaric cooling or progressive burial of a troctolitic pluton. Coronas in the Eifel xenoliths consist of radially arranged orthopyroxene with interstitial plagioclase, amphibole and Ti-magnetite surrounding garnet (Fig. 7-4a), which Okrusch et al. (1979) interpreted as being due to a retrograde reaction occurring at lower P and T than the peak metamorphism. Coronas in the Chudleigh province xenoliths, coupled with variable modal mineralogy at relatively constant bulk composition, allow estimation of derivation depths as well as P-T history for the xenolith suite (Kay and Kay, 1983; Rudnick et al., 1986). Two corona structures are observed:

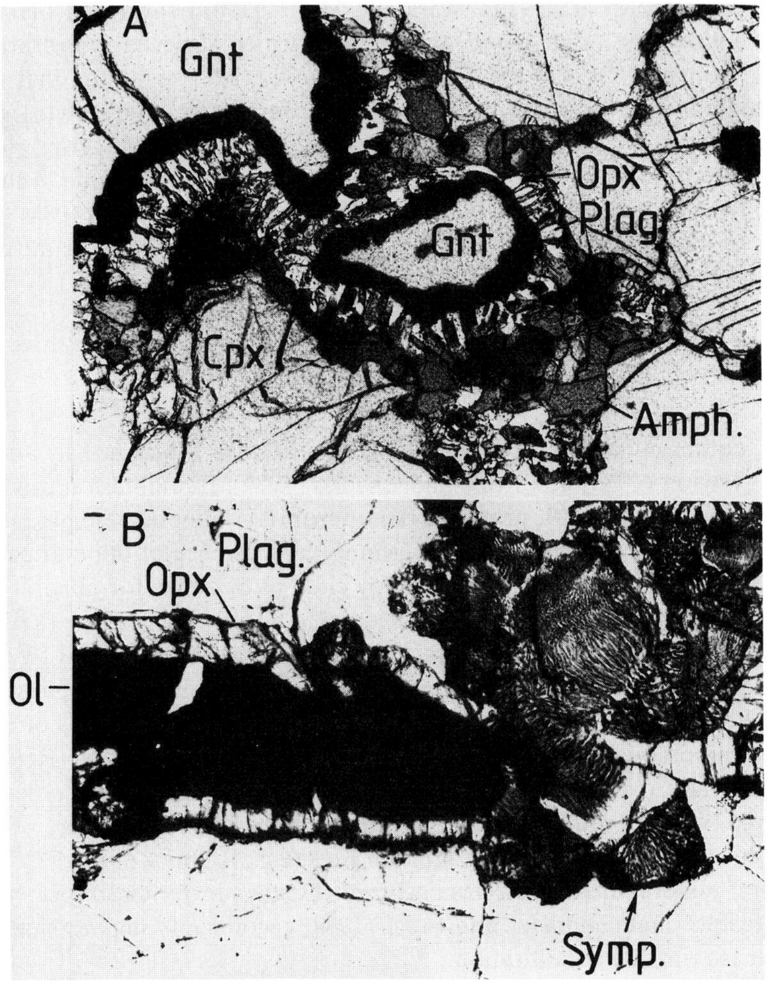

Fig. 7-4. Coronas in granulite facies xenoliths. (A) Radially aligned orthopyroxene (*Opx*) and plagioclase (*Plag*) surround garnet (*Gnt*) (with thin, dark kelyphite rim) and are in turn surrounded by amphibole (*Amph*). Clinopyroxene and ilmenite occur in the groundmass. Xenolith from the Eifel, West Germany. (B) Olivine (*Ol*) altered to iddingsite is rimmed by orthopyroxene (*Opx*), which is in turn rimmed by symplectic spinel-pyroxene (*Symp*), all set in a plagioclase groundmass. Xenolith from the Chudleigh volcanic province, Queensland, Australia.

(1) olivine rimmed by orthopyroxene that is in turn rimmed by pyroxene-spinel symplectite (Fig. 7-4b) and (2) spinel rimmed by garnet. Both reactions were interpreted by Rudnick et al. (1986) to reflect isobaric cooling of mafic cumulates in the deep crust. In addition, xenoliths from this suite with similar bulk composition display widely variable mineralogies, representing almost the entire gabbro to eclogite transition (Green and Ringwood, 1971; Ito and Kennedy, 1971; Irving, 1974; Herzberg, 1978; Wood, 1987), and were presumably derived from a large depth range (~20 to 40 km). Mineral compositions change progressively across this transition with plagioclase becoming more sodic with depth.

Equilibration temperatures estimated for lower crustal xenoliths generally range from 700 to 1000°C. Figure 7-5 shows granulite xenolith equilibration temperatures calculated using either the Wells (1977) calibration of the two-pyroxene thermometer or the Ellis and Green (1979) calibration of the garnet-clinopyroxene thermometer, and assuming $Fe^{2+} = Fe_{total}$ (these are therefore maximum temperatures). Temperatures are taken or calculated from the references given in the figure caption. Estimated uncertainty in these temperatures is ±50°C (Ellis and Green, 1979). In spite of this rather large uncertainty (on the scale of the diagram), the data in Figure 7-5 show that systematic temperature differences occur between xenolith suites. Xenoliths from the Colorado Plateau, South Australia, the Man Shield in northwest Africa, Uganda and the Siberian platform exhibit the lowest equilibration temperatures,

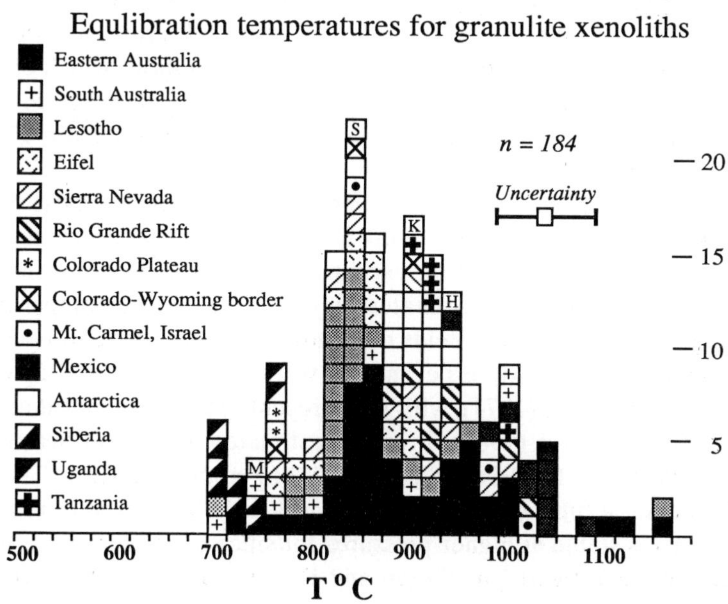

Fig. 7-5. Equilibration temperatures for granulite xenoliths calculated from either the Wells (1977) two-pyroxene thermometer or the Ellis and Green (1979) garnet-clinopyroxene thermometer, assuming $Fe^{2+} = \Sigma Fe$. Sources of data are: eastern Australia - Arculus et al. (1988); Griffin et al. (1987) Rudnick and Taylor (1987, 1990); Stolz (1987); Wass and Hollis (1983); South Australia - Arculus et al. (1988); Lesotho - Griffin et al. (1979); Eifel - Okrusch et al. (1979); Loock et al. (1990); Sierra Nevada - Dodge et al. (1988, 1986); Rio Grande Rift - Warren et al. (1979), Padovani and Carter (1977a), Baldridge et al. (Baldridge, 1979); Colorado Plateau - Ehrenberg and Griffin (1979); Broadhurst (1986), Colorado/Wyoming kimberlites - Bradley and McCallum (1984); Mexico - Nimz et al. (1986); Mt. Carmel, Israel - Esperança et al. (1988); Antarctica - Berg et al. (1989); Siberia - Shatsky et al. (1990); Uganda - Thomas and Nixon (1987); Tanzania - Jones et al. (1983). In some cases only average temperatures are reported and are shown as a box with a letter in it. M = Man Shield (Toft et al., 1989); S = Scotland (Hunter et al., 1984); K = Kimberly, South Africa (Van Calsteren et al., 1986); and H = Hoggar, Algeria (Leyreloup et al., 1982).

generally falling below 850°C, whereas xenoliths from the Rio Grande and Antarctic Rifts, Mexico, Hoggar in Algeria, Kimberly in South Africa, eastern Australia, Israel, Tanzania and the Sierra Nevada exhibit the highest equilibration temperatures, generally falling above 850°C, with the majority falling above 900°C. Suites exhibiting intermediate average equilibration temperatures include Lesotho, Eifel, Arizona and Scotland.

The temperatures recorded by cation exchange thermometers may represent either the *in situ* lower crustal temperature at the time of xenolith entrainment or a paleotemperature (if the xenolith has cooled below the blocking temperature of cation diffusion). It should be possible to test which scenario is true by isotopic analyses of mineral phases (e.g., Harte et al., 1981). There is a rough correlation between xenolith equilibration temperatures and crustal heat flow for the xenolith suites shown in Figure 7-5. Xenoliths erupted within stable continental platforms where heat flow is low (e.g., Lesotho, Siberia and the Man Shield, west Africa), generally exhibit low equilibration temperatures (Kimberly is a notable exception to this) and isotopic disequilibrium between mineral phases at the time of eruption (Harte et al., 1981; Shatsky et al., 1990). The temperatures recorded by these xenoliths, although at the lower end of the temperature histogram (Fig. 7-5), lie well above a shield geotherm and presumably reflect blocking temperatures for cation exchange. Xenoliths erupted within regions that are characterized by very high heat flow (e.g., continental rift zones or areas of extensive Tertiary magmatism with heat flow on the order of 90 to >100 mW m^{-2}), generally show high equilibration temperatures (e.g., the Rio Grande Rift, Mexico, Antarctica, eastern Australia) and are in isotopic equilibrium at the time of eruption (Reid et al., 1989; McDonough et al., 1990). An exception to this is the Eifel xenoliths, which erupted near a rift yet have moderate equilibration temperatures (Okrusch et al., 1979; Loock et al., 1990) and show isotopic disequilibrium between mineral phases (Stosch and Lugmair, 1984; Loock et al., 1990).

Compared with temperature estimates, fewer pressure estimates exist for lower crustal xenoliths. This is because many xenoliths do not contain appropriate assemblages for estimating pressures on the basis of cation exchange barometry. Because most lower crustal xenoliths are mafic, rough pressure estimates may be obtained by comparing the observed mineral assemblage to phase stability fields determined for the gabbro to eclogite transition. These fields are based on the early experimental work of Green and Ringwood (1971), Ito and Kennedy (1971) and Irving (1974), with more recent thermodynamical constraints placed by Herzberg (1978) and Wood (1987). Pressure estimates for garnet-bearing lower crustal xenoliths range from 0.8 to 1.4 GPa, with most falling above 0.9 GPa. The appearance of garnet in a mafic rocks is a function of pressure, as well as bulk composition. It can be seen on Figure 7-6 that garnet is stable at higher pressures for silica-oversaturated compositions, such as the quartz tholeiite investigated by Green and Ringwood (1967, 1971) and Wood (1987), but appears at lower pressures in silica-saturated compositions, such as the olivine tholeiite composition investigated by Irving (1974).

6. Physical properties

To date few laboratory determinations of the physical properties of granulite xenoliths have been published. The reason for this is probably due to the small sample size and variable degrees of freshness of many xenoliths. Nevertheless, it should be possible to extrapolate measurements made on surface granulites to xenoliths provided the rocks have similar mineralogies.

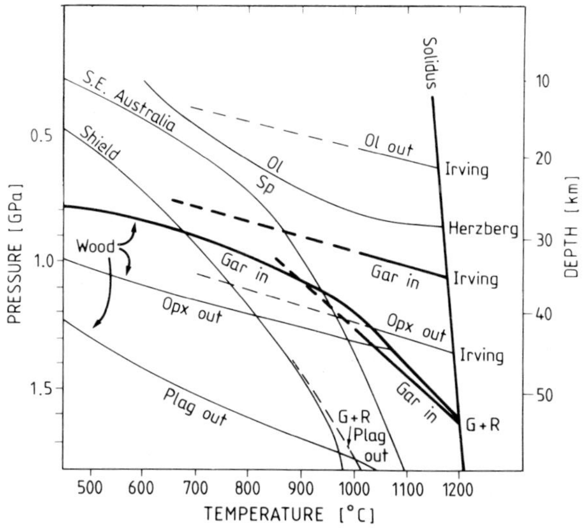

Fig. 7-6. *P-T* diagram showing phase stability fields for mafic compositions based on work of Irving (1974), Herzberg (1978), Green and Ringwood (1971) (G + R) and Wood (1987). Also shown are postulated geotherms for eastern Australia (from Griffin et al., 1984) and a stable shield region (Jones et al., 1983). Quartz tholeiite solidus is from Green and Ringwood (1971).

6.1. Resistivity measurements

Within the last decade, d.c. resistivity, magnetotelluric and geomagnetic variation techniques have documented regions within the deep crust that contain abnormally low resistivities (ρ) (see Jones, this volume). Normal lower crust typically has ρ = 100–1000 Wm, whereas anomalous lower crust is defined as having ρ < 100 Wm, and often as much as an order of magnitude lower (Haak and Hutton, 1986). Low resistivities can be produced in a variety of ways: interconnected graphite, magnetite, sulfides or aqueous fluids have all been suggested as causes for low resistivities (Haak and Hutton, 1986; Jones, this volume). However, because of the rarity of sufficient quantities of these mineral phases in lower crustal samples, saturation with aqueous saline fluids is generally appealed to in electrical studies (Shankland and Ander, 1983; Gough, 1986; Haak and Hutton, 1986).

No resistivity measurements have been reported for lower crustal xenoliths, however, Lee et al. (1983) measured the conductivities of several gneisses from the Lewisian complex of Scotland under a variety of *P-T* and fluid saturation conditions (275°C and 0.35 GPa, equivalent to ~11 km depth), and these results may be compared with observations on xenoliths. Lee et al. (1983) found that ρ increased with increasing pressure in saturated samples, which they attributed to successive cutting off of conduction paths as a result of the elastic deformation of the pore walls. As expected, the presence of fluid, increasing salinity and increasing temperature caused ρ to decrease. One unexpected result of this study was the observation that mafic rocks yielded consistently lower ρ than felsic rocks. This observation has been used to suggest that low-ρ regions within the lower crust may be explained simply by the presence of mafic rock types, thereby overcoming the apparent conflicting evidence from petrology and geoelectric studies regarding the volatile content of the lower crust (Yardley, 1986; see below). However, it is probably significant that the mafic rocks measured by Lee et al. (1983) all contain large amounts of amphibole. Amphibole is a much better conductor than quartz or feldspar (the other modally abundant phases in the gneisses). In addition, laboratory measurements of the conductivity of pyroxenes and garnets, which would be volumetrically important minerals in dry mafic granulites, show them to be highly resistive ($\rho > 500,000\ \Omega$m at $T = 500°$C) (Duba et al., 1973; Parkhomenko, 1982). Thus, the enhanced conductivity of the mafic samples relative to the felsic samples may be primarily due to the presence of a hydrous phase and not due to the compositional differences. It is also noteworthy that none of the measured ρ's reach the low values measured in deep crustal ρ anomalies, even when saturated with 0.5 *M* NaCl solution (the salinity of sea water). Extrapolation to higher temperatures, as might exist in the lower crust, would cause ρ to decrease, but this decrease is offset by the effects of increasing pressure, which closes pore spaces. Therefore, these measured ρ's may be representative of these rocks in the deep crust (i.e., $\geq$20 km depth).

It is apparent from the above discussion that, barring the presence of interlocking graphite, magnetite or sulfide, a free, highly saline fluid phase is required to account for the deep crustal low-resistivity layers. However, several problems exist with postulating the presence of free H_2O in the lower crust. First, a free saline aqueous fluid cannot exist metastably in the lower crust, as implied by Gough (1986), because it would react with the rocks to form hydrous minerals (Yardley, 1986). However, most lower crustal samples (xenoliths or granulites from surface terrains) have anhydrous mineralogies. In addition, fluid inclusions in granulite xenoliths are CO_2-rich; no H_2O fluid inclusions have been reported. Secondly, the low density of H_2O means it would rise in the crust unless trapped by impermeable rocks at some deep level (Haak and Hutton, 1986). The effects of CO_2 saturation on the resistivity of rocks is poorly known. Kay and Kay (1981) calculated that in order to explain resistivity layers of 10–100 Ωm at crustal levels of 26 km ($\sim$400°C) a solute-bearing CO_2 fluid phase would need to have a resistivity of 0.01–0.001 Ωm, which they considered very unlikely for a non-polar solvent such as CO_2.

Resistivity profiles from the Rhenish Massif may help to reconcile the apparent discrepancies between observed mineralogy of lower crustal xenoliths and the presence of resistivity anomalies. Low-ρ layers ($\sim$10 Ωm), which are between 1 and $\sim$5 km thick, occur at 15 km depth in the north and dip to 30 km depth in the south of the Rhenish Massif, western Germany (Jödicke et al., 1983). The Eifel basalts erupt through the southern portion of the Rhenish Massif and carry lower crustal xenoliths containing significant quantities of amphibole (Okrusch et al., 1979). Isotopic investigations suggest that this amphibole resulted from reaction of originally dry granulites with a mantle-derived fluid sometime after 125 Ma ago (the age of the last mineral equilibration) (Stosch and Lugmair, 1984). The experimental data of Lee et al. (1983) suggest that simply the presence of amphibole cannot account for the observed low resistivities, however, it is possible that these hydrous phases document the passage of aqueous solutions through the lower crust. These solutions may have been subsequently trapped at slightly higher levels, giving rise to the low-resistivity layers currently observed in the Rhenish Massif.

6.2. Seismic velocities

In contrast to resistivity measurements, several laboratory P-wave determinations have been made on xenoliths. Figure 7-7 shows a histogram of V_p for the lower crust based on seismic refraction profiles compiled by Christensen (1982) with newer data sources given in the figure caption. Also shown are 1.0-GPa room temperature ultrasonic velocity determinations for a variety of possible lower crustal rock types, including samples from both

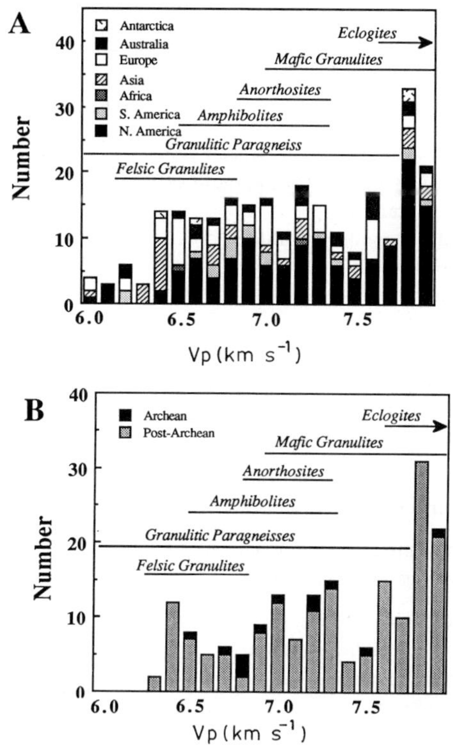

Fig. 7-7. Histogram of lower crustal P-wave velocities as compiled by Christensen (1982); more recent data are from Drummond (1982), Finlayson (1982), Kaila et al. (1979), Luosto and Korhonen (1986) and Srivastava et al. (1984). (A) All data for layers greater than 20 km, keyed to locality; (B) Data for the lowest crustal layer, keyed to crustal age.

terrains and xenoliths. Note that these velocities will decrease by up to 0.5 km s^{-1} when corrected for *in situ* lower crustal temperatures [the velocity temperature derivative used is -5×10^{-4} km s^{-1} $°C^{-1}$, Christensen (1979)]. Table 7-2 tabulates the results of these ultrasonic studies and data sources.

Padovani et al. (1982) measured ultrasonic velocities of hydrated crustal xenoliths from the Colorado Plateau at 0.6 GPa confining pressure. All the samples were characterized by variable hydrothermal alteration to greenschist facies minerals. Amphibolites yielded V_p = 6.86 to 6.91 km s^{-1}, whereas a mafic granulite and eclogite gave V_p = 6.73 and 7.74 km s^{-1}, respectively. The velocities of the amphibolites and the mafic granulite are lower than those of hydrous mafic granulites measured by Manghnani et al. (1974) and Fountain (1976). Likewise, the velocities are lower than those observed at comparable depths in the crust of the Colorado Plateau and lower than velocities calculated using single-crystal data for the original amphibolite- to granulite-facies minerals. Padovani et al. (1982) suggested that sealing of microcracks in the xenoliths by greenschist facies minerals, formed due to fluid infiltration from the entraining kimberlite, lowered the measured V_p's. They concluded that the greenschist facies hydration seen in the xenoliths is solely a product of transport by the kimberlite and does not reflect the *in situ* lower crustal environment.

Ultrasonic P- and S-wave velocities were measured by Jackson and Arculus (1984) at 0.4 GPa for a suite of lower crustal xenoliths from Calcutteroo, South Australia. Measured

TABLE 7-2

Room temperature ultrasonic P-wave velocity measurements for lower crustal rock types

Reference	Rock type	Pressure (GPa)	Measured V_p (km/sec)	V_p * 1.0 GPa
Manghnani et al. (1974)	Felsic to intermediate granulites	1.0	6.5–6.7	6.5–6.7
	Mafic granulites		7.0–7.3	7.0–7.3
	Hydrous mafic granulites		7.1–7.3	7.1–7.3
	Anorthosites		6.9–7.1	6.9–7.1
	Eclogites		7.9–8.6	7.9–8.6
Fountain (1974)	Mafic granulites	1.0	7.3–7.4	7.3–7.4
Christensen and Fountain (1975)	Felsic granulites	1.0	6.5–6.8	6.5–6.8
	Mafic granulite		7.5	7.5
	Hydrous mafic granulites		7.0–7.4	7.0–7.4
Fountain (1976)	Granulite facies Paragneiss	1.0	7.0–7.7	7.0–7.7
	Hydrous mafic granulites		7.1–7.6	7.1–7.6
Evans (1980)	Granulite facies Paragneisses	0.4	5.7–7.4	5.8–7.5
	Hydrous mafic granulites		6.6–7.0	6.7–7.1
Padovani et al. (1982) [1]	Paragneiss	0.6	6.47	6.60
	Hydrous mafic granulite		6.73	6.84
	Amphibolites		6.8–6.9	6.9–7.1
	Eclogite		7.74	8.03
Chroston and Evans (1983)	Hydrous mafic granulites	1.0	6.5–7.0	6.5–7.0
Jackson and Arculus (1984)	Mafic granulite [1]	0.4	6.72	6.8
	Mafic granulites		7.5–7.9	7.6–8.0
	Paragneiss		7.16	7.26
	Eclogite1		7.24	7.34
	Eclogites		7.5–7.9	7.6–8.0
Padovani et al. (1986)	Granulite facies paragneiss	0.6	7.2	7.3
Reid et al. (1989)	Paragneiss xenoliths	0.0	6.7–7.3	7.2–7.9
Fountain et al., 1990	Metatonalite	0.6	6.3–6.5	6.4–6.6
	Paragneiss		6.4	6.5
	Mafic granulite		7.0–7.4	7.1–7.5
	Anorthosite		7.2	7.3
Kern and Schenk (1985)	Mafic granulite	0.6	6.3–6.8	6.3–6.8
	Felsic granulite		6.1–6.7	6.2–6.7
	Paragneiss		6.4–7.6	6.4–7.7
Chroston and Brooks (1989)	Felsic to intermediate granulites [2]	1.0	6.3–6.6	6.3–6.6

* Calculated when not given in reference using a pressure derivative of 0.1 km/s GPa (Jackson et al., 1990).
[1] Altered samples. [2] Measured under water-saturated conditions.

velocities ranged from 7.2 to 8.0 km s^{-1} for mafic granulites and eclogites and 7.2 km s^{-1} for a felsic paragneiss. One highly altered mafic garnet granulite gave a velocity of 6.7 km s^{-1}. Jackson and Arculus concluded that their measurements are broadly consistent with V_p measurements made on surface granulites by Manghnani et al. (1974) (see also Table 7-2). One intriguing result in the Jackson and Arculus data is the high velocity observed for the paragneiss xenolith (7.2 km s^{-1}). Similar, high velocities have been measured for paragneisses from the Ivrea zone (Fountain, 1976) and paragneiss xenoliths from Kilbourne Hole, New Mexico (Reid et al., 1989) (Table 7-2) and are probably controlled by the presence of high velocity phases such as garnet (V_p = 8.0–8.6 km s^{-1}) and Al-silicates [kyanite or sillimanite, $V_p > 8$ km s^{-1} (Christensen, 1982)]. These observations suggest that lower crustal P-wave velocities greater than 7 km s^{-1} may not be due strictly to mafic granulites; metasedimentary granulites may also be important lower crustal constituents in these regions.

6.3. Heat production

The mafic composition of most lower crustal xenoliths causes their concentrations of heat producing elements to be low. Histograms of K, Th and U contents for the available literature data are shown in Figure 7-8. Felsic granulite and paragneiss xenoliths contain the greatest amount of K, Th and U, whereas mafic granulites are generally depleted in these elements.

The K_2O contents of mafic granulite xenoliths overlap those of basalts, but are lower, on average, than K_2O in most continental basalts (BVSP, 1981). Kimberlite-hosted mafic granulites have higher average K_2O contents than those carried by alkali basalts (Fig. 7-8). This suggests that either the lower crust beneath areas where kimberlites erupt has a higher K_2O content, or that the K_2O contents of the xenoliths are enhanced through some form of post-entrainment contamination. Griffin et al. (1979) suggested that the K_2O contents of most of the mafic granulites they studied are accounted for by the primary mineral phases. However, the low modal abundances of biotite and K-feldspar in these rocks (Griffin et al., 1979, table 1) requires another site for the K, possibly on grain boundaries or in the prevalent sericite that formed in the plagioclase feldspars.

Mafic granulite xenoliths contain low abundances of Th and U (Fig. 7-8), overlapping those found in mid-ocean ridge basalts (Jochum et al., 1983), but lower than those typical of continental basalts (BVSP, 1981). These low abundances may be due to depletion during granulite facies metamorphism, but it is also likely that they are, to some extent, primary features and due to the cumulate nature of many xenoliths (Van Calsteren et al., 1986). Some felsic and paragneiss granulites also show quite low concentrations of Th and U, suggesting these elements have been depleted during high-grade metamorphism of these rocks.

Heat production for lower crustal xenoliths is 0.28 μW m^{-3} [using average values of K_2O = 0.79 (n = 350), Th = 1.6 (n = 86) and U = 0.38 (n = 55)) or 0.075 μW m^{-3} (using median values of K_2O = 0.4, Th = 0.3 and U = 0.08) (from Rudnick and Presper (1990), also see Table 7-4]. These values include data from kimberlite-hosted xenoliths and are thus likely to be too high. Even so, they are considerably lower than the 0.4 μW m^{-3} determined for several surface outcrops of granulites (Ashwal et al., 1987; Fountain et al., 1987; Pinet and Jaupart, 1987), and are attributable to the higher proportion of mafic rocks in granulite xenolith suites.

7. Chemical composition

Whole-rock chemical compositions have been determined for a large number of samples (336 analyses in the literature)(Rudnick and Presper, 1990). In principle, the chemical composition of a xenolith can yield information on what type of rock it is, and the important processes by which it was formed or modified. Small sample sizes may hinder obtaining a representative sample, particularly if the xenolith is mineralogically layered (see Rudnick and

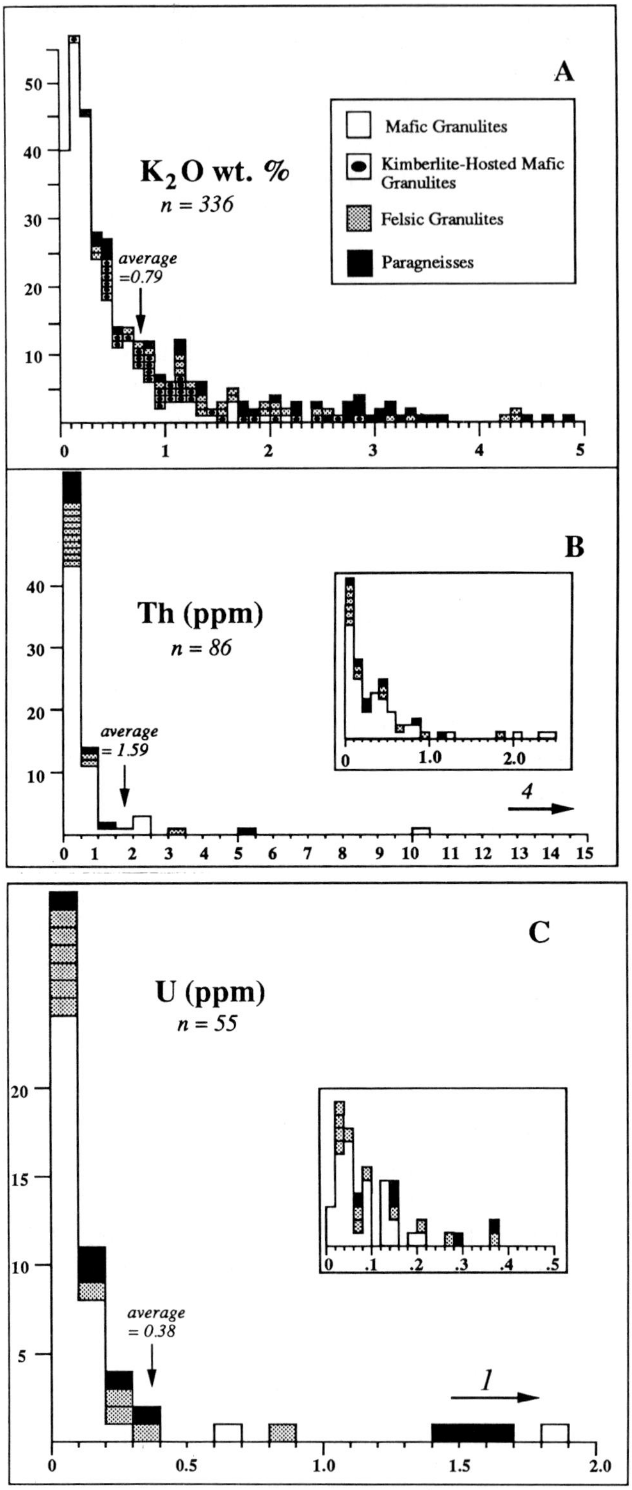

Fig. 7-8. Histogram of heat-producing elements in granulite facies xenoliths. Key to symbols shown in part (A). Average values taken from Table 7-4.

Taylor, 1987). Nevertheless, this does not appear to be an important problem in using xenolith chemistry to distinguish the origin of many samples.

The mafic composition of most lower crustal xenoliths (Fig. 7-2) may arise from a variety of processes. These rocks could represent basaltic melts and/or cumulates which crystallized either on or near the Earth's surface, and were subsequently buried to deep crustal levels; they could represent basaltic melts and/or cumulates which crystallized deep within the crust and underwent granulite facies metamorphism during isobaric cooling (often referred to as basaltic underplating); or they could represent mafic residua left after extraction of a partial melt during granite genesis (restites). Distinguishing cumulates from restites may be difficult in the absence of igneous textures; some chemical criteria have been discussed by Rogers and Hawkesworth (1982), Rudnick et al. (1986) and Rudnick and Taylor (1987), but each xenolith suite must be evaluated separately. No steadfast guidelines apply to all xenoliths. Most mafic xenoliths have been interpreted to represent basalts and their cumulates; possible mafic restite has only been described in the McBride suite of northern Queensland (see below and Rudnick and Taylor, 1987) and the North Hessian xenoliths of Germany (Mengel, 1990).

The normative mineralogies of mafic granulite xenoliths cover the basalt spectrum ranging from nepheline-normative to hypersthene-normative to quartz-normative varieties. It is likely that this reflects the range of parental basalt compositions which gave rise to these granulites, although cumulates will not reflect the same degree of silica saturation as their parental melts. Transition-element abundances are highly variable in mafic granulite xenoliths: Ni and Cr contents range from >1000 ppm and >2000 ppm, respectively, in high-Mg granulites to $\leq$ 10 ppm in more evolved, low-Mg mafic granulites. Concentrations of these elements are generally low in compositionally intermediate to felsic metaigneous granulites and paragneisses, although some intermediate and felsic granulite xenoliths contain unusually high Ni and Cr contents (Fig. 7-9) (50–120 ppm Ni and 50–110 ppm Cr)(Leyreloup et al., 1977; Rudnick and Taylor, 1987; Stolz and Davies, 1989).

Incompatible trace elements also show highly variable concentrations in granulite xenoliths, as exemplified in plots of Mg# [100 Mg/(Mg + ΣFe)] vs. La and Ba (Fig. 7-10). La in mafic xenoliths ranges from <1 ppm to 40 ppm and is generally higher in the low-Mg# samples. The Eifel xenoliths exhibit the highest La concentrations of any mafic xenoliths, and La concentration appears to correlate with modal amphibole content (Loock et al., 1990). Paragneiss and intermediate to felsic granulite xenoliths have high La contents, generally above 10 ppm. Ba concentrations range from $\sim$5 ppm to >1000 ppm in mafic granulites, with kimberlite-hosted samples showing very high values (Fig. 7-10). This enrichment in Ba seen in the kimberlite-hosted xenoliths is attributed to secondary processes and is not an *in situ* lower crustal feature (see section 2.3). As with La, Ba contents increase with decreasing Mg#, and paragneiss and intermediate to felsic granulite xenoliths generally exhibit the highest Ba contents ($\geq$300 ppm).

Rb and Sr concentrations of lower crustal xenoliths are shown in Figure 7-11. The relatively good correlation between Rb and Rb/Sr ratio reflects little variation in Sr contents (Fig. 7-11b) relative to the large variation of Rb contents (Fig. 7-11a) for xenoliths spanning a range of compositions. Mafic xenoliths exhibit the lowest Rb concentrations and Rb/Sr ratios and felsic/intermediate xenoliths and paragneiss xenoliths exhibit the highest concentrations and Rb/Sr ratios, which are similar to upper crustal values. Here again, kimberlite-hosted xenoliths have been distinguished. They show higher average Rb concentrations than other mafic granulite xenoliths (akin to their higher K_2O contents) and plot at higher Rb values for a given Rb/Sr ratio. This may be simply due to Sr addition, with Rb content held constant, or

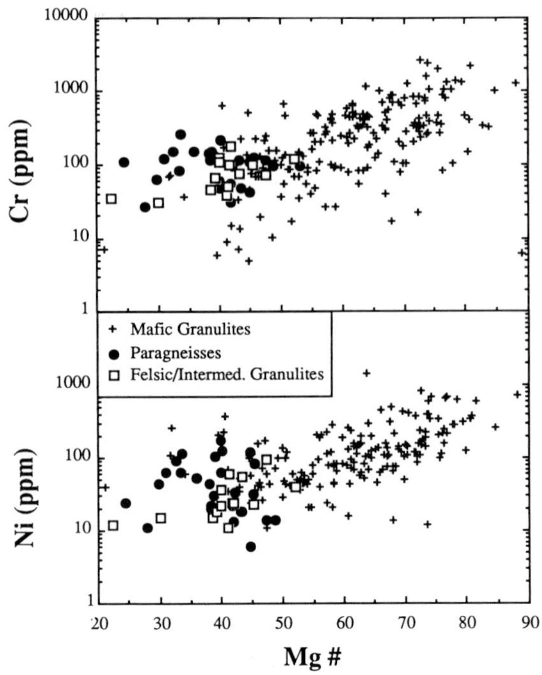

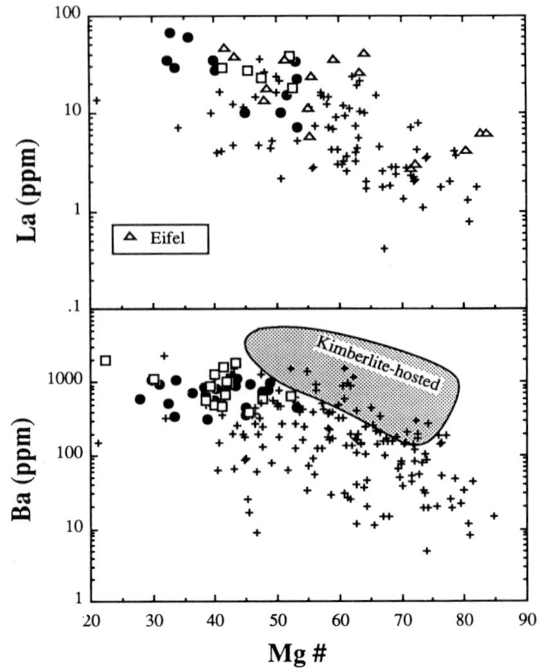

Fig. 7-9 (left). Mg# versus Cr (top) and Ni (bottom) for granulite xenolith data from the literature. Mg# is calculated as 100 Mg/(Mg+ΣFe). Felsic/intermediate xenoliths are designated as those with $SiO_2 > 55$ wt.%, whereas mafic xenoliths have $SiO_2 < 55$ wt.%.

Fig. 7-10 (right). Mg# versus La (top) and Ba (bottom) for granulite xenolith data from the literature. Symbols as in Fig. 7-9. Mafic Eifel xenoliths are shown as triangles in upper plot. Field of kimberlite-hosted mafic granulites shown in bottom plot.

it could be due to both Rb and Sr addition, as long as the Rb/Sr ratio of the added material was less than or equal to the original Rb/Sr ratio of the rock. The lower crust, as sampled by xenoliths, has a Rb/Sr ratio significantly lower than that of the upper crust, but similar to the lower crustal estimate of Taylor and McLennan (1985). Low Rb and Rb/Sr in the lower crust may be the result of Rb depletion caused by granulite facies metamorphism and/or the cumulate origin of many mafic granulite facies xenoliths.

8. Isotopic studies

A large number of isotopic studies have recently been made on lower crustal xenolith suites (see Table 7-1). These studies allow characterization of the isotopic composition of the lower continental crust, delineation of the main processes by which the lower crust forms and determination of its age. The age of the lower crust may be one of the most important parameters in leading to an understanding of lower crustal growth processes.

8.1. Sr and Nd

Present-day Sr and Nd isotopic compositions of lower crustal xenoliths are shown in Figure 7-12. Data sources are listed in the figure caption and paragneiss and felsic/intermediate

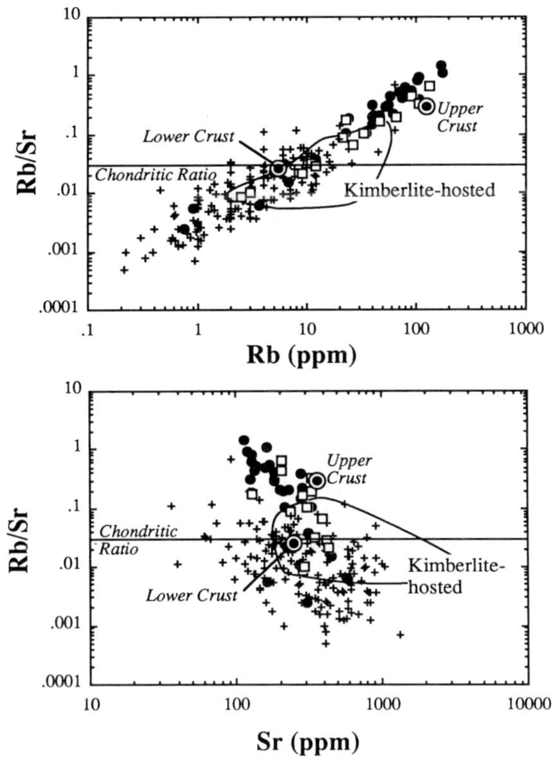

Fig. 7-11. Rb versus Rb/Sr (upper) and Sr versus Rb/Sr (lower) for granulite xenolith data from the literature. Symbols as in Fig. 7-9. Also shown are the chondritic Rb/Sr ratio, Taylor and McLennan's (1985) upper and lower crustal estimates and the field for mafic, kimberlite-hosted granulites.

metaigneous xenoliths are distinguished as in previous diagrams. It is clear from Figure 7-12 that there is not one isotopic composition that is characteristic of the lower crust. Xenolith isotopic compositions range from very depleted, MORB-like ratios, to low ϵ_{Nd}, low $^{87}Sr/^{86}Sr$ ratios (similar to the Lewisian granulites), to evolved isotopic ratios, having high $^{87}Sr/^{86}Sr$ and negative ϵ_{Nd} (similar to many upper crustal rocks). The latter are mainly paragneiss xenoliths. Such radiogenic Sr values require the rocks to have had high Rb/Sr ratios during much of their history. The upper crustal Rb/Sr ratios observed in many of these paragneiss xenoliths (Fig. 7-11) suggest that little metamorphism-induced Rb depletion has occurred in them. Consequently the high $^{87}Sr/^{86}Sr$ in these rocks is independent of when they underwent granulite facies metamorphism and may be a characteristic feature of these lower crustal rock types. Most lower crustal xenoliths have negative ϵ_{Nd} values, indicating they possessed a time-integrated LREE enrichment.

The range of isotopic compositions found within a given xenolith suite is highly variable. Some suites show only a limited range of $^{87}Sr/^{86}Sr$ and ϵ_{Nd} [e.g, the Anakies and Gloucester suites (O'Reilly et al., 1988)], whereas others span nearly the entire range shown in Figure 7-11 [e.g., McBride xenoliths (Rudnick, 1990); Massif Central xenoliths (Downes and Leyreloup, 1986)]. These data indicate the importance of determining lower crustal characteristics for a given region before attempting to model the effects of lower crustal assimilation on magmas that traverse the crust. For example, the use of Lewisian Sr and Nd isotopic compositions for the lower crust (Weaver and Tarney, 1980) may only be valid in northern Scotland or other

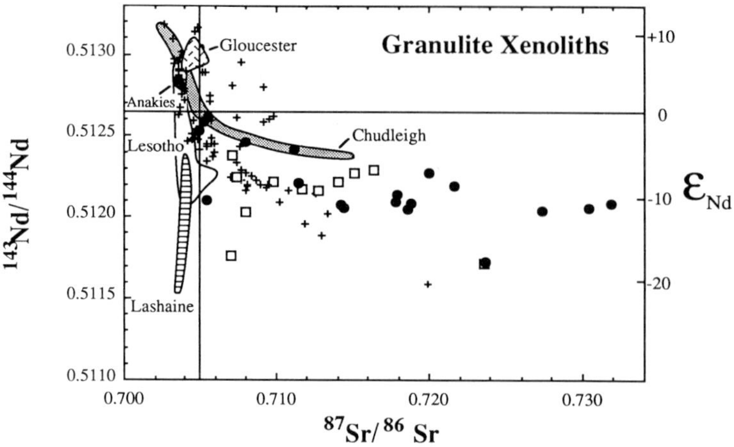

Fig. 7-12. $^{87}Sr/^{86}Sr$ versus $^{143}Nd/^{144}Nd$ diagram for granulite xenoliths. Data from the literature (listed in Table 7-1). Symbols as in Fig. 7-9 with several fields of mafic granulites outlined: Chudleigh, north Queensland; Lashaine, Tanzania; Lesotho, south Africa; Anakies, Victoria, Australia; Gloucester, SE Australia.

regions which can be shown to be underlain by Archean crust that was depleted early on. Such an isotopic signature is not a viable composition for the lower crust in Phanerozoic regions.

The isotopic ratios of mafic granulite xenoliths correlate to some degree with the age of the crust through which the xenoliths were erupted: xenoliths having the most negative ϵ_{Nd} occurring in the oldest crustal regions. The only xenoliths that fall into the lower left-hand quadrant in Figure 7-12 (indicating time-integrated Rb depletion and LREE enrichment) are the Lesotho and Lashaine suites, which erupt near the edge of the Archean craton in south Africa (Rogers and Hawkesworth, 1982), and within a mid-Proterozoic fold belt in east Africa (Cohen et al., 1984), respectively. Other mafic granulite xenoliths that were erupted within Proterozoic crust, such as the Mexican xenoliths (Ruiz et al., 1988a,b), Calcutteroo in south Australia (McCulloch et al., 1982), and the McBride province xenoliths from north Queensland, Australia (Stolz and Davies, 1989; Rudnick, 1990), have elevated $^{87}Sr/^{86}Sr$, indicating that their low Rb/Sr ratios are relatively recent features. These observations may be explained in one of two ways. First, recent granulite facies metamorphism may have lowered the Rb/Sr ratio, leaving the measured $^{87}Sr/^{86}Sr$ unsupported. Secondly, if the rocks are relatively young underplated basaltic cumulates, in which low Rb/Sr ratios are a primary feature, the evolved isotopic compositions may have resulted from mixing of the mantle-derived magma with an isotopically evolved lower crust. In the case of the north Queensland mafic xenoliths, zircon ion microprobe dates indicate the rocks were intruded and metamorphosed 300 Ma ago (Rudnick and Williams, 1987), so their evolved isotopic characteristics are likely the result of crust–mantle mixing (Rudnick, 1989).

Sr and Nd isotopic compositions may be used to help define the processes by which the lower crust formed. Several xenolith suites have Sr and Nd isotopic compositions that fall on mixing arrays. In one case, the Chudleigh xenoliths of north Queensland, isotopic compositions correlate with major- and trace-element compositions (see below), indicating the xenoliths formed as basaltic cumulates derived from an underplated magma that underwent assimilation as it crystallized. Other suites showing mixing arrays may have formed in a similar fashion [e.g., McBride xenoliths (Rudnick, 1990) and the Massif Central xenoliths (Downes et al., 1990)].

8.2. Pb

The Pb isotopic characteristics of lower crustal xenoliths are of special interest because the lower crust has been hypothesized to be the unradiogenic Pb reservoir needed to mass balance Pb from the radiogenic upper crust and depleted mantle reservoirs so that the bulk earth plots on the geochron (Davies, 1984; Newsom et al., 1986). [The geochron is the line in $^{206}Pb/^{204}Pb$ vs. $^{207}Pb/^{204}Pb$ space that all Pb's in the earth that have evolved from the Earth's initial Pb isotopic composition (i.e., troilite Pb from the Cañon Diablo iron meteorite) with a single U/Pb ratio over 4.57 Ga should fall on].

Available Pb isotopic analyses for lower crustal xenoliths are plotted in Figure 7-13. An interesting feature of these Pb data is that most granulite xenoliths, in sharp contrast to many granulite facies terrains, have radiogenic Pb isotopic compositions and plot near the fields of modern basalts and upper crustal rocks. This is true even though their Nd model ages are Proterozoic. It is therefore likely that these rocks became granulites (and underwent U-depletion) only recently, during the Paleozoic or later (Rudnick and Goldstein, 1990). In addition, several well-studied suites show that their Pb isotopic compositions, as well as Sr and Nd isotopes, reflect mixing between a mantle-derived basaltic magma and surrounding continental lower crust. In all cases studied, this pre-intrusion lower crust had upper crustal-like Pb isotopic compositions. These observations led Rudnick and Goldstein (1990) to

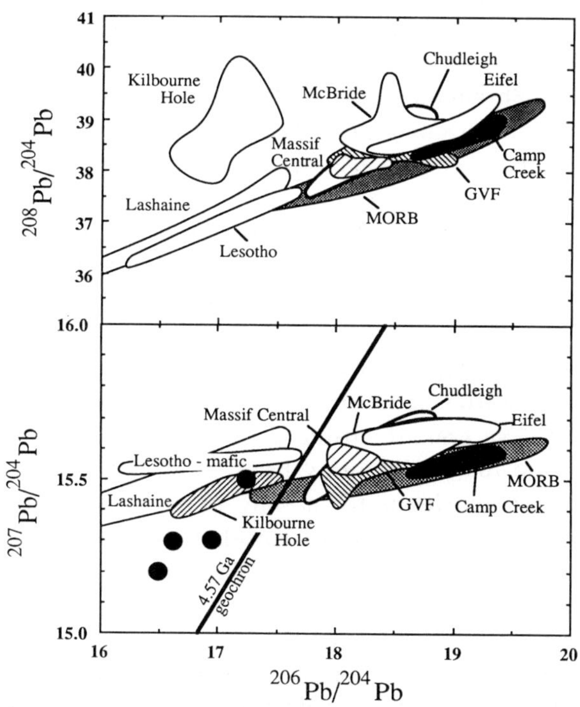

Fig. 7-13. Fields of various granulite xenoliths shown on $^{206}Pb/^{204}Pb$ versus $^{207}Pb/^{204}Pb$ (lower) and $^{208}Pb/^{204}Pb$ (upper) plots. Dots represent model lower crustal compositions from Rudnick and Goldstein (1990). Data sources: Lesotho, van Calsteren et al. (1988); Lashaine: Cohen et al. (1984); Massif Central: Vidal and Postaire (1985); Kilbourne Hole: Reid et al. (1989); GVF = Geronimo Volcanic Field: Kempton et al. (1990); Eifel: Rudnick and Goldstein (1990); Camp Creek: Esperança et al. (1988); Chudleigh and McBride, N. Queensland: Rudnick and Goldstein (1990).

suggest that even though some crustal sections may have grown primarily in the Proterozoic, subsequent orogenic events have caused lower crustal re-working and "rejuvenation" of the Pb isotopes through intrusion of mantle-derived magmas and tectonic interleaving of upper crustal slices (similar to the plumbotectonics model of Doe and Zartman, 1979). Thus unradiogenic Pb can develop in the lower crust only in regions that have remained stable for very long times (i.e., since the mid-Proterozoic). This may explain the Pb characteristics of xenoliths from Kilbourne Hole, Lesotho and Lashaine. Xenoliths from these latter two suites are the only samples that show evidence for long-term Rb and U, depletion, as seen by their low $^{87}Sr/^{86}Sr$ and unradiogenic Pb isotopes.

If the Pb isotopic composition of the lower crust is a function of tectonothermal age, as implied in the Rudnick and Goldstein model, then the Pb isotopic composition of the bulk lower crust can be calculated using the data for granulite terrains and xenoliths and tectonothermal age maps of the crust. The resulting lower crust compositions fall to the left of the geochron (Fig. 7-13), but are too radiogenic to cause the bulk earth to fall on the geochron, requiring the existence of yet another unradiogenic Pb reservoir to solve the Pb mass balance problems.

8.3. Age determinations

Ages of lower crustal xenoliths may provide some of the most important information regarding how and when the lower crust grew, especially in relation to the upper crust and regional tectonics. Several methods exist for determining xenolith ages. These include whole-rock Rb-Sr, Sm-Nd and Pb-Pb isochrons, Nd model ages, Sr, Nd, Pb and K-Ar measurements of mineral separates, and U-Pb zircon analyses. Table 7-3 summarizes the ages that have been ascribed to various lower crustal xenolith suites and the methods used to determine them. By far the most meaningful xenolith ages are obtained by mineral separate analyses or single-zircon analyses.

Whole-rock Pb, Sr and Nd isochrons suffer from the problem of demonstrating a unique initial ratio for the xenoliths within a suite, with no subsequent modification of the parent/daughter ratios. Xenoliths will fall on a whole-rock Sm-Nd isochron only if they are cogenetic and have not experienced crustal assimilation. No cases have yet been described where this occurs. The only suite shown in Table 7-3 that comes close to forming a Sm-Nd isochron is Lesotho, but even these data scatter about an isochron that is, in part, defined by a felsic granulite at the low $^{147}Sm/^{144}Nd$ end. If mixing does occur, however, it may be possible to determine an age based on the mixing array itself, as shown by Rudnick et al. (1986). Using this method, isotopic compositions for a suite of xenoliths that are thought to be genetically related on the basis of trace-element correlations are back-calculated to earlier times until the data lie on a reasonable mixing array. This time is then assumed to represent the crystallization age of the xenoliths.

Nd model ages are, at best, mantle-extraction ages and, at worst, a mixed age (see Arndt and Goldstein, 1987). The possibility of underplated mafic magmas being derived from an enriched mantle source or the magma mixing with pre-existing crust, makes Nd model ages of doubtful value in determining crystallization age for the protoliths of many mafic granulite xenoliths. For example, the most isotopically evolved xenoliths from the Chudleigh suite have Proterozoic Nd model ages, however, the xenoliths formed by basalt–crust mixing in the Cenozoic (see section 9.1). In cases such as this, the Nd model age may place some constraints on the age of the crust being assimilated.

TABLE 7-3

Age determinations for lower crustal xenoliths

Locality	Rock type	Method(s)	Age (Ga)	Comments	Ref.
Partan Craig,	Felsic granulite	Sm-Nd min.	0.36	eruption age	1
Scotland	Felsic granulite	U-Pb zircon	>2.0	upper intercept, highly discordant	2
Lesotho,	Mafic cumulates	Sm-Nd WR	1.4	high MSWD	3
S Africa	& Felsic granulite	Nd model age	1.8	T_{DM} age felsic gran.	3
	Mafic granulite	K-Ar biotite	1.0	cooling age	4
	Mafic granulite	K-Ar amph.	0.7	cooling age	4
	? Granulite	U-Pb zircon	1.0–1.5	discordant	5
Calcutteroo,	Mafic granulites,	Sm-Nd WR	2.2	T_{DM} age felsic gran.	6
S. Australia	& Felsic granulite				
Eifel,	Mafic granulites	Sm-Nd min.	0.175	cooling age?	7
Germany		Sm-Nd WR	1.5	4 point isochron	7
		Pb, Sr, Nd WR	~0.45	intrusion age based on mixing array	8
Mexico	Paragneiss	U-Pb zircon	0–0.8		9
	Mafic granulite		0.001		
	Felsic granulite	(ion probe)	0.03–1.4		
Kilbourne	Paragneiss	Rb-Sr, Sm-Nd min.	0.037	eruption age	10
Hole, NM	Paragneiss	Rb-Sr, Sm-Nd layers	1.56	metamorphic age?	10
	Paragneiss	Rb-Sr WR	1.7	metamorphic age	10
	Paragneiss	U-Pb zircon	1.4	upper intercept, highly discordant	11
Chudleigh,	Mafic cumulates	Sr, Nd WR	<0.1	intrusion age based on mixing array	12
Queensland					
McBride,	Paragneiss, mafic	U-Pb zircon	0.2–0.3	metamorphic age, intrusion age	13
Queensland	granulites, felsic	(ion probe)	1.7	protolith age	
	granulites				
Massif Central,	Paragneisses,	Sr, Nd WR	~0.35	mixing age	14
France					
Chinese Peak,	Mafic granulite	U-Pb zircon	0.18	slightly discordant, protolith age	15
Sierra Nevada					
Camp Creek,	Amphibolites,	Nd model age	1.2–1.9	T_{DM} ages amphibolites, eclogites	16
Arizona	eclogites,	Pb-Pb WR	1.1	protolith age or mixing line	
	mafic granulite	Sm-Nd mineral	0.03	eruption age	

Key to references: [1] van Breemen and Hawkesworth, 1980; [2] Halliday et al., 1984; [3] Rogers and Hawkesworth, 1982; [4] Harte et al., 1981; [5] Davis, 1976; [6] McCulloch et al., 1982; [7] Stosch and Lugmair, 1984; Loock et al., 1990; [8] Rudnick and Goldstein, 1990; [9] Rudnick and Cameron, 1991; [10] Padovani and Hart, 1981; Reid et al., 1989; [11] Davis and Grew, 1977; [12] Rudnick et al., 1986; [13] Rudnick and Williams, 1987; [14] Downes et al., 1989; [15] Dodge et al., 1986; [16] Esperança et al., 1988.

A Pb-Pb whole-rock isochron has been described for the xenoliths of Camp Creek, Arizona (Esperança et al., 1988). Esperança et al. found that amphibolite, eclogite and granulite xenoliths plot about a line with positive slope in $^{206}Pb/^{204}Pb$ vs. $^{207}Pb/^{204}Pb$ space, which they interpreted as an 1.1 Ga isochron. However, similar lines may be formed through mixing. The observation that the xenoliths plot within the field of modern MORB suggests either that the rocks originally crystallized in the Proterozoic and Th and U concentrations were not altered by later metamorphism or that the line represents a mixing trend for relatively young mantle-derived mafic rocks that underplated the crust.

Ages determined for mineral separates date the time of last equilibration or cooling below a given isotopic system's blocking temperature (van Breeman and Hawkesworth, 1980; Harte et al., 1981; Padovani and Hart, 1981). This information is useful in determining the thermal and metamorphic history of a given crustal segment. In areas where the host magmatism was accompanied by high heat flow, mineral equilibration ages are indistinguishable from eruption ages (van Breemen and Hawkesworth, 1980; Padovani and Hart, 1981).

Perhaps the best way of dating igneous crystallization and intrusion ages for granulite xenoliths is through single-zircon U-Pb isotopic analyses. Zircons are particularly resilient minerals, even during granulite facies metamorphism (e.g., Black et al., 1986) and their growth or Pb-loss during metamorphic events makes it possible to determine the geological history of a sample through single-grain zircon analyses. The events thus documented for a lower crustal xenolith may then be related to the regional geologic history and tectonism (see following section and Rudnick and Williams, 1987).

9. Case studies

Due to the variable lithologies present in many granulite xenolith suites and the equally variable origins of the rocks, each suite should be examined separately, using a variety of techniques, in order to understand the likely processes by which the rocks, and hence the lower crust, formed. In the following sections I review petrological, geochemical and isotopic features of three xenolith suites, selected because they are erupted from diverse tectonic settings and because I have first-hand knowledge of their chemical characteristics. These examples highlight the variability of lower crustal compositions and they focus attention on the role of basaltic underplating in the formation of the lower crust.

9.1. The Chudleigh province, north Queensland — A Paleozoic fold belt

Lower crustal xenoliths from the Chudleigh volcanic province in north Queensland, Australia, were described by Kay and Kay (1983), Rudnick et al. (1986) and Rudnick and Taylor (1991). These xenoliths are unusual for their large size (up to 50 cm) and freshness. The <1 Ma basalts that carry the xenoliths erupt just to the east of a large fault that separates Paleozoic rocks of the Tasman fold belt to the east from Proterozoic rocks of the Georgetown Inlier to the west. All granulite xenoliths have anhydrous mineralogies and mafic compositions, with plagioclase-rich samples the most abundant type. Despite nearly constant major-element composition, the mineralogy of the plagioclase-rich xenoliths varies from olivine-bearing assemblages through the gabbro-granulite transition to plagioclase-garnet-clinopyroxene assemblages. These mineralogies indicate a range of equilibration depths between 20 and 40 km. Coronal structures, described in section 5 and shown in Figure 7-4b, indicate isobaric cooling of the rocks in the deep crust prior to entrainment in the host basalts.

The REE patterns of the xenoliths correlate with lithology: plagioclase-rich xenoliths are LREE-enriched with large positive Eu anomalies, pyroxene-rich xenoliths are LREE-depleted with no Eu anomalies and xenoliths transitional between these two types have

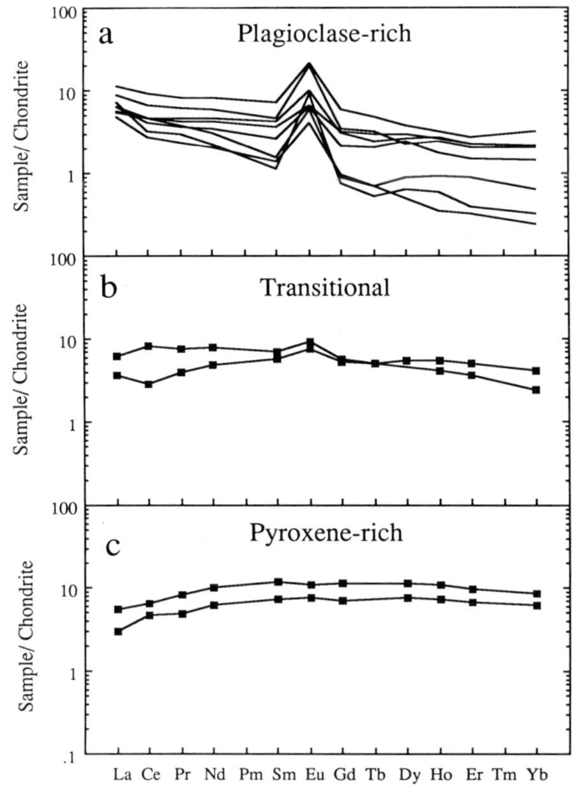

Fig. 7-14. REE patterns for Chudleigh xenoliths, north Queensland, Australia (from Rudnick et al., 1986).

intermediate LREE abundances and slight positive Eu anomalies (Fig. 7-14). These are features consistent with a cumulate origin for the xenoliths. Good positive correlations are observed between compatible trace elements and Mg# and good negative correlations occur between incompatible trace elements and Mg#. In addition, moderately incompatible elements, such as Y, Zr, Hf and V show strong negative correlations with Al_2O_3 content. These correlations are interpreted to reflect chemical variations produced during crystal accumulation from an evolving basaltic melt, or group of melts. The Mg#–trace element correlations reflect the composition of the coexisting melt, whereas the Al_2O_3–trace element correlations reflect the proportions of cumulate plagioclase to clinopyroxene. The very low HREE abundances in some of the more evolved, low-Mg# cumulates (<0.5 times chondrite) limits the amount of trapped melt to less than 2%. Thus, the elemental correlations cannot be attributed to variable amounts of trapped melt.

The Sr and Nd isotopic compositions of these rocks vary from MORB-like ratios (0.7023, +10) to evolved ratios (0.715, −6) and plot along a mixing trend on the $^{87}Sr/^{86}Sr$ vs. ϵ_{Nd} diagram (see the Chudleigh field in Fig. 7-12). Perhaps the most definitive chemical relationships observed for these rocks are the excellent correlations observed between Mg# (hence compatible and incompatible trace elements) and Sr and Nd isotopic compositions (Fig. 7-15). Such correlations require the parental melt of the crystal cumulates to have assimilated an isotopically evolved crustal component as it differentiated in the deep crust. These tight correlations, coupled with the highly variable Sm/Nd ratios, can be also used

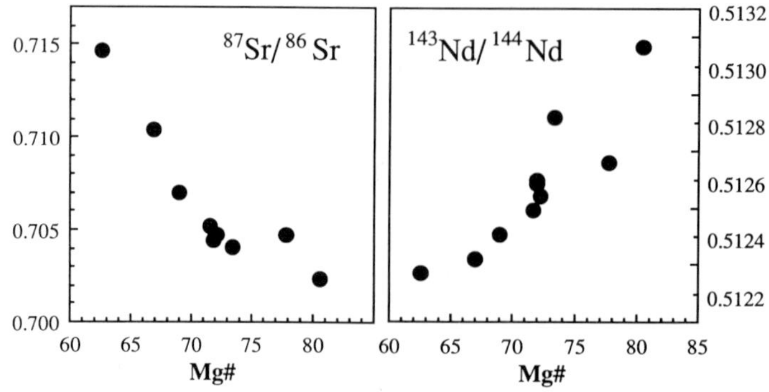

Fig. 7-15. Mg# versus $^{87}Sr/^{86}Sr$ and $^{143}Nd/^{144}Nd$ for Chudleigh xenoliths (from Rudnick et al., 1986).

to place constraints on the intrusion age of the magma. As the present-day isotopic ratios are back-calculated to earlier times, the correlations fall apart. Therefore, the xenoliths must be younger than ~100 Ma in order to maintain the trends. It is likely that these xenoliths represent cumulates derived from basaltic magmas which intruded the lower crust during an earlier phase of Cenozoic magmatism in this region of north Queensland.

9.2. The McBride province, north Queensland — A Proterozoic inlier

Granulite facies xenoliths from the McBride volcanic province, north Queensland, Australia have been described by Kay and Kay (1983), with subsequent petrological, geochemical, Sr and Nd isotopic and U-Pb zircon ion probe studies by Rudnick and Taylor (1987), Rudnick and Williams (1987), Stolz (1989), Stolz and Davies (1989) and Rudnick (1989). The <3 Ma host basalts that carried the xenoliths erupted within the Proterozoic Georgetown Inlier. The McBride xenoliths contrast markedly with those from the Chudleigh province: they are smaller ($\leq$ 10 cm), felsic to intermediate lithologies make up a significant (~20%) proportion of the suite, mineralogical layering is common, textures are generally well-equilibrated and many contain hydrous phases. Equilibration pressures, where they can be estimated, lie at 0.8 GPa and higher.

Good element-element correlations are not observed in this suite. The major- and trace-element compositions of the xenoliths are interpreted as reflecting a diversity of origins for these rocks. Likely protoliths include: pyroxene-rich mafic cumulates, underplated mafic melts, mafic restites, calc-alkaline granitic magmas, metapelites and possibly a mafic cumulate from a granite. In some cases mineralogical layering and small sample sizes have combined to cause an unrepresentative sampling, which is readily discerned in the whole-rock chemistry.

Another difference between the McBride xenoliths and those from the Chudleigh province is that many of the former contain zircon. U-Pb ion microprobe ages for these zircons reveal the chronology of geological events in the deep crust of this region. All analyzed samples contain a Permo-Triassic zircon population, with zircons plotting along concordia between 200 and 300 Ma. Two samples (a metapelite and a mafic restite) contain a Proterozoic zircon population (~1570 Ma), most of which are discordant and plot along a chord to 300 Ma. Finally, several samples also contain xenocrystic zircons. One mafic granulite contains several concordant or near-concordant xenocrysts with ages between 500 and 1200 Ma and one felsic meta-igneous granulite contains a single 2200 Ma zircon.

These zircon ages correlate well with the regional surface geology. The ~1570 Ma zircon populations correspond to the timing of high-grade metamorphism and granite intrusion in the Georgetown Inlier. It is, therefore, likely that these samples formed at, or near the Earth's surface at this time. The Permo-Triassic ages recorded by the zircons correspond to the intrusion and eruption age of voluminous, high-level, calc-alkaline granites and ash-flow tuffs. The exact tectonic setting for this igneous activity is not resolved, although a continental arc setting has been suggested. The last phase of this igneous activity occurred at 270 Ma. Thus, the 200–300 Ma zircon ages, found in all samples, record the initial crystallization of most samples and subsequent granulite facies metamorphism of all samples. The scatter of ages to 200 Ma is interpreted as due to slow cooling in the lower crust following the granulite facies metamorphism. These ages also verify that the granulite xenoliths were derived from the present-day lower crust since they were metamorphosed at 300–200 Ma, the same time which the ash-flows were erupted at the surface, and these ash flows have been preserved, little erosion has occurred since then.

One feature the McBride xenoliths share with the Chudleigh xenoliths is their large variation in Sr and Nd isotopic compositions. These range from depleted mantle-like $^{87}Sr/^{86}Sr$ ratios and ϵ_{Nd} values (0.7039 and +6, respectively) to more evolved compositions (0.724 and −15, respectively). The isotopic compositions broadly correlate with lithology: the metasedimentary xenolith has the highest $^{87}Sr/^{86}Sr$ and lowest ϵ_{Nd}, whereas the mafic pyroxene-rich cumulates have the lowest $^{87}Sr/^{86}Sr$ and highest ϵ_{Nd}. The present-day isotopic ratios follow a mixing array, except for the felsic meta-igneous samples which plot at higher $^{87}Sr/^{86}Sr$ for a given ϵ_{Nd}. At 300 Ma, the time of granulite facies metamorphism and basaltic intrusion, all samples lay along this mixing array.

The combined petrographic, geochemical, U-Pb zircon ages and Sr and Nd isotopic compositions of the McBride xenoliths lead to the following lower crustal history. At ~300 Ma ago, Proterozoic rocks sitting at or near the Earth's surface were tectonically emplaced into the lower crust. This was accompanied by intrusion of mantle-derived basaltic melts that heated the crust, causing partial melting and mixed with the crust. Many of the granitic melts thus produced rose in the crust, intruding at higher levels or erupting as ash flows. Some of these melts, however, remained behind and underwent granulite facies metamorphism upon cooling. There is no direct evidence for Cenozoic basaltic underplating in this xenolith suite. Most of the mafic granulites formed as basaltic underplates in the Paleozoic, and at least one is a Proterozoic mafic restite. Thus, the McBride xenolith suite may be somewhat typical of lower crust formed in Phanerozoic continental margins.

9.3. The Eifel, western Germany — A Cenozoic rift

Granulite facies xenoliths from the Eifel, western Germany, have been described by Okrusch et al. (1979) and Voll (1983), with geochemical and isotopic data provided by Stosch and Lugmair (1984), Stosch et al. (1986), Rudnick and Goldstein (1990) and Loock et al. (1990). The Eifel xenoliths are carried by young (<1 Ma) alkali basalts that erupted within an uplifted Paleozoic block north of, and associated with, the Rheingraben. The xenoliths are 1–10 cm across and are often finely banded. Equilibration conditions are in the range of 750–900°C, 0.6–1.2 GPa. Amphibole is abundant in many of the Eifel xenoliths and occurs in several morphologies, including as secondary rims on clinopyroxene and as lamellae within clinopyroxene. Coronas of orthopyroxene-plagioclase and amphibole surround garnets in some samples (Fig. 7-4a) indicating garnet breakdown due to pressure decrease. In addition,

Ca increase on orthopyroxene rims and Ca depletion on clinopyroxene rims reflect a late heating event.

The Eifel xenoliths have mafic compositions and show a considerable spread in Mg# (from 42 to 84). Rough negative correlations between incompatible elements and Mg# and positive correlations between compatible trace elements and Mg# are interpreted as reflecting a range of bulk compositions characterized by cumulates at the high Mg# end and fractionated melts at the low Mg# end. K_2O, Rb and La contents correlate with modal amphibole content; as noted earlier, these xenoliths display some of the highest La contents of any mafic granulites yet measured.

Sr and Nd isotopic compositions of these xenoliths vary from near bulk-earth to more evolved compositions (0.703, +1.5 to 0.7115, −8.6). Like the trace-element–Mg# variations, Sr and Nd isotopic compositions correlate roughly with Mg#: xenoliths with the highest Mg# have the lowest $^{87}Sr/^{86}Sr$ and highest ϵ_{Nd}, whereas those with the lowest Mg# are the only samples that exhibit high $^{87}Sr/^{86}Sr$ and low ϵ_{Nd}.

The variable to high amphibole content in these lower crustal xenoliths is a distinctive feature of this suite. Some amphibole is clearly secondary (forming rims around clinopyroxene) and other appears primary. All amphiboles have similar compositions, irrespective of their morphology. Carefully prepared mineral separates for one amphibole-rich granulite xenolith, measured by Stosch and Lugmair (1984), show that the minerals were not in Sm-Nd isotopic equilibrium at the time of eruption and, significantly, the amphibole does not fall on the 170 Ma isochron defined by the garnet, clinopyroxene and plagioclase. The amphibole also has the lowest $^{87}Sr/^{86}Sr$ ratio of any mineral in the rock but has the highest Rb/Sr ratio. These features are interpreted to reflect relatively recent amphibole growth in the granulites, possibly due to infiltration by a hydrous, mantle-derived fluid.

The original crystallization age of these xenoliths is not clear. Stosch and Lugmair (1984) and Stosch et al. (1986) suggested a ~1.5-Ga age based on Nd model ages for granulites having the lowest amphibole contents, which presumably experienced the least amount of recent metasomatic alteration of Sr and Nd isotopic compositions. New whole-rock Sr and Nd results show that the old T_{DM} ages are not restricted to amphibole-poor varieties (Loock et al., 1990) and radiogenic Pb compositions for all samples (Rudnick and Goldstein, 1990) suggest that these rock have only recently become depleted in U. If the low U content of these samples is a primary feature (due to a cumulate origin), then it is likely that the rocks are Paleozoic or younger. Rudnick and Goldstein (1990) suggested an age of ~450 Ma based on the presence of mixing trends between Sr, Nd and Pb isotopes at that time. A minimum age for these rocks is defined by the mineral isochrons of 126 and 170 Ma (Loock et al., 1990).

The information gained from study of granulite facies xenoliths from the Eifel suggests the following lower crustal history. Intrusion and crystallization of mantle-derived mafic melts occurred during the Paleozoic or earlier. This may have been closely followed by granulite facies metamorphism. Uplift at a later date caused garnet breakdown, and was later followed by metasomatism of the granulites by mantle-derived hydrous fluids sometime after 170 Ma. Finally, heating associated with the Quaternary rifting caused Ca-enrichment of rims on orthopyroxenes and partial resetting of the Sm-Nd mineral isochrons. The rift-associated heating was not intense enough to cause total isotopic resetting of the minerals in the granulites prior to eruption of the Eifel basalts ~0.5–1 Ma ago.

10. A lower crustal composition from xenoliths

Various estimates of the bulk composition of the lower crust have been made on the basis of averages of analyses from granulite terrains (Weaver and Tarney, 1984; Shaw et al., 1986), average analyses of xenoliths from selected suites (Dupuy et al., 1979; Rudnick and Taylor, 1987) and on the basis of crustal evolution models (Taylor and McLennan, 1985). These model compositions are shown in Table 7-4. The average compositions of lower crustal xenoliths, on a suite-by-suite basis, are given in Table 7-5. These data demonstrate that some degree of lithological diversity occurs between xenolith suites: several suites have andesitic average compositions, but the majority are mafic. Lithological diversity within suites is somewhat masked by taking an average composition. However, the fact that most suites have mafic averages means that very few xenoliths from the averaged suites can have evolved compositions.

The average composition of all lower crustal xenoliths is shown in the last column of Table 7-4. Although heavily weighted toward the Massif Central locality (which represents 24% of the total published analyses), this average is not significantly different from the average obtained from locality averages (last column of Table 7-5). The reason for this is that the Massif Central is not the only xenolith locality containing a significant number of non-mafic xenoliths: Hoggar, in Algeria, the Mexican localities and Kilbourne Hole, New Mexico, also have andesitic averages. (Note, however, that the average composition of the latter suite, based on only 3 analyses of paragneiss xenoliths, is probably too evolved since no data for the numerous mafic granulites which occur there are published.)

The mafic composition of both xenolith averages reflects the dominance of these lithologies in lower crustal xenolith suites. This composition contrasts markedly with these models based on granulite terrain data (Table 7-4). Thus concentrations of compatible elements (e.g., the transition metals) are high, and concentrations of incompatible elements (e.g., LREE, Rb, Ba, Zr, Pb, Th and U) are low in the xenolith average relative to the terrain-based averages. The middle to heavy REE, Y and Sr are similar between the Canadian shield terrains and xenoliths, but HREE and Y are lower, and Sr higher, in the Lewisian-based lower crustal estimate of Weaver and Tarney (1984). The lower crustal model of Taylor and McLennan (1985) is most similar to the xenolith compositions, although the xenolith average has considerably higher alkali and alkaline earth elements and lower Cu than this model.

11. Representativeness of xenoliths as lower crust

The well established compositional differences between xenoliths and granulite terrains lead naturally to the question: how representative are granulite xenoliths of the present-day lower crust? Do the dominantly mafic compositions reflect sampling of earlier mafic intrusions associated with the very magmatic activity which brought the xenoliths to the surface, thus biasing the average composition to mafic lithologies? Are felsic xenoliths underrepresented due to disintegration and melting in the host? Finally, how well do xenoliths reflect lower crust formed throughout geologic time?

The question of whether mafic granulites are overrepresented in xenolith populations because they are earlier manifestations of the volcanism that carried the xenoliths to the surface can be answered by careful study of each xenolith suite. A number of lines of evidence suggest that mafic granulite xenoliths are not cogenetic with their hosts. These include: (1) the observation that most xenoliths exhibit metamorphic textures and mineralogy, thus

TABLE 7-4

Estimates of the bulk composition of the lower continental crust

	Terrain averages		Model	Xenolith averages		
	Lewisian Granulites	Canadian Shield		Massif Central	McBride Province	All xenoliths [1]
	Weaver and Tarney (1984)	Shaw et al. (1986)	Taylor and McLennan (1985)	Dupuy et al. (1979)	Rudnick and Taylor (1987)	Rudnick and Presper (1989)
SiO_2	61.2	61.2	54.4	56.3	50.3	50.5
TiO_2	0.5	0.8	0.9	1.1	1.3	1.0
Al_2O_3	15.6	14.9	16.1	17.1	16.6	16.5
FeO	5.3	7.2	10.6	7.9	12.1	9.0
MnO	0.1	0.12	0.2	0.11	0.22	0.15
MgO	3.4	4.1	6.3	5.0	8.0	7.8
CaO	5.6	5.2	8.5	5.5	9.0	9.7
Na_2O	4.4	3.1	2.8	2.1	1.6	2.4
K_2O	1.0	1.6	0.34	1.42	0.45	0.79
P_2O_5	0.18	0.13	–	0.16	–	0.23
Mg#	53.2	50.4	51.1	52.7	53.8	60.7
Sc		19	36	25	33	34
V		139	285	167	217	214
Cr	88	144	235	220	217	308
Co		39	35	29	31	51
Ni	58	89	135	61	141	137
Cu		38	90	22	29	53
Rb	11	41	5.3	27	12	17
Sr	569	317	230	266	196	403 *
Y	7	30	19		28	20
Zr	202	169	70	174	70	84
Nb	5	14	6		13	11
Cs			0.10	1.3	0.07	1.0
Ba	757	458	150	545	212	384 *
La	22		11	26	12	13
Ce	44		23	57	28	28
Nd	18		13		16	15
Sm	3.3		3.2	5.8	4.1	3.6
Eu	1.2		1.2	1.5	1.4	1.3
Tb	0.43		0.59		0.79	0.58
Yb	1.2		2.2	3.0	3.2	1.9
Hf	3.6	4	2.1	4.6	3.3	2.5
Pb	13	23	4		3	5
Th	0.42	7.7	1.1	5.8	0.5	1.6
U	0.05	1.1	0.3	0.6	0.2	0.4
K/Rb	754	324	532	436	311	394
Rb/Sr	0.019	0.129	0.023	0.102	0.061	0.042
Th/U	8.4	7.0	3.7	9.7	2.5	4.0
$(La/Yb)_N$	12.2	3.3	5.8	2.5	4.6	
μ	0.3	3.5	5.4		4.8	5.8

* These averages exclude kimberlite-hosted xenoliths. [1] Data base for xenolith average includes 336 analyses for major elements. μ calculated using Pb isotopic ratios from lower crust model of Zartman and Haines (1988).

have experienced recrystallization prior to eruption (Frey and Prinz, 1978); (2) chemical analyses of many xenoliths rule out a cogenetic relationship with their host; and (3) isotopic compositions of xenoliths are often distinct from those of the host basalt or kimberlite, ruling out a cogenetic relationship [an exception to this general rule are the Anakies xenoliths, SE Australia (O'Reilly et al., 1988)].

If mafic xenoliths are not cogenetic with their host basalts, they may, nevertheless, be related to an earlier phases of igneous activity that shortly preceded the eruption of the host. Such rocks may not necessarily have chemical or isotopic compositions similar to the host basalts and may well have developed metamorphic mineral assemblages and textures due to re-equilibration in a hot lower crustal environment over a relatively short time interval. An example of this type of xenolith suite is found in the Chudleigh xenoliths (see section 9.1). Determining the ultimate origin of mafic xenoliths requires age information, which is often difficult to obtain (see section 8.3). With the data available to date, it is clear that some xenoliths suites do represent young basaltic additions, possibly associated with the host-basalt volcanism; however, many are demonstrably older than the host-related volcanism (see Tables 7-1 and 7-3).

Non-mafic granulites generally constitute a small proportion of lower crustal xenolith suites. Is the low abundance of non-mafic granulites due to the disintegration of felsic to intermediate xenoliths in the hot basaltic hosts? The answer to this question is probably no. Although some felsic granulite xenoliths are extensively melted, as the sillimanite-bearing felsic granulites from Victoria, Australia (McDonough, 1990), the host basalts that carry these rocks show no indication of assimilation of felsic crustal material. They have high Ni and Cr contents, high Mg# (Frey et al., 1978) and primitive Sr, Nd and Pb isotopic compositions (McDonough et al., 1985), thereby severely limiting the amount of felsic lower crustal material that could have been assimilated.

It is an interesting and significant feature that lower crustal xenolith suites are primarily confined to Post-Archean, non-cratonic locations. Only recently have granulite xenoliths been reported from within Archean cratons. These are: sapphirine-quartz granulites from the Kaapvaal craton, South Africa (Dawson and Smith, 1987), mafic granulites from the Man Shield, Liberia (Toft et al., 1989), mafic granulites from the Siberian platform (Shatsky et al., 1983, 1990) and mixed lithology granulites from the Bearpaw Mountains, Montana (Collerson et al., 1988, 1992). Therefore, in the xenolith population we have, from a temporal point of view, a non-representative sampling of the lower crust. This is unfortunate, as the majority of granulite facies terrains are Archean. Until more data become available for the few xenolith occurrences within Archean cratons (particularly geochronological data), it is difficult to reach firm conclusions regarding the reasons for the compositional differences between Archean granulite terrains and lower crustal xenoliths.

12. Conclusions

The studies on lower crustal xenoliths, summarized above, allow some general conclusions to be drawn about the lower crust as sampled by xenoliths.

(1) Most granulite facies xenoliths have equilibration pressures above 0.8 GPa, based on cation exchange geobarometry. However, many xenoliths do not contain garnet, so their pressures cannot be determined, and it is probable that some of these rocks were derived from shallower depths.

TABLE 7-5

Average compositions of lower crustal xenolith by locality

	1	2	3	4	5	6	7	8	9	10	11	12	13	14	15	16	17	18	19	20	$\overline{X}$
SiO_2	54.93	50.44	48.31	49.63	47.22	55.69	47.97	49.70	49.87	54.53	48.65	49.70	49.86	46.70	46.21	50.52	45.36	53.15	59.08	49.84	50.4
TiO_2	1.01	0.65	0.86	0.68	1.13	1.08	0.33	1.27	0.99	1.20	-	0.82	1.31	1.37	0.95	1.02	1.22	0.88	1.11	1.23	1.0
Al_2O_3	16.75	17.17	15.95	19.26	17.42	16.93	17.53	17.67	13.61	17.34	17.11	17.56	16.40	17.64	14.28	14.94	13.09	18.50	21.80	15.70	16.8
FeO*	8.51	10.15	9.59	7.64	9.11	8.04	4.27	9.85	8.57	7.66	6.71	8.19	11.24	9.65	9.67	11.75	11.55	6.26	10.46	10.95	9.0
MnO	0.14	0.15	0.16	0.14	0.20	0.10	0.09	0.15	0.14	0.13	-	0.15	0.22	0.16	0.17	0.16	0.20	0.11	0.15	0.17	0.15
MgO	5.80	8.15	8.08	5.45	8.09	5.47	11.61	5.79	9.31	5.74	10.59	9.86	8.21	7.94	10.76	7.27	11.33	5.47	2.15	7.14	7.7
CaO	6.98	10.88	10.40	12.24	12.28	5.50	15.97	9.08	12.87	8.31	14.53	10.65	9.47	10.45	12.39	9.77	14.01	9.39	1.08	9.58	10.3
Na_2O	2.82	2.56	2.94	3.22	2.44	2.20	1.74	2.84	2.02	2.87	0.75	2.61	2.01	2.73	2.06	1.56	1.11	3.41	0.78	2.82	2.27
K_2O	0.82	0.39	0.99	1.23	0.33	1.48	0.11	1.00	0.55	1.00	0.19	0.19	0.44	0.48	0.27	0.90	0.35	1.14	3.16	1.10	0.81
P_2O_5	0.18	0.16	0.16	0.15	0.40	0.17	0.03	0.43	0.25	0.13	-	0.37	0.25	0.39	0.15	0.08	0.53	0.17	0.05	0.06	0.22
n	12	9	19	17	16	72	2	14	6	17	3	21	28	18	50	10	19	22	3	5	336
Mg#	51.6	59.2	60.2	57.1	61.0	48.3	82.3	50.3	61.3	55.8	73.6	67.6	54.5	58.0	65.7	51.2	62.9	57.4	26.8	53.5	57.9

n = number of analyses averaged.

1 Hoggar, Algeria (Leyreloup et al., 1982)
2 Fort Portal, Uganda (Thomas and Nixon, 1987)
3 Lesotho, south Africa (Griffin et al., 1979; Rogers and Hawkseworth, 1982)
4 Man Shield, Liberia (Toft et al., 1989)
5 Eifel, West Germany (Stosch et al., 1986)
6 Massif Central, France (Leyreloup et al., 1977; Dostal et al., 1980)
7 Erebus Volcanic Field, Antarctica (Kyle et al., 1987)
8 Chinese Peak, Sierra Nevada, U.S.A. (Dodge et al., 1986)
9 Big Creek, Sierra Nevada, U.S.A. (Dodge et al., 1988)
10 Mexico (Roberts et al., 1989)
11 Pali Aike, Chile (Selverstone and Stern, 1983)
12 Chudleigh, Queensland, Australia (Kay and Kay, 1983; Rudnick et al., 1986)
13 McBride, Queensland (Kay and Kay, 1983; Rudnick and Taylor, 1987; Stolz and Davies, 1989)
14 Central Queensland (Griffin et al., 1987)
15 SE Australia (Wilkinson and Taylor, 1980; Arculus et al., 1988; O'Reilly et al., 1988)
16 South Australia (Arculus et al., 1988)
17 Camp Creek, AZ, U.S.A. (Esperança et al., 1988)
18 Geronimo, NM (Kempton et al., 1990)
19 Kilbourne Hole, NM, U.S.A. (Padovani and Carter, 1977; Reid et al., 1989)
20 Colorado/Wyoming border (Bradley and McCallum, 1986).

TABLE 7-5 (continued)

	1	2	3	4	5	6	7	8	9	10	11	12	13	14	15	16	17	18	19	20	$\overline{X}$
Sc	–	29	–	–	34	–	–	–	–	–	–	–	38	–	34	44	46	18	28	–	34
V	180	211	198	–	–	170	–	–	–	–	–	104	222	316	237	281	362	141	–	–	220
Cr	137	–	285	–	166	241	738	226	437	–	–	301	390	76	587	375	343	114	115	–	302
Co	48	38	77	–	35	40	–	–	–	–	–	–	33	–	62	65	82	–	–	–	53
Ni	95	–	103	–	–	70	194	71	360	–	–	154	199	88	234	124	141	77	–	–	147
Cu	17	–	54	–	–	23	28	–	–	–	–	31	80	109	74	58	92	20	–	–	53
Zn	–	–	–	77	–	–	91	24	–	–	–	–	53	107	104	105	94	105	67	–	83
Ga	–	–	–	–	–	–	–	12	–	–	–	–	–	–	–	15	15	–	–	–	14
Rb	–	10	18	–	2	32	–	–	–	17	–	1	10	4	6	37	9	31	94	–	21
Sr	522	277	863	–	589	298	243	513	302	383	–	426	235	608	420	180	187	715	119	–	405
Y	42	12	11	–	–	18	6	–	–	–	–	8	26	18	15	27	24	19	–	–	19
Zr	124	194	43	–	–	97	18	130	72	105	–	27	175	38	41	83	71	98	268	–	99
Nb	10.3	–	53	–	–	7.0	2.5	–	–	–	–	0.7	11.7	3.8	7.3	60.0	–	7.5	–	–	12
Cs	–	–	–	–	–	–	–	–	–	–	–	0.2	0.1	–	4.7	0.1	–	0.8	0.1	–	1.0
Ba	515	292	2144	–	–	648	24	729	567	–	–	152	218	119	238	885	66	435	–	–	502
La	17.3	5.84	8.33	–	19.9	16.6	–	–	–	15.1	–	2.28	12.6	–	4.60	15.0	–	15.1	42.1	–	14.6
Ce	35.2	12.1	18.7	–	45.8	34.9	–	–	–	32.3	–	4.92	2.92	–	11.4	33.9	–	32.4	72.3	–	29.9
Nd	–	6.58	9.47	–	26.0	–	–	–	–	18.5	–	3.71	3.41	–	8.44	16.5	17.6	17.9	32.7	–	15.6
Sm	3.21	1.55	2.54	–	5.27	4.08	–	–	–	4.09	–	1.12	1.25	–	2.37	4.72	4.29	3.73	8.93	–	3.79
Eu	1.47	0.71	1.13	–	1.51	1.39	–	–	–	1.32	–	0.86	–	–	0.77	1.39	–	1.31	1.63	–	1.23
Gd	–	1.89	–	–	–	–	–	–	–	4.75	–	1.26	4.11	–	2.46	5.40	–	–	9.14	–	4.14
Tb	0.54	–	0.47	–	0.71	0.54	–	–	–	0.70	–	0.16	0.86	–	0.37	1.40	–	0.55	1.70	–	0.73
Ho	–	0.48	0.68	–	–	–	–	–	–	0.87	–	0.30	1.05	–	0.53	1.11	–	–	–	–	0.72
Er	–	1.37	–	–	–	–	–	–	–	2.56	–	0.78	3.62	–	1.45	3.10	–	–	–	–	2.15
Yb	1.56	1.30	1.77	–	2.00	1.64	–	–	–	2.09	–	0.69	3.13	–	1.30	2.87	–	1.00	7.52	–	2.24
Lu	0.24	0.20	0.27	–	0.31	–	–	–	–	0.32	–	–	0.40	–	–	–	–	0.15	1.06	–	0.37
Hf	2.21	–	1.63	–	2.33	–	–	–	–	2.88	–	0.54	6.39	–	0.76	2.38	–	2.44	7.24	–	2.88
Pb	–	–	–	–	–	–	3.00	–	–	4.16	–	1.12	3.86	–	3.87	8.11	–	9.67	13.8	–	5.9
Th	0.77	–	–	–	2.20	–	2.00	–	–	2.85	–	0.06	0.68	–	0.44	6.09	–	1.74	10.9	–	2.77
U	–	–	–	–	–	–	–	–	–	–	–	0.03	0.25	–	0.11	0.52	–	0.85	0.48	–	0.37

(2) The majority of lower crustal xenoliths are mafic with compositions corresponding to a variety of basaltic melts and their cumulates.

(3) The presence of small amounts of paragneiss xenoliths in many granulite xenolith suites requires that tectonic interleaving (burial of supracrustals) occurred in many areas.

(4) Sr, Nd and Pb isotopic compositions of lower crustal xenoliths are highly variable, spanning the range observed in supracrustal rocks and oceanic basalts, and including low $^{87}Sr/^{86}Sr$, low ϵ_{Nd}, unradiogenic Pb types, which are characteristic of some Archean granulite terrains.

(5) Some xenolith suites are relatively young underplated mafic material, associated with earlier phases of the host basalt magmatism, however, many are demonstrably older and attest to the importance of basaltic underplating throughout the Phanerozoic and possibly Proterozoic times.

Although much new data and information have recently become available on granulite facies xenoliths, questions regarding their origin and significance persist. Some of the more outstanding questions include:

(1) Are granulite terrains or granulite xenoliths more representative of the present-day lower continental crust, especially in Archean regions? Are the compositional differences between granulite terrains and xenoliths due to their age differences or other reasons.

(2) To what degree do density differences between mafic lithologies found in granulite xenoliths versus the felsic to intermediate lithologies that characterize granulite terrains control sampling? Can dense, mafic bodies be gravitationally separated from the lower crust and sink into the lithospheric mantle? Does the comparatively low density of granulite terrains promote their uplift to the surface compared with denser lithologies?

Answers to these and other questions will hopefully be forthcoming as more and more lower crustal xenolith suites are discovered and studied.

Acknowledgments

I thank K. Collerson, K. Mezger, D. Fountain, S. McLennan, S. Harley, W. McDonough, J. Stone and G. Wörner for comments on earlier versions of this paper.

References

Aoki, K.-I., 1971. Petrology of mafic inclusion from Itinome-gata, Japan. Contrib. Mineral. Petrol., 30: 314–331.

Arculus, R.J., Ferguson, J., Chappell, B.W., Smith, D., McCulloch, M.T., Jackson, I., Hensel, H.D., Taylor, S.R., Knutson, J. and Gust, D.A., 1988. Trace element and isotopic characteristics of eclogites and other xenoliths derived from the lower continental crust of southeastern Australia and southwestern Colorado Plateau, U.S.A. In: D.C. Smith (Editor), Eclogites and Eclogite-facies Rocks. Elsevier, Amsterdam, pp. 335–386.

Arculus, R.J. and Smith, D., 1979. Eclogite, pyroxenite and amphibolite inclusions in the Sullivan Buttes latite, Chino Valley, Yavapai County, Arizona. In: F.R. Boyd and H.O.A. Meyer (Editors), The Mantle Sample: Inclusions in Kimberlites and Other Volcanics. Am. Geophys. Union, Washington, D.C., pp. 309–317.

Arndt, N.T. and Goldstein, S.L., 1987. Use and abuse of crust-formation ages. Geology, 15: 893–895.

Ashwal, L.D., Morgan, P., Kelley, S.A. and Percival, J.A., 1987. Heat production in an Archean crustal profile and implications for heat flow and mobilization of heat-producing elements. Earth Planet. Sci. Lett., 85: 439–450.

Baldridge, W.S., 1979. Mafic and ultramafic inclusion suites from the Rio Grande Rift (New Mexico) and their bearing on the composition and thermal state of the lithosphere. J. Volcanol. Geotherm. Res., 6: 319–351.

Berg, J.H., Moscati, R.J. and Herz, D.L., 1989. A petrologic geotherm from a continental rift in Antarctica. Earth Planet. Sci. Lett., 93: 98–108.

Best, M.G. and Brimhall, W.H., 1974. Late Ceonozoic alkalic basaltic magmas in the western Colorado Plateau and the Basin and Range transition zone, U.S.A., and their bearing on mantle dynamics. Geol. Soc. Am. Bull., 85: 1677–1690.

Black, L.P., Williams, I.S. and Compston, W., 1986. Four zircon ages from one rock: the history of a 3930 Ma-old granulite from Mount Sones, Enderby Land, Antarctica. Contrib. Mineral. Petrol., 94: 427–437.

Bohlen, S.R., 1987. Pressure-temperature-time paths and a tectonic model for the evolution of granulites. J. Geol., 95: 617–632.

Bradley, S.D. and McCallum, M.E., 1984. Granulite facies and related xenoliths from Colorado–Wyoming kimberlite. In: J. Kornprobst (Editor), Kimberlites II: The Mantle and Crust–Mantle Relationships. Elsevier, Amsterdam, pp. 205–218.

Broadhurst, J.R., 1986. Mineral reactions in xenoliths from the Colorado Plateau; implications for lower crustal conditions and fluid composition. In: J.B. Dawson, D.A. Carswell, J. Hall and K.H. Wedepohl (Editors), The Nature of the Lower Continental Crust. Geol. Soc. Spec. Publ., 24: 331–349.

BVSP, 1981. Basaltic Volcanism on the Terrestrial Planets. Pergamon, New York, N.Y., 1286 pp.

Christensen, N.I., 1979. Compressional wave velocities in rocks at high temperatures and pressures, critical thermal gradients and crustal low-velocity zones. J. Geophys. Res., 84: 6849–6857.

Christensen, N.I., 1982. Seismic velocities. In: R.S. Carmichael (Editor), Handbook of Physical Properties of Rocks, Volume II. CRC Press, Boca Raton, Fl., pp. 1–228.

Christensen, N.I. and Fountain, D.M., 1975. Constitution of the lower continental crust based on experimental studies of seismic velocities in granulite. Geol. Soc. Am. Bull., 86: 227–236.

Chroston, P.N. and Broods, S.G., 1989. Lower crustal seismic velocities from Lofoten–Vesterålen, north Norway. Tectonophysics, 157: 251–269.

Chroston, P.N. and Evans, C.J., 1983. Seismic velocities of granulites from the Seiland petrographic province, N. Norway: implications for Scandinavian lower continental crust. J. Geophys., 52: 14–21.

Cohen, R.S., O'Nions, R.K. and Dawson, J.B., 1984. Isotope geochemistry of xenoliths from East Africa: implications for development of mantle reservoirs and their interaction. Earth Planet. Sci. Lett., 68: 209–220.

Collerson, K.D., Hearn, B.C., MacDonald, R.A., Upton, B.F. and Park, J.G., 1988. Granulite xenoliths from the Bearpaw mountains, Montana: constraints on the character and evolution of lower continental crust. Terra Cognita, 8: 270.

Collerson, K.D., Upton, B.G.J., Hearn, B.C., Jr., Macdonald, R., Thomas, J.E., Harmon, R.S. and Stewart, D.B., 1992. Archean and early Proterozoic lithosphere: new xenolithic evidence from the Wyoming craton. Nature, in review.

Davies, G.F., 1984. Geophysical and isotopic constraints on mantle convection: an interim synthesis. J. Geophys. Res., 89: 6017–6040.

Davis, G.L., 1976. The ages and uranium contents of zircons from kimberlites and associated rocks. Carnegie Inst. Washington Yearb., 77: 631–635.

Davis, G.L. and Grew, E.S., 1977. Age of zircon from a crustal xenolith, Kilbourne Hole, New Mexico. Carnegie Inst. Washington Yearb., 77: 897–898.

Dawson, J.B., 1977. Sub-cratonic crust and upper-mantle models based on xenolith suites in kimberlite and nephelinitic diatremes. J. Geol. Soc., 134: 173–184.

Dawson, J.B. and Smith, J.V., 1987. Reduced sapphirine granulite xenoliths from the Lace kimberlite, South Africa; implication for the deep structure of the Kaapvaal Craton. Contrib. Mineral. Petrol., 95: 376–383.

Dodge, F.C.W., Lockwood, J.P. and Calk, L.C., 1988. Fragments of the mantle and crust from beneath the Sierra Nevada batholith: Xenoliths in a volcanic pipe near Big Creek, California. Geol. Soc. Am. Bull., 100: 938–947.

Dodge, F.W., Calk, L.C. and Kistler, R.W., 1986. Lower crustal xenoliths, Chinese Peak lava flow, central Sierra Nevada. J. Petrol., 27: 1277–1304.

Doe, B.R. and Zartman, R.E., 1979. Plumbotectonics I, The Phanerozoic. In: H.L. Barnes (Editor), Geochemistry of Hydrothermal Ore Deposits. Wiley, New York, NY, pp. 22–70.

Domenick, M.A., Kistler, R.W., Dodge, F.W. and Tatsumoto, M., 1983. Nd and Sr isotopic study of crustal and mantle inclusions from the Sierra Nevada and implications for batholith petrogenesis. Geol. Soc. Am. Bull., 94: 713–719.

Dostal, J., Dupuy, C. and Leyreloup, A., 1980. Geochemistry and petrology of meta-igneous granulitic xenoliths in Neogene volcanic rocks of the Massif Central, France — implications for the lower crust. Earth Planet. Sci. Lett., 50: 31–40.

Downes, H., Dupuy, C. and Leyreloup, A.F., 1990. Crustal evolution of the Hercynian belt of western Europe: evidence from lower crustal granulitic xenoliths (French Massif Central). Chem. Geol., 83: 209–231.

Downes, H. and Leyreloup, A.F., 1986. Granulitic xenoliths from the French massif central — petrology, Sr and Nd isotope systematics and model age estimates. In: J.B. Dawson, D.A. Carswell, J. Hall and K.H. Wedepohl (Editors), The Nature of the Lower Continental Crust. Geol. Soc. Spec. Publ., 24: 319–330.

Drummond, B.J., 1982. Seismic constraints on the chemical composition of the crust of the Pilbara craton, northwest Australia. Rev. Bras. Geociências, 12: 113–120.

Duba, A., Boland, J.N. and Ringwood, A.E., 1973. The electrical conductivity of pyroxene. J. Geol., 81: 727–735.

Duebendorfer, E.M. and Frost, B.R., 1988. Retrogressive dissolution of garnet: effect on garnet-biotite geothermometry. Geology, 16: 875–877.

Dupuy, C., Leyreloup, A. and Vernieres, J., 1979. The lower continental crust of the Massif Central (Bournac, France) — with special references to REE, U and Th composition, evolution, heat-flow production. In: L.H. Ahrens (Editor), Origin and Distribution of the Elements. Pergamon Press, Oxford, pp. 401–415.

Eberz, G.W., Clarke, D.B., Chatterjee, A.K. and Giles, P.S., 1991. Chemical and isotopic composition of the lower crust beneath the Meguma Lithotectonic Zone, Nova Scotia: evidence from granulite facies xenoliths. Contrib. Mineral. Petrol., 109: 69–88.

Edwards, A.C., Lovering, J.F. and Ferguson, J., 1979. High pressure basic inclusions from the Kayrunnera kimberlitic breccia pipe in New South Wales, Australia. Contrib. Mineral. Petrol., 69: 185–192.

Ehrenberg, S.N. and Griffin, W.L., 1979. Garnet granulite and associated xenoliths in minette and serpentinite diatremes of the Colorado Plateau. Geology, 7: 483–487.

Ellis, D.J. and Green, D.H., 1979. An experimental study of the effect of Ca upon garnet-clinopyroxene Fe-Mg exchange equilibria. Contrib. Mineral. Petrol., 71: 13–22.

Emery, J.V., Arculus, R.J. and Henderson, C.E., 1985. A possible mineral host for the unusual Sr- and Ba-enrichments in some lower crustal granulites and eclogites: Strontiobarytes. EOS Trans. Am. Geophys. Union, 66: 387.

England, P.C. and Thompson, A.B., 1984. Pressure-temperature-time paths of regional metamorphism I. Heat transfer during the evolution of regions of thickened continental crust. J. Petrol., 25: 894–928.

Esperança, S. and Garfunkel, Z., 1986. Ultramafic xenoliths from the Mt. Carmel area (Karem Maharal Volcano), Israel. Lithos, 19: 43–49.

Esperança, S., Carlson, R.W. and Shirey, S.B., 1988. Lower crustal evolution under central Arizona: Sr, Nd and Pb isotopic and geochemical evidence from the mafic xenoliths of Camp Creek. Earth Planet. Sci. Lett., 90: 26–40.

Essene, E.J., 1982. Geologic thermometry and barometry. In: J.M. Ferry (Editor), Characterization of Metamorphism through Mineral Equilibria. Mineral. Soc. Am., Rev. Mineral., pp. 153–206.

Evans, C.J., 1980. The seismic velocities of the Ox Mountain granulites of Ireland and the implications for the interpretation of the crustal structure of north Britain. Geophys. J. R. Astron. Soc., 63: 417–426.

Finlayson, D.M., 1982. Geophysical differences in the lithosphere between Phanerozoic and Precambrian Australia. Tectonophysics, 84: 287–312.

Fountain, D.M., 1974. Seismic velocities in granulite facies rocks and rocks from the Ivrea–Verbano and Strona–Ceneri zones: a study of continental crustal composition. PhD Thesis, Univ. Washington, 178 pp.

Fountain, D.M., 1976. The Ivrea–Verbano and Strona–Ceneri zones, northern Italy: a cross-section of the continental crust — new evidence from seismic velocities of rock samples. Tectonophysics, 33: 145–165.

Fountain, D.M., Salisbury, M.H. and Furlong, K.P., 1987. Heat production and thermal conductivity of rocks from the Pikwitonei–Sachigo continental cross section, central Manitoba: implications for the thermal structure of Archean crust. Can. J. Earth Sci., 24: 1583–1594.

Fountain, D.M., Salisbury, M.H. and Percival, J.A., 1990. Seismic structure of the continental crust based on rock velocity measurements from the Kapuskasing uplift. J. Geophys. Res., 95: 1167–1186.

Francis, D.M., 1976. Corona-bearing pyroxene granulite xenoliths and the lower crust beneath Nunivak island, Alaska. Can. Mineral., 14: 291–298.

Frey, F.A. and Prinz, M., 1978. Ultramafic inclusions from San Carlos, Arizona: petrologic and geochemical data bearing on their petrogenesis. Earth Planet. Sci. Lett., 38: 129–176.

Frey, F.A., Green, D.H. and Roy, S.D., 1978. Integrated models of basalt petrogenesis: a study of quartz tholeiites to olivine melilitites from south eastern Australia utilizing geochemical and experimental petrological data. J. Petrol., 19: 463–513.

Frost, B.R. and Chacko, T., 1989. The granulite uncertainty principle: limitations on thermobarometry in granulites. J. Geol., 97: 435–450.

Garvie, O.C. and Robinson, D.N., 1984. The formation of kelyphite and associated sub-kelyphitic sculptured surfaces on pyrope from kimberlite. In: J. Kornprobst (Editor), Kimberlites I: Kimberlites and Related Rocks. Elsevier, Amsterdam, pp. 371–382.

Gough, D.I., 1986. Seismic reflectors, conductivity, water and stress in the continental crust. Nature, 323: 143–144.

Grapes, R.H., 1986. Melting and thermal reconstitution of pelitic xenoliths, Wehr Volcano, east Eifel, West Germany. J. Petrol., 27: 343–396.

Green, D.H. and Ringwood, A.E., 1967. An experimental investigation of the gabbro to eclogite transformation and its petrological applications. Geochim. Cosmochim. Acta, 31: 767–833.

Green, D.H. and Ringwood, A.E., 1971. A comparison of recent experimental data on the gabbro-garnet granulite-eclogite transition. J. Geol., 80: 277–288.

Griffin, W.L. and O'Reilly, S.Y., 1986. The lower crust in eastern Australia: xenolith evidence. In: J.B. Dawson, D.A. Carswell, J. Hall and K.H. Wedepohl (Editors), The Nature of the Lower Continental Crust. Geol. Soc. Spec. Publ., 24: 363–374.

Griffin, W.L. and O'Reilly, S.Y., 1987. The composition of the lower crust and the nature of the continental Moho — xenolith evidence. In: P.H. Nixon (Editor), Mantle Xenoliths. Elsevier, Amsterdam, pp. 413–430.

Griffin, W.L., Carswell, D.A. and Nixon, P.H., 1979. Lower-crustal granulites and eclogites from Lesotho, southern Africa. In: F.R. Boyd and H.O.A. Meyer (Editors), The Mantle Sample: Inclusions in Kimberlites. Am. Geophys. Union, Washington, D.C., pp. 59–86.

Griffin, W.L., Wass, S.Y. and Hollis, J.D., 1984. Ultramafic xenoliths from Bullenmerri and Gnotuk maars, Victoria, Australia: petrology of a subcontinental crust–mantle transition. J. Petrol., 25: 53–87.

Griffin, W.L., Sutherland, F.L. and Hollis, J.D., 1987. Geothermal profile and crust–mantle transition beneath east-central Queensland: volcanology, xenolith petrology and seismic data. J. Volcanol. Geotherm. Res., 31: 177–203.

Haak, V. and Hutton, R., 1986. Electrical resistivity in continental lower crust. In: J.B. Dawson, D.A. Carswell, J. Hall and K.H. Wedepohl (Editors), The Nature of the Lower Continental Crust. Geol. Soc. Spec. Publ., 24: 35–49.

Haggerty, S.E. and Toft, P.B., 1985. Native Iron in the continental lower crust: petrological and geophysical implications. Science, 229: 647–649.

Halliday, A.N., Aftalion, M., Upton, B.G.J., Aspen, P. and Jocelyn, J., 1984. U-Pb isotopic ages from a granulite-facies xenolith from Partan Craig in the Midland Valley of Scotland. Trans. R. Soc. Edinburgh: Earth Sci., 75: 71–74.

Halliday, A.N., Stephens, W.E., Hunter, R.H., Menzies, M.A., Dickin, A.P. and Hamilton, P.J., 1985. Isotopic and chemical constraints on the building of the deep Scottish lithosphere. Scott. J. Geol., 21: 465–491.

Harley, S.L., 1989. The origin of granulites: a metamorphic perspective. Geol. Mag., 126:

Harley, S.L. and Black, L.P., 1987. The Archaean geological evolution of Enderby Land, Antarctica. In: R.G. Park and J. Tarney (Editors), Evolution of the Lewisian and Comparable Precambrian High Grade Terrains. Geol. Soc. Spec. Publ., 27: 285–296.

Harte, B., Jackson, P.M. and Macintyre, R.M., 1981. Age of mineral equilibria in granulite facies nodules from kimberlites. Nature, 291: 147–148.

Hayob, J.L., Essene, E.J., Ruiz, J., Ortega-Gutiérrez, F. and Aranda-Gómez, J.J., 1989. Young high-temperature granulites form the base of the crust in central Mexico. Nature, 342: 265–268.

Herzberg, C.T., 1978. Pyroxene geothermometry and geobarometry: experimental and thermodynamic evalutation of some subsolidus phase relations involving pyroxenes in the system CaO-MgO-Al_2O_3-SiO_2. Geochim. Cosmochim. Acta, 42: 945–957.

Hunter, R.H., Upton, B.G.J. and Aspen, P., 1984. Meta-igneous granulite and ultramafic xenoliths from basalts of the Midland Valley of Scotland: petrology and mineralogy of the lower crust and upper mantle. Trans. R. Soc. Edinburgh, 75: 75–84.

Irving, A.J., 1974. Geochemical and high pressure experimental studies of garnet pyroxenite and pyroxene granulite xenoliths from the Delegate basaltic pipes, Australia. J. Petrol., 15: 1–40.

Ito, K. and Kennedy, G.C., 1971. An experimental study of the basalt-garnet granulite-eclogite transition. In: J.G. Heacock (Editor), The Structure and Physical Properties of the Earth's Crust. Am. Geophys. Union, Washington, D.C., pp. 303–314.

Jackson, I. and Arculus, R.J., 1984. Laboratory wave velocity measurements on lower crustal xenoliths from Calcutteroo, South Australia. Tectonophysics, 101: 185–197.

Jackson, I., Rudnick, R.L., O'Reilly, S.Y. and Bezant, C., 1990. Measured and calculated elastic wave velocities for xenoliths from the lower crust and upper mantle. Tectonophysics, 173: 207–210.

Jagoutz, E., Carlson, R.W. and Lugmair, G.W., 1980. Equilibrated Nd-unequilibrated Sr isotopes in mantle xenoliths. Nature, 286: 708–710.

James, D.E., Padovani, E.R. and Hart, S.R., 1980. Preliminary results on the oxygen isotopic composition of the lower crust, Kilbourne Hole Maar, New Mexico. Geophys. Res. Lett., 7: 321–324.

Jochum, K.P., Hofmann, A.W., Ito, E., Seufert, H.M. and White, W.M., 1983. K, U, and Th in mid-ocean ridge basalt glasses and heat production, K/U and K/Rb in the mantle. Nature, 306: 431–436.

Jödicke, H., Untiedt, J., Olgemann, W., Schulte, L. and Wagenitz, V., 1983. Electrical conductivity structure of the crust and upper mantle beneath the Rhenish Massif. In: K. Fuchs et al. (Editors), Plateau Uplift. Springer Verlag, Berlin, pp. 288–362.

Jones, A.P., Smith, J.V., Dawson, J.B. and Hansen, E.C., 1983. Metamorphism, partial melting, and K-metasomatism of garnet-scapolite-kyanite granulite xenoliths from Lashaine, Tanzania. J. Geol., 91: 143–166.

Kaila, K.L., Reddy, P.R., Krishna, V.C., Roy Chowdhury, K., Tewari, H.C., Murty, P.R.K. and Tripathi, K.M., 1979. Curstal investigations in India from deep seismic soundings. Geophys. Res. Bull., 17: 273–292.

Kalamarides, R.I., Berg, J.H. and Hank, R.A., 1987. Lateral isotopic discontinuity in the lower crust: an example from Antarctica. Science, 237: 1192–1195.

Kay, R.W. and Kay, S.M., 1981. The nature of the lower continental crust: inferences from geophysics, surface geology, and crustal xenoliths. Rev. Geophys. Space Phys., 19: 271–297.

Kay, S.M. and Kay, R.W., 1983. Thermal history of the deep crust inferred from granulite xenoliths, Queensland, Australia. Am. J. Sci., 283-A: 486–513.

Kempton, P.D. and Harmon, R.S., 1992. Oxygen-isotope evidence for large-scale hybridization of the lower crust during magmatic underplating. Geochim. Cosmochim. Acta, 55: 971–986.

Kempton, P.D., Harmon, R.S., Hawkesworth, C.J. and Moorbath, S., 1990. Petrology and geochemistry of lower crustal granulites from the Geronimo volcanic field, southeastern Arizona. Geochim. Cosmochim. Acta, 54: 3401–3426.

Kern, H. and Schenk, V., 1985. Elastic wave velocitites in rocks from a lower crustal section in southern Calabria (Italy). Phys. Earth Planet. Inter., 40: 147–160.

Kuno, H., 1967. Mafic and ultramafic nodules from Itinome-gata, Japan. In: P.J. Wiley (Editor), Ultramafic and Related Rocks. John Wiley and Sons, New York, N.Y., pp. 337–342.

Kyle, P.R., Wright, A. and Kirsch, I., 1987. Ultramafic xenoliths in the late Cenozoic McMurdo volcanic group, western Ross Sea embayment, Antarctica. In: P.H. Nixon (Editor), Mantle Xenoliths. John Wiley and Sons, New York, N.Y., pp. 287–294.

Lee, C.D., Vine, F.J. and Ross, R.G., 1983. Electrical conductivity models for the continental crust based on laboratory measurements on high-grade metamorphic rocks. Geophys. J. R. Astron. Soc., 72: 353–371.

Leeman, W.P., Menzies, M.A., Matty, D.J. and Embree, G.F., 1985. Strontium, neodymium and lead isotopic compositions of deep crustal xenoliths from the Snake Rive Plain: evidence for Archean basement. Earth Planet. Sci. Lett., 75: 354–368.

Leyreloup, A., Bodinier, J.L., Dupuy, C. and Dostal, J., 1982. Petrology and geochemistry of granulite xenoliths from Central Hoggar (Algeria) — Implications for the lower crust. Contrib. Mineral. Petrol., 79: 68–75.

Leyreloup, A., Dupuy, C. and Andriambololona, R., 1977. Catazonal xenoliths in French Neogene volcanic rocks: constitution of the lower crust. 2. Chemical composition and consequences of the evolution of the French Massif Central Precambrian crust. Contrib. Mineral. Petrol., 62: 283–300.

Loock, G., 1988. Zur Geochemie und Petrologie der granulitischen Unterkrustenauswürflinge der Eifel. Diplomarbeit Thesis, Univ. Köln, 124 pp.

Loock, G., Seck, H.A. and Stosch, H.-G., 1990. Granulite facies lower crustal xenoliths from the Eifel, West Germany: petrological and geochemical aspects. Contrib. Mineral. Petrol., 105: 25–41.

Lovering, J.F. and White, A.J.R., 1964. The significance of primary scapolite in granulitic inclusions from deep-seated pipes. J. Petrol., 5: 195–218.

Lovering, J.F. and White, A.J.R., 1969. Granulitic and eclogitic inclusions from basic pipes at Delegate, Australia. Contrib. Mineral. Petrol., 21: 9–52.

Luosto, U. and Korhonen, H., 1986. Crustal structure of the Baltic shield based on off-Fennolora refraction data. Tectonophysics, 128: 183–208.

Manghnani, M.H., Ramananantoandro, R. and Clark, S.P., Jr., 1974. Compressional and shear wave velocities in granulite facies rocks and eclogites to 10 kbar. J. Geophys. Res., 79: 5427–5446.

Mazzucchelli, M., 1983. The upper zone of the Ivrea–Verbano layered complex (Italian Western Alps). Neues Jahrb. Mineral. Abh., 146: 101–116.

McCulloch, M.T., Arculus, R.J., Chappell, B.W. and Ferguson, J., 1982. Isotopic and geochemical studies of nodules in kimberlite have implications for the lower continental crust. Nature, 300: 166–169.

McDonough, W.F., 1990. Constraints on the composition of the continental lithospheric mantle. Earth Planet. Sci. Lett., 101: 1–18.

McDonough, W.F., McCulloch, M.T. and Sun, S.S., 1985. Isotopic and geochemical systematics in Tertiary–Recent basalts from southeastern Australia and implications for the evolution of the sub-continental lithosphere. Geochim. Cosmochim. Acta, 49: 2051–2067.

McDonough, W.F., Rudnick, R.L. and McCulloch, M.T., 1991. The chemical and isotopic composition of the lower eastern Australian lithosphere: a review. In: B. Drummond (Editor), The Nature of the Eastern Australian Lithosphere. Geol. Soc. Aust., Spec. Publ., 17: 163–188.

Mehnert, K.R., 1975. The Ivrea zone. A model of the deep crust. Neues Jahrb. Mineral. Abh., 125: 156–199.

Mengel, K., 1990. Crustal xenoliths from Tertiary volcanics of the northern Hessian depression. Petrological and chemical evolution. Contrib. Mineral. Petrol., 104: 8–26.

Mengel, K. and Wedepohl, K.H., 1983. Crustal xenoliths in Tertiary volcanics from the northern Hessian depression. In: K. Fuchs et al. (Editors), Plateau Uplift. Springer-Verlag, Berlin, pp. 332–335.

Meyer, H.O.A. and Brookins, D.G., 1976. Sapphirine, sillimanite, and garnet in granulite xenoliths from Stockdale kimberlite, Kansas. Am. Mineral., 61: 1194–1202.

Mezger, K., Hanson, G.N. and Bohlen, S.R., 1988. U-Pb systematics of garnet: dating the growth of garnet in the late Archean Pikwitonei granulite domain at Cauchon and Natawahunan Lakes, Manitoba, Canada. Contrib. Mineral. Petrol., 101: 136–148.

Mittlefehldt, D.W., 1984. Genesis of clinopyroxene-amphibole xenoliths from Birket Ram: trace element and petrologic constraints. Contrib. Mineral. Petrol., 88: 280–287.

Mittlefehldt, D.W., 1986. Petrology of high pressure clinopyroxenite series xenoliths, Mount Carmel, Israel. Contrib. Mineral. Petrol., 94: 245–252.

Newsom, H.E., White, W.M., Jochum, K.P. and Hofmann, A.W., 1986. Siderophile and chalcophile element abundances in oceanic basalts, Pb isotope evolution and growth of the Earth's core. Earth Planet. Sci. Lett., 80: 299–313.

Newton, R.C., 1983. Geobarometry of high-grade metamorphic rocks. Am. J. Sci., 283-A: 1–28.

Newton, R.C. and Perkins, D., III, 1982. Thermodynamic calibration of geobarometers based on the assemblages garnet-plagioclase-orthopyroxene (clinopyroxene)-quartz. Am. Mineral., 67: 203–222.

Nimz, G.J., Cameron, K.L., Cameron, M. and Morris, S.L., 1986. Petrology of the lower crust and upper mantle beneath southeastern Chihuahua, Mexico. Geof. Int., 25: 85–116.

Nixon, P.H., 1987a. Kimberlitic xenoliths and their cratonic setting. In: P.H. Nixon (Editor), Mantle Xenoliths. John Wiley and Sons, New York, N.Y., pp. 215–240.

Nixon, P.H., 1987b. Mantle Xenoliths. John Wiley and Sons, New York, 844 pp.

O'Reilly, S.Y., Griffin, W.L. and Stabel, A., 1988. Evolution of Phanerozoic eastern Australian lithosphere: Isotopic evidence for magmatic and tectonic underplating. In: M.A. Menzies and K.G. Cox (Editors), Oceanic and Continental Lithosphere: similarities and differences. J. Petrol. Spec. Vol., pp. 89–108.

Okrusch, M., Schröder, B. and Schnütgen, A., 1979. Granulite-facies metabasite ejecta in the Laacher See area, Eifel, West Germany. Lithos, 12: 251–270.

Oliver, G.J.H., 1977. Feldspathic hornblende and garnet granulites and associated anorthosite pegmatites from Doubtful Sound, Fiordland, New Zealand. Contrib. Mineral. Petrol., 65: 111–121.

Oliver, G.J.H., 1980. Geology of the granulite and amphibolite facies gneisses of Doubtful Sound, Fiordland, New Zealand. N.Z. J. Geol. Geophys., 23: 27–41.

Omerod, D.S., Hawkesworth, C.J., Rogers, N.W., Leeman, W.P. and Menzies, M.A., 1988. Tectonic and magmatic transition in the western great basin, U.S.A. Nature, 333: 349–353.

Owen, J.V., Greenough, J.D., Hy, C. and Ruffman, A., 1988. Xenoliths in a mafic dyke at Popes Harbour, Nova Scotia: implications for the basement to the Meguma Group. Can. J. Earth Sci., 25: 1464–1471.

Padovani, E.R. and Carter, J.L., 1977a. Aspects of the deep crustal evolution beneath south central New Mexico. In: J.G. Heacock (Editor), The Earth's Crust. Am. Geophys. Union, Washington, D.C., pp. 19–55.

Padovani, E.R. and Carter, J.L., 1977b. Non-equilibrium partial fusion due to decompression and thermal effects in crustal xenoliths. In: H.J.B. Dick (Editor), Magma Genesis. Oreg. Dept Geol. Mineral. Ind. Bull., Portland, Oreg., pp. 43–57.

Padovani, E.R., Hall, J. and Simmons, G., 1982. Constraints on crustal hydration below the Colorado Plateau from Vp measurements on crustal xenoliths. Tectonophysics, 84: 313–328.

Padovani, E.R. and Hart, S.R., 1981. Geochemical constraints on the evolution of the lower crust beneath the Rio Grande rift. In: Processes of Planetary Rifting. Lunar Planet. Sci. Inst., Houston, Texas, pp. 149–152.

Parkhomenko, E.I., 1982. Electrical resistivity of minerals and rocks at high temperature and pressure. Rev. Geophys. Space Phys., 20: 193–218.

Pinet, C. and Jaupart, C., 1987. The vertical distribution of radiogenic heat production in the Precambrian crust of Norway and Sweden: geothermal implications. Geophys. Res. Lett., 14: 260–263.

Reid, M.R., Hart, S.R., Padovani, E.R. and Wandless, G.A., 1989. Contribution of metapelitic sediments to the composition, heat production, and seismic velocity of the lower crust of southern New Mexico. Earth Planet. Sci. Lett., 95: 367–381.

Roberts, S. and Ruiz, J., 1989. Geochemical zonation and evolution of the lower crust in Mexico. J. Geophys. Res., 94: 7961–7974.

Rogers, N.W., 1977. Granulite xenoliths from Lesotho kimberlites and the lower continental crust. Nature, 270: 681–684.

Rogers, N.W. and Hawkesworth, C.J., 1982. Proterozoic age and cumulate origin for granulite xenoliths, Lesotho. Nature, 299: 409–413.

Rudnick, R.L., 1990. Nd and Sr isotopic compositions of lower crustal xenoliths from North Queensland, Australia: implications for Nd model ages and crustal growth processes. Chem. Geol., 83: 195–208.

Rudnick, R.L. and Cameron, K.L., 1991. Age diversity of the deep crust in northern Mexico. Geology, 19: 1197–1200

Rudnick, R.L. and Goldstein, S.L., 1990. The Pb isotopic compositions of lower crustal xenoliths and the evolution of lower crustal Pb. Earth Planet. Sci. Lett., 98: 105–207.

Rudnick, R.L. and Presper, T., 1990. Geochemistry of intermediate- to high-pressure granulites. In: D. Vielzeuf and P. Vidal (Editors), Granulites and Crustal Evolution. Kluwer, Dordrecht, pp. 523–550.

Rudnick, R.L. and Taylor, S.R., 1987. The composition and petrogenesis of the lower crust: a xenolith study. J. Geophys. Res., 92: 13,981–14,005.

Rudnick, R.L. and Taylor, S.R., 1991. Petrology and geochemistry of lower crustal xenoliths from northern Queensland and inferences on lower crustal composition. In: B. Drummond (Editor), The Australian Lithosphere. Geol. Soc. Aust., Spec. Publ., 17: 189–208.

Rudnick, R.L. and Williams, I.S., 1987. Dating the lower crust by ion microprobe. Earth Planet. Sci. Lett., 85: 145–161.

Rudnick, R.L., McDonough, W.F., McCulloch, M.T. and Taylor, S.R., 1986. Lower crustal xenoliths from Queensland, Australia: evidence for deep crustal assimilation and fractionation of continental basalts. Geochim. Cosmochim. Acta, 50: 1099–1115.

Ruiz, J., Patchett, P.J. and Arculus, R.J., 1988a. Nd-Sr isotope composition of lower crustal xenoliths — evidence for the origin of mid-Tertiary felsic volcanics in Mexico. Contrib. Mineral. Petrol., 99: 36–43.

Ruiz, J., Patchett, P.J. and Ortega-Gutierrez, F., 1988b. Proterozoic and Phanerozoic basement terranes of Mexico from Nd isotopic studies. Geol. Soc. Am. Bull., 100: 274–281.

Rutter, M.J., 1987. The nature of the lithosphere beneath the Sardinian continental block: Mantle and deep crustal inclusions in mafic alkaline lavas. Lithos, 20: 225–234.

Schulze, D.J. and Helmstaedt, H., 1979. Garnet pyroxenite and eclogite xenoliths from the Sullivan Buttes latite, Chino valley, Arizona. In: F.R. Boyd and H.O.A. Meyer (Editors), The Mantle Sample: Inclusions in Kimberlites and Other Volcanics. Am. Geophys. Union, Washington, D.C., pp. 318–329.

Selverstone, J. and Stern, C.R., 1983. Petrochemistry and recrystallization history of granulite xenoliths from the Pali-Aike volcanic field, Chile. Am. Mineral., 68: 1102–1111.

Shankland, T.J. and Ander, M.E., 1983. Electrical conductivity, temperatures, and fluids in the lower crust. J. Geophys. Res., 88: 9475–9484.

Shatsky, V.S., Sobolev, N.V. and Pavlyuchenko, V.S., 1983. Fassaïte-garnet-anorthite xenolith from the Udachnaya kimberlite pipe, Yakutia. Doklady Akad. Nauk SSSR, 272: 188–192.

Shatsky, V., Rudnick, R.L. and Jagoutz, E., 1990. Mafic granulites from Udachnuya pipe, Yakutia: Samples of Archean lower crust? In: N.V. Sobolev (Editor), Deep Seated Magmatism and Evolution of Lithosphere of Siberian Platform. Siberia Branch U.S.S.R. Acad. Sci., Mirny.

Shaw, D.M., Cramer, J.J., Higgins, M.D. and Truscott, M.G., 1986. Composition of the Canadian Precambrian shield and the continental crust of the earth. In: J.B. Dawson, D.A. Carswell, J. Hall and K.H. Wedepohl (Editors), The Nature of the Lower Continental Crust. Geol. Soc. Spec. Publ., 24: 257–282.

Spera, F.J., 1980. Aspects of magma transport. In: R.B. Hargraves (Editor), Physics of Magmatic Processes. Princeton Univ. Press, Princeton, N.J., pp. 265–323.

Srivastava, H.N., Verma, R.K., Verma, G.S. and Chaudgury, H.M., 1984. Crustal studies of the Koyna region using explosion data from deep seismic soundings. Tectonophysics, 110: 61–72.

Stolz, A.J., 1987. Fluid activity in the lower crust and upper mantle: mineralogical evidence bearing on the origin of amphibole and scapolite in ultramafic and mafic granulite xenoliths. Mineral. Mag., 51: 719–732.

Stolz, A.J. and Davies, G.R., 1989. Metasomatized lower crustal and upper mantle xenoliths from north Queensland: chemical and isotopic evidence bearing on the composition and source of the fluid phase. Geochim. Cosmochim. Acta, 53: 649–660.

Stosch, H.-G. and Lugmair, G.W., 1984. Evolution of the lower continental crust: granulite facies xenoliths from the Eifel, West Germany. Nature, 311: 368–370.

Stosch, H.-G., Lugmair, G.W. and Seck, H.A., 1986. Geochemistry of granulite facies lower crustal xenoliths: implications for the geological history of the lower continental crust underneath the Eifel, West Germany. In: J.B. Dawson, D.A. Carswell, J. Hall and K.H. Wedepohl (Editors), The Nature of the Lower Continental Crust. Geol. Soc. Spec. Publ., 24: 331–350.

Sutherland, F.L. and Hollis, J.D., 1982. Mantle-lower crust petrology from inclusions in basaltic rocks in eastern Australia –an outline. J. Volcanol. Geotherm. Res., 14: 1–29.

Tanaka, T. and Aoki, K.-I., 1981. Petrogenetic implications of REE and Ba data on mafic and ultramafic inclusions from Itinome-gata, Japan. J. Geol., 89: 369–390.

Taylor, S.R. and McLennan, S.M., 1985. The Continental Crust: its Composition and Evolution. Blackwell, Oxford, 312 pp.

Thomas, C.W. and Nixon, P.H., 1987. Lower crustal granulite xenoliths in carbonatite volcanoes of the Western Rift of East Africa. Mineral. Mag., 51: 621–633.

Toft, P.B., Hills, D.V. and Haggerty, S.E., 1989. Crustal evolution and the granulite to eclogite transition in xenoliths from kimberlites in the West African Craton. Tectonophysics, 161: 213–231.

Upton, B.G.J., Aspen, P. and Chapman, N.A., 1983. The upper mantle and deep crust beneath the British Isles: evidence from inclusion suites in volcanic rocks. J. Geol. Soc. London, 140: 105–122.

van Breeman, O. and Hawkesworth, C.J., 1980. Sm-Nd isotopic study of garnets and their metamorphic host rocks. Trans. R. Soc. Edinburgh, 71: 97–102.

Van Calsteren, P.W., Kempton, P.D. and Hawkesworth, C.J., 1988. Depletion of U in the lower crust: evidence from granulite xenoliths from southern Africa. Chem. Geol., 70: 74.

Van Calsteren, P.W.C., Harris, N.B.W., Hawkesorth, C.J., Menzies, M.A. and Rogers, N.W., 1986. Xenoliths from southern Africa: a perspective on the lower crust. In: J.B. Dawson, D.A. Carswell, J. Hall and K.H. Wedepohl (Editors), The Nature of the Lower Continental Crust. Geol. Soc. Spec. Publ., 24: 351–362.

Vidal, P. and Postaire, B., 1985. Étude par la méthode Pb-Pb de roches de haut grade métamorphique impliquées dans la chaîne Hercynienne. Chem. Geol., 49: 429–449.

Vielzeuf, D., 1983. The spinel and quartz associations in high grade xenoliths form Tallante (S.E. Spain) and their potential use in geothermometry and barometry. Contrib. Mineral. Petrol., 82: 301–311.

Voll, G., 1983. Crustal xenoliths and their evidence for crustal structure underneath the Eifel volcanic district. In: K. Fuchs et al. (Editors), Plateau Uplift, The Rhenish Shield — a Case History. Springer-Verlag, Berlin, pp. 336–342.

Warren, R.G., Kudo, A.M. and Keil, K., 1979. Geochemistry of lithic and single-crystal inclusions in basalts and a characterization of the upper mantle–lower crust in the Engle Basin, Rio Grande Rift, New Mexico. In: R.E. Riecker (Editor), Rio Grande Rift: Tectonics and Magmatism. Am. Geophys. Union, Washington, D.C., pp. 393–415.

Wass, S.Y. and Hollis, J.D., 1983. Crustal growth in south-eastern Australia — evidence from lower crustal eclogitic and granulitic xenoliths. J. Metamorph. Geol., 1: 25–45.

Weaver, B.L. and Tarney, J., 1980. Continental crust composition and nature of the lower crust: constraints from mantle Nd-Sr isotope correlation. Nature, 286: 342–346.

Weaver, B.L. and Tarney, J., 1984. Empirical approach to estimating the composition of the continental crust. Nature, 310: 575–577.

Wells, P.R.A., 1977. Pyroxene thermometry in simple and complex systems. Contrib. Mineral. Petrol., 62: 129–139.

Wever, T. and Sadowiak, P., 1989. A relationship between crustal thickness and mean crustal velocity. Tectonophysics, 170: 159–163.

White, R.S., 1988. The Earth's crust and lithosphere. In: M.A. Menzies and K.G. Cox (Editors), Oceanic and Continental Lithosphere: Similarities and Differences. J. Petrol., Spec. Vol., pp. 1–10.

Wilkinson, J.F.G., 1975. An Al-spinel ultramafic-mafic inclusion suite and high pressure megacrysts in an analcimite and their bearing on basaltic magma fractination at elevated pressures. Contrib. Mineral. Petrol., 53: 71–104.

Wilkinson, J.F.G. and Taylor, S.R., 1980. Trace element fractionation trends of thoeiitic magma at moderate pressure: evidence from an Al-spinel ultramafic-mafic inclusion suite. Contrib. Mineral. Petrol., 75: 225–233.

Windrim, D.P. and McCulloch, M.T., 1986. Nd and Sr isotopic systematics of central Australian granulites: chronology of crustal development and constraints on the evolution of lower continental crust. Contrib. Mineral. Petrol., 94: 289–303.

Wood, B.J., 1987. Thermodynamics of multicomponent systems containing several solid solutions. In: I.S.E. Carmichael and H.P. Eugster (Editors), Thermodynamic Modelling of Geological Materials: Minerals, Fluids and Melts. Mineral. Soc. Am., Washington, D.C., pp. 71–95.

Wörner, G., Schmincke, H.-U. and Schreyer, W., 1982. Crustal xenoliths from the Quaternary Wehr volcano (East Eifel). Neues Jahrb. Mineral. Abh., 144: 29–55.

Yardley, B.W.D., 1986. Is there water in the deep continental crust? Nature, 323: 111.

Zartman, R.E. and Haines, S.M., 1988. The plumbotectonic model for Pb isotopic systematics among major terrestrial reservoirs — a case for bi-directional transport. Geochim. Cosmochim. Acta, 52: 1327–1339.

Zashu, S., Kaneoka, I. and Aoki, K.-I., 1980. Sr isotope study of mafic and ultramafic inclusions from Itinome-gata, Japan. Geochem. J., 14: 123–128.

Chapter 8

Exposed crustal cross sections as windows on the lower crust

JOHN A. PERCIVAL, DAVID M. FOUNTAIN and MATTHEW H. SALISBURY

1. Introduction

The Earth's continental crust is a dynamic entity. Although the history of continental growth can be traced back at least 3800 Ma, the record is incomplete and many questions remain, not only regarding its evolution, but also its fundamental properties and structure. In particular, there has been a general lack of understanding of the character and behavior of the lower continental crust because of its inaccessibility. In this paper we follow the usage that the lower continental crust is the bottom half and thus has a variable thickness and maximum depth. Certain properties appear common to the lower crust, such as P-wave velocity above 6.5 km s^{-1} (Percival and Berry, 1987), but the upper boundary rarely coincides exactly with this velocity contour. Much of the lower crust may be granulite-grade metamorphic rock, but it is doubtful that the upper boundary is everywhere an amphibolite-granulite isograd.

Significant advances in our understanding of the lower crust have been made over the last decade through the study of exposed crustal cross sections (e.g., Fountain and Salisbury, 1981) and many other avenues of research. Since Fountain and Salisbury's (1981) recognition that common features characterize cross sections, much attention has been drawn to these exposures as windows on the lower continental crust. New data, acquired as a consequence of the development of techniques such as accurate geothermobarometry, precise geochronology and reflection seismic profiling, permit a more comprehensive analysis of the character, formation and uplift history of these terranes on a crustal scale. In several locations, exposures have been geophysically traced to their normal position in the deep crust, allowing extrapolation of surface observations over wide areas.

Figure 8-1 shows the general map and section character of a crustal cross section exposed in a compressional uplift, independent of the width of the metamorphic transition. Common features include increasing metamorphic grade and paleopressure through 15–30 km of section, commonly from greenschist to granulite facies, and concomitant changes in rock properties: a trend toward more mafic compositions with higher density and seismic velocity and lower heat production with depth. Certain characteristics are specific to individual situations and depend on uplift history, such as the paired gravity anomalies associated with overthrust settings. Structural character varies to some extent with the timing of uplift; ductile structures related to the unroofing process may be imposed during synmetamorphic exhumation whereas early structures may be preserved if uplift results from a late, superposed event. In contrast to isolated high-grade terranes and xenoliths, lower crust exposed in cross

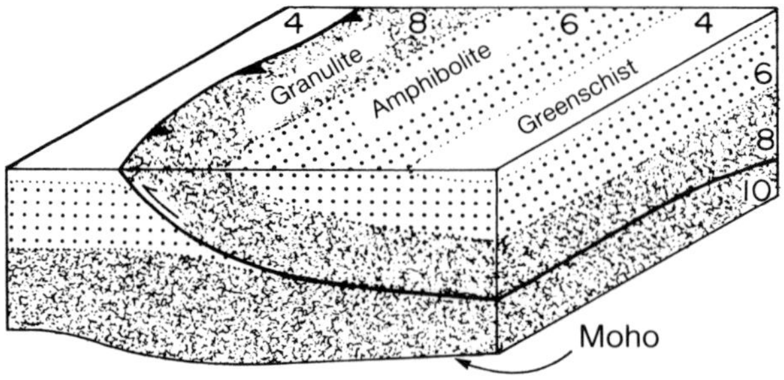

Fig. 8-1. Generalized map and section views of an exposed crustal cross section. Crustal thickness may vary from 20 to 50 km; lateral width of metamorphic transition: 20–150 km. Numbers refer to vertical and lateral distribution of paleopressure (after Percival, 1988).

sections (Fig. 8-1) remains coupled to upper crustal levels. Understanding of the geological evolution of these shallow levels may enhance interpretation of cryptic events observed in the deeper levels.

However, only some properties of the lower crust survive the tortuous path to the surface; features such as structure, rock and mineral composition, cooling history and seismic velocity and reflectivity may be preserved during uplift, whereas others, such as electrical conductivity and ambient fluid, are generally lost. Thus our picture of the lower crust is a collage based on fragmentary evidence from a wide spectrum of observations.

A critical assessment is required to recognize exposures of the lower crust within a much larger population of high-grade terranes with similar characteristics. Notably, pressures and temperatures of formation are comparable [4–12 kbar; 700–900°C (Newton and Perkins, 1982)]; similar rock types may form at identical *P-T* conditions in several different environments along discrete burial and uplift paths. Hence, other types of information must be employed to derive inferences on origin and history (e.g., Percival, 1989a). Whereas the origin of many high-grade terranes is attributed to an early structural (e.g., Ellis, 1987) or magmatic (e.g., Bohlen, 1987) thickening event and subsequent rebound, granulite terranes at the base of exposed crustal cross sections (e.g., Fountain and Salisbury, 1981) were thrust from the deep crust after establishment of the gross crustal structure. Only through integrated studies of regional and detailed geology and geophysics can valid assessments be made.

This paper presents a survey of terranes that have been proposed as crustal cross sections, in order to assess the character, origin and relationships of the contained lower crust. Crustal sections should ideally be studied in conjunction with geophysical and xenolith studies, to provide links between depth and the surface. However, owing to natural limitations, ideal situations are rarely encountered.

Individual cross sections are grouped below according to the type of mechanism that exposes the terrane. Although the tectonic setting at the time of formation is more relevant to the origin of the terrane, it is a less useful parameter for classification of the terranes which are overwhelmingly magmatic in character. Four different mechanisms may transport crustal sections to the surface. In addition to the compressional uplifts documented by Fountain and Salisbury (1981), extensional (e.g., core complexes), transpressional and impactogenic environments have subsequently been recognized. A separate category, referred to as wide, oblique transitions, describes gradational (>100 km wide) greenschist to granulite progressions, where the uplift mechanism is problematic.

2. Compressional uplifts

Thrust faults that carry deep crust to the surface, occur both at plate margins and in intracratonic settings. The depth of detachment, which depends on crust and mantle rheology at the time of deformation, determines the maximum depth exposed. Mantle detachments, exposing Moho, are inferred in the Ivrea (Berckhemer, 1969) and Kohistan (Coward et al., 1986) sections. Figure 8-2 shows a generalized cross section through a compressional uplift; a single fault is depicted although thrust systems are also common. Normal faults form as a common consequence of relaxation following uplift (Burchfiel and Royden, 1985; Dewey, 1988). The age of uplift with respect to crust formation varies greatly; the Kohistan arc was obducted while still near its metamorphic peak (Chamberlain et al., 1989), whereas the Kapuskasing uplift was exposed long after crust formation and high-grade metamorphism (Percival et al., 1988). Ambient temperature, related to timing of exhumation, therefore determines the nature of structures imposed during uplift.

2.1. Kapuskasing uplift, Ontario, Canada

Geophysical anomalies associated with the northeast-striking Kapuskasing uplift transect the easterly subprovince structure of the Superior Province (e.g., Card and Ciesielski, 1986) over a distance of almost 500 km (Fig. 8-3). Interpretations of the origin of the structure have evolved from Garland's (1950) initial model of thinning of the granitic upper crustal layer, to ideas of a Proterozoic rift-related structure (Innes et al., 1967; Bennett et al., 1967; Burke and Dewey, 1973), to a transcurrent fault zone of early Proterozoic age (Watson, 1980). More recently, an intracratonic, basement-uplift style thrust has been proposed to explain the oblique crustal cross section exposed in part of the area (Percival and Card, 1983, 1985; Percival and McGrath, 1986). This latter interpretation has significant implications for the properties and origin of the contemporary lower crust beneath the Superior Province and is undergoing rigorous tests as part of the Canadian LITHOPROBE program.

High-grade rocks of the Kapuskasing uplift comprise three geologically and geophysically distinct segments, from south to north, the Chapleau, Groundhog River and Fraserdale–Moosonee blocks (Percival and McGrath, 1986). In the south, a continuous transition is exposed from the greenschist-facies Michipicoten greenstone belt of the Wawa subprovince, through amphibolite-facies tonalitic orthogneiss of the Wawa gneiss terrane, to a heteroge-

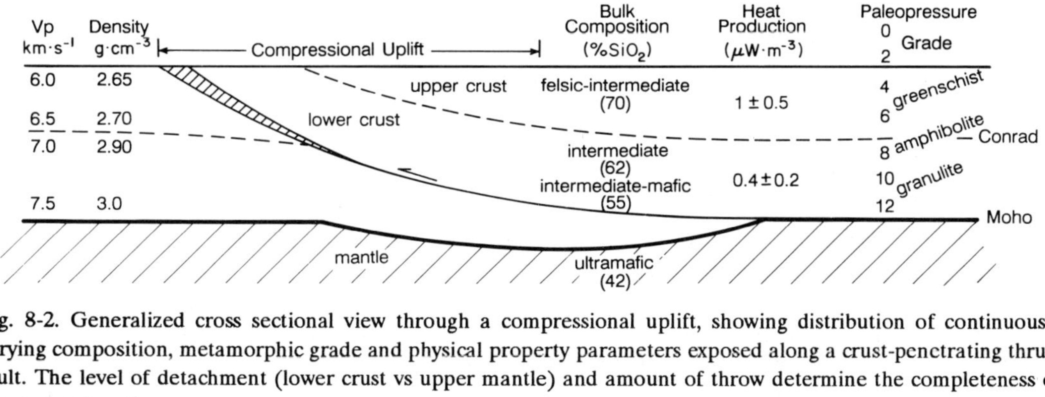

Fig. 8-2. Generalized cross sectional view through a compressional uplift, showing distribution of continuously varying composition, metamorphic grade and physical property parameters exposed along a crust-penetrating thrust fault. The level of detachment (lower crust vs upper mantle) and amount of throw determine the completeness of the exposed section.

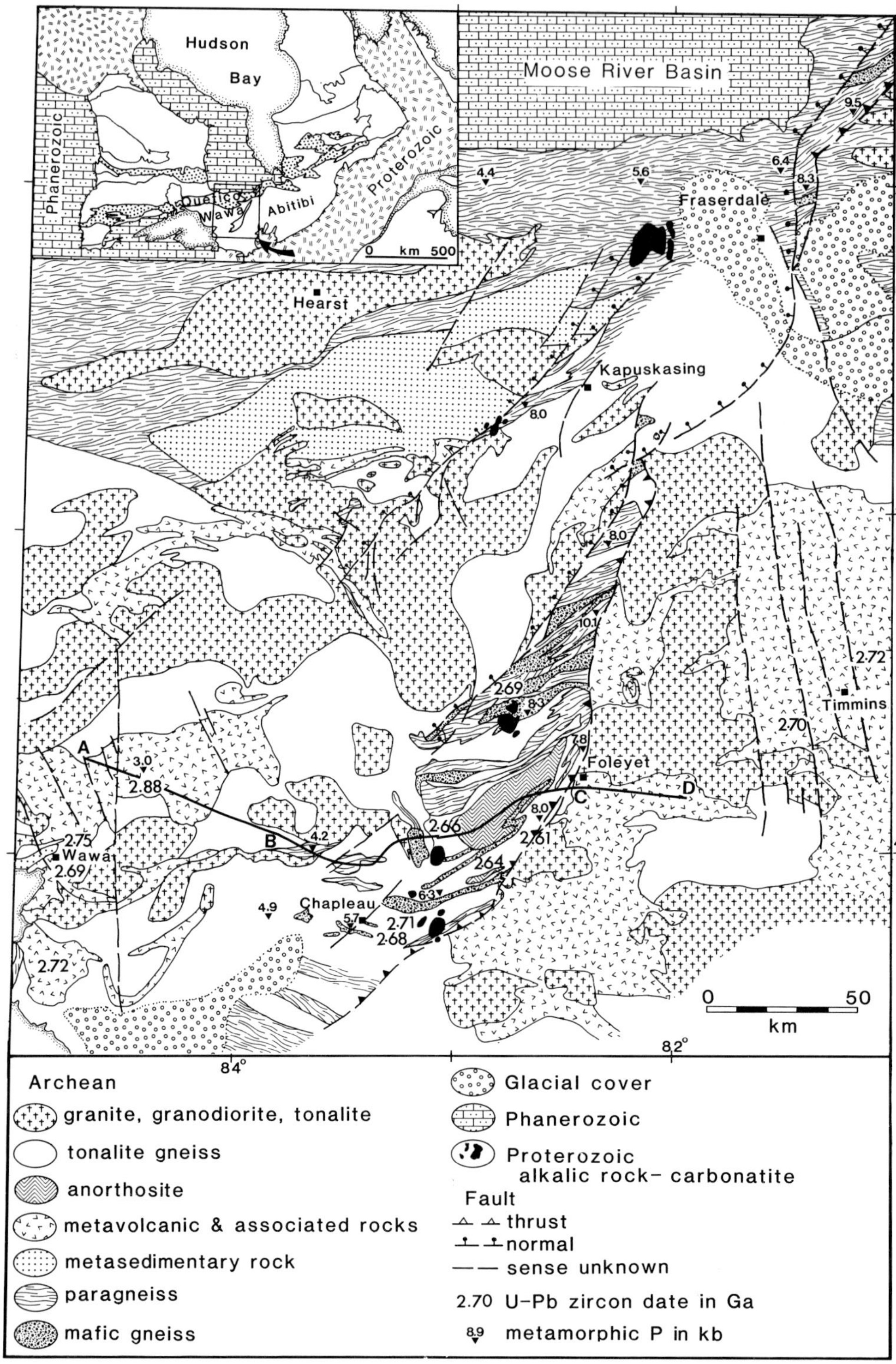

Fig. 8-3. (a) Geology of the Kapuskasing uplift of the central Superior Province (inset) (after Percival, 1990). The oblique cross section through the Abitibi-Wawa greenstone terrane exposes an unbroken transition from low-grade rocks of the Michipicoten belt, through medium-grade tonalitic gneisses, into high-grade gneisses of the Kapuskasing zone, along section line ABCD.

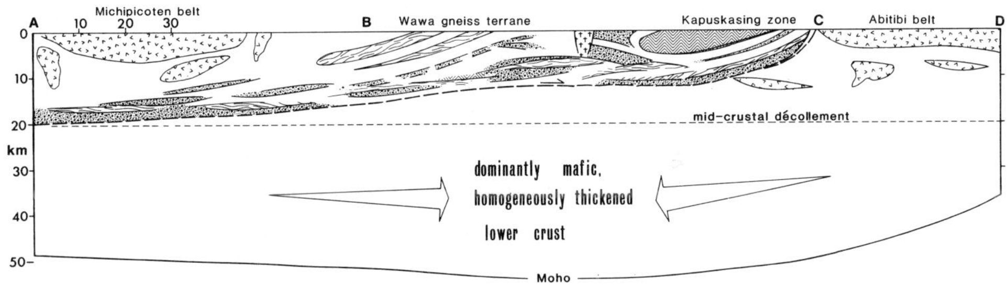

Fig. 8-3 (continued). (b) Cross section along line ABCD, constrained by surface observations and seismic reflection and refraction data. Shortening along the Ivanhoe Lake fault is estimated to be at least 55 km.

neous gneiss sequence in the upper amphibolite and granulite facies of the Chapleau block of the Kapuskasing uplift. Across this lithological and metamorphic transition, metamorphic pressure increases from 2–3 kbar in rocks of the greenschist facies (Studemeister, 1983), to 5–6 kbar in the amphibolite facies, based on hornblende geobarometry (Hammarstrom and Zen, 1986) of tonalites, to 7–9 kbar in rocks of the granulite facies, based on garnet-pyroxene-plagioclase-quartz barometry (Percival, 1983). A moderately-dipping sequence of tonalitic, dioritic and mafic rocks, paragneiss and anorthosite, interlayered on a km scale, makes up the Chapleau block (Thurston et al., 1977; Percival and Card, 1985). It is separated from low-grade rocks of the Abitibi belt to the east by a ductile and brittle, northwest-dipping thrust, the Ivanhoe Lake fault (Percival and Card, 1983; Cook, 1985; Bursnall, 1989; Percival et al., 1989; Geis et al., 1990). Over part of its length, the western boundary is a complex lithological, structural and metamorphic transition that may represent an exposed example of the Conrad mid-crustal seismic discontinuity (Percival, 1986), based on density and seismic velocity contrasts (Percival and Fountain, 1989). Tonalites of the Wawa gneiss terrane have isotropic P-wave velocities averaging 6.55 km s^{-1} whereas layered granulites of the Kapuskasing zone have lithologically-dependent velocities ranging from 6.55 to 7.4 km s^{-1}, with variable anisotropy and an average of 6.9 km s^{-1} (Fountain et al., 1990).

In a zone several tens of kilometers west of and structurally above this boundary, Moser (1988) reported evidence of subhorizontal, synmagmatic extensional lozenge structures in tonalitic gneiss; recent work indicates similar structural style within the high-grade Kapuskasing zone (Bursnall, 1989; Moser, 1989). Although interpretation of the exact timing and regional significance of these structures depends upon geochronology in progress, their form resembles the subhorizontal, discontinuous reflectors present at mid- to deep crustal levels on many seismic reflection profiles (e.g., Oliver et al., 1983; Barazangi and Brown, 1986a; Reston, 1988).

The temporal evolution of the upper 30 km of crust in the central Superior Province is documented through precise U-Pb zircon dates of rock units at various structural levels. The main volcanic accumulations and coeval plutons in the Wawa and Abitibi areas range from ~2750 to 2696 Ma (Krogh and Turek, 1982; Turek et al., 1984; Mortensen, 1987) and are cut by widespread post-tectonic plutons of 2680 ± 10 Ma (Frarey and Krogh, 1986). Tonalitic rocks underlying the greenstone belts and exposed in the Wawa gneiss terrane have ages in the range 2707–2677 Ma (Percival and Krogh, 1983). Their origin is related to a garnet-bearing mafic source (Rudnick and Taylor, 1986), probably either subducted oceanic slabs or mafic lower crust. Supracrustal rocks of the underlying high-grade Kapuskasing zone have no direct dates but tonalite associated with the Shawmere complex has a minimum age of 2765 Ma, slightly older than the main volcanic piles. Model Nd-Sm ages on the supracrustal rocks suggest mantle extraction times in the 2750–2700 Ma range (McNutt and Dickin, 1989). Percival and Card (1985) concluded that the crustal column was initiated with development of the Kapuskasing supracrustal sequence and Shawmere anorthosite. The Wawa–Abitibi volcanic sequences were deposited on this substrate and late syn-volcanic tonalites intruded as thick sheets approximately between the two supracrustal packages. Regional shortening and crustal thickening was followed by injection of late granite-granodiorite plutons.

The present >53-km-thick crust beneath parts of the Chapleau block is thicker by 10–15 km than that of the adjacent regions (Fig. 8-3b; Boland et al., 1988). Velocity contours are subhorizontal below ~20 km depth, but the 6.6-km s^{-1} contour dips ~20°W to extend from near the surface beneath the Kapuskasing zone, to typical regional depths of about 20 km. The Moho downwarp may have developed by ductile flow of the lower crust below a 20-km décollement horizon in response to the shortening that caused brittle uplift of the high-velocity Kapuskasing granulites to higher levels in the crust (Percival and Green, 1988; Percival et al., 1989). Beneath 20 km, velocities in the range 6.8–7.6 km s^{-1} (Boland and Ellis, 1989) can be modeled using homogeneous, dominantly garnet-rich, possibly geochemically depleted mafic compositions (Fountain et al., 1990), based on extrapolation of compositional trends observed at surface and the featureless character of reflection profiles in this depth range.

Recent seismic reflection profiles of the Kapuskasing–Wawa area vary in character on the local and regional scales. Reflective sequences can be traced from at least 12 km depth to within 600 m of the surface in several locations (Fig. 8-3b; Geis et al., 1990). A prominent lamellar zone about 1.5 km thick within the Kapuskasing zone probably corresponds to layered mafic, tonalitic and anorthositic rocks near the base of the Shawmere anorthosite complex (Percival et al., 1989; Green et al., 1990). Thinner, weaker reflective zones from the Wawa gneiss terrane could arise from ductile, high-strain zones in tonalite (Moser, 1989). A zone of branching, gently west-dipping reflections can be traced upward to surface in the Ivanhoe Lake fault zone through a zone of high-angle truncations of short reflection packages, suggestive of listric fault geometry.

Measurements of heat flow and heat generation in the area of the Groundhog River block indicated low values for the highest grade sampled (Cermak and Jessop, 1971), although no granulites were intersected owing to non-coincidence of geophysical anomalies and granulites. Further south, Ashwal et al. (1987) reported variable, but generally low values of heat production from all levels of the crustal section, which they attributed to the mafic character of the greenstone belt crust. In a test of whether the lower crust, known to be electrically conductive at normal (20–40 km) depth, is also conductive where it reaches the surface, Woods and Allard (1986) and Kurtz et al. (1989) measured conductivity on a regional scale in the Kapuskasing area. The lack of a regional anomaly associated with the Kapuskasing structure (Mareschal, 1990) suggests that lower crustal conductivity stems from either: (1) the presence of fluids at depth which escaped from the Kapuskasing rocks during uplift (Woods and Allard, 1986; Bailey et al., 1989); (2) thin, grain-boundary graphite films such as reported by Frost et al. (1989) and observed in Kapuskasing rocks (Mareschal et al., 1992), that cracked during exhumation; or (3) the conductive layer is deeper than the Kapuskasing décollement. Magnetic susceptibility is variable throughout the Wawa–Kapuskasing section, independent of structural level (Shive and Fountain, 1988); susceptibility levels in even the most magnetic rocks of the Chapleau block are insufficient to account for magnetic levels deduced for the deep crust from aerial surveys.

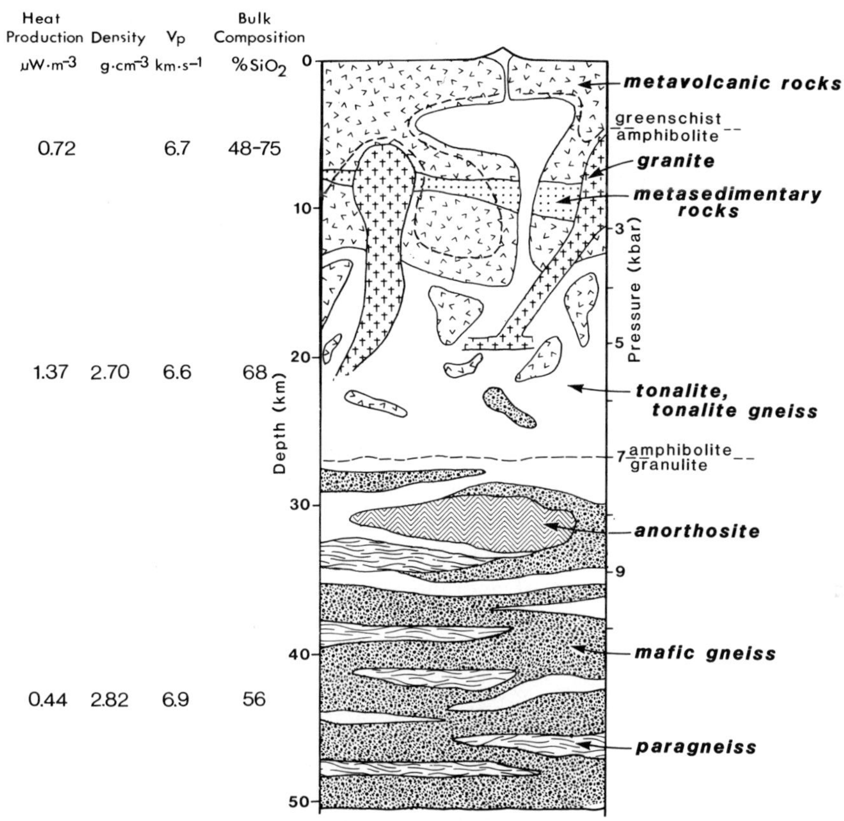

Fig. 8-4. Composite crustal profile through the central Superior Province, based on depth relations observed in the Kapuskasing uplift. The three major depth units correspond to the Michipicoten belt, Wawa gneiss terrane and Kapuskasing structural zone. Bulk composition and other physical properties are characteristic of each terrane type and vary systematically as a function of depth; references can be located in the text.

In an attempt to define depth-related changes in crustal composition, Truscott and Shaw (1986, 1990) determined weighted average compositions for the middle and lower parts of the crustal profile. In comparison with published estimates of the composition of the lower continental crust, the Chapleau block is relatively mafic (~56 wt.% SiO_2; Truscott, 1989), with similar levels of lithophile elements (Shaw et al., 1988). Properties are summarized in a schematic section in Figure 8-4.

Profiles of radiometric age through the crustal section show decreasing apparent age with increasing paleopressure and decreasing crustal level (Corfu, 1987; Percival et al., 1988), in the U-Pb zircon (2700–2610 Ma) and sphene systems (2690–2490 Ma; Percival and Krogh, 1983; Krogh et al., 1986; 1988), the K-Ar and $^{40}Ar/^{39}Ar$ hornblende (2690–2480 Ma) and biotite (2500–2000 Ma) systems (Hunt and Roddick, 1987; Archibald et al., 1986; Lopez Martinez and York, 1990) and the Rb-Sr biotite system (2500–1930 Ma; Z.E. Peterman, pers. commun., 1985). Furthermore, the age gap between systems with different blocking temperatures widens with depth. The age–depth pattern can be interpreted in terms of progressively slower cooling with depth, down to temperatures in the 300°C range, because rapid (fault-related) uplift of rocks above 300°C would have caused instantaneous cooling, with a resultant plateau in the distribution of biotite ages. The concept of late uplift of a cold slab is consistent with the brittle nature of some of the faults in the Ivanhoe Lake zone; local ductile structures (Bursnall, 1989) may have formed earlier or deeper (e.g., Sibson, 1977). Paleomagnetic results (Costanzo-Alvarez and Dunlop, 1988) support a post-2550 Ma age of uplift.

The metamorphic environment of the basal granulites of the section (25–30 km; 700–800°C (Percival, 1983)) is consistent with conditions within a magmatic arc; carbonic fluid inclusions (Rudnick et al., 1984)) could have been produced at depth within a magmatic arc, as indicated by the high proportion of igneous rocks throughout the section and the evidence for prolonged residence in the lower crust. The thermal effects of the long magmatic history may be reflected in the metamorphism and growth of new zircon for about 100 Ma after major magmatism (Krogh et al., 1988), in the manner modeled by Wells (1981). Percival and Card (1985) proposed a tectonic model in which supracrustal rocks of the Kapuskasing zone were progressively buried to lower crustal depths by volcanic accumulations and subsequent intraplated tonalitic intrusions. Many authors have considered Archean greenstone belts to be analogous to contemporary island arcs; however, the 40 ± 5-km average crustal thickness of greenstone belt crusts is anomalous with respect to modern 20-km-thick intra-oceanic arcs. Either accretion of arcs causes structural and magmatic thickening and subsequent intracrustal melting, or thicker, more felsic arcs were produced in the Archean owing to subduction of warmer oceanic crust of average younger age than at present (Abbott and Hoffman, 1984; Martin, 1986).

2.2. Ivrea Zone, Southern Alps, Italy

The Ivrea–Verbano zone (IVZ) of northern Italy (Fig. 8-5), together with the adjacent Strona–Ceneri zone (SCZ), was the first region to be interpreted as an exposed cross section of the continental crust (e.g., Berckhemer, 1969; Mehnert, 1975; Fountain, 1976). Metamorphic grade increases across the IVZ from upper amphibolite facies in the southeast to granulite facies in the northwest (R. Schmid, 1967; Zingg, 1980, 1983) where the zone is truncated by the Insubric Line, a major fault that marks the southern limit of Alpine metamorphism and deformation (e.g., S. Schmid et al., 1987). The eastern boundary with the SCZ, in the upper greenschist and amphibolite facies (Reinhard, 1964; Boriani and Sacchi, 1973; Boriani et al., 1982/1983), is the Pogallo fault zone in the north and the Cossato–Mergozzo–Brissago line (CMBL) in the south (Boriani, 1970; Boriani and Sacchi, 1974).

Pronounced Bouguer gravity and magnetic anomalies (Weber et al., 1949; Vecchi, 1969; Kaminski and Menzel, 1968; Albert, 1974; Froidevaux and Guillaume, 1979; Guillaume and Guillaume, 1980) characterize the IVZ. Early seismic refraction studies (Ansorge, 1968; Giese, 1968; German Research Group for Explosion Seismology, 1968), in conjunction with potential field data, indicated that the IVZ was underlain by rocks with high velocity and density that dip to the southeast and merge with the lower crust under the Po basin; recent geophysical investigations (e.g., Lanza, 1982; Wagner, 1984; Kissling, 1984) support this general view. Seismic reflection data over the southern part of the Ivrea gravity anomaly (Bayer et al., 1987; Nicolas et al., 1990) revealed a seismically transparent zone beneath the IVZ and a group of east-dipping reflectors at 3–15 seconds under the Po plain.

The backbone of the IVZ is a chain of ultramafic bodies (peridotite, hornblendite, pyroxenite) with associated granulite facies mafic gneiss and paragneiss that extends from Baldissero in the south to Finero in the north. The peridotite bodies are commonly interpreted as samples of continental mantle based on their high pressures and temperatures of emplacement (8–11 kbar and 850°C) and their composition (e.g., Garuti et al., 1978/1979; Shervais, 1979;

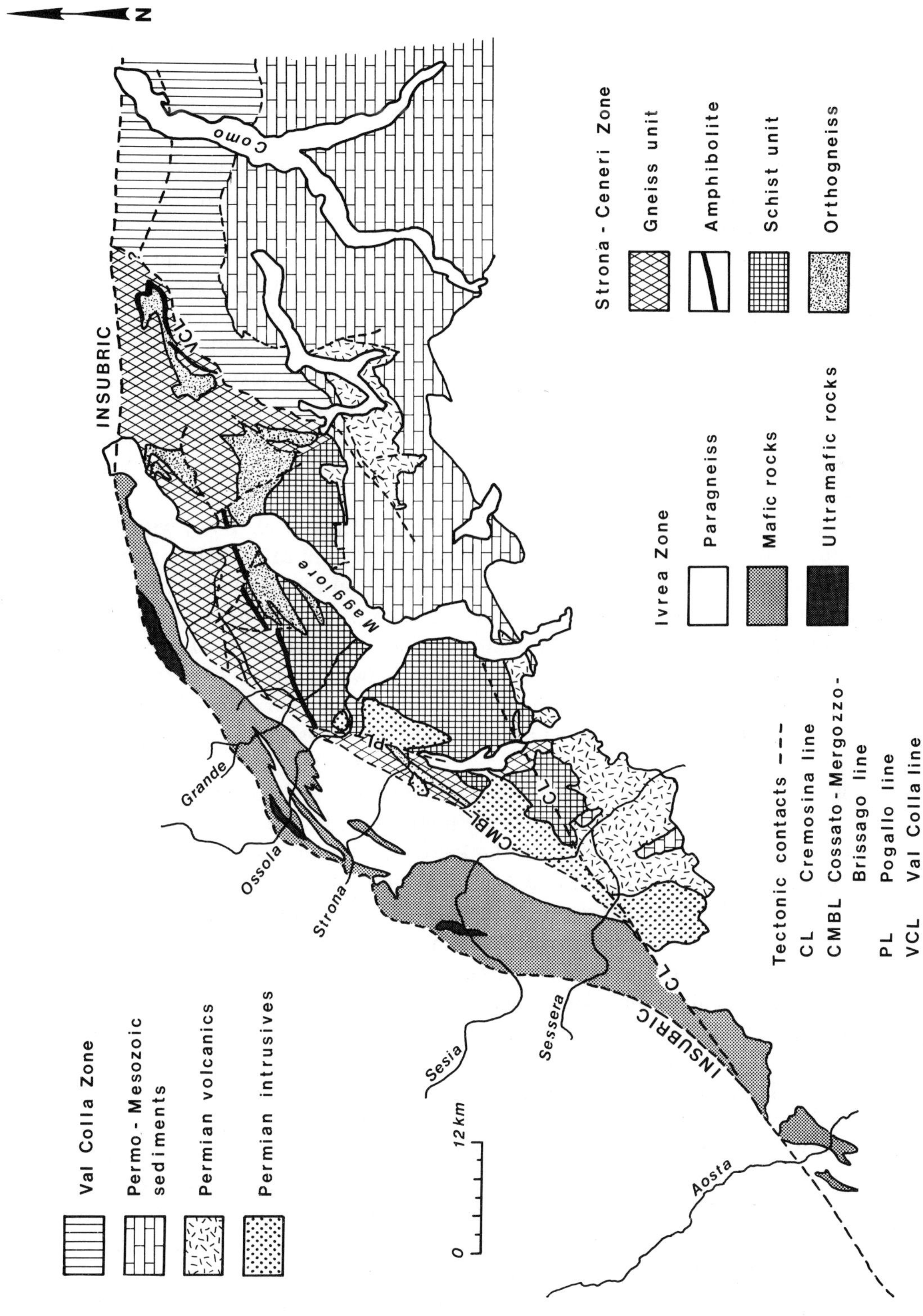

Fig. 8-5. Geology of the Ivrea-Verbano zone of the southern Alps (after Zingg, 1983).

Rivalenti et al., 1981). The age of the peridotites is not known although websterite dikes and gabbroic pods in the Balmuccia peridotite are dated by Sm-Nd at 270 Ma (Voshage et al., 1988). Metamorphic conditions for paragneiss and mafic gneiss (R. Schmid and Wood, 1976; Sills, 1984) reached 8–11 kbar and 700–800°C. Peridotite mylonites occur within and at the contacts of the ultramafic bodies (e.g., Shervais, 1978/1979; Brodie, 1980). High-temperature extensional shear zones in mafic gneisses in Valle d'Ossola developed prior to 280–300 Ma ago (Brodie and Rutter, 1985, 1987; Brodie et al., 1989).

The higher structural levels of the IVZ vary along the strike of the terrane. High-grade paragneiss and marble intercalated with amphibolite (e.g., R. Schmid, 1967; Bertolani, 1968; Mehnert, 1975), commonly with MORB geochemical signatures (Sills and Tarney, 1984; Mazzucchelli and Sienna, 1986; Cumming et al., 1987), dominate the central IVZ. The paragneisses are compositionally similar to graywackes (Sills and Tarney, 1984); higher-grade equivalents, depleted in granitic components, are interpreted as restites (Sighinolfi and Gorgoni, 1978; R. Schmid, 1978/1979). Geothermobarometric data for paragneiss from Val Strona indicate peak metamorphic conditions of 750°C at 6 kbar (Sills, 1984). The paragneisses have Proterozoic crustal residence ages (Köppel and Grünenfelder, 1971; Köppel, 1974; Ben Othman et al., 1984; Voshage et al., 1987; Cumming et al., 1987), monazite dates of 275 Ma (Köppel, 1974; Köppel and Grünenfelder, 1978/1979) and Sm-Nd mineral isochron dates of 245 Ma (Polvé, 1983).

In the southern IVZ, gabbros emplaced at 6 kbar (Rivalenti et al. 1984) with preserved magmatic textures (Rivalenti et al., 1975; Mazzucchelli, 1983) and calc-alkaline affinities (Pin and Sills, 1986) constitute the higher levels. Their subvertical magmatic layering led Zingg (1983) to infer that the IVZ was subhorizontal during gabbro emplacement. Gabbro intrudes and contains xenoliths of high-grade, deformed paragneiss (Rivalenti et al., 1984; Pin and Sills, 1986). A diorite unit in the upper contact zone between this 6–7-km-thick gabbroic unit and paragneiss in the Val Sesia region gave a U-Pb zircon date of 285 Ma (Pin, 1986); Sm-Nd mineral isochrons on gabbro (Ben Othman et al., 1984; Voshage et al., 1987) support a Hercynian age of intrusion.

The age of peak metamorphism in the IVZ is debated. On the basis of a 478-Ma Rb-Sr whole-rock isochron, Hunziker and Zingg (1980) inferred Ordovician peak metamorphism, suggesting persistent high temperatures until the Permo-Carboniferous to account for Rb-Sr isochrons from thin slabs. Others (Pin and Vielzeuf, 1983; Ben Othman et al., 1984; Pin, 1986; Fountain, 1986) relate the gabbros to the peak metamorphism and thus favor a Permo-Carboniferous age. However, because the gabbros intruded late in the metamorphic history (Zingg et al., 1990) they cannot be used to directly date the peak metamorphism. Mineral dates for the IVZ [see Hodges and Fountain (1984) and Handy (1987) for reviews] record progressive cooling from the high temperatures of the Permo-Carboniferous to temperatures around 300°C in the Jurassic (Fig. 8-6).

In the SCZ, major rock types include paragneiss, orthogneiss and mica schist in the amphibolite facies, with minor amphibolite and ultramafic rocks. Based on a 473-Ma Rb-Sr whole-rock isochron and 450-Ma U-Pb monazite dates, Hunziker and Zingg (1980) inferred a Caledonian age for metamorphism in the SCZ whereas Rb-Sr mica dates suggest a Hercynian age (Boriani et al., 1982/1983, 1985). The SCZ was intruded by post-metamorphic granites of 270–276 Ma age (Köppel, 1974; Köppel and Grünenfelder, 1978/1979; Hunziker and Zingg, 1980) that are compositionally equivalent to volcanic rocks to the east (Hunziker, 1974). South Alpine supracrustal rocks, commonly in fault contact with the SCZ, include Permo-Carboniferous (300–260 Ma) calc-alkaline basalt, basaltic andesite, dacite and rhyolite (Stille

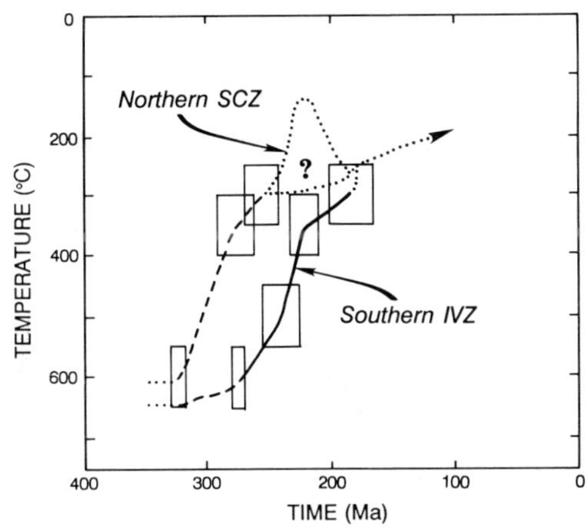

Fig. 8-6. Thermal history of southern Alpine deep crustal exposures, based on geochronological and P-T constraints. Ivrea-Verbano zone (after Handy, 1987).

and Buletti, 1987) and clastic rocks as well as early Mesozoic carbonates, deposited along the rifted Tethyan margin (Winterer and Bosellini, 1981).

The northeasterly structural trends of the SCZ are discordant to the more northerly trends of the IVZ along the kilometer-wide Pogallo ductile fault zone, containing at its eastern margin a mylonite zone up to 100 m thick (Boriani, 1970, 1971; Boriani and Sacchi, 1973). Handy (1986, 1987) determined an oblique-sinistral shear sense in present-day coordinates and inferred that deformation mechanisms change from low to high temperature from south to north. Current interpretations of the Pogallo fault zone involve an early Mesozoic low-angle normal fault subsequently rotated into a steep orientation during emplacement of the IVZ (Hodges and Fountain, 1984; Handy, 1986, 1987; S. Schmid et al., 1987). The sense of shear is the same as that inferred from high-temperature shear zones within the IVZ (Brodie and Rutter, 1987). Alternatively, the Pogallo fault zone and the CMBL have been interpreted as late Paleozoic strike-slip faults (Boriani and Sacchi, 1974; Boriani et al., 1977) that displaced the SCZ relative to the IVZ after their emplacement in the upper crust.

Although many aspects of the geological evolution of the IVZ remain enigmatic and controversial, the terrane can be viewed as an example of lower continental crust that evolved in convergent and extensional continental margin environments (Fountain, 1986; Zingg et al., 1990). Various isotopic data (e.g., Köppel, 1974; Hunziker and Zingg, 1980) indicate that high temperatures prevailed during the Hercynian, either as a continuation of an Ordovician thermal episode (Hunziker and Zingg, 1980) or as a separate event. This Hercynian event overlaps the emplacement of the deep-level gabbroic bodies, intrusion of shallow-level granites in the SCZ and eruption of the calc-alkaline volcanic suite in the southern Alpine domain. These magmatic additions at all crustal levels may have occurred in a Permian Andean margin setting (Stille and Buletti, 1987). The high-temperature extensional event (Brodie and Rutter, 1987) may correspond to crustal dilation during this arc stage or the collapse of the margin upon relaxation of convergence as suggested for other orogenic belts (e.g., Coney and Harms, 1984; Royden and Burchfiel, 1987). The cooling patterns of the IVZ and SCZ, along with the kinematic data for the Pogallo fault zone suggest that the sequence was further extended during Late Triassic–Middle Jurassic development of the Tethyan continental margin (Hodges and Fountain, 1984; Handy, 1986, 1987; S. Schmid et al., 1987).

Rift basins developed in the upper crust (Winterer and Bosellini, 1981) in response to the extension; Handy (1987) speculated that the thickness of crystalline crust represented by the IVZ and SCZ was reduced to about 20 km. Initial uplift of the IVZ occurred in this extensional phase and was followed by a second uplift stage during Late Cretaceous convergence (S. Schmid et al., 1987). Neogene backfolding and backthrusting (Hodges and Fountain, 1984; S. Schmid et al., 1987) during Alpine shortening led to development of low-grade mylonites and folds near the Insubric Line (e.g., R. Schmid, 1967; Brodie and Rutter, 1987; S. Schmid et al., 1987).

Laboratory studies of the physical properties of the IVZ and SCZ sequences have been used to model the geophysical nature of the lower continental crust. P-wave velocities of rocks from the IVZ (Fountain, 1976; Burke and Fountain, 1990) are variable but generally high (>6.9 km s^{-1}). Velocities of SCZ rocks are consistently in the 6.2-km s^{-1} range. The velocity change across the SCZ to IVZ transition is similar to that at the Conrad discontinuity. Within the section, the change corresponds to an increase in metamorphic grade and a general change in composition with depth. Synthetic reflection seismograms (Hale and Thompson, 1982; Fountain, 1986; Hurich and Smithson, 1987) predicted that lower crust as represented by the IVZ should be reflective because of high reflection coefficients associated with alternating high- and low-velocity layers; use of Brodie and Rutter's (1987) reconstruction of pre-Alpine geometry (Rutter and Brodie, this volume; Fig. 8-9) accentuates this conclusion (Burke and Fountain, 1990). Heat production measurements (Höhndorf, 1975; Dostal and Capredi, 1979; Galson, 1983) show that IVZ mafic gneisses have very low values but high-grade paragneisses are only slightly lower than the higher-level SCZ gneisses; the IVZ shows considerable lateral and vertical variation in heat production. Galson (1983) concluded that data from the SCZ/IVZ section does not support general crustal models for the variation of heat production with depth (e.g., Lachenbruch, 1970). Magnetic susceptibility is extremely variable for IVZ rocks (Wasilewski and Fountain, 1982; Wagner, 1984; Wasilewski and Warner, 1988; Belluso et al., 1990). Some high-grade mafic rocks have high susceptibilities but Shive (1990) concluded that the overall level of susceptibility of IVZ rocks is insufficient to account for the IVZ magnetic anomaly or the long-wavelength magnetic anomalies arising from the lower crust in general.

2.3. Calabrian massif, southern Italy

High-grade metamorphic rocks exposed in the Calabrian massif (e.g., Paglionico and Piccarreta, 1978; Moresi et al., 1978/1979; Schenk, 1980, 1984) are regarded as the base of a nearly intact cross section of the continental crust (e.g., Schenk and Schreyer, 1978). This granulite-facies terrane, located in the Serre and Sila between the Apennines and Maghrebides, was emplaced along the easterly-trending Curinga–Girifalco line and is overlain by amphibolite facies gneisses, low-grade phyllites and Mesozoic and Tertiary sedimentary rocks (Fig. 8-7). In contrast to the IVZ, the Calabrian massif lacks the distinctive geophysical anomalies that link the exposed terrane to the present-day lower crust; the region is characterized by a magnetic low (Schwarz, 1978) and a Bouguer gravity minimum between highs over the Ionian and Tyrrhenian Seas (Görler and Giese, 1978). A P-wave velocity inversion under the terrane has been used as evidence that low-density crust was thrust beneath the Calabrian arc to induce uplift of the high-grade rocks (Schütte, 1978; Görler and Giese, 1978). Xenoliths of rocks similar to those exposed in the Calabrian massif were recovered from the nearby Lipari Island volcano (Honnorez and Keller, 1968).

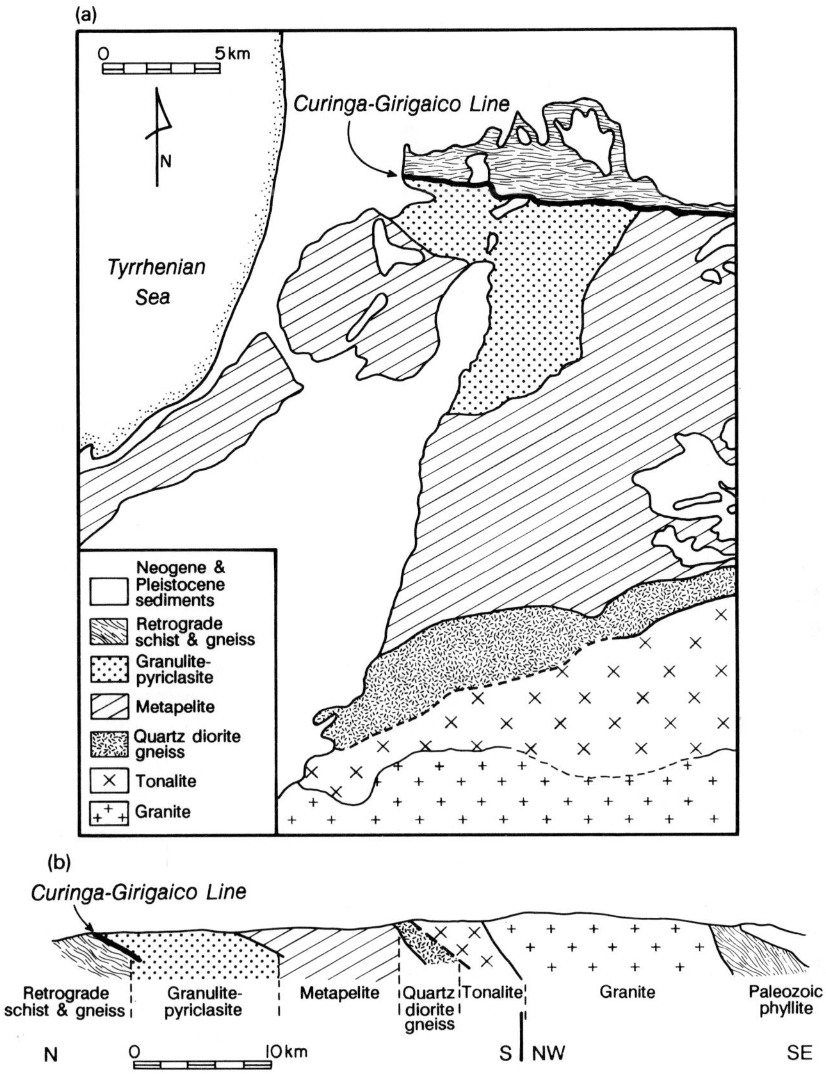

Fig. 8-7. Geology (a) and cross section (b) of the Calabrian massif of southern Italy (after Schenk, 1984).

The high-grade terrane exposed in the Serre can be divided into two units. The structurally lower unit is composed of granulite-facies mafic and felsic gneiss with thin intercalations of ultramafic rock. Mafic rocks of variable composition comprise pyroxene-plagioclase±garnet assemblages. The ultramafic suite includes spinel peridotite, pyroxenite and hornblendite, interpreted as cumulates (Schenk, 1984). Geothermobarometric data indicate that the terrane experienced peak metamorphic conditions of ca. 7–8 kbar and 760–850°C, followed by a pressure drop of 1.5–2 kbar and subsequent isobaric cooling (Schenk, 1984, 1989).

The higher structural unit is dominated by various metapelitic rocks (aluminous paragneiss, cordierite-sillimanite-biotite gneiss and garnet-biotite gneiss) with minor marble, calc-silicate gneiss and mafic gneiss. Metamorphic pressure ranges from about 7.5 kbar at the base to about 5.5 kbar at the top (Schenk, 1984, 1989). Maccarrone et al. (1983) suggested that some of the metapelites experienced anatexis and melt removal. The high-grade terrane is overlain by a unit of quartz diorite gneiss that dips to the south below a 230–295-Ma post-metamorphic tonalitic-granodioritic-granitic complex (Borsi et al., 1976; Schenk, 1980). These plutons intrude a Lower Carboniferous phyllite sequence that is overlain by Mesozoic and Tertiary sedimentary rocks. Restoring the surface geology of the Serre (Fig. 8-7) to its pre-uplift orientation (Schenk, 1981; Kern and Schenk, 1985, 1988) yields a 20–25-km-thick crustal column, consistent with pressures recorded by granulites at the base.

The termination of granulite facies metamorphism is estimated at about 295 Ma on the basis of U-Pb zircon and monazite dates (Schenk, 1980, 1984). This date coincides approximately with intrusion ages for higher-level tonalites and granites. Schenk (1984) postulated that the terrane was uplifted 5–7 km at the end of the Hercynian orogeny as continental crust was thrust beneath the Calabrian terrane (e.g., Görler and Giese, 1978). Rb-Sr biotite dates (Schenk, 1980) decrease from approximately 135 Ma in the upper part of the metapelite unit to about 110 Ma in the mafic granulite unit indicating that the deep level rocks cooled through the 300°C isotherm during the Early Cretaceous.

The Calabrian section has also been used to model the geophysical character of the deep crust. Laboratory-determined seismic velocities of Calabrian rocks (Kern and Schenk, 1985, 1988) are generally similar to those of the IVZ/SCZ section, although mafic rocks from the IVZ generally have higher velocities. Velocities increase downward through the Calabrian section from <6.5 to 7 km s^{-1} but are lower in the deepest mafic units. The major velocity change that occurs at the contact between the high-level plutonic complex and the metapelite unit is similar in magnitude to that reported for mid-crustal discontinuities.

The Calabria and Ivrea sections bear a number of important similarities (Moresi et al., 1978/79; Schenk, 1981) and a few distinct differences. Lithologically, the exposed lower crust in both areas is dominated by metapelites and mafic gneisses. The metapelite-rich Calabrian section (Maccarrone et al., 1983) is most comparable to the Val d'Ossola and Val Strona sections of the IVZ (R. Schmid, 1967; Mehnert, 1975) rather than the more mafic Val Sesia sequence. Simple cooling curves for the Calabrian and Ivrea sections (Schenk, 1984) offer further comparisons: both terranes experienced high-grade metamorphism in the lower crust and magmatism in the upper crust during the Hercynian event. However, the early compressional uplift in Calabria contrasts with the early extensional uplift postulated for Ivrea. Biotite dates are Mesozoic for both areas, but slightly younger in Calabria. It thus appears that the geological history of these terranes, 1200 km apart, involved similar pre-Hercynian and Hercynian evolution but divergence in post-Hercynian time.

2.4. Kohistan arc, Pakistan

The Kohistan arc of Pakistan represents up to 40 km of section through a deformed, intruded island-arc sequence exposed in the Himalayan collision zone. As such it provides many clues to the nature of growth of continental crust.

Measuring about 600 by 100 km, the dominantly magmatic Kohistan complex is bounded to the north and south by major faults, the Northern and Indus sutures respectively (Coward et al., 1982). Two stages of growth, separated by major deformation, include a Late Jurassic–Late Cretaceous island-arc building stage and Late Cretaceous–Eocene, calc-alkaline Andean batholith emplacement (Coward et al., 1986). At least two phases of deformation are recognized in the older island-arc sequence, making imprecise the reconstruction of a crustal cross section at the close of arc magmatism. However, metamorphic grade generally guides recognition of different crustal levels.

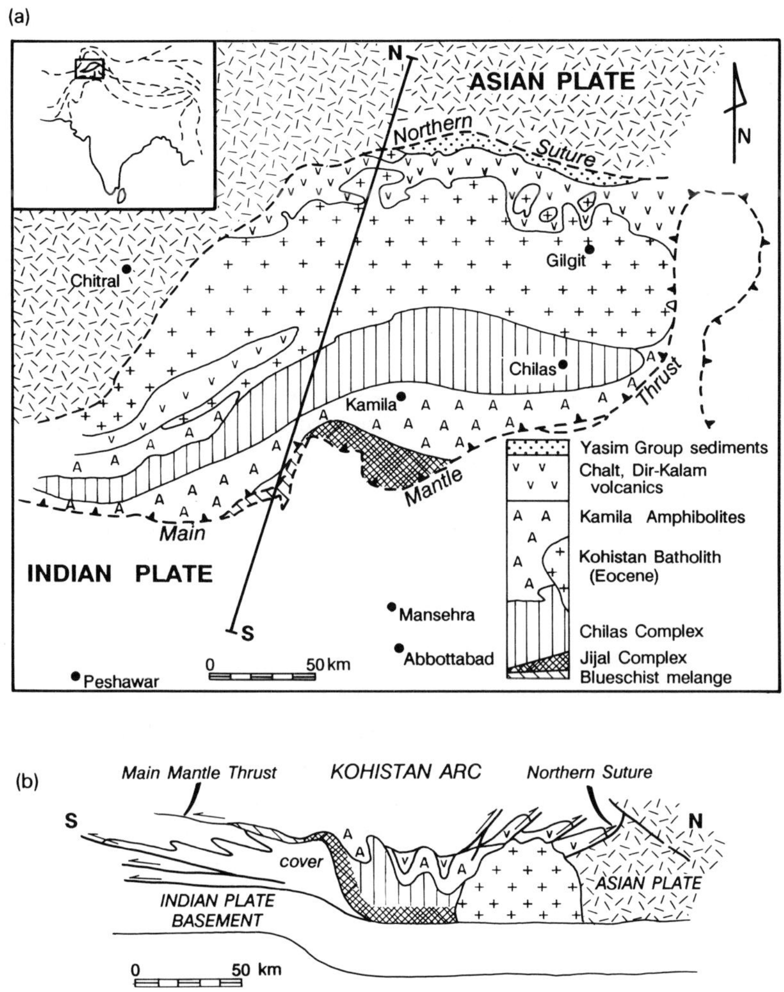

Fig. 8-8. (a) Generalized geological map of the Kohistan arc of Pakistan. The composite crustal section includes high-level volcano-sedimentary sequences (Chalt Group), the intermediate-level Kohistan batholith, the lower crustal Chilas complex and lowermost Jijal-Patan complex. (b) Schematic cross section NS (after Coward et al., 1986), showing the folded Kohistan arc as an allochthonous terrane between the Asian and Indian plates.

From the base, a reconstructed section through the Late Jurassic–Cretaceous Kohistan column (Fig. 8-8) consists of: (1) a discontinuous basal blueschist mélange, overlain by (2) the Jijal–Patan complex, high-pressure garnet-granulites and ultramafic rocks, possibly representing uppermost mantle; (3) the Chilas complex, two-pyroxene mafic granulites with locally preserved igneous layering; (4) the Kamila amphibolites, deformed mafic igneous and sedimentary rocks; (5) the Chalt volcanics, greenschist-amphibolite subaerial and submarine volcanic and sedimentary rocks; and (6) the Yasin Group, consisting of slates, turbidites, volcaniclastics and limestones (Coward et al., 1986). Gneissic tonalites of similar age occur within the section. The Eocene Kohistan batholith, occupying the axis of the arc, consists of plutons of gabbro, diorite, tonalite and granite that cut Cretaceous structures. A swarm of late pegmatites represents the latest intrusive phase. Although the polarity of subduction is disputed (Bard, 1983; Windley, 1983; Coward et al., 1986; 1987) the calc-alkaline Andean plutonism is thought to relate to continued subduction of Tethys following accretion of the Kohistan arc onto a continental mass, probably Asia (Treloar et al., 1989a,b).

The stratiform, layered Chilas complex, about 10–15 km thick, was metamorphosed to granulite facies at pressures of 12–14 kbar and temperatures of 670–790°C (Jan and Howie, 1981; Bard, 1983). Although generally accepted as lower arc crust, the nature and timing of its granulite-facies metamorphism must be considered in a crustal reconstruction; if the high pressures were imposed during an early, intra-arc event (Bard, 1983), a 40-km-thick island-arc crust is implied, whereas if they accompanied Eocene magmatism (e.g., Chamberlain et al., 1989), the crustal thickness relates to an Andean setting. As a consequence of igneous layering and thrust imbrication, the Chilas complex has a subhorizontal fabric, that probably extends laterally for at least 350 km. This type of lithological continuity contrasts with the heterogeneity noted in most exposed sections and in many seismic reflection profiles, although continuous planar reflections characterize the lower crust in some areas (Oliver et al.,1983).

A reconstructed velocity section through the Kohistan sequence based on laboratory measured P-wave velocities indicates that the upper crust of the section is typified by velocities between 6.2 and 6.4 km s^{-1} and a lower crust with P-wave velocities between 6.4 and 6.7 km s^{-1} (Chroston and Simmons, 1989). This structure is similar to that of volcanic arcs and some sections of continental crust.

Although imperfect as a section through an island arc, the Kohistan sequence may be a typical example of an oceanic fragment accreted to an Andean margin. Thus the deformation, related to the accretion process, and late magmatism, a product of continued subduction, constitute parts of the process of conversion of oceanic to continental crust; had it not been arrested by collision, magmatic thickening could have proceeded to typical Andean values. The resultant section, with a garnet-granulite mantle, and pyroxene-granulite mafic lower crust, overlain by calc-alkaline plutons, resembles many continental cross sections (e.g., Fountain and Salisbury, 1981), particularly those of Archean age.

2.5. Central Australia

Central Australia hosts two major east–west-trending Proterozoic granulite terranes, the Arunta and Musgrave Ranges, separated by the Phanerozoic Amadeus basin (Figs. 8-9 and 8-10, respectively). Significant geophysical anomalies over the terranes, such as large-amplitude Bouguer gravity anomalies and pronounced teleseismic P-wave residuals (Lambeck and Penney, 1984) support a model of crustal structure (Forman and Shaw, 1973) in which the lower crust and upper mantle have been elevated along crustal-scale thrust faults. The Redbank zone (e.g., Bell, 1978), a 7–10-km-thick mylonite zone that limits the Arunta block on the south, is a prominent north-dipping seismic reflector (Goleby et al., 1989). The south-dipping Woodroffe thrust (e.g., Bell, 1978) forms the northern boundary to the eastern Musgrave block.

Combined seismic reflection data (Goleby et al., 1989) and kinematic analysis (Teyssier, 1985) demonstrate that the Arunta granulites were thrust to the south along the Redbank fault zone onto the Ormiston nappes and Amadeus basin. An array of north-dipping faults that transect the Arunta block are interpreted as components of a crustal-scale duplex system (Teyssier, 1985). The age of major uplift was inferred by Teyssier (1985) to be at about 350 Ma (Alice Springs orogeny), based on K-Ar mica dates ranging from 350 to 400 Ma (Shaw et al., 1980); Rb-Sr dates point to uplift at earlier times (Marjoribanks and Black, 1974; Stewart et al., 1984).

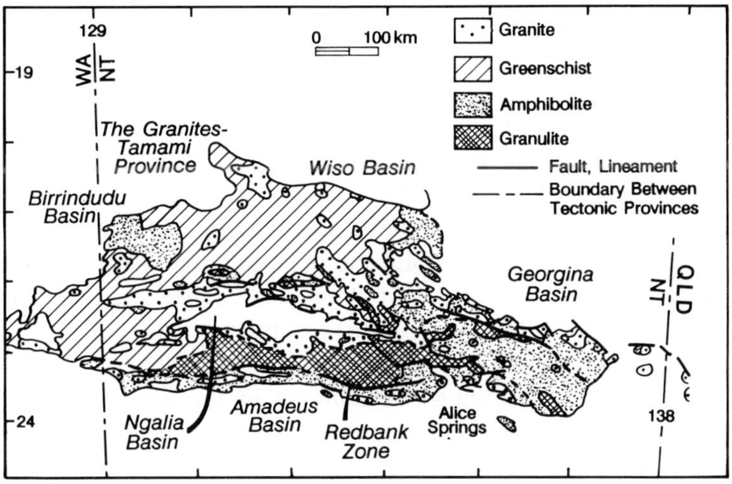

Fig. 8-9. Geology of the Arunta block, central Australia (after Stewart et al., 1984).

High-grade metamorphic rocks are lithologically similar in the two terranes. The sequence of quartzofeldspathic gneiss, intercalated with minor mafic gneiss, meta-pelitic gneiss, marble, calc-silicate gneiss and banded iron formation, is interpreted as a metamorphosed supracrustal succession dominated by silicic and mafic volcanic rocks (Stewart et al., 1984). Peak metamorphic conditions, based on assemblages and geobarometry, are estimated at 6.5–9 kbar and 600–900°C in the Arunta block (Warren, 1983; Windrim and McCulloch, 1983; Glikson, 1986) and 12 kbar and 760°C in the Musgrave block (Maboko et al., 1989). Complex overgrowth and corona textures suggest isobaric cooling (Warren and Stewart, 1988; Maboko et al., 1989), possibly in the interval 1100–350 Ma although the polymetamorphic history complicates interpretation (Clarke et al., 1990).

Layered intrusions (e.g., Giles, Kalka complexes), consisting of pyroxenite, norite, olivine gabbro, and anorthosite, are an important lithological component of both granulite terranes (Goode and Moore, 1975; Glikson, 1986; Gray and Goode, 1989). The sequence of growth

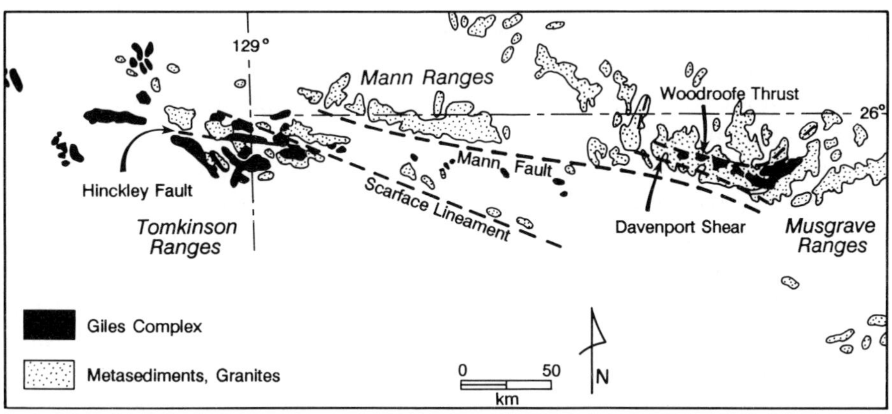

Fig. 8-10. General geology of the Musgrave Range, central Australia (after Goode and Moore, 1975).

of igneous and metamorphic minerals in several bodies in the Tomkinson range of the western Musgrave terrane indicate that the intrusions were emplaced (Goode and Moore, 1975) and deformed (Etheridge, 1975) under deep crustal conditions, either during or immediately following the granulite-facies metamorphic peak (Moore and Goode, 1978; Gray and Goode, 1989). Higher structural levels of the Arunta block consist of amphibolite- to greenschist-facies aluminous metasedimentary rocks, amphibolite and post-metamorphic granite, granodiorite, diorite and tonalite (Stewart et al., 1984).

Geochronological investigations of the Arunta and Musgrave blocks are largely confined to Rb-Sr whole-rock isochrons on samples representing large areas (Arriens and Lambert, 1969; Marjoribanks and Black, 1974; Iyer et al., 1976; Gray and Compston, 1978; Black et al., 1983; Windrim and McCulloch, 1983). The spectrum of resulting dates (1100–1700 Ma) have been interpreted to reflect various metamorphic episodes. Dating of events in polymetamorphic terranes by Rb-Sr whole-rock techniques is tenuous in the absence of control by high-resolution U-Pb zircon data on cross-cutting intrusive bodies. Sm-Nd model ages for gneisses from the Arunta block suggest a mantle fractionation event at 2000 Ma to form the protoliths (Windrim and McCulloch, 1983; Black and McCulloch, 1984).

Reconstruction of crustal sections for the Arunta and Musgrave blocks may not be possible owing to their dissection by uplift-related faults. Also, the indication of polymetamorphic history and lack of precise geochronology for the terranes inhibit assessment of the relative importance of crustal growth and reworking. This situation particularly inhibits assessment of the role of mafic intrusions as a metamorphic heat source.

2.6. Kasila Group, west Africa

A compressional uplift mechanism has been proposed to explain the occurrence of a linear belt of Archean granulites, the Kasila Group of the west African craton (Williams, 1988). Mafic rocks, metasediments, iron formation and leucogabbro, all metamorphosed to granulite facies (10 kbar, ~850°C) constitute an inferred basal section (Williams, 1978); however, Proterozoic cover prohibits observations of higher structural levels. Intense strain recorded in the Kasila rocks is attributed to syn-metamorphic thrusting and uplift (Williams, 1988).

2.7. Summary — compressional uplifts

Many common depth-related features characterize the compressional uplifts. Foremost is a downward decreasing proportion of supracrustal rocks relative to intrusive bodies. The intrusions vary from cross-cutting plugs at high levels to tabular sheets as they change in composition from felsic to intermediate, mafic and anorthositic. Metamorphic grade ranges from greenschist to granulite.

Restite compositions have been recognized in the lower crustal parts of only the Ivrea and Calabria profiles although migmatitic textures implying anatectic conditions are universal. There appears to be no unique setting for crustal construction of terranes exposed by this mechanism; accretionary prism, magmatic arc and rift settings are represented. Compressional uplifts generally occur at plate boundaries; the exposed sections commonly bear the imprint of earlier, boundary-related deformation, such as high-temperature extensional shear zones, or collision-related folds and intrusions. The Kapuskasing and central Australian uplifts occur in intracratonic settings, hundreds to thousands of kilometers from plate boundaries active at the time of uplift. Transmission of stress to the craton interior and cause of localization of uplifts remain outstanding problems. Uplift of rocks from mantle depths is rare and appears restricted to cross sections in plate boundary settings (Ivrea, Kohistan).

The highest metamorphic pressures in an uplift provide a guide to the depth of décollement; values in the 7–9-kbar range are common. Perhaps this 20–25-km level corresponds to the temperature of a brittle-ductile transition at the time of compression. Alternatively, the high density of the crust below this level might energetically favor descent rather than ascent and cause the crust to delaminate to form wedge structures (e.g., Oxburgh, 1972). Lower crustal thickening, as observed at Kapuskasing (Boland et al., 1988; Percival et al., 1989) and inferred in other sections, may be a manifestation of this process.

3. Wide, oblique transitions

Several low- to high-grade transitions included in Fountain and Salisbury's (1981) compilation of exposed cross sections are here referred to as wide, oblique transitions. These exhibit many similarities to the narrow, compressional uplifts described above, including systematic variation in metamorphic grade and pressure (~3–8 kbar), associated with decreasing proportions of supracrustal rocks. Notable differences include the laterally wide metamorphic transitions (>100 km), consistently steeply-oriented structures independent of structural level and lack of a well-defined uplift mechanism. Some of the examples cited below may be transitional to large high-grade terranes, where uplift is related to a premetamorphic thickening event (*e.g.*, Percival, 1990).

3.1. Dharwar craton, southern India

A strike-parallel transect (Fig. 8-11) through the Archean Dharwar craton begins in the north in greenschist-facies supracrustal belts and terminates about 200 km to the south in granulite-facies charnockite (Pichamuthu, 1965; Hansen et al., 1984a; Raase et al., 1986). Although the craton contains components ranging from >3400 Ma to 2500 Ma, the metamorphism responsible for the greenschist (3 kbar) to granulite (8 kbar) transition occurred about 2500 Ma ago.

Two main sequences of supracrustal rocks have been recognized in the schist belts. The older Sargur Group of >3000 Ma age consists of metapelite, quartzite, iron formation, calc-silicate, mafic metavolcanic rocks and gabbro-anorthosite intrusions, whereas the 3000–2500 Ma Dharwar Supergroup contains mafic to intermediate volcanic and volcaniclastic rocks, conglomerate, sandstone, iron formation, greywacke and shale (Swami Nath and Ramakrishnan, 1981). Dharwar supracrustals are dominant in the lower grade northern part of the craton; fragments of the Sargur Group occur as inclusions in gneisses to the south, generally at higher grade. The Peninsular gneiss complex of amphibolite to granulite-facies tonalitic gneiss and migmatite underlies much of the craton and contains components as old as 3400 Ma (Beckinsale et al., 1980; Buhl, 1987).

Although metamorphic grade is relatively low in the schist belts, their structural and evolutionary history is complex. For example, the Kolar belt, containing important gold deposits, is the site of a suture between four terranes with distinctive structural and isotopic histories (Balakrishnan and Rajamani, 1987; Krogstad et al., 1988).

Supracrustal rocks become less abundant to the south as grade increases to amphibolite facies in Peninsular gneiss. The boundary between the amphibolite and granulite facies, also

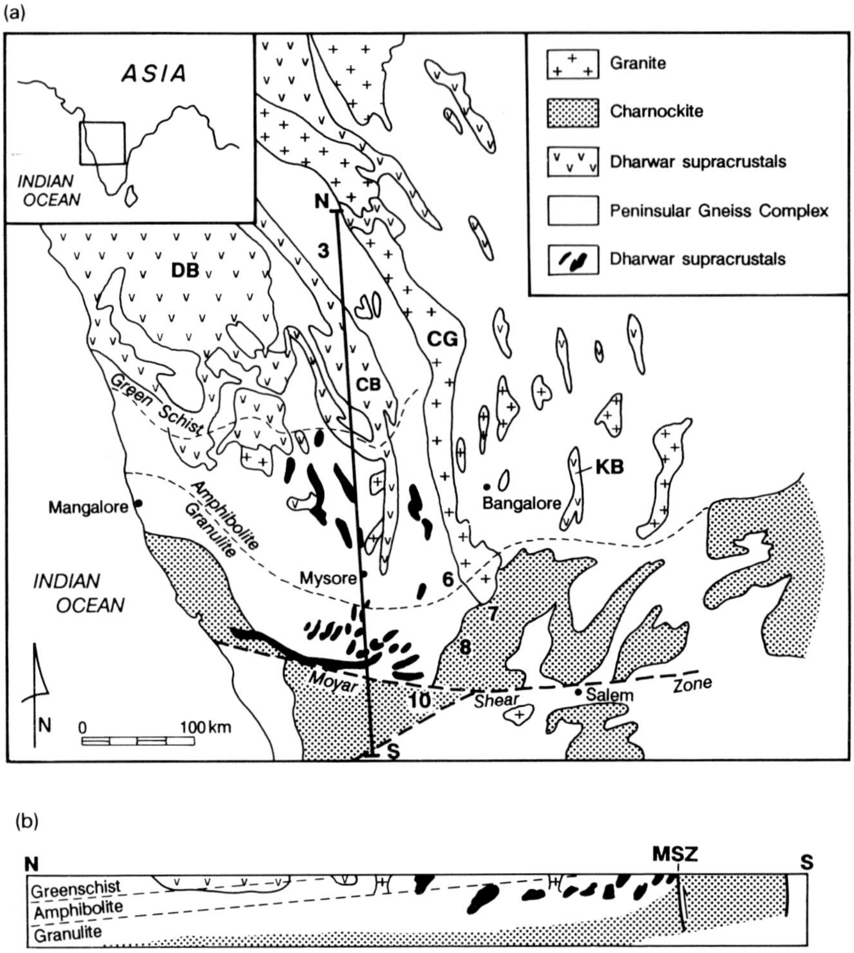

Fig. 8-11. (a) Geology of part of the Dharwar craton (after Janardhan et al., 1982). Numbers refer to metamorphic pressure in kbar, from Harris and Jayaram (1982), Hansen et al. (1984a,b) and Raith et al. (1988). *DB* = Dharwar belt; *CB* = Chitradurga belt; *KB* = Kolar belt; *CG* = Closepet granite. (b) Longitudinal N-S cross section showing inferred depth relationships beneath the craton and granulite zone, reworked at 2500 Ma, in the south.

known as the charnockite front, is one of the most thoroughly studied yet controversial metamorphic transitions anywhere. Early debate centered on the problem of igneous vs metamorphic origin for charnockites; recently, the relative importance of volatile fluxing and partial melting to produce incipient charnockite textures has been disputed [see Frost and Frost (1987) and Ashwal (1988) for reviews].

Several quarries near the amphibolite–granulite boundary display textures indicating transformation of Peninsular gneiss to massive charnockite in patches and along discrete minor shear zones and fingers (Pichamuthu, 1961; Janardhan et al., 1982). In these zones, the background grey, medium-grained, hornblende-biotite tonalite gneiss becomes brown, coarse-grained orthopyroxene charnockite with distinct chemistry and new monazite, dated by U-Pb at 2520 Ma (Buhl, 1987). Detailed geochemical traverses have documented changes in major and minor element composition (Condie et al., 1982; Hansen et al., 1987; Stahle et al., 1987; Jiang et al., 1988) as well as in the composition of fluid inclusions (Hansen et al., 1984b,

1987) and their isotopic character (Jackson et al., 1988). Virtually all results point to the influx of dehydrating fluids, probably dominantly CO_2, along channelways to produce the incipient charnockite textures. Massif charnockites to the south have been interpreted as a pervasively CO_2-flushed terrane (Harris et al., 1982) and Newton et al. (1980) suggested that similar processes may account for granulite-facies metamorphism in general.

To the north of the transition, the biotite±hornblende-bearing Closepet granite of 2520 Ma age (Grew and Manton, 1984; Buhl, 1987) forms a linear north-striking batholith of about 40 by 500 km, with thin veins and dikes extending into the transition zone. Friend (1981, 1985) proposed that H_2O, driven off the charnockites ahead of the CO_2 front, caused partial melting of amphibolite-facies rocks to produce the granite. Newton (1988) suggested that H_2O and associated lithophile elements flushed from the lower crust by CO_2 passage produced the granite metasomatically.

The Moyar shear zone separates the structurally coherent greenschist to granulite transition from the 10-kbar (Raith et al., 1983, 1988) massif charnockites to the south. Although this region has previously been interpreted as the deepest part of the crustal section, a Rb-Sr model age (Buhl, 1987) suggests that it has a short pre-history before its 2500 Ma metamorphism, in contrast to the Peninsular gneiss protoliths to the north. The region south of the Moyar shear zone probably had an independent history and, therefore, does not represent deep levels of the Dharwar craton.

Tectonic models for the Dharwar craton must incorporate the evidence for CO_2 streaming, the oblique exposure and a prominent set of dominantly dextral, N–S-trending shear zones (Newton, 1987). These constraints led Newton (1990) to suggest an event at about 2500 Ma in which the Dharwar craton was obliquely loaded from the south by a second continent. While explaining many elements of the geology, this model does not account for the decreasing proportion of supracrustal rocks with increasing paleodepth. This distribution may have been established prior to the 2500-Ma metamorphism. Newton's model implies that the charnockites spent only a brief period of their history at 8 kbar (i.e., in the lower crust). Therefore, although the inferred fluid-driven processes are instructive, the charnockites cannot be considered to be analogous to the contemporary lower crust beneath the Dharwar craton (Newton, 1990).

3.2. Pikwitonei granulite domain, Manitoba, Canada

Like the Kapuskasing uplift, the Pikwitonei region of the northwestern Superior Province (Fig. 8-12) has been interpreted as the base of an oblique cross section through a greenstone belt–gneiss terrane (Weber, 1980; Fountain and Salisbury, 1981). Narrow, east-striking volcano-sedimentary belts of the Cross Lake area decrease in width and continuity as they increase in metamorphic grade westward. In greenschist-facies rocks, pressures are about 3 kbar (Mezger et al., 1989a) whereas pressure increases westward from 7 to 10 kbar (Arima and Barnett, 1984; Paktunc and Baer, 1986; Mezger et al., 1989a, 1990; Mezger, this volume) across the orthopyroxene isograd, which defines the eastern Pikwitonei boundary (Hubregtse, 1980). Farther west, granulites are retrogressed to amphibolite facies in a 25-km-wide zone adjacent to the Proterozoic Thompson belt (Weber and Scoates, 1978; Bleeker, 1990).

The Pikwitonei domain consists dominantly of felsic rock types, including large areas of tonalite and granodiorite, with minor paragneiss, calc-silicate, anorthosite and mafic rocks with locally preserved pillows (Weber, 1980). Orthopyroxene, cordierite and sillimanite are

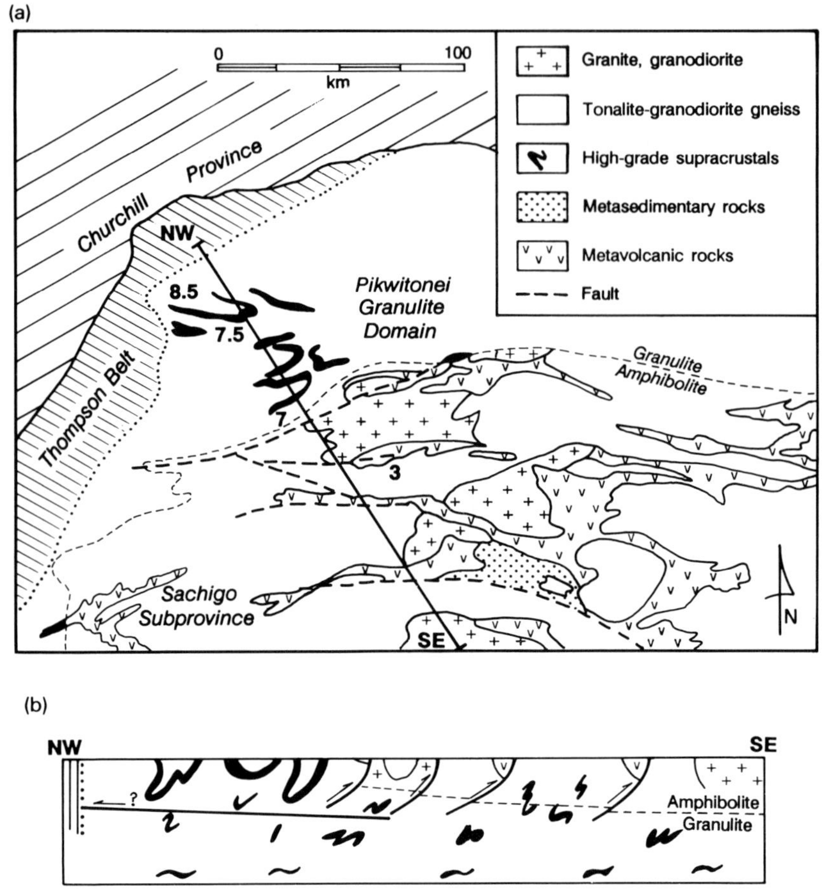

Fig. 8-12. (a) Geology of the Pikwitonei granulite domain and surrounding terranes (after Hubregtse, 1980). The oblique section includes greenstone belts in the amphibolite facies, tonalitic gneisses and the dominantly felsic Pikwitonei granulites. Numbers represent metamorphic pressure (kbar) from Mezger et al. (1989a,b). Dotted line marks the extent of Proterozoic reworking to amphibolite facies. (b) NW-SE cross section (modified after Weber, 1980), showing the wide, oblique transition from amphibolite to granulite facies.

widespread; sapphirine is rare. Foliation and gneissosity dip steeply; two structural phases have been dated by analyzing zircon from leucosome at 2690 and 2637 Ma (Heaman et al., 1986). Mezger et al. (1989a, 1990) quantified stages of the counterclockwise *P-T*-time path by U-Pb dating of garnet and concluded that several events are probable, at 2744–2734, 2700–2687, 2660–2637 and 2605–2591 Ma. The metamorphic path was interpreted to result from mineral growth during magmatic underplating (Bohlen and Mezger, 1989).

Most of the crustal section in the 3–10-kbar interval exposed in the Cross Lake–Pikwitonei region is felsic and has a magmatic origin. Intrusions in the deepest part of the section have similar age (Heaman et al., 1986; Mezger et al., 1989a), density and heat production values (Fountain et al., 1987) to those at higher levels. Metamorphism at deep levels was long-lived; the youngest activity post-dates that recorded in the high-level greenstone terranes (Mezger et al., 1989a;b). This pattern resembles that observed in the Kapuskasing uplift. The uplift mechanism for the Pikwitonei domain has been widely discussed; simple compressional models (Weber, 1980; Fountain and Salisbury, 1981), extensional (Weber and Mezger, 1990), complex transcurrent (Green et al., 1985) and transpressional (Bleeker, 1990) mechanisms have been suggested.

3.3. Western gneiss terrane, Yilgarn block, Australia

The metamorphic character and crustal structure of the Archean Yilgarn block of south-western Australia led to an interpretation that the block is an oblique crustal section (Glikson and Lambert, 1976). In the eastern part of the block, north-striking metavolcanic rocks of the eastern Goldfields Province have preserved supracrustal features and greenschist-facies metamorphic assemblages (Gee et al., 1981). Smaller, sinuous belts of the Southern Cross Province to the west form septa within granitoid masses and have attained greenschist–amphibolite-facies conditions. Farther west in the Western Gneiss Terrane (WGT), supracrustal rocks of dominantly sedimentary origin reach amphibolite and granulite conditions, corresponding to 5–6.5 kbar, 750–840°C (Gee et al., 1981). In contrast to the relatively simple upright structures in the greenstone belts, early subhorizontal foliations are overprinted by steep structures in the gneisses.

The low- to high-grade transition corresponds to marked changes in gross crustal structure determined by seismic refraction studies. In the east, the 32-km-thick crust consists of an upper low-velocity (6.13 km s^{-1}) layer and a lower intermediate-velocity (6.72 km s^{-1}) horizon (Mathur et al., 1977). This package thins to $\sim$14 km to the west and is underlain by a 32-km-thick layer of high-velocity (7.49 km s^{-1}) material. The thickness variations have alternatively been attributed to features inherited from the Archean (Gee et al., 1981) or a result of mafic magmas underplated during the break-up of Gondwana (Glikson and Lambert, 1976).

Several models have been proposed to account for the evolution of the Yilgarn Block. Based on field relations and early geochronological work, Gee et al. (1981) concluded that the WGT represents sialic basement to the greenstone association. This conflicts with the interpretation of Glikson and Lambert (1976) in which the greenstone belts formed at the same time as the deep crustal levels of the WGT. Neither model is supported by Sm-Nd studies that yield 3200-Ma model ages from the WGT and 2700-Ma ages with no indication of crustal pre-history from the east (McCulloch et al., 1983). It is, therefore, probable that new crust was added from west to east in a lateral accretionary environment.

3.4. Summary — wide, oblique transitions

Wide transitions show many similar features to compressional uplifts, chiefly a 3–8-kbar metamorphic transition. However, in considering whether high-grade terranes are representative of the contemporary lower crust, it is necessary also to consider the crustal evolution. In the Dharwar craton, the oblique section crosses a 2500-Ma metamorphic transition that is apparently unrelated to crustal formation. The deepest rocks spent only a short period of their history under 8 kbar conditions and cannot be extrapolated beneath lower-grade terranes. Similarly, the Western Gneiss Terrane can similarly not be viewed as the basement or lower crust beneath lower-grade greenstone terranes on the basis of Sm-Nd results. Of the examples described, only the Pikwitonei has characteristics that allow extrapolation beneath the greenstone terrane. It differs, however, from the compressional uplifts in its steep structural attitudes and lack of increasingly mafic composition with depth.

4. Impactogenic uplifts

Of all the mechanisms that expose the lower crust, uplift resulting from meteorite impact is perhaps the simplest in concept, yet neither of the examples described below is without controversy. In theory, hypervelocity impact excavates a crater, causing rebound of deep crustal material along major fault networks, leading to symmetrical cross sections. Frictional heating in the mantle may generate basic magma which intrudes the crater. In nature, the evidence is rarely unequivocal.

4.1. Vredefort dome, South Africa

Although the origin of the Vredefort structure remains enigmatic (e.g., Dietz, 1961; Schreyer and Abraham, 1978; Simpson, 1981; Hart et al., 1981a), there is general agreement that a 14-km section is exposed through the Archean crust (Slawson, 1976), possibly by the processes that also produced shock metamorphic features (Dietz, 1961; Lilly, 1981). In the Vredefort dome, regionally flat-lying strata of early Proterozoic age are turned into a steep collar surrounding a core of Archean rocks that increases in metamorphic grade from amphibolite at the vertical unconformity, to granulite toward the centre [see Nicolaysen (1990) and Hart et al. (1990a) for reviews]. Dominated by felsic compositions, the interpreted section (Slawson, 1976; Hart et al., 1981b, 1990a) consists of an outer granite gneiss unit (trondhjemite, granodiorite, biotite-rich gneiss, granite gneiss) in the amphibolite facies, underlain by a leucocratic gneiss containing numerous mafic, ultramafic and metasedimentary inclusions, all metamorphosed to upper amphibolite and granulite facies. The boundary between the outer granite gneiss unit and the higher-grade rocks is characterized by highly brecciated rocks and pseudotachylite (Hart et al., 1990a). Borehole and geophysical data suggest that mafic and ultramafic rocks are common in the covered lower portion of the section (Hart, 1990b). Hart et al. (1990a,b) infer that ultramafic rocks are dominant at the unexposed bottom of the section. Along with the grade change, progressive geochemical depletion in Rb/Sr ratio (Slawson, 1976), U, Th and K content (Hart et al., 1981a) and heat production (Schneider et al., 1987) through the section support a model of increasing paleodepth toward the centre of the structure. Age data from various whole-rock isotopic systems reveals a complex evolution with granulite precursors as old as 3800 Ma, reset by high-grade metamorphism at about 3000 and 2800 Ma at different structural levels (Hart et al., 1981b).

Uplift of the dome probably occurred about 2000 Ma ago (Nicolaysen et al., 1963). Shock features such as shatter cones suggest an impactogenic origin; however, detailed studies (e.g., Schreyer and Abraham, 1978; Simpson, 1981) have revealed inconsistencies in the relative ages of structural and metamorphic events that preclude this simple interpretation.

4.2. Levack gneiss complex, Superior Province, Ontario

At the southern edge of the Superior Province, the Levack gneiss complex (Fig. 8-13; Card and Innes, 1981; Card et al., 1984) forms the northern collar of the Sudbury structure of 1849 Ma age (Krogh et al., 1982). North of Sudbury, outliers of Huronian (2480–2200 Ma) strata are tilted into variable orientations, indicating post-Huronian deformation, possibly related to the Sudbury event (Schneider et al., 1987). The Levack complex may thus represent the basal part of a crustal cross section, whose origin could be similar to that envisaged for

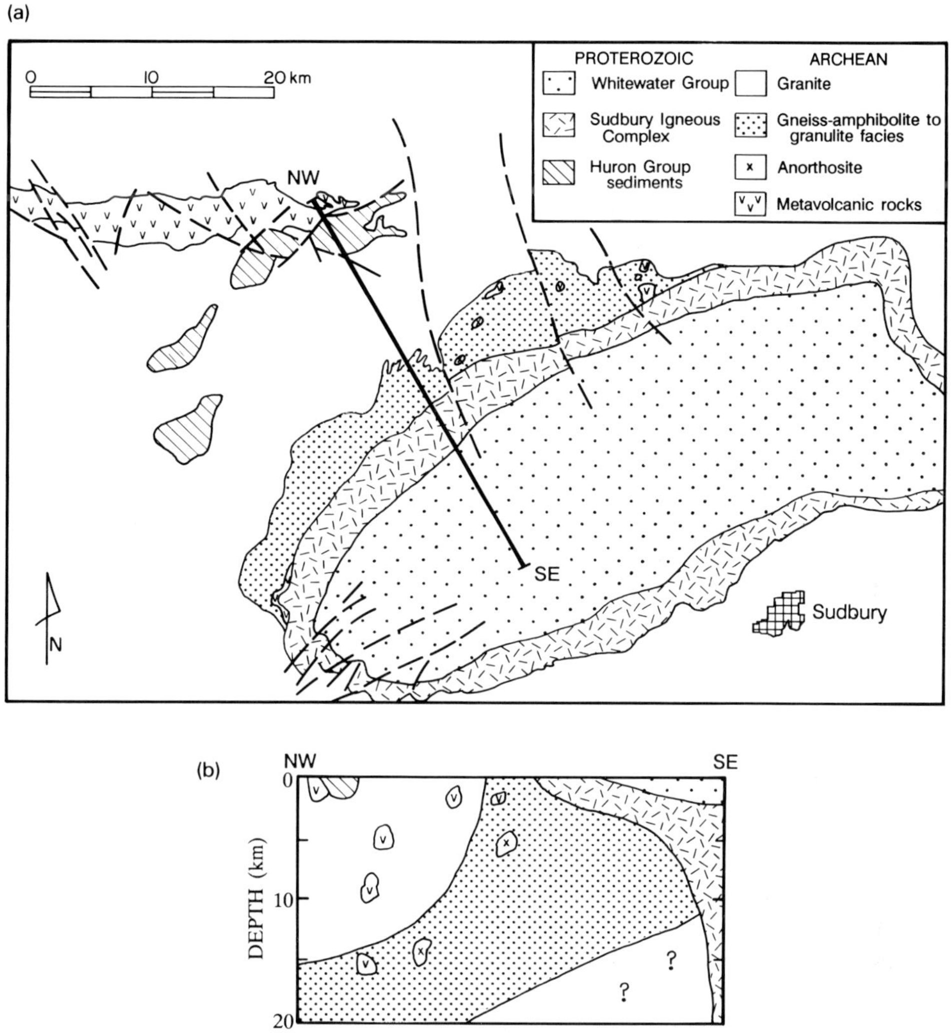

Fig. 8-13. (a) Generalized geological map of the Sudbury area showing position of the Levack gneiss complex at the northern collar of the Sudbury structure (after Card et al., 1984). The oblique section through greenstones and granites into granulites may expose up to 10 km of section. (b) Cross section NW-SE.

the Vredefort structure. Metamorphic grade increases from greenschist facies in the Benny greenstone belt of the Abitibi subprovince, through a 10-km-wide region, occupied mainly by orthogneiss and massive granite, to granulite facies in the 5-km-wide Levack gneiss complex, where retrogression to greenschist facies is widespread and possibly related to faults. The complex consists mainly of hornblende-bearing tonalitic orthogneiss with locally-preserved two-pyroxene assemblages, along with minor paragneiss, mafic gneiss and small anorthositic bodies (Card and Innes, 1981), yielding low heat productivity on average (Schneider et al., 1987). Tonalitic gneiss from the complex has a U-Pb zircon age of 2707 Ma whereas leucosome layers, presumably dating from the time of metamorphic crystallization, are 2650 Ma (Krogh et al., 1984). The lithological and chronological sequences are similar to those of the Kapuskasing uplift, 150 km to the northwest, however gneissic layering and foliations are vertical in the Levack structure. No barometric estimates are available for the depth of origin of the Levack complex. However, the lateral distance between the closest Huronian outlier and the granulites is on the order of 10 km; assuming vertical dips on Archean isobars, post-Huronian uplift could be of similar magnitude.

4.3. Summary — impactogenic uplifts

Based on a population of two impactogenic uplifts in Archean cratons, both controversial, common depth-related features appear to be: (1) decreasing heat production; (2) increasing metamorphic grade, and below the amphibolite-granulite transition: (3) gneissic intrusive rocks of dominantly felsic composition, containing (4) variable proportions of supracrustal relicts. Shock metamorphic features, including faults and alteration zones, obscure deep-crustal attributes.

5. Transpressional uplifts

The fourth and perhaps most complex mechanism responsible for exposing lower crust is transpressional uplift. By this mechanism, uplift results from a vertical component of motion along large transcurrent faults. Exposures may be especially deep if uplift occurs in two stages, as when a transcurrent fault cuts a terrane previously exposed by a different mechanism.

5.1. Fiordland, South Island, New Zealand

The Fiordland complex of southwestern New Zealand has a two-stage uplift history involving early extensional and later transpressional movements. In the Late Cretaceous (?), granulite facies orthogneisses of Early Cretaceous age were elevated in an extensional metamorphic core complex; the complex is currently undergoing tectonic elevation at about 1 cm yr^{-1} by transpressional motion along the Alpine fault and being rapidly eroded by one of the highest levels of rainfall in the world (10 m yr^{-1}).

The Fiordland complex (Fig. 8-14) is characterized by the largest positive gravity anomaly in the world (+200 mGal; Reilly, 1965) suggesting that the Moho is close to the surface (8–25 km depending on assumptions; Woodward, 1972; Oliver and Coggon, 1979). Refraction studies over the complex confirm that high-velocity (7.0–7.3 km s^{-1}) lower crustal rocks extend from the surface to 15-km depth but failed to detect the Moho (Priestly and Davey, 1983).

As can be seen in Figure 8-14, the complex is bounded to the west by the Alpine fault, along which dextral motion has displaced volcanic and sedimentary rocks of the Nelson and Westland terranes 400 km to the northeast while reverse motion thrust the Pacific plate, including Fiordland, over the Indian plate. The upper levels of the complex are cut by a 100–300-m-thick, flat to east-dipping ductile shear zone (Gibson et al., 1988) and are overlain by amphibolite-grade ortho- and paragneisses grading upward and eastward through granitic plutons and unmetamorphosed volcanics cut by a west-dipping normal fault on the east. The calc-alkaline granulites in the core of the complex, metamorphosed at pressures of 9–12.4 kbar (Newton and Perkins, 1982) and temperatures of 650–700°C have since suffered retrograde, amphibolite-facies metamorphism near the ductile shear zone. The granulites are interpreted to represent the lower crust (Oliver and Coggon, 1979), whereas the transposed and extended amphibolite-grade rocks above the shear zone, metamorphosed at 5–9 kbar, are derived from mid-crustal levels.

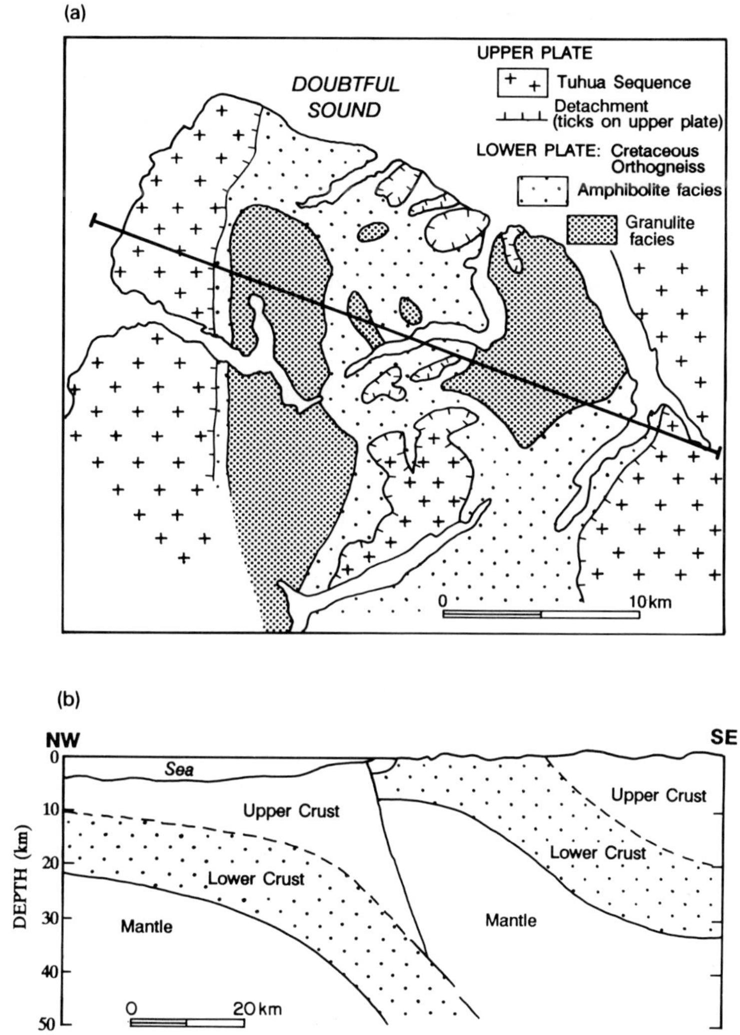

Fig. 8-14. (a) Geology of the Fiordland area (modified after Oliver, 1980 and Gibson et al., 1988). Cretaceous orthogneisses in the granulite facies are exposed in the lower plate of a metamorphic core complex, according to Gibson et al. (1988). (b) Cross section NW-SE (after Oliver and Coggon, 1979), showing transpressional elevation of the Fiordland sequence adjacent to the Alpine fault.

The high-grade unit within the core of the complex consists of two-pyroxene granulites with relict primary igneous to gneissic textures overprinted by rare eclogitic (Blattner, 1976) and garnet reaction zones (Bradshaw, 1989a) near plagioclase-rich veins and dikes. These are found in thin concordant sills and stocks invading and enclosing metasedimentary and metaplutonic rafts, all cut by gabbronorite, anorthosite and ultramafic dikes and sills (Fig. 8-14). The overprinting relationships, along with U-Pb, K-Ar and Rb-Sr dates (Mattinson et al., 1986) and geobarometric studies suggest that the granulite protoliths crystallized 120–130 Ma ago in a magmatic arc environment at about 20-km depth and were then loaded to as much as 40 km in a collisional environment, where garnet crystallized during peak metamorphism, possibly in response to fluid influx from the mantle (Bradshaw, 1989a,b). According to Gibson et al. (1988), these events were followed by rapid extensional unroofing of the granulites to depths equivalent to the 400°C isotherm about 90 Ma ago, marking initial rifting of the Tasman Sea, a view disputed by Bradshaw and Kimbrough (1989).

Although the rocks in the core of the complex almost certainly represent lower continental crust, the age and crustal stratigraphy remain problematic. In particular, the section above the detachment is allochthonous with respect to the Fiordland rocks, although Pb and Sr isotopic data from gabbronorites above and below the detachment indicate similar ages (Mattinson et al., 1986). A thick section of the middle crust (up to 9 km) appears to be missing and other detachments may be present at depth. The exposures at Doubtful Sound are nevertheless instructive. They clearly expose the lower crust of a young magmatic arc consisting dominantly of granulite-facies orthogneiss of mafic to andesitic composition.

5.2. Tehachapi complex, Sierra Nevada, California

The Sierra Nevada batholith of California, a 500 by 100 km continental-margin igneous complex of Late Triassic to Late Cretaceous age, is exposed at a variety of structural levels ranging from caldera level to granulite facies (Sams and Saleeby, 1988), with the deepest levels in the 300 km^2 Tehachapi Complex in the south (Fig. 8-15), adjacent to the San Andreas and Garlock faults (Ross, 1985). Most of the batholith consists of calc-alkaline plutons of diorite to granite, crystallized at pressures of $\sim$1 to 6 kbar (Ague and Brimhall, 1988a) from variably crustally contaminated, slab-derived magmas (Ague and Brimhall, 1988b).

Considering only plutons and volcanic rocks of 100 $\pm$ 5 Ma age, Sams and Saleeby (1988) constructed a generalized cross section from near surface to about 25 km depth. Uppermost in the section are high-level intrusions and extrusive equivalents. The main exposed mass of the batholith, crystallized at 1–5 kbar, consists of I-type granitoid rocks with evidence of variable contamination by metasedimentary or granitoid crust. At the deepest structural levels, the mafic Tehachapi Complex consists of mafic and felsic gneiss, amphibolite, diorite, tonalite, granulite and rare metasediments. Metamorphic orthopyroxene and/or garnet is present in mafic igneous and sedimentary compositions; hypersthene of probable igneous origin occurs in tonalite (Ross, 1985). Depth estimates from geobarometry (Sharry, 1982; Ague and Brimhall, 1988a) range from 25 to 30 km and suggest that the complex represents a sample of the root zone of the batholith.

Uplift of the southern Sierra Nevada batholith is bracketed between $\sim$80 and 40 Ma (Sams and Saleeby, 1988), suggesting high rates of uplift and erosion soon after formation. The cause of the uplift may be southward-directed thrusting or transpressional motion along the San Andreas–Garlock fault system.

With the exception of minor metasedimentary inclusions, most of the observed crustal profile in the Sierra Nevada has a magmatic origin, increasing in mafic content, crystallization temperature and anhydrous compositions with depth. An underlying cratonal base is implied by the evidence for magma contamination. Thus the substrate on which the arc was built may consist of an assemblage of continental North America and marginal sediments (Domenick et al., 1983), highly modified by the passage of magmas (Ague and Brimhall, 1988b).

Saleeby (1990; Fig. 8-15b) suggested a mechanism for convecting high-level supracrustal rocks to granulite depths in an active plutonic environment. Local convection cells have upward flow in mantle-derived plutons, with descending margins recycling volcanic and sedimentary material.

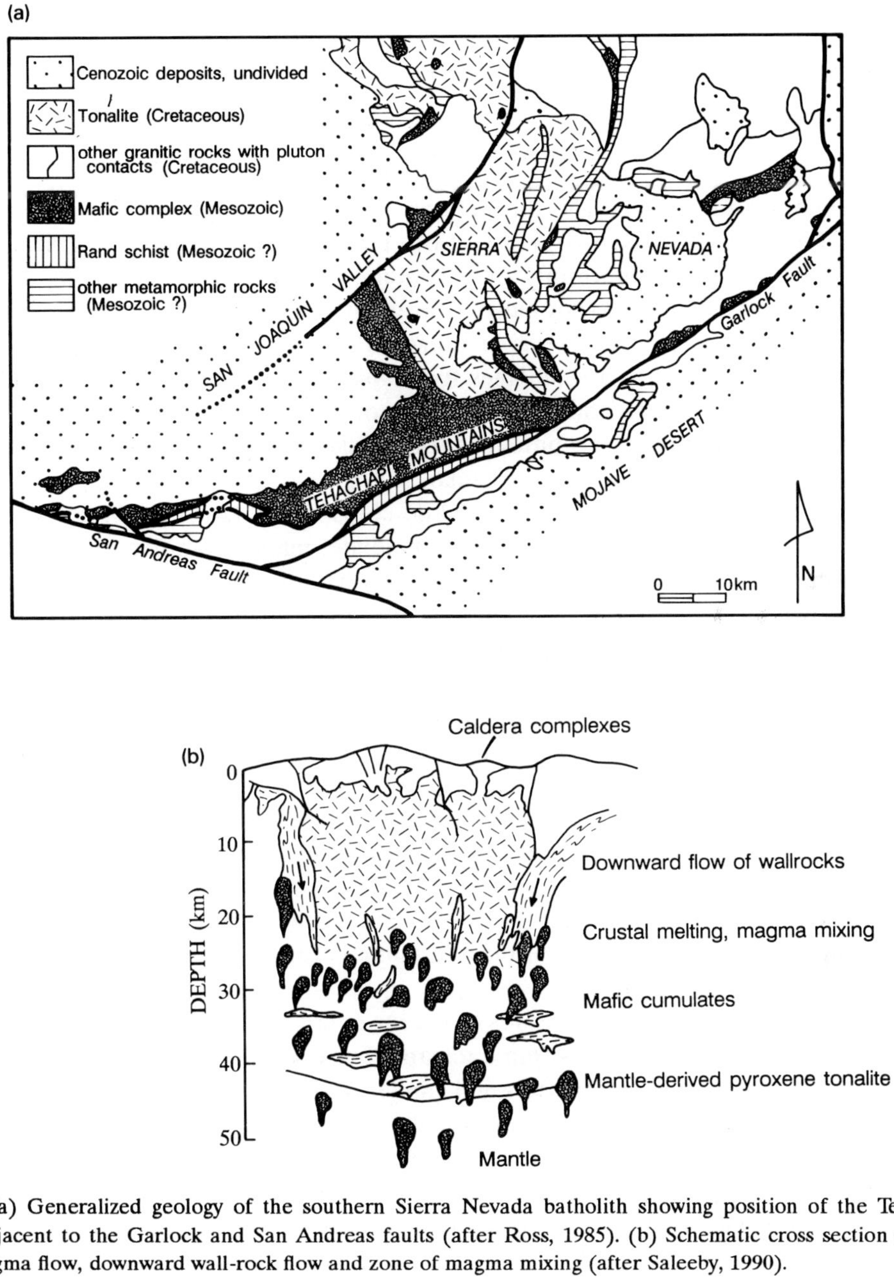

Fig. 8-15. (a) Generalized geology of the southern Sierra Nevada batholith showing position of the Tehachapi complex adjacent to the Garlock and San Andreas faults (after Ross, 1985). (b) Schematic cross section showing upward magma flow, downward wall-rock flow and zone of magma mixing (after Saleeby, 1990).

5.3. Summary — transpressional uplifts

Magmatic rocks are the dominant components of the transpressional uplifts described. The geological history of the Sierra Nevada is relatively simple, involving Cretaceous intrusion and subsequent transpressional uplift to expose the roots in the Tehachapi Complex adjacent to the San Andreas–Garlock fault system. In contrast, the plutonic rocks of Fiordland were intruded in the Cretaceous, metamorphosed to high pressure, partly exhumed, further uplifted in an extensional core complex and finally exposed by transpression along the Alpine fault. Both complexes were elevated along currently active faults; their exposure may be a transitory phenomenon.

6. Characteristics of lower crust in cross sections

Several features are common to all sections, regardless of age, tectonic setting or uplift history. By definition, the basal parts of sections record high metamorphic grade. Somewhat surprising, however is the uniformity of maximum pressure values, implying similar uplift magnitudes. Where reliable geobarometry is available, metamorphic pressure at the base of the section is consistently in the 7–9 kbar range although values as high as 12–14 kbar have been inferred on the basis of assemblages for Kohistan and Doubtful Sound. Metamorphic temperatures derived from geothermometers are commonly in the range 700–850°C and may be minimum values owing to the potential for diffusional re-equilibration in slowly cooled terranes (e.g., Perkins, 1988; Frost and Chacko, 1989).

All terranes consist dominantly of magmatic components that vary in composition from generally felsic at high structural level to intermediate, mafic or anorthositic at depth. Most sections contain only minor supracrustal relics; thick paragneiss units in the lower parts of the Ivrea and Calabria sections are exceptional. Widespread migmatitic textures in supracrustal rocks indicate partial melting at the prevailing metamorphic temperatures.

Cross-cutting intrusive phases are commonly recognized at middle and high crustal levels; where dated, the age range is on the order of tens of millions of years. This history, in conjunction with evidence for supracrustal relics, suggests residence in a magmatic arc environment, superimposed on a pre-existing basement or supracrustal sequence. Tectonic settings suggested for the early history of various sections include accretionary prisms, island arcs and rifts. In light of the present 30–40 km crustal thickness of most of the terranes examined, in contrast to <20 km values for mature island arcs, a continental arc (Andean-type) environment is probable during the late magmatic phase of crustal development. If cross sections represent a random sample of the continental crust, then much of the continents may have developed by similar processes.

Lithological and metamorphic features of the lower crust persist over large distances where more than one section is exposed within a terrane. For example, Ivrea and Calabria, 1200 km apart, both have deep level paragneisses and similar timing of structural-metamorphic events. Three deep crustal exposures in the Superior Province, located 1400 km apart across strike, show similar thick tonalitic upper crust, formed at similar times, and later, high-grade metamorphism at depth. The large across-strike areal extent of coeval magmatism in the Superior Province has no apparent analogue in recent settings. Magmatic arcs such as the Sierra Nevada batholith were active over <50 km widths at any time. Widening of the arc occurs by subsequent magmatism in different locations. It is difficult to explain the widespread, synchronous magmatism in the Superior Province as the product of a single arc. However, simultaneously active arcs, subsequently amalgamated, is a feasible process.

Integrated heat production values for the deep parts of the Pikwitonei, Kapuskasing and Ivrea sections commonly occupy the range 0.3–0.5 $\mu W\ m^{-3}$, similar to that in many high-grade terranes. Such values are higher than those predicted for lower crust in models of crustal heat production based on surface heat flow measurements. The lowermost crust, not exposed in most cross sections, may be inferred to be more depleted in heat-producing elements than the parts of the lower crust sampled by cross sections.

Several characteristics distinguish compressional uplifts from wide, oblique transitions. Although metamorphic conditions overlap for the two types, the wide transitions lack the large, layered igneous bodies and depth-correlated, increasingly mafic bulk composition common in compressional uplifts. Compositional layering on the km scale that characterizes the deep parts of compressional uplifts occurs only sporadically in the high-grade parts of wide oblique transitions. In many compressional uplifts foliation and layering is inferred on the basis of igneous layering to have been subhorizontal prior to uplift into orientations dipping 30–90°. In contrast, vertical foliation and lithological units, striking at high angles to the metamorphic gradient, can be traced across the wide oblique transitions. These features suggest fundamental differences in crustal structure, composition and uplift history between compressional uplifts and wide oblique transitions.

7. Implications for the nature of the lower crust

Fragments of the lower continental crust attain the surface by a variety of mechanisms including compression, transpression and rebound following meteorite impact. These events generally followed an earlier period of uplift due to compression, extension or simple regional erosion. In the following section, some of the results outlined above are used to address major issues of the lower crust.

A great deal of controversy has been engendered by the observation of *deep subhorizontal structures* on seismic reflection profiles, attributed to some combination of lower crustal fluids, mafic sills, shear zones or lithological layering [see Barazangi and Brown (1986a,b) and Matthews and Smith (1987) for reviews]. Although deep fluids, inferred on the basis of electrical conductivity anomalies, may contribute to the reflection character of the deep crust, they are not a unique cause; prominent shallow reflections arise from non-conductive high-grade terranes which have lost their ambient fluids (Green et al., 1990). In the Kapuskasing example, shallow, lamellar reflections can be projected a short distance to interlayered mafic and felsic rock types with velocity-anisotropy characteristics which also generate synthetic reflections (Fountain et al., 1990). Cumulate-layered mafic complexes like the Chilas could give rise to lamellar reflective zones provided that the scale of layering and compositional contrasts are appropriate.

The subhorizontal orientation which characterizes the reflections from the deep crust and restored cross sections has diverse origins. For example, gently-dipping shear zones in the Ivrea zone are attributed to subhorizontal extension (Brodie and Rutter, 1987). Similar low-angle detachments or distributed strain zones characterize deep parts of the Doubtful Sound section. Subhorizontal rolling foliation in parts of the Kapuskasing uplift is ascribed to synmagmatic extension (Moser, 1989), possibly controlled by structural level. Horizontal cumulate layering of the Chilas complex probably originated in an oceanic setting and was subsequently deformed during accretion. It may be significant that lower crust in all of the sections studied contains evidence of polyphase deformation, attributed to major tectonic events that shaped the crust and to long residence under ductile conditions.

Extreme ductility in the lower crust during much of its history has hindered resolution of events and correlation with structures recognized at higher crustal levels, although significant advances have been made over the last decade. The dominant structural fabrics are generally subhorizontal layering and gneissosity that probably result from some combination of processes including 1) sill-like igneous intrusion, 2) compressional low-angle shear, and 3) extensional collapse (e.g., Sandiford, 1989) or spreading.

The *nature of the Moho* is a long-standing question which exposed cross sections may illuminate. For example, ultramafic rocks in the Ivrea zone have been proposed as upper mantle material, carried up and emplaced with the Ivrea section. The mafic-ultramafic contact, which may represent an exposed example of the Moho, is at least locally a high-strain zone (Brodie, 1980). Some section may have been eliminated along this structure if it has a similar origin to the extensional structures at higher structural levels within the Ivrea zone. Ultramafic sills in gabbroic rocks at Doubtful Sound may constitute part of a transitional crust–mantle interface as inferred for eastern Australia by Griffin and O'Reilly (1987). The garnet granulite-grade rocks of the Jijal–Patan complex of Kohistan may represent an exposed example of sub-arc mantle. The cumulates may result from fractionation of mafic magmas ponded at the base of the crust; mafic and ultramafic cumulates would become part of the geophysically-defined mantle.

Mid-crustal discontinuities in lithology and/or physical properties are common. In both the Ivrea zone and Kapuskasing uplift these boundaries have been interpreted as exposed examples of Conrad-type discontinuities. They correspond to increases in density and velocity at diffuse contact zones between felsic rocks and underlying layered lithological sequences that also generally overlap the amphibolite-granulite transition. If this feature is common and accounts for Conrad-type discontinuities elsewhere, it would imply a crustal density filter mechanism (e.g., Glazner and Ussler, 1988) that ponds magmas at different structural levels according to their composition, density, temperature and viscosity. In the Ivrea zone the stratification results partly from ponding of mafic magmas at deep levels whereas in the Kapuskasing section the boundary relates to the base of a tonalitic layer ponded in the mid crust.

Deep crustal expressions of extensional events have been recognized in several sections. In the southern Alps, depositional basins can be correlated temporally and kinematically with deep-crustal extensional shear zones exposed in the Ivrea zone (Handy, 1986, 1987; S. Schmid et al., 1987). Similarly, Cretaceous basins in New Zealand are temporally related to uplift and unroofing of the postulated core complex at Doubtful Sound (Gibson et al., 1988).

Following the inference that much of the continental crust developed in an active, marginal environment, the potential for *an early stacking history* of Cordilleran-style terranes prior to magmatic arc development must be considered. However, evidence for different histories in upper and lower crust may be difficult to document. To date, geochronological evidence is lacking for older supracrustal rocks in the upper crust structurally overlying younger supracrustal rocks in the lower crust. Examples of younger zircon dates in deeper levels than in higher-level granite-greenstone belts occur in the Kapuskasing and Pikwitonei sections, however both of these record metamorphic events subsequent to possible terrane amalgamation.

The *heat source for metamorphism* and granite production remains a relevant problem, considering hypotheses that postulate mafic, underplated magmas (Furlong and Fountain, 1986; Bohlen, 1987; Bohlen and Mezger, 1989; Fountain, 1989). Layered, mafic-ultramafic complexes, common in compressional uplifts and in Doubtful Sound, are obvious candidates. The gabbros in the Ivrea zone, intruded ~290 Ma ago in the late stages of the high-grade metamorphism (310–290 Ma), are possible contributing heat sources, given the geochronological uncertainty. Although underplating has been inferred at Calabria to account for metamorphic heating during anatectic melt extraction (Schenk, 1984), only minor mafic bodies are present. In the Kapuskasing zone, the Shawmere anorthosite complex is pre-metamorphic. The mafic-ultramafic complexes in the Musgrave Range are either syn- or post-metamorphic. No general trend is apparent among the sections. Intrusions of more felsic composition have been implicated in high-grade metamorphic processes in the Pikwitonei and Kapuskasing sections and may be responsible for some metamorphic heating in the Kohistan arc. At Kapuskasing, the heat source may have been partly in tonalites intruded at intermediate crustal levels.

The composition and origin of *deep-crustal metamorphic fluids* has received considerable attention, particularly in the Adirondack, Dharwar and Doubtful Sound granulite terranes. A common observation, first noted by Touret (1971), of high-density CO_2-filled fluid inclusions in granulites, leads to several possible interpretations: 1) the inclusions represent the peak metamorphic fluid which is carbonic owing to extraction of water by granitic melts or influx of CO_2 (e.g., Touret, 1971; Frost and Frost, 1987); or 2) the inclusions were trapped during post-metamorphic uplift (e.g., Lamb et al., 1988). At the amphibolite-granulite transition in the Dharwar craton, textural, mineralogical and geochemical evidence point to infiltration of CO_2 along structurally-controlled pathways. Conversely, high-grade metamorphism in the Adirondacks was fluid-absent and thus carbonic fluids in inclusions are late (Lamb et al., 1988). In the Kapuskasing zone volatiles are partitioned into water-bearing granitoid melts (Percival and Fountain, 1989; Percival, 1989b) and CO_2-bearing fluids (Rudnick et al., 1984). Regardless of their origin, the carbonic fluids trapped in inclusions testify to the presence of a CO_2-rich phase at some stage of the metamorphic history of deep crustal rocks. If this fluid precipitates grain-boundary films of graphite, as postulated by Frost et al. (1989), the phenomenon of enhanced electrical conductivity in the deep crust can be readily understood.

Models for the origin of *vertical geochemical zonation* within the continental crust take three main forms. At one end of the spectrum, Taylor and McLennan (1981) proposed, on the basis of geochemical evidence, that a bulk andesitic crust differentiated by mobilization and intrusion into an upper crustal granitic component, leaving a residual mafic lower crust. Alternatively, the broad differences could result from intrusion to density-determined levels of mantle or slab-derived magmas of different composition. Also possible are processes whereby mafic magmas of mantle origin are incorporated into the deep crust, at the same time providing heat to drive granitic melts off pre-existing units. Although pelitic rocks in the Ivrea and Calabria sections show geochemical signs of depletion, geochemical data are insufficient from other terranes to generalize. However, based on the presence of widespread migmatitic textures at the deepest exposed levels of most terranes and the probability of higher degrees of fusion at deeper, hotter levels, geochemical depletion is probably a significant process (see Rudnick, this volume).

Lower crustal composition is probably incompletely represented by the basal parts of cross sections. Several authors have emphasized differences between the generally intermediate compositions of exposed high-grade terranes, including basal cross sections, and the mafic composition of xenolith suites derived from the lowermost crust (e.g., Rudnick et al., 1988; Bohlen and Mezger, 1989; Truscott and Shaw, 1990), stressing the likelihood of magmatic underplating. An inference drawn from this observation is that high-grade exposures are samples of the middle crust; true lowermost crust can only be observed in xenoliths, possibly being too dense for tectonic exhumation. However, several occurrences of detached fragments of high-pressure granulite, with no link to the deep crust, are known [e.g., Bergen arc (Austrheim, 1987); Furua complex (Coolen, 1982)]. Xenoliths sampling the lowermost crust of eroded high-grade terranes are rare (e.g., Collerson, 1989), and therefore it is difficult to construct complete crustal columns. Furthermore, in the Ivrea zone, where a relatively complete sample is available through thin crust, possibly including the Moho, the proportion of mafic compositions increases downwards but is variable along strike. Evaluating the bulk composition of the crust beneath exposed levels remains a challenge for multidisciplinary earth science study.

A global inventory of exposed cross sections contains only a handful of examples where the upper and lower crust can be linked structurally, chronologically and geophysically. The few hundred square kilometers of lower crust thus exposed represents an exceedingly small, and to variable extent, altered, sample of the lower half of the continental crust. Unique opportunities are nevertheless provided by these windows to examine the variation of igneous, metamorphic and structural processes and their geophysical expression with depth. Relationships established in such sections can be applied to the interpretation of regions where only deep-crustal geophysics and an understanding of the tectonic history from surface geology are available.

Acknowledgments

Klaus Mezger's comments on an early draft prompted clarification of several important points. DMF wishes to acknowledge support from the National Science Foundation (grants EAR-8418350, EAR-8720798, EAR-9003956 and EAR-9017894) for research on deep crustal terranes. Geological Survey of Canada Publication number 13490.

References

Abbott, D.H. and Hoffman, S.E., 1984. Archean plate tectonics revisited 1: Heat flow, spreading rate, and the age of subducting oceanic lithosphere and their effects on the origin and evolution of continents. Tectonics, 3: 420–448.

Ague, J.J. and Brimhall, G.H., 1988a. Magmatic arc asymmetry and distribution of anomalous plutonic belts in the batholiths of California: Effects of assimilation, crustal thickness, and depth of crystallization. Geol. Soc. Am. Bull., 100: 912–927.

Ague, J.J. and Brimhall, G.H., 1988b. Regional variations in bulk chemistry, mineralogy, and the compositions of mafic and accessory minerals in the batholiths of California. Geol. Soc. Am. Bull., 100: 891–911.

Albert, G., 1974. Die magnetische Anomalie der Ivreazone. J. Geophys., 40: 283–301.

Ansorge, J., 1968. Die Struktur der Erdkruste an der Westflanke der Zone von Ivrea. Schweiz. Mineral. Petrogr. Mitt., 48: 247–254.

Archibald, D.A., Farrar, E. and Hanes, J.A., 1986. An $^{40}Ar/^{39}Ar$ study of the eastern boundary of the Kapuskasing structural zone: Ivanhoe Lake cataclastic zone and contiguous Abitibi greenstone belt. Geol. Assoc. Can., Prog. Abstr., 11: 41–42.

Arima, M. and Barnett, R.L., 1984. Sapphrine bearing granulites from the Sipiwesk Lake area of the late Archean Pikwitonei granulite terrain, Manitoba, Canada. Contrib. Mineral. Petrol., 88: 102–112.

Arriens, P.A. and Lambert, I.B., 1969. On the age and strontium isotopic geochemistry of granulite-facies rocks from the Fraser Range, western Australia, and the Musgrave Ranges, central Australia. Spec. Publ. Geol. Soc. Aust., 2: 377–388.

Ashwal, L.D., 1988. Workshop on the Deep Continental Crust of South India. In: L.D. Ashwal (Editor), Lunar Planet. Inst. Tech. Rept. 88-06, Houston, p. 388.

Ashwal, L.D., Morgan, P., Kelley, S.A. and Percival, J.A., 1987. Heat production in an Archean crustal profile and implications for heat flow and mobilization of heat-producing elements. Earth Planet. Sci. Lett., 85: 439–450.

Austrheim, H., 1987. Eclogitization of lower crustal granulites by fluid migration through shear zones. Earth Planet. Sci. Lett. 81: 221–232.

Bailey, R.C., Craven, J.A., Macnae, J.C. and Polzer, B., 1989. Imaging deep fluids in the Archean crust. Nature, 340: 136–138.

Balakrishnan, S. and Rajamani, V., 1987. Geochemistry and petrogenesis of granitoids around the Kolar schist belt, south India: Constraints for the evolution of the crust in the Kolar area. J. Geol., 95: 219–240.

Barazangi, M. and Brown, L.D. (Editors), 1986a. Reflection Seismology: A Global Perspective. Am. Geophys. Union, Geodyn. Ser. 13, 311 pp.

Barazangi, M. and Brown, L.D. (Editors), 1986b. Reflection Seismology: The Continental Crust. Am. Geophys. Union, Geodyn. Ser. 14, 339 pp.

Bard, J.P., 1983. Metamorphism of an obducted island arc: example of the Kohistan sequence (Pakistan) in the Himalaya collided range. Earth Planet. Sci. Lett., 65: 133–144.

Bayer, R., Cazes, M., Dal Piaz, G.V., Damotte, B., Elter, G., Gosso, G., Hirn, A., Lanza, R., Lombardo, B., Mugnier, J.-L., Nicolas, A., Nicolich, R., Polino, R., Roure, F., Sacchi, R., Scarascia, S., Tabacco, I., Taponnier, P., Tardy, M., Taylor, M., Thouvenot, F., Torreilles, G. and Villien, A., 1987. Premiers résultats de la traversée des Alpes occidentales par sismique reflexion verticale (Programme ECORS-CROP). C. R. Acad. Sci. Paris, 305, Ser. II: 1461–1470.

Beckinsale, R.D., Drury, S.A. and Holt, R.W., 1980. 3,360-Myr old gneisses from the South Indian craton. Nature, 283: 469–470.

Bell, T.H., 1978. Progressive deformation and reorientation of fold axes in a ductile mylonite zone: The Woodroffe thrust. Tectonophysics, 44: 285–320.

Belluso, E., Biino, G. and Lanza, R., 1990. New data on the rock magnetism in the Ivrea–Verbano Zone (Northern Italy) and its relationships to the magnetic anomalies. Tectonophysics, 182: 79–89.

Ben Othman, D., Polvé, M. and Allègre, C.J., 1984. Nd-Sr isotopic composition of granulites and constraints on the evolution of the lower crust. Nature, 307: 510–515.

Bennett, G., Brown, D.D., George, P.T. and Leahy, E.J., 1967. Operation Kapuskasing. Ont. Dept. Mines, Misc. Pap. 10, 98 pp.

Berckhemer, H., 1969. Direct evidence for the composition of the lower crust and the Moho. Tectonophysics, 8: 97–105.

Bertolani, M., 1968. La petrografia della Valle Strona (Alpi occidentali italiane). Schweiz. Mineral. Petrogr. Mitt., 48: 695–732.

Black, L.P. and McCulloch, M.T., 1984. Sm-Nd ages of the Arunta, Tennant Creek, and Georgetown Inliers of northern Australia. Aust. J. Earth Sci., 31: 49–60.

Black, L.P., Shaw, R.D. and Stewart, A.J., 1983. Rb-Sr geochronology of Proterozoic events in the Arunta inlier, Central Australia. B.M.R. J. Aust. Geol. Geophys., 8: 129–137.

Blattner, P., 1976. Replacement of hornblende by garnet in granulite-facies assemblages near Milford Sound, New Zealand. Contrib. Mineral. Petrol., 55: 155–190.

Bleeker, W., 1990. New structural-metamorphic constraints on Early Proterozoic oblique collision along the Thompson nickel belt, northern Manitoba, Canada. In: J.F. Lewry and M.R. Stauffer (Editors), The Early Proterozoic Trans-Hudson Orogen. Geol. Assoc. Can. Spec. Pap., 37: 57–73.

Bohlen, S.R., 1987. Pressure-temperature-time paths and a tectonic model for the evolution of granulites. J. Geol., 95: 617–632.

Bohlen, S.R. and Mezger, K., 1989. Origin of granulite terranes and the formation of the lowermost continental crust. Science, 244: 326–329.

Boland, A.V. and Ellis, R.M., 1989. Velocity structure of the Kapuskasing uplift, northern Ontario, from seismic refraction studies. J. Geophys. Res., 94: 7189–7204.

Boland, A.V., Ellis, R.M., Northey, D.J., West, G.F., Green, A.G., Forsyth, D.A., Mereu, R.F., Meyer, R.P., Morel-à-l'Huissier, P.A., Buchbinder, G.G.R., Asudeh, I. and Haddon, R.A.W., 1988. Seismic delineation of upthrust Archaean crust in Kapuskasing, northern Ontario. Nature, 335: 711–713.

Boriani, A., 1970. The "Pogallo line" and its connection with the metamorphic and the anatectic phases of "Massiccia dei Laghi" between the Ossola Valley and Lake Maggiore (Northern Italy). Boll. Soc. Geol. Ital., 89: 415–433.

Boriani, A., 1971. Blastomylonites and recrystallized mylonites. Rend. Soc. Ital. Mineral. Petrol., 27: 35–47.

Boriani, A. and Sacchi, R., 1973. Geology of the junction between the Ivrea–Verbano and Strona–Ceneri Zones (southern Alps). Mem. Ist. Geol. Mineral. Padova, 28: 1–36.

Boriani, A. and Sacchi, R., 1974. The "Insubric" and other tectonic lines in the Southern Alps (NW Italy). Mem. Soc. Geol. Ital., 13: 1–11.

Boriani, A., Bigioggero, B. and Origoni Giobbi, E., 1977. Metamoprhism, tectonic evolution and tentative stratigraphy of the 'Serie dei Laghi'. Geological map of the Verbania area (northern Italy). Mem. Ist. Geol. Mineral. Padova, 32: 1–25.

Boriani, A., Columbo, A. and Macera, P., 1985. Radiometric geochronology of Central Alps. Rend. Soc. Ital. Mineral. Petrol., 40: 139–186.

Boriani, A., Origoni Giobbi, E. and Del Moro, A., 1982/1983. Composition, level of intrusion and age of the "Serie dei Laghi" orthogneisses (northern Italy–Ticino, Switzerland). Rend. Soc. Ital. Mineral. Petrol., 38: 191–205.

Borsi, S., Hieke-Merlin, O., Lorenzoni, S., Paglionico, A. and Zanettin-Lorenzoni, E., 1976. Stilo unit and "dioritic-kinzigitic" unit in le Serre (Calabria, Italy). Geological, petrological, geochronological characters, Boll. Soc. Geol. Ital., 95: 219–245.

Bradshaw, J.Y., 1989a. Early Cretaceous vein-related garnet granulite in Fiordland, southwest New Zealand: A case for infiltration of mantle-derived CO_2-rich fluids. J. Geol., 97: 697–717.

Bradshaw, J.Y., 1989b. Origin and metamorphic history of an Early Cretaceous polybaric granulite terrain, Fiordland, southwest New Zealand. Contrib. Mineral. Petrol., 103: 346–360.

Bradshaw, J.Y. and Kimbrough, D.L., 1989. Comment on "Age constraints on metamorphism and the development of a metamorphic core complex in Fiordland, southern New Zealand". Geology, 17: 380–381.

Brodie, K.H., 1980. Variations in mineral chemistry across a shear zone in phlogopite peridotite. J. Struct. Geol., 2: 265–272.

Brodie, K.H. and Rutter, E.H., 1985. On the relationship between deformation and metamorphism with special reference to the behaviour of basic rocks. In: A.B. Thompson and D.C. Rubie (Editors), Metamorphic Reactions: Kinetics, Textures and Deformation. Adv. Phys. Geochem., 4, Springer, Berlin, pp. 138–179.

Brodie, K.H. and Rutter, E.H., 1987. Deep crustal extensional faulting in the Ivrea Zone of northern Italy. Tectonophysics, 140: 193–212.

Brodie, K.H., Rex, D. and Rutter, E.H., 1989. On the age of deep-crustal extensional faulting in the Ivrea zone, northern Italy. In: M.P. Coward, D. Dietrich and R.G. Park (Editors), Alpine Tectonics. Geol. Soc. Spec. Publ., 45: 203–210.

Buhl, D., 1987. U-Pb und Rb-Sr Altersbestimmungen und Untersuchungen zum Strontiumisotopenaustausch an Granuliten Sudindiens. Dissertation, University of Munster, FRG.

Burchfiel, B.C. and Royden, L.H., 1985. North–south extension within the convergent Himalayan region. Geology, 13: 679–682.

Burke, K. and Dewey, J.F., 1973. Plume-generated triple junctions: Key indicators in applying plate tectonics to old rocks. J. Geol., 81: 406–433.

Burke, M. and Fountain, D.M., 1990. Seismic properties of rocks from extended continental crust — New laboratory measurements from the Ivrea zone. Tectonophysics, 182: 119–146.

Bursnall, J.T., 1989. Structural sequence fron the southeastern part of the Kapuskasing Structural Zone in the vicinity of Ivanhoe Lake, Ontario. Geol. Surv. Can. Pap., 89-1C: 405–411.

Card, K.D. and Ciesielski, A., 1986. Subdivisions of the Superior Province of the Canadian Shield. Geosci. Can., 13: 5–13.

Card, K.D. and Innes, D.G., 1981. Geology of the Benny area, District of Sudbury. Ont. Geol. Surv., Rep. 206, 117 pp.

Card, K.D., Gupta, V.K., McGrath, P.H. and Grant, F.S., 1984. The Sudbury structure: Its regional geological and geophysical setting. In: E.G. Pye, A.J. Naldrett and P.E. Giblin (Editors), The Geology and Ore Deposits of the Sudbury Structure. Ont. Geol. Surv., Spec. Vol., 1: 431–446.

Cermak, V. and Jessop, A.M., 1971. Heat flow, heat generation and crustal temperature in the Kapuskasing area of the Canadian Shield. Tectonophysics, 11: 287–303.

Chamberlain, C.P., Zeitler, P.K. and Jan, M.Q., 1989. The dynamics of the suture between the Kohistan island arc and the Indian plate in the Himalaya of Pakistan. J. Metamorph. Geol., 7: 135–149.

Chroston, P.N. and Simmons, G., 1989. Seismic velocities from the Kohistan Arc, northern Pakistan. J. Geol. Soc. London, 146: 971–979.

Clarke, G.L., Collins, W.J. and Vernon, R.H., 1990. Successive overprinting granulite facies metamorphic events in the Anmatjira Range, central Australia. J. Metamorph. Geol., 8: 65–88.

Collerson, K.D., 1989. The character and evolution of lower continental crust: A review. Geol. Assoc. Can., Prog. Abst., 14: A101.

Condie, K.C., Allen, P. and Narayana, B.L., 1982. Geochemistry of the Archean low- to high-grade transition zone, southern India. Contrib. Mineral. Petrol., 81: 157–167.

Coney, P.J. and Harms, T.A., 1984. Cordilleran metamorphic core complexes: Cenozoic extensional relics of Mezozoic compression. Geology, 12: 550–554.

Cook, F.A., 1985. Geometry of the Kapuskasing structure from a Lithoprobe pilot reflection survey. Geology, 13: 368–371.

Coolen, J.J.M.W., 1982. Carbonic fluid inclusions in granulites from Tanzania: a comparison of geobarometric methods based on fluid density and mineral chemistry. Chem. Geol., 37: 59–77.

Corfu, F., 1987. Inverse age stratification in the Archean crust of the Superior province: evidence for infra- and subcrustal accretion from high-resolution U-Pb zircon and monazite ages. Precambrian Res., 36: 259–275.

Costanzo-Alvarez, V. and Dunlop, D.J., 1988. Paleomagnetic evidence for post-2.55-Ga tectonic tilting and 1.1 Ga reactivation in the southern Kapuskasing zone, Ontario, Canada. J. Geophys. Res., 93: 9126–9136.

Coward, M.P., Jan, M.Q., Rex, D., Tarney, J., Thirlwall, M. and Windley, B.F., 1982. Geotectonic framework of the Himalaya of N. Pakistan. J. Geol. Soc. London, 139: 299–308.

Coward, M.P., Windley, B.F., Broughton, R.D., Luff, I.W., Petterson, M.G., Pudsey, C.J., Rex, D.C. and Asif Khan, M., 1986. Collision tectonics in the NW Himalayas, In: M.P. Coward and A.C. Ries (Editors), Collision Tectonics. Geol. Soc. Spec. Publ., 19: 203–219.

Coward, M.P., Butler, R.W.H., Khan, M.A. and Knipe, R.J., 1987. The tectonic history of Kohistan and its implications for Himalayan structure. J. Geol. Soc. London, 144: 377–391.

Cumming, G.L., Köppel, V. and Ferrario, A., 1987. A lead isotope study of the northeastern Ivrea zone and the adjoining Ceneri zone (N Italy): evidence for a contaminated subcontinental mantle. Contrib. Mineral. Petrol., 97: 19–30.

Dewey, J.F., 1988. Extensional collapse of orogens. Tectonics, 7: 1123–1139.

Dietz, R.S., 1961. Vredefort ring structure: Meteorite impact scar? J. Geol., 69: 499–516.

Domenick, M.A., Kistler, R.W., Dodge, F.C.W. and Tatsumoto, M., 1983. Nd and Sr isotopic study of crustal and mantle inclusions from the Sierra Nevada and implications for batholith petrogenesis. Geol. Soc. Am. Bull., 94: 713–719.

Dostal, J. and Capredi, S., 1979. Rare earth elements in high-grade metamorphic rocks from the western Alps. Lithos, 12: 41–49.

Ellis, D.J., 1987. Origin and evolution of granulites in normal and thickened crusts. Geology, 15: 167–170.

Etheridge, M., 1975. Deformation and recrystallization of orthopyroxene from the Giles Complex, Central Australia. Tectonophysics, 25: 87–114.

Forman, D.J. and Shaw, R.D., 1973. Deformation of the crust and mantle in central Australia. Aust. Bur. Miner. Resour., Geol. Geophys., Bull. 144, 20 pp.

Fountain, D.M., 1976. The Ivrea–Verbano and Strona–Ceneri zones, northern Italy: a cross section of the continental crust — new evidence from seismic velocities. Tectonophysics, 33: 145–166.

Fountain, D.M., 1986. Implications of deep crustal evolution for seismic reflection seismology. In: M. Barazangi and L. Brown (Editors), Reflection Seismology: The Continental Crust. Am. Geophys. Union, Geodyn. Ser., 14: 1–7.

Fountain, D.M., 1989. Growth and modification of lower continental crust in extended terrains: the role of extension and magmatic underplating. In: R.F. Mereu, S. Mueller and D.M. Fountain (Editors), Lower Crust: Properties and Processes. Am. Geophys. Union, Monogr., 51: 287–299.

Fountain, D.M. and Salisbury, M.H., 1981. Exposed cross sections through the continental crust: Implications for crustal structure, petrology and evolution. Earth Planet. Sci. Lett., 5: 263–277.

Fountain, D.M., Salisbury, M.H. and Furlong, K.P., 1987. Heat production and thermal conductivity of rocks from the Pikwitonei–Sachigo continental cross section, central Manitoba: implications for the thermal structure of Archean crust. Can. J. Earth Sci., 24: 1583–1594.

Fountain, D.M., Salisbury, M.H. and Percival, J.A., 1990. Seismic structure of the continental crust based on rock velocity measurements from the Kapuskasing uplift. J. Geophys. Res., 95: 1167–1186.

Frarey, M.J. and Krogh, T.E., 1986. U-Pb zircon ages of late internal plutons of the Abitibi and eastern Wawa subprovinces, Ontario and Quebec. Current Research, Part A, Geol. Surv. Can. Pap., 86-1A: 43–48.

Friend, C.R.L., 1981. Charnockite and granite formation and influx of CO_2 at Kabbaldurga. Nature, 294: 550–552.

Friend, C.R.L., 1985. Evidence for fluid pathways through Archaean crust and the generation of the Closepet granite, Karnataka, south India. Precambrian Res., 27: 239–250.

Froidevaux, P. and Guillaume, A., 1979. Contribution à l'analyse structural des Alpes Liguro–Piemontaises par l'étude du champ magnétique terrestre. Tectonophysics, 54: 139–157.

Frost, B.R. and Chacko, T., 1989. The granulite uncertainty principle: limtations on thermobarometry in granulites. J. Geol., 97: 435–450.

Frost, B.R. and Frost, C.D., 1987. CO_2, melts and granulite metamorphism. Nature, 327: 503–506.

Frost, B.R., Fyfe, W.S., Tazaki, K. and Chan, T., 1989. Grain-boundary graphite in rocks and implications for high electrical conductivity in the lower crust. Nature, 340: 134–136.

Furlong, K.P. and Fountain, D.M., 1986. Continental crustal underplating: Thermal considerations and seismic-petrologic consequences. J. Geophys. Res., 91: 8285–8294.

Galson, D.A., 1983. Heat Production and Temperature of the Alpine Crust. Ph.D. Diss., Cambridge, 171 pp.

Garland, G.D., 1950. Interpretations of gravimetric and magnetic anomalies on traverses in the Canadian Shield in northern Ontario. Dom. Observ. Ottawa, Publ. 16, 57 pp.

Garuti, G., Rivalenti, G., Rossi, A. and Sinigoi, S., 1978/1979. Mineral equilibria as geotectonic indicators in the ultramafics and related rocks of the Ivrea–Verbano Basic Complex (Italian Western Alps): pyroxenes and olivine. Proc. 2nd Symp. Ivrea–Verbano. Mem. Soc. Geol. Ital., 33: 147–160.

Gee, R.D., Baxter, J.L., Wilde, S.A. and Williams, I.R., 1981. Crustal development in the Archaean Yilgarn Block, Western Australia. Spec. Publ. Geol. Soc. Aust., 7: 43–56.

Geis, W.T., Cook, F.A., Green, A.G., Percival, J.A., West, G.F. and Milkereit, B., 1990. Thin thrust sheet formation of the Kapuskasing structural zone revealed by LITHOPROBE seismic reflection data. Geology, 18: 513–516.

German Research Group of Explosion Seismology, 1968. Topographie des Ivrea–Korpers abgeleitet aus seismichen und gravimetrischen Daten. Schweiz. Mineral. Petrogr. Mitt., 48: 235–246.

Gibson, G.M., McDougall, I. and Ireland, T.R., 1988. Age constraints on metamorphism and development of a metamorphic core complex in Fiordland, southern New Zealand. Geology, 16: 405–408.

Giese, P., 1968. Die Struktur der Erdkruste im Bereich der Ivrea-Zone. Ein Vergleich verschiedener, seismischer Interpretationen und der Versuch einer petrographisch-geologischen Deutung. Schweiz. Mineral. Petrogr. Mitt., 48: 261–284.

Glazner, A.F. and Ussler, W., 1988. Trapping of magma at midcrustal density discontinuities. Geophys. Res. Lett., 15: 673–675.

Glikson, A.Y., 1986. An upthrusted early Proterozoic basic granulite-anorthosite suite and anatectic gneisses, south-west Arunta block, central Australia: Evidence on the nature of the lower crust. Trans. Geol. Soc. S. Afr., 89: 263–283.

Glikson, A.Y., 1987. Regional structure and evolution of the Redbank–Mount Zeil thrust zone: a major lineament in the Arunta Inlier, central Australia. B.M.R. J. Aust. Geol. Geophys., 10: 89–107.

Glikson, A.Y. and Lambert, I.B., 1976. Vertical zonation and petrogenesis of the early Precambrian crust in western Australia. Tectonophysics, 30: 55–89.

Goleby, B.R., Shaw, R.D., Wright, C., Kennett, B.L.N. and Lambeck, K., 1989. Geophysical evidence for 'thick-skinned' crustal deformation in central Australia. Nature, 337: 325–330.

Goode, A.D.T. and Moore, A.C., 1975. High pressure crystallization of the Ewarara, Kalka and Gosse Pile intrusions, Giles complex, central Australia, Contrib. Mineral. Petrol., 51: 77–97.

Görler, K. and Giese, P., 1978. Aspects of the evolution of the Calabrian arc. In: H. Closs, D. Roeder and K. Schmidt (Editors), Alps, Apennines, Hellenides. Schweizerbart, Stuttgart, pp. 374–388.

Gray, C.M. and Compston, W., 1978. A rubidium-strontium chronology of the metamorphism and prehistory of central Australian granulites. Geochim. Cosmochim. Acta, 42: 1735–1747.

Gray, C.M. and Goode, A.D.T., 1989. The Kalka layered intrusion, Central Australia. A strontium isotopic history of contamination and magma dynamics. Contrib. Mineral. Petrol., 103: 35–43.

Green, A.G., Hajnal, Z. and Weber, W., 1985. An evolutionary model of the western Churchill Province and western margin of the Superior Province in Canada and the north-central United States. Tectonophysics, 116: 281–322.

Green, A.G., Milkereit, B., Percival, J., Davidson, A., Parrish, R., Cook, F., Geis, W., Cannon, W., Hutchinson, D., West, G. and Clowes, R., 1990. Origin of deep crustal reflections: Seismic profiling across high grade metamorphic terranes in Canada. Tectonophysics, 173: 627–638.

Grew, E.S. and Manton, W.I., 1984. Age of allanite from Kabbaldurga quarry, Karnataka. J. Geol. Soc. India, 25: 193–195.

Griffin, W.L. and O'Reilly, S.Y., 1987. Is the continental Moho the crust–mantle boundary? Geology, 15: 241–244.

Guillaume, A. and Guillaume, S., 1980. Nouvelles cartes des anomalies de la pesanteur dans les Alpes occidentales. C.R. Acad. Sci. Paris, 290, Sér. D: 163–166.

Hale, L.D. and Thompson, G.A., 1982. The seismic reflection character of the continental Mohorovicic discontinuity. J. Geophys. Res., 87: 4625–4635.

Hammarstrom, J.M. and Zen, E-an, 1986. Aluminum in hornblende: An empirical igneous geobarometer. Am. Mineral., 71: 1297–1313.

Handy, M.R., 1986. The structure and rheological evolution of the Pogallo Fault zone, a deep crustal dislocation in the southern Alps of northwestern Italy (Prov. Novara). Ph.D. Thesis University of Basel, 327 pp.

Handy, M.R., 1987. The structure, age and kinematics of the Pogallo fault zone; Southern Alps, northwestern Italy. Eclogae Geol. Helv., 80: 593–632.

Hansen, E.C., Newton, R.C. and Janardhan, A.S., 1984a. Pressures, temperatures and metamorphic fluids across an unbroken amphibolite-facies to granulite-facies transition in southern Karnataka, India In: A. Kröner, A.M. Goodwin and G.N Hanson (Editors), Archean Geochemistry. Springer-Verlag, Berlin, pp. 161–181.

Hansen, E.C., Newton, R.C. and Janardhan, A.S., 1984b. Fluid inclusions in rocks from the amphibolite-facies gneiss to charnockite progression in southern Karnataka, India: Direct evidence concerning the fluids of granulite metamorphism. J. Metamorph. Geol., 2: 249–264.

Hansen, E.C., Janardhan, A.S., Newton, R.C., Prame, W.K.B.N. and Ravindra Kumar, G.R., 1987. Arrested charnockite formation in southern India and Sri Lanka. Contrib. Mineral. Petrol., 96: 225–244.

Harris, N.B.W. and Jayaram, S., 1982. Metamorphism of cordierite gneisses from the Bangalore region of the Indian Archaean. Lithos, 15: 89–97.

Harris, N.B.W., Holt, R.W. and Drury, S.A., 1982. Geobarometry, geothermometry, and late Archean geotherms from the granulite facies terrain of South India. J. Geol., 90: 509–528.

Hart, R.J., Nicolaysen, L.O. and Gale, N.H., 1981a. Radioelement concentrations in the deep profile through Precambrian basement of the Vredefort structure. J. Geophys. Res., 86: 10,639–10,652.

Hart, R.J., Welke, H.J. and Nicolaysen, L.O., 1981b. Geochronology of the deep profile through Archean basement at Vredefort, with implications for early crustal evolution. J. Geophys. Res., 86: 10,663–10,680.

Hart, R.J., Andreoli, M.A.G., Tredoux, M. and De Wit, M.J., 1990a. Geochemistry across an exposed section of Archaean crust at Vredefort, South Africa: with implications for mid-crustal discontinuities. Chem. Geol., 82: 21–50.

Hart, R.J., Andreoli, M.A.G., Smith, C.B., Otter, M.L. and Durrheim, R., 1990b. Ultramafic rocks in the centre of the Vredefort structure (South Africa): Possible exposure of upper mantle? Chem. Geol., 83: 233–248.

Heaman, L.M., Machado, N., Krogh, T.E. and Weber, W., 1986. Preliminary U-Pb zircon results from the Pikwitonei granulite terrain, Manitoba. Geol. Assoc. Can., Progr. Abstr., 11: 79.

Hodges, K.V. and Fountain, D.M., 1984. Pogallo Line, South Alps, northern Italy: An intermediate crustal level, low-angle normal fault. Geology, 12: 151–155.

Höhndorf, A., 1975. Radioaktive Wärmeproduktion in Gesteinen der Ivrea-Zone. Schweiz. Mineral. Petrogr. Mitt., 55: 89–102.

Honnorez, J. and Keller, J., 1968. Xenolithe in vulkanischen Gesteinen der Äolischen Inseln. Geol. Rundsch., 57: 719–736.

Hubregtse, J.J.M.W., 1980. The Archean Pikwitonei granulite domain and its position at the margin of the northwestern Superior Province (Central Manitoba). Man. Miner. Res., Publ. GP80-3, 16 pp.

Hunt, P.A. and Roddick, J.C., 1987. A compilation of K-Ar ages. Geol. Surv. Can. Pap., 87-2: 143–210.

Hunziker, J.C., 1974. Rb-Sr and K-Ar age determination and the Alpine tectonic history of the Western Alps. Mem. Ist. Geol. Mineral. Univ. Padova, 31, 54 pp.

Hunziker, J.C. and Zingg, A., 1980. Lower Paleozoic amphibolite to granulite facies metamorphism in the Ivrea zone, southern Alps, northern Italy. Schweiz. Mineral. Petrogr. Mitt., 60: 181–213.

Hurich, C.E. and Smithson, S.B., 1987. Compositional variation and the origin of deep crustal reflections. Earth Planet. Sci. Lett., 85: 416–426.

Innes, M.J.S., Goodacre, A.K., Weber, J.R. and McConnell, R.K., 1967. Structural implications of the gravity field in Hudson Bay and vicinity. Can. J. Earth Sci., 4: 977–993.

Iyer, S.S., Woodford, P.J. and Wilson, A.F., 1976. Rb-Sr isotopic studies of a polymetamorphic granulite terrane, Strangways Range, central Australia. Lithos, 9: 211–224.

Jackson, D.H., Mattey, D.P. and Harris, N.B.W., 1988. Carbon isotope compositions of fluid inclusions in charnockites from southern India. Nature, 333: 167–170.

Jan, M.Q. and Howie, R.A., 1981. The mineralogy and geochemistry of the metamorphosed basic and ultrabasic rocks of the Jijal complex, Kohistan, NW Pakistan. J. Petrol., 22: 85–126.

Janardhan, A.S., Newton, R.C. and Hansen, E.C., 1982. The transformation of amphibolite facies gneiss to charnockite in southern Karnataka and and northern Tamil Nadu, India. Contrib. Mineral. Petrol., 79: 130–149.

Jiang, J., Clayton, R.N. and Newton, R.C., 1988. Fluids in granulite facies metamorphism: A comparative oxygen isotope study on the south India and Adirondack high-grade terrains. J. Geol., 96: 517–533.

Kaminski, W. and Menzel, H., 1968. Zur deutung der Schwere-anomalie des Ivrea-Korpers. Schweiz. Mineral. Petrogr. Mitt., 48: 255–260.

Kern, H. and Schenk, V., 1985. Elastic wave velocities in rocks from a lower crustal section in southern Calabria (Italy). Earth Planet. Sci. Lett., 40: 147–160.

Kern, H. and Schenk, V., 1988. A model of velocity structure beneath Calabria southern Italy, based on laboratory data. Earth Planet. Sci. Lett., 87: 325–337.

Kissling, E., 1984. Three-dimensional gravity model of the northern Ivrea-Verbano Zone. Matér. Géol. Suisse, Géophys., 21: 53–61.

Köppel, V., 1974. Isotopic U-Pb ages of monazites and zircons from the crust–mantle transition and adjacent units of the Ivrea and Ceneri zones (southern Alps, Italy). Contrib. Mineral. Petrol., 43: 55–70.

Köppel, V. and Grünenfelder, M., 1971. A study of inherited and newly formed zircons from paragneisses and granitised sediments of the Strona–Ceneri Zone (Southern Alps). Schweiz. Mineral. Petrogr. Mitt., 51: 385–409.

Köppel, V. and Grünenfelder, M., 1978/1979. Monazite and zircon U-Pb ages from the Ivrea and Ceneri zones (Abst.). Mem. Sci. Geol., 33: 257.

Krogh, T.E. and Turek, A., 1982. Precise U-Pb zircon ages from the Gamitagama greenstone belt, southern Superior Province. Can. J. Earth Sci., 19: 859–867.

Krogh, T.E., McNutt, R.H. and Davis, G.L., 1982. Two high-precision U-Pb zircon ages for the Sudbury Nickel irruptive. Can. J. Earth Sci., 19: 723–728.

Krogh, T.E., Davis, D.W. and Corfu, F., 1984. Precise U-Pb zircon and baddeleyite ages for the Sudbury area. In: E.G. Pye, A.J. Naldrett and P.E. Giblin (Editors), The Geology and Ore Deposits of the Sudbury Structure. Ont. Geol. Surv., Spec. Vol., 1: 431–446.

Krogh, T.E., Corfu, F. and Percival, J.A., 1986. U-Pb zircon and sphene ages ages from the Kapuskasing zone in the Chapleau–Agawa Bay region. Geol. Assoc. Can., Program Abstr., 11: 91.

Krogh, T.E., Heaman, L.M. and Machado, N., 1988. Detailed U-Pb chronology of successive stages of zircon growth at medium and deep levels using parts of single zircon and titanite grains. 1988 KSZ Lithoprobe Workshop, Univ. Toronto, 243 pp.

Krogstad, E.J., Hanson, G.N. and Rajamani, V., 1988. U-Pb ages and Sr, Pb and Nd isotope data for gneisses near the Kolar schist belt: Evidence for the juxtaposition of discrete Archean terranes, In: L.D. Ashwal (Editor), Workshop on the Deep Continental Crust of South India. Lunar Planet. Inst. Tech. Rept., 88-06: 84–86.

Kurtz, R.D., Macnae, J.C. and West, G.F., 1989. A controlled-source, time-domain electromagnetic survey over an upthrust section of Archean crust in the Kapuskasing structural zone. Geophys. J. Int., 99: 195–203.

Lachenbruch, A.H., 1970. Crustal temperature and heat production: implications of the linear heat-flow relation. J. Geophys. Res., 75: 3291–3300.

Lamb, W.M., Valley, J.W. and Brown, P.E., 1987. Post-metamorphic CO_2-rich fluid inclusions in granulites. Contrib. Mineral. Petrol., 96: 485–495.

Lambeck, K. and Penney, C., 1984. Teleseismic travel time anomalies and crustal structure in central Australia. Earth Planet. Sci. Lett., 34: 46–56.

Lanza, R., 1982. Models for interpretation of the magnetic anomaly of the Ivrea body. Geol. Alpine, 58: 85–94.

Lilly, P.A., 1981. Shock metamorphism in the Vredefort collar: evidence for internal shock sources. J. Geophys. Res., 86: 10,689–10,700.

Lopez Martinez, M. and York, D., 1990. A comparative $^{40}Ar/^{39}Ar$ study of the Kapuskasing structural zone. Can. J. Earth Sci., 27: 787–793.

Maboko, M.A.H., McDougall, I. and Zeitler, P.K., 1989. Metamorphic P-T path of granulites in the Musgrave Ranges, central Australia. In: J.S. Daly, R.A. Cliff and B.W.D. Yardley (Editors), Evolution of Metamorphic Belts. Geol. Soc. Spec. Publ., 43: 303–307.

Maccarrone, E., Paglionico, A., Piccarreta, G. and Rottura, A., 1983. Granulite-amphibolite facies metasediments from the Serre (Calabria, Southern Italy): Their protoliths and the processes controlling their chemistry. Lithos, 16: 95–111.

Mareschal, M., 1990. Electrical conductivity: The story of an elusive parameter, and of how it possibly relates to the Kapuskasing Uplift (Lithoprobe, Canada). In: M.H. Salisbury and D.M. Fountain (Editors), Exposed Cross-Sections of the Continental Crust. Kluwer, Dordrecht, pp. 453–468.

Mareschal, M., Fyfe, W.S. and Percival, J., 1992. Graphite films in Kapuskasing gneisses relieve the lower-crustal conductivity paradox (abstract). EOS, Trans. Am. Geophys. Union, 73: 314.

Marjoribanks, R.W., 1975. The structural and metamorphic geology of the Ormiston area, central Australia. Bur. Min. Res, Australia Rec., 13.

Marjoribanks, R.W. and Black, L.P., 1974. Geology and geochronology of the Arunta complex north of Ormiston Gorge, Central Australia. J. Geol. Soc. Aust., 21: 291–299.

Martin, H., 1986. Effect of steeper Archean geothermal gradients on geochemistry of subduction-zone magmas. Geology, 14: 753–756.

Mathur, S.P., Moss, F.J. and Branston, J.C., 1977. Seismic and gravity investigations along the geotraverse, Western Australia. Bur. Miner. Res. Geol. Geophys. Aust., Bull. 191

Matthews, D.H. and Smith, C. (Editors), 1987. Deep Seismic Profiling of the Continental Lithosphere. Geophys. J. R. Astron. Soc. Spec. Issue 89, 447 pp.

Mattinson, J., Kimbrough, D. and Bradshaw, J., 1986. Western Fiordland orthogneiss: Early Cretaceous arc magmatism and granulite-facies metamorphism, New Zealand. Contrib. Mineral. Petrol., 92: 383–392.

Mazzucchelli, M., 1983. The upper zone of the Ivrea–Verbano layered complex (Italian Western Alps). Neues Jahrb. Mineral. Abh., 146: 101–116.

Mazzucchelli, M. and Siena, F., 1986. Geotectonic significance of the meta-basites of the kinzigitic series, Ivrea–Verbano zone (western Italian Alps). Tscherm. Min. Petr. Mitt., 35: 99–116.

McCulloch, M.T., Collerson, K.D. and Compston, W., 1983. Growth of Archaean crust within the Western Gneiss Terrain, Yilgarn Block, Western Australia. J. Geol. Soc. Aust., 30: 155–160.

McNutt, R.H. and Dickin, A.P., 1989. Rb/Sr and Sm/Nd studies of four units in the vicinity of the Kapuskasing structural zone. Geol. Assoc. Can., Program. Abstr., 14: A124.

Mehnert, K., 1975. The Ivrea zone, a model of the deep crust. Neues Jahrb. Mineral. Abh., 125: 156–199.

Mezger, K., 1992. Temporal evolution of regional granulite terranes: Implications for the formation of the lowermost continental crust. In: R.W. Kay, S.J. Arculus and D.M. Fountain (Editors), Lower Continental Crust. Elsevier, Amsterdam, this volume.

Mezger, K., Hanson, G.N. and Bohlen, S.R., 1989a. U-Pb systematics of garnet: dating the growth of garnet in the Late Archean Pikwitonei granulite domain at Cauchon and Natawahunan Lakes, Manitoba, Canada. Contrib. Mineral. Petrol., 101: 136–148.

Mezger, K., Hanson, G.N. and Bohlen, S.R., 1989b. High-precision U-Pb ages of metamorphic rutile: application to the cooling history of high-grade terranes. Earth Planet. Sci. Lett., 96: 106–118.

Mezger, K., Bohlen, S.R. and Hanson, G.N., 1990. Metamorphic history of the Archean Pikwitonei granulite domain and the Cross Lake Subprovince, Superior Province, Manitoba, Canada. J. Petrol., 31: 483–517.

Moore, A.C. and Goode, A.D.T., 1978. Petrography and origin of granulite-facies rocks in the western Musgrave Block, central Australia, J. Geol. Soc. Aust., 25: 341–358.

Moresi, M., Paglionico, A., Piccarreta G., and Rottura, A., 1978/1979. The deep crust in Calabria (Polia-Copanello Unit): a comparison with Ivrea Verbano zone. Mem. Ist. Geol. Mineral. Padova, 33: 233–242.

Mortensen, J.K., 1987. Preliminary U-Pb zircon ages for volcanic and plutonic rocks of the Noranda–Lac Abitibi area, Abitibi subprovince, Quebec. Geol. Surv. Can. Pap., 87-1A: 581–590.

Moser, D., 1988. Structure of the Wawa gneiss terrane near Chapleau, Ontario. Geol. Surv. Can. Pap., 88-1C: 93–99.

Moser, D., 1989. Mid-crustal structures of the Wawa gneiss terrane near Chapleau, Ontario. Geol. Surv. Can. Pap., 89-1C: 215–224.

Newton, R.C., 1987. Petrologic aspects of Precambrian granulite facies terrains bearing on their origins. In: A. Kröner (Editor), Proterozoic Lithosphere Evolution. Am. Geophys. Union, Geodyn. Ser., 17: 11–26.

Newton, R.C., 1988. Closepet granite of south India: Origin by K-metasomatism melting. Geol. Soc. Am., Abstr. Program, 20: A204.

Newton, R.C., 1990. The late Archean high-grade terrain of South India and deep structure of the Dharwar craton of south India. In: M.H. Salisbury and D.M. Fountain (Editors), Exposed Cross-Sections of the Continental Crust. Kluwer, Dordrecht, pp. 305–326.

Newton, R.C. and Perkins, D., 1982. Thermodynamic calibration of geobarometers based on the assemblage garnet-plagioclase-orthopyroxene (clinopyroxene)-quartz. Am. Mineral., 67: 203–222.

Newton, R.C., Smith, J.V. and Windley, B.F., 1980. Carbonic metamorphism, granulites and crustal growth. Nature, 292: 144–146.

Nicolas, A., Hirn, A., Nicolich, R., Polino, R. and ECORS-CROP Working Group, 1990. Lithospheric wedging in the western Alps inferred from the ECORS-CROP traverse. Geology. 18: 587–590.

Nicolaysen, L.O., 1990. The Vredefort structure: an introduction and guide to recent literature. Teconophysics, 171: 1–6.

Nicolaysen, L.O., Burger, A.J. and van Niekerk, C.B., 1963. The origin of the Vredefort structure in light of new isotopic data (abstr.). Int. Union Geophys. Geod., v. 13.

Oliver, G.J.H., 1980. Geology of the granulite and amphibolite facies gneisses of Doubtful Sound, Fiordland, New Zealand. N. Z. J. Geol. Geophys., 23: 27–41.

Oliver, G.J.H. and Coggon, J.H., 1979. Crustal structure of Fiordland, New Zealand. Tectonophysics, 54: 253–292.

Oliver, J.E., Brown, L.D. and Cook, F.A., 1983. COCORP and the continental crust. J. Geophys. Res., 88: 3329–3347.

Oxburgh, E.R., 1972. Flake tectonics. Nature, 239: 202–204.

Paglionico, A. and Piccarreta, G., 1978. History and petrology of a fragment of deep crust in the Serre (Calabria, Southern Italy). Neues Jahrb. Mineral. Monatsh., 9: 385–396.

Paktunc, A.D. and Baer, A.J., 1986. Geothermobarometry of the northwestern margin of the Superior province: Implications for its tectonic evolution. J. Geol., 94: 381–394.

Percival, J.A., 1983. High-grade metamorphism in the Chapleau-Foleyet area, Ontario. Am. Mineral., 68: 667–686.

Percival, J.A., 1986. A possible exposed Conrad discontinuity in the Kapuskasing uplift, Ontario In: M. Barazangi and L.D. Brown (Editors), Reflection Seismology: The Continental Crust. Am. Geophys. Union, Geodyn. Ser., 14: 135–141.

Percival, J.A., 1988. Deep geology out in the open. Nature, 335: 671.

Percival, J.A., 1989a. Granulite terranes and the lower crust of the Superior Province. In: R.F. Mereu, S. Mueller and D.M. Fountain (Editors), Properties and Processes of Earth's Lower Crust. Am. Geophys. Union, Geophys. Monogr. 51, Int. Union Geod. Geophys. Vol. 6: 301–310.

Percival, J.A., 1989b. Melt-induced fluid pumping and the source of CO_2 in granulites. In: D. Bridgwater (Editor), Fluid Movements — Element Transport and the Composition of the Deep Crust. Kluwer, Dordrecht, pp. 61–69.

Percival, J.A., 1990. Archean tectonic settings of granulite terranes of the Superior Province, Canada: A view from the bottom. In: D. Vielzeuf and Ph. Vidal (Editors), Granulites and Crustal Evolution. Kluwer, Dordrecht, pp. 171–193.

Percival, J.A. and Berry, M.J., 1987. The lower crust of the continents. In: K. Fuchs and C. Froidevaux (Editors), Composition, Structure and Dynamics of the Lithosphere–Asthenosphere System. Am. Geophys. Union, Geodyn. Ser., 16: 33–59.

Percival, J.A. and Card, K.D., 1983. Archean crust as revealed in the Kapuskasing uplift, Superior Province, Canada. Geology, 11: 323–326.

Percival, J.A. and Card, K.D., 1985. Structure and evolution of Archean crust in central Superior Province, Canada. In: L.D. Ayres, P.C. Thurston, K.D. Card and W. Weber (Editors), Evolution of Archean Supracrustal Sequences. Geol. Assoc. Can., Spec. Pap., 28: 179–192.

Percival, J.A. and Fountain, D.M., 1989. Metamorphism and melting at an exposed example of the Conrad discontinuity, Kapuskasing uplift, Canada. In D. Bridgwater (Editor), Fluid Movements — Element Transport and the Composition of the Deep Crust. Kluwer, Dordrecht, pp. 51–60.

Percival, J.A. and Green, A.G., 1988. Towards a balanced crustal-scale cross section of the Kapuskasing uplift. 1988-89 Kapuskasing Workshop, Univ. Toronto, pp. 233–234.

Percival, J.A. and Krogh, T.E., 1983. U-Pb zircon geochronology of the Kapuskasing structural zone and vicinity in the Chapleau-Foleyet area, Ontario. Can. J. Earth Sci., 20: 830–843.

Percival, J.A. and McGrath, P.H., 1986. Deep crustal structure and tectonic history of the northern Kapuskasing uplift of Ontario: An integrated petrological-geophysical study. Tectonics, 5: 553–572.

Percival, J.A., Parrish, R.R., Krogh, T.E. and Peterman, Z.E., 1988. When did the Kapuskasing zone come up? 1988 KSZ Lithoprobe Workshop, Univ. Toronto, 43–47.

Percival, J.A., Green, A.G., Milkereit, B., Cook, F.A., Geis, W. and West, G.F., 1989. Seismic reflection profiles across deep continental crust exposed in the Kapuskasing uplift structure. Nature, 342: 416–420.

Perkins, D., 1988. Geothermometry and geobarometry: Not as precise as we want and unlikely to get any better (Abstr.). Terra Cognita, 8: 265.

Pichamuthu, C.S., 1961. Transformation of Peninsular gneiss into charnockite in Mysore State, India. J. Geol. Soc. India, 2: 46–49.

Pichamuthu, C.S., 1965. Regional metamorphism and charnockitization in Mysore State, India. Ind. Mineral., 6: 119–126.

Pin, C., 1986. Datation U-Pb sur zircons à 285 M.a. du complexe gabbro-dioritique du Val Sesia–Val Mastallone et age tardi-hercynien du metamorphsime granulitique de la zone Ivrea–Verbano (Italie). C.R. Acad. Sci. Paris, 303, Sér. II: 827–830.

Pin, C. and Sills, J., 1986. Petrogenesis of layered gabbros and ultramafic rocks from Val Sesia, the Ivrea zone, northwest Italy: trace element and isotope geochemistry. In: J.B. Dawson, D.A. Carswell, J.Hall, and K.H. Wedepohl (Editors), The Nature of the Lower Continental Crust. Geol. Soc. Spec. Publ., 24: 231–249.

Pin, C. and Vielzeuf, D., 1983. Granulites and related rocks in Variscan median Europe: a dualistic interpretation. Tectonophysics, 93: 47–74.

Polvé, M. 1983. Les isotopes du Nd et Sm dans les lherzolite orogéniques: contribution à la détermination de la structure et de la dynamique de manteau supérieur, Thèse, Université de Paris.

Priestly, K. and Davey, F., 1983. Crustal structure of Fiordland, New Zealand, from seismic refraction measurements. Geology, 11: 660–663.

Raase, P., Raith, M., Ackermand, D. and Lal, R.K., 1986. Progressive metamorphism of mafic rocks from greenschist to granulite facies in the Dharwar craton of south India. J. Geol., 94: 261–282.

Raith, M., Raase, P., Ackermand, D. and Lal, R.K., 1983. Regional geothermobarometry in the granulite facies terrane of south India. Trans. R. Soc. Edinburgh Earth Sci., 73: 221–244.

Raith, M., Hengst, C., Nagel, B., Bhattacharya, A. and Srikantappa, C., 1988. Metamorphic conditions in the Nilgiri granulite terrane and the adjacent Moyar and Bhavani shear zones: A reinterpretation. J. Geol. Soc. India, 31: 112–113.

Reilly, W., 1965. Gravity map of New Zealand. N. Z. Dept. Sci. Indust. Res., scale 1:4 000 000.

Reinhard, M., 1964. Über das Grundgebirge des Sottoceneri im Süd-Tessin und die darin auftretenden Ganggesteine. Beitr. Geol. Karte Schweiz., NF 117: 89 pp.

Reston, T.J., 1988. Evidence for shear zones in the lower crust offshore Britain. Tectonics, 7: 929–946.

Rivalenti, G., Garuti, G. and Rossi, A., 1975. The origin of the Ivrea–Verbano basic formation (Western Italian Alps) — Whole rock geochemistry. Boll. Soc. Geol. Ital., 94: 1149–1186.

Rivalenti, G., Garuti, G., Rossi A. and Sinigoi, S., 1981. Chromian spinel in the Ivrea–Verbano basic complex, western Alps, Italy. Tscherm. Mineral. Petrogr. Mitt., 29: 33–53.

Rivalenti, G., Rossi, A., Siena F. and Sinigoi, S., 1984. The layered series of the Ivrea Verbano igneous complex, Western Alps, Italy. Tscherm. Mineral. Petrogr. Mitt., 33: 77–99.

Ross, D.C., 1985. Mafic gneiss complex (batholithic root?) in the southernmost Sierra Nevada, California. Geology, 13: 288–291.

Royden, L.H. and Burchfiel, B.C., 1987. Thin-skinned N–S extension within the convergent Himalayan region: gravitational collapse of a Miocene topographic front. In: M.P. Coward, J.F. Dewey and P.L. Hancock (Editors), Continental Extensional Tectonics. Geol. Soc. Spec. Publ., 28: 611–619.

Rudnick, R.L., 1992. Xenoliths — samples of the lower continental crust. In: R.W. Kay, R.J. Arculus and D.M. Fountain (Editors), Lower Continental Crust. Elsevier, Amsterdam, this volume.

Rudnick, R.L. and Taylor, S.R., 1986. Geochemical constraints on the origin of Archean tonalitic-trondhjemitic rocks and implications for lower crustal composition. In: J.B. Dawson, D.A. Carswell, J.Hall, and K.H. Wedepohl (Editors), The Nature of the Lower Continental Crust. Geol. Soc. Spec. Publ., 24: 179–191.

Rudnick, R.L., Ashwal, L.D. and Henry, D.J., 1984. Fluid inclusions in high-grade gneisses of the Kapuskasing structural zone, Ontario: Metamorphic fluids and uplift/erosion path. Contrib. Mineral. Petrol., 87: 399–406.

Rudnick, R.L., Arndt, N.T. and Presper, T., 1988. Geochemistry of granulite terrains and granulite xenoliths: Relevance to present day lower crustal composition (abstr.). Terra Cognita, 8: 271.

Saleeby, J.B., 1990. Progress in tectonic and petrogenetic studies in an exposed cross section of young (100 Ma) continental crust, southern Sierra Nevada, California. In: M.H. Salisbury and D.M. Fountain (Editors), Exposed Cross-Sections of the Continental Crust. Kluwer, Dordrecht, pp. 137–158.

Sams, D.B. and Saleeby, J.B., 1988. Geology and petrotectonic significance of crystalline rocks of the southernmost Sierra Nevada, California. In: W.G. Ernst (Editor), Metamorphic and Crustal Evolution of the Western U.S. Prentice-Hall, Englewood Cliffs, N.J., pp. 865–893.

Sandiford, M., 1989. Horizontal structures in granulite terrains: A record of mountain building or mountain collapse? Geology, 17: 449–452.

Schenk, V., 1980. U-Pb and Rb-Sr radiometric dates and their correlation with metamorphic events in the granulite-facies basement of the Serre, southern Calabria (Italy). Contrib. Mineral. Petrol., 73: 23–38.

Schenk, V., 1981. Synchronous uplift of the lower crust of the Ivrea zone and of southern Calabria and its possible consequences for the Hercynian orogeny in southern Europe. Earth Planet. Sci. Lett., 56: 305–320.

Schenk, V., 1984. Petrology of felsic granulites, metapelites, metabasics, ultramafics and metacarbonates from southern Calabria (Italy): Prograde metamorphism, uplift and cooling of a former lower crust. J. Petrol., 25: 255–298.

Schenk, V., 1989. *P-T-t* path of the lower crust in the Hercynian fold belt of southern Calabria. In: J.S. Daly, R.A. Cliff and B.W.D. Yardley (Editors), Evolution of Metamorphic Belts. Geol. Soc. London, Spec. Publ., 43: 337–342.

Schenk, V. and Schreyer, W., 1978. Granulite-facies metamorphism in the Northern Serre, Calabria, Southern Italy. In: H. Closs, D. Roeder and K. Schmidt (Editors), Alps, Apennines, Hellenides. Schweizerbart, Stuttgart, pp. 341–346.

Schmid, R., 1967. Zur Petrographie und Struktur der Zone Ivrea–Verbano zwischen Valle d'Ossala und Val Grande (Prov. Novara, Italien). Schweiz. Mineral. Petrogr. Mitt., 47: 935–1117.

Schmid, R., 1978/1979. Are the metapelites in the Ivrea Zone restites? Mem. Ist. Geol. Mineral. Univ. Padova, 33: 67–69.

Schmid, R. and Wood, B.J., 1976. Phase relationships in granulitic metapelites from the Ivrea–Verbano Zone (Northern Italy). Contrib. Mineral. Petrol., 54: 255–279.

Schmid, S.M., Zingg, A. and Handy, M.R., 1987. The kinematics of movements along the Insubric Line and the emplacement of the Ivrea zone. Tectonophysics, 135: 47–66.

Schneider, R.V., Roy, R.F. and Smith, A.R., 1987. Investigations and interpretations of the vertical distribution of U, Th and K: South Africa and Canada. Geophys. Res. Lett., 14: 264–267.

Schreyer, W. and Abraham, K., 1978. Symplectitic cordierite-orthopyroxene-garnet assemblages as products of contact metamorphism of pre-existing basement granulites in the Vredefort structure, South Africa, and their relations to pseudotachylite. Contrib. Mineral. Petrol., 68: 53–62.

Schütte, K.-G., 1978. Crustal structure in southern Italy. In: H. Closs, D. Roeder and K. Schmidt (Editors), Alps, Apennines, Hellenides. Schweizerbart, Stuttgart, pp. 315–320.

Schwarz, G., 1978. Magnetic measurements in northern Calabria. In: H. Closs, D. Roeder and K. Schmidt (Editors), Alps, Apennines, Hellenides. Schweizerbart, Stuttgart, pp. 321–323.

Sharry, J., 1982. Minimum age and westward continuation of the Garlock fault zone, Tehachapi Mountains, California. Geol. Soc. Am., Abstr. Program, 14: 23.

Shaw, D.M., Truscott, M.G., Gray, E.A. and Middleton, T.A., 1988. Boron and lithium in high-grade rocks and minerals from the Wawa–Kapuskasing region, Ontario. Can. J. Earth Sci., 25: 1485–1502.

Shaw, R.D., Stewart, A.J. and Black, L.P., 1984. The Arunta Inlier: a complex ensialic mobile belt in central Australia. Part 2: tectonic history. Aust. J. Earth Sci., 31: 457–484.

Shervais, J.W., 1978/1979. Ultramafic and mafic layers in the Alpine-type Lherzolite massif of Balmuccia, NW Italy. Proc. 2nd Symp. Ivrea Verbano, Mem. Sci. Geol., 33: 135–145.

Shervais, J.W., 1979. Thermal emplacement model for the Alpine Lherzolite massif at Balmuccia, Italy. J. Petrol., 20: 795–820.

Shive, P.N., 1990. The Ivrea zone and lower crustal magnetization. Tectonophysics, 182: 161–167.

Shive, P.N. and Fountain, D.M., 1988. Magnetic mineralogy in an Archean crustal cross section: Implications for crustal magnetization. J. Geophys. Res., 93: 12,177–12,186.

Sibson, R.H., 1977. Fault rocks and fault mechanisms. J. Geol. Soc. London, 133: 191–213.

Sighinolfi, G.P. and Gorgoni, C., 1978. Chemical evolution of high-grade metamorphic rocks — anatexis and remotion of material from granulite terrains. Chem. Geol., 22: 157–176.

Sills, J.D., 1984. Granulite facies metamorphism in the Ivrea Zone, N.W. Italy. Schweiz. Mineral. Petrogr. Mitt., 64: 169–191.

Sills, J.D. and Tarney, J., 1984. Petrogenesis and tectonic significance of amphibolite interlayered with metasedimentary gniesses in the Ivrea zone, Southern Alps, Northwest Italy. Tectonophysics, 107: 187–206.

Simpson, C., 1981. Occurrence and orientation of shatter cones in Pretoria Group quartzites in the collar of the Vredefort 'Dome': Impact origin precluded. J. Geophys. Res., 86: 10,701–10,706.

Slawson, W.F., 1976. Vredefort dome: A cross-section of the upper crust. Geochim. Cosmochim. Acta, 40: 117–121.

Stahle, H.J., Raith, M., Hoernes, S. and Delfs, A., 1987. Element mobility during incipient granulite formation at Kabbaldurga, southern India. J. Petrol., 28: 803–843.

Stille, P. and Buletti, M., 1987. Nd-Sr isotopic characteristics of the Lugano volcanic rocks and constraints on the continental crust formation in the South Alpine domain (N Italy-Switzerland). Contrib. Mineral. Petrol., 96: 140–150.

Stewart, A.J., Shaw, R.D. and Black, L.P., 1984. The Arunta Inlier: a complex ensialic mobile belt in central Australia. Part 1: stratigraphy, correlations and origin. Aust. J. Earth Sci., 31: 445–455.

Studemeister, P.A., 1983. The greenschist facies of an Archean assemblage near Wawa, Ontario. Can. J. Earth Sci., 20: 1409–1420.

Swami Nath, J. and Ramakrishnan, M., 1981. Early Precambrian supracrustals of southern Karnataka. Geol. Surv. Ind. Mem., 112: 1–38.

Taylor, S.R. and McLennan, S.M., 1981. The composition and evolution of the continental crust: rare earth element evidence from sedimentary rocks. Philos. Trans. R. Soc. London, Ser. A, 301: 381–399.

Teyssier, C., 1985. A crustal thrust system in an intracratonic tectonic environment. J. Struct. Geol., 7: 689–700.

Thurston, P.C., Siagusa, G.N. and Sage, R.P., 1977. Geology of the Chapleau area, Districts of Algoma, Sudbury and Cochrane. Ont. Div. Mines Geol., Rep. 157, 293 pp.

Touret, J., 1971. Le facies granulite en Norvege meridionale. II: Les inclusions fluids. Lithos, 4: 423–436.

Treloar, P.J., Rex, D.C., Guise, P.G., Coward, M.P., Searle, M.P., Windley, B.F., Petterson, M.G., Jan, M.Q. and Luff, I.W., 1989a. K-Ar and Ar-Ar geochronology of the Himalayan collision in NW Pakistan: Constraints on the timing of suturing, deformation, metamorphism and uplift. Tectonics, 8: 881–909.

Treloar, P.J., Williams, M.P. and Coward, M.P., 1989b. Metamorphism and crustal stacking in the North Indian plate, north Pakistan. Tectonophysics, 165: 167–184.

Truscott, M.G., 1989. Models of intermediate and lower crust composition: comparing the Kapuskasing structural zone, Ontario. Geol. Assoc. Can. Program Abstr., 14: A102.

Truscott, M.G. and Shaw, D.M., 1986. Preliminary estimate of lower continental crust composition in the Kapuskasing structural zone. Geol. Assoc. Can. Program Abstr., 11: 138.

Truscott, M.G. and Shaw, D.M., 1990. Average composition of lower and intermediate continental crust, Kapuskasing structural zone, Ontario. In: M.H. Salisbury and D.M. Fountain (Editors), Exposed Cross Sections of the Continental Crust. Kluwer, Dordrecht, pp. 421–436.

Turek, A., Smith, P.E. and Van Schmus, W.R., 1984. U-Pb zircon ages and the evolution of the Michipicoten greenstone belt, Wawa, Ontario. Can. J. Earth Sci., 21: 457–464.

Vecchia, O., 1969. La Zone Cuneo–Ivrea–Locarno, element fondamental des Alpes, geophysique et geologie. Schweiz. Mineral. Petrogr. Mitt., 48: 215–226.

Voshage, H., Hunziker, J.C., Hofmann, A.W. and Zingg, A., 1987. A Nd and Sr isotopic study of the Ivrea zone, Southern Alps, N. Italy. Contrib. Mineral. Petrol., 97: 31–42.

Voshage, H., Sinigoi, S., Mazzucchelli, M., Demarchi, G., Rivalenti, G. and Hofmann, A.W., 1988. Isotopic constraints on the origin of ultramafic and mafic dikes in the Balmuccia peridotite (Ivrea Zone). Contrib. Mineral. Petrol., 100: 261–267.

Wagner, J.-J., 1984. Geophysical studies of the Ivrea zone — a review. Matér. Géol. Suisse Géophys., 21: 13–19.

Warren, R.G. and Hensen, B.J., 1989. The *P-T* evolution of the Proterozoic Arunta Block, central Australia, and implications for tectonic evolution. In: J.S. Daly, R.A. Cliff and B.W.D. Yardley, (Editors), Evolution of Metamorphic Belts. Geol. Soc. Spec. Publ., 43: 349–355.

Warren, R.G., 1983. Metamorphism and tectonic evolution of granulites, Arunta block, central Australia. Nature, 305: 300–303.

Warren, R.G. and Stewart, A.J., 1988. Isobaric cooling of Proterozoic high-temperature metamorphites in the northern Arunta block, central Australia: Implications for tectonic evolution. Precambrian Res., 40/41: 175–198.

Wasilewski, P. and Fountain, D.M., 1982. The Ivrea zone as a model for the distribution of magnetization in the continental crust. Geophys. Res. Lett., 9: 333–336.

Wasilewski, P. and Warner, R.D., 1988. Magnetic petrology of deep crustal rocks — Ivrea Zone, Italy. Earth Planet. Sci. Lett., 87: 347–361.

Watson, J., 1980. The origin and history of the Kapuskasing structural zone, Ontario, Canada. Can. J. Earth Sci., 17: 866–876.

Weber, E.K., Gassmann, F., Niggli, E. and Roethlisberger, H., 1949. Die magnetische Anomalie westlich von Locarno. Schweiz. Mineral. Petrogr. Mitt., 29: 492–508.

Weber, W., 1980. The Pikwitonei granulite domain: edge of the Superior Province craton. EOS, Trans. Am. Geophys. Union, 61: 386–387.

Weber, W. and Mezger, K., 1990. An oblique cross section of Archean continental crust at the northwestern margin of Superior Province, Manitoba, Canada. In: M.H. Salisbury and D.M. Fountain (Editors), Exposed Cross-Sections of the Continental Crust. Kluwer, Dordrecht, pp. 327–341.

Weber, W. and Scoates, R.F.J., 1978. Archean and Proterozoic metamorphism in the northwestern Superior Province and along the Churchill–Superior boundary, Manitoba. Metamorphism in the Canadian Shield. Geol. Surv. Can. Pap., 78-10: 5–16.

Wells, P.R.A., 1981. Thermal models for the magmatic accretion and subsequent metamorphism of continental crust. Earth Planet. Sci. Lett., 46: 253–265.

Williams, H.R., 1978. The Archean geology of Sierra Leone. Precambrian Res., 6: 251–268.

Williams, H.R., 1988. The Archean Kasila Group of western Sierra Leone: Geology and relations with adjacent granite-greenstone terrane. Precambrian Res., 38: 201–213.

Windley, B.F., 1983. Metamorphism and tectonics of the Himalya. J. Geol. Soc. London, 140: 849–865.

Windrim, D.P. and McCulloch, M.T., 1983. Nd and Sr isotopic systematics of central Australian granulites: chronology of crustal development and constraints on the evolution of lower continental crust. Contrib. Mineral. Petrol., 94: 289–303.

Winterer, E.L. and Bosellini, A., 1981. Subsidence and sedimentation on Jurassic passive continental margin, southern Alps, Italy. Bull. Am. Assoc. Pet. Geol., 65: 394–421.

Woods, D.V. and Allard, M., 1986. Reconnaissance electromagnetic induction study of the Kapuskasing structural zone: implications for lower crustal conductivity. Phys. Earth Planet. Inter., 42: 135–142.

Woodward, D.J., 1972. Gravity anomalies in Fiordland, southwest New Zealand. N. Z. J. Geol. Geophys., 15: 22–32.

Zingg, A., 1980. Regional metamorphism in the Ivrea zone (Southern Alps, N. Italy): field and microscopic investigations. Schweiz. Mineral. Petrogr. Mitt., 60: 153–179.

Zingg, A., 1983. The Ivrea and Strona–Ceneri zones (Southern Alps, Ticino and N-Italy) — A review. Schweiz. Mineral. Petrogr. Mitt., 63: 361–392.

Zingg, A., Handy, M., Hunziker, J. and Schmid, S.M., 1990. Tectonometamorphic history of the Ivrea zone and its relationship to the crustal evolution of the southern Alps. Tectonophysics, 182: 169–192.

Chapter 9

Magmas as tracers of lower crustal composition: an isotopic approach

G. LANG FARMER

1. Introduction

Although there now exists a considerable body of data on the chemical and petrologic features of exposed lower crustal terranes (Dawson et al., 1986; Park and Tarney, 1987), many fundamental questions remain regarding the chemical composition of the lower continental crust. For example, no consensus exists as to whether the lower crust is dominantly felsic or mafic in composition (Weaver and Tarney, 1984; Kay and Kay, 1986), whether it is everywhere depleted in heat producing elements (Taylor and McLennan, 1985), or whether fundamental chemical differences exist between lower crust produced in the Archean, Proterozoic, and Phanerozoic (Tarney and Weaver, 1987). One method of investigating present-day lower crustal compositions involves analyses of the crustally-derived xenoliths entrained in young continental igneous rocks (cf. Rudnick, this volume). Although such studies have the advantage of dealing directly with portions of the lower crust, xenoliths represent small-scale and possibly biased samples. An alternative method of investigating the lower crust is provided by exposed continental igneous rocks that crystallized from magmas derived either wholly or in part from the deep crust. Such rocks represent remote probes of the lower crust on a large (km) scale and can provide important constraints regarding the chemical nature of the lower continental crust. In this chapter the basic techniques involved in applying isotopic data from continental igneous rocks towards inferring lower crustal compositions are reviewed and the advantages and disadvantages of this approach are outlined through a discussion of case studies, with particular emphasis on data from the western U.S.

2. General techniques

Magmas that form from crustal anatexis, or that originate in the upper mantle and subsequently rise through and interact with the continental crust, are composed at least in part of material that originally resided in the deep crust. As a result, igneous rocks now exposed at the Earth's surface represent remote samples of the lower crust and should provide information regarding the compositions of their deep crustal sources. Unfortunately, such information is not always easily extracted. The major problem is that the chemical compositions of continental igneous rocks do not necessarily record the compositions of their crustal sources, even for rocks that represent the products of crustal anatexis. Instead, magmatic major and trace element compositions are strongly influenced by such parameters as the degree of partial melting of the crustal source, the residual mineralogy of the source(s), and the extent and nature of closed system magmatic differentiation (Fig. 9-1). As a result, it can be difficult to unambiguously determine the chemical characteristics of the crust from the elemental composition of the continental igneous rocks, although important constraints regarding crustal compositions and magmatic evolutionary paths have been made in this manner (Henderson, 1984).

An alternative approach is to infer aspects of the source chemical compositions from igneous rock isotopic compositions. The advantage of using isotopic data is that magmas will generally acquire the isotopic compositions of their source(s), and can, for the heavier elements (>Si, atomic weight = 28 g/mole), retain these isotopic compositions through subsequent closed system differentiation. Therefore, many of the processes that can modify magma chemical compositions will not affect their isotopic characteristics (Fig. 9-1), in which case continental igneous rocks can contain accurate information regarding deep crustal isotopic compositions. But the isotopic information must be translated into a measure of crustal composition. The art of making such a translation is the subject of this chapter.

The use of igneous rock isotopic data to infer crustal compositions ultimately depends upon the fact that the isotopic compositions of a select group of elements in the Earth vary as a result of radioactive decay. These elements include Pb, Sr, Nd, Ca, Hf, Os, Ce, and Ba (Faure, 1986), all of which have radiogenic isotopes contributed by the decay of a long-lived parent isotope. For example, ^{147}Sm decays by α-decay to ^{143}Nd with a half-life of 109 Ga. The ^{147}Sm decay results in variations in the ^{143}Nd/^{144}Nd ratio of the Earth, where ^{144}Nd is a non-radiogenic reference isotope of Nd.

In general, temporal changes in the isotopic composition of any element due to radioactive decay can be described by the equation:

$$D/R = (D/R)_0 + (P/R)\,(e^{\lambda_P t} - 1) \tag{1}$$

where D/R is the present-day ratio of the number of atoms of the daughter isotope (D) to the atoms of reference isotope (R), $(D/R)_0$ is the initial daughter/reference isotope ratio, P/R is the present-day ratio of the parent isotope to the daughter, t is time, and λ is the decay constant for the parent isotope, P (Faure, 1986). This expression illustrates that the rate of change in the D/R ratio for a given decay pair (i.e.^{147}Sm-^{143}Nd, ^{187}Re-^{187}Os, ^{238}U-^{206}Pb, ^{235}U-^{207}Pb, ^{87}Rb-^{86}Sr, etc.) depends primarily on the rate of decay of the parent, time, and the P/R ratio. The P/R ratio is, in turn, proportional to the weight ratio of the parent element to the daughter element (hereafter referred to as the P/D ratio; Faure, 1986).

The application of igneous rock isotopic compositions as probes of crustal compositions, however, relies not just on the fact that long-lived radioactive isotopes exist in the Earth but also on the fact that different domains within continental crust can evolve different elemental isotopic compositions. These domains ultimately develop because variations exist in the weight ratios of the parent to daughter elements (P/D) within the crust. The variations in P/D ratios arise, in turn, because the parent and daughter elements in a given decay pair have different physical and chemical properties and respond differently to different geologic processes. The result is a geochemical *fractionation* between the parent and daughter element, with the magnitude of the fractionation depending on the process and on the elements involved in the decay pair. For example, the range of Sm/Nd ratios in the crust is fairly small, because the parent and daughter are both light rare earth elements (DePaolo, 1988). In contrast, the variation of crustal Rb/Sr ratios is large, reflecting the fact that the parent and daughter elements belong to different chemical families (Faure, 1986).

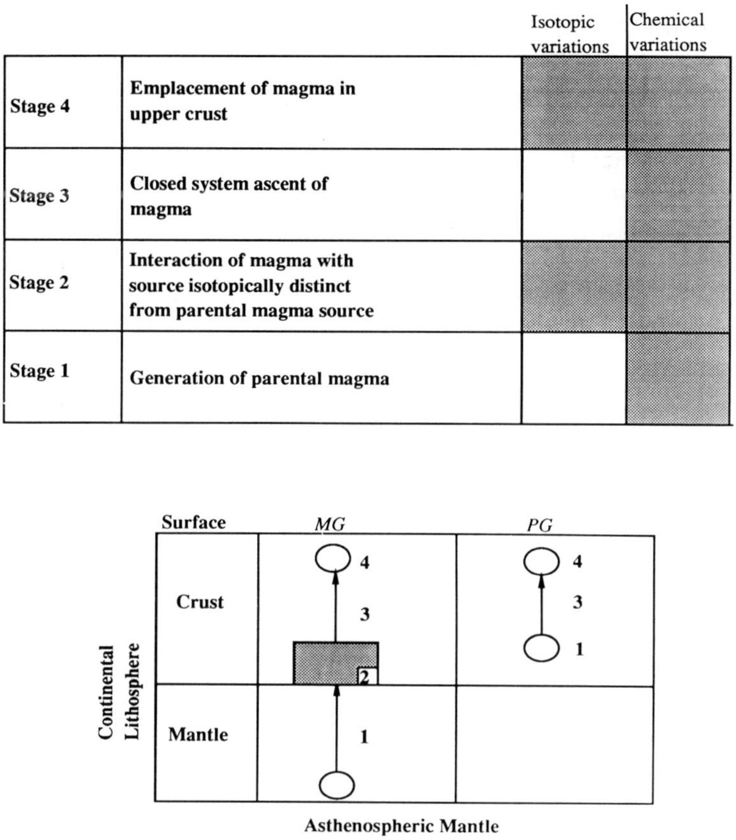

Fig. 9-1. Idealized stages of the chemical and isotopic evolution of continental magmas (after Farmer and DePaolo, 1983). In this figure, an idealized evolutionary framework for continental magmas is presented in which the magma evolution is divided in four stages. Filled boxes at right in upper diagram depict stages in which either chemical or isotopic variations in the magma may take place. Note that magma isotopic variations occur only in stages 2 and 4, while chemical variations can occur in all stages. Stage 1 represents the initial magma generation, which can occur either by melting within the continental crust or underlying mantle (lower diagram). The chemical composition of this parental magma will not match the source, except in the special case of complete melting, and will otherwise depend on pressure, temperature, source mineralogy, and extent of melting. The magma and its source are likely, however, to have the same isotopic compositions. Upon separating from the parental source, the magma may then rise and encounter a foreign reservoir, defined as material chemically and isotopically distinct from the parental source. Open system interaction with this foreign reservoir represents stage 2, and produces further shifts in the magma chemical composition as well as shifts in its isotopic composition. Stage 3 is reached if the magma ceases to interact with the foreign reservoir, but continues to differentiate in a closed system. Some magmas may finally reach stage 4 during emplacement at high levels in the crust. At this stage, material exchange between crust and magma can occur via circulating hydrothermal fluids or by wall rock assimilation.

Two examples of how the above terminology might apply to the generation of metaluminous (MG) and peraluminous (PG) granites are shown in lower diagram. In the example shown, the MG parental magmas are generated in the lithospheric mantle (stage 1), interact with the lower continental crust (stage 2), rise through the mid-crust without crustal interaction (stage 3), and are finally emplaced in the upper crust (stage 4). The PG magmas, on the other hand, are shown as being generated by anatexis in the mid-crust (stage 1), with the magmas rising without crustal interaction into the upper crust (stage 3) where they ultimately crystallize (stage 4).

Several general processes can produce variations in crustal P/D ratios, the most important of which involve the magmatic processes responsible for the initial extraction of the continental crust from the upper mantle, and the magmatic and/or metamorphic processes that subsequently differentiate the crust internally (Ben Othman et al., 1984). The extraction of the crust from the upper mantle is of critical importance, because the chemical differentiation of magmas derived from the mantle can result in variations in P/D ratios that correlate with the magma bulk composition. For example, magmatic differentiation processes, such as crystal fractionation, generally result in igneous rocks where Rb/Sr and U/Pb increase, and Sm/Nd decreases, with increasing SiO_2 (Faure, 1986). Therefore the isotopic composition developed over time by a given crustal segment can reflect its bulk composition. It is important to emphasize that without the P/D fractionations associated with crust formation no internal isotopic differences would develop within the crust, and the isotopic compositions of the crust and mantle, and magmas derived thereof, would remain indistiguishable from one another (Fig. 9-2). It should also be noted that crustal segments with the same P/D ratios, but of substantially different ages, can also have distinctly different isotopic compositions.

Any relationship between crustal composition and P/D ratios that is initially present in the crust can be modified by subsequent internal differentiation processes. For example, crustal material may be transfered from the lower to upper crust, either via fluids or magmas produced during high-grade metamorphism (Brown and Fyfe, 1970; Frost et al., 1989; Wickham and Taylor, 1987). This internal differentiation can accompany crust formation (Moorbath, 1984), or lag by some significant time interval (Ben Othman et al., 1984). During internal differentiation, various parent elements (which, not coincidently, are also the major heat producing elements in the crust) may be preferentially transfered to the upper crust, leaving a lower crust with low Rb/Sr, U/Pb, and Th/Pb ratios (Fig. 9-2). For any given crustal bulk composition, such depletions can significantly alter the isotopic evolution of the lower crust (Fig. 9-2; Moorbath and Taylor, 1986). A few of the P/D ratios, however, have been shown to be largely unaffected by the high grade metamorphism. In particular, the Sm/Nd ratio seem to be only mildly affected by even high-grade metamorphism (Cohen et al., 1988; DePaolo et al., 1982; Jacobsen and Wasserburg, 1978) although some examples of REE mobility have been described (McCulloch, 1984; Whitehouse, 1988; Windrim et al., 1984). Therefore, crustal Nd isotopic compositions may still evolve in response to the original Sm/Nd ratios imposed during crust formation, and may still be related to the crustal bulk compositions. Such a relationship, however, cannot be presumed and must be tested on a case by case basis (see example from western U.S. presented below).

Crustal P/D ratios can also be affected by surficial processes. The Rb/Sr of clastic sediments, for example, can be significantly higher than their crustal sources (Faure, 1986). In this case, the effect on the crustal isotopic evolution is the opposite of high-grade metamorphism, with more radiogenic Sr isotopic compositions being produced for fine-grained clastic sediments relative to unweathered igneous rocks. Other P/D ratios, such as U/Pb, also increase in clastic sediments (Doe and Zartman, 1979), but Sm/Nd ratios are not greatly different between clastic sediments and sources (Taylor and McLennan, 1985). If incorporated into the lower crust, metasedimentary rocks that had significant residence times in the upper crust could represent relatively radiogenic Sr and Pb reservoirs.

The above considerations suggest that isotopic heterogeneities will develop in the continental crust primarily in response to variations in P/D ratios imposed during magmatic and/or metamorphic processes. Magmas derived from the lower crust that retain the isotopic compositions of their crustal sources can therefore contain important indirect information regarding the lower crustal composition or metamorphic grade. But this technique of inferring crustal compositions is not without limitations. First, the crust must be old enough for its

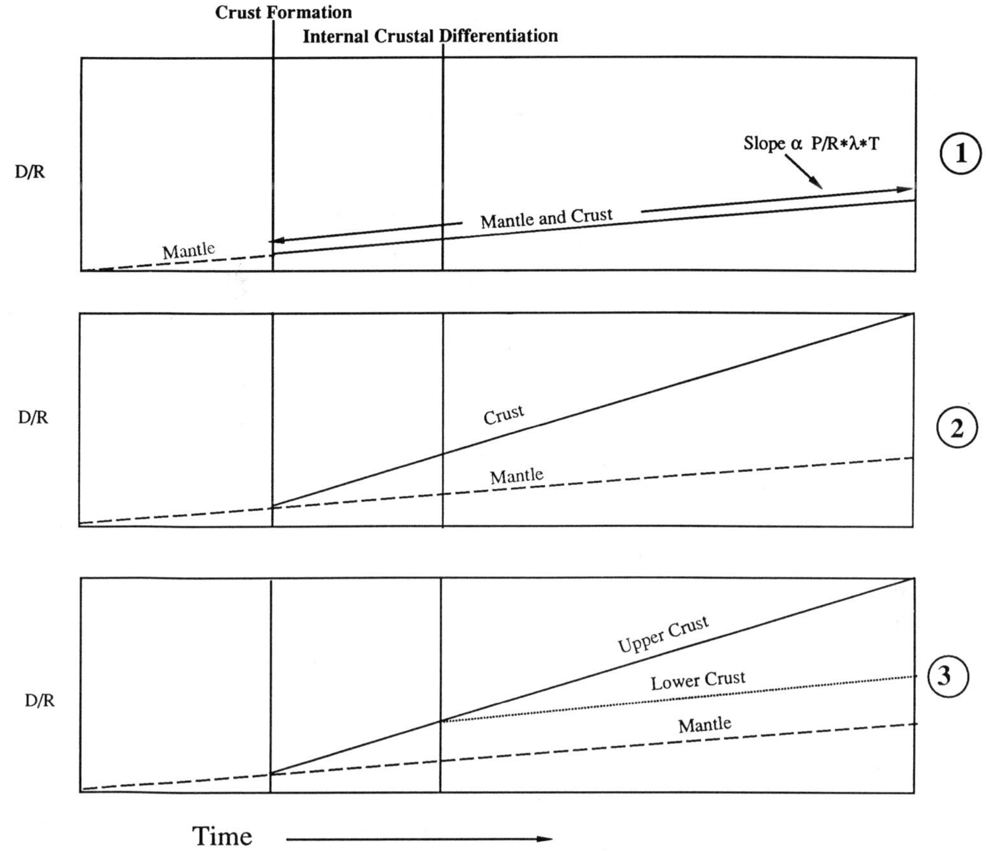

Fig. 9-2. Hypothetical growth curves for daughter/reference (D/R) isotope ratios through time for three different cases.

Case 1: in this case the relative chemical properties of the parent and daughter elements are such that continental crust is formed with the same parent/reference isotope ratio (P/R) as the mantle from which the crust was derived. The P/R ratio is also unmodified during any internal differentiation process accompanying or post-dating crust formation. As a result, the crust and mantle evolve the same D/R ratios through time. As a result, no information regarding crustal compositions can be gained from measuring the isotopic composition of this daughter element in younger continental igneous rocks.

Case 2: In this case, the parent and daughter elements are fractionated from one another during initial crust formation but not during subsequent internal differentiation. As a result, the crust can evolve isotopic compositions distinct from the upper mantle, and can become internally inhomogeneous isotopically if igneous processes produced a spread of P/R ratios at the time of crust formation. Note, however, that a long time period is required for the isotopic compositions of crust and mantle to reflect the differences in their P/R ratios. Decay pairs following this scenario are Sm-Nd and, possibly, K-Ca (see text) and La-Ce.

Case 3: In this case, parent and daughter element are fractionated both during crustal formation and internal differentiation. The isotopic evolution of the crust can be extremely complex, depending on the relative time of the external and internal differentiation of the crust. Decay pairs following this scenario include Rb-Sr, U-Pb, and Th-Pb.

isotopic compositions to have diversified in response to any internal P/D variations (Fig. 9-2). Secondly, continental magmas may sample several compositionally and isotopically distinct crustal segments, and it may not always be possible to isolate the isotopic charateristics of each crustal source from the isotopic compositions of the crystallized igneous rock. Thirdly, only the most fusible segments of the crust may be involved in magma genesis, and inferences regarding crustal compositions from igneous rock isotopic compositions may apply only to select portions of the crust. Fourthly, the crustal magmas must inherit the isotopic compositions their sources had prior to melting, and the source isotopic compositions must have remained unaffected by the metamorphism that preceded crustal anatexis or assimilation (Wickham and Taylor, 1987). Finally, it must also be recognized that the precise crustal depth from which a given continental igneous rocks was either derived, or at which their parental magmas underwent crustal interaction, can rarely be precisely defined. Therefore although continental igneous rocks provide information regarding the isotopic compositions of deep crust, the "deep crust" actually corresponds merely to crust at some deeper level in the continent and not necessarily the lowermost crust.

Given the above difficulties, it is clear that isotopic data from continental igneous rocks cannot be indiscriminately applied towards investigating the chemical makeup of the deep crust. Nevertheless, there are a growing number of studies that have successfully used this approach to indirectly infer aspects of crustal compositions. Most of the applications involve granitic rock isotopic data and such studies will be stressed in the following discussion, although several studies have demonstrated that isotopic and chemical data from continental volcanic rocks can yield similar information (Thompson et al., 1982; Stille and Buletti, 1987; Pichavant et al., 1988). One of the most extensive data sets is from Mesozoic and younger granitic rocks in the western U.S. and these data constitute the case study discussed in detail below.

3. Case example — Western U.S.

The western portion of the U.S. was the site of extensive Mesozoic and Tertiary magmatism (Snyder et al., 1976; Coney and Reynolds, 1977; Barton et al., 1988), much of it corresponding to the subduction-related granite batholiths that formed directly at the paleocontinental margin. The batholith granites, however, are poor indicators of the pre-batholith crustal compositions. The available isotopic data from the Mesozoic batholith granites in the western U.S. suggest that they were derived from a source comprised of pre-existing crust and voluminous amounts of mafic magmas injected into the lower crust presumably from the underlying mantle (DePaolo, 1981c). A similar mixed crustal source has been postulated for other continental margin batholiths (Hildreth and Moorbath, 1988). The effect of the addition of large amounts of mafic material is to pervasively alter the crustal chemical and isotopic compositions. Therefore, although the batholith granites may record the isotopic characteristics of the mixed lower crust, they provide little unambiguous information regarding the pre-magmatism crustal isotopic compositions.

But unlike other circum-Pacific regions there has been extensive continental interior magmatism in the western U.S., at distances up to 1,500 km from the inferred position of the convergent plate margin at the continental margin. Because of the lower magma volumes involved, the injection of mafic magmas may not have greatly altered the crust in which case the igneous rocks resulting from this magma may more faithfully record the pre-magmatism crustal isotopic compositions. Furthermore, the continental interior magmas primarily sampled Precambrian continental lithosphere where there is most likely to be a wide range of internal radiogenic isotopic compositions.

Among the continental interior igneous rocks, there are examples of both peraluminous and metaluminous granites (Farmer and DePaolo, 1983, 1984). The interpretation of the isotopic characteristics of these two granite types requires different approaches and are dealt with separately below.

3.1. Peraluminous granites

Muscovite-bearing, peraluminous granites (PG) of Late Cretaceous to Early Tertiary age are exposed in linear belt inboard but parallel to the continental margin batholiths in the western U.S. (Miller and Bradfish, 1980). These granites have been interpreted to be the products of the anatexis of quartzofeldspathic crust (Miller, 1985) and so their initial isotopic compositions should provide direct information regarding the isotopic compositions of their parental crustal sources.

Field relationships suggest that these granites are intrusive with respect to their immediate wall rocks and so must have been generated at greater paleo-crustal depths than that represented by the current levels of exposure (Lee et al., 1984; Haxel et al., 1981). This conclusion is supported by the fact that, where data are available, the peraluminous granites and their wall rocks have distinctly different isotopic compositions. For example, in the Great Basin and eastern Washington, metasedimentary wall rocks generally have much higher $^{87}Sr/^{86}Sr$ than the PG (Farmer and DePaolo, 1983; Asmerom et al., 1988a; Farmer, unpublished data). Therefore, the granites could not be derived from the anatexis of their immediate wall rocks (see Kistler et al., 1981, for a possible exception), although limited interaction between wall rock and magma may have taken place near the pluton boundaries (Lee and van Loenen, 1971; Lee and Christiansen, 1983). Because the peraluminous granites were generally emplaced at depths between 5 and 15 km (Kistler et al., 1981; Barton, 1987), the granites must have been derived from some greater crustal depth. However, it is unlikely that these magmas could have risen great distances from their source, given their relatively high water contents (Miller, 1985) and, in some cases, their unevolved trace element characteristics (e.g., light rare earth element enrichments, the lack of Eu anomalies; Barton, 1987). On this basis, it is probable that the peraluminous granites represent mid-crustal melting (Haxel et al., 1984) and so their isotopic compositions should provide information regarding the crustal compositions at this depth.

The most striking aspect of the PG isotopic characteristics is the fact that their Nd isotopic compositions vary systematically as a function of position in the southwest U.S., with PG in the northern Great Basin and southeastern California having lower ϵ_{Nd} values (-16 to -19; ϵ_{Nd} values defined in Fig. 9-3 caption) than PG in southern Arizona (-10 to -12; Fig. 9-3). In contrast, the Sr isotopic compositions are fairly uniform throughout the western U.S., ranging, with some exceptions, from 0.715 to 0.725 (Fig. 9-5a). Only a few common Pb determinations are available for the PG (Miller, 1985; Wooden et al., 1988), but these data reveal that the PG in southeastern California have variable, but relatively non-radiogenic, Pb isotopic compositions ($^{206}Pb/^{204}Pb \sim 17$–$19$, $^{207}Pb/^{204}Pb \sim 15.5$–$15.7$, $^{208}Pb/^{204}Pb \sim 38.5$–$39.5$). The Ca isotopic compositions of PG in the northern Great Basin, on the other hand, are the most radiogenic of the few continental rocks yet analyzed ($\epsilon_{Ca} = +3$ to $+5$, Fig. 9-4; Marshall and DePaolo, 1989).

Based on the PG isotopic data alone, only a few broad generalizations can be made regarding their crustal sources. For example, the low ϵ_{Nd} values require an ancient light rare earth element enriched source, as expected for rocks derived from Precambrian continental crust (DePaolo, 1988). But the low $^{206}Pb/^{204}Pb$ and $^{207}Pb/^{204}Pb$ ratios, combined with relatively high $^{208}Pb/^{204}Pb$ ratios, suggest a source depleted in U, but not in Th, relative to Pb (Wooden

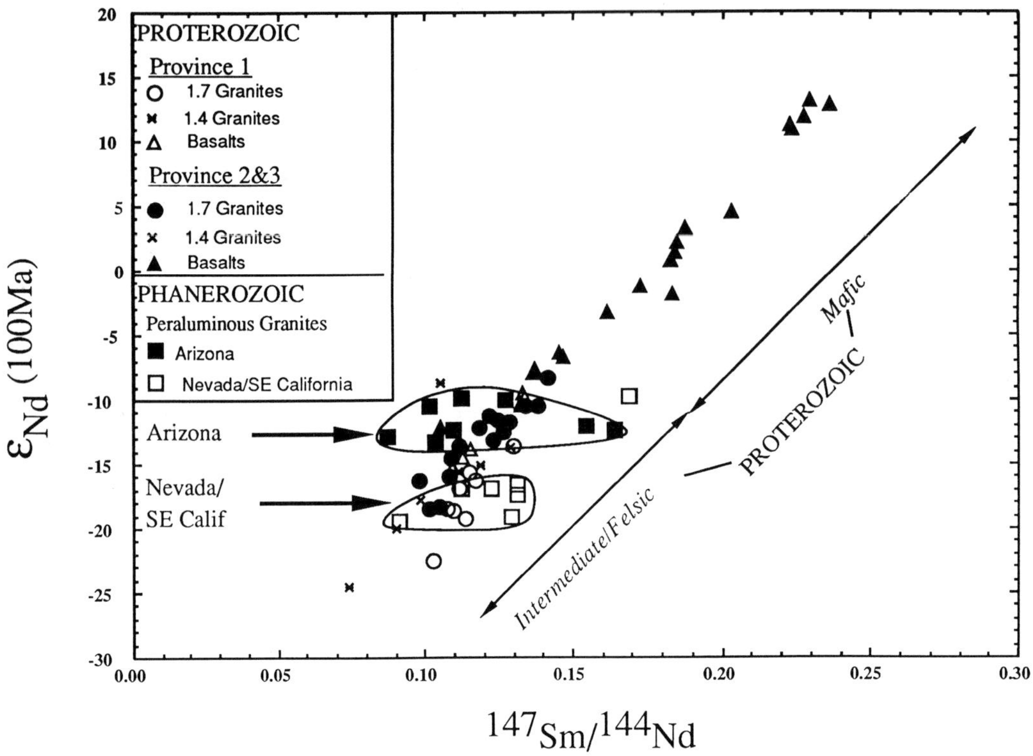

Fig. 9-3. ϵ_{Nd} (100 Ma) vs. $^{147}Sm/^{144}Nd$ for Late Cretaceous to Early Tertiary granitic rocks and Proterozoic continental crust from the southwestern U.S. The ϵ_{Nd} (100 Ma) values are defined as: $[(^{143}Nd/^{144}Nd)_{sample}/(^{143}Nd/^{144}Nd)_{CHUR}) - 1] \times 10^4$, where $(^{143}Nd/^{144}Nd)_{sample}$ is the sample Nd isotopic compositions measured at 100 Ma, and $(^{143}Nd/^{144}Nd)_{CHUR}$ is the equivalent value for a reference reservoir, CHUR (chondritic uniform reservoir; DePaolo, 1988). Peraluminous granite isotopic data from Farmer and DePaolo (1983,1984). Proterozoic crust data from DePaolo (1981a), Nelson and DePaolo (1984), and Bennett and DePaolo (1987). Province 1 (Bennett and DePaolo, 1987) refers to Proterozoic crust (1.7 to 1.4 Ga in age) with Nd model ages between 2.0 and 2.2 Ga which is present in the Great Basin and southeastern California. Provinces 2 and 3 exist within the Proterozoic crust in Colorado, New Mexico, and Arizona and have younger Nd model ages from 1.6 to 2.0 Ga. Felsic to intermediate composition crust in Province 1 and Provinces 2 and 3 also have discretely different ϵ_{Nd} (100 Ma) values (-16 to -19 vs. -10 to -12) which are matched by the spatial variations in the ϵ_{Nd} (100 Ma) of younger peraluminous granites (PG). The PG isotopic compositions from Province 1, and Provinces 2 and 3, are predominately within the circled regions. The similarity between the PG and Proterozoic crust Nd isotopic compositions suggest that the peraluminous granites were derived from felsic to intermediate composition crust at depth within the Proterozoic craton. Note that the peraluminous granites match the Proterozic ϵ_{Nd} values, but not $^{147}Sm/^{144}Nd$, due to fractionation of Sm from Nd during crustal melting and/or differentiation of the magmas parental to the PG.

et al., 1988). Therefore, the data suggest a low μ ($^{238}U/^{204}Pb$) crustal source, but the bulk composition of this source cannot be estimated. Furthermore, these data do not in themselves allow a determination of whether the low μ represents an original characterisitic of the source, or whether this apparent U-depletion was imposed during internal differentiation of the crust.

One approach towards extracting further information on the source composition from the PG isotopic data is to compare the PG isotopic compositions to the values for exposed, weakly metamorphosed, Proterozoic crust. The comparison is useful because most of the PG are exposed in regions underlain by Proterozoic crust and rocks of similar age (but of unknown

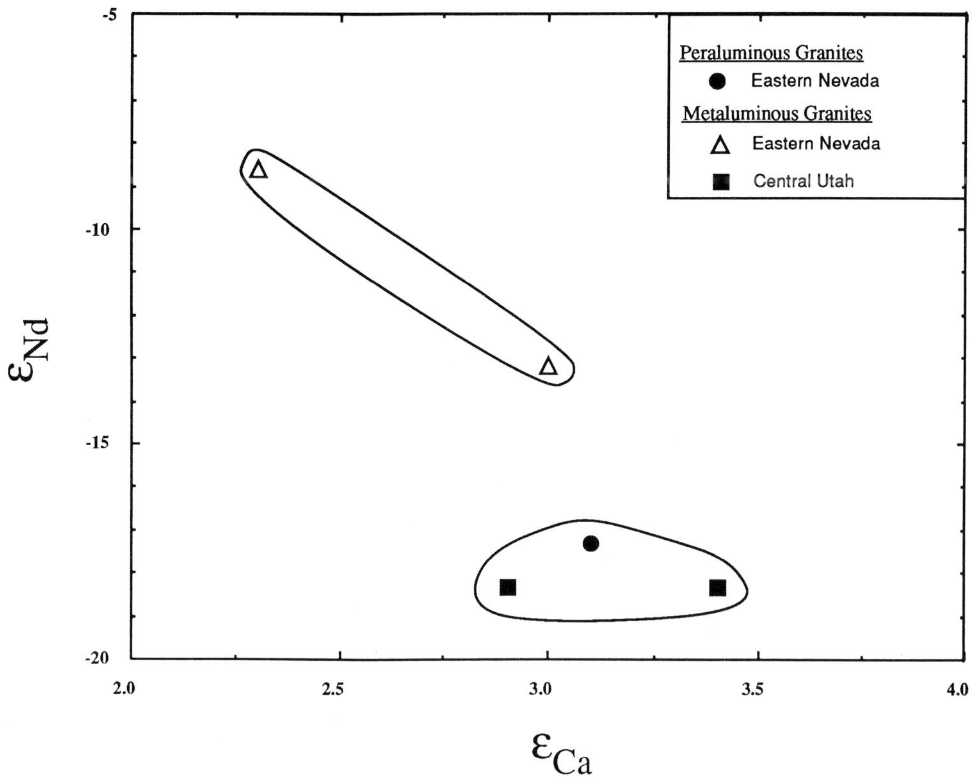

Fig. 9-4. Initial ϵ_{Nd} vs. initial ϵ_{Ca} for metaluminous and peraluminous granites in the northeastern Great Basin. Ca isotopic data from Marshall and DePaolo (1989).

composition) are likely source regions for the granites. Furthermore, the Proterozoic crust isotopic data provides a test of whether the crustal P/D ratios and isotopic compositions do generally vary with crustal bulk composition, a prerequisite for the use of isotopic data from the younger igneous rocks to determine crustal compositions. The exposed Proterozoic crust also provides a template against which the isotopic compositions inferred for the lower crust can be compared, in order to assess the effects of internal differentiation processes on the composition of the deep crust.

The available data suggest that there is a good correlation between the Proterozoic crust ϵ_{Nd} (at 100 Ma, the approximate average age of the peraluminous granites in this region) and $^{147}Sm/^{144}Nd$ values, and its bulk composition (Fig. 9-3). In general, ϵ_{Nd} and $^{147}Sm/^{144}Nd$ both decrease with increasing SiO_2. In addition, felsic Proterozoic crust in the Great Basin and southeastern California, and Proterozoic crust in southern Arizona and New Mexico, have discretely different ϵ_{Nd} values (Bennett and DePaolo, 1987). This observation is significant because the ϵ_{Nd} values for felsic Proterozoic crust and for the PG in each region are coincident (Fig. 9-3). Therefore, it is probable that the peraluminous granites throughout the western U.S. were derived from deep-seated felsic to intermediate compositions rocks, and that this crustal source must have been the same age as, and had similar Sm/Nd ratios to, the exposed Precambrian crust.

A similar conclusion can be reached on the basis of the available Ca isotopic data although a direct comparison of the PG and Proterozoic crust Ca isotopic compositions cannot be

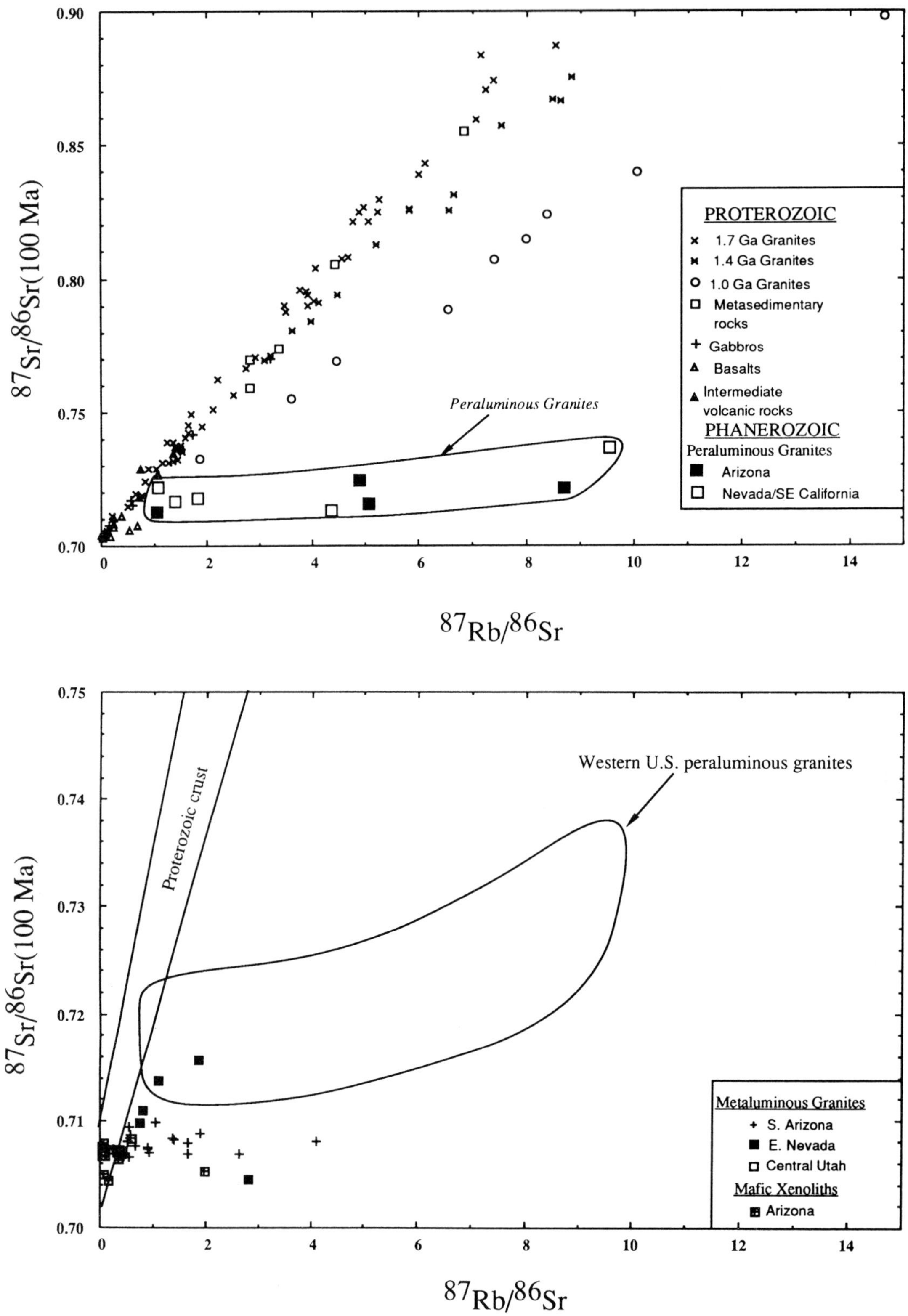
PROTEROZOIC
1.7 Ga Granites
1.4 Ga Granites
1.0 Ga Granites
Metasedimentary rocks
Gabbros
Basalts
Intermediate volcanic rocks
PHANEROZOIC
Peraluminous Granites
Arizona
Nevada/SE California
Peraluminous Granites
$^{87}Sr/^{86}Sr(100 Ma)$
$^{87}Rb/^{86}Sr$
Proterozoic crust
Western U.S. peraluminous granites
Metaluminous Granites
S. Arizona
E. Nevada
Central Utah
Mafic Xenoliths
Arizona

made due to the lack of data from the latter. Nevertheless, Marshall and DePaolo (1989) argued that the high initial ϵ_{Ca} values for the PG (Fig. 9-4) requires a high K/Ca for their crustal source. The time integrated K/Ca ratio for the crustal sources can be calculated using equation (1), with the assumption that the crustal source was 1.6 Ga old at the time the PG were generated and that the crust was originally derived from an island-arc source ($\epsilon_{Ca} = 1$). This calculation yields a crustal K/Ca ratio of about 1. As with Sm/Nd, there is generally a good correlation between K/Ca and crustal bulk composition (LeMaitre, 1976) with a K/Ca ratio of 1 roughly corresponding to a granodiorite composition ($SiO_2 = 65\%$; Marshall and DePaolo, 1989).

The constraints imposed by the Nd and Ca isotopic data preclude the possibility that the PG were derived from mafic crustal sources, as has been proposed to account for the relatively high Na contents for the PG in the southwestern U.S (White et al., 1986). Instead, both the Na contents and the ϵ_{Nd} values for the PG require an ancient, mildly peraluminous, quartzofeldspathic source, such as granitic rocks or immature sedimentary rocks derived thereof (cf. Miller, 1985).

Although both the Nd and Ca isotopic data are consistent with an intermediate to felsic composition, mid-crustal source for the PG, the Sr isotopic compositions of exposed Proterozoic rocks of this compositions do not correspond to the PG Sr isotopic compositions. As shown in Figure 9-5, there is a good correlation between $^{87}Sr/^{86}Sr$ (at 100 Ma), $^{87}Rb/^{86}Sr$, and rock bulk composition for the exposed Proterozoic rocks. In general, basaltic crustal rocks have the lowest $^{87}Sr/^{86}Sr$ and Rb/Sr, and both values increase with increasing SiO_2. The $^{87}Sr/^{86}Sr$ ratios of the PG (0.710–0.730) are considerably lower than the values for most of the exposed intermediate to felsic crustal rocks. Given the requirement that the PG be derived from silicic crust of the same age as the exposed crust, it must be concluded that the crustal source for the PG has undergone a depletion of Rb relative to Sr. Furthermore, this lowering must have occurred during internal differentiation processes operating near the time of crust formation, in order for the crust to have had significantly lower $^{87}Sr/^{86}Sr$ ratios than their nearer surface equivalents 100 Ma ago (Fig. 9-2). Therefore the PG isotopic data reveal crust which has differentially lost Rb relative to Sr, but not Sm relative to Nd, must exist at depth within the Proterozoic basement throughout the southwest U.S., and in portions of the northwest U.S. as well (Asmerom et al., 1988a). Loss of U relative to Pb apparently accompanied the Rb depletion, given the nonradiogenic Pb isotopic compositions and low U contents (Keith and Reynolds, 1980) of many of the PG.

Fig. 9-5. (a) $^{87}Sr/^{86}Sr$ (100 Ma) vs. $^{87}Rb/^{86}Sr$ for Proterozoic basement rocks and Phanerozoic peraluminous granites in the southwestern U.S. Data compiled from many sources (Bickford et al., 1969; Hedge, 1970; Taggart and Brookins, 1975; McCallum and Hedge, 1976; Bryant and Hedge, 1978; Brookins et al., 1980; Wobus and Hedge, 1980; Farmer and DePaolo, 1983, 1984).

(b) Same as (a), but with the inclusion of metaluminous granite data from Farmer and DePaolo (1983, 1984) and Anthony and Titley (1988), and mafic xenolith data from Esperanca et al. (1988). Included with the Arizona MG compilation are isotopic analyses of contemporaneous volcanic rocks reported by Anthony and Titley (1988). Proterozoic crust array is for 100 Ma, all MG $^{87}Sr/^{86}Sr$ are initial values. The MG in southern Arizona have lower $^{87}Sr/^{86}Sr$ ratios than any of the peraluminous granites in this region, while MG and PG in eastern Nevada have overlapping $^{87}Sr/^{86}Sr$ values. The southern Arizona MG are interpreted to have been derived either by direct melting of mafic crust similar isotopically to mafic xenoliths found in this region or by mixing of mafic mantle-derived magmas and anatectic melts of the mafic crust. MG in eastern Nevada probably represent mixtures between low $^{87}Sr/^{86}Sr$ mafic magmas and felsic Proterozoic basement.

The time-integrated Rb/Sr of this crustal source can be explicitly calculated by solving for the P/R ratio in equation (1) and using recent estimates of the initial Sr isotopic compositions for the Proterozoic crust in the western U.S. (Bennett et al., 1988). For example, in eastern California and the Great Basin initial Sr isotopic compositions for exposed Proterozoic crust range from 0.7027 to 0.7120, which, combined with the PG initial values (at 100 Ma) and the assumption that that the source Rb/Sr ratio has not changed since 1.6 Ga ago, yields a source $^{87}Rb/^{86}Sr$ of 0.2 to 0.6 (or Rb/Sr ratios from about 0.07 to 0.21). These are maximum estimates of the source Rb/Sr ratios, since the present day Rb/Sr ratios would have to be even lower if the internal differentiation process that shifted the source Rb/Sr took place at any significant time interval after crust formation. However, the estimated source Rb/Sr ratios are not unreasonable given that similar Rb/Sr ratios are commonly observed for intermediate composition portions of exposed deep crustal terranes (Shaw et al., 1986). Compared to the exposed intermediate to felsic compositions Proterozoic rocks the sources of the PG have undergone about a 2–10 fold decrease in Rb/Sr. It is also important to note that the $^{87}Rb/^{86}Sr$ ratios of the PG (= 1–9) are generally much higher than the values calculated for their sources. The increase in Rb/Sr is most likely the result of the preferential retention of Sr by residual feldspar in the crustal source, or feldspar crystal fractionation from the PG magmas. Fractionation between source and PG Sm/Nd ratios must also have occurred, as evidenced by the differences between the PG Sm/Nd ratios and the values for the felsic crust from which the PG were derived (Fig. 9-3). Both observations reemphasize the difficulties involved in determining source elemental ratios from the chemical compositions of crustally derived magmas, given that the P/D ratios of the PG do not match the source region values inferred from their isotopic compositions. It is also important to note that the range of measured PG Sm/Nd and Rb/Sr ratios overlap those for a wide variety of Precambrian crustal rocks, including both felsic and mafic varieties. Obviously the PG have not preserved the relationship between P/D ratios and bulk composition present within the Precambrian crust itself.

Implicit in all of the above discussion is the assumption that the PG isotopic compositions faithfully reflect the pre-anatexis isotopic compositions of the sources. In other words, the assumption is that no process took place in the source to alter the crustal isotopic compositions just prior to or during crustal melting. Such an alteration could decouple the source chemical and isotopic compositions, in which case the PG isotopic compositions could no longer be used to estimate crustal compositions. The assumption that the PG accurately reflect the crustal isotopic compositions is reasonable if element exchange between magma and residual or unmelted solids occurred only through solid or static liquid state diffusion. In this case the characterisitic elemental diffusion lengths are only on a meter scale (Margaritz and Hofmann, 1978). But if a circulating fluid phase was present at or just prior melting, then this fluid could have transferred material from one isotopically distinct source to another. Such interaction could then have erased any isotopic contasts that may have arisen between crustal rocks of different bulk compositions. An example of such volatile fluxing has been recognized in the Pyrenees (Wickham and Taylor, 1987; Bickle et al., 1988). Here major hydrothermal circulation was apparently induced in conjunction with rising crustal temperatures prior to the extension-related crustal anatexis that led to Hercynian granite formation. In this case, the fluxing was characterized by major lowering in $\delta^{18}O$ values and decrease in $^{87}Sr/^{86}Sr$ of pelitic country rocks. As pointed out by Bickle et al. (1988), this alteration decoupled the isotopic compositions of the crustal rocks, and magmas derived from them, from their pre-alteration bulk compositions.

Detailed oxygen isotopic data sets for peraluminous granites and older crustal rocks in the western U.S. are just becoming available (Wickham and Peters, 1988; Solomon and Taylor, 1989). As with the European Hercynian rocks, the $\delta^{18}O$ values of peraluminous granites in the Great Basin are low relative to the values expected for pelitic sedimentary rocks, the latter representing one proposed source for these granites (Miller, 1985). However, the PG $\delta^{18}O$, as well as their ϵ_{Nd} values, are identical to the values for exposed, and unaltered, Precambrian granitic rocks in southeastern California (Bennett et al., 1988). Therefore although major volatile fluxing of the PG source regions cannot be excluded, such a process is not required to account for the PG oxygen isotopic characteristics.

In summary, the PG isotopic data, when coupled with a similar data set for exposed and unmetamorphosed Proterozoic crustal rocks, provides some important constraints regarding mid-crustal compositions. First, the Nd isotopic data reveal that there is intermediate to felsic composition crust at mid-crustal depths throughout much of the western U.S. The PG data do not require that the mid-crust be exclusively felsic in composition, but only that such material be present in sufficient volume to produce the PG during crustal anatexis. The PG sources are also the same age(s) as exposed Proterozoic crust, and which has experienced no significant shift in Sm/Nd since the time of crust formation. But relative to the exposed crust, the PG sources have had long-lived low U/Pb and Rb/Sr ratios, which could reflect ancient U and Rb depletion events in the mid-crust.

3.2 Metaluminous granites

Another class of young granitic rocks found in the continental interior of the western U.S. is metaluminous in composition (molecular $Al_2O_3/(Na_2O + K_2O + CaO) < 1.00$). The metaluminous granites (MG) range mainly from Jurassic to mid-Tertiary in age (Barton et al., 1988), and differ from the peraluminous granites both in their bulk compositions and their isotopic characteristics. Specifically, the MG have extremely variable isotopic compositions both within and between the various Proterozoic crustal provinces. For example, metaluminous granites that intruded Proterozoic crust in the northern Great Basin span a range of correlated Nd and Sr isotopic compositions from $\epsilon_{Nd} = +1$ to -19, and from $^{87}Sr/^{86}Sr = 0.704$ to 0.716 (Farmer and DePaolo, 1983, 1984; Miller et al., 1989), with the highest ϵ_{Nd} and lowest $^{87}Sr/^{86}Sr$ being associated with the oldest (Jurassic) granitic rocks (Fig. 9-6).

Similar, roughly hyperbolic arrays, of Sr and Nd isotopic data have been observed in metaluminous granitic rock suites worldwide and yet their origin remains enigmatic (McCulloch and Chappell, 1982; Gray, 1984; Juteau et al., 1986; Downes and Duthou, 1988). Several different mechanisms for producing the MG isotopic arrays have been proposed, each of which has different implications regarding whether or not the MG isotopic compositions can be used to extract information regarding the compositions of the deep crust. For example, each granite on such an array could have been derived from a separate, isotopically distinct, crustal source, with the crustal isotopic compositions being themselves correlated. In this case, each MG retains information regarding the composition of a specific crustal segment, as with the peraluminous granites. Most workers, however, interpret hyperbolic arrays of Sr and Nd isotopic compositions as mixing curves. The most commonly invoked model suggests that the mixing takes place between high ϵ_{Nd}, low $^{87}Sr/^{86}Sr$, mafic magmas derived from LREE depleted mantle and ancient low ϵ_{Nd} felsic continental crust (Farmer and DePaolo, 1983, 1984; Anthony and Titley, 1988). In this model, mixing is considered to have occurred during the interaction between mafic magmas and pre-existing crust as these magmas enter

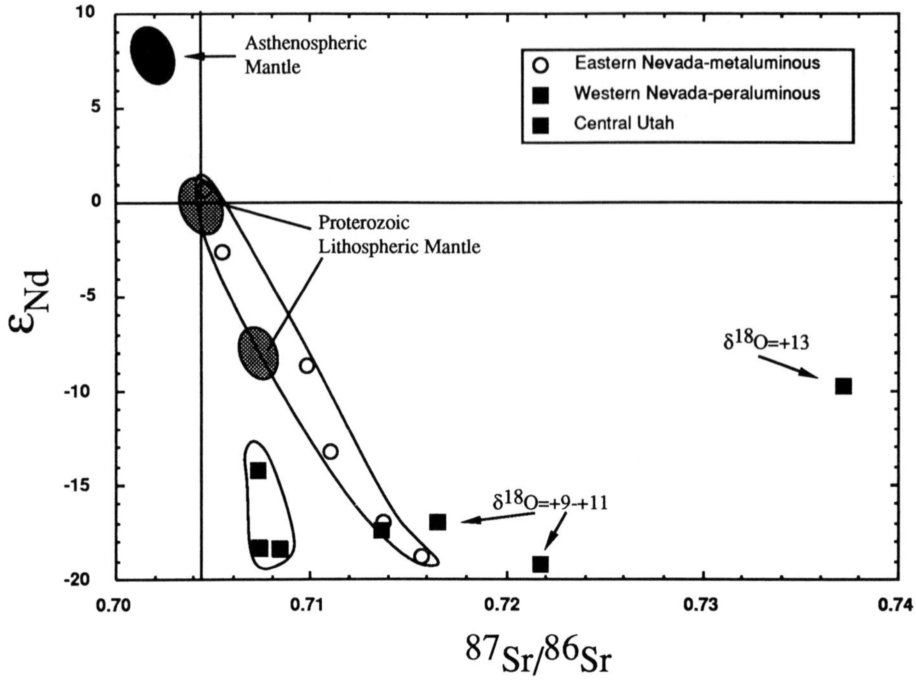

Fig. 9-6. Initial Nd and Sr isotopic compositions for metaluminous and peraluminous granites in the northern Great Basin, Nevada and Utah. Granite data from Farmer and DePaolo (1983, 1984). Estimated isotopic compositions for lithosphere and asthenosphere derived basaltic magmas from Menzies et al.(1983), Phelps et al. (1983), Alibert et al. (1986), Perry et al. (1987), Perry et al. (1988), Farmer et al. (1989). The $\delta^{18}O$ values are from Kistler et al. (1981) and Lee and Christiansen (1983). The high $\delta^{18}O = +13$ PG was probably derived from anatexis of miogeoclinal sedimentary rocks, whereas other PG are interpreted to have been derived from Proterozoic basement underlying the geocline (Farmer and DePaolo, 1983).

the lower crust from the underlying mantle (stage 2 evolution, as defined in Fig. 9-1). With increasing intermixed proportions of the crustal component, the magma isotopic composition more closely approaches the isotopic compositions of the crustal end member. One prediction of this model is that the isotopic compositions of the MG should reflect in part the isotopic characteristics of the deepest crust, given that the crustal temperatures should be greatest here and the crust should be more easily incorporated by the rising magmas (Farmer and DePaolo, 1983).

If mixing of mantle derived magmas and the lowermost continental crust does occur, the difficulty becomes one of determining the isotopic compositions of crust involved in any resulting metaluminous granites. The main problem is that even though each MG in a given suite may represent mixing with the same crust, none of the MG need have acquired the isotopic compositions of this crustal end member. Furthermore, depending on the mechanism and physical conditions of mixing, the MG need not even define an isotopic array that either passes through, or that can be extrapolated to, the crustal isotopic compositions. This problem is best expressed in mixing models that involve concomittant assimilation and fractional crystallization (AFC) by the magmas rising through the crust (Taylor, 1980; DePaolo, 1981b). In such AFC models, the trajectory of mixing lines on an ϵ_{Nd} vs. $^{87}Sr/^{86}Sr$ plot depends on a variety of parameters, including the bulk partition coefficients for Nd and

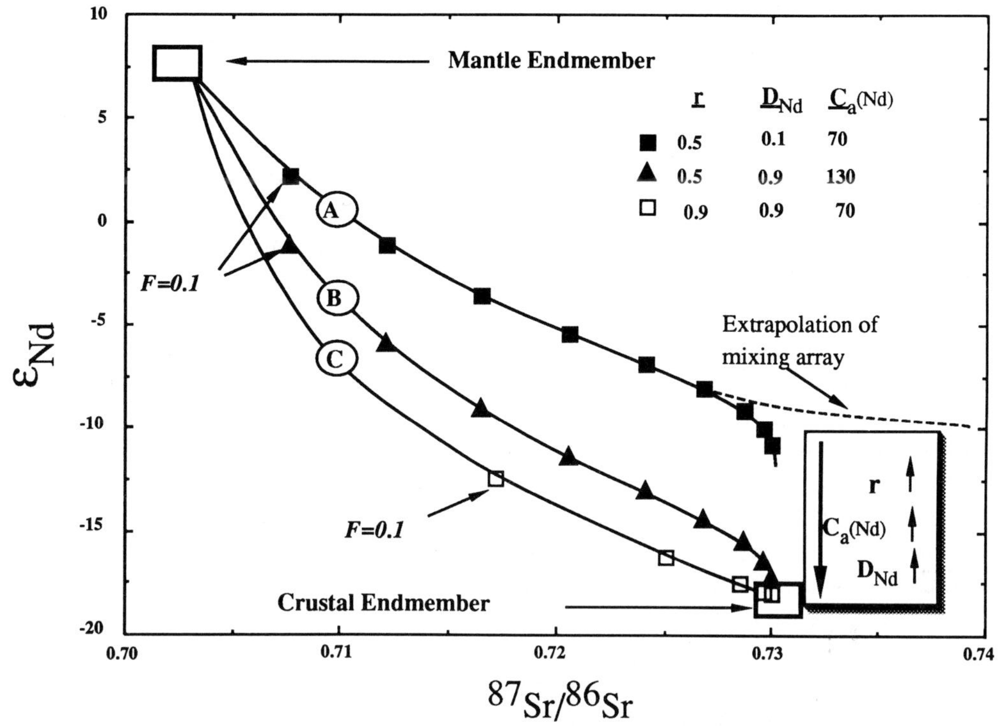

Fig. 9-7. Model assimilation-fractional crystallization (AFC) curves between hypothetical mantle (ϵ_{Nd} = +8, $^{87}Sr/^{86}Sr$ = 0.7035) and crustal (ϵ_{Nd} = −18, $^{87}Sr/^{86}Sr$ = 0.730) end members. Three different mixing arrays are shown (labeled A-C), which correspond to varying r, D_{Nd}, and C_a(Nd) (Nd concentration of crustal assimilant). In all models, the initial magma Sr and Nd concentrations are 100 ppm and 30 ppm, respectively, the crustal assimilant Sr concentration is 150 ppm, and D_{Sr} = 2.

Sr in the fractionating mineral phases (D_{Nd}, D_{Sr}), the rate of assimilation relative to the rate of fractional crystallization (r), and the Sr and Nd concentrations in the original magma and the crustal assimilant (Fig. 9-7). However, for the case of basaltic magmas entering the lower crust only a limited set of AFC parameters are likely. For example, the magmas are unlikely to assimilate large amounts of crust relative to the original magma, unless the crust was previously heated to near solidus temperatures. As a result, low r values ($<\sim 0.5$) will generally be favored for AFC processes in the lower crust, because such low r values imply low relative amounts of assimilated crust (Farmer and DePaolo, 1983). In addition, it is likely that the magma and the crustal assimilant will have similar Nd concentrations (20–90 ppm; DePaolo, 1988). And finally, recent studies have demonstrated that for a wide variety of initial mafic magma and assimilant compositions, the fractionating crystalline phases (plagioclase, olivine, pyroxene) will have low D_{Nd} (<1) and high D_{Sr} (>1) (Nielsen, 1989).

The AFC parameters outlined above will produce a granite isotopic array similar to the curve labeled "A" on Figure 9-7. The pertinent feature of this array is that as crystallization proceeds (as measured by the parameter F, which is the ratio of the final to the original magma mass), crustal assimilation initially drives the magmas towards the crustal isotopic composition along a regular hyperbolic path. But because the original magmatic Sr, but not Nd, is rapidly flushed from the magma and replaced by crustal Sr, the magma attains the crustal Sr isotopic composition well before it reaches the crustal Nd isotopic value. Therefore, with increasing crystallization the magma Sr isotopic composition remains fixed, but the ϵ_{Nd} values continue to decrease until the magma is completely crystallized (Fig. 9-7). Such inflections in MG isotopic arrays have not yet been observed in nature. So if MG ever evolve under AFC conditions similar to those used in this model, then their F values must have been in the range where the MG isotopic compositions follow a regular hyperbolic array. Unfortunately, in this situation an extrapolation of the hyperbolic array towards lower ϵ_{Nd} and higher $^{87}Sr/^{86}Sr$ (Fig. 9-7) will not pass through the Nd isotopic composition of the crustal end member. As a result, inferences regarding the crustal isotopic and chemical compositions based on such extrapolations would be seriously in error, as pointed out by many other workers (James, 1981; Taylor and Sheppard, 1986; Johnson et al., 1990). Of course as r, D_{Nd}, or the crustal Nd concentration increases the AFC mixing arrays do shift downwards towards lower ϵ_{Nd} and eventually intersect both the crustal Sr and Nd isotopic compositions (curves "B" and "C", Fig. 9-7). In this case, the mixing array could yield accurate information regarding the crustal isotopic characteristics. But for a given set of MG it is no simple matter to determine whether or not the operating AFC parameters allowed the MG to follow a trend that leads to the crustal isotopic compositions. Therefore, if a suite of metaluminous granites formed by AFC processes, then it may not always be possible to use their isotopic compositions to draw conclusions regarding crustal compositions.

However, mafic magmas and crust need not mix by AFC processes. One likely alternative is magma mixing, in which case the interaction between mantle magma and crust can be modeled as a two-component mixing (as long as both the mantle and crust derived magmas have uniform and discretely different isotopic compositions). The critical aspect of such simple mixing is that it generates hyperbolic Nd and Sr isotopic arrays that always pass directly through the isotopic compositions of the mixing end members (Fig. 9-8; Gray, 1984; Juteau et al., 1986). The shape (i.e. curvature) of the hyperbola varies, however, and is controlled by the relative ratios of Sr to Nd in each end member, as measured by the parameter W, where $W = (Sr/Nd)_{mantle}/(Sr/Nd)_{crust}$ (DePaolo and Wasserburg, 1979; Juteau et al., 1986). Most importantly, the covariation between the Nd and Sr isotopic compositions in the mixed magmas will be preserved through later closed system magmatic differentiation which could erase evidence of two component mixing from the trace-element compositions of the crystallized granites (Juteau et al., 1986). If magma mixing has taken place, some firm constraints regarding the isotopic compositions of the end members can be extracted from the shape of the MG mixing hyperbola. As shown in Figure 9-8, the Sr isotopic composition of the mantle end member is defined by the x asymptote (x_o), while the Nd isotopic composition of the the crustal end member is constrained by the y asymptote (y_o). Therefore if a given MG isotopic array can be demonstrated to be the result of two component mixing, then some isotopic charateristics of the crustal end member can be extracted from the shape and bounding limits of the mixing array itself.

Obviously, the task of extracting information regarding crustal compositions from MG isotopic data is greatly complicated by the uncertainties regarding whether a suite of MG represent the products of crustal anatexis or mixing between mantle and crustal-derived material. These problems are manifested in the MG in the western U.S. In southern Arizona, for example, Mesozoic and younger metaluminous granites define an isotopic array from ϵ_{Nd} values of about -2 to -9 and $^{87}Sr/^{86}Sr$ from 0.707 to 0.709 (Fig. 9-9; Farmer and DePaolo, 1984; Anthony and Titley, 1988; Asmerom et al., 1988b). One possible interpretation of this array is that the MG represent the interaction between a high ϵ_{Nd}, low $^{87}Sr/^{86}Sr$ magma, either derived from the upper mantle or from mafic Proterozoic crust, and intermediate to felsic

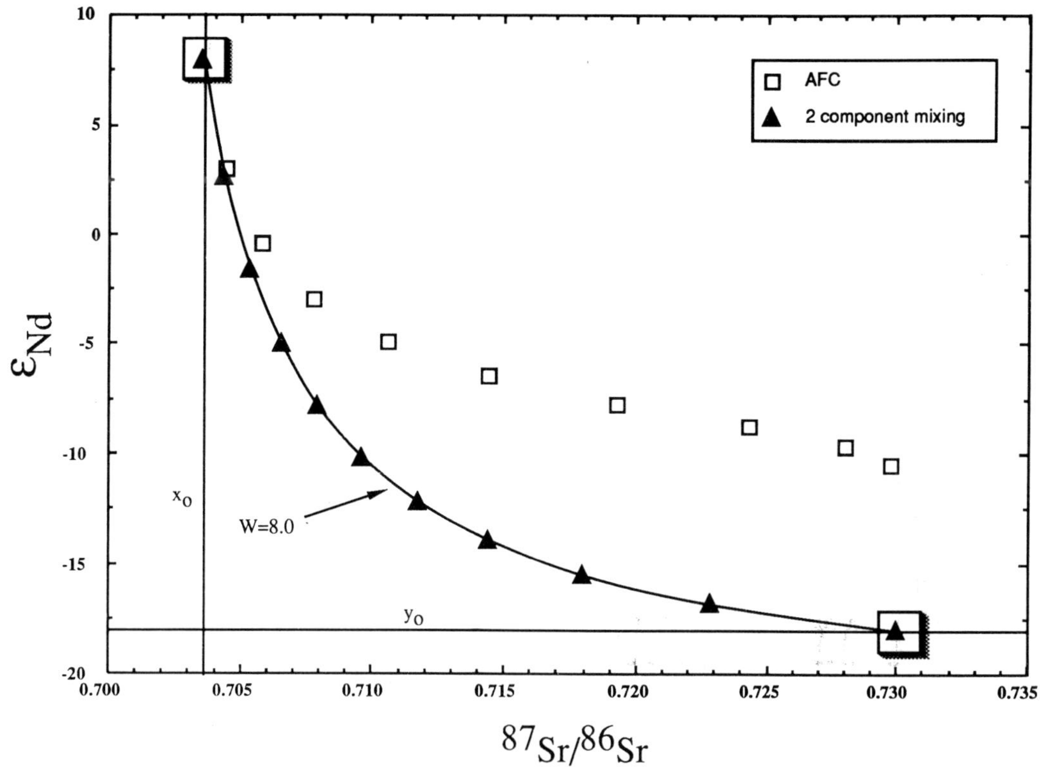

Fig. 9-8. Comparison of 2-component mixing and assimilation-fractional crystallization models. In both models, the same crustal and mantle end members were chosen (crust: $^{87}Sr/^{86}Sr = 0.730$, [Sr] = 150 ppm, $\epsilon_{Nd} = -18$, [Nd] = 70 ppm, mantle: $^{87}Sr/^{86}Sr = 0.7035$, [Sr] = 500 ppm, $\epsilon_{Nd} = +8$, [Nd] = 30 ppm). As in Fig. 9-7, the AFC curve does not pass through crustal end member, due to the low D_{Nd} (0.1) and r (0.5) values used in model. In contrast, the hyperbolic 2-component mixing curves always pass through both the crustal and mantle end members. The curvatures of the hyperbolae produced by mixing, however, will depend on the relative Sr/Nd ratios in the two end members. With the parameters used in this example, $W = (Sr/Nd)_{mantle}/(Sr/Nd)_{crust} = 8.0$, which produces a strongly concave-up hyperbola similar to those observed in metaluminous granite suites worldwide. X_o and Y_o refer to the asymptotes of the hyperbola.

crust (or anatectic melts of such crust) similar to the source of the PG in southern Arizona. However, both Farmer and DePaolo (1984) and Anthony and Titley (1988) noted that neither the MG Sr or Nd isotopic compositions overlap the values for peraluminous granites in this region, with the MG having discretely higher ϵ_{Nd}, and lower $^{87}Sr/^{86}Sr$, values (Fig. 9-9). While this observation could be explained as being the natural consequence of AFC interaction between the magmas and the intermediate composition crust (Fig. 9-7), it is significant to note that the MG isotopic compositions cluster at a specific set of isotopic compositions ($\epsilon_{Nd} \sim -9$ and $^{87}Sr/^{86}Sr \sim 0.7070$; Fig. 9-9). To account for this grouping of isotopic compositions, Anthony and Titley (1988) proposed that many of the MG were actually derived largely from a crustal source with slightly higher ϵ_{Nd} and lower $^{87}Sr/^{86}Sr$ ratios than the PG source. Based on a comparison with the exposed Proterozoic rocks (Fig. 9-3), an $\epsilon_{Nd} = -9$ corresponds to a mafic composition crust ($^{147}Sm/^{144}Nd \sim 0.13$–$0.14$; Anthony and Titley, 1988). If the lowest ϵ_{Nd} MG in southern Arizona do represent melts of mafic crust, then the observed trend towards higher ϵ_{Nd} and lower $^{87}Sr/^{86}Sr$ could represent interaction between a high ϵ_{Nd}

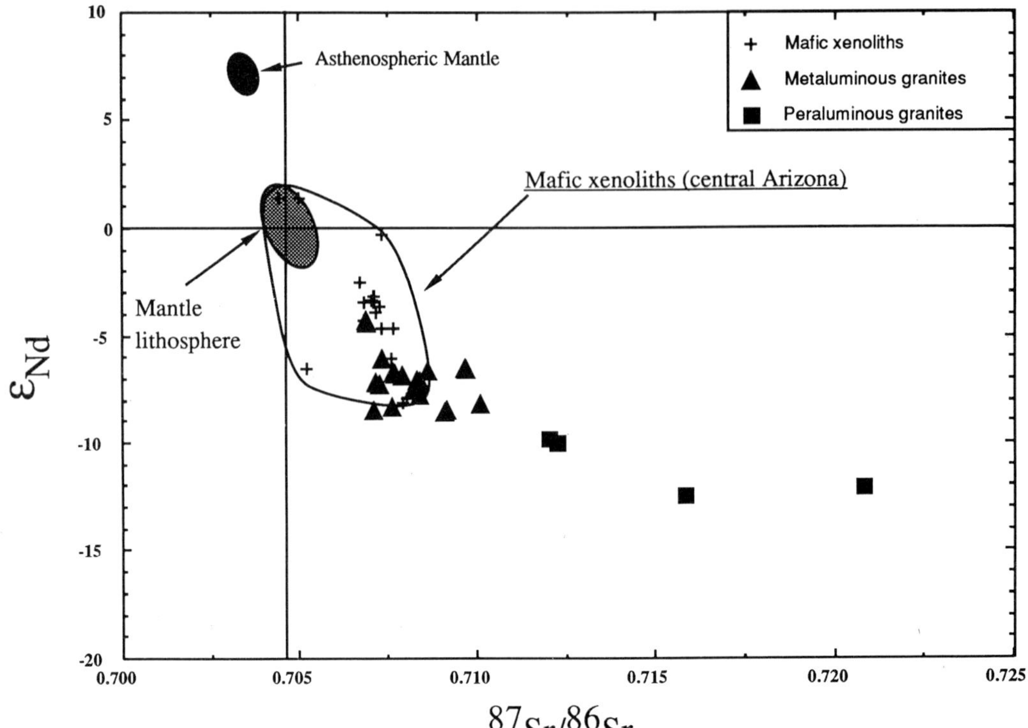

Fig. 9-9. Initial Nd and Sr isotopic compositions for metaluminous and peraluminous granites in southern Arizona. Granite data from Farmer and DePaolo (1984) and Anthony and Titley (1988). Mantle isotopic compositions estimated as in Fig. 9-6. Xenolith isotopic data from Esperanca et al. (1988).

mantle-derived (?) magma ($\epsilon_{Nd} > -2$) and the mafic lower crust. Because the isotopic array formed by the MG passes through the inferred mafic crustal end member, then the interaction might be best modeled as a two-component mixing process (Fig. 9-8).

Therefore, according to Anthony and Titley (1988) the MG isotopic data require that at least some portion of the lower crust beneath southern Arizona be mafic in composition. Interestingly, recent interpretations of seismic reflection profiles in western Arizona have been interpreted as requiring a mafic lower crust in this region which is overlain by a mid-crust of intermediate composition mid-crustal depths (McCarthy et al., 1990). The mafic lower crust and intermediate composition mid-crust could represent the dominant sources of the MG and PG, respectively.

Isotopic studies of lower crustal xenoliths entrained in mid-Tertiary volcanics from central Arizona seem to support models that involve mafic crust in the generation of the MG (Esperanca et al., 1988). The xenolith population contains a large component of mafic amphibolites and eclogites which have a range of Nd and Sr isotopic compositions (calculated at 100 Ma) (Figs. 5b, 9). In general, the eclogites have somewhat higher ϵ_{Nd} and lower $^{87}Sr/^{86}Sr$ than the amphibolites. Esperanca et al. (1988) have interpreted the xenoliths as Proterozoic age mafic lower continental crust, whereas other workers (Schulze and Helmstaedt, 1990) have suggested that eclogite xenoliths may be largely represent samples of the mantle lithosphere. In either case, the range of xenolith and MG ϵ_{Nd} and $^{87}Sr/^{86}Sr$ values overlap, which supports models in which the MG represent magmas (or mixtures of magmas) derived solely from continental lithosphere sources both in the lower crust and lithospheric mantle.

A somewhat different situation is presented by the MG isotopic data in the northeastern Great Basin of Nevada and Utah. Here two distinct isotopic arrays are observed for Mesozoic and younger metaluminous granites (Fig. 9-6). In central Utah, a region underlain by crystalline Proterozoic basement, mid-Tertiary granites show a restricted range of low ϵ_{Nd} values (-14 to -18) and low $^{87}Sr/^{86}Sr$ (0.707–0.708). A few hundred kilometers to the west within the Cordilleran miogeocline, Jurassic to mid-Tertiary metaluminous granites in eastern Nevada have a wide range of correlated ϵ_{Nd} and $^{87}Sr/^{86}Sr$ values that trend towards the same ϵ_{Nd} values as those observed in central Utah (-18), but towards considerably higher $^{87}Sr/^{86}Sr$ ratios (0.714–0.716; Fig. 9-6). Furthermore, the MG isotopic array passes directly through the isotopic compositions observed for peraluminous granites in this region, unlike the situation observed for the MG in southern Arizona. The simplest interpretation of the eastern Nevada granite data is that it represents two component mixing between either asthenosphere or mantle lithosphere-derived mafic magmas, and magmas produced from the anatexis of felsic or intermediate composition crust. Alternatively, the higher ϵ_{Nd}, lower $^{87}Sr/^{86}Sr$ MG could represent the anatexis of more mafic crustal sources although there is no xenolith isotopic data set available to test this hypothesis.

The MG in central Utah must also have been derived from Proterozoic felsic crust, given that their Nd isotopic compositions are also identical to PG in eastern Nevada. But the lower $^{87}Sr/^{86}Sr$ of the central Utah granites requires a source with an even lower time-integrated Rb/Sr than that estimated for the PG crustal source. Again assuming an initial crustal Sr isotopic composition equivalent to that determined for Proterozoic crust in southeastern California, the maximum source Rb/Sr is about 0.07, which is about 3 times lower than the maximum value determined for the PG source. Central Utah MG also tend to have lower $\delta^{18}O$ ratios (6.5–9‰ vs. 9–13‰; Solomon and Taylor, 1988) and less radiogenic Pb isotopic compositions (Fig. 9-10a,b) than those in eastern Nevada (Stacey et al., 1968; Stacey and Zartman, 1978). These observations all suggest that the central Utah MG were derived from a deep-seated, Proterozoic age, felsic crustal source that is at a significantly higher metamorphic grade than felsic crust beneath eastern Nevada (Farmer and DePaolo, 1983; Christiansen et al., 1988; Solomon and Taylor, 1989). The high-grade (granulite?) metamorphism, and the possible attendent Rb and U depletions, must have occurred early in the history of this crustal segment in order to account for its very unradiogenic Sr and Pb isotopic compositions (Fig. 9-2). However, the metamorphism apparently did not affect the crustal K/Ca ratio, given that metaluminous granites in central Utah and peraluminous granites in eastern Nevada have the same Ca isotopic compositions (Fig. 9-4; Marshall and DePaolo, 1989). Although the K/Rb of the high-grade crust cannot be explicitly calculated from the available isotopic data, the apparent decrease in Rb/Sr, combined with the lack of any shift in K/Ca, suggests that the felsic crust experienced increasing K/Rb with increasing metamorphism, a feature common in exposed granulite terranes (Rudnick et al., 1985).

There are several other important implications of the MG data worth emphasizing. First, Proterozoic crust in central Utah can be demonstrated, on the basis of the MG isotopic data, to contain low Rb/Sr and low U/Pb, possibly high metmamorphic grade (granulite?), intermediate to felsic composition lower crust. In contrast, there is no evidence from the MG in southern Arizona that similar intermediate composition lower crust is present here, even though both regions are underlain by cratonal segments of the Proterozic crust. The significance of this observation in terms of models for the origin and accretionary tectonics of

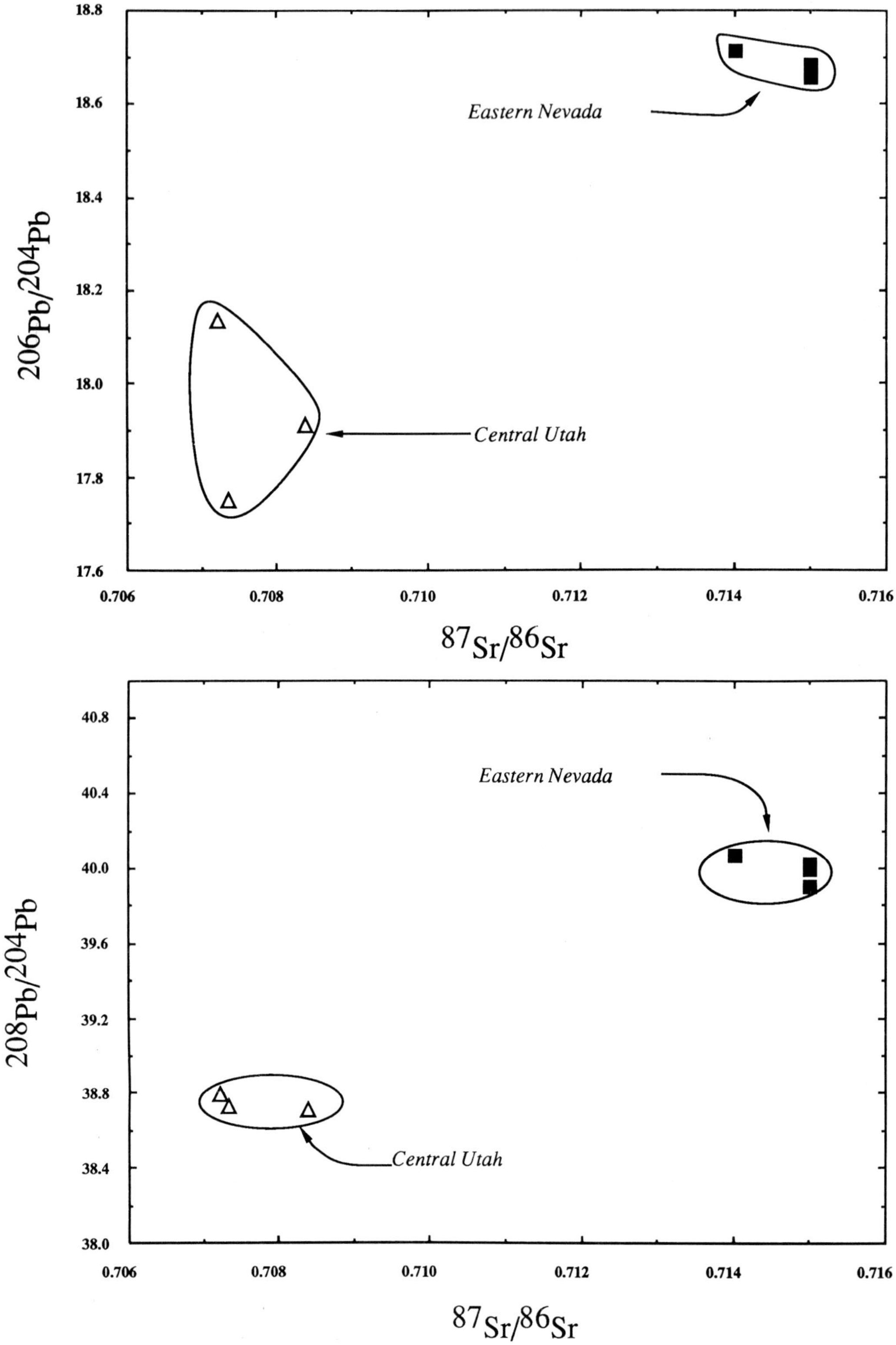

Fig. 9-10. $^{206}Pb/^{204}Pb$ vs. initial $^{87}Sr/^{86}Sr$ (a) and $^{208}Pb/^{204}Pb$ vs. initial $^{87}Sr/^{86}Sr$ (b) for metaluminous granites in central Utah and eastern Nevada. Data from Stacey et al. (1968), Stacey and Zartman (1978), and Farmer and DePaolo (1983).

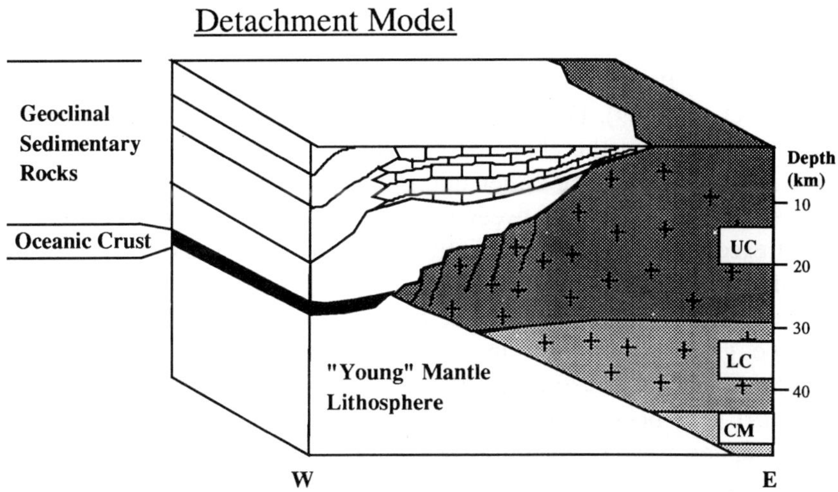

Fig. 9-11. Detachment model (Lister et al., 1986) for Late Proterozoic rifting in the northern Great Basin. Shown is a possible cross-section of the northern Great Basin in the Late Triassic. The detachment model refers to the mechanism of rifting and is used to account for the relative distribution of high grade (granulite) Proterozoic lower crust (*LC*) and lower grade upper crust (*UC*) in this region, as deduced from the metaluminous granite isotopic data. In this model, Late Proterozoic rifting is assumed to take place along a eastward dipping low angle normal fault. With the removal of the footwall (or lower plate) of the normal fault during rifting, only the hanging wall remains with North America. And because of the disposition of the original detachment fault, the leading (western) edge of the remaining Precambrian basement consists of only of upper crust. The space originally occupied by the lower crust, and any associated lithospheric mantle (*CM*), is replaced by upwelling asthenospheric mantle (which subsequently cools to produce a relatively young lithospheric mantle). The rootless Proterozoic upper crust is the source tapped by both metaluminous and peraluminous granites in eastern Nevada. Further to the east the Proterozoic crust thickens and eventually crust that was largely unaffected by rifting is encountered. The lower crust preserved beneath central Utah is considered to be the source of the MG in this region.

the Proterozoic crust is not clear, but the MG isotopic data reveal nevertheless that there are important regional differences in the average composition of Proterozoic lower crust in the western U.S.

Secondly, significant variations in the makeup of cratonal vs. non-cratonal portions of the Proterozoic basement also seem to exist. For example, granulite grade felsic lower crust appears to exist in the craton of central Utah, but not at the rifted margin of the Proterozoic crust in eastern Nevada. To account for the apparent lack of granulite grade, felsic lower crust in eastern Nevada, Farmer and DePaolo (1983) suggested that the crustal basement underlying the miogeoclinal sedimentary rocks in this region either was never severely Rb or U depleted, or that such lower crust, and the mantle lithosphere, associated with the Precambrian craton were physically removed during Late Proterozoic continental rifting (Fig. 9-11) Although the detachment rift model shown in Fig. 9-11 is speculative, it demonstrates how the constraints provided by the MG isotopic data can be used to investigate models of crustal tectonic processes.

4. Other examples

Extensive isotopic data sets are available for Phanerozoic granitic rocks worldwide, particularly from granites in Australia (McCulloch and Chappell, 1982; Gray, 1984; Hensel et al.,

1985), Alaska (Blum et al., 1987; Arth et al., 1989), Europe (Bernard-Griffiths et al., 1985; Juteau et al., 1986; Downes and Duthou, 1988; Liew and Hofmann, 1988), Canada (Ashwal et al., 1986), Antartica (Borg et al., 1987, 1990), and Asia (Gariepy et al., 1985; Deniel et al., 1987; Xianhua et al., 1989). Few of these studies, however, have been designed with the specific intent of determining the chemical compositions of unexposed deep crust. Instead, these studies have concentrated on either determining the sources of the granitic magma or on mapping subsurface boundaries between discrete age provinces within the crust.

An important exception are the isotopic studies of the Caledonian (515–390 Ma) granites in Scotland and northern England, as summarized by Harmon et al. (1984). As in the western U.S., the Caledonian granites include both peraluminous and metaluminous granites, and much of the granite plutonism in northern Britain occurred within a segment of preexisting Precambrian crust. However, the pre- and syn-tectonic peraluminous granites (the older granites) have been generally interpreted as anatectic melts of Late Proterozoic metasedimentary upper crust (Harmon et al., 1984), and so the isotopic characteristics of these granites yield few contraints regarding the nature of the deep crust in northern Britain. The Sr and Pb isotopic compositions of the younger metaluminous granites, however, were much lower than those of the metasedimentary upper crust at the time of granite formation (Fig. 9-12). Therefore, although some of the granites have clearly interacted with the upper crust (Fig. 9-12), most workers have suggested that the deeper lithosphere was the parental source of the younger granites (Harmon et al., 1984). In much of Scotland, this deeper component in the Caledonian granites apparently consists of both mantle and lower crustal sources (Halliday, 1981, 1984). For example, the Pb, Nd and Sr isotopic compositions of one of the best studied Caldedonian granites (the Etive Complex) span between the values expected for the oceanic mantle 400 Ma years ago, and the values measured for Archean (Lewisian) granulites exposed in northernmost Scotland (Fig. 9-12). Many workers have interpreted these data, as well as isotopic data from Tertiary volcanic rocks in this region, as evidence that Archean crust underlies much of Scotland, even in regions where only Proterozic crust is exposed (Dickin, 1981; Frost and O'Nions, 1985). Other workers have suggested, largely on the basis of the Caledonian granite Pb isotopic compositions, that mid-Proterozoic crust may also underlie this region (Clayburn et al., 1983; Clayburn, 1988). However, regardless of whether the deep crustal component in the Caledonian granites in Scotland is Archean (2.7 Ga) or mid-Proterozoic (1.0 Ga), the crust clearly had time-integrated Rb/Sr and U/Pb lower than, and Th/U higher than, the exposed crust. Therefore, as in the western U.S., broad tracts of the lower crust in northern Britain may have experienced ancient Rb and U depletions.

5. Conclusions

The above examples serve to point out both the advantages and disadvantages of dealing with the isotopic compositions of igneous rocks as probes of deep crustal compositions. In the western U.S., the most stringent constraints regarding the composition of the deep crust were achieved by a direct comparison of the isotopic compositions of younger peraluminous granites and exposed, relatively unmetamorphosed, older crust. Some of the important observations drawn from this comparison were:

(1) The crustal P/D ratios (Sm/Nd, Rb/Sr, etc.) of exposed Proterozoic crust in the western U.S. do covary with crustal bulk composition, and that these different crustal compositions have evolved discretely different isotopic compositions.

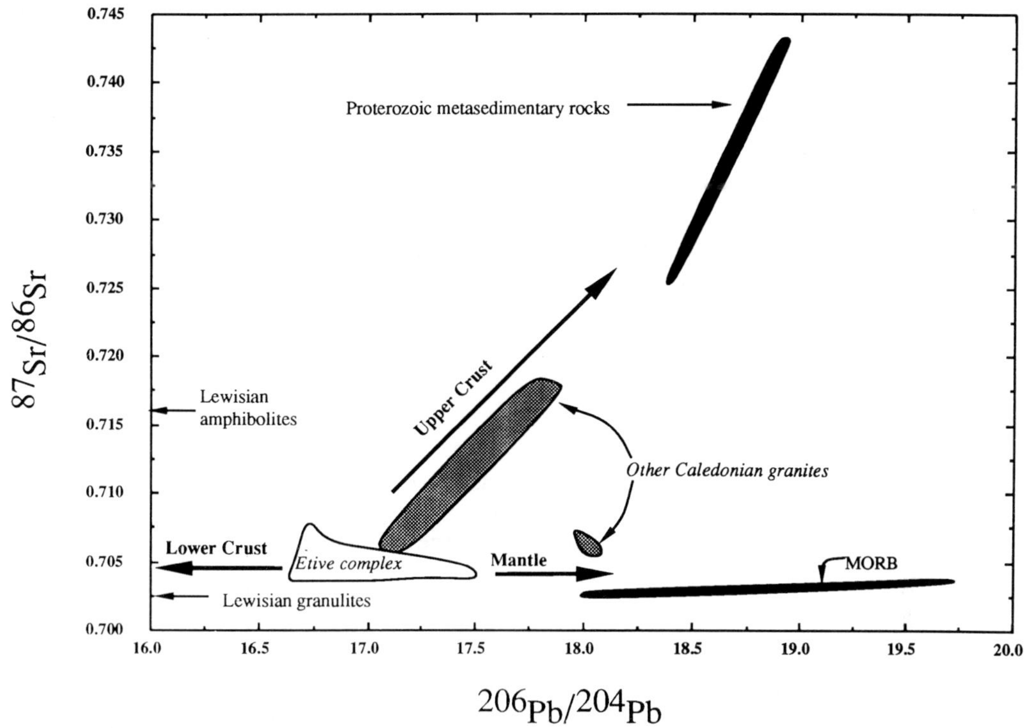

Fig. 9-12. Pb and Sr isotopic data from younger Caledonian granites and basement rocks in northern Great Britain. Isotopic compositions for both granites, crust, and upper mantle source similar to that of mid-ocean ridge basalts (MORB) plotted at the approximate time of granite formation (400 Ma). Modified from Clayburn (1988). The data suggest that the younger granites may represent mixtures between material derived from at least three different sources including the upper and lower crusts, and the upper mantle. Although the lower crustal source may include both Archean and mid-Proterozoic components, this source clearly has low $^{87}Sr/^{86}Sr$ and low $^{206}Pb/^{204}Pb$ ratios relative to the values for exposed upper crustal rocks in this region. Therefore, it is probable that this lower crust has had long term low U/Pb and Rb/Sr ratios.

(2) Peraluminous granites throughout the western U.S. have been derived from ancient felsic to intermediate composition crust, based on the similarity between felsic crust and PG Nd isotopic compositions. This similarity also suggests that felsic mid-crust within the Proterozoic craton of the western U.S. is of the same age and has the same Sm/Nd ratio as the exposed felsic crust. This result supports the claim that Sm/Nd is largely unaffected by internal differentiation processes within the crust.

(3) Large tracts of the felsic middle crust in the western U.S. apparently experienced, near the time of crust formation, a major depletion in Rb relative to Sr and U relative to Pb.

Firm conclusions regarding crustal compositions from the metaluminous granites in the western U.S. are not as forthcoming, due to uncertainties regarding the mode of origin of these rocks. But a comparison of the isotopic compositions of the MG, PG, and Proterozoic crustal rocks suggests:

(1) that major regional differences in the bulk composition of the Proterozoic lower crust may exist, with the lower crust in southern Arizona being dominantly mafic in composition, while further to the north in central Utah an important component of intermediate to felsic composition lower crust is required.

(2) highly Rb and U depleted lower crust exists within the cratonal portions of the Proterozoic crust in central Utah (as also observed in northern Britain), but is apparently absent in the deepest portions of the Proterozoic basement found directly to the west. The differences in the makeup of the lower crust across the Great Basin could have been imposed during Late Proterozoic rifting, and illustrate that the composition of the lowermost crust can be affected by tectonic processes operating well after crust formation.

It should be reemphasized that in both the western U.S. and Britain, the comparison between the isotopic compositions of the granite and exposed crust provided the key element in interpreting the granite data. Without such a comparison, the granite isotopic data would provide few firm constaints regarding the composition of the deep crust. But under appropriate circumstances, granite isotopic data can provide a useful alternative methods of studying the deep continental crust.

Acknowledgments

The manuscript has been greatly improved by the careful editing and reviews of D. Fountain and C. Frost.

References

Alibert, C., Michard, D. and Albarede, F., 1986 Isotope and trace element geochemistry of Colorado Plateau volcanics. Geochim. Cosmochim. Acta, 50: 2735–2750.

Anthony, E.Y. and Titley, S.R., 1988 Progressive mixing of isotopic reservoirs during magma genesis at the Sierrita porphyry copper deposit, Arizona: inverse solutions. Geochim. Cosmochim. Acta, 52: 2235–2249.

Arth, J.G., Zmuda, C.C., Foley, N.K., Criss, R.E., Patton, W.W.J. and Miller, T.P., 1989. Isotopic and trace element variations in the Ruby and Angayucham terranes. J. Geophys. Res., 94: 15491–15955.

Ashwal, L.D., Wooden, J.L. and Emslie, R.F., 1986. Sr, Nd, and Pb isotopes in Proterozoic intrusives astride the Grenville Front in Labrador: Implications for crustal contamination and basement mapping. Geochim. Cosmochim. Acta, 50: 2571–2585.

Asmerom, Y., Ikramuddin, M. and Kinart, K., 1988a. Geochemistry of Late Cretaceous granitoids from northeastern Washington: Implications for genesis of two-mica Cordilleran granites. Geology, 16: 431–435.

Asmerom, Y., Damon, P.E., Patchett, P.J. and Shafiqullah, M., 1988b. Geochemical data from the southern Cordillera of North America: Implications for crust–mantle interaction during early Mesozoic arc magmatism. EOS, Trans. Am. Geophys. Union, 69: 518.

Barton, M.D., 1987. Lithophile-element mineralization associated with Late Cretaceous two-mica granites in the Great Basin. Geology, 15: 337–340.

Barton, M.D., Battles, D.A., Bebout, G.E., Capo, R.C., Christiensen, J.N., Davis, S.R., Hanson, R.B., Michelsen, C.J. and Trim, H.E., 1988. Mesozoic contact metamorphism in the western United States. In: W. G. Ernst, editor, Metamorphism and Crustal Evolution of the Western United States. Rubey Volume VII, pp. 110–178.

Ben Othman, D., Polve, M. and Allegre, C.J., 1984. Nd-Sr isotopic composition of granulities and constraints on the evolution of the lower continental crust. Nature, 307: 510–515.

Bennett, V.C. and DePaolo, D.J., 1987 Proterozoic crustal history of the western United States as determined by neodymium isotopic mapping. Geol. Soc. Am. Bull., 99: 674–685.

Bennett, V.C., DePaolo, D.J. and Smith, B.M., 1988. Nd, O, and Sr isotopic contrasts in the Proterozoic crust across the Colorado river region. Geol. Soc. Am. Abstr. Programs, 20: 142.

Bernard-Griffiths, J., Peucat, J.J., Sheppard, S. and Vidal, P., 1985. Petrogenesis of Hercynian leucogranites from the southern Armorican Massif: Contribution of REE and isotopic (Sr, Nd, Pb and O) geochemical data to the study of source rock characteristics and ages. Earth Planet. Sci. Lett., 74: 235–250.

Bickford, M.E., Wetherill, G.W., Barker, F. and Lee-Hu, C., 1969. Precambrian Rb-Sr chronology in the Needle Mountains, southwestern Colorado. J. Geophys. Res., 74: 1660–1687.

Bickle, M.J., Wickham, S.M., Chapman, H.J. and Taylor, J., H.P., 1988. A strontium, neodymium and oxygen isotope study of hydrothermal metamorphism and crustal anatexis in the Trois Seigneurs Massif, Pyrennes, France. Contrib. Mineral, Petrol., 100: 399–417.

Blum, J.D., Blum, A.E., Davis, T.E. and Dillon, J.T., 1987. Petrology of cogenetic silica-saturated and oversaturated plutonic rocks in the Ruby geanticline of north-central Alaska. Can. J. Earth Sci., 24: 159–169.

Borg, S.G., DePaolo, D.J. and Smith, B.M., 1990. Isotopic structure and tectonics of the Central Transantarctic Mountains. J. Geophys. Res., 95: 6647–6667.

Borg, S.G., Stump, E., Chappell, B.W., McCulloch, M.T., Wyborn, D., Armstrong, R.L. and Holloway, J.R., 1987. Granitoids of northern Victoria Land, Antarctica: Implications of chemical and isotopic variations to regional crustal structure and tectonics. Am. J. Sci., 287: 127–169.

Brookins, D.G., Bolton, W.R. and Condie, K.C., 1980. Rb-Sr isochron ages of four Precambrian igneous rock units from south-central New Mexico. Isochron/West, 29: 31–36.

Brown, G.C. and Fyfe, W.S., 1970. The production of granitic melts during ultrametamorphism. Contrib. Mineral. Petrol., 28: 310–318.

Bryant, B. and Hedge, C.E., 1978. Granite of Rosalie Peak, a phase of the 1700-million-year-old Mount Evans pluton, Front Range, Colorado. J. U.S. Geol. Surv., 6: 447–451.

Christiansen, E.H., Funkhouser, M.J. and Howell, K.A., 1988. Petrogenesis of rare-metal granites from depleted crustal source, an example from the Cenozoic of western Utah, USA. In: R.P. Taylor (Editor), Recent Advances in the Geology of Granite-Related Mineral Deposites. Spec. Vol. Can. Inst. Min. Metal., pp. 53–57.

Clayburn, J.A.P., 1988. The crustal evolution of Central Scotland and the nature of the lower crust: Pb, Nd and Sr isotope evidence from Caledonian granites. Earth Planet. Sci. Lett., 90: 41–51.

Clayburn, J.A.P., Harmon, R.S., Pankhurst, R.J. and Brown, J.F., 1983. Sr, O, and Pb isotope evidence for origin and evolution of Etive Igneous Complex, Scotland. Nature, 303: 492–497.

Cohen, A.S., O'Nions, R.K., Siegenthaler, R. and Griffin, W.L., 1988. Chronology of the pressure–temperature history recorded by a granulite terrain. Contrib. Mineral. Petrol., 98: 303–311.

Coney, P.J. and Reynolds, S.J., 1977. Cordilleran Benioff zones. Nature, 270: 403–405.

Dawson, J.B., Carswell, D.A., Hall, J. and Wedepohl, K.H., 1986. The nature of the lower continental crust. Geol. Soc. Spec. Pub. 24: 394pp.

Deniel, C., Vidal, P., Fernandez, A., LeFort, P. and Peucat, J.J., 1987. Isotopic study of the Manaslu granite (Himalaya, Nepal): Inferences on the age and source of Himalayan leucogranites. Contrib. Mineral. Petrol., 96: 78–92.

DePaolo, D.J., 1981a. Nd in the Colorado Front Range and implications for crust formation and mantle evolution in the Proterozoic. Nature, 291: 193–196.

DePaolo, D.J., 1981b. Trace element and isotopic effects of combined wall-rock assimilation and fractional crystallization. Earth Planet. Sci. Lett., 53: 189–202.

DePaolo, D.J., 1981c. A neodymium and strontium isotopic study of Mesozoic calc-alkaline granitic batholiths of the Sierra Nevada and Peninsular ranges, California. J. Geophys. Res., 86: 10470–10488.

DePaolo, D.J., 1988. Neodymium Isotope Geochemistry, An Introduction. Springer-Verlag, Berlin, 187 pp.

DePaolo, D.J., Manton, W.I., Grew, E.S. and Halpern, M., 1982. Sm-Nd, Rb-Sr and U-Th-Pb systematics of granulite facies rocks from Fyfe Hills, Enderby Land, Antarctica. Nature, 298: 614–618.

DePaolo, D.J. and Wasserburg, G.J., 1979. Petrogenetic mixing models and Nd-Sr isotopic patterns. Geochim. Cosmochim. Acta, 43: 615–627.

Dickin, A.P., 1981. Isotope geochemistry of Tertiary igneous rocks from the Isle of Skye, NW Scotland. J. Petrol., 22: 155–189.

Doe, B.R. and Zartman, R.E., 1979. Plumbotectonics, The Phanerozoic. In: H.L. Barnes (Editor), Geochemistry of Hydrothermal Ore Deposits. Wiley-Interscience, New York, N.Y., pp. 22–70.

Downes, H. and Duthou, J., 1988. Isotopic and trace element arguments for the lower-crustal origin of Hercynian granitoids and pre-Hercynian orthogneisses, Massif Central (France). Chem. Geol., 68: 291–308.

Esperanca, S., Carlson, R.W. and Shirey, S.B., 1988. Lower crustal evolution under central Arizona: Sr, Nd, and Pb isotopic and geochemical evidence from the mafic xenoliths of Camp Creek. Earth Planet. Sci. Lett., 70: 26–40.

Farmer, G.L. and DePaolo, D.J., 1983. Origin of Mesozoic and Tertiary granite in the western United States and implications for pre-Mesozoic crustal structure. I. Nd and Sr isotopic studies in the geocline of the northern Great Basin. J. Geophys. Res., 88: 3379–3401.

Farmer, G.L. and DePaolo, D.J., 1984. Origin of Mesozoic and Tertiary granite in the western United States and implications for pre-Mesozoic crustal structure. 2. Nd and Sr isotopic studies of unmineralized and Cu- and Mo-mineralized granite in the Precambrian craton. J. Geophys. Res., 89: 10141–10160.

Farmer, G.L., Perry, F.V., Semken, S., Crowe, B., Curtis, D. and DePaolo, D.J., 1989. Isotopic evidence on the structure and origin of subcontinental lithospheric mantle in southern Nevada. J. Geophys. Res., 94: 7885–7898.

Faure, G., 1986. Principles of Isotope Geology. John Wiley and Sons, New York, N.Y., 589 pp.

Frost, C.D. and O'Nions, R.K., 1985. Caledonian magma genesis and crustal recycling. J. Petrol., 26: 515–544.

Frost, R.B., Frost, C.D. and Touret, J.L.R., 1989. Magmas as a source of heat and fluids in granulite metamorphism. In: D. Bridgewater (Editor), Fluid Movements — Element Transport and the Composition of the Deep Crust. Kluwer Academic Publishers, Dordrecht, pp. 1–18.

Gariepy, C., Allegre, C.J. and Hua Xu, R., 1985. The Pb-isotopic geochemistry of granitoids from the Himalaya–Tibet collision zone: Implications for crustal evolution. Earth Planet. Sci. Lett., 74: 220–234.

Gray, C.M., 1984. An isotopic mixing model for the origin of granitic rocks in southeastern Australia. Earth Planet. Sci. Lett., 70: 47–60.

Halliday, A.N., 1981. On the sources of uranium in some Scottish Caledonian granites. Mineral. Mag., 44: 437–442.

Halliday, A.N., 1984. Coupled Sm-Nd and U-Pb systematics in late Caledonian granites and the basement under northern Britain. Nature, 307: 229-233.

Harmon, R.S., Halliday, A.N., Clayburn, J.A.P. and Stephens, W.E., 1984. Chemical and isotopic systematics of the Caledonian intrusions of Scotland and Northern England: A guide to magma source region and magma–crust interaction. Philos. Trans. R. Soc. London, Ser. A, 310: 709–742.

Haxel, G.B., Tosdal, R.M., May, D.J. and Wright, J.E., 1984. Latest Cretaceous and early Tertiary orogenesis in south central Arizona: Thrust faulting, regional metamorphism, and granitic plutonism. Geol. Soc. Am. Bull., 95: 631–653.

Hedge, C.E., 1970. Whole rock Rb-Sr age of the Pikes Peak batholith, Colorado. U.S. Geol. Surv., Prof. Pap., 700-B: B86–B89.

Henderson, P. (Editor), 1984. Rare Earth Element Geochemistry. Elsevier, Amsterdam, 510 pp.

Hensel, H.D., McCulloch, M.T. and Chappell, B.W.,1985. The New England batholith: Constraints on its derivation from Nd and Sr isotopic studies of granitoids and country rocks. Geochim. Cosmochim. Acta, 49: 369–384.

Hildreth, W. and Moorbath, S., 1988. Crustal contributions to arc magmatism in the Andes of central Chile. Contrib. Mineral. Petrol., 89: 455–489.

Jacobsen, S.B. and Wasserburg, G.J., 1978. Interpretation of Nd, Sr and Pb isotope data from Archean migmatites in Lofoten–Vesteralen, Norway. Earth Planet. Sci. Lett., 41: 245–253.

James, D.E., 1981. The combined use of oxygen and radiogenic isotopes as indicators of crustal contamination. Annu. Rev. Earth Planet. Sci., 9: 311–344.

Johnson, C.M., Lipman, P.W. and Czamanske, G.K., 1990. H, O, Sr, Nd, and Pb isotope geochemistry of the Latir volcanic field and cogenitic intrusions, New Mexico, and relations between evolution of a continental magmatic center and modification of the lithosphere. Contrib. Mineral. Petrol., 104: 99–124.

Juteau, M., Michael, A. and Albarede, F., 1986. The Pb-Sr-Nd isotope geochemistry of some recent circum-Mediterranean granites. Contrib. Mineral. Petrol., 92: 331–340.

Kay, R.W. and Kay, S.M., 1986. Petrology and geochemistry of the lower continental crust: An overview. In: J.B. Dawson, D.S. Carswell, J. Hall and K.H. Wedepohl (Editors), The Nature of the Lower Continental Crust. Geol. Soc. Spec. Publ., 24: 147–159.

Keith, S.B. and Reynolds, S.J. 1980. Geochemistry of Cordilleran metamorphic core complexes. In: P.J. Coney and S.J. Reynolds (Editors), Cordilleran Metamorphic Core Complexes and their Uranium Favorability. U.S. Dept. of Energy Open-file Rpt. GJBx-256(80), 321 pp.

Kistler, R.W., Ghent, E.D. and O'Neil, J.R., 1981. Petrogenesis of garnet two-mica granites in the Ruby Mountains, Nevada. J. Geophys. Res, 86: 10591–10606.

Lee, D.E. and Christiansen, E.H., 1983. The granite problem as exposed in the southern Snake Range, Nevada. Contrib. Mineral. Petrol., 83: 99–116.

Lee, D.E., Kistler, R.W., Friedman, I. and VonLoenen, R.E., 1981. Two-mica granites of northeastern Nevada. J. Geophys. Res., 86: 10,607–10,616.

Lee, D.E. and van Loenen, R.E., 1971. Hybrid granitoid rocks of the southern Snake Range, Nevada. U.S. Geol. Surv., Prof. Pap., 668: 48 pp.

LeMaitre, R.W., 1976. The chemical variability of some common igneous rocks. J. Petrol.,17: 589–637.

Liew, T.C. and Hofmann, A.W., 1988. Precambrian crustal components, plutonic associations, plate environment of the Hercynian Fold Belt of Central Europe: Indications from a Nd and Sr isotopic study. Contrib. Mineral. Petrol., 98: 129–138.

Lister, G.S., Etheridge, M.A. and Symonds, P.A., 1986. Detachment faulting and the evolution of passive continental margins. Geology, 14: 246–250.

Margaritz, M. and Hofmann, A.W., 1978. Diffusion of Sr, Ba and Na in obsidian. Geochim. Cosmochim. Acta, 42: 595–605.

Marshall, B.D. and DePaolo, D.J., 1989. Calcium isotopes in igneous rocks and the origin of granite. Geochim. Cosmochim. Acta, 53: 917–922.

McCallum, M.E. and Hedge, C.E., 1976. Rb-Sr ages of granitic rocks in the Rawah Batholith, northern Colorado. Isochron/West, 17: 33–37.

McCarthy, J., Okaya, D.A. and Goodwin, E.B., 1990. Geophysical character of the middle and lower crust in the southwestern US: Results from seismic profiling. Geol. Soc. Am. Abstr. Programs, 22: 65–66.

McCulloch, M.T. and Chappell, B.W., 1982. Nd isotopic characteristics of S- and I-type granites. Earth Planet. Sci. Lett., 58: 51–64.

McCulloch, M.T.a.B., L.P., 1984. Sm-Nd isotopic systematics of Enderby Land and evidence for the redistribution of Sm and Nd during metamorphism. Earth Planet. Sci. Lett., 71: 46–58.

Menzies, M.A., Leeman, W.P. and Hawkesworth, C.J., 1983. Isotope geochemistry of Cenozoic volcanic rocks reveals mantle heterogeneity below western USA. Nature, 303: 205–209.

Miller, C.F., 1985. Are strongly peraluminous magmas derived from pelitic sedimentary sources? J. Geol., 93: 673–689.

Miller, C.F. and Bradfish, L.J., 1980. An inner Cordilleran belt of muscovite-bearing plutons. Geology, 8: 412–416.

Miller, D.M., Wooden, J.L. and Wright, J.E., 1989. Mantle-derived Late Jurassic plutons emplaced during possible regional extension of the crust, northwest Utah and northeast Nevada. Geol. Soc. Am. Abstr. Programs, 21: 117.

Moorbath, S., 1984. Origin of granulites. Nature, 312: 290.

Moorbath, S. and Taylor, P.N., 1986. Geochronology and related isotope chemistry of high-grade metamorphic rocks from the lower continental crust. In: J.B. Dawson, D.A. Carswell, J. Hall and K.H. Wedepohl (Editors), The Nature of the Lower Continental Crust. Geol. Soc. Spec. Publ., 24: 211–220.

Nelson, B.K. and DePaolo, D.J., 1984. 1,700 Myr greenstone volcanic successions in southwestern North America and isotopic evolution of Proterozoic mantle. Nature, 311: 143–146.

Nielsen, R.L., 1989. Phase equilibria constraints on liquid lines of descent generated by paired assimilation and fractional crystallization: Trace elements and Sr and Nd isotopes. J. Geophys. Res., 94: 787–794.

Park, R.G. and Tarney, J., 1987. Evolution of Lewisian and Comparable Precambrian High Grade Terrains. Geol. Soc. Spec. Pub., 27: 315 p.

Perry, F., Baldridge, W.S. and DePaolo, D.J., 1987. Role of asthenosphere and lithosphere in the genesis of late Cenozoic basaltic rocks from the Rio Grande Rift and adjacent regions of the southwestern United States. J. Geophys. Res., 92: 9193–9213.

Perry, F.V., Baldridge, W.S. and DePaolo, D.J., 1988. Chemical and isotopic evidence for lithospheric thinning beneath the Rio Grande rift. Nature, 332: 432–434.

Phelps, D.W., Gust, D.A. and Wooden, J.L., 1983. Petrogenesis of the mafic feldspathoidal lavas of the Raton-Clayton volcanic field, New Mexico. Contrib. Mineral. Petrol., 84: 182–190.

Pichavant, M., Kontak, D.J., Briqueu, Herrera, J.V. and Clark, A.J., 1988. The Miocene–Pliocene Macusani volcanics, SE Peru. Contrib. Mineral. Petrol., 100: 325–338.

Rudnick, R.L., McLennan, S.M. and Taylor, S.R., 1985. Large ion lithophile elements in rocks from high-pressure granulite facies terrains. Geochim. Cosmochim. Acta, 49: 1645–1655.

Schulze, D.J. and Helmstaedt, H., 1990. Garnet pyroxenites and eclogites from Chino Valley, Arizona: Lower crust or upper mantle? Geol. Soc. Am. Abstr. Programs, 22: 81.

Shaw, D.M., Cramer, J.J., Higgins, M.D. and Truscott, M.G., 1986. Composition of the Canadian Precambrian shield and the continental crust of the Earth. In: J.B. Dawson, D.A. Carswell, J. Hall and K.H. Wedepohl (Editors), The Nature of the Lower Continental Crust. Geol. Soc. Spec. Publ., 24: pp. 275–282.

Snyder, W.S., Dickinson, W.R. and Silberman, M.L., 1976. Tectonic implications of space-time patterns of Cenozoic magmatism in the western United States. Earth Planet. Sci. Lett., 32: 91–106.

Solomon, G.C. and Taylor, J., H.P., 1989. Isotopic evidence for the origin of Mesozoic and Cenozoic granitic plutons in the northern Great Basin. Geology, 17: 591–594.

Stacey, J.S. and Zartman, R.E., 1978. A lead and strontium isotopic study of igneous rocks and ores from the Gold Hill mining district, Utah. Utah Geol., 5: 1–15.

Stacey, J.S., Zartman, R.E. and Nkomo, I.T., 1968. A lead isotope study of galenas and selected feldspars from mining districts in Utah. Econ. Geol., 63: 796–814.

Stille, P. and Buletti, M., 1987. Nd-Sr isotopic characteristics of the Lugano volcanic rocks and constraints on the continental crust formation in the South Alpine domain (N-Italy–Switzerland). Contrib. Mineral. Petrol., 96: 140–150.

Taggart, J.E. and Brookins, D.G., 1975. Rb-Sr whole rock age determinations for Sandia Granite and Cibola Gneiss, New Mexico. Isochron/West, 12: 5–8.

Tarney, J. and Weaver, B.L., 1987. Geochemistry of the Scourian complex: Petrogenesis and tectonic models. In: R. G. Park and J. Tarney,editors, Geol. Soc. Spec. Publ., 27: pp. 45–56.

Taylor, H.P., Jr. and Sheppard, S.M.F., 1986. Igneous rocks: I. Processes of isotopic fractionation and isotope systematics. In: J.W. Valley, H.P. Taylor, Jr. and J.R. O'Neil (Editors), Reviews in Mineralogy, 21: 227–272.

Taylor, H.P., Jr., 1980. The effects of assimilation of country rocks by magmas on $^{18}O/^{16}O$ and $^{87}Sr/^{86}Sr$ systematics in igneous rocks. Earth Planet. Sci. Lett., 36: 359–362.

Taylor, S.R. and McLennan, S.M., 1985. The Continental Crust: Its Composition and Evolution. Oxford, Blackwell Scientific Publishers, 312 pp.

Thompson, R.N., Dickin, A.P., Gibson, I.L. and Morrison, M.A., 1982. Elemental fingerprints of isotopic contamination of Hebridean Paleocene mantle-derived magmas by Archaen sial. Contrib. Mineral. Petrol., 79: 159–168.

Weaver, E.B. and Tarney, J., 1984. Major and trace element composition of the continental lithosphere. In: H.N. Pollack and V.R. Murthy (Editors), Phys. Chem. Earth, 17: 39–68.

White, A.J.R., Clemens, J.D., Hollaway, J.R., Silver, L.T., Chappell, B.W. and Wall, V.J., 1986. S-type granites and their probable absence in southwestern North America. Geology, 14: 115–118.

Whitehouse, M.J., 1988. Granulite facies Nd-isotopic homogenization in the Lewisian complex of northwest Scotland. Nature, 331: 705–707.

Wickham, S.M. and Peters, T.M., 1988. Fluid and melt transport in anatectic environments. Geol. Soc. Am. Abstr. Programs, 20: A304.

Wickham, S.M. and Taylor, J., H.P., 1987. Stable isotope constraints on the origin and depth of penetration of hydrothermal fluids associated with the Hercynian regional metamorphism and crustal anatexis in the Pyrennes. Contrib. Mineral. Petrol., 95: 255–268.

Windrim, D.P., McCulloch, M.T., Chappell, B.W. and Cameron, W.E., 1984. Nd and Sr isotopes and REE in central Australian granulites: A case study of rare earth element mobility. Earth Planet. Sci. Lett., 70: 27–39.

Wobus, K.A. and Hedge, C.E., 1980. Rb-Sr isochron age of Precambrian plutons of the San Pedro Mountains, north-central New Mexico. Isochron/West, 27: 19–25.

Wooden, J.L., Stacey, J.S., Howard, K.A., Doe, B.R. and Miller, D.M., 1988. Pb isotopic evidence for the formation of Proterozoic crust in the southwestern United States. In W.G. Ernst (Editor), Metamorphism and Crustal Evolution of the Western United States., Rubey Vol. VII. Prentice-Hall, New York, N.Y., pp. 69–86.

Xianhua, L., Tatsumoto, M., Premo, W.R. and Xuntang, G., 1989. Age and origin of the Tanghu Granite, southeast China: Results from U-Pb single zircon and Nd isotopes. Geology, 17: 395–399.

Chapter 10

Fluids in the deep crust — Petrological and isotopic evidence

STEPHEN M. WICKHAM

1. Introduction

Fluid transport in crustal rocks is a fundamental process controlling the geochemical evolution of the continental crust (e.g., Fyfe et al., 1978; Ferry, 1980; Etheridge et al., 1983; Newton, 1989; Wickham, 1989a). Because it is mostly inaccessible to direct observation, this process is poorly understood and difficult to study, even within the upper crust. We have even less understanding of fluid transport within the enigmatic and remote lower crust, for which we have a restricted number of samples exposed to study at the surface, in which multiple deformational and thermal overprints often seriously complicate the mineralogical and petrographic evidence.

Deep crustal fluids probably mostly comprise complex mixtures of various volatile components (the most important of which are H_2O and CO_2 but which also include other species such as CH_4, N_2, SO_2, and the rare gases) and also various dissolved species. Some of these fluids are highly reactive with silicate minerals, particularly at high pressures and temperatures, and can promote melting. Because they are supercritical fluids and have very low viscosities, they are also potentially highly mobile, and therefore theoretically able to transport dissolved components over considerable distances. They are therefore theoretically capable of altering the chemical and isotopic composition of large regions of the crust. Understanding the scale of these chemical transport processes is seriously hindered by our very incomplete knowledge of the solubility of common ionic species in C–O–H fluids at elevated P and T.

Aside from chemical components, fluids also transport heat efficiently and therefore play a very significant role in regulating the temperature field within the crust (e.g., Norton and Taylor, 1979). Inasmuch as the stability of mineral assemblages, and mineral solubilities, are governed by temperature and by fluid composition, and because reactions involving breakdown of hydrous silicates or melting can in turn influence the chemical composition and physical properties of the fluid phase, there is a very complex interplay between these various parameters. Understanding the behaviour of fluids in the lower crust is a first-order geological problem, but one that necessarily involves a complete understanding of the various physical and chemical processes involved and their influence on each other. In the study of fluid or melt transport and its influence on the mineral assemblages that form the lower crust we are still at an embryonic stage.

This chapter will review evidence from various lines of investigation of the nature of lower crustal fluids including studies of phase equilibria, fluid inclusions, and isotope systematics, and also fluid dynamical constraints. Although observations and analytical measurements are for the most part easy enough to make, all of the data provided by these techniques are beset with problems of interpretation that in some cases make them inadequate to constrain fluid transport. These problems vary with the various techniques, and they must be well understood in order that significant conclusions may be drawn. This situation is changing rapidly as a result of the current intense research activity in this area, with a number of significant discoveries having been made during the eighteen months prior to the time of writing (1989). Although we will not be primarily concerned with melting, which is also covered in a separate chapter (see Kay et al., this volume) it is impossible to divorce the behaviour of lower crustal fluids from phase equilibria involving melting, or from melt transport itself. However, in this chapter we will only be concerned with relatively low melt-fraction systems containing less than ~20% melt.

2. Equilibrium thermodynamic calculations

2.1. Overview

The most extensive samples of the deep crust that are available for us to study are amphibolite and granulite facies terranes that are either exposed in craton interiors (e.g., the Archean rocks of the Superior Province in Ontario, Canada — see Percival et al., this volume) or in large tectonic slices brought up adjacent to major fault zones (e.g., the Ivrea Zone in the Southern Alps, Zingg, 1983). Because pressure estimates from mineral assemblages in these ancient terranes fully encompass the range of pressures that are today experienced

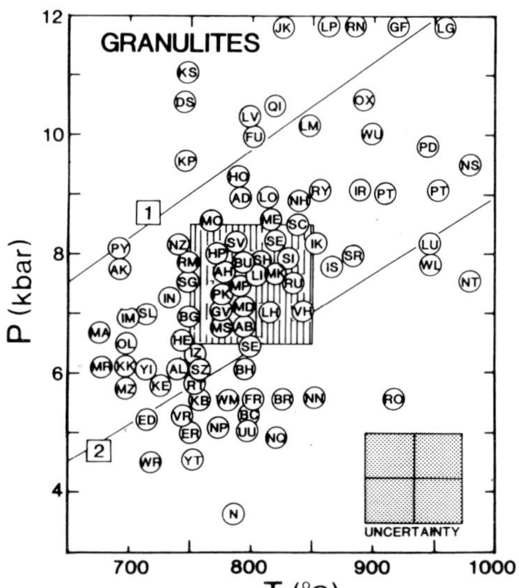

Fig. 10-1. Pressure–temperature conditions for granulite terranes worldwide (after Harley, 1989). Each circle denotes the average conditions within a particular terrane as determined by geothermobarometry. The letters within each circle correspond to acronyms for each terrane (see Harley, 1989, table 1 for list). Typical uncertainties of the P–T estimates are shown by the shaded box. The lines labelled (1) and (2) divide the diagram into what are commonly called high-, medium-, and low-pressure granulites. The box with vertical ruled lines refers to Bohlen's (1987) "average granulite conditions."

within the lower parts of normal thickness continental crust (see Fig. 10-1), they may be considered to be bulk samples of the lower crust or of the mid-levels of overthickened crust (Ellis, 1987). However, regardless of the validity of this interpretation, these rocks contain a wealth of information concerning phase equilibria, pressure–temperature relationships, and geochemistry in the mineral assemblages and their compositions and reaction relationships. It is important to realize that the recrystallisation processes observed in high-grade gneiss terranes predominantly relate to orogenic events, during which pressures and temperatures are perturbed from their normal values. Between orogenies, temperatures in the lowest part of normal thickness crust will not normally exceed 600°C, and little recrystallization will be occurring. Extensive reviews of studies that estimate pressure and temperature conditions within these terranes have been recently made by Newton (1985, 1989), Mezger (this volume), and Harley (1989), whose recent compilation is shown in Figure 10-1. However, since the same mineral equilibria used for calculating pressures and temperatures often involve hydrous phases and aqueous and carbonaceous fluids, it is also possible to calculate C–O–H fluid fugacities (in particular water activities) for these same assemblages (e.g., Lamb and Valley, 1984, 1985, 1988; Newton, 1985, 1989; Edwards and Essene, 1988; Chacko et al., 1988). These calculations place important constraints on the behaviour of fluids in lower crustal metamorphism and the petrogenesis of granulite facies rocks.

2.2. Fluid fugacity calculations

Unlike volcanic rocks, and xenoliths transported to the surface therein, regional outcrops of lower crustal rocks have cooled relatively slowly during their transport to the surface. Although this process is still fast enough to preserve high-grade mineral assemblages, it means that the outcrops that we see today represent a time-integrated picture of lower crustal recrystallization processes with mineral assemblages continually trying to equilibrate to new P, T, or X_{fluid} conditions. In some sense this is an advantage to the petrologist in that it allows us (e.g., using garnet zoning profiles and appropriate geothermobarometers) to reconstruct the pressure–temperature–time history (P–T–t path) of individual portions of the lower crust (e.g., England and Richardson, 1977). However, the situation is complicated by the possible presence of earlier thermal events (polymetamorphism) and the extreme difficulty of accurately dating specific generations of mineral growth (see Mezger, this volume). Granulite petrology makes use of both well-equilibrated mineral assemblages to quantify pressure and temperature conditions and also disequilibrium features (e.g., zoned mineral composition) in order to extract the temporal or kinematic information that is vital to petrogenetic interpretation. The question of thermodynamic equilibrium and when or where to assume it is probably one of the most important and difficult questions facing metamorphic petrologists. In order to address this problem, some researchers are now turning to the study of the kinetics of metamorphic reactions (e.g., Joesten, 1986; Lasaga, 1986) and to quantifying diffusion rates of important components such as oxygen in silicate minerals using isotopic tracers (e.g., Jiang et al., 1988; Elphick et al., 1989).

This preamble is necessary because it underlines some of the assumptions involved in extracting P–T–t and fluid fugacity data from partially equilibrated metamorphic rocks. Calculation of fluid fugacities first requires that pressures and temperatures are accurately determined and that appropriate mineral assemblages exist, and that these are in equilibrium. Inasmuch as H_2O (and to a lesser extent CO_2) is extremely reactive with crustal rocks at elevated P and T, and also extremely mobile, and may strongly enhance diffusion rates, its presence is likely to speed up reaction rates (at least locally) but also to leave disequilibrium textures and assemblages, particularly if fluid moves rapidly and non-pervasively (e.g., along discrete cracks) through a given rock volume. Fluid fugacity calculations are thus fraught with difficulties in assuming equilibrium, determining accurate pressures and temperatures, and making simplifying assumptions related to the mixing properties of chemical components in minerals and fluids at high pressures and temperatures. More importantly, these calculations relate only to the fugacities that existed during the appropriate equilibration event for the reactions in question. This time period might be virtually instantaneous, or it might involve a much longer period of time (10^2–10^3 years?) depending on the thermal history of the assemblage. Inasmuch as most fluid fugacities are calculated at elevated temperatures above ambient (normal) deep crustal temperatures, simple thermal models (e.g., England and Thompson, 1984) dictate that these temperatures will be transient and are unlikely to have been experienced for longer than 10^3 years during a particular orogenic event. This important point must be carefully addressed before calculated fluid fugacities may be generalized in space and/or time, for example, to infer fluid composition, fluid transport or the presence or absence of fluid over an entire terrain for the duration of a given metamorphic event. It is an even larger step to extrapolate further to the point where these data are used to describe the ambient state of the lower crust. These uncertainties detract from the utility of water activity calculations, and to some extent account for the relatively few detailed studies that there have been of fluid fugacity variation in high-grade terrains thought to be representative of the lower crust. Nevertheless, these calculations place some useful constraints on the nature of deep crustal fluids.

Some recent studies of a_{H_2O} and other fluid fugacities illustrate some of the problems involved in these calculations. Edwards and Essene (1988) studied samples of the Major Paragneiss, a mostly pelitic metasedimentary unit, collected from the amphibolite–granulite transition observed regionally in the northwest Adirondack Mountains, New York (see also Engel and Engel, 1960; Valley et al., 1983; Lamb and Valley, 1988). These samples contained assemblages that allowed peak metamorphic pressures and temperatures to be estimated and revealed a somewhat systematic variation across the terrane (650 to 760°C based on two feldspar thermometry and 5.4 to 8.0 kbar based on anorthite–grossular–sillimanite–quartz barometry), although some of the temperature estimates may have been affected by retrograde resetting. Using these pressure and temperature estimates, Edwards and Essene then estimated water activities in 15 samples using the reactions:

$$\text{muscovite} + \text{quartz} = \text{K-feldspar} + \text{sillimanite} + H_2O \quad (1)$$

$$\text{annite} + \text{sillimanite} + 2\,\text{quartz} = \text{almandine} + \text{K-feldspar} + H_2O \quad (2)$$

and in one sample:

$$\text{annite} + 3\,\text{quartz} = 3\,\text{ferrosilite} + \text{K-feldspar} + H_2O \quad (3)$$

All of these rocks are aluminous pelites and in many samples both reactions (1) and (2) could be used to calculate a_{H_2O}; in biotite + sillimanite + quartz + K-feldspar + garnet assemblages a_{H_2O} could be calculated from (2) and a maximum a_{H_2O} from (1). Using more than one reaction greatly improves the certainty of estimates because the estimates of a_{H_2O} can be cross-compared.

Errors in the calculations for each reaction on its own are difficult to assess, and include uncertainties in peak temperature and pressure estimates, and activity-composition models

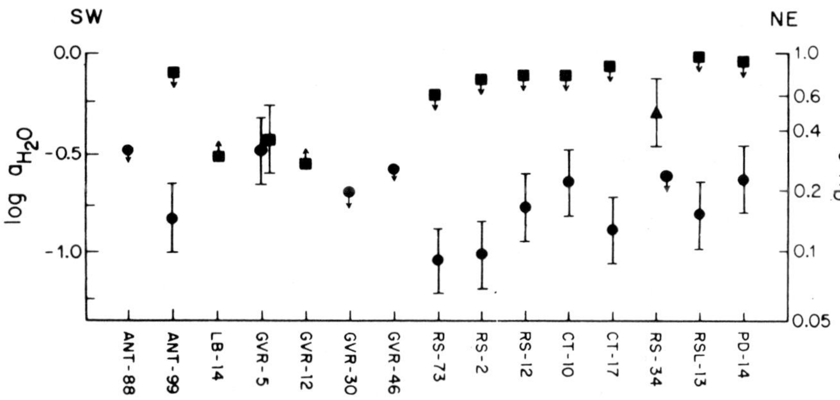

Fig. 10-2. Calculation of water activities across the amphibolite–granulite transition in the northwest Adirondacks (after Edwards and Essene, 1988). Samples PD-14, RSL-13, RS-34, CT-17, and RS-12 are from the granulite grade part of the terrane (i.e., higher grade than "orthopyroxene in" in mafic rocks — see Edwards and Essene (1988), figs. 2 and 11). a_{H_2O} is calculated from the muscovite–quartz breakdown reaction (squares), the biotite–quartz breakdown reaction (triangle), and the biotite–sillimanite–quartz breakdown reaction (circles). Maxima and minima calculated when only partial assemblages were present are denoted by arrows.

for garnet, biotite and feldspar. Edwards and Essene estimated an error in a_{H_2O} of $\pm 50\%$ for each reaction, but this figure is probably considerably reduced when more than one reaction is used. The results from 15 samples are shown in Figure 10-2 and all have $a_{H_2O} < 1$ and range from 0.08 to 0.50 in the better-constrained samples. The values show no obvious trend and do not correlate with metamorphic grade, although considering the magnitude of the errors, subtle trends may not be resolved by the calculated values. Nevertheless, these authors concluded that the amphibolite–granulite transition in the Adirondacks was probably a manifestation of increasing temperature and not of decreasing water activity.

Some of Edwards and Essene's samples contained graphite and if a_{H_2O} is known, this allows calculation of f_{CO_2}, f_{CH_4}, f_{CO}, f_{H_2}, and f_{O_2} if $P_{fluid} = P_{total}$ (see also Lamb and Valley, 1984, 1985). Although the thermodynamic properties of mixed fluids containing these components are not known in detail, these calculations again offer important constraints on metamorphic fluid composition, and in this study implied a fluid mostly composed of CO_2 and H_2O (assuming that f_{O_2} was $\sim$1 log unit below QFM). The study concluded that the water activity data could not be simply explained by any fluid–rock interaction scenario and that a_{H_2O} was probably buffered locally. The estimation of fluid composition is predicated on the assumption that a fluid phase was present in the first place; an alternative possibility is that the rocks equilibrated under vapour-absent conditions. This underscores another problem in calculating fluid composition from mineral assemblages, namely the difficulty in proving that water activity corresponds to fluid present or fluid absent conditions (see Thompson, 1983).

Lamb and Valley (1984, 1985), in an assessment of C–O–H fluid fugacities calculated from mineral equilibria in another part of the Adirondack granulite terrane and elsewhere, noted that CO_2 fluid infiltration under lithostatic pressure should result in precipitation of graphite, provided the oxygen fugacity of the rock system is low enough. For example, at typical granulite pressures and temperatures and with f_{O_2} slightly below QFM, addition of CO_2 sufficient to grow 10 volume% orthopyroxene via the reaction:

$$\text{tremolite} = 2\,\text{diopside} + 3\,\text{enstatite} + \text{quartz} + H_2O$$

requires the precipitation of 1.5 volume% graphite. However, they also suggested that f_{O_2} and f_{H_2O} values calculated for some granulite assemblages from the Adirondacks and elsewhere imply fluid absent conditions. This point is illustrated in Figure 10-3, which shows a plot of f_{H_2O} vs. f_{CO_2} at 800°C, 8 kbar and curves denoting the stability of graphite at various fluid pressures. The outer curve denotes $P_{fluid} = P_{lithostatic} = 8$ kbar and graphite activity = 1. Above (outside) this curve, graphite is not stable, and inside it, it would only be stable at $P_{fluid} < P_{lithostatic}$. Lamb and Valley (1984) noted that the f_{H_2O} and f_{O_2} estimates of a number of assemblages from several granulite terranes plot inside the graphite stability curve of Figure 10-3 (for appropriate P and T), implying that $P_{fluid} < P_{lithostatic}$ whether or not graphite was present. This requires conditions to be fluid absent when these assemblages equilibrated. This and the absence of graphite in these same rocks implies that CO_2 infiltration could only have occurred in minute quantities, not nearly large enough to buffer the CO_2/H_2O ratio in the fluid phase to a constant high value and promote dehydration reactions.

These calculations place important constraints on fluid composition in these granulites, provided that oxygen fugacities can be well constrained (as noted by Lamb and Valley, 1984, 1985, 1988) and temporally linked to calculated water activities. In this respect, Frost and Chacko (1989) have reanalyzed the Fe–Ti oxide f_{O_2} determinations for some of the same

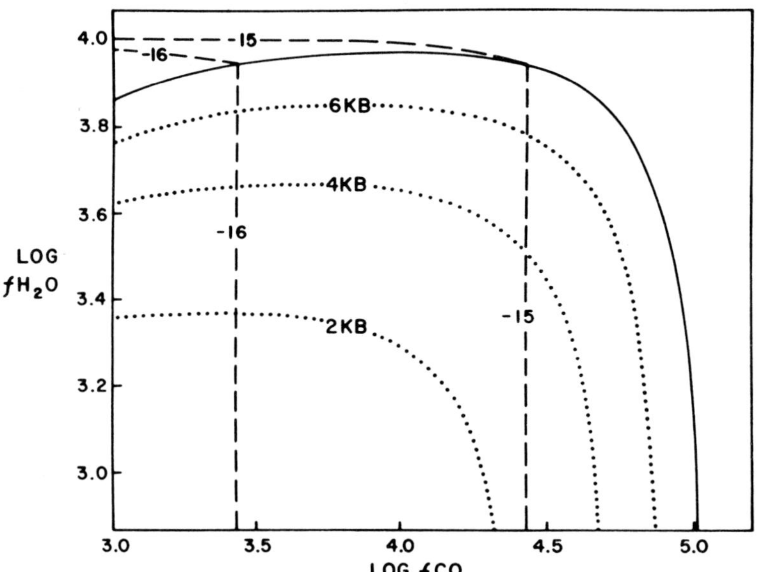

Fig. 10-3. log f_{H_2O} vs. log f_{CO_2} at 800°C and at lithostatic pressure (P_L) = 8 kbar [after Lamb and Valley (1985) and Lamb et al. (1987)]. The solid curve denotes the conditions when graphite is present (i.e., graphite saturation, $a_{graphite} = 1$) and the fluid pressure (P_F) is lithostatic ($P_F = P_L$). Above this curve, $a_{graphite} < 1$ and $P_F = P_L$. Below this curve, the dotted lines show the position of the graphite saturation curve at lower P_F shown for 6 kbar, 4 kbar, and 2 kbar. For graphite-absent rocks ($a_{graphite} < 1$), these contours represent an upper limit to P_F. Thus, while $P_L = 8$ kbar any value of f_{H_2O} and f_{CO_2} plotting inside the 8 kbar curve requires P_F to be less than P_L, implying fluid-absent conditions (Lamb and Valley, 1984, 1985). The dashed lines are contours of f_{O_2}, and allow a combination of f_{H_2O} and f_{O_2} determinations from mineral equilibria to be plotted onto this diagram. A review of oxygen fugacity data from several granulite terranes led Lamb and Valley (1984) to suggest that conditions did indeed fall below the graphite saturation curve for the appropriate pressure and that the metamorphism in these cases was fluid absent. This conclusion has been disputed by Frost and Chacko (1989), who suggest that the f_{O_2} values have suffered some resetting since peak metamorphic conditions, and that a CO_2 fluid was in equilibrium with these rocks, at least during the cooling history.

set of data [using an updated version of the iron-titanium oxide thermometer (Anderson and Lindsley, 1988)] and come to a different conclusion, namely that these oxide samples crystallized on (or re-equilibrated on) the graphite saturation surface, and were therefore probably in equilibrium with a CO_2-rich fluid during cooling. Furthermore, these authors have suggested that this indicates some resetting of Fe-Ti oxides during cooling due to oxyexsolution reaction with CO_2 on the graphite saturation surface, and that oxide f_{O_2} determinations may not apply to peak metamorphic conditions. The determination of peak temperature f_{O_2} values in deep crustal rocks is therefore critical to these phase equilibrium constraints on the nature of deep crustal fluids and remains a most important research area.

Another recent example of a *regional* study of water activities is that of Chacko et al. (1988) for the khondalite belt in southern India. This terrane provides the most compelling evidence of any for large-scale CO_2 mobility in the deep crust (e.g., Hansen et al., 1987) and has given rise to intense debate over the question of whether the granulites of this region have formed in response to CO_2 infiltration (see section (4.2) below). Although it is not the intention of this chapter to address this question exclusively, this terrane has been more thoroughly studied than many, and the water activity dataset of Chacko et al. (1988) is one of the most comprehensive of its type. These workers calculated a_{H_2O} using the two reactions:

$$2\,\text{phlogopite} + 6\,\text{quartz} = 3\,\text{enstatite} + 2\,\text{K-feldspar} + 2\,H_2O$$

and

$$\text{phlogopite} + 2\,\text{quartz} + \text{sillimanite} = \text{pyrope} + \text{K-feldspar} + H_2O$$

and well-determined P and T values (5.5 kbar and 750°C) which were assumed to be uniform across the area studied (Chacko et al., 1987). They found that for 13 charnockite samples and 13 khondalite samples collected from localities encompassing an area of some 5000 km^2, a_{H_2O} was 0.25 ± 0.06 (1σ) (see Fig. 10-4). The uncertainties on these calculated values are probably <50% (T. Chacko, pers. commun., 1989), but it is not the absolute value that is of interest so much as the astonishing uniformity across the entire region. Ruling out the unlikely event of a coincidence, this uniformity in such a large number of widely scattered samples strongly supports the validity of the calculations and lessens the likelihood of equilibration problems, retrograde overprints, etc. as discussed above. The scale of the uniformity is most easily explained by buffering of the fugacities of volatile species (and perhaps also temperature) at fixed values on a regional scale. If a_{H_2O} was internally buffered, then much more variation would be expected within the various lithological units sampled, some of which have very different mineral assemblages and also (presumably) different fluid transport properties.

Several other studies have been made of a_{H_2O} variation in amphibolite and granulite terranes, some documenting local variation on km to metre scales (e.g., Waters and Whales, 1984; Lamb and Valley, 1988) and some revealing regional gradients in a_{H_2O}, such as Phillips (1980), who showed that X_{H_2O} in the fluid phase decreased steadily from 1.0 to 0.35 in the amphibolite to granulite transition in the Willyama Complex of southeast Australia. To some extent the discovery of regionally constant a_{H_2O} in a metamorphic terrane is a more powerful constraint on metamorphic processes than the documentation of seemingly random variation. There are a variety of reasons why a_{H_2O} values should be variable regionally, and only one of these is due to local buffering of fluid composition at the metamorphic peak, the others being related to the uncertainties outlined above. On the other hand, regionally constant values remove some of these uncertainties and suggest buffering over a wide area. This would be most easily promoted by infiltration of a pervasive fluid or melt phase.

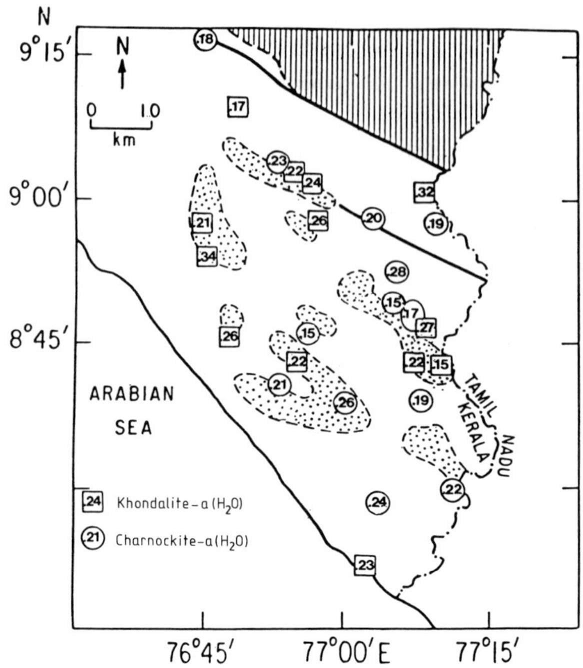

Fig. 10-4. a_{H_2O} for charnockite gneisses (circles) and khondalites (squares) from a wide area within the Kerala Khondalite Belt, calculated from the biotite–quartz and biotite–sillimanite–quartz breakdown reactions at a uniform pressure of 5.5 kbar and 750°C, after Chacko et al. (1988).

Despite the many uncertainties attached to phase equilibrium calculations, further careful documentation of fluid fugacities and f_{O_2} in other terranes is certainly worthwhile, particularly where textures and other considerations prove that the mineral assemblages used to make the calculations have equilibrated simultaneously, and where retrograde effects are taken into careful consideration (e.g., Lamb and Valley, 1988; Frost and Chacko, 1989). In particular, it is important to acquire better estimates of peak metamorphic temperatures in granulite facies rocks, for example by integration of exsolution lamellae, one of the "fossil thermometers" of Frost and Chacko (1989). To date there are relatively few places where these constraints can be unambiguously applied, and the present results can only be used to rule out CO_2 infiltration at a few restricted locations. This deficiency should spur on others to make more of these calculations on a systematic, regional scale, and to compare these data with other petrological and geochemical observations.

3. Fluid inclusions

The measurement of fluid inclusion abundances and their densities and chemical characteristics in high-grade gneiss terranes has been extremely influential in developing models of fluid behaviour in the lower crust. In particular, the discovery by Touret (1971) of high-density, almost pure CO_2 fluid inclusions in Scandinavian granulites, followed by their observation in many other granulite terranes worldwide (e.g., Hansen et al., 1984; Hollister, 1988) became an important foundation upon which models of large-scale CO_2 infiltration into the lower crust were constructed (e.g., Newton et al., 1980). Fluid inclusion studies have also in some cases been used to refute these models (e.g., Lamb et al., 1987) by maintaining that the CO_2-rich fluids were introduced into these rocks long after the metamorphic peak, and that the inclusions are therefore irrelevant to the high-grade metamorphic conditions.

Although potentially providing a powerful constraint on fluid–rock interaction, there are clearly a number of uncertainties in both making and interpreting fluid inclusion measurements. Some of these are solely related to the preservation of the inclusion itself, such as leakage, necking down, deformational overprinting or selective removal of some inclusion components by crystallization on, or diffusion into, the host crystal (see Hollister, 1988). Others relate to petrographic observation and the identification of successive generations of inclusion. These problems require that very detailed, painstaking studies involving a large number of measurements be made before any broad conclusion can be drawn (e.g., Touret and Hartel, 1990), and this accounts for the rather small number of comprehensive studies that have been thus far made (e.g., Hollister et al., 1979; Touret, 1985). Another fundamental problem is in relating fluid inclusion composition to mineral equilibria in the same rocks and in relating both of these to geochronology and the overall thermal history. Thus there is still confusion as to whether fluid inclusion compositions observed in some terranes represent high-grade fluids that the rocks equilibrated with, or whether they are merely a retrograde phenomenon (e.g., Lamb et al., 1987; Morrison and Valley, 1988b). Touret (1985) and Touret and Hartel (1990) cite three essential conditions that must be satisfied in order to extract meaningful data from fluid inclusion studies: (1) selection of favourable samples in order to establish an inclusion chronology; (2) very careful observation and comparison of fluid inclusion compositions with phase equilibria in the same thin section; and (3) comparison of fluid inclusion data with other petrological or tectonic constraints. When these conditions are well satisfied, the inclusion data can then provide important information about lower crustal fluids and their behaviour during metamorphism.

In high-grade gneiss terranes, granulites commonly contain CO_2-rich, high-density fluid inclusions as opposed to those found in amphibolite facies assemblages which are usually more water-rich (see review of Hollister, 1988). In some particular cases the CO_2 contents are extremely high (over 20% by volume in garnet, quartz and feldspar from some southern Indian charnockites — see Fig. 10-5) representing a bulk CO_2 content sometimes as high as several weight percent in the whole rock. Enrichments in carbon in fluid inclusions from orthopyroxene-bearing charnockites from India (including samples from the same locality as the sample shown in Fig. 10-5) have been more quantitatively measured by analysis of fluid inclusions in quartz and garnet from adjacent amphibolite–charnockite samples (e.g., Jackson et al., 1988a,b) that show carbon contents increasing by several fold in the charnockites over the precursor amphibolites (see section (4.3) below). Unfortunately, there is presently very little analytical information that documents the variation of absolute carbon abundance in the silicate rocks of the lower crust, either in India or elsewhere, and it would be highly desirable to expand the current dataset. This is central to the whole issue of CO_2 fluxes in these rocks, inasmuch as it would provide quantitative documentation of regional CO_2 enrichment, and place strong budgetary constraints on carbon sources. This information would greatly enhance the significance of both fluid inclusion and other petrological studies. Budgetary calculations on CO_2 fluxes have been attempted by Wickham (1988) and Stone et al. (1988), who used thermodynamic constraints on CO_2-rock ratios in various granulites, based on estimates (e.g., Frost and Frost, 1987; Frost et al., 1989) of the amount of CO_2 necessary to dilute the fluid phase to give appropriate water activities, to fix upper limits to the typical CO_2 fluxes required (see Fig. 10-6). Evidence of abso-

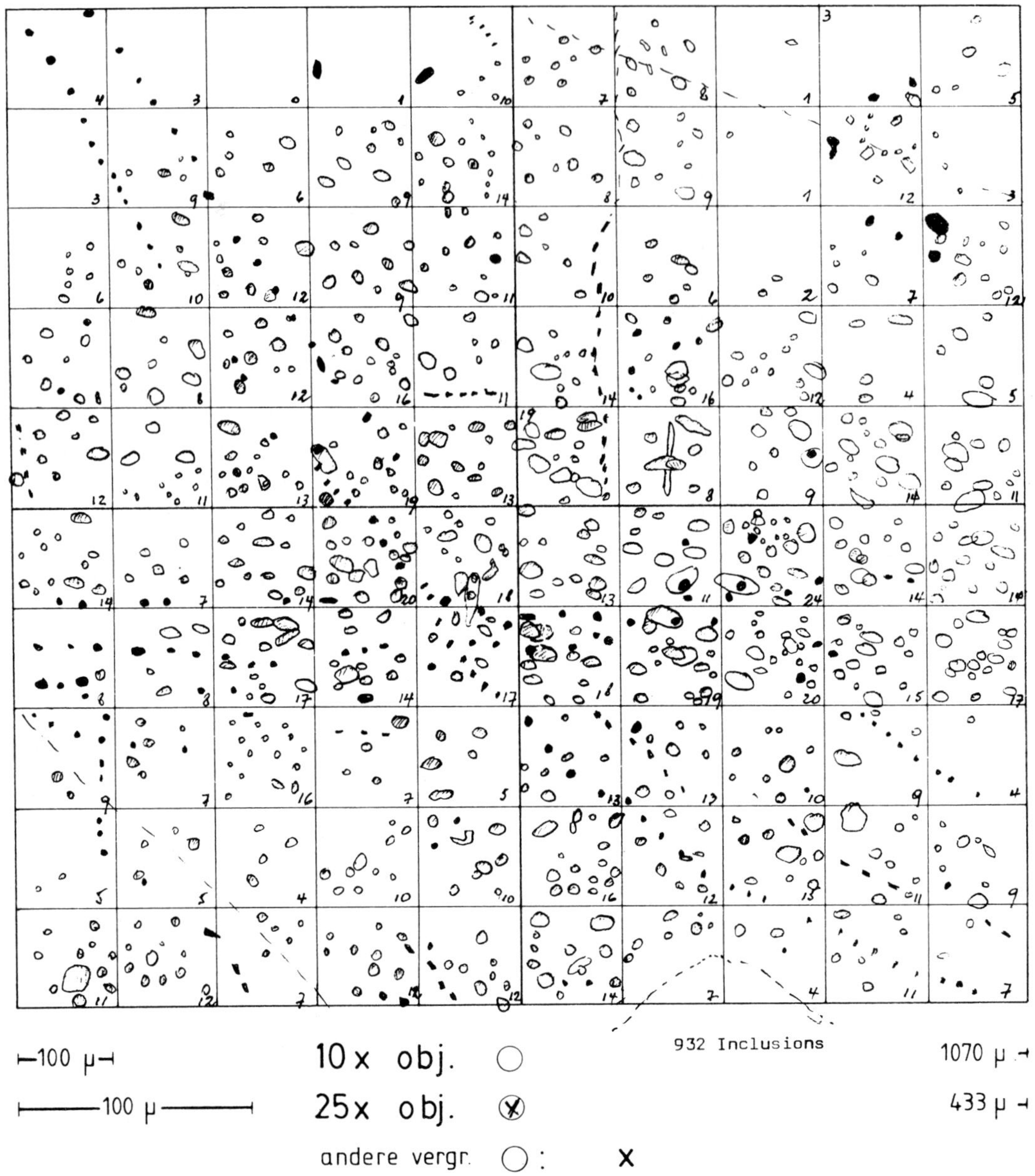

Fig. 10-5. Hand drawing of fluid inclusions in quartz from the Doddabetta Quarry charnockite locality, India (data courtesy of J.L.R. Touret and R. Sinclair, unpublished). The diagram shows the extremely high abundance (10–15 volume %) of fluid inclusions in quartz (which are just as abundant in garnet and feldspar from the same locality) in these rocks, most of which are very rich in CO_2. This corresponds to an abundance of 3–5 wt.% in these phases (Touret and Hansteen, 1988), an estimate that has been independently corroborated by step-heating decrepitation measurements (Jackson et al., 1988b). The lower scale bar represents 100 μm (25x objective).

lute CO_2 enrichment in certain crustal rocks would place these estimates on much firmer ground.

Recent studies of CO_2-rich fluid inclusions in Adirondack granulites (Lamb et al., 1987; Morrison and Valley, 1988b) has led to the suggestion that the CO_2 was trapped during

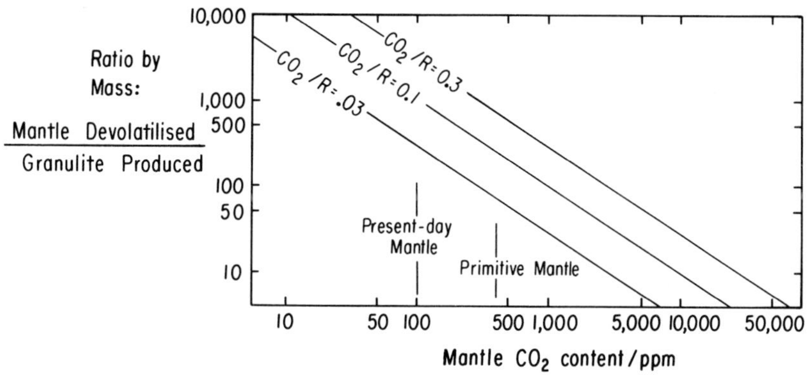

Fig. 10-6. Mass balance relationships illustrating the mass of mantle that must be decarbonated in order to produce a unit mass of granulite by the CO_2 infiltration mechanism (after Stone et al., 1988). The diagram shows a range of plausible CO_2/rock ratios required to convert a (hydrated) amphibolite to granulite (see text and Lamb and Valley, 1984; Frost and Frost, 1987). Estimates of mantle CO_2 content (e.g., Harris, 1981; Pineau and Javoy, 1983; Des Marais and Moore, 1984) for the present-day MORB source and for a more primitive mantle (Stone et al., 1988) indicate participation of unreasonably large mantle volumes in the production of thick crustal granulite sequences, whether in the Archean or at the present day. Derivation of the CO_2 required to produce 5 km of granulite from a 100-km-thick lithospheric reservoir requires such a reservoir to carry an average CO_2 content on the order of thousands of ppm.

a post-metamorphic retrograde event. This was based partly on petrologic evidence that calculated peak metamorphic fluid compositions were not as CO_2-rich as the fluid inclusions present in the same rocks, and partly on textural relationships between the inclusions and low-temperature retrograde calcite–chlorite–sericite veinlets. Such retrograde veins are commonplace in many granulite terranes (J. Touret, pers. commun., 1988) and although they are so small (<1 mm) as to have escaped attention in many cases, they may constitute a significant bulk mass fraction (Morrison and Valley, 1988b). There is no question that these veins formed at low temperatures (at least <400°C) and indicate the former presence of a CO_2-rich fluid at these low temperatures. However, it is far from certain *when* the CO_2 was originally introduced into the rocks (or where it came from). Stable isotope data (Morrison and Valley, 1988b) reveal relatively homogeneous $\delta^{18}O$ and $\delta^{13}C$ in vein calcite ($\delta^{18}O$ = +11 to +15, $\delta^{13}C$ = −4 to 0), intermediate between primary igneous values and those of Adirondack marbles. This could reflect derivation of the vein carbonate from the marble and partial exchange with the anorthosite (Taylor, 1969) or it could reflect derivation of the calcite from external sources and partial exchange with both marbles and anorthosite (Morrison and Valley, 1988b). Some high-grade granulites are extremely rich in high-density syn-metamorphic CO_2 inclusions (e.g., Touret and Hansteen, 1988; see Fig. 10-5) and also contain tiny retrograde carbonate veinlets. In these cases, it is quite likely that the veinlets represent CO_2 liberated from the high-pressure fluid inclusions as they ruptured during uplift. In this case the carbonate need not be derived from a hypothetical external source, but could merely represent an internal redistribution of CO_2-rich fluid. If this interpretation is correct, then it implies that the estimates of CO_2 contents in some granulites based solely on fluid inclusions are minima (even though in some cases these can be as high as several wt.% CO_2; see Fig. 10-5), and that some high-grade gneiss terranes could contain even more primary (i.e., high-grade) CO_2 than these earlier estimates. If this interpretation applied to the Adirondacks, it would also explain why Fe-Ti oxide equilibria throughout the region still give high-grade metamorphic temperature estimates >600°C (Bohlen et al., 1985) rather than being reset by a later-stage lower-temperature alteration event associated with the formation of the low-grade carbonate veinlets.

It has recently been proposed that fluid immiscibility in C–O–H–Cl systems might be responsible for the CO_2-rich inclusions observed in granulites (Touret, 1985; Henry, 1988; Thompson, 1989). At high Cl-concentration and high P and T, separation of such fluid into a Cl-rich aqueous brine and a CO_2-rich phase is predicted both by theoretical calculations (Bowers and Helgeson, 1983) and some natural observations (Trommsdorff and Skippen, 1986). High mobility and/or reactivity of the brine relative to the CO_2 would lead to escape of the former from the system (either by intergranular flow, or by reaction with appropriate phases), leaving behind the less mobile CO_2. At present, insufficient data exist to test this imaginative hypothesis further, for example by documenting the Cl-content required for phase separation to occur under lower crustal conditions, and correlating this with observations of natural rocks.

4. Stable isotope constraints on deep crustal fluids

4.1. Overview

Although oxygen isotope systematics have been widely studied in igneous rock systems (e.g., Taylor, 1968; Taylor and Sheppard, 1986) and provide powerful constraints on shallow-level hydrothermal processes and on the origin of different magma types, they have only recently been extensively used in the study of metamorphic processes (see Valley, 1986 for a review). However, because oxygen is the major component in both rocks and in most metamorphic fluids, it is potentially a very powerful tracer of fluid flow, particularly in metamorphic terranes that contain a variety of interlayered lithologies of initially contrasting isotopic composition, such as metasedimentary and igneous rocks. In this situation, measurement of the oxygen isotopic composition of adjacent layers indicates whether the isotopic homogenization expected to be associated with high fluxes of pervasive metamorphic fluid has occurred. Situations where complete homogenization has not occurred can be even more useful because in this case the isotopic profiles across lithological contacts can provide quantitative constraints on fluxes and rock porosities (Bickle and McKenzie, 1987). If fluid flow across lithological boundaries has not occurred, then the pre-metamorphic (depositional) isotopic contrast between the lithologies should be preserved revealing a sharp compositional step, perhaps showing diffusive broadening over a few metres or less (e.g., Bickle and McKenzie, 1987). Measurement of these diffusive (as opposed to advective) isotopic profiles can also provide information regarding fluid composition and rock porosity.

As yet there have been relatively few systematic, detailed studies of this type in terranes thought to be representative of the deep crust, but enough data exist to draw some important preliminary conclusions, and some of these are discussed below (e.g., Shieh and Schwarcz, 1974; Wilson and Baksi, 1983; Valley and O'Neil, 1984; Valley, 1986; Baker and Fallick, 1988). As with phase equilibria and fluid inclusions, one of the drawbacks is a lack of information constraining the exact timing of the fluid flow event under investigation. Stable isotope ratios reveal only very indirect evidence about *when* they were established or whether they have been reset. As pointed out earlier, deep crustal terranes now exposed at the surface often have complex polymetamorphic thermal histories, and in many cases the present-day oxygen isotope ratios may be the result of multiple superimposed equilibration events involving several styles of fluid–rock interaction. As with the petrological evidence, the isotope systematics reveal a *time-integrated* picture of these terranes.

Another important caveat concerns the use of stable isotope profiles to constrain metamorphic fluid budgets, as mentioned above. These data constrain fluid movement across lithological boundaries, but say little about fluid movement along strike within individual layers. Inasmuch as effective permeability probably varies enormously in high-grade metamorphic rocks, in space as well as in time, particularly in lithologies in which reactions involving significant volume changes may occur, one might expect *a priori* that fluid flow in deep crustal rocks would be strongly channelled. Deformation resulting in tilting of lithological units might therefore promote episodic gravity-driven fluxes along particular, more permeable layers, with little or no movement across lithological boundaries. Isotope profiles made across lithological boundaries would offer little constraint on this sort of scenario, because they might locally show major isotopic anomalies, even though some layers had seen significant fluid fluxes. Isotope data should thus be interpreted with caution where isotope profiles show across-strike anomalies, but along-strike homogeneity.

Until comparatively recently, stable isotope studies of high-grade metamorphic terranes have been reconnaissance in nature, and have provided general information on the bulk isotopic compositions of these terranes and the lithologies within them, but only crude information constraining fluid fluxes (e.g., Shieh and Schwarcz, 1974; Valley and O'Neil, 1984; Wickham and Taylor, 1985). Garlick and Epstein (1967) were first to discover that there was a general tendency for metasediment $\delta^{18}O$ values to decrease at higher metamorphic grades within individual terranes, and this was attributed by them to result from progressive equilibration at higher grades and deeper structural levels with lower-^{18}O primitive mantle-derived igneous rocks. In some deep crustal terranes this explanation is probably correct, although at shallower levels, interaction with pore or surface fluids during prograde metamorphism has also been postulated as a mechanism capable of causing ^{18}O-depletion (Wickham and Taylor, 1985). In other terranes, the isotopic heterogeneity of the high-grade rocks has been explained in terms of a lack of equilibration because metamorphic conditions were fluid-absent or fluid infiltration was limited (e.g., Valley and O'Neil, 1984; Baker, 1988).

In the isotopically homogenized high-grade terranes, the exact *mechanism* of large-scale isotopic exchange required to transform amphibolite and granulite facies lithologies to relatively homogeneous $\delta^{18}O$ values often less than +10 and sometimes close to +6 (e.g., Shieh and Schwarcz, 1974; Wilson and Baksi, 1983; Wickham et al., 1987; Wickham and Peters, 1990) is not at all clear. It requires large-scale mobility of oxygen, but whether this occurs by flushing with H_2O- or CO_2-rich fluids, silicate melts or by pervasive circulation of small quantities of fluid is a subject of intense debate (e.g., Valley and O'Neil, 1984; Wickham and Taylor, 1985; Hansen et al., 1987; Thompson, 1989). The solution to this question lies at the centre of the general debate over the nature of deep crustal fluids.

Stable isotope studies are allowing us to approach solutions to these problems, but there is still a very long way to go. Detailed oxygen isotope profiles across lithological contacts can reveal homogeneity or asymmetrically displaced profiles that are the signature of advective fluid motion, and also reveal the fluid transport direction, and give an indication of the scale over which advective transport has operated (Bickle and McKenzie, 1987). These sorts of studies are just beginning to be made (e.g., Baker et al., 1989; Wickham and Peters, 1990), although at the time of writing there have been very few studies that are specifically related to granulites or to lower crustal fluids (e.g., Valley and O'Neil, 1984; Wickham and Taylor, 1987; Baker, 1988).

4.2. Deep crustal CO_2 fluxes

Research into the petrology of the lower crust has been dominated over the last decade by the concept of large CO_2 fluxes accompanying and promoting granulite facies metamorphism, and also causing redistribution of trace elements (most notably a depletion of incompatible elements) in the lower crust. In the original formulation of this hypothesis, Newton et al. (1980) postulated that large CO_2 fluxes derived from enriched regions of the mantle which had been carbonated by earlier subduction of supracrustal carbonate sediments, infiltrate and traverse the lower crust in association with heating. This hypothesis was based primarily on the observation of the amphibolite–granulite transition in southern India, where progressive charnockitization is manifested by the breakdown of biotite to orthopyroxene, corresponding to an isograd that can be mapped continuously over a huge distance (Fig. 10-7). This process is accompanied by a change from H_2O-dominated fluid inclusions to CO_2-dominated inclusions (Newton, 1985; Hansen et al., 1987). Within the granulites the concentration of CO_2-rich inclusions can sometimes be astonishingly high (Fig. 10-5), representing (at least locally) an apparent increase in absolute CO_2 abundance (Touret and Hansteen, 1988).

The main features of the CO_2-flushing hypothesis and its ramifications have been detailed many times elsewhere and will not be repeated here. One of its most important consequences is the destabilization of hydrous phases, and the resulting dehydration of large volumes of the lower crust. It implies that large CO_2 fluxes have characterized the evolution of the

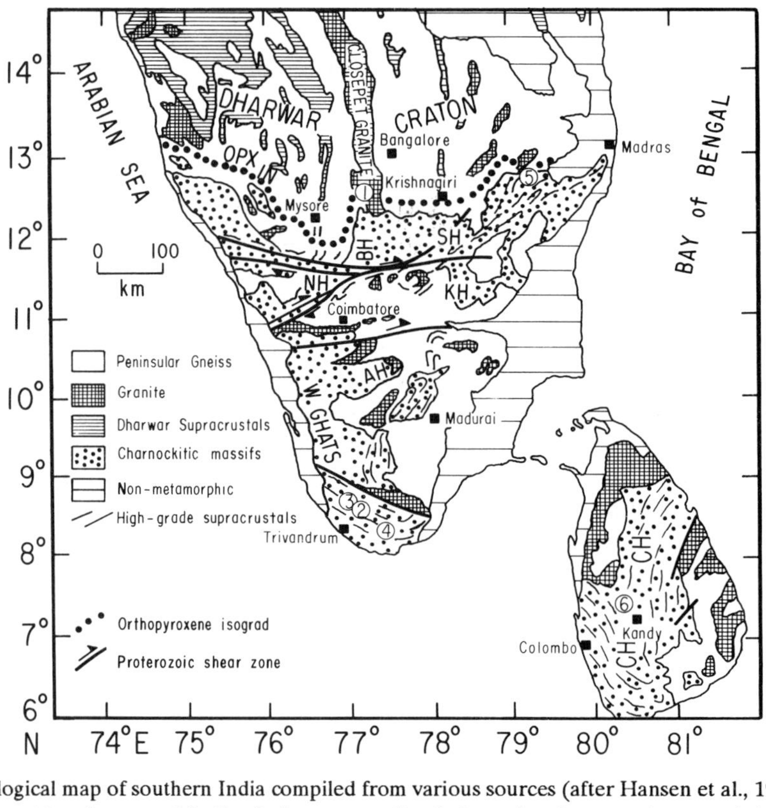

Fig. 10-7. Geological map of southern India compiled from various sources (after Hansen et al., 1987, fig. 1) showing the regional transition from amphibolite facies to granulite facies going from north to south in peninsular India. Note the regionally defined "orthopyroxene in" isograd (bold dotted line).

lower crust of some regions and that large volumes of the lower crust behave chemically as open systems, at least periodically. However, the hypothesis has recently been challenged by many petrologists and geochemists for a variety of reasons, the most important of which are: (1) such clear-cut examples of a progressive but regional transition from amphibolite to granulite accompanied by a change from H_2O- to CO_2-rich inclusions (as seen in India) have not been convincingly documented elsewhere; (2) the dehydration of granulites can be achieved by other mechanisms such as vapour absent dehydration melting and the extraction of this melt containing dissolved H_2O (Fyfe, 1973; Powell, 1983a, 1983b; Waters, 1987), or by dehydration of terranes during earlier metamorphic episodes (Valley and O'Neil, 1982; McClelland and Husain, 1986) or intrusion and crystallization of CO_2-rich magmas (Frost and Frost, 1987); (3) estimates of water activity in some (but not all) terranes vary considerably (Edwards and Essene, 1988; Lamb and Valley, 1988; Waters, 1987), implying that fluid compositions were locally, not regionally, buffered; (4) at low f_{O_2}, infiltration of CO_2 should result in graphite precipitation which is not observed in several granulite terranes (Lamb and Valley, 1984, 1985); and (5) oxygen isotope studies of some granulite terranes have revealed steep isotopic gradients adjacent to metamorphosed carbonate units, and other isotopic heterogeneity (e.g., $\delta^{13}C$ of CO_2 in cordierite) indicating that these samples at least have not been subjected to extensive fluid infiltration (e.g., Valley and O'Neil, 1984; Wickham and Taylor, 1987; Baker, 1988; Vry et al., 1988).

Some of these objections disprove the CO_2 flooding hypothesis in its purest sense (i.e., that CO_2 flooding was absolutely pervasive) but the issue is not clear-cut and it remains uncertain whether more selective modes of CO_2 transport have occurred or whether granulites and most of the deep crust remain fluid-absent for most of the time, even during high-grade metamorphism. This is because, like the CO_2 flooding hypothesis, none of the objections listed above has been rigorously established on a regional scale. Geological exposures of amphibolite–granulite transitions (see point (1) above) are not necessarily as straightforward as the transition observed in southern India nor are individual outcrops as well exposed, and it still remains to be seen just how common this patchy charnockitization is in other terranes. Furthermore, granulite facies metamorphism can occur within a variety of different tectonic settings, and these may each have their own distinct petrological or geochemical characteristics (Harley, 1989). Partial melting textures (see point (2) above) are a common feature of high-grade gneisses and are consistent with estimated pressure–temperature conditions, but melt extraction at low melt-fraction from these rocks is not predicted by fluid dynamical theory (McKenzie, 1985; Wickham, 1987) because melt viscosity of typical minimum melt granitic liquids is too high (but see section 5.3 below). Nor has it been convincingly demonstrated that deep crustal terranes have been *regionally* depleted in a minimum melt (granitic) component (see also section 5). If dehydration of the lower crust is not achieved by removal of hydrous melts and the melt simply remains where it forms, then dehydration reactions should reverse as melts cool and crystallize and release water (e.g., Lamb et al., 1987; Lamb and Valley, 1988), as has been demonstrated over laboratory time scales by Peterson and Newton (1989a). As noted in section (2), there have been only a very few regional-scale determinations of water activity (point (3)) in deep crustal rocks, far too few to establish any general rules. The same is true for correlations between f_{O_2} and the presence or absence of graphite (point (4)). To date there have been few detailed oxygen isotope studies of granulites (point (5)), although the situation is now changing, and this area probably has the most promise both for resolving the CO_2-flushing controversy, and for documenting the behaviour of metamorphic fluids in the lower crust.

4.3. Oxygen isotope studies of granulites

A relatively early stable isotope study of the Adirondack amphibolite and granulite facies terrane by Taylor (1969) revealed that the anorthosite bodies of this region have considerably higher $\delta^{18}O$ than most other anorthosites worldwide (+8.9 to +11.2 in the Adirondacks, cf. +6 to +7 for most others). In contrast to many anorthosites, however, those in the Adirondacks have been subjected to a complex metamorphic history after their emplacement, during the ~1100–1000 Ma Grenville Orogeny, and Taylor (1969) concluded that the ^{18}O-enrichment in the Adirondack anorthosite resulted from syn-metamorphic isotopic exchange with surrounding ^{18}O-rich country rocks such as the carbonate-rich metasediments that are locally common in this area.

This interpretation has been challenged by Valley and coworkers (e.g., Morrison and Valley, 1988a) who have ascribed the high $\delta^{18}O$ of the anorthosite as being due to assimilation of high-^{18}O rocks at the magmatic stage or derivation from ^{18}O-rich crustal rocks (see also Anderson and Morin, 1969). The isotopic heterogeneity observed in the Adirondack marbles, both regionally (Taylor, 1969; Valley and O'Neil, 1984) and locally (Valley, 1986), was interpreted as evidence of limited fluid transport during granulite metamorphism (Morrison and Valley, 1988a). This conclusion has been locally supported by other isotopic studies (e.g., Valley and O'Neil, 1982; Jiang et al., 1988) and has been integrated with the phase equilibrium argument of Lamb and Valley (1984) for peak-metamorphic fluid-absent conditions and lack of CO_2 infiltration, and the assertion of Lamb et al. (1987) and Morrison and Valley (1988) that CO_2-rich fluid inclusions and tiny carbonate veinlets reflect late-stage post-peak metamorphic fluid infiltration. Figure 10-8 shows the variation in $\delta^{18}O$ values across the contact between a carbonate-rich marble xenolith and the Marcy anorthosite, metamorphosed to granulite facies (Valley and O'Neil, 1984). The preservation of such sharp isotopic contrasts makes it unlikely that there was an intense flux of CO_2 (or H_2O) across this lithological boundary at any stage during the metamorphic history, because this would tend to homogenize the $\delta^{18}O$ values (as observed, for example, at the contact between amphibolite facies pelites and carbonates in the Pyrenees (Wickham and Taylor, 1987)). It does not rule out more selective flow geometries involving channelling of fluid within specific lithologies (Valley and O'Neil, 1984). However, these types of isotopic profiles are extremely useful because they may be used to place upper limits on total fluid flux across lithological contacts (e.g., Baker et al., 1989).

The data referred to above make it unlikely that pervasive equilibration with externally derived CO_2 occurred within the Adirondack granulite terrane. However, this extreme formulation of the CO_2 infiltration hypothesis is easy to disprove, and it is much harder to validate the converse hypothesis that metamorphism has been largely vapour-absent. In fact, the interpretation of Taylor (1969) does not require pervasive fluid flushing — simply that *exchange* of oxygen at some stage of the Adirondack metamorphic-anatectic history occurred between the anorthosite and volatiles derived from the local country rocks. This could have taken place at quite low fluid/rock ratio, with the fluid simply transporting oxygen from country-rock to anorthosite and not in itself representing the high-^{18}O reservoir. However, Morrison and Valley (1988a) present material balance arguments refuting the likelihood of exchange with local carbonates as a cause of ^{18}O-enrichment on account of the paucity of carbonate available. These viewpoints have been discussed elsewhere (e.g., Taylor, 1969; Morrison and Valley, 1988a) and will probably continue to be reiterated for some time to come. Considering the likelihood of a strongly heterogeneous permeability structure within

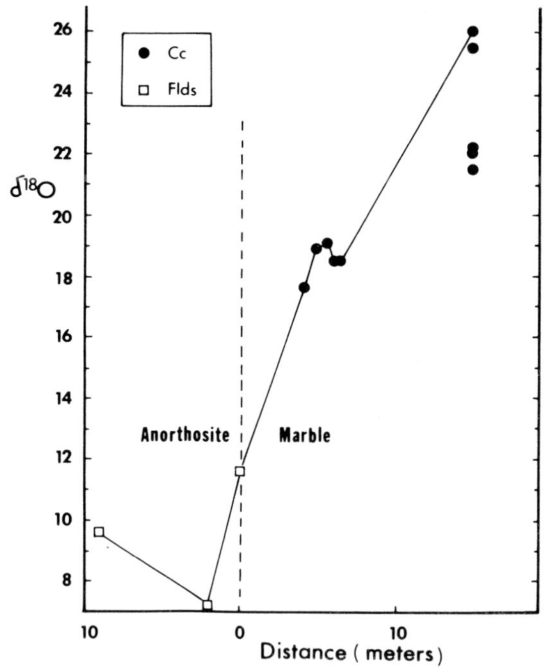

Fig. 10-8. Oxygen isotopic composition of calcite and feldspar from a traverse across a marble xenolith within anorthosite, Adirondack Mountains, New York (after Valley and O'Neil, 1984; Valley, 1986). The steep gradient in $\delta^{18}O$ across the contact zone between marble and anorthosite suggests that the values have mainly been set by diffusive processes and preclude a pervasive large flux of oxygen-bearing fluid such as CO_2 or H_2O across this zone during the metamorphism. On the other hand, the data do not preclude advection of fluid parallel to lithological contacts.

a lithologically varied high-grade gneiss terrane such as the Adirondacks (especially during metamorphic and deformational events), the isotopic and petrological data so far published from this huge area probably make generalized conclusions premature. In order to quantify fluid budgets or identify fluid pathways in terranes thought to be representative of the deep crust, we need to make more detailed isotopic studies than have thus far been made, both in the Adirondacks and elsewhere (see below).

A contrasting dataset was recently reported by Baker and Fallick (1988) from the Lofoten–Vesteralen granulites of northern Norway (cf. Valley and O'Neil, 1984; Baker, 1988), where marbles from the granulite facies and from near the amphibolite–granulite transition show large isotopic shifts to much lighter carbon and oxygen isotopic composition ($\delta^{13}C \sim -7‰$ and $\delta^{18}O \sim +10$) compared to fairly normal sedimentary values in amphibolite facies precursors (see Fig. 10-9). Because of the magnitude of the $\delta^{18}O$ and $\delta^{13}C$ shifts and the limited oxygen and carbon loss through decarbonation in these samples, these shifts are almost certainly related to open system behaviour and exchange with an external oxygen and carbon reservoir. It is possible that CO_2 may have been introduced into some of these rocks by associated syn-metamorphic mangerite–charnockite magmatic intrusions. No study of $^{18}O/^{16}O$ systematics in the silicate rocks has yet been reported from this interesting area, and although the stable isotope data provide a case for large-scale advection of fluid at deep (9–12 kbar) crustal levels, further interpretation must await more isotopic and petrological data.

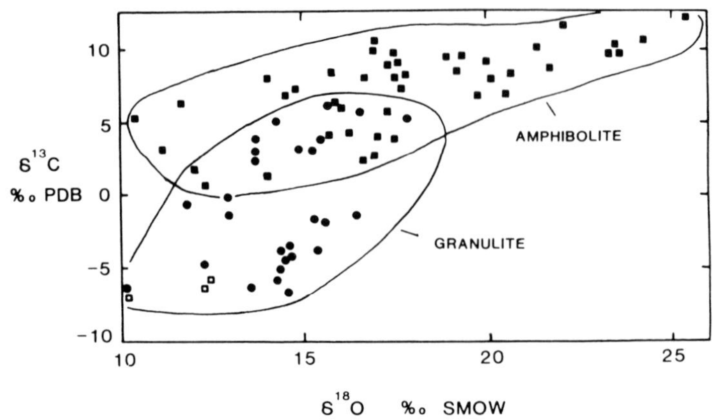

Fig. 10-9. $\delta^{18}O$ vs. $\delta^{13}C$ in calcite from amphibolite and granulite facies marbles of the Lofoten–Vesteralen gneiss terrane in Norway (after Baker and Fallick, 1988). Filled squares are from pods with dolomite + diopside stable, filled circles are from pods with dolomite + diopside unstable, and open squares are calcite-poor samples with no dolomite. The data define a clear isotopic shift to lower $\delta^{13}C$ and $\delta^{18}O$ going from the amphibolite to the granulite facies.

A further extension of the use of stable isotopes in the study of deep crustal fluids is in the isotopic analysis of fluid contained within fluid inclusions. Recent advances in the application of laser microprobe techniques, both for liberating the volatile contents of specific inclusions and making *in situ* measurements such as Raman spectroscopic identification of phases, suggest that this will be a promising area for future research. Some of the problems associated with the study of fluid inclusions have been outlined in section 3, and stable isotope analyses of these materials is rendered especially difficult because of the difficulty in analyzing individual inclusions (because of their small size) or single generations of inclusion (because several generations can often be intermixed within the same grain). However, the chemical and isotopic composition of inclusions can provide an extremely powerful constraint on fluid–rock interaction in the deep crust when combined with detailed petrographic studies.

An excellent recent study of this type is that of Jackson et al. (1988a,b) in which quartz and garnet were separated from both amphibolite and closely adjacent granulite facies gneisses from southern India and analyzed for both bulk carbon content and also for carbon isotopic composition. The carbon was present as CO_2 in fluid inclusions within these phases. The data indicate a bulk two- to fourfold enrichment in carbon in most granulite (orthopyroxene-bearing) samples compared with their amphibolite precursors, coupled with distinct shifts in carbon isotopic composition to lighter values. This clearly indicates an enrichment in CO_2 associated with the conversion of amphibolite to granulite at the partially-charnockitized localities studied. However, the lack of additional data at this stage from other areas, coupled with the difficulty of making a time-consuming study of this type, make any generalization based on these few results premature.

The model of carbonic metamorphism proposed by Newton et al. (1980) requires granulite metamorphism (at least in the well-known southern India terrane) to result from the pervasive flooding of all rocks with mantle-derived CO_2. As noted above, this interpretation probably does not apply in the Adirondacks (e.g., Valley and O'Neil, 1984; Jiang et al., 1988) and *in its purest form* probably does not apply anywhere *on a regional scale*, since it is very unlikely that *all* lithologies will be sufficiently permeable to be infiltrated on the appropriate time-scale and length-scale. However, the question of whether fluid flow is channelized in the lower crust, with fluid being preferentially transported along some layers while others remain almost impermeable is still very open. Detailed isotopic profiles measured across carbonate/silicate contacts are useful because a pre-metamorphic sedimentary oxygen isotope anomaly can easily be inferred. However, to date there have been very few detailed studies of the variation in silicate mineral oxygen isotopic composition across the boundaries between different types of silicate lithology from any high-grade gneiss terrane (e.g., Valley and O'Neil, 1984). The Adirondacks data preclude significant fluid transport across the boundary between carbonate and silicate units (at localities such as the one shown in Fig. 10-8) during granulite facies metamorphism, but do not constrain the presence or absence of significant fluid fluxes elsewhere, particularly within the silicate rocks of the same terrane.

Existing oxygen isotope data from various amphibolite and granulite facies rocks suggest that relatively unreactive (but very ductile) pure calcite marbles may be extremely (anomalously?) impermeable under high-grade conditions as compared with lithologies within which devolatilization is taking place (Rumble and Spear, 1983; Nabelek et al., 1988; Wickham and Peters, 1990). For example, Baker et al. (1989) have convincingly shown that in the Miocene metamorphic sequence on Naxos the thick and relatively pure marble units that make up about 50% of the terrane have been infiltrated by less than half the total volume of fluid released by devolatilization of the surrounding schists, proving that metamorphic fluids were channelized preferentially within the silicate units. This study also rules out pervasive infiltration of mantle CO_2 as postulated by others (e.g., Kreulen, 1988) in this terrane. Conversely, in terranes where marbles are much less pure, allowing carbonate–silicate reaction to cause substantial decarbonation, homogeneous, flat oxygen isotope profiles across carbonate-rich layers demonstrate that fluid infiltration has been intense and pervasive (e.g., Wickham and Taylor, 1985; 1987). Detailed profiles of individual phases such as quartz from adjacent, contrasting silicate lithologies within these areas and their analysis in terms of fluid transport theory (e.g., Bickle and McKenzie, 1987) will in future reveal the scale of advective fluid movements and help us to understand the nature of fluid transport in the various common lithologies, and the magnitude of fluid fluxes in the deep crust.

5. Melting and fluid transport

5.1. Overview

Most high-grade gneiss terranes show evidence for melting, at least on a small scale, and any discussion of deep crustal fluids must also consider the behaviour of silicate melts in this type of environment. Although precise definition is difficult, we will follow the normal petrological practice in this discussion and refer to melts as liquids that are relatively rich in aluminosilicate components and relatively poor in volatiles (i.e., less than about 30% dissolved H_2O and CO_2), and to *fluids* as being dominantly composed of volatiles, principally H_2O and CO_2, with relatively small amounts of dissolved silicates. Obviously this distinction may become blurred in some cases (e.g., pegmatites, which may contain roughly equal proportions of both) where the normally wide miscibility gap between fluids and melts may break down.

Deep crustal pressure–temperature conditions, even in anorogenic regions, are often close to rock melting temperatures, and are well in excess of most water-saturated solidi during granulite facies metamorphism (Harley, 1989; see Fig. 10-1). This makes melting of a variety of common crustal rock types likely, at least to a small extent. Inasmuch as the solubility of some common volatiles (particularly H_2O) in these liquids is high, especially at elevated deep crustal pressures (Holloway, 1973), melting and fluid transport are inevitably intimately connected. In particular, the coincidence of the water-saturated and dry solidi for common crustal rock types with the typical P–T range of granulite metamorphism (Fig. 10-1), espoused by some as being symptomatic of a link between melting and the dehydration of the lower crust (e.g., Powell, 1983a,b; Clemens, 1988), ensures the potential for widespread melting and crystallization. It should be remembered that small-scale melting (i.e., $<30\%$) will favour chemical fractionation by differential motion of the melt with respect to the residue, whereas large-scale melting (i.e., $>50\%$ melt) will lead to bulk mobilisation and intrusive (diapiric?) ascent of large regions of the lower crust without necessarily causing chemical fractionation (Wickham, 1987). In the deep crust, where P–T conditions remain close to lower crustal solidi, there is great potential for volatiles and other incompatible components to dissolve and be released by melts as they form and crystallize. Thus, fluid transport may be directly linked to melt transport.

5.2. Experimental studies

The solubility of volatiles in silicate melts under deep crustal conditions has been addressed by numerous experimental studies, mostly working in chemically simple systems. The variation in the solubility of H_2O in compositionally simple silicate melts has been documented over a range of P and T using infrared spectroscopic techniques (Stolper, 1982). Experimental studies in haplogranitic systems have shown that CO_2 solubility is less than 1 wt.% at pressures below 10 kbar and that addition of CO_2 to coexisting mixed H_2O-CO_2 vapour raises the vapour saturated solidus temperature (Eggler and Kadik, 1979). However, recent experiments of Peterson and Newton (1989a,b, 1990) have confirmed the earlier preliminary results of Wendlandt (1981), showing that when MgO is added to the system, the solidus is actually lowered at high X_{CO_2} relative to the water-saturated system (see Fig. 10-10). This result has profound implications for a variety of important lower crustal processes, and is probably caused by an interaction of MgO and CO_2 in the melt.

The Peterson and Newton (1989a,b, 1990) experiments in the system KMASH-CO_2 (melting the assemblages phlogopite–quartz-vapour and phlogopite–quartz–sanidine-vapour in the pressure range 3–10 kbar) place the debate over the behaviour of CO_2 in the lower crust in a new light. First, the likelihood of the presence of a free CO_2 fluid in the lower crust is greatly reduced because the new data imply that biotite- and quartz-bearing compositions would melt in the presence of CO_2 at typical granulite facies temperatures (i.e., above 650°C). Second, the presence of a CO_2 vapour and CO_2 enrichment of the lower crust does not exclude melting. It would in fact promote it. Third, the solubility of CO_2 in the melt would be high. Peterson and Newton (1989b, 1990) report syenitic or lamprophyric compositions in glass from their experiments and volatile contents probably in excess of 10 wt.% and possibly as high as 40 wt.% (R.C. Newton, pers. commun., 1989). Thus CO_2-rich silicate melt, even if it were only present in tiny quantities, could be an important intermediary transporting medium for CO_2 (Frost and Frost, 1987) rather than the vapour phase itself. Inasmuch as the transport properties of melts vary enormously and in view of the potential for melts to transport dissolved volatiles, it is therefore pertinent to review quickly the principal physical controls on these properties.

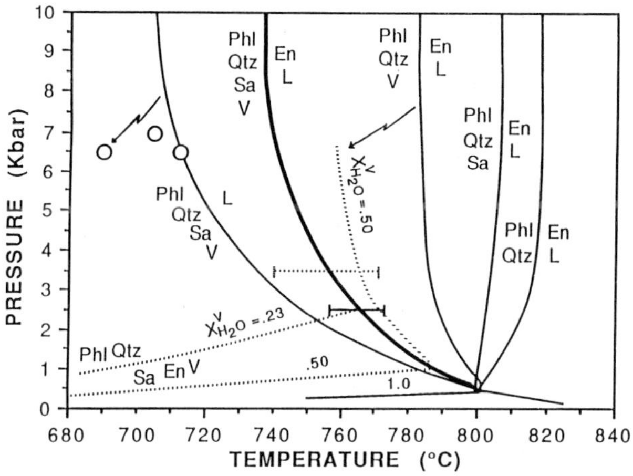

Fig. 10-10. Univariant melting in the KMASH and KMASH–CO_2 systems (after Peterson and Newton, 1989b). The bold curve denotes the incongruent melting of quartz + phlogopite + sanidine + vapour (Wendlandt, 1981) at $CO_2/H_2O + CO_2$ in the vapour equal to 0.5, with bold bracket from Peterson and Newton (1989a). Light curves denote reaction boundaries in the pure H_2O system (see Peterson and Newton, 1989a). Steep dashed curve is reaction phlogopite + quartz + vapour = enstatite + liquid in the H_2O–CO_2 (KMASH–CO_2) system at $CO_2/H_2O + CO_2$ in the vapour equal to 0.5 (with dashed reversal bracket). The arrow denotes the displacement of this reaction to lower temperatures compared with the pure H_2O curve. Circles denote runs in which quartz + phlogopite + sanidine + vapour melted congruently at $CO_2/H_2O + CO_2 = 0.5$ in the vapour phase, with arrow denoting displacement relative to the pure H_2O curve. A comparison with Fig. 10-1 shows how relevant these equilibria are to granulite facies conditions. For further details see Peterson and Newton (1989b, 1990).

5.3. Fluid and melt transport properties

Under typical lower crustal pressure–temperature conditions, at relatively low melt- or fluid-fraction (i.e., <30%), melts and fluids will be distributed in the intergranular pore space between individual silicate grains. Their mobility is fundamentally controlled by whether or not the individual drops of liquid are interconnected, a condition that is governed by the dihedral angle, the angle formed between the liquid and two adjacent grains (Beeré, 1975). Liquid interconnectedness and its relevance to geological systems has been discussed by others (e.g., McKenzie, 1984; Jurewicz and Watson, 1985) and will only be briefly outlined here. The critical condition for interconnectedness at all porosities is that the dihedral angle should be less than 60° (Beeré, 1975). Of the relatively small number of such measurements that have so far been made, *all* silicate melts ranging from basalts (Waff and Bulau, 1979) to rhyolites (Jurewicz and Watson, 1985) to carbonatites (Hunter and McKenzie, 1988) distributed within silicate matrices have had $\theta < 60°$ (see Table 1) and are at least theoretically capable of extraction and movement relative to their solid matrix even at very low porosities. Water also appears to have $\theta < 60°$ and most probably remains interconnected where it exists as a free phase (Watson and Laporte, 1989). Indeed, the fact that rocks progressively lose their water with increasing metamorphic grade and become more dehydrated supports this idea, because efficient removal of a pervasive, internally generated, water-rich fluid would be very difficult unless it was interconnected.

On the other hand, Watson and Brenan (1987) and Watson and Laporte (1989) have measured the dihedral angle for CO_2 in olivine, quartz, and feldspar matrices respectively and found it to be greater than 60°, implying that pervasive intergranular flow of pure CO_2

at low porosities within silicate rocks is unlikely; of course, this does not preclude CO_2 becoming interconnected at higher porosities [e.g., above 8% (Hunter and McKenzie, 1988)]. Presumably there must be a crossover between these two types of behaviour in mixed H_2O and CO_2 fluids but it has yet to be identified. These data put important constraints on hypotheses involving CO_2-flushing. If the fluid was pure CO_2, then it would presumably have to flow in discrete cracks in order to enter and pass through silicate lithologies, and because of the lack of connectedness would have difficulty in penetrating and equilibrating pervasively with the rocks it traversed (i.e., the regions between the cracks). In effect, equilibration would be limited by the spacing of the cracks and diffusion through the low-porosity rock between. However, a mixed H_2O-CO_2 fluid might still be interconnected and therefore able to flow pervasively, even at quite high X_{CO_2}.

For interconnected liquids (fluids or melts) calculations show that the efficiency with which they can be extracted from crustal rocks is strongly dependent on viscosity (McKenzie, 1985; Wickham, 1987) and this can vary enormously for the materials under consideration. Silicate melts range in viscosity from $\sim10^{-3}$ Pa s (carbonatite) through $\sim10^{1}$ Pa s (basalt) to $\sim10^{4}$ Pa s (water-saturated rhyolite) and $\sim10^{11}$ Pa s (dry rhyolite); in general, composition (in particular volatile content) and temperature are the principal controls (Shaw, 1972). This huge range of viscosity means that even if the liquid is interconnected (which appears to be the case for most geological fluids and melts, with the important exception of pure CO_2), the rate at which it will separate from its solid matrix and flow through the mantle or crust is extremely dependent on viscosity. Thus carbonatites and basalts can quickly and easily separate from their sources and move through hundreds or thousands of kilometres within the mantle at extremely low melt-fraction and over geologically rapid time-scales, whereas more silica-rich melts (which are likely to dominate within the continental crust), even if they are relatively low viscosity through virtue of being hot or volatile-rich, can only segregate, even in the least viscous cases, on a small (cm or m) scale (Wickham, 1987, 1989; cf. Clemens and Vielzeuf, 1987). For normal viscosity silicic melts (i.e., 10^5–10^8 Pa s), segregation will be even more sluggish.

Thus, an important conclusion of compaction theory is that removal of a small fraction of silica-rich, (granitic), water-undersaturated melt from the lower crust will be so sluggish as to be virtually impossible to achieve over a reasonable geological time-scale. This conclusion is in direct conflict with the theory that low melt-fraction segregation is an important process leading to dehydration and chemical differentiation of the lower crust (e.g., Fyfe, 1973; Clemens and Vielzeuf, 1987). However, as mentioned above, this latter issue is still very much open; it remains to be demonstrated on a regional scale that large volumes of the middle and lower crust have been stripped of a minimum melt fraction, and this is certainly not *required* as a granite- or granulite-forming process (see Wickham, 1989; cf. Miller et al., 1988).

The huge viscosity variation in deep crustal melts and fluids places stringent limits on melt and fluid transfer processes in the lower crust. In particular it limits intergranular movement of crustal melts at low melt-fraction unless they are (1) volatile-rich and/or (2) relatively poor in silica and/or (3) very hot (>1000°C). It also suggests that H_2O-rich fluids, where present, should be highly mobile, whereas pure CO_2 should not be mobile at all as a pervasive intergranular phase.

5.4. Implications for melt and fluid motion

In the deep crust, the amount of fluid that can be generated internally by devolatilization of H_2O- or CO_2-bearing minerals is generally small — probably just a few weight percent H_2O and substantially less CO_2 unless the terrane is unusually rich in metasedimentary carbonate (see below). This is because most volatiles will already have been lost from the rocks during the earlier stages of prograde metamorphism to amphibolite facies at higher structural levels in the crust. Thus, at deep levels, only relatively small *internally generated* fluxes are likely, in the main corresponding to prograde breakdown of biotite and amphibole (i.e., the amphibolite–granulite transition).

The only *major* source for a deep crustal fluid flux is the mantle (including material emplaced within the mantle by tectonic processes such as subduction). However, it is extremely difficult to place accurate constraints on the possible magnitude of volatile fluxes from the mantle, mainly because of uncertainties in the subducted fluid budget. Stone et al. (1988) reviewed the constraints on fluxes from the main mantle reservoirs and concluded that the convecting (MORB source) mantle is an inadequate source for the CO_2 required to generate granulites by CO_2 infiltration. This conclusion has also been arrived at by O'Nions and Oxburgh (1988) through helium isotope studies. However, the mantle *lithosphere* is a potentially rich source for both CO_2 and H_2O because its longevity means that it can have been repeatedly enriched during previous subduction or underthrusting events, or by infiltration of CO_2-rich melts derived from deeper mantle levels. This enrichment is attested to by the carbonate and alkali metasomatism observed in some peridotite xenoliths (e.g., Nixon, 1987) and by carbonatite or kimberlite magmatism (e.g., Dawson et al., 1986) indicating the presence of substantial CO_2-enriched volumes at depth within the mantle.

There is thus a large potential source of CO_2 (and possibly also H_2O) of very loosely constrained magnitude within the mantle lithosphere. There is also a potentially significant source of CO_2 in metasedimentary carbonate in the deep crust which might be mobilized during major magmatic underplating events (Wickham, 1988a,b; see also Baker and Fallick, 1988). These source considerations, coupled with the fluid dynamical constraints outlined earlier, suggest the following implications for deep crustal fluid flow.

(1) H_2O-rich fluid will enter, flow through, and leave the lower crust relatively easily, but in most cases at amphibolite or granulite pressure–temperature conditions (i.e., $T > 600°C$, $P > 5$ kbar) it will not be in stable equilibrium with common silicate rock-types and will promote melting. Because of this, most H_2O transport at deep levels probably occurs with the H_2O dissolved in silicate melt (although a free H_2O-rich vapour phase may be generated for short periods of time when hydrous minerals undergo subsolidus dehydration reactions). Of course, under vapour-absent conditions, such minerals can dehydrate without producing a vapour phase when the rock undergoes dehydration melting (Thompson, 1982). Because the total water content of deep crustal rocks is not high, and H_2O fluxes from the mantle not likely to be large, transport of free H_2O in the deep crust over large distances is in general not likely to occur.

(2) Because it has been considered relatively inert and less reactive with silicate minerals than H_2O (but see (5.2) above), CO_2 has been viewed as a better candidate for traversing deep crustal rocks as a free fluid phase (particularly in view of the potential for replenishment of the mantle CO_2 reservoir by subduction). However, the limited experimental data on its textural distribution in silicates suggests that it should be relatively immobile (Watson and Brenan, 1987). This preliminary constraint tends to exclude the likelihood of *pervasive*

intergranular flow of CO_2-rich fluid, although it does not rule out flow in fractures.

(3) Because they are interconnected at all porosities, silicate melts, spanning all compositions from rhyolite to basalt to carbonatite, can potentially separate from their solid matrices and traverse the lower crust as intergranular liquids. They are equally capable of carrying dissolved volatiles with them, but their separation behaviour is *highly* dependent on their viscosity. Compaction theory (McKenzie, 1985) predicts that silicic melts, unless they contain high H_2O contents or are very hot (>1000°C), will be too viscous to separate except on a small (cm to m) scale. This small-scale separation is possibly represented by the layered migmatites often observed in high-grade gneiss terranes. It is thus very unlikely that intergranular flow of silicic liquid with small H_2O and CO_2 concentration (i.e., <5 wt.%) occurs in the lower crust over length scales of more than centimetres or at the most metres.

(4) Basalts or carbonatite liquids, on the other hand, have the physical properties that would allow them to move within the lower crust by pervasive intergranular flow. However, they are unlikely to be *internally* generated within the lower crust for compositional reasons, and if introduced from outside (i.e., from below) will probably be transported along discrete conduits, such as fracture zones. Hitherto, CO_2 has been regarded as almost insoluble in silicic melts (e.g., Holloway, 1973) and therefore unable to influence viscosity (as it does dramatically in the case of carbonatite liquids). However, experiments of Wendlandt (1981) and more recently Peterson and Newton (1989) (see section 5.2) have shown that CO_2-rich fluids can actually lower the solidus of biotite-bearing rocks, and that the liquid produced during the early stages of melting has a very high dissolved CO_2 content. This means that the liquid probably has a low viscosity (perhaps similar to that of basalt) even at high silica contents, which raises the possibility that it would be highly mobile and could both separate from and flow pervasively through large volumes of lower crustal rocks.

5.5. CO_2-rich melting and an alternative model of granulite petrogenesis

The high volatile content of a CO_2-saturated, mafic, silica-rich (granitic) liquid almost certainly implies that it has a low viscosity (see Fig. 10-11) that is probably considerably less than that of H_2O-undersaturated granitic liquid. Thus this liquid could behave in the crust in the same way as basalts and carbonatites in the mantle. It would be capable of separating from its solid matrix at very low melt fraction (<1%?) and of undergoing pervasive intergranular flow over very large, kilometric length scales (McKenzie, 1985). Since incompatible elements would have a high solubility in these liquids, it would probably be highly reactive and capable of stripping the lower crust of incompatible trace elements. This could account for the widespread LIL depletions that have been observed in some granulite terranes (e.g., Weaver and Tarney, 1983). It would also be capable of imposing a low and uniform water activity over wide regions

This raises the possibility of an alternative model of granulite petrogenesis. There has been much recent discussion as to whether dehydration of granulites is promoted by pervasive flooding of the lower crust with CO_2 derived from the mantle (Newton et al., 1980) or by removal of water in silicate melts formed by dehydration melting (Fyfe, 1973) — see also section (4.2). The experimental results suggest that a third scenario is possible. CO_2 introduced in fluids could induce melting to produce small amounts of CO_2-rich, syenitic or lamprophyric liquid capable of strongly concentrating incompatible elements and H_2O. Such liquids would be rich enough in volatiles to have a low viscosity and would spontaneously separate from the residual solid material at very low melt fraction (probably less than 1

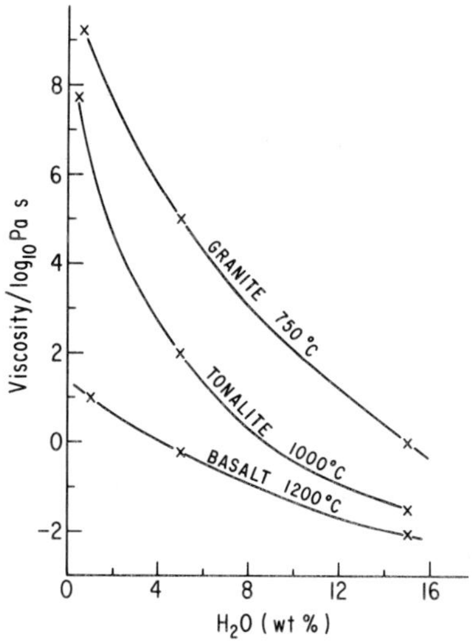

Fig. 10-11. The variation of viscosity with increasing volatile (H_2O) content for three common silicate liquids at appropriate magmatic temperatures (basalt at 1200°C, tonalite at 1000°C, and granite at 750°C). Viscosities were calculated using the empirical method of Shaw (1972), using the major-element analyses listed in Cox et al. (1979), Appendix 2. This model neglects any possible pressure effects, and is based on experiments with liquids containing 6.2 wt.% H_2O. However, the relatively smooth increase in water solubility at higher pressures and temperatures reported by Stolper (1982) probably justifies the extrapolation involved. The graph shows the dramatic decrease in viscosity that occurs with increasing water content, especially for the more silicic compositions. Increasing the CO_2 content of the liquid would probably have a similar effect, resulting in a profound lowering of viscosity, although not necessarily as pronounced as with H_2O. Unfortunately, there are too few viscosity data available for CO_2-rich silicate melts to make quantitative predictions in this respect.

volume% melt (McKenzie, 1985)). They would be capable of rapid, pervasive flow (again at low melt-fraction) over large distances — at least through silicate lithologies — and would therefore be capable of achieving profound, pervasive, and regional geochemical effects. In this scenario, the agent of CO_2 transport and H_2O removal would be a CO_2-rich (syenitic?) liquid, not a CO_2-rich fluid, nor a CO_2-poor haplogranitic liquid. The mechanism of CO_2 transport would be similar to that proposed by Frost and Frost (1987), but the CO_2 concentration in the melt would be much higher, and the amount of melt required could be tiny (<1 volume%), so that it might be undetectable petrographically. Such a carbonic liquid would have a low water activity and could impose this on rock systems with which it equilibrated. It would thus promote dehydration of biotite or amphibole, and would also be capable of removing H_2O and other elements from the lower crust. It would also release CO_2 vapour when it crystallized. This would readily explain the petrological characteristics of some granulites, particularly in southern India, including the very high CO_2 fluid inclusion contents observed locally (e.g., Doddabetta — see Fig. 10-5), the CO_2 inclusions commonly observed in migmatite leucosomes (Olsen, 1987; Hollister, 1988), and the aureoles of granulite observed around syenitic and charnockitic plutons in some terranes (e.g., Frost and Frost, 1987). Although the agent of granulite formation would be a silicate melt rather than a CO_2 fluid, it would be a melt of quite different composition than that assumed by previous models (e.g., Powell, 1983a,b) inasmuch as it would be syenitic, low-viscosity, and CO_2-rich, as opposed to granitic, viscous, CO_2-poor, and H_2O-undersaturated.

6. Conclusions

We have reached an exciting stage in the evolution of our understanding of deep crustal fluids. The last decade has seen the development and application of many petrological and geochemical techniques, which have led to the discovery of numerous new phenomena. Direct evidence for the presence of mineralizing fluids in crystalline basement at 8–10 km depth has been obtained from the Kola Peninsula superdeep borehole (Kozlovsky, 1984, p. 268). Much evidence both supporting and denying the existence of large fluid or melt fluxes at deep crustal levels has been amassed. Supporting evidence includes field and petrologic evidence (e.g., Newton, 1985), stable isotope studies (e.g., Baker and Fallick, 1988; Wickham and Taylor, 1987), experimental petrology (e.g., Peterson and Newton, 1989b), fluid inclusion studies (e.g., Touret and Hansteen, 1988), resistivity surveys (e.g., Bailey et al., 1989), and equilibrium thermodynamic calculations (e.g., Chacko et al., 1988). Other data have been interpreted to reflect only very limited fluid involvement (e.g., Lamb and Valley, 1984, 1988; Valley and O'Neil, 1984; Clemens and Vielzeuf, 1987; Lamb et al., 1987; Baker, 1988) by identifying deep crustal regions exhibiting isotope heterogeneity, or from detailed phase equilibrium and fluid inclusion studies.

Helium isotope systematics have over the last decade provided us with an especially powerful way to identify present-day discharge of mantle-derived fluids, although this technique is of course of no use in studying old terranes. Extensive $^3He/^4He$ measurements from Europe, Asia, and North America all indicate a significant present-day flux of mantle helium through the continental crust within regions of presently active volcanism and/or active extensional tectonics (O'Nions and Oxburgh, 1988; Stone et al., 1988). These same localities also often show high present-day CO_2 discharge [e.g., in the Massif Central in France (Matthews et al., 1987)] and tend to confirm the existence, at least locally within the lower crust or the mantle lithosphere, of substantial CO_2 sources. Whether the CO_2 is derived from crystallising magmas or decarbonating metamorphic rocks is presently undecided. However, if mantle He also tracks deep level CO_2 discharge, the data available today suggest that its release is geographically extremely heterogeneous. This heterogeneity probably echoes (albeit on a much larger scale) the likely permeability structure of the crust through which these volatiles have passed.

The clear message from the observations and interpretations described in this chapter is that there is a pressing need for more geochemical studies, and more integration of research techniques. In particular, the application of fluid dynamical theory (e.g., Bickle and McKenzie, 1987) to the petrological and isotopic study of high-grade gneiss terranes (e.g., Harris and Bickle, 1989) needs much development and offers much potential. Part of the problem today is in the paucity of detailed isotopic and geochemical data available from petrologically well-documented terranes. In particular detailed isotopic profiling (e.g., Valley and O'Neil, 1984) (principally for oxygen) across silicate/silicate lithological contacts has been almost entirely neglected to date, despite the certainty that there is a wealth of information pertaining to deep crustal fluid transport that is waiting to be collected. Our present understanding is so vague that we cannot even identify with certainty which lithologies will be permeable and which impermeable during metamorphism, let alone determine the nature and scale of fluid transport within whole terranes. Detailed geochemical and isotopic mapping combined with petrological and fluid dynamical analysis of high-grade metamorphic rocks (e.g., Labotka et al., 1988; Wickham and Peters, 1990) is now beginning to define (both qualitatively and quantitatively) metamorphic permeabilities, and to identify the main pathways of fluid flow.

The next decade should bring many more of these types of study, and should see a new expansion of our knowledge and understanding of the behaviour of fluids in the deep crust.

Acknowledgements

This review benefited from discussions with John Stone, Dave Waters, Mike Bickle, Hugh Taylor, Bob Newton, Jacques Touret, Tom Chacko, Jon Peterson, Ron Frost, Vic McGregor, Richard Arculus, and many others. I am grateful to Simon Harley, Jacques Touret, and Daniel Vielzeuf for sending me various reprints and preprints, to John Valley, Roger Powell, and David Fountain for reviews, to the editors for inviting me to write this chapter, and to James Eason for typing the manuscript. This work was supported by NSF grant EAR 8720097.

References

Andersen, D.J. and Lindsley, D.H., 1988. Internally consistent solution models for Fe-Mg-Mn-Ti oxides. Am. Mineral., 73: 714–726.

Anderson, A.T., Jr. and Morin, M., 1969. Two types of massif anorthosites and their implications regarding the thermal history of the crust. In: Y.W. Isachsen (Editor), Origin of Anorthosite and Related Rocks. New York State Mus. Sci. Ser., Mem. 18: 57–69.

Bailey, R.C., Craven, J.A., Macnae, J.C. and Polzer, B.D., 1989. Imaging of deep fluids in Archaean crust. Nature, 340: 136–138.

Baker, A.J., 1988. Stable isotope evidence for limited fluid infiltration of deep crustal rocks from the Ivrea Zone, Italy. Geology, 16: 492–495.

Baker, A.J. and Fallick A.E., 1988. Evidence for CO_2 infiltration in granulite facies marbles from Lofoten–Vesteralen, Norway. Earth Planet. Sci. Lett., 91: 132–140.

Baker J., Bickle M.J., Buick, I.S., Holland, T.J.B. and Matthews, A., 1989. Isotopic and petrological evidence for the infiltration of water-rich fluids during the Miocene M2 metamorphism on Naxos, Greece. Geochim. Cosmochim. Acta, 53: 2037–2057.

Beeré, W., 1975. A unifying theory of the stability of penetrating liquid phases and sintering pores. Acta Metall., 23: 131–138.

Bickle, M.J. and McKenzie, D., 1987. The transport of heat and matter by fluids during metamorphism. Contrib. Mineral. Petrol., 95: 384–392.

Bohlen, S.R., 1987. Pressure–temperature–time paths and a tectonic model for the evolution of granulites. J. Geol., 95: 617–632.

Bohlen, S.R., Valley, J.W. and Essene, E.J., 1985. Metamorphism in the Adirondacks I. Petrology, pressure, and temperature. J. Petrol., 26, 971–992.

Bowers, T.S. and Helgeson, H.C., 1983. Calculation of the thermodynamic and geochemical consequences of non-ideal mixing in the system H_2O–CO_2–NaCl on phase relations in geologic systems: equation of state for H_2O–CO_2–NaCl fluids at high pressures and temperatures. Geochim. Cosmochim. Acta, 47: 1247–1275.

Chacko, T., Ravindra Kumar, G.R. and Newton, R.C., 1987. Metamorphic P–T conditions in the Kerala (south India) Khondalite Belt: a granulite-facies supracrustal terrain. J. Geol., 95: 343–358.

Chacko, T., Ravindra Kumar, G.R. and Peterson, J.W., 1988. Water activities in the Kerala khondalite belt. J. Geol. Soc. India, 31: 16–18.

Clemens, J.D., 1987. The granulite–granite connexion. Terra Cognita, 8: 257.

Clemens, J.D. and Vielzeuf, D., 1987. Constraints on melting and magma production in the crust. Earth Planet. Sci. Lett., 86: 287–306.

Cox, K.G., Bell, J.D. and Pankhurst, R.J., 1979. The Interpretation of Igneous Rocks. George, Allen and Unwin, London, 450 pp.

Dawson, J.B., Carswell, D.A., Hall, J. and Wedepohl, K.H. (Editors), 1986. The Nature of the Lower Continental Crust. Geol. Soc. London, Spec. Publ. 24.

Des Marais, D.J. and Moore, J.G., 1984. Carbon and its isotopes in mid-oceanic basaltic glasses. Earth Planet. Sci. Lett., 69: 43–57

Edwards, R.L. and Essene, E.J., 1988. Pressure, temperature, and C-O-H fluid fugacities across the amphibolite–granulite transition, northwest Adirondack Mountains, New York. J. Petrol., 29: 39–72.

Eggler, D.H. and Kadik A.A., 1979. The system $NaAlSi_3O_8$-H_2O-CO_2 to 20 kbar pressure: I. Compositional and thermodynamic relations of liquids and vapours coexisting with albite. Am. Min., 64: 1036–1048.

Ellis, D.J., 1987. Origin and evolution of granulites in normal and thickened crusts. Geology, 15: 167–170.

Elphick, S.C., Graham, C.M. and Dennis P.F., 1989. An ion microprobe study of anhydrous oxygen diffusion in anorthite: a comparison with hydrothermal data and some geological implications. Contrib. Mineral. Petrol., 100: 490–495.

Engel, A.E.J. and Engel, C.G., 1960. Progressive metamorphism and granitization of the Major Paragneiss, N.W. Adirondack Mts., N.Y. Part II, mineralogy. Geol. Soc. Am. Bull., 71: 1–58.

England, P.C. and Richardson, S.W., 1977. The influence of erosion upon mineral facies of rocks from different metamorphic environments. Q. J. Geol. Soc. London, 134: 201–213.

England, P.C. and Thompson, A.B., 1984. Pressure–temperature–time paths of regional metamorphism, I. Heat transfer during the evolution of regions of thickened continental crust. J. Petrol., 25: 894–928.

Etheridge, M.A., Wall, V.J. and Vernon, R.H., 1983. The role of the fluid phase during regional metamorphism and deformation. J. Metamorph. Geol., 1: 205–226.

Ferry J.M., 1980. A case study of the amount and distribution of heat and fluid during metamorphism. Contrib. Mineral. Petrol., 71: 373–385.

Ferry, J.M., 1986. Infiltration of aqueous fluid and high fluid:rock ratios during greenschist facies metamorphism: a reply. J. Petrol., 27: 695–714.

Frost, B.R. and Chacko, T., 1989. The granulite uncertainty principle: limitations on thermobarometry in granulites. J. Geol., 97: 435–450.

Frost, B.R. and Frost, C.D., 1987. CO_2, melts, and granulite metamorphism. Nature (London), 327: 503–506.

Frost, B.R., Frost, C.D. and Touret, J.L.R., 1989. Magmas as a source of heat and fluids in granulite metamorphism. In: D. Bridgwater (Editor), Fluid Movements — Element Transport and the Composition of the Deep Crust, Kluwer Academic Publishers, Dordrecht, Holland, pp. 1–18.

Fyfe, W.S., 1973. The granulite facies, partial melting, and the Archean crust. Philos. Trans. R. Soc. London, Ser. A, 273: 457–461.

Fyfe, W.S., Price, N.J. and Thompson, A.B., 1978. Fluids in the Earth's Crust. Elsevier, Amsterdam–New York, 338 pp.

Garlick, G.D. and Epstein, S., 1966. Oxygen isotope ratios in coexisting minerals of regionally metamorphosed rocks. Geochim. Cosmochim. Acta, 31: 181–214.

Hansen, E.C., Janardhan, A.S., Newton, R.C., Prame, W.K.B.N. and Ravindra Kumar, G.R., 1987. Arrested charnockite formation in southern India and Sri Lanka. Contrib. Mineral. Petrol., 96: 225–244.

Harley, S.L., 1989. The origin of granulites: a metamorphic perspective. Geol. Mag., 126: 215–247.

Harris, D.M., 1981. The concentration of CO_2 in submarine tholeiitic basalts. J. Geol., 89: 689–701.

Harris, N.B.W. and Bickle, M.J., 1989. Advective fluid transport during charnockite formation; an example from southern India. Earth Planet. Sci. Lett., 93: 151–156.

Henry, D.J., 1988. Cl-rich minerals in Archean granulite facies ironstones from the Beartooth Mountains, Montana, USA: implications for fluids involved in granulite metamorphism. J. Geol. Soc. India, 32: 43–45.

Hollister, L.S., 1988. On the origin of CO_2-rich fluid inclusions in migmatites. J. Metamorph. Geol., 6: 467–474.

Hollister, L.S., Burruss, R.C., Henry, D.L. and Hendel, E.M., 1979. Physical conditions during uplift of metamorphic terrains as recorded by fluid inclusions. Soc. Mineral. Fr. Bull. 102: 555–561.

Holloway, J.R., 1973. The system pargasite–H_2O–CO_2: a model for melting of a hydrous mineral with a mixed-volatile fluid: I. Experimental results to 8 kbar. Geochim. Cosmochim. Acta, 37: 651–666.

Hunter, R. and McKenzie, D., 1989. The equilibrium geometry of carbonate melts in rocks of mantle composition. Earth Planet. Sci. Lett., 92: 347–356.

Jackson, D.H., Mattey, D.P. and Harris, N.B.W., 1988a. Carbon isotope compositions of fluid inclusions in charnockites from southern India. Nature (London), 333: 167–170.

Jackson, D.H., Mattey, D.P., Santosh, M. and Harris, N.B.W., 1988b. Carbon stable isotope analysis of fluid inclusions by stepped heating. Mem. Geol. Soc. India, 11: 149–158.

Jiang, J., Clayton, R.N. and Newton, R.C., 1988. Fluids in granulite facies metamorphism: a comparative oxygen isotope study on the south India and Adirondack high-grade terrains. J. Geol., 96: 517–533.

Joesten, R., 1986. The role of magmatic reaction, diffusion and annealing in the evolution of coronitic microstructure in troctolite gabbro from Risor, Norway. Mineral. Mag., 50: 445–457.

Jurewicz, S.R. and Watson, E.B., 1985. The distribution of partial melt in a granitic system: the application of liquid phase sintering theory. Geochim. Cosmochim. Acta, 49: 1109–1121.

Koslovsky, Y.A., 1984. The Superdeep Well of the Kola Peninsula. Springer-Verlag, New York.

Kreulen, R., 1988. High integrated fluid/rock ratios during metamorphism at Naxos: evidence from carbon isotopes of calcite in schists and fluid inclusions. Contrib. Mineral. Petrol., 98: 28–32.

Labotka, T.C., Nabelek, P.I. and Papike, J.J., 1988. Fluid infiltration through the Big Horse limestone member in the Notch Peak contact-metamorphic aureole, Utah. Am. Mineral., 73: 1302–1324.

Lamb, W.M. and Valley, J.W., 1984. Metamorphism of reduced granulites in a low-CO_2, vapour-free environment. Nature, 312: 56–58.

Lamb, W.M. and Valley, J.W., 1985. C-O-H fluid calculations and granulite genesis. In: A.C. Tobi and J.L.R. Touret (Editors), The Deep Proterozoic Crust in the North Atlantic Provinces. D. Reidel, Dordrecht, Holland, pp. 119–131.

Lamb, W.M. and Valley, J.W., 1988. Granulite facies amphibole and biotite equilibria, and calculated peak metamorphic water activities. Contrib. Mineral. Petrol., 100: 349–360.

Lamb, W.M., Valley, J.W. and Brown, P.E., 1987. Post-metamorphic CO_2-rich fluid inclusions in granulites. Contrib. Mineral. Petrol., 96: 485–495.

Lasaga, A., 1986. Metamorphic reaction rate laws and development of isograds. Mineral. Mag., 50: 359–373.

Matthews, A., Fouillac, C., Hill, R., O'Nions, R.K. and Oxburgh, E.R., 1987. Mantle-derived volatiles in continental crust: the Massif Central of France. Earth Planet. Sci. Lett., 85: 117–128.

McKenzie, D., 1984. The generation and compaction of partially molten rock. J. Petrol., 25: 713–765.

McKenzie, D., 1985. The extraction of magma from the crust and mantle. Earth Planet. Sci. Lett., 74: 81–91.

McLelland, J.M. and Husain, J., 1986. Nature and timing of anatexis in the eastern and southern Adirondack Highlands. J. Geol., 94: 17–25.

Miller, C.F., Watson, E.B. and Harrison, T.M., 1988. Perspectives on the source, segregation and transport of granitoid magmas. Trans. R. Soc. Edinburgh: Earth Sci., 79: 135–156.

Morrison, J. and Valley, J.W., 1988a. Contamination of the Marcy Anorthosite Massif–Adirondack Mountains, N.Y.: petrologic and isotopic evidence. Contrib. Mineral. Petrol., 98: 97–108.

Morrison, J. and Valley, J.W., 1988b. Post-granulite facies fluid infiltration in the Adirondacks. Geology, 16: 513–516.

Nabelek, P.I., Labotka, T.C, O'Neil, J.R. and Papike, J.J., 1984. Contrasting fluid/rock interaction between the Notch Peak granitic intrusion and argillites and limestones in western Utah: evidence from stable isotopes and phase assemblages. Contrib. Mineral. Petrol., 86: 25–34.

Newton, R.C., 1985. Temperature, pressure, and metamorphic fluid regimes in the amphibolite facies to granulite facies transition zones. In: A.C. Tobi and J.L.R. Touret (Editors), The Deep Proterozoic Crust in the North Atlantic Provinces. D. Reidel, Dordrecht, Holland, pp. 75–104.

Newton, R.C., 1989. Metamorphic fluids in the deep crust. Annu. Rev. Earth Planet. Sci., 17: 385–412.

Newton, R.C., Smith, J.V. and Windley, B.F., 1980. Carbonic metamorphism, granulites, and crustal growth. Nature (London), 288: 45–50.

Nixon, P.H. (Editor), 1987. Mantle Xenoliths. Wiley and Sons, Chichester, 844 pp.

Norton, D. and Taylor, H.P., Jr., 1979. Quantitative simulation of the hydrothermal systems of crystallizing magmas on the basis of transport theory and oxygen isotope data: an analysis of the Skaergaard intrusion. J. Petrol., 20: 421–486.

Olsen, S.N., 1987. The composition and role of the fluid in migmatites: a fluid inclusion study of the Front Range rocks. Contrib. Mineral. Petrol., 96: 104–120.

O'Nions, R.K. and Oxburgh, E.R., 1988. Helium, volatile fluxes and the development of the continental crust. Earth Planet. Sci. Lett., 90: 331–347.

Peterson, J.W. and Newton, R.C., 1989a. Reversed experiments on biotite-quartz-feldspar melting in the system KMASH: implications for crustal anatexis. J. Geol., 97: 465–485.

Peterson, J.W. and Newton, R.C., 1989b. CO_2-enhanced melting of biotite-bearing rocks at deep-crustal pressure–temperature conditions. Nature (London), 340: 378–380.

Peterson, J.W. and Newton, R.C., 1990. Experimental biotite-quartz melting in the KMASH–CO_2 system and the role of CO_2 in the petrogenesis of granites and related rocks. Am. Mineral., 75: 1029–1042..

Phillips, G.N., 1980. Water activity changes across an amphibolite–granulite facies transition, Broken Hill, Australia. Contrib. Mineral. Petrol., 75: 377–386.

Pineau, F. and Javoy, M., 1983. Carbon isotopes and concentrations in mid-ocean ridge basalts. Earth Planet. Sci. Lett., 62: 239–257.

Powell, R., 1983a. Processes in granulite facies metamorphism. In: M.P. Atherton and C.D. Gribble (Editors), Migmatites, Melting, and Metamorphism. Shiva Publishing Ltd., Nantwich, Cheshire, UK, pp. 127–139.

Powell, R., 1983b. Fluids and melting under upper amphibolite conditions. Q. J. Geol. Soc. London, 140: 629–637.

Rumble, D., III and Spear, F.S., 1983. Oxygen-isotope equilibration and permeability enhancement during regional metamorphism. Q. J. Geol. Soc. London, 140: 619–628.

Shaw, H.R., 1972. Viscosities of magmatic silicate liquids: an empirical method of prediction. Am. J. Sci., 272: 876–893.

Shieh, Y.N. and Schwarcz, H.P., 1974. Oxygen isotope studies of granite and migmatite, Grenville Province of Ontario, Canada. Geochim. Cosmochim. Acta, 38: 21–45.

Stolper, E.M., 1982. Water in silicate glasses: an infrared spectroscopic study. Contrib. Mineral. Petrol., 81: 1–17.

Stone, J.O.H., Wickham, S.M. and O'Nions, R.K., 1988. Helium and carbon in crustal fluids: constraints on granulite production by mantle-derived CO_2. EOS, 69.

Taylor, H.P., Jr., 1968. The oxygen isotope geochemistry of igneous rocks. Contrib. Mineral. Petrol., 19: 1–71.

Taylor, H.P., Jr., 1969. Oxygen isotope studies of anorthosite with particular reference to the origin of bodies in the Adirondack Mountains, New York. In: Y.W. Isachsen (Editor), Origin of Anorthosite and Related Rocks. New York State Mus. Sci. Ser., Mem. 18: 111–134.

Taylor, H.P., Jr. and Sheppard, S.M.F., 1986. Igneous rocks: I. Processes of isotopic fractionation and isotope systematics. In: J.W. Valley, H.P. Taylor, Jr. and J.R. O'Neil (Editors), Stable Isotopes in High-Temperature Geological Processes. Reviews in Mineralogy, 16: 227–271.

Thompson, A.B., 1982. Dehydration melting of pelitic rocks and the generation of H_2O-undersaturated granitic liquids. Am. J. Sci., 282: 1567–1595.

Thompson, A.B., 1983. Fluid-absent metamorphism. Q. J. Geol. Soc. London, 140: 533–547.

Thompson, A.B., 1990. Heat, fluids and melting in the granulite facies. In: D. Vielzeuf and Ph. Vidal (Editors), Granulites and Crustal Evolution. Kluwer, Dordrecht, pp. 37–57.

Touret, J.L.R., 1971. Le facies granulite en Norvège Méridionale. II. Les inclusions fluides. Lithos, 4: 423–436.

Touret, J.L.R., 1985. Fluid regime in southern Norway: the record of fluid inclusions. In: A.C. Tobi and J.L.R. Touret (Editors), The Deep Proterozoic Crust in the North Atlantic Provinces. D. Reidel, Dordrecht, Holland, pp. 517–550.

Touret, J.L.R. and Hansteen, T.H., 1988. Geothermobarometry and fluid inclusions in a rock from the Doddabetta charnockite complex, southwest India. Rend. Soc. Ital. Mineral. Petrol., 43: 65–82.

Touret J.L.R. and Hartel, T.H.D., 1990. Synmetamorphic fluid inclusions in granulites. In: D. Vielzeuf and Ph. Vidal (Editors), Granulites and Crustal Differentiation. Kluwer, Dordrecht, pp. 397–417.

Trommsdorff, V. and Skippen, G., 1986. Vapour loss ("boiling") as a mechanism for fluid evolution in metamorphic rocks. Contrib. Mineral. Petrol., 94: 317–322.

Valley, J.W., 1986. Stable isotope geochemistry of metamorphic rocks. In: J.W. Valley, H.P. Taylor, Jr. and J.R. O'Neil (Editors), Stable Isotopes in High-Temperature Geological Processes. Reviews in Mineralogy, 16: 445–489.

Valley, J.W. and O'Neil, J.R., 1982. Oxygen isotope evidence for shallow emplacement of Adirondacks anorthosite. Nature (London), 300: 497–500.

Valley, J.W. and O'Neil, J.R., 1984. Fluid heterogeneity during granulite facies metamorphism in the Adirondacks: stable isotope evidence. Contrib. Mineral. Petrol., 85: 158–173.

Valley, J.W., McLelland, J., Essene, E.J. and Lamb, W.M., 1983. Metamorphic fluids in the deep crust: evidence from the Adirondacks. Nature (London), 301: 226–228.

Valley, J.W., Taylor, H.P., Jr. and O'Neil, J.R. (Editors), 1986. Stable Isotopes in High-Temperature Geological Processes. Reviews in Mineralogy, Volume 16, 570 pp.

Vry, J., Brown, P.E., Valley, J.W. and Morrison, J., 1988. Constraints on granulite genesis from carbon isotope compositions of cordierite and graphite. Nature (London), 332: 66–68.

Waff, H.S. and Bulau, J.R., 1979. Equilibrium fluid distribution in an ultramafic partial melt under hydrostatic stress conditions. J. Geophys. Res., 84: 6109–6114.

Waters, D.J., 1988. Partial melting and the formation of granulite facies assemblages in Namaqualand, South Africa. J. Met. Geol., 6: 387–404.

Waters, D.J. and Whales, C.J., 1984. Dehydration melting and the granulite transition in metapelites from southern Namaqualand, South Africa. Contrib. Mineral. Petrol., 88: 269–275.

Watson, E.B. and Brenan, J.M., 1987. Fluids in the lithosphere, I. Experimentally-determined wetting characteristics of CO_2–H_2O fluids and their implications for fluid transport, host-rock physical properties, and fluid inclusion formation. Earth Planet. Sci. Lett., 85: 497–515.

Watson, E.B. and Laporte, D., 1989. Textural aspects of partially molten granitoids: an overview of experimental rationale, results and possible implications for migmatite development. Abstracts, 28th International Geological Congress, Washington, D.C., 3: 338.

Weaver, B.L. and Tarney, J., 1983. Major and trace element composition of the continental lithosphere. Phys. Chem. Earth, 15: 39–68.

Wendlandt, R.F., 1981. Influence of CO_2 on melting of model granulite facies assemblages: a model for the genesis of charnockites. Am. Mineral., 66: 1164–1174.

Wickham, S.M., 1987. The segregation and emplacement of granitic magmas. Q. J. Geol. Soc. London, 144: 281–297.

Wickham, S.M., 1988a. Underplating, anatexis, and assimilation of metacarbonate; a possible source for large CO_2 fluxes in the deep crust. J. Geol. Soc. India, 31: 162.

Wickham, S.M., 1988b. Evolution of the lower crust. Nature (London), 333: 119–120.

Wickham, S.M., 1989. Melt transfer at low melt-fraction and its role in the geochemical modification of the crust. Abstract, 28th International Geological Congress, Washington, D.C. 3: 356.

Wickham, S.M., 1990. Isotopic modification of the continental crust; implications for the use of isotope tracers in granite petrogenesis. In: J.R. Ashworth and M. Brown (Editors), High Temperature Metamorphism and Crustal Anatexis. Unwin Hyman, London, pp. 124–148.

Wickham, S.M. and Peters, M.T., 1990. An oxygen isotope discontinuity in high-grade rocks of the East Humboldt Range, Nevada. Nature (London), 345: 150–153.

Wickham, S.M. and Taylor, H.P., Jr., 1985. Stable isotopic evidence for large-scale seawater infiltration in a regional metamorphic terrane: the Trois Seigneurs Massif, Pyrenees, France. Contrib. Mineral. Petrol., 91: 122–137.

Wickham, S.M. and Taylor, H.P., Jr., 1987. Stable isotope constraints on the origin and depth of penetration of hydrothermal fluids associated with Hercynian regional metamorphism and crustal anatexis in the Pyrenees. Contrib. Mineral. Petrol., 95: 255–268.

Wickham, S.M., Taylor, H.P., Jr. and Snoke, A.W., 1987. Fluid–rock–melt interaction in metamorphic core complexes — a stable isotope study of the Ruby Mountains–East Humboldt Range, Nevada. Geol. Soc. Am. Abstr., 19: 463.

Wilson, A.F. and Baksi, A.K., 1983. Widespread ^{18}O depletion in some Precambrian granulites of Australia. Precambrian Res., 23: 33–56.

Zingg, A., 1983. The Ivrea and Strona–Ceneri Zones (northern Alps, Ticino and N. Italy) — A review. Schweiz. Min. Pet. Mitt., 63: 361–392.

Chapter 11

Magma genesis and crustal processing

R.W. KAY, S. MAHLBURG KAY and R.J. ARCULUS

1. Introduction

Mafic and felsic magmas have played essential roles in the growth and differentiation of the crust. The addition of mass from the mantle to the crust by the ascent of mantle-derived basalt across the Moho, and the redistribution of mass within the crust by the ascent of crust-derived granitoids, are obvious and recurring examples of this igneous connection. Less obvious but perhaps no less important is the interplay between igneous and tectonic events, especially in the lower crust, which is the focus of this chapter. The plan of this chapter is first to review some processes and models that relate to crustal magmatism in a generic way, then to consider some examples of mafic and then felsic magmas in the continental crust, and finally to review alternative viewpoints on the question: how has andesitic crust been extracted from the mantle?

2. Magmas in the crust: processes and models

Mantle-derived magmas encountering the continental crust are in strong physical and chemical disequilibrium with crustal materials, and are the primary cause of many structural and thermal events in the crust. In the deep crust, the processes are largely unobserved, yet they can be inferred using petrological and geochemical analyses of rocks and magmas that originate in the lower crust as outlined in the later section. On the theoretical side, a number of process-oriented studies have combined physical, phase equilibria, mass, and thermal balance constraints with realistic boundary conditions in attempts to model the behavior of magmas in the crust.

Assume that a given basalt magma batch has formed within and segregated from asthenospheric mantle (see McKenzie and Bickle, 1988), and has risen through overlying denser, cooler lithospheric mantle to reach the crust–mantle boundary. Further rise may occur by three mechanisms: porous flow, hydraulic fracturing (magma-fracting), and diapirism. Heat loss to the surrounding rocks causes crystallization, and unless upward rise precedes crystallization and fractionation, the magma will no longer have the chemistry appropriate to equilibrium with the mantle, but rather will follow a compositional path appropriate to equilibrium or fractional crystallization in the crust. Furthermore, a batch of magma need not be a closed system: it can react with or dissolve wall-rock minerals with which it is not at equilibrium; it can partially melt the wall rocks, and wall-rock melt and mantle melt (or several mantle melts themselves) can mix. The physics of each of these processes has recently been the topic of some fundamental advances, several of which are outlined below. The fate of a magma in the crust is, for all these advances, not predictable at present.

2.1. Magma migration by crustal fusion: thermal constraints

In rising through the crust the thermal energy that a magma expends may easily freeze it. A calculation by Ahern et al. (1981) illustrates (Fig. 11-1) the extreme case: a melt layer rises through a cooler overlying crust by melting (dissolving) it, a process akin to zone refining. A major heat source for melting is the latent heat of crystallization of liquidus minerals of the melt, which largely equals the heat necessary for fusion at the temperature of the melt (Bowen, 1928). But the heat expended on raising the temperature of the overlying crust to that of the melt is large, and increasingly so at shallower (cooler) crustal levels. Therefore, even as the melt rises and becomes more silicic and cooler, its total remaining mass decreases. Ahern et al. (1981) calculate that, for a one dimensional model, the rise of a molten layer from the base of the crust is generally limited to less than twice its height.

The thermal energy expended by a rising magma on overlying rocks may be less than that indicated in Ahern et al.'s model provided that the overlying rock does not melt (or is not dissolved), and provided that the solid breaks off and sinks in sufficiently large blocks that the time scale of block heating is less than the time scale of magmatic rise. Any size less than 10's to 100's of meters will heat up in 10^2 to 10^4 yr ($D = 10^{-2}$ cm^2/s). Practically, this means that a 1-km melt layer rising at 10 cm/yr would have to stope blocks exceeding 100 m in order to avoid "thermal death". Thus, only if large blocks do break off and sink rapidly, will the magma rise a significant distance in the crust.

2.2. Crustal melting

It is the conclusion of Clemens and Vielzeuf (1987) that for most crust-derived magmas, melting is due to heat sources transferred from the underlying mantle, most often basaltic

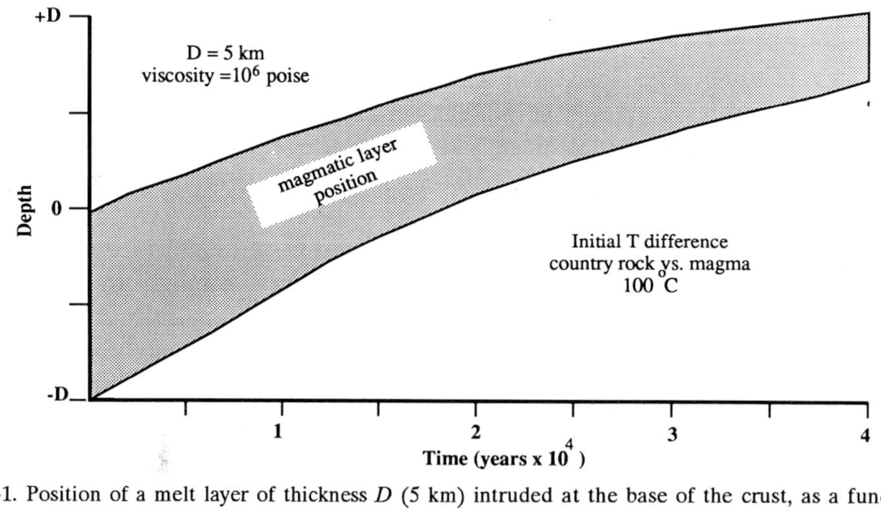

Fig. 11-1. Position of a melt layer of thickness D (5 km) intruded at the base of the crust, as a function of time. Migration is by assimilation of overlying crust. Initial temperature difference between magma and country rock is assumed to be 100°C with a magma viscosity of 3 x 10^6 poise (after Ahern et al., 1981).

intrusions. Typically, the evidence for basaltic intrusion is indirect: for example, observation of the effects of intrusion, such as the spawning of large volumes of crust-derived melt that intrude shallower crustal levels, and extrude as voluminous high-T rhyolites and dacites.

Clemens and Vielzeuf (1987) emphasize that partial melting of the crust occurs most frequently under fluid-absent conditions, and therefore is controlled by the breakdown of hydrous phases. The released water acts as a flux for the granitic fraction of the source rocks, which melts at a temperature that is lower than that of a water-free source, but higher than than for a water- saturated melting condition. In contrast, Wickham (1987) argues for introduction of external water into Pyreneean crust in order to account for the large percentage of melting at relatively low temperature. The introduction of CO_2 may also enhance melting by inducing breakdown of hydrous minerals like biotite (e.g., Peterson and Newton, 1989). The melting conditions for some common crustal materials are indicated in Figure 11-2.

Patchett (1980) and Huppert and Sparks (1988) have examined specific models for melting of crust overlying a basaltic sill. As shown in Figure 11-3, for a relatively hot silicic crust, a 500 m basalt sill produces an overlying crustal melt zone up to 300 m thick within a hundred years. Basalt flowing continuously in a thinner conduit produces the same effect. In Figure 11-3, the crustal melt is hot (≈950°C) and represents a high-percentage partial melt that is convecting and therefore homogeneous. Basalt intrusions are unable to traverse this molten region, and extensive mixing and assimilation may occur between basalt and lower crustal rocks, as invoked by many authors (e.g., Ewart et al., 1971; Arculus and Johnson, 1981; Thompson, 1982; Kay et al., 1987; Hildreth and Moorbath, 1988, Voshage et al., 1990). However, the intrusion of a basaltic sill or dike into cool crust or dry, previously melted crust results in much less melting, and there is a greater probability that the basalt will traverse the entire crust. This is because the density of most basaltic magmas is marginally less than that of the majority of non-sedimentary continental crust rock types, granite being an exception (Fig. 11-4). In the absence of crustal melt layers with density less than that of basalt, there are no barriers to the rise of a basalt magma batch to very high levels in the crust (see Glazner and Ussler, 1989). On the other hand, vertically interconnected magma batches rise as long as the integrated density of magma is less then rock (see below), but only provided that the volume and flow rate of the magma is sufficient for the batch to escape "thermal death": solidification due to heat loss.

The existence of huge, gabbroic-anorthositic layered intrusions that were emplaced into the crust at relatively shallow levels (<10 km depth) (e.g., Bushveld and Stillwater) attests to the development of arrest horizons of mafic magmas closer to the surface than to the Moho (e.g., Glazner and Ussler, 1989). In the case of the Bushveld, it has been suggested (Irvine and Sharpe, 1982) that widespread and voluminous anatexis of lower crustal components occurred, with the production of the Al-rich basaltic andesite that, upon rising in the crust, fractionated abundant plagioclase, now observed in anorthositic layers. As an alternative to basalt intrusion, it may be noted that crustal melting caused by radioactive heating has been advocated by several workers. The Himalayan example (Pinet and Jaupart, 1987) is the best documented, with the resulting leucogranites carrying the high U and Th signature (up to 28 ppm U and 10 ppm Th in the Manaslu granite) of their source. Crustal granitoids commonly have much lower U and Th contents and much higher crystallization temperatures than the Himalayan examples. At present, the accumulation of radioactivity-generated heat is not a primary cause for high percentage (as opposed to low percentage or migmatitic) crustal melting; earlier in Earth history, this mechanism becomes increasingly more plausible. Crustal

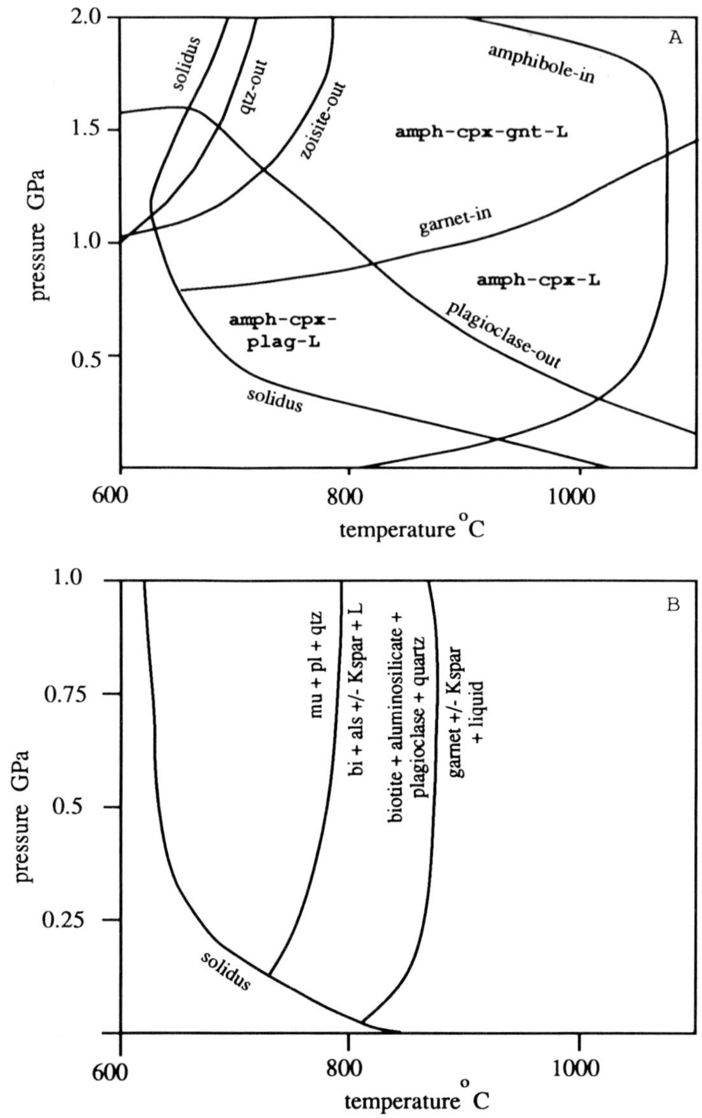

Fig. 11-2. *P–T* stability diagrams for mafic and pelitic compositions under crustal conditions: (Note 1 GPa = 10 kbar ≈ 35 km depth). (A) Composite for tholeiitic basalt with 5 wt.% H_2O after Green (1982). Note that in a broad supersolidus regime, melt coexists with amphibole–clinopyroxene ± plagioclase ± garnet, and is of trondhjemitic-tonalitic composition. (B) For fluid-absent melting of pelite with Mg/(Mg + Fe) = 0.5 after Vielzeuf and Holloway (1988). Note the most important melting reaction, where the volume of melt exceeds the critical melt fraction (see Fig. 12-4) of 0.3, involves the breakdown of biotite. Abbreviations: *qtz* = quartz; *mu* = muscovite; *pl* or *plag* = plagioclase, *bi* = biotite; *als* = aluminosilicate; *Kspar* = K-rich feldspar; *L* = liquid.

thickening can also cause crustal melting, by bringing fusable crust to higher temperature, but only after a time lag of several tens of millions of years (e.g., Zen, 1988; Patino Douce et al., 1990). However, thickening alone of materials displaced from a normal geothermal gradient is not sufficient to attain the temperatures of many of the silicic (and especially of the less silicic) magmas that erupt in areas of thickened crust like the Central Andes (England and Thompson, 1984).

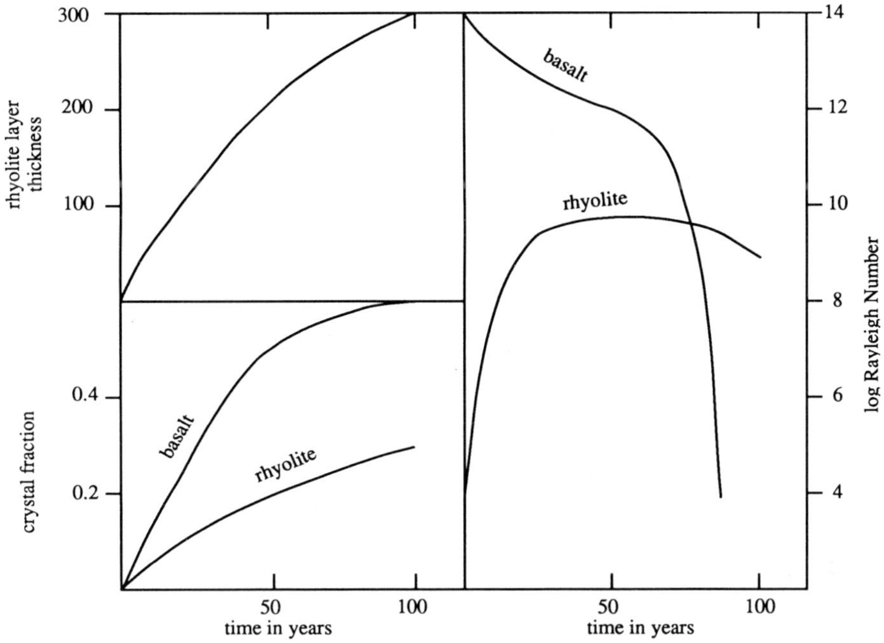

Fig. 11-3. Variations in the properties as a function of time of a ryholite melt layer, and underlying a 500-m-thick basalt layer with an initial temperature of 1200°C, assuming the rhyolite is generated by heat transferred from the basalt to continental crust with a melting temperature of 850°C at an initial T of 500°C (after Huppert and Sparks, 1988): top left is the increase in thickness of the rhyolite layer; bottom left is the variation of crystal content of the basalt and rhyolite; right is the Rayleigh Number of the different melt layers (at Rayleigh Numbers exceeding about 2000 convection occurs).

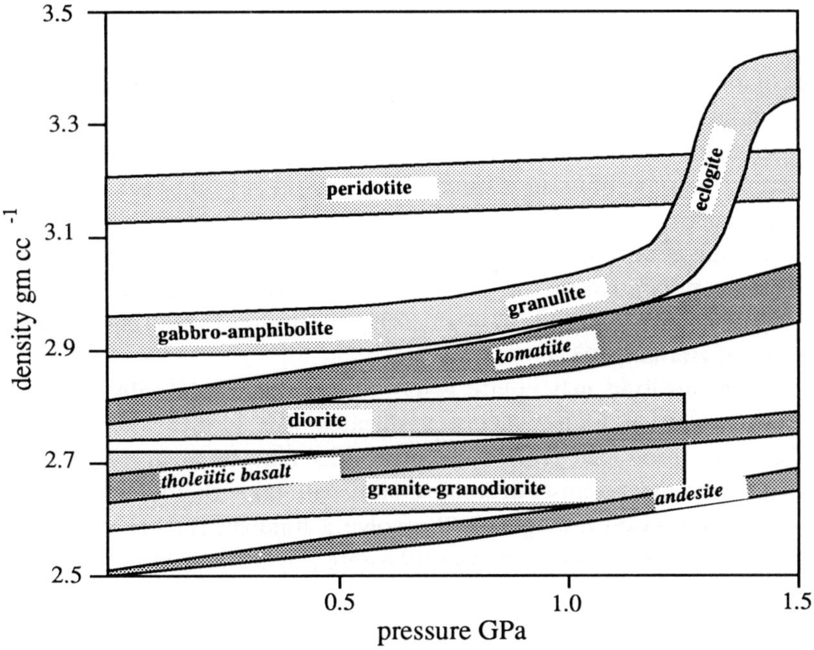

Fig. 11-4. Densities of rocks and magmas (labelled in italics) as a function of pressure, calculated for lower crustal temperatures of 600–1000°C after Herzberg et al. (1983).

2.3. The buoyancy of magmas in the crust

Magmas that are buoyant rise, but whereas basalt is invariably less dense than peridotitic and eclogitic lithosphere it is much closer in density to the feldspathic and particularly quartz-rich rocks that are common in the mid to upper crust (e.g., Ryan, 1987; Glazner and Ussler, 1989). Generally, though, magma rise must be considered using the concept of hydraulic head at the top of a magmatic column. For example, in a column that extends from the mantle through the crust, density deficits of basalt melt relative to peridotite counterbalance density excesses against granitic crustal rocks to create a pressure head at the top of the column, permitting the rise of basaltic magmas in the upper crust. Trans-lithospheric fracturing (Turcotte, 1990) and interconnectedness is required for such an ascent mechanism. Nevertheless, density interfaces (especially the Moho) remain likely sites for magmatic stagnation, as outlined by Herzberg et al. (1983) — see Figure 11-4. As magma crystallizes, it changes density. For example, an olivine tholeiite originating in the mantle, may undergo olivine and pyroxene crystallization in the crust with the residual magma becoming increasingly less dense, especially if crustal assimilation occurs, than the mantle-derived basalt. Although it has probably lost any hydraulic head with loss of vertical interconnectedness to a magma column extending into the mantle, this residual magma may evolve to a density less than that of the surrounding rocks. It seems certain that partial melts of crust are less dense than their equilibrium coexisting solids (restite) and density will not bar their segregation and ascent, although other factors (viscosity, water saturation) may.

2.4. Segregation of crustal melts and fluids

Once partial melting is sufficient to establish an intergranular melt network, melt will flow through the rock, and by porosity reduction, melt segregation will occur. The viscosity of silicic crustal melts is much higher than that of basalt, and segregation is much slower (e.g., McKenzie, 1985). Many partly molten crustal rocks flow as mushes (Fig. 11-5) without anything more than local segregation. With basaltic underplating mechanisms like that outlined above, the melting can reach high percentages very quickly and the residual crystals (restite) no longer form an interconnected matrix. Segregation of melt from this fluidized melt–crystal mixture then amounts to a question of crystal settling rate, which of course is slow in silicic magmas.

Because fluid fluxing is so important in crustal melting, the flow of crustal fluids (H_2O, CO_2) is a very significant control on the distribution of crustal melts. The migration of H_2O by porous flow is much faster than the flow of CO_2, due to lower wetting angles of aqueous solutions (Watson and Brenan, 1987).

Migration of a hydrous fluid may create H_2O-rich regions, specifically, those with higher H_2O than that contained in high-T hydrous silicates like biotite and hornblende (e.g., Wickham, 1987). Melts produced from such protoliths may be near H_2O saturation and incapable of rising very far because decompression will take them into a H_2O-saturated condition. Once H_2O-saturated a magma will exsolve a fluid on further decompression and will have a temperature too low to remain molten (e.g., Sykes and Holloway, 1987). Figure 11-6 shows some decompression paths.

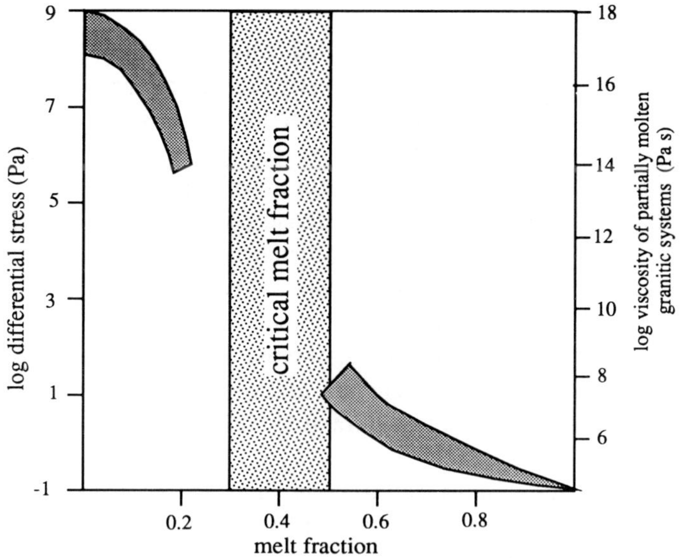

Fig. 11-5. Variation in strength and effective viscosity of partially molten granitic systems at a strain rate of 10^{-5} S^{-1} and 0.3 GPa confining pressure. An essentially solid-behaving system at low melt fraction (shaded region, top left) will behave more like a liquid at high melt fraction (shaded region, bottom right). A critical melt fraction regime for which no data exists separates the solid-like and liquid like regimes. (after Wickham, 1987).

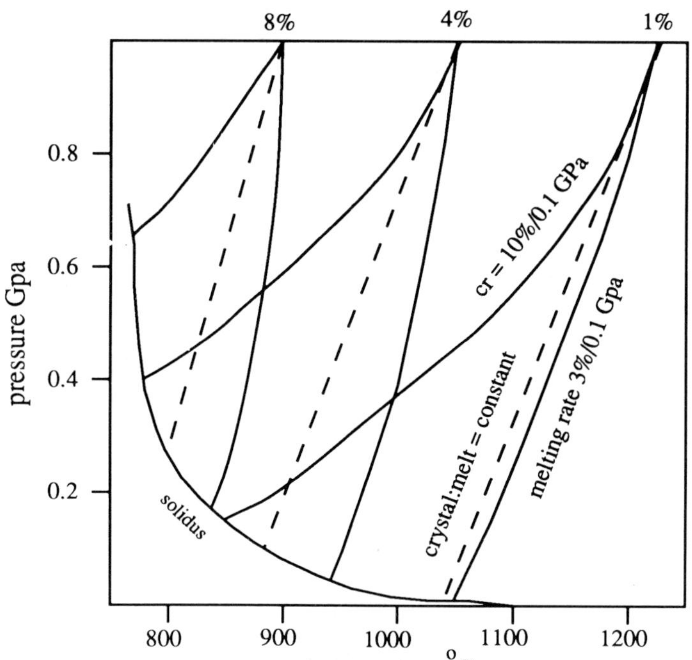

Fig. 11-6. Representative ascent trajectories for a model granitic system after Sykes and Holloway (1987), for bulk H_2O contents of 1, 4, and 8 wt.%. For each water content, 3 cases are shown, as labelled only for 1% H_2O: *cr* (crystallization rate) of 10% per 0.1 GPa, constant crystal-melt ratio, and melting rate of 3% per 0.1 GPa.

2.5. Magma mixing

Compositionally dissimilar magmas mix with difficulty, most frequently due to contrasting physical properties, particularly viscosity and density (cf. Sparks and Marshall, 1986; Kouchi and Sunagawa, 1985), but sometimes due to immiscibility, as in some Fe and K-rich magmas. Various models of Huppert, Sparks, and co-workers (e.g., Huppert et al., 1982) have been proposed to explore the conditions prevailing following the intrusion of dense hot mafic magma into a magma chamber containing less dense cooler magma, including cases where crystallizing magmas reach vapor saturation. McBirney et al. (1985) have examined the role of sidewall crystallization in generating felsic magmas that flow upward and form physically isolated caps over mafic magma chambers. In some plutons, convection has incompletely mixed mafic intrusions into felsic magma chambers. Frost and Mahood (1987) outline constraints on mixing of mafic and felsic magmas, and conclude that homogenization is likely only if the compositional differences is less than 10% SiO_2, or if the mafic magma is greater than half the total mass.

Evidence for magma mixing, or in general the behavior of magma chambers as open systems, comes from both petrographic and geochemical studies. Petrographic observations, including mineral zoning and incompatable mineral assemblages, are commonly interpreted as evidence of mixing immediately before eruption. Often, however, geochemical studies indicate mixing even in the absense of mineralogical evidence. Models are easy enough to formulate but often the characteristics of mantle-derived and crust-derived components are difficult to define (e.g., Powell, 1984).

3. Magmatic underplating

Magmatic underplating of the continental crust occurs at plate boundaries and within plates, wherever basaltic magmas rise to the top of the mantle. Evidence for underplating is almost entirely indirect, and comes from a diverse set of observations and deductions (Furlong and Fountain, 1986; Bohlen and Mezger, 1989). Some specific examples are outlined in this section.

3.1. Underplating at rifted continental margins: seismic evidence

The crust of rifted continental margins thins in the transition zone between continent and ocean. The thinned zone is often (in so-called magmatic margins) the site of basaltic volcanism, temporally coincident with rifting. Recent seismic profiling has identified these buried basaltic extrusions (seaward dipping reflectors) as well as large volumes of high velocity lower crust, thought to be gabbroic intrusives. White et al. (1987) explain the contrast between "magmatic" (e.g., the Hatton Bank) and amagmatic (e.g., the Bay of Biscay) rifted margins by invoking, for magmatic margins, higher potential temperature (e.g., temperature at the same pressure, with no melting) in the mantle. The ratio of intrusive (underplated) to extrusive magmatic rocks is at least 4 at the Hatton Bank margin. For magmatic margins, the intrusion of large volumes of basalt invalidates the simple relation between stretching factor (beta), crustal thickness, and crustal inflation and subsidence during rifting.

3.2. Underplating of non-rift crust: evidence of silicic volcanism

Some crustal regions that have undergone little extension have experienced periods of voluminous and widespread crust-derived silicic volcanism that demand a magnitude of thermal input that only mantle-derived magma intruded at deep crustal level can provide. The heating of mantle under large continental masses (supercontinents), shown theoretically by Gurnis (1988), may explain the largest occurrences of these "anorogenic" volcanic episodes. Kay et al. (1989), describe a Gondwana analogue to the well known Proterozoic examples of mid-continental North America (Bickford, 1988). The late Paleozoic to Jurassic rhyolitic magmatism at the Gondwana margin is not accompanied by basalt, but the existence of the basaltic underplating of equal or greater mass to the erupted silicic volcanic rocks follows from the heat source requirement. Presumably the crust in the region, which (from radiogenic isotope constraints) was formed only up to several hundred million years previously by assembly of terranes that comprised Gondwana, contained much fuseable rock that formed a molten barrier against basaltic rise.

3.3. Underplating in continental interiors: thermal evolution evidence

In the lithosphere, heat is transported more quickly by rising magmas than by conduction. As well, thermal anomalies that are caused by insertion of hot magma into cooler crust are shorter-lived than those caused by diffusion of the thermal pulses through the lithosphere (e.g., Sass and Lachenbrach, 1979). Therefore, the thermal evolution of crust recorded by geothermometry of dated minerals, and by crustal uplift and subsidence rates, offers a powerful constraint on the distribution of crustal thermal anomalies. For example, in a study of Tertiary metamorphic core complexes of the western United States Gans (1987) argues for basaltic underplating on the basis of thermal models and crustal thickness in localized regions of crustal extension (see Fig. 11-7).

The currently active Yellowstone Plume of western North America represents another clear example of crustal reprocessing by basaltic underplating. A progression of early-erupted, silicic melt from crustal magma chambers has been followed by extrusion of basalt (i.e., the basalt of the Snake River Plain) (Brott et al., 1981). Note that this sequence is the reverse of that observed in the Archean Yilgarn block of Western Australia, which has also been attributed to the impingement of a hot-spot beneath continental crust (Campbell and Hill, 1988); a puzzle in the case of Yellowstone is the ability of late-rising basalt to traverse the crust, and presumably implies the absence, at this evolutionary stage, of a silicic melt layer of lower density than the basalt.

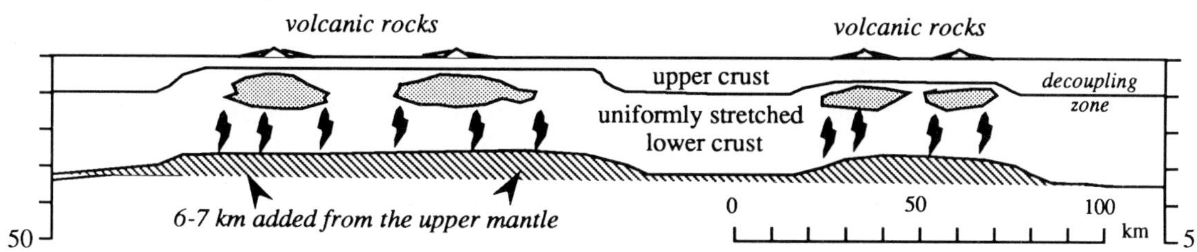

Fig. 11-7. A schematic and generalized cross-section of the eastern part of the Great Basin of the western United States, that shows a two-layer, open-system model for crustal stretching with underplating by basalts derived from the underlying upper mantle (after Gans, 1987).

3.4. Underplating at convergent margins: thermal and compositional arguments

Magmas of a wide compositional spectrum erupt from centers spaced along the magmatic arcs that parallel oceanic trenches. The deeper levels of extinct arcs, where exposed [e.g., Kohistan (Jan and Howie, 1981); Peninsular Terrane, Alaska (DeBari and Coleman, 1989)], contain abundant rocks with basaltic composition or, in the case of crystal cumulates, with basaltic sources. Rock units representing the crust–mantle boundary are rarely exposed, so arguments favoring underplating of entire crustal sections are often conjectural, based on the inability of a conductive geotherm to transport sufficient heat through the lithosphere to account for the high heat flow without wholesale melting both crust and mantle (Stern, 1987; Sass and Lachenbruch, 1979). In contrast, thick (kilometer to ten kilometer) underplates of mantle-derived magma furnish sufficient heat for both high geothermal gradients and crustal melting (e.g., Hildreth and Moorbath, 1988; Ingebritsen et al., 1989). Bohlen (1987) has noted that the last leg of pressure–temperature paths of some granulite (high-grade metamorphic) terranes indicates an isobaric cooling path that is consistant with thermal relaxation of thickened crust underplated with basalt at a convergent margin.

In arcs which are built on oceanic crust, the common andesitic composition volcanic and plutonic rocks are thought to originate by deep crustal differentiatiation (e.g., Gill, 1981; Conrad and Kay, 1984), in combination with assimilation of and mixing with melts derived from the arc crust itself (Kay et al., 1990). Mass balance (e.g., Kay and Kay, 1985) indicates that a large volume, perhaps often twice that of the complementary andesitic extrusives, lies in the deep crust. Additionally, some mantle-derived magmas crystallize entirely within the crust. The recognition of sections of oceanic arc among ophiolites (e.g., Pearce et al., 1984) allows direct observation of magmatic underplating of a mafic crust. Similarly, the famous Ivrea section (e.g., Voshage et al., 1990) in which the youngest major mafic units intruded a crust with much fusable silicic material at the former Moho, (i.e., underplated) appear to be arc-related (e.g., Pin and Sills, 1986). Finally, magmatic underplating has been claimed for the deeper parts of the Sierra Nevada (e.g., Gromet and Silver, 1987; Dodge et al., 1988) and the Andes (e.g., Hildreth and Moorbath, 1988; Kay et al., 1991).

3.5. Xenolithic evidence

The magmatic underplating hypothesis is supported by the common occurrence of basaltic composition xenoliths that crystallized at lower crustal depths, as reviewed by Kay and Kay (1981), Griffin and O'Reilly (1987), Arculus et al., (1987) and Rudnick (this volume). The basic xenoliths include examples from diverse tectonic regimes: active island arcs (e.g., Japan, Aleutians) extinct arcs (Sierra Nevada, Chudleigh province, Queensland), rifted continental margins (eastern Australia) and continental rifts (Rio Grande, Rheingraben). Both cognate (derived from the magma that transported them) and accidental xenoliths occur. Often, the chemical and isotopic composition and the metamorphic history of the accidental basaltic xenoliths is consistent with their being underplated magmatic material from an earlier tectonic cycle.

4. Anatexis and granitoid melt generation

Among those considered in this paper, this topic has been the most thoroughly explored in the literature, from field, experimental and geochemical viewpoints. Some extensive contemporary reviews of granitoid genesis are available (e.g., Pitcher, 1987; and Transactions of the Royal Society of Edinburgh, 79, 1988). We will use several examples to emphasize the end-member tectonic contexts of granitoid formation in continental crust. Not treated explicitly here are migmatities, most recently, a term used for rocks with granitoid melt segregated or irregularly mixed on a meter scale. Very generally, partial melts of a wide range of hydrated crustal compositions (basalt to pelagic sediment) are siliceous (>60% SiO_2). In these melts (intrusive granitoids and extrusive andesites, dacites and rhyolites) differences in the felsic minerals (e.g., proportion and composition of quartz and alkali feldspars) are less than those in the mafic minerals, which are often interpreted (particularly in the granitoids) as unmelted remainders of the source rocks (restites — Chappell et al. 1987; Sawka and Chappell, 1987). A "letter" classification of granitoids (A, I, M, S) based partly on the identity of mafic minerals (hornblende, biotite, cordierite, garnet, etc.) calls attention to the primacy of source rock differences (anhydrous, igneous, mantle, and sedimentary) in determining granitoid type. Diagnostic isotopic and trace element distinctions also exist between the melts of different source rocks and tectonic environments.

For models that invoke partial melting, amphibolite through garnet granulite facies materials are plausible protoliths of all these granitoid types. Diagnostic isotopic and trace element distinctions exist between the granitoid melts of different source rocks and tectonic enviroments. Between different granitoid types, the hydrous mineral breakdown reactions that trigger melting are distinct. In the case of "S-type" granitoids, the most significant melting reaction in a volumetric sense is the breakdown of biotite in metapelites through (Vielzeuf and Holloway, 1988; Fig. 2b):

$$\text{biotite} + \text{aluminosilicate} + \text{plagioclase} + \text{quartz} = \text{garnet} \pm \text{K-feldspar} + \text{melt}$$

In I-type granitoid protoliths, amphibole breakdown reactions predominate.

Importantly, not all intermediate and silicic melts that intrude shallower crustal levels are crustal melts. In particular, Brown et al. (1984) note a class of granitoids with mantle-like geochemical and isotopic characteristics, which are associated with mantle-derived gabbros (basalts), and may have segregated from gabbroic crystal-liquid mushes (a process that is observed in the Palisades diabase — Shirley, 1987). The evidence for a crustal origin for all or, (in mixed magmas), part of a granitoid is best approached on a case-by-case basis, using appropriate process-independent mass tracers. In some cases, in particular where the existing lower crust is young and geochemically similar to subsequent intrusions, melting of pre-existing crust and differentiation are very difficult to distinguish (e.g., Kay et al., 1990).

4.1. Himalayan leucogranites: crustal melting in a collision zone

Miocene leucogranites intrude the upper thrust slices along the entire length of the Himalayan collision zone. The lower thrust slices, exposing rocks that were up to 30 km deep, contain migmatitic gneisses; the leucogranites segregated from rocks like these (Deniel et al., 1987). The gneisses were originally sediments and the leucogranites are peraluminous, tourmaline and muscovite-bearing (broadly, S-type, even given the arguments of White et al., 1986). The high and extremely variable $^{87}Sr/^{86}Sr$ ratios in the leucogranites reflect heterogeneities in their metasedimentary sources. The areal distribution requires production of numerous isotopically distinct batches of magma of limited bulk compositional range that segregated and rose in the crust, but did not intermix. Thermal models of these granities are consistent with an origin by self-heating (to 600°–650°) by radioactive decay of radiogenic element-rich sedimentary protoliths that were trapped in crustal sites with low thermal conductivity during thrusting. Among restite fragments in the plutons, none represents a mafic intrusive that might have served as a heat source. In this way the Himalayan leucogranites do not resemble muscovite-bearing leucogranites from Kodiak Island Alaska (Hill et al., 1981) which have trace-element and isotopic characteristics analogous to their sedimentary protoliths, oceanic sediment being subducted at the Aleutian trench south of Kodiak (Kay and Kay, 1988), but which also have some oxygen isotope characteristics of oceanic basalt that may have served as a heat source for the melting. The Kodiak granites have lower radioactive heat production than the Himalayan ones.

4.2. Crustal melting at the Taupo (New Zealand) and Trois Seigneurs (Pyrenees) rift zones

The Taupo rift zone is an active back-arc basin that developed within continental lithosphere. It lies west of a volcanic line associated with subduction of the Pacific plate at the Hikurangi trench. The rift is filled with several kilometers of rhyolitic to andesitic pyroclastic material; a drill hole intersects quartz diorite at about 2.5 km (Stern, 1987). Mesozoic greywacke forms mountain ranges on the rift flanks. Andesites and basalts have erupted from the volcanic centers south and east of the Taupo volcanic zone, yet rhyolites and shallow intrusive equivalents predominate (over 94% of the total; Cole, 1981) within the Taupo zone itself. Cole (1981) favored an origin of the rhyolites by partial melting of greywacke and argillite (e.g., Conrad et al., 1988) like that exposed on the flanks of the rift, a view not held by Blattner and Reid (1982) (see below). The rhyolites are only slightly peraluminous (some contain biotite, but none has muscovite, cordierite or garnet). Amphibole-bearing rhyolites crystallized at lower temperatures (735–780°C) than amphibole-free rhyolites (860°–890°C) (Ewart et al., 1971). Both formed at higher temperature than the Himalayan leucogranites.

Cole (1981) favors the introduction of water to flux the argillites and greywackes and increase the melting percentage to yield the voluminous Taupo rhyolites. Water bound in metamorphic minerals is not sufficient (cf. Clemens and Vielzeuf, 1987). The argument for introduction of water is more fully developed in the Pyreneean rift zone, setting of the Trois Seigneurs leucogranite (Wickham, 1987). At Trois Seigneurs, Bickel et al. (1988) postulate several crustal magma sources, with the main granitoid mass (a biotite granite) derived by melting of two metasedimentary units by invading mafic magma from the mantle. Components from both mantle and crust are found in the biotite granite. In contrast, a late granodiorite phase of the Trois Seigneurs was probably derived from a hydrated mafic igneous lithology at mid to lower crustal depth. Bickle et al. (1988) attribute the homogenization of Sr and O isotope ratios in large volumes of crust to fluid and melt migration during rifting. The original Sr and O isotope ratios in sediment are substantially lowered, reducing differences between peraluminous crustal magma sources and the mantle. This lowering of Sr and O isotope ratios in sedimentary sources is also necessary at Taupo, where a high-temperature hydrothermal zone overlies a mafic intrusion, to explain the data of Blattner and Reid (1982), who maintain that sediment is not the source, but that the Taupo rhyolites are deep crustal melting of a basaltic source, or fractionation from basalt.

4.3. Crustal melting under dry conditions: alkaline, anorogenic, anhydrous (A-type) granitoids

Many orogenic zones, that have been the site of voluminous, syntectonic I and S-type magmatism, have been invaded by post-tectonic silicic magmas. In southeastern Australia, the source of plutons formed by these magmas appears to be lower crust that has melted under hot, anhydrous conditions (Collins et al., 1982) Generally, the silicic magmas are both intrusive (A-type granitoids) and extrusive (Bailey and MacDonald, 1987). Often the silicic melts are peralkaline, F-rich, and of low crystallinity at the time of intrusion or extrusion. Their low volume and association with alkaline basaltic magmatism means that alternative origins such as extreme basaltic differentiation or liquid immiscibility may be possible. But some of the variability in, for example, Sr isotope ratios and in mineralogy more likely reflect a crustal source (e.g., Bailey and MacDonald, 1987). The near absence of water in the magmas mirrors its low abundance in the source rock: either the water was removed by an earlier magma that did not melt out all the granite fraction of the rock, and left it enriched in F and Cl (in residual biotites and amphiboles), or water pressure was reduced by invasion of a CO_2-rich fluid, usually taken to be mantle-derived (see Wickham, 1988, for discussion of a crustal source). A likely source for the heat for melting and for the CO_2 (especially when it is associated with mantle-derived He), is alkaline basaltic magmas, which are CO_2-rich. Migration of CO_2 through rock over large distances seems unlikely (Watson and Brenan, 1987) but local migration of solute-rich (low water activity) F and Cl-bearing carbonic solutions may enrich the crustal sources in F and Cl, which act as fluxes. The ability of CO_2 to lower melting temperatures of biotite-bearing granitoids (Peterson and Newton, 1989) is an important fact linking CO_2 transport to migration of alkaline (including syenitic) silica-rich magmas. Whatever the processes of volatile transfer are, they occur only on a local scale in places like the east African Rifts, where peralkaline felsic lavas erupted in close proximity have come from distinct sources (metasomatized sialic, basic and ultrabasic rocks — Bailey and MacDonald, 1987).

5. Continental growth, recycling, and the lower crust

Based on solar and meteoritic abundances, it is clear that a major proportion of the terrestrial inventory of some large ionic radius, mantle-incompatible elements such as Cs, Rb, K, Ba, Th, and U resides in the uppermost continental crust, despite the trivial weight percentage ($<0.2\%$) of the whole Earth represented by this region (Taylor and McLennan, 1985; Anderson, 1989). An efficient process of continental crust extraction is indicated, with concomitant disposal of the residues from this extraction. Some form of subduction zone process, albeit different perhaps in character from that presently operating, should be considered (e.g., Fyfe, 1978).

But, if the continental mass has been extracted from the mantle, it may seem that the knowledge of magmatic processes as applied to mafic and felsic magmas in the crust and mantle, would be sufficient to answer the question: how is the crust extracted from the mantle? On the contrary, major questions continue to cloud understanding of crustal growth: has the composition of crustal additions from the mantle changed since the Archean? At various times in Earth history, how much greater has been the rate of mass addition from mantle to crust than the rate of mass addition from crust to mantle (crustal recycling)? In particular, is andesitic mean continental crust composition created directly out of the mantle or is it best understood as the net product of exchange of compositionally different mass between mantle to crust?

Some simple boundary conditions apply to these fundamental questions:

(1) Despite the attractiveness of island arc magmatism as a mechanism for continental crustal formation, the magmatic flux from the mantle into present-day intra-oceanic arcs is

overwhelmingly basaltic (e.g., Arculus, 1981); extraction of large volumes of granodioritic upper crust from this protolith requires a second stage: either differentiation or (intracrustal) melting. Both processes leave a mafic-ultramafic residue (e.g., Kay and Kay, 1985).

In a tectonic regime where upward redistribution of granitoids occurs with retention in the continental lithosphere of the restites, we might expect to find a volumetrically predominant portion of the continental crust, and specifically the lower part, to be composed of mafic-to-ultramafic residues from granitoid extraction. Despite the widespread occurrence of mafic xenoliths derived from the lower crust (see Rudnick, this volume), the evidence from exposed crustal cross sections (Percival et al., this volume) is that the lower part of the continental crust does not consist predominantly of these materials.

(2) There has been a decrease in the production rate of continental crust over time. In addition, there is a difference in the relative abundance of continental granitoid types with age (Taylor and McLennan, 1985). Naturally, this applies only to surviviors of an initial population. Generally, sodic granitoid magmas ("trondhjemite-tonalite") with strongly fractionated rare earth element (REE) abundances (chondrite-normalized La/Yb >10) are predominant in Archean terranes, whereas relatively potassic granitoids with less fractionated REE become volumetrically significant in Proterozoic and younger terranes. Even given this generalization, significant present-day diversity occurs that correlates strongly with tectonic regime (e.g., in the Andes, Kay et al., 1991). Hydrated mafic protoliths (garnet amphibolite) are the most likely sources of the trondhjemite-tonalite suite (Barker and Arth, 1976), whereas feldspar-bearing protoliths, often with recycled supracrustal components, are important in the genesis of the potassic granitoids (Chappell, 1984). Neither suite can be a direct melt of mantle peridotite, but as the trondhjemite-tonalite suite can be produced by melting of a mafic source at mantle pressures (e.g., Rapp, 1990), models involving direct derivation of andesite crust from the mantle can be entertained. In such models the mafic residues are located in the mantle, not the lower crust.

5.1. Andesitic crust from the mantle

Arculus and Ruff (1990) have argued that the critical difference between early Archean and younger subduction zones relates to the thermal structures involved (see also Martin, 1986 and Kay, 1978). At present, the dT/dP along the upper surfaces of a subducted slab are such that dehydration occurs through reactions such as:

amphibole = H_2O-rich fluid + garnet + pyroxene $\pm$ other non-hydrous phases

The ingress of the H_2O-rich fluid into the overlying, hotter peridotite wedge triggers melting and production of a spectrum of basaltic magmas in island- and continental arcs.

In the event that higher temperatures at lower or equivalent pressures were achieved in Archean subduction zones, dehydration melting would have occurred through reactions such as:

amphibole = SiO_2-rich melt + pyroxene + olivine $\pm$ garnet and other non-hydrous phases

Experimental studies have documented the trondhjemite-tonalite-like character of the magmas produced during these types of dehydration melting processes (Helz, 1976; Johnston and Wyllie, 1988; Rapp et al., 1991). It is emphasized that a major secular change in the chemistry of melts associated with subduction zones is implied. Furthermore, a residue disposal mechanism through recycling into the mantle is clearly appropriate for the Archean situation, and retention of the residues from granitoid extraction within the lower continental crust is not expected.

5.2. Lower crustal delamination

One of the mechanisms by which individual continental blocks can grow is through the accretion of smaller crustal masses (terranes). But several observations indicate that the continental mass is not simply the sum of unassembled parts: Severe shortening of upper crustal levels shown in balanced cross sections of collisional margins implies a large thickness of similarly shortened underlying lower crustal and mantle lithosphere. In recent collision zones (Alps, Himalayas) this lithosphere may be present (e.g., Laubscher, 1988) but it does not appear to exist in old inactive collision zones and orogens (e.g., Drummond, 1981). During assembly of continents, are there losses of the mafic as well as the ultramafic parts of continental lithosphere?

Following collision of a number of indivdual terranes, the crust can continue to grow as basaltic magmas from the mantle invade continental lithosphere, including lower crust (e.g., Nelson, 1991). Through geologic time, a "basification" of the continental crust (Arculus and Ruff, 1990) including a decrease in the concentration of elements like K (Kay, 1980), as well as a smearing of the crust–mantle boundary (Griffin and O'Reilly, 1987) is implied. In some cases, extension within continental plates results in large volumes of basaltic lower crust being formed, as in magmatic rifted margins and in the Basin and Range of the western United States (Fig. 11-7). This crust appears to have lower production rate but a higher preservation probability (it is likely never to be exposed and eroded) than, for example, oceanic arc crust, which is vulnerable to erosion after it is welded to the main continental masses by collision.

In summary, the composition of many of the pre-assembly pieces of the continental mosaic was basaltic and post-assembly processes only contributes more basalt. How then could the bulk composition of the continents be andesitic? This question also applies to smaller scale examples, where basaltic oceanic terranes fuse to form new andesitic crust [e.g., British Columbia, an example used by Kay and Kay (1988)].

The problems outlined above are those of crustal growth from the mantle since the Archean, when direct production of voluminous sodic granitoids through melting of subducted, hydrated mafic crust faltered. They can be broken into two parts — crustal construction and crustal destruction (together these concepts incorporate the concept of crustal recycling: e.g., Kay and Kay, 1991). Much emphasis has been placed on construction through terrane accretion (e.g., Hoffman, 1988) and direct production from the mantle. Destruction of upper crust by erosion and its possible return to the mantle by subduction is also a popular topic (e.g., Hart and Gülen, 1989). Here we focus on mechanisms for destruction of lower crust: lower crustal erosion, delamination, or hidden subduction (e.g., Kay and Kay, 1986).

5.2.1. Tectonic setting for lower crustal delamination

In the asthenosphere, large masses with only small positive density contrasts sink; the subduction of oceanic plates is the most familiar example. Turcotte (1989) calculates that a structural unit comprised of the bottom 15 km of 35-km-thick continental crust (with an in situ density of 2.8 g/cm^3) together with an underlying 180-km-thick mantle lithosphere (with an in situ density of 3.3 g/cm^3) is denser (due to thermal contraction) than mantle

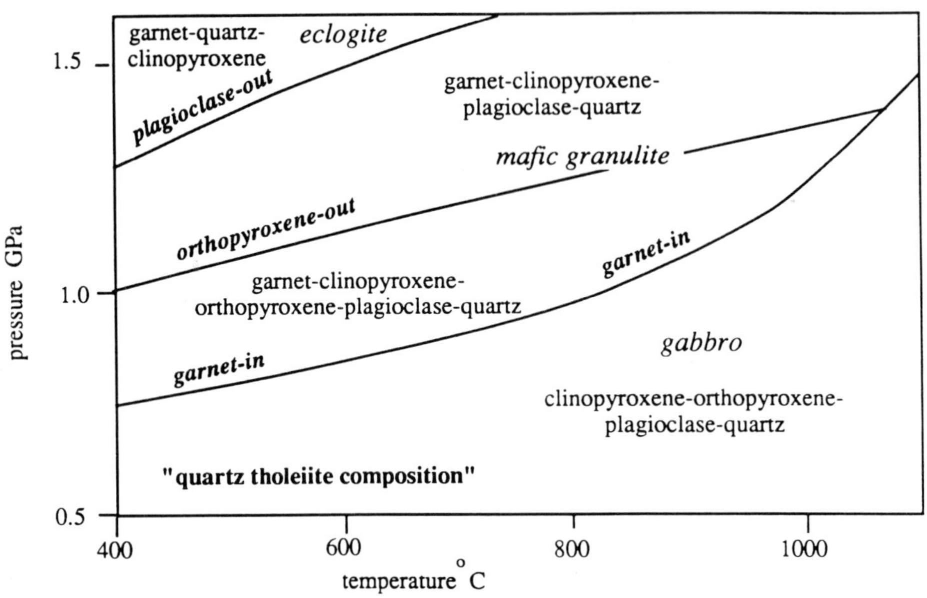

Fig. 11-8. Transition under lower continental crustal conditions of dry basalt (quartz tholeiite composition) from gabbro through granulite to quartz eclogite (after Wood, 1987).

asthenosphere, and therefore may sink into asthenosphere (delaminate). The lower crust in this calculation is assumed to be basaltic in composition and gabbroic in mineralogy: with normal geothermal gradients, Wood (1987) has shown that at depths less than 35 km the equilibrium mineralogy of mafic rocks is essentially garnet-free (Fig. 11-8), with a density is about 2.8 g/cm^3. Even given its negative buoyancy, initiating delamination of this thick mantle lithosphere with attached gabbroic lower crust is a formidable obstacle. There is no evidence that the lower parts of thickened (that is old) lithospheric plates with 35 km crust have spontaneously delaminated. Rather, lithospheric delamination that includes lower crust, if it happens, appears to occur [for reasons outlined below, following Kay and Kay (1991)] in regions where crustal doubling by collision has occurred and in regions of thickened crust that develop by compression in some continental volcanic arcs. Both regions have much thinner lithosphere and thicker crust than used in Turcotte's example. It is perhaps significant that both of these regions have relatively thin lithosphere, although at first this may seem to detract (from density considerations) from the possibility of any lower crustal delamination. This is because, when combined with thin mantle lithosphere, gabbroic lower crustal thicknesses in excess of only a very few kilometers are sufficient to make the whole lithosphere buoyant. This buoyancy is enhanced if the lithosphere is refractory (Fe, Al-poor) and therefore less dense than asthenosphere (e.g., Oxburgh and Parmentier, 1977).

The key to making lithosphere non-buoyant is that any mafic lower crust in an abnormally thick crustal section (e.g., 50 km), especially with cool crust (e.g., 600°C at the Moho) will have significantly higher density than gabbro, due to the formation of garnet and aluminous pyroxene and amphibole as equilibrium metamorphic minerals. If the garnet-forming reactions proceed to completion — with less dense plagioclase reacting out of mafic composition rocks in favor or more dense garnet and aluminous pyroxene or amphibole — the resulting rocks (eclogite or garnet-amphibolite) are denser than peridotite. The garnet-forming reactions are relatively temperature insensitive (Fig. 11-8), so that even lower crust whose temperature is greater that the solidus will, if melt is drained from it, be denser than mantle. The thin lithospheric mantle peridotite plus lower crust will now be denser than peridotitic mantle lithosphere alone. We deduce that delamination as a mechanism for recycling lower crust into the mantle becomes possible only where crust is abnormally thick, and mantle lithosphere is thin, which occurs in a limited number of tectonic regimes and within those, only over limited time spans. The mechanism is enhanced if the deepest part of a thick crust melts and the melt migrates to shallower crustal levels, for the garnet-bearing residue is denser than the unmelted rock (the less dense, feldspathic components melt preferentially).

Once the dense lower crustal residue has sunk into immediately underlying asthenosphere, its complement, granitoid melt that has migrated to shallower crustal levels, represents the only remaining observable evidence for the presence of deep crustal lithologies that were dense enough to delaminate. Thus we are drawn to ask whether post-orogenic (or post-thickening) magmatism, often extension-related, contains evidence for equilibrium with a dense residue.

5.2.2. Lower crustal delamination: nature of evidence to test the model

Lower crustal delamination involves spliting of crust, perhaps at a mid-crustal asthenosphere level (e.g., Turcotte et al. 1984), and sinking of lower crust with any underlying lithospheric mantle, into mantle asthenosphere. A more restrictive type of lithospheric delamination involves only mantle lithosphere, and this is specifically or by implication, the type of delamination envisioned by Bird (1979 and subsequent publications). Many subsequent effects, including replacement of cool by hot mantle, melting of the mantle to release basaltic magmas, and crustal heating, would be observed when mantle lithosphere delaminates, whether lower crust delaminates or not. Observation of these effects is not sufficient to indicate lower crustal delamination.

As the entire delamination process occurs at depth, with its successful completion resulting in sinking of the dense lithosphere even deeper, evidence for all types of delamination is indirect. It is conceivable that even indirect evidence for delamination is entirely absent. For instance, during lower crustal delamination, if melting of lower crust does not occur, or if the melt, once formed, does not segregate and migrate into crust that does not delaminate, then no evidence of the delamination event exists in the remaining lithosphere. Tracking the delaminated crust in the mantle using tomographic techniques may be impossible, due to inadequate spatial resolution.

If melts do segregate from the lower crustal layer as it delaminates, the trace element signatures of these melts should be distinctive (e.g., Kay and Kay, 1986, 1991). In a general way, Martin (1987) has outlined major-element and trace-element models relevant to extraction of intermediate to high SiO_2 granitoids from mafic crust at pressures corresponding to the deepest levels of tectonically overthickened crust. Most important, the minerals in Martin's calculated residue for extraction of dacitic (68% SiO_2) melt (25% by mass) from basalt (50% SiO_2) are, in weighted sum, denser than peridotitic mantle (see also Rapp, 1990). Garnet, amphibole and oxide phases are sufficiently dense to overbalance the remaining plagioclase. Incidentally, from a mass balance point of view, fractionation of these dense phases from basalt at lower crustal pressure is indistinguishable from partial melting.

Extreme fractionation of the rare earth element (REE) abundance pattern is the most diagnostic characteristic of partial melts extracted from a garnet–amphibole or garnet–

clinopyroxene-rich rock. This same fractionation is generated by melting olivine-bearing and olivine-free rocks, (olivine is essentially REE-free). Therefore, from the REE data alone, melting of garnet peridotite and garnet-clinopyroxenite are indistinguishable. Choice as to alternative source rocks rests on an assessment of major elements and elements like Ni and Cr, which are enriched in melts that are derived from mantle peridotite.

Melts associated with thickened crust in the Andes and the Alps including the "appinite suite" of intrusives, do meet many of the geochemical attributes expected for melts in equilibrium with high pressure mafic crustal rocks (Kay and Kay, 1991). But even in these cases, some of the andesitic magmas have high Ni and Cr, and may therefore be presumed to come from associated (and somewhat deeper?) ultramafic sources (e.g., Venturelli et al., 1984).

5.2.3. Lower crustal delamination: signature recognized in mantle-derived magmas?

The requirement of a complementary mafic residue for the bulk upper-crustal granodiorite composition, that does not appear to reside in general in the lower part of the continental crust, should prompt a search for evidence of the existence of this material within the mantle. Although their large volume should make the residues easily observable in the crust, any evidence for recycling of the residues into the mantle is likely to be obsured by the overwhelming mass of the latter. Only in the case of distinctive isotopic or trace element characteristics, could we expect to see any convincing evidence within mantle-derived basalts of the presence of deeply recycled, lower continental crust. Three anticipated characteristics are: low $^{87}Sr/^{86}Sr$ inherited through time from the low Rb/Sr of granitoid residues; high $^{143}Nd/^{144}Nd$ as a result of elevated Sm/Nd of melt-depleted, possibly garnet-dominated residues; unradiogenic $^{206}Pb/^{204}Pb$, $^{207}Pb/^{204}Pb$, and $^{208}Pb/^{204}Pb$ due to low U/Pb and Th/Pb of residual materials. Although the first two of these characterisitics are typical of ocean ridge and some hot-spot basalts, the third is not (e.g., Hart and Gülen, 1989) and in essence constitutes the "Pb paradox" (mixtures of Pb from continental upper crust and the upper mantle do not lie on the geochron). On balance, it is commonly assumed that the mantle source regions of oceanic basalts represents the complement after extraction by some means unspecified, of the upper continental crust (e.g., Hofmann, 1988). But the lack of an unradiogenic Pb reservoir as a sufficient mass balance of the relatively radiogenic upper continental crust and the majority of ridge and hot-spot basalts, means that we are at present uncertain of the fate of any putative delaminated lower continental crust within the Earth.

6. Concluding remarks

The growth of continental crust and its geochemical relationship with the mantle is an area of vigorous research. It is fair to say that only within the last decade has the early enthusiasm for the andesite model of crustal genesis begun to wane, despite the continued presentation by many introductory geology texts of this appealingly simple, conceptual approach. It would appear that an appreciation for the significance of secular changes in the style of subduction zone magmatism, and the consequences for upper crustal production is only beginning to be realized. Most importantly, it seems likely that due to crust–mantle mass exchange including sediment subduction and lower crustal delamination, no simple mass balance relationship exists whereby the sum of lower plus upper continental crust can be equated to a single, primary mantle-or subducted lithosphere-derived magma type.

The failure of a purely igneous solution to the genesis of continental crust also requires recognition of the important role of surface processes on Earth. For example, in the absence of weathering and hydration of potential granitoid protoliths, it is doubtful whether granitic magmas would be produced in any significant quantity at all (Campbell and Taylor, 1983). In addition, the loss of Mg^{2+} from continents to the mantle via the intermediate agents of weathering, H_2O-transport and removal in subducted oceanic crust is also critical in terms of crustal evolution (Albarede and Michard, 1986).

Nevertheless, igneous processes have clearly been important in the realm of ultrametamorphism, and the formation of volumetrically extensive granulite terrains that were once at lower continental crustal conditions (Powell, 1983; Bohlen, 1987). We have endeavored to show in this chapter that injection of mantle-derived mafic magmas at the base of or within the lower crust has been an important bulk crustal addition and modifier of crustal structure and chemistry, at least from the Proterozoic onwards.

Some quantitative and conceptual problems that need to be resolved include:

(1) the reality of "tectospheric roots" to continents and their relationship to crustal genesis (Jordan, 1988);

(2) the nature of the Moho and crust–mantle interfaces (Hale and Thompson, 1982; Griffin and O'Reilly, 1987);

(3) mechanisms and secular changes of crustal growth (e.g., Fyfe, 1978; Kröner, 1985);

(4) mode of sequestration of early-formed (4.5–4.0 Ga old) crust and subsequent recycling into the continental crust.

References

Ahern, J.L., Turcotte, D.L. and Oxburgh, E.R., 1981. On the upward migrationof an intrusion. J. Geol., 89: 421–432.

Albarede, F. and Michard, A., 1986. Transfer of continental Mg, S, O, and U to the mantle through hydrothermal alteration of the oceanic crust. Chem. Geol., 57: 1–15.

Anderson, D.L., 1989. Where on Earth is the crust? Physics Today, March, 1989: 38–46.

Arculus, R.J., 1981. Island arc magmatism in relation to the evolution of the crust and mantle. Tectonophysics, 75: 113–133.

Arculus, R.J. and Johnson, R.W., 1981. Island arc magma sources: a geochemical assessment of the roles of slab-derived components and crustal contamination. Geochem. J., 15: 109–133.

Arculus, R.J. and Ruff, L.J., 1990. Genesis of continental crust: evidence from island arcs, granulites and exospheric processes. In: D. Vielzeuf and P. Vidal (Editors), Granulites and Crustal Differentiation. NATO ASI Series, Kluwer, Amsterdam, pp. 1–18.

Arculus, R.J., Ferguson, J., Chappell, B.W., Smith, D., McCulloh, M.T., Jackson, I., Hensel, H.D., Taylor, S.R., Knutson, J. and Gust, D.A., 1987. Trace element and isotopic characteristics of eclogites and other xenoliths derived from the lower continental crust of southeastern Australia and southwestern Colorado Plateau, U.S.A. In: D.C. Smith (Editor), Eclogite and Eclogite-facies Rocks. Elsevier, Amsterdam, pp. 335–386.

Bailey, D.K. and MacDonald, R., 1987. Dry peralkaline felsic liquids and carbon dioxide flux through the Kenya rift zone. Geochem. Soc. Spec. Publ., 1: 91–105.

Barker, F. and Arth, J.G., 1976. Generation of trondhjemitic-tonalitic liquids and Archean bimodal trondhjemite-basalt suites. Geology, 4: 596–600.

Bickford, M.E., 1988. The formation of continental crust: Part 1. A review of some principles; Part 2. An application to the Proterozoic evolution of southern North America. Geol. Soc. Am. Bull., 100: 1375–1391.

Bickle, M.J., Wickham, S.M., Chapman, H.J. and Taylor, H.P. Jr., 1988. A strontium, neodymium and oxygen isotope study of hydrothermal metamorphism and crustal anatexis in the Trois Seigneurs Massif, Pyrenees, France. Contrib. Mineral. Petrol., 100: 399–417.

Bird, P., 1979. Continental delamination and the Colorado Plateau. J. Geophys. Res., 84: 7561–7571.

Blattner, P. and Reid, F., 1982. The origin of lavas and ignimbrites of the Taupo Volcanic Zone, New Zealand, in the light of oxygen isotope data. Geochim. Cosmochim. Acta, 46: 1417–1429.

Bohlen, S.R., 1987. Pressure–temperature–time paths and a tectonic model for the evolution of granulites. J. Geol., 95: 617–632.

Bohlen, S.R. and Mezger, K., 1989. Origin of granulite terrains and the formation of the lowermost continental crust. Science, 244: 326–329.

Bowen, N.L., 1928. The Evolution of the Igneous Rocks. Princeton Univ. Press, Princeton, N.J., 334 pp.

Brott, C.A., Blackwell, D.D. and Ziagos, J.P., 1981. Thermal and tectonic implications of heat flow in the eastern Snake River Plain, Idaho. J. Geophys. Res., 86: 11,709–11,734.

Brown, G.C., Thorpe, R.S. and Webb, P.C., 1984. The geochemical characteristics of granitoids in constrasting arcs and comments on magma sources, J. Geol. Soc. London, 141: 413–426.

Campbell, I.H. and Taylor, S.R., 1983. No water, no granites — no oceans, no continents. Geophys. Res. Lett., 10: 1061–1064.

Campbell, I.H. and Hill, R.I., 1988. A two-stage model for the formation of the granite-greenstone terrains of the Kalgoorlie-Norseman area, Western Australia. Earth Planet. Sci. Lett., 90: 11–25.

Chappell, B.W., 1984. Source rocks of I- and S-type granites in the Lachlan Fold Belt of southeastern Australia. Philos. Trans. R. Soc. London, Ser. A, 310: 693-707.

Chappell, B.W., White, A.J.R. and Wyborn, D., 1987. The importance of residual source material (restite) in granite petrogenesis. J. Petrol., 28: 1111-1138.

Clemens, J.D. and Vielzeuf, D., 1987. Constraints on melting and magma production in the crust. Earth Planet. Sci. Lett., 86: 287-306.

Cole, J.W., 1981. Genesis of lavas of the Taupo volcanic zone, North Island, New Zealand. J. Volcanol. Geotherm. Res., 10: 317-337.

Collins, W.J., Beams, S.D., White, A.J.R. and Chappell, B.W., 1982. Nature and origin of A-type granites with particular reference to Southeastern Australia. Contrib. Mineral. Petrol., 80: 189-200.

Conrad, W.K. and Kay, R.W., 1984. Ultramafic and mafic inclusions from Adak Island: crystallization history, and implications for the nature of primary magmas and crustal evolution in the Aleutian Arc. J. Petrol., 25: 88-125.

Conrad, W.S., Nicholls, I.A. and Wall, V.J., 1988. Water-saturated and undersaturated melting of metaluminous and peraluminous model crustal compositions at 10 kbar: The origin of rhyoltic magmas in the Taupo Volcanic Zone, New Zealand, and other occurrences. J. Petrol., 29: 765- 804.

DeBari, S.M. and Coleman, R.G., 1989. Examination of the deep levels of an island arc, evidence from the Tonsina ultramafic-mafic assemblage, Tonsina, Alaska. J. Geophys. Res., 94: 4373–4391.

Deniel, C., Vidal, P., Fernandez, A., Le Fort, P. and Peucat, J.J., 1987. Isotopic study of the Manaslu granite (Himalaya, Nepal): inferences on the age and source of Himalayan leucogranites. Contrib. Mineral. Petrol. 96: 78–92.

Dodge, F.C.W., Lockwood, J.P. and Calk, L.C., 1988. Fragments of the mantle and crust from beneath the Sierra Nevada batholith: xenoliths in a volcanic pipe near Big Creek, California. Geol. Soc. Amer. Bull., 100: 938–947.

Drummond, B.J., 1981. Crustal structure of the Precambrian terrains of northwest Australia from seismic refraction data. B.M.R. J. Geol. Geophys., 6: 123–141.

England, P.C. and Thompson, A.B., 1984. Pressure–temperature–time paths of regional metamorphism I — Heat transfer during the evolution of regions of thickened continental crust. J. Petrol., 25: 894–928.

Ewart, A., Green, D.C., Carmichael, I.S.E. and Brown, F.H., 1971. Voluminous low temperature rhyoltic magmas in New Zealand, Contrib. Mineral. Petrol., 33: 128–144.

Frost, T.P. and Mahood, G.A., 1987. Field, chemical, and physical constraints on mafic-felsic magma interactions in the Lamarch Granodiorite, Sierra Nevada, California. Geol. Soc. Am. Bull., 99: 272–291.

Furlong, K.P. and Fountain, D.M., 1986. Continental crustal underplating: thermal considerations and seismic-petrologic consequences. J. Geophys. Res., 91: 8285–8294.

Fyfe, W.S., 1978. The evolution of the Earth's crust: modern plate tectonics to ancient hot spot tectonics. Chem. Geol., 23: 89–114.

Gans, P.B., 1987. An open-system, two-layer crustal stretching model for the Eastern Great Basin. Tectonics, 6: 1–12.

Gill, J.B., 1981. Orogenic Andesites and Plate Tectonics. Springer-Verlag, Berlin, 392 pp.

Glazner, A.F. and Ussler, W., 1989. Crustal extension, crustal density, and the evolution of Cenozoic magmatism in the Basin and Range of the Western United States. J. Geophys. Res., 94: 7952–7960.

Green, T.H., 1982. Anatexis of mafic crust and high pressure crystallization of andesite. In: R.S. Thorpe (Editor), Andesites: Orogenic Andesites and Related Rocks. Wiley and Sons, Chichester, pp. 456–488.

Griffin, W.L. and O'Reilly, S.Y., 1987. The composition of the lower crust and the nature of the continental Moho — xenolith evidence. In: P.H. Nixon (Editor), Mantle Xenoliths. Wiley and Sons, Chichester, pp. 413–431.

Gromet, L.P. and Silver, L.T., 1987. REE variations across the Peninsular Ranges batholith: implications for batholithic petrogenesis and crustal growth in magmatic arcs. J. Petrol., 28: 75–126.

Gurnis, M., 1988. Large-scale mantle convection and the aggregation and dispersal of supercontinents. Nature, 332: 695–699.

Hale, L.D. and Thompson, G.A., 1982. The seismic reflection character of the continental Mohorovicic discontinuity. J. Geophys. Res., 87: 4625–4635.

Hart, S.R. and Gülen, L., 1989. Crust/mantle recycling at subduction zones: Proceeding of the NATO Advanced Research Workshop on 'Crust/Mantle Recycling at Convergence zones'. NATO ASI Series 258, 279 pp.

Helz, R.J., 1976. Phase relations of basalts in their melting ranges at P_{H_2O} = 5 kb. Part II. Melt compositions. J. Petrol., 17: 139–193.

Herzberg, C.T., Fyfe, W.S. and Carr, M.J., 1983. Density constraints of the formation of the continental Moho and crust. Contrib. Mineral. Petrol., 84: 1–5.

Hildreth, W. and Moorbath, S., 1988. Crustal contributions to arc magmatism in the Andes of central Chile. Contrib. Mineral. Petrol., 98: 455–489.

Hill, M., Morris, J and Whelan, J., 1981. Hybrid granites indruding the accretionary prism, Kodiak, Shumagin and Sanak Islands, Southwest Alaska. J. Geophys. Res., 86: 10,569–10,590.

Hoffman, P.F., 1988. United Plates of America, the birth of a craton: early Proterozoic assembly and growth of Laurentia. Annu. Rev. Earth. Planet. Sci., 16: 543–603.

Hofmann, A.W., 1988. Chemical differentiation of the Earth: the relationship between mantle, continental crust, and oceanic crust. Earth. Planet. Sci. Lett., 90: 297–314.

Huppert, H.E. and Sparks, R.S.J., 1988. The generation of granitic magmas by intrustion of basalt into continental crust. J. Petrol., 29: 599–624.

Huppert, H.E., Sparks, R.S.J. and Turner, J.S., 1982. Effects of volatiles on mixing in calc-alkaline magma systems. Nature, 297: 554–557.

Ingebritsen, S.E., Sherrod, D.R. and Mariner, R.H., 1989. Heat flow and hydrothermal circulation in the Cascade Range, North-Central Oregon. Science, 243: 1458–1462.

Irvine, J.N. and Sharpe, M.R., 1982. Source-rock compositions and depths of origin of Bushveld and Stillwater magmas. Carnegie Inst. Washington Yb., 81: 294–303.

Jan, M.Q. and Howie, R.A., 1981. The mineralogy and geochemistry of the metamorphosed basic and ultrabasic rocks of the Jijal Complex, Kohistan, NW Pakistan. J. Petrol., 22: 85–126.

Johnston, A.D. and Wyllie, P.J. (1988. Constraints on the origin of Archean trondhjemites based on phase relationships of Nuk gneiss with H_20 at 15 kbar. Contrib. Mineral. Petrol., 100: 35–46.

Jordan, T.H., 1988. Structure and formation of the continental tectosphere. J. Petrol., Special Lithosphere Issue, pp. 11–37.

Kay, R.W., 1980. Volcanic arc magma genesis: implications for element recycling in the crust–upper mantle system. J. Geol., 88: 497–522.

Kay, R.W., 1978. Aleutian magnesian andesites: melts from subducted Pacific Ocean crust. J. Volcanol. Geotherm. Res., 4: 117–132.

Kay, R.W. and Kay, S. Mahlburg, 1981. The nature of the lower continental crust: inferences from geophysics, surface geology and crustal xenoliths. Rev. Geophys. Space Phys., 19: 271–297.

Kay, R.W. and Kay, S. Mahlburg, 1986. Petrology and geochemistry of the lower continental crust: an overview. In: J.B. Dawson, D.A. Carswell, J. Hall and K.H. Wedepohl (Editors), The Nature of the Lower Continental Crust. Geol. Soc. Spec. Publ., 24: 147–159.

Kay, R.W. and Kay, S. Mahlburg, 1988. Crustal recycling and the Aleutian Arc. Geochim. Cosmochim. Acta, 52: 1351–1359.

Kay, R.W. and Kay, S. Mahlburg, 1991. Creation and destruction of lower continental crust. Geol. Rundsch., 80: 259–278.

Kay, S. Mahlburg, Maksaev, V., Moscoso, R., Mpodozis and Nasi, C., 1987. Probing the evolving Andean lithosphere: Mid-late Tertiary magmatism in Chile (29°–30°30′S) over the modern zone of subhorizontal subduction. J. Geophys. Res., 92: 6173–6189.

Kay, S. Mahlburg and Kay, R.W., 1985. Role of crystal cumulates and the oceanic crust in the formation of the lower crust of the Aleutian arc. Geology, 13: 461–464.

Kay, S. Mahlburg, Kay, R.W., Perfit, M.R. and Citron, G., 1990. Calc-alkaline plutonism in an oceanic arc: the Aleutian arc, Alaska. In: S. Mahlburg Kay and C. Rapela (Editors), Plutonism from Alaska to Antarctica. Geol. Soc. Am., Spec. Pap., 241: 233–255.

Kay, S. Mahlburg, Mpodzis, C., Ramos, V. and Munizaga, F., 1991. Magma source variations for late Tertiary magmatism associated with a shallowing subduction zone and a thickening crust in the central Andes, 28°–33°S. In: R.S. Harmon and C. Rapela (Editors), Andean Magmatism and its Tectonic Setting. Geol. Soc. Am., Spec. Paper, 265: 113–137.

Kay, S. Mahlburg, Ramos, V., Mpodozis, C. and Sruoga, P., 1989. Late Paleozoic to Jurassic silicic magmatism at the Gondwana margin: analogy to the Middle Proterozoic in North America? Geology, 17: 324–328.

Kouchi, A. and Sunagawa, I., 1985. A model for mixing basaltic and dacitic magmas as deduced from experimental data. Contrib. Mineral. Petrol., 89: 17–23.

Kröner, A., Evolution of the Archean continental crust, 1985. Annu. Rev. Earth Planet. Sci., 13: 69–106.

Laubscher, H., 1988. Material balance in Alpine orogeny. Geol. Soc. Am. Bull., 100: 1313–1328.

Martin, H., 1986. Effect of steeper Archean geothermal gradient on geochemistry of subduction-zone magmas. Geology, 14: 753–756.

Martin, H., 1987. Petrogenesis of Archaean trondhjemities, tonalites, and granodiorites from eastern Finland: major and trace element geochemistry. J. Petrol., 28: 921–953.

McBirney, A.R., Baker, B.H. and Nilson, R.H., 1985. Liquid fractionation. Part I: Basic principles and experimental simulations. J. Volcanol. Geotherm. Res., 24: 1–24.

McKenzie, D., 1985. The extraction of magma from the crust and mantle. Earth Planet. Sci. Lett., 74: 81–91.

McKenzie, D. and Bickle, J., 1988. The volume and composition of melt generated by extension of the lithosphere. J. Petrology 29: 625–679.

Nelson, K.D., 1991. A unified view of craton evolution motivated by recent deep seismic reflection and refraction results. Geophys. J. Int., 105: 25–36.

Oxburgh, E. and Parmentier, E., 1977. Compositional and density stratification in oceanic lithosphere — causes and consequences. J. Geol. Soc. London, 133: 343–355.

Patchett, J., 1980. Thermal effects of basalt on continental crust and crustal contamination of magmas. Nature, 283: 559–561.

Patino Douce, A.E., Humphreys, E.D. and Johnston, A.D., 1990. Anatexis and metamorphism in tectonically thickened continental crust exemplified by the Sevier hinterland, western North America. Earth Planet. Sci. Lett., 97: 290–315.

Pearce, J.A., Lippard, S.J. and Roberts, S., 1984. Characteristics and tectonic significance of supra-subduction zone ophiolites. In: B.P. Kokelaar and M.F. Howells (Editors), Marginal Basin Geology. Geol. Soc. Spec. Publ., 16: 77–94.

Peterson, J.W. and Newton, R.C., 1989. CO_2-enhanced melting ofbiotite-bearing rocks at deep-crustal pressure–temperature conditions. Nature, 340: 378–380.

Pin, C. and Sills, J.D., 1986. Petrogenesis of layered gabbros and ultramafic rocks from Val Sesia, the Ivrea Zone, NW Italy: trace element and isotope geochemistry. In: J.B. Dawson, D.A., Carswell, J. Hall and K.H. Wedepohl (Editors), The Nature of the Lower Continental Crust. Geol. Soc. Spec. Publ., 24: 231–249.

Pinet, C. and Jaupart, C., 1987. A thermal model for the distribution in space and time of the Himalayan granites. Earth Planet. Sci. Lett. 84: 87–99.

Pitcher, W.S., 1987. Granites and yet more granites forty years on. Geol. Rundsch., 76: 51–79.

Powell, R., 1983. Processes in granulite-facies metamorphism. In: M.P. Atherton and C.D. Gribble (Editors), Migmatities, Melting and Metamorphism. Shiva, Cheshire, pp. 127–139.

Powell, R., 1984. Inversion of the assimilation and fractional crystallizaton (AFC) equations; characterization of contaminants from isotope and trace element relationships in volcanic suites. J. Geol. Soc. London, 141: 447–452.

Rapp, R., 1990. Vapor-absent partial melting of amphibolite/eclogite at 8–32 kbar: implications for the origin and growth of the continental crust. Ph.D. Thesis, Rensselaer Polytechnic Institute, Troy, N.Y., 316 pp.

Rapp, R.P., Watson, E.B. and Miller, C.F., 1991. Partial melting of amphibolite/eclogite and the origin of archean trondhjemites and tonalites. Precambrian Res.: 51: 1–26.

Ryan, M.P., 1987. Neutral bouyancy and the mechanical evolution of magmatic systems. Geochem. Soc., Spec. Publ., 1: 259–287.

Sass, J.H. and Lachenbruch, A.H., 1979. Thermal regime of the Australian continental crust. In: M. McElhinney (Editor), The Earth: its Origin, Structure, and Evolution. Academic Press, London, pp. 301–351.

Sawka, W.N. and Chappell, B.W., 1987. Mafic xenoliths from I- and S-type granitoids: evidence for variations in deep crustal radioactive heat production. Geophys. Res. Lett.,, 14: 303–306.

Shirley, D.N., 1987. Differentiation and compaction in the Palisades Sill, New Jersey. J. Petrol., 28: 835–865.

Sparks, R.S.J. and Marshall, L.A., 1986. Thermal and mechanical constraints on mixing between mafic and silicic magmas. J. Volcanol Geotherm. Res., 29: 99–124.

Stern, T.A., 1987. Asymmetric back-arc spreading, heat flux and structure associated with the Central Volcanic Region of New Zealand. Earth Planet. Sci. Lett., 85: 265–276.

Sykes, M.L. and Holloway, J.R., 1987. Evolution of granitic magmas during ascent: A phase equilibrium model. Geochem. Soc. Spec. Publ., 1: 447–461.

Taylor, S.R. and McLennan, S.M., 1985. The Continental Crust: Its Composition and Evolution. Blackwell, London, 312 pp.

Thompson, R.N., 1982. Magmatism of the British Tertiary volcanic province. Scott. J. Geol., 18: 49–107.

Turcotte, D.L., 1989. Dynamics of recycling. In: S. Hart and L. Gülen (Editors), Crust/Mantle Recycling at Convergence Zones. NATO ASI Ser., 258: 245–257.

Turcotte, D.L., 1990. On the role of laminar and turbulent flow in buoyancy driven magma fractures. In: M.P. Ryan (Editor), Magma Transport and Storage. Wiley, Chihester, pp. 103–112.

Turcotte, D.L., Liu, J.Y. and Kulhawy, F.H., 1984. The role of an intracrustal asthenosphere on the behavior of major strike-slip faults. J. Geophys. Res., 89: 5801–5816.

Venturelli, G., Thorpe, R.S., Dal Piaz, G.V., Del Moro, A. and Potts, P.J., 1984. Petrogenesis of calc-alkaline, shoshonitic and associated ultrapotassic Oligocene volcanic rocks from the Northwestern Alps, Italy. Contrib. Mineral. Petrol., 86: 209–220.

Vielzeuf, D. and Holloway, J.R., 1988. Experimental determination of the fluid-absent melting relations in the pelitic system. Consequences for crustal differentiation. Contrib. Mineral Petrol., 98: 257–276.

Voshage, H., Hofmann, A.W., Mazzucchelli, M., Rivalenti, G., Sinigoi, S., Raczek, I. and Demarchi, G., 1990. Isotopic evidence from the Ivrea Zone for hybrid lower crust formed by magmatic underplating. Nature, 347: 731–736.

Watson, E.B. and Brenan, J.M., 1987. Fluids in the lithosphere, 1. Experimentally-determined wetting characterisitics of CO_2-H_2O fluids and their implications for fluid transport, host-rock physical properties, and fluid inclusion formation. Earth Planet. Sci. Lett. 85: 497–515.

White, A.J.R., Clemens, J.D., Holloway, J.R., Silver, L.T., Chappell, B.W. and Wall, V.J., 1986. S-type granites and their probable absence in southwestern North America. Geology, 14: 115–118.

White, R.S., Spence, G.D., Fowler, S.R., McKenzie, D.P., Westbrook, G.K. and Bowen, A.N., 1987. Magmatism at rifted continental margins. Nature 330: 439–444.

Wickham, S.M., 1987. The segregation and emplacement of granitic magmas. J. Geol. Soc. London, 144: 281–297.

Wickham, S.M., 1988. Evolution of the lower crust. Nature, 333: 119–120.

Wood, B.J., 1987. Thermodynamics of multicomponent systems containing several solid solutions. In: I.S.E. Carmichael and H.P. Eugster (Editors), Reviews in Mineralogy, 17. Mineral. Soc. Am., pp. 71–95.

Zen, E. An, 1988. Thermal modelling of stepwise anatexis in a thrust-thickened sialic crust. Trans. R. Soc. Edinburgh Earth Sci., 79: 223–235.

Chapter 12

Temporal evolution of regional granulite terranes: Implications for the formation of lowermost continental crust

K. MEZGER *

1. Introduction

The lower crust (depth range $\geq$15 km) is a poorly understood component of the continental crust. Due to its inaccessibility, its petrogenesis and evolution are fairly obscure. Only a few cross sections through the continental crust may expose parts of the lowermost ($\geq$25 km) crust and possibly the Mohorovicic discontinuity (e.g., Doubtful Sound, New Zealand; Ivrea Zone, northern Italy; Fountain and Salisbury, 1981) and thus allow a direct study. Since these cross sections expose only small parts (if any) of the lowermost crust most information about the deep continental crust has to be obtained indirectly. Different approaches have been used to obtain information on the chemical and temporal evolution of the lower crust.

Important information on the composition and metamorphic grade of the lower crust may be obtained by the study of xenoliths brought up in volcanic eruptions (e.g., Kay and Kay, 1981, 1983; Stosch and Lugmair, 1984; Rudnick and Williams, 1987; Rudnick and Taylor 1987; Rudnick, this volume). Many xenoliths contain mineral assemblages that allow the determination of the pressure of equilibration using well calibrated geobarometers and thus establish unambiguously their lower crustal origin (Bohlen and Mezger, 1989; Rudnick, this volume). Although lower crustal xenoliths are widespread, their geologic context and thus the temporal evolution of the lower crust can rarely be reconstructed. The important question of how representative xenoliths are for the composition and evolution of the lower crust is difficult to answer (see Rudnick, this volume).

Crustal melts are also important samples of the lower crust. These melts are typically of granitic composition (most of them are close to minimum melts) and may sample large areas of the crust, but only certain rock-types. Therefore, crustal melts are of a much more uniform composition than their source rocks and obscure the possible variety in rock composition in the deep crust. Unless crustal melts contain restite components, they can provide little information on the possible diversity in the major-element composition of the lower crust. As a consequence, direct estimates of the bulk composition of the lower crust generally cannot be made by studying crustal melts. Such melts, however, may be valuable tracers for studying the isotopic composition and thus the geochemical evolution of the lower crust through time on a large scale (e.g., Nelson and DePaolo, 1985).

* Current address: Max-Planck-Institut für Chemie, Saarstraße 23, D-6500 Mainz, Germany.

Estimates of the average composition of the lower crust have also been made by studying the composition of the upper crust via sediments and assuming a bulk composition (andesite) for the total crust (e.g., Taylor and McLennan, 1985). This model assumes that andesites are the major building blocks of continental crust and that the net flux of material from the mantle to the crust is of andesitic composition. In general, andesites are produced along active continental margins and if they are responsible for crustal growth, the continental mass would grow dominantly by lateral accretion of island arcs type crust. Crustal growth due to magmatism in rift settings or above hot-spots would have to be considered insignificant.

It is also possible to obtain information on the lowermost continental crust by studying granulites. Since granulites are among the deepest buried continental rocks, granulite terranes (particularly the Lewisian granulites of northern Scotland) have been used in some models as representatives for the composition and evolution of the lower crust (e.g., Weaver and Tarney 1984; Tarney and Weaver 1987). This latter approach is discussed in more detail in this chapter. However, most granulite terranes may not represent the lowermost continental crust but rather middle to lower crust (e.g., Bohlen and Mezger, 1989). Nevertheless, by studying granulite terranes it might be possible to obtain important information on the lowermost continental crust that underlies these terranes. The discussion in this chapter will concentrate on the temporal evolution of regional granulite terranes and their possible significance for the temporal evolution of the lowermost continental crust.

2. Characteristics of regional granulite terranes

2.1. General characteristics

In the context of this discussion only granulites with an areal extent of more than 1000 km^2 will be considered. Such granulite terranes occur in all shield areas around the world (e.g., Perkins and Newton, 1981; Newton and Perkins, 1982; Newton, 1987; Bohlen and Mezger, 1989; Harley, 1989) and are Archean to Proterozoic in age. The oldest regional granulites are described from the Napier Complex in Enderby Land, Antarctica (e.g., Black et al., 1986a,b) and western Greenland (e.g., Moorbath et al., 1986). The youngest regional granulite terranes currently recognized are the Khondalite belt of south India and the western part of Sri Lanka which exhibit crystallization ages of ca. 600 Ma (Kröner et al., 1987). The apparent absence of younger regional granulites may be due to the slow exhumation rates for terranes that eventually expose granulites on a regional scale, rather than fundamental differences in the processes that generate and modify continental crust. With respect to their lithologies, regional granulite terranes range from dominantly meta-igneous [e.g., Pikwitonei (Manitoba), Nilgiri Hills (south India)] to dominantly metasedimentary (e.g., Khondalite Belt). Some terranes show a continuum from amphibolite to granulite facies [e.g., Pikwitonei, Namaqualand (South Africa)], others are fault bounded blocks that show only granulite facies rocks [Nilgiri Hills, Furua Complex (Tanzania)]. Some terranes show systematic regional variations in pressure [e.g., Central Gneiss Belt, Ontario (Anovitz and Essene, 1990); Pikwitonei, Manitoba (Mezger et al., 1990a); Napier Complex, Antarctica (Harley, 1985a, 1989), others show uniform pressure distributions [e.g., Adirondacks, New York (Bohlen et al., 1985); Furua Complex (Coolen, 1980)].

The variability in composition and metamorphic conditions reflects the complexity in the possible chemical and tectonic evolution of regional granulite terranes. In detail, every terrane may have some of its own characteristics, and it is necessary to evaluate which characteristics are of general significance and which are unique to a given terrane. Only then can a decision be made whether a specific granulite terrane or granulites in general, can be used as models for the lower crust.

2.2. Pressure–temperature conditions

Over the past ten years a variety of new geobarometers and thermometers have been experimentally calibrated and others have been refined using better thermodynamic data (Essene, 1989). Many of these barometers and thermometers can be used to obtain precise pressures and temperatures, particularly in granulite grade rocks because they involve anhydrous phases that are important rock-forming minerals in high-grade rocks. Improved thermodynamic data and solution models particularly for anhydrous phases are now available that allow the use of experimentally calibrated geothermometers and geobarometers well outside the composition used in the experiments (e.g., Ganguly and Saxena, 1984; Davidson and Lindsley, 1985; Fuhrman and Lindsley, 1988; Powell and Holland, 1988; Essene, 1989).

Care has to be taken in using geothermometers and geobarometers to determine peak metamorphic conditions in high-grade terranes. Diffusion studies indicate that major elements in many rock-forming minerals can diffuse quite readily under upper-amphibolite to granulite facies conditions (Freer, 1981). This can lead to widespread resetting of pressures and temperatures following peak metamorphism and could mean that pressures and temperatures obtained using geothermometers and geobarometers do not represent the peak metamorphic conditions, but rather conditions at some unspecified time during retrogression (Frost and Chacko, 1989). Therefore, in order to obtain reliable information on the conditions during peak metamorphism it is important to apply a variety geothermometers and geobarometers that rely on the partitioning of different cations with different diffusivities. Only if agreement can be obtained among the different geothermometers and geobarometers it is likely that they record the peak metamorphic conditions. In addition, these geothermobarometric results should be confirmed by independent evidence given by the presence and absence of characteristic mineral assemblages. Additional support for the validity of the geothermobarometric results may be obtained from critical mineral assemblages. The regional occurrence of sapphirine + quartz, spinel + quartz, orthopyroxene + sillimanite + quartz and metamorphic pigeonite in the Napier Complex are strong support that this terrane has experienced temperatures in excess of 900°C as recorded by geothermometers (Sheraton et al., 1980; Grew, 1980, 1982; Ellis, 1980; Ellis et al., 1980; Harley, 1985a, 1987a; Sandiford and Powell, 1986b). The absence of pigeonite in Fe-rich rocks containing orthopyroxene + clinopryroxene and the occurrence of grossular + quartz rather than plagioclase + wollastonite, or the abundance of garnet + sillimanite assemblages instead of spinel + quartz in the Adirondacks and the Pikwitonei granulite domain can be used as an indication that these terranes did not experience temperatures in excess of 850°C. This supports the suggestion that the temperatures obtained from a variety of geothermometers in these terranes are close to the peak metamorphic conditions and not substantially reset (Bohlen et al., 1985; Mezger et al., 1990a). The lack of extensive resetting despite the high temperatures reached during granulite facies metamorphism may be typical for granulites and probably is a direct result of the vapor-absent conditions during the peak of metamorphism and for some time thereafter.

A compilation of pressure and temperature data shows that a majority of regional granulite terranes exhibit inferred peak metamorphic pressures of 7.5 ± 1 kbar and temperatures of 800 ± 50°C (Newton and Perkins, 1982; Bohlen and Mezger, 1989; Harley, 1989) (Fig. 12-1).

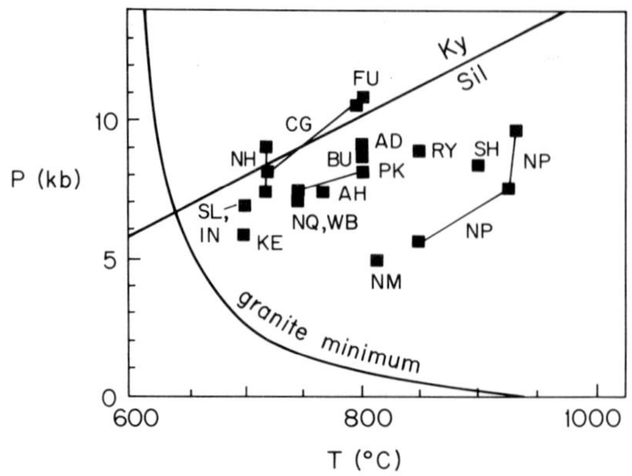

Fig. 12-1. P–T diagram showing the peak metamorphic conditions for selected regional granulite terranes. Also shown are the granite melting curves for $X(H_2O) = 1$ after Huang and Wyllie (1975). Squares connected by lines give range for areas that show a significant variation in peak P–T conditions. *AD* = Aldan Shield, Siberia, USSR (Perchuck et al., 1986); *AH* = Adirondack Highlands, New York, USA (Bohlen et al., 1985); *BU* = Buksefjorden, Greenland (Wells, 1979, 1980; Newton and Perkins, 1982); *CG* = Central Gneiss Belt, Ontario, Canada (Anovitz and Essene, 1990); *FU* = Furua Complex, Tanzania (Coolen, 1980); *IN* = Inajärvi, Finland (Hörmann et al., 1980); *KE* = Kerala Khondalite Belt, Kerala, India (Chacko et al., 1987); *NH* = Nilgiri Hills, South India (Raith et al., 1988); *NM* = Namaqualand, South Africa (Waters, 1985, 1986); *NP* = Napier Complex, Enderby Land, Antarctica (Ellis, 1980; Harley, 1985a; Sandiford, 1985); *NQ* = New Quebec, Canada (Newton, 1987); *PK* = Pikwitonei Granulite Domain, Manitoba, Canada (Mezger et al., 1990a); *RY* = Rayner Complex, Enderby Land, Antarctica (Ellis, 1983; Black et al., 1987); *SH* = Sri Lanka Highlands (Sandiford et al., 1988; Schenk et al., 1988, 1990); *SL* = Sri Lanka Lowlands (Sandiford et al., 1988); *WB* = Wheat Belt, Australia (Newton, 1987).

However, there are some notable exceptions, like the Napier Complex discussed in detail below. The pressures and temperatures shown in Figure 12-1 are also consistent with the occurrence of sillimanite as the typical Al_2SiO_5-phase in the majority of the regional granulite terranes. Only a few regional granulite terranes contain kyanite as a characteristic Al_2SiO_5-phase [Furua Complex (Coolen, 1980); Central Gneiss Belt (Anovitz and Essene, 1990)].

2.3. Fluids in granulite terranes

The composition of the fluid phase during metamorphism is of central importance in the discussion of the genesis of granulite terranes and one of the most controversial topics (e.g., Touret, 1971, 1987; Newton et al., 1980; Newton, 1985, 1989; Frost and Frost, 1987; Valley et al., 1990; Wickham, this volume). In lower and medium grade non-carbonate rocks a_{H_2O} is typically close to unity (e.g., Ferry and Burt, 1982). However, in granulites a_{H_2O} has to be significantly lower than unity in order to prevent large-scale melting. Rocks of granitic composition (sensu strictu) are known from most granulite terranes, but they are typically scarce. The preservation of these granites requires that a_{H_2O} must be significantly less than unity at least during peak metamorphism, but may have been high during the early stages of metamorphism. Further evidence for low a_{H_2O} comes from a combination of thermobarometry with the stability fields of major rock-forming minerals. Although biotite, hornblende and cordierite can be stable under granulite facies pressure–temperature conditions, their hydroxyl sites characteristically have $X_{OH} \ll 1$. In many cases hydrous minerals are replaced by equivalent anhydrous assemblages such as orthopyroxene + K-feldspar, garnet + K-feldspar, orthopyroxene + clinopyroxene + plagioclase, garnet + K-feldspar + sillimanite/kyanite, orthopyroxene + sillimanite $\pm$ quartz or sapphirine + quartz. If appropriate mineral assemblages with hydrous and non-hydrous phases are present, it is possible to estimate a_{H_2O} (e.g., Edwards and Essene 1988). These calculations show that a_{H_2O} is typically less than 0.4 [see Harley (1989) for a summary].

In granulite facies rocks, CO_2-rich fluid inclusions that are consistent with having been trapped during peak metamorphism are widespread. This observation has led to the suggestion that a H_2O-dominated fluid had been diluted by CO_2 during high-grade metamorphism (e.g., Touret, 1971, 1987; Janardhan et al., 1979, 1982; Hansen et al., 1984). The increased CO_2/H_2O ratio could be the result of influx of CO_2 derived from an external source or be caused by the removal of a granitic melt that dissolved most of the H_2O and left the CO_2 behind (e.g., Touret, 1987).

The transformation of amphibole-bearing orthogneisses to charnockites can be observed in-situ in several superb outcrops in southern India. The newly formed charnockite forms patches within an amphibole- and biotite-bearing country rock. The charnockite has dominantly CO_2-rich fluid inclusions whereas the amphibole- and biotite-bearing country rock has dominantly H_2O-rich fluid inclusions. This observation lead to the suggestion that CO_2 flooding from deep-seated sources, possibly the mantle, is a mechanism that can lead to the reduction of a_{H_2O} and the transformation of amphibolite to granulite facies rocks during vapor-present conditions (Newton et al., 1980; Janardhan et al., 1979, 1982; Hansen et al., 1984). However, this model of bulk flooding by CO_2-rich fluids has been shown to be problematic for several reasons ranging from the difficulty to identify a deep source that could yield sufficient CO_2 to the lack of homogenization of oxygen and carbon isotopes in granulites (Newton, 1989; Wickham, this volume). New experimental data seem to indicate that CO_2 might readily dissolve in the silicate melt during partial melting of phlogopite-bearing rocks (Peterson and Newton, 1989). This experimental finding would imply that the presence of CO_2-rich fluid inclusions cannot be used as an argument for the extraction of a granitic melt (e.g., Touret, 1987), unless melting occurred under conditions where biotite was stable. In addition, if CO_2 indeed enhances partial melting, most regional granulites must have formed under conditions of low a_{H_2O} and low a_{CO_2} or under vapor-absent conditions, otherwise regional granulites would have to show evidence for widespread and high degrees of partial melting, which is not observed. More commonly evidence for small degrees of partial melting by the break-down of hydrous phases under vapor-absent conditions can be observed under high-grade conditions (Waters and Whales, 1984; Vielzeuf and Holloway, 1988). Based on phase equilibria and stable isotope evidence vapor-absent metamorphism has been suggested for a variety of regional granulite terranes (e.g., Lamb and Valley, 1984; Waters and Whales, 1984; Powers and Bohlen, 1985; Valley et al., 1990). It is still possible that CO_2-rich fluids play an important role in granulite terranes, however, they may be more of local importance (Frost and Frost, 1987), and the charnockitization phenomenon observed in well described amphibolite to granulite transition zones from southern India may be restricted to this area and may be due to local production of a CO_2-rich fluid (e.g., Stähle et al., 1987).

2.4. Heat sources and tectonic setting

Pressure and temperature estimates from regional granulite terranes indicate that the geothermal gradient during metamorphism was higher than the gradient for stable continental crust (Perkins and Newton, 1981; Newton and Perkins, 1982; Newton, 1987; Harley, 1989; Fig. 12-1). To raise the geothermal gradient, heat must be added to the crust, or be generated internally by radioactive decay. If the extra heat needed to increase the geothermal gradient is generated within the crust, heat production must be higher than heat loss by conduction or convection. External heat has to be added to the crust by underplating and intrusion of magmas, because large fluid/rock ratios (i.e. >5–10) would be needed in order for an influx of hot fluids to cause a significant temperature change within a metamorphic terrane (e.g., Barnicoat, 1987; Peacock, 1987; Brady, 1988). Influx of such large quantities of fluids would lead to an abundance of high-variance mineral assemblages and a homogenization of most pre-metamorphic variation in the oxygen and carbon isotopic ratios on a regional scale (Valley et al., 1983; Valley and O'Neil, 1984; Vry et al., 1988). Since most rocks have mineral assemblages preserved that buffer fluid activities (Powell, 1983) and show a correlation of stable isotope ratios with lithology, the addition of heat by the influx of hot fluids on a regional scale must be considered insignificant. However, on a local scale the influx of external fluids can have a significant control on the thermal regime (Chamberlain and Rumble, 1988), mineral assemblages and stable isotope signatures (e.g., Valley et al., 1983; Valley and O'Neil, 1984).

Since input of external heat seems to be only possible through the addition of mantle-derived melts, granulite-facies conditions can be achieved in a variety of geologic settings such as in a rift environment, above a mantle hot-spot or along an active continental margin (e.g., Wells, 1980; Hodges et al., 1982; Harley, 1985a; Wickham and Oxburgh, 1985; Sandiford and Powell, 1986a; Bohlen, 1987, 1991). Some unusual high-pressure granulites may form in tectonic settings dominated by thrusting that may even lead to a double-thickened continental crust (Ellis, 1987). In such a setting granulite facies conditions may be achieved just by heating due to radioactive decay and without extensive input of externally derived heat.

Although all these tectonic settings are possible sites for granulite-facies metamorphism (e.g., Bohlen, 1991), it is not at all clear what the significance of these different settings might be with respect to the generation of stable deep continental crust. The major problem so far has been to deduce the tectonic setting from the fragmentary information preserved in rocks from granulite terranes. Once the tectonic setting for the evolution of a large number of regional granulites is established, it will be possible to evaluate the tectonic settings that lead to the formation of these granulite terranes and possibly the lowermost continental crust.

3. Evolution of regional granulite terranes: examples

The lower crust is expected to have formed by a complex interplay of magmatism, metamorphism and deformation and thus only a detailed study of the geologic history of a section of deep crust may reveal the processes and driving forces that led to its formation (e.g., Dewey, 1986). By reconstructing the absolute time frame for the igneous and metamorphic history of these granulite terranes it is possible to establish constraints on the tectonic environment in which these once deeply buried granulite terranes formed and how the underlying lowermost continental crust evolved through time. In this chapter, the discussion on the temporal evolution of regional granulite terranes and their implication for the genesis of the lowermost continental crust will be restricted to the evolution of terranes where the uplift was apparently unrelated to the processes that lead to the formation and stabilization of the crust, or where the available information indicates that the time-integrated uplift rates were slow (i.e., less than ca. 100 m/Ma). Thus these terranes may have been lower continental crust for some time of their history. For these reasons the granulite terranes of the Napier

Complex (Enderby Land, Antarctica), Pikwitonei granulite domain (NW Superior Province, Canada), and the Adirondack terrane (Grenville Province, USA) were selected for the detailed discussion. In addition, these regional granulite terranes are the only examples where sufficient geochronologic data and reliable geobarometric and geothermometric information are available to allow a quantitative reconstruction of the major part of their geologic history. Of particular importance in these reconstructions are the determination of variations in the conditions of metamorphism as a function of time and the timing of metamorphism, magmatism and deformation. This information is essential in order to be able to elucidate the processes involved in the chemical and temporal evolution of the deep continental crust. These three regional granulite terranes have spent several hundred Ma at a depth corresponding to middle to lower parts (20–30 km) of stable continental crust and may have been lower crust for a substantial time of their history. Their current exposure at the Earth's surface may be accidental and unrelated to their genesis.

3.1. Napier Complex (Enderby Land, Antarctica)

3.1.1. Geologic setting

The Archean Napier Complex is located at the northern margin of Enderby Land, Antarctica (Fig. 12-2). Its southern margin is bound by the Proterozoic Rayner Complex (Black et al., 1987). The Napier Complex is dominated by orthogneisses of trondhjemitic, tonalitic, granodioritic and granitic composition (Sheraton and Black, 1983) with interlayered pelitic, quartzitic and ferruginous metasedimentary rocks. The regional occurrence of the equilibrium assemblages sapphirine–orthopyroxene–quartz ± osumilite, spinel–quartz and orthopyroxene–sillimanite–quartz in metapelites and the preservation of pigeonite (now inverted and exsolved) in ironstones indicate that the terrane experienced metamorphic temperatures possibly as high as 1000°C during the highest grade metamorphism (e.g., Sheraton et al., 1980; Grew, 1980, 1982; Ellis, 1980; Ellis et al., 1980; Harley, 1985a, 1987a; Sandiford and Powell, 1986b). The Napier Complex thus represents one of the highest temperature exposed granulite terranes. The temperatures do not show a significant regional variation; however,

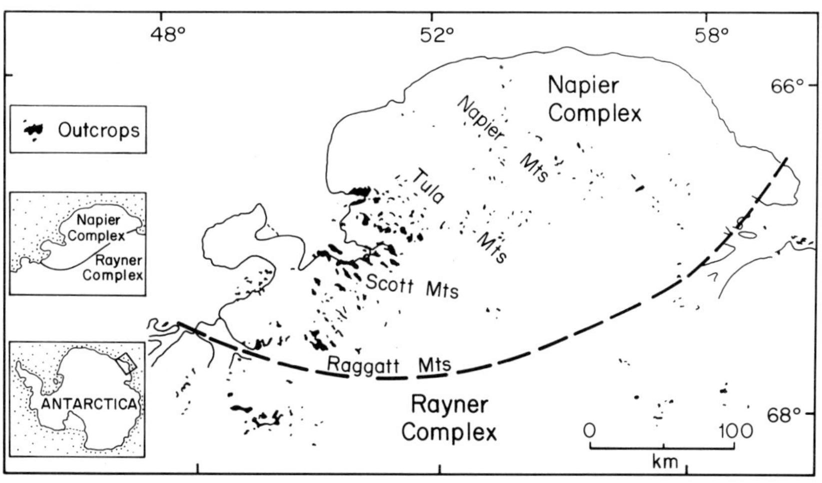

Fig. 12-2. Location map of the Napier Complex, Antarctica (after Black et al., 1987). A geologic map for this area is currently not available due to the paucity of exposure.

TABLE 12-1

Generalized geological history of the Napier Complex (after Harley, 1985a, 1987b)

Ages (in Ma)	Events
3927–3700	intrusion of orthogneisses, deposition of layered series
>3100	intrusion of many orthogneisses
ca. 3072	granulite facies (>950°C/6–10 kbar) with disturbance of most isotopic systems. Flat-lying isoclinal folds, layer-parallel foliation, elongation lineations
>2900	granulite facies (850–900°C, 6–10 kbar), asymmetric tight folds followed by the emplacement of mafic/ultramafic dikes and cooling to 800–700°C/5–7 kbar.
ca. 2463	granulite–upper amphibolite facies (700–750°C/7–9 kbar), formation of upright close to open folds, large basins and domes, weak vertical deformation, localized intense deformation, contemporaneous shearing. Followed by the intrusion of mafic dikes.
ca. 1000	amphibolite facies metamorphism, local overprint by Rayner Complex metamorphism.
ca. 960	rapid uplift to 3–4 kbar and 650–700°C
ca. 540	weak greenschist facies overprint at 3–4 kbar

the pressures increase systematically from the northeast to southwest across the terrane (Fig. 12-1). The current exposure of Napier Complex is interpreted as an oblique cross-section through part of the Archean lower continental crust (Sandiford, 1985). Structural studies indicate that the area was deformed and metamorphosed during three major tectonothermal events, each of which led to the formation of a distinct fabric and is also recorded by mineral reactions (coronas, pseudomorphs) (e.g., Harley, 1985a,b) (Table 1).

3.1.2. Geologic history

A U-Pb zircon age of 3927 ± 10 Ma from a metatonalite indicates that continental crust was formed in the Napier Complex during the early Archean (Black et al., 1986b) (Table 1). The supracrustal units of the layered series (metasediments with intercalated metavolcanics) were probably deposited at the same time. The protoliths of many orthogneisses intruded between 3927 Ma and 3072 Ma. However, none of these have been dated so far and their crustal history has been obscured by the intense recrystallization during the pervasive granulite facies metamorphism around 3100 Ma. During this metamorphism temperatures of 900–1025°C at pressures of 8–10 kbar were reached on a regional scale (Harley, 1985a, 1987a; Sandiford and Powell, 1986b). This metamorphism was accompanied by extensive deformation and magmatism and led to the disturbance of most isotopic systems (Black et al., 1983, 1986a; McCulloch and Black, 1984; Black and McCulloch, 1987; Harley and Black, 1987). Following this high-grade metamorphism, the terrane cooled isobarically and experienced a second granulite facies metamorphism at about 2950 Ma. This event was characterized by folding and extensive recrystallization at temperatures of 850–900°C (Harley, 1985a; Black et al., 1986a). Subsequently, minor mafic to ultramafic dikes intruded. At about the same time the terrane cooled and corona and exsolution textures developed that record metamorphic conditions of 800–700°C at 5–7 kbar.

A third regional metamorphism that reached upper amphibolite to granulite facies conditions (700–650°C/5–8 kbar) affected the area around 2500 Ma. This metamorphism is associated with locally intense deformation and led to variable recrystallization to fine grain sizes and local formation of hydrous minerals (biotite, amphibole). During this metamorphic episode, mafic dikes, which show granulite facies assemblages, intruded (Harley, 1985a). After this metamorphism and until ca. 1200 Ma, several generations of mafic dikes intruded while the terrane was still at ca. 8 kbar (Kuehner and Green, 1987).

Metamorphism in the adjacent Rayner Complex at ca. 1100 Ma led to local overprinting at amphibolite grade in the Napier Complex. Mineral assemblages in chilled margins of mafic dikes indicate that the terrane stayed at mid-crustal levels (>5–6 kbar) until at least 960 Ma. At about this time the terrane was uplifted rapidly to 3–4 kbar. At about 540 Ma greenschist facies metamorphism that affected large parts of East Antarctic also led to a weak overprint in the Napier Complex (Sandiford, 1985; Harley, 1987b).

3.1.3. Possible tectonic setting

The metamorphic temperatures at ca. 3100 Ma were uniformly high (>950°C) (Harley, 1985a, 1987a,b; Sandiford and Powell, 1986b; Grew, 1980, 1982) although the pressure variations indicate that the terrane now exposes a crustal depth range of 10–13 km. This indicates that the temperatures were apparently buffered, possibly by extensive partial melting. The high temperatures and the apparently buffered temperatures over 10–13 km indicate that convective heat flow, most likely in the form of advecting early- to syn-metamorphic magmas, dominated the thermal regime. Isotopic evidence indicates that magmas intruded just prior to and during this tectonothermal event (Harley and Black, 1987). This regime of high heat flux and abundant magmatic intrusions, the magmatic and possibly minor tectonic thickening, the development of a pronounced subhorizontal fabric and the isobaric cooling path are interpreted to indicate that the tectonic setting of this high-grade metamorphism was possibly in an extensional environment at the base of ca. 30-km-thick continental crust and most of the heat was supplied by underplated and intruding magmas (Sandiford and Powell, 1986a; Harley and Black, 1987; Harley, 1987b). If this interpretation is correct, it would imply that parts of the Napier Complex represent a rare example of currently exposed rocks that were actual lowermost crust before ca. 3100 Ma. However, subsequent to this tectonothermal episode these lowermost crustal rocks resided in the middle crust as a result of underplating of magmas that led to the formation of this unusual high-temperature granulite terrane.

In contrast, Ellis (1987) suggested that the high-grade metamorphism in the Napier Complex was the result of double thickening of continental crust by thrusting. In this model the rocks would have an early high-pressure/low-temperature history and pass through the stability field of kyanite. After the upper half of the thick crust was eroded, the lower half equilibrated under granulite-grade conditions, and then cooled along an nearly isobaric path. The supporting argument for this model is the lack of mafic rocks that could have intruded around 3100 Ma ago. Felsic intrusives, which include a distinct group of I-type granitoids with a strong depletion in heavy rare earth elements, are widespread. This depletion is interpreted as an indication of garnet in the residue and thus high pressures (8–15 kbar) (Ellis, 1987). This model implies rapid syn-orogenic uplift, therefore breakdown reactions from high- to low-pressure assemblages, such as orthopyroxene + plagioclase coronas around garnet should be preserved. Even if the rocks equilibrated at high temperatures as suggested by Ellis (1987) the low a_{H_2O} that prevented the rocks from melting would also have retarded complete recrystallization. The apparent lack of such decompression textures casts some doubt on the validity of the crustal thickening model for the metamorphism around 3100 Ma in the Napier Complex. In addition, the heavy rare earth element depletion is not a conclusive argument for the presence of garnet in the source, because the source may have been already depleted in the heavy rare earth elements. Crustal extension may not necessarily lead to intrusion of mafic magmas as shown in the case of the Adirondacks (see below). For these reasons the model involving crustal extension seems to be the better explanation for the high temperature metamorphism in the Napier Complex.

During the second high-grade episode at ca. 2900 Ma inclined folds developed, possibly indicating a transition to a compressional tectonic regime. Since the terrane cooled nearly isobarically following this episode the compression did not lead to a substantial crustal thickening and the terrane remained in isostatic equilibrium. The tectonothermal episode from 3100 to 2900 Ma led to the stabilization of the continental crust in the Napier Complex. Later metamorphic episodes were associated only with weak compression at ca. 2463 Ma and marginal remobilization at ca. 1000 Ma. The currently exposed area stayed at lower to middle crustal conditions for ca. 2000 Ma. The uplift of this lower crust may have been caused by underplating of magmas and rocks of the adjacent Rayner Complex (Ellis, 1983; Harley, 1985a, b). Therefore uplift of the Napier Complex was unrelated to the processes that formed and stabilized the crust (Harley, 1987b; Sandiford, 1985).

3.2. Pikwitonei granulite domain (Superior Province)

3.2.1. Geologic setting

The Cross Lake subprovince, the Pikwitonei granulite domain and the Thompson mobile belt (Fig. 12-3) represent the northwestern part of the Archean Superior Province, bordered by the younger Churchill Province to the north and west and the Archean Berens plutonic belt to the south (Ermanovics and Davison, 1976; Corfu and Ayres, 1984). In the northwestern Superior Province, the greenschist- to amphibolite-facies tonalite–greenstone terrane of the Cross Lake subprovince grades into the granulite-facies Pikwitonei domain toward the western margin of the Superior Province. The boundary between the Pikwitonei granulite domain and the Cross Lake subprovince is defined by the orthopyroxene isograd. The transition is the result of prograde metamorphism and does not show any obvious structural or lithological breaks (Weber and Scoates, 1978; Hubregtse, 1980; Weber and Loveridge, 1981). The Pikwitonei–Cross Lake area is a rare example of an Archean terrane where greenstone belts can be traced continuously from greenschist through granulite facies. The northwestern margin of the Superior Province has been interpreted as an uplifted and tilted part of Archean crust (Weber, 1980; Fountain and Salisbury, 1981; Green et al., 1985; Mezger et al., 1990a). A transect from Utik Lake in the east to the Burntwood River in the west shows an increase in metamorphic grade from middle-amphibolite to granulite facies and exposes an oblique cross-section through approximately 20 km of Archean continental crust (Fig. 12-4).

The greenstone belt at Utik Lake reached conditions of the middle-amphibolite facies and consists largely of ultramafic to mafic metavolcanic rocks with minor metapelitic units and iron formations. The greenstone belt is surrounded by tonalitic gneisses to the southeast and by large bodies of undeformed postkinematic pink granites to the northeast. At Cauchon Lake the orthopyroxene isograd marks the transition from amphibolite to granulite facies. The narrow and discontinuous greenstone belts consist of metamorphosed volcanic rocks (ul-

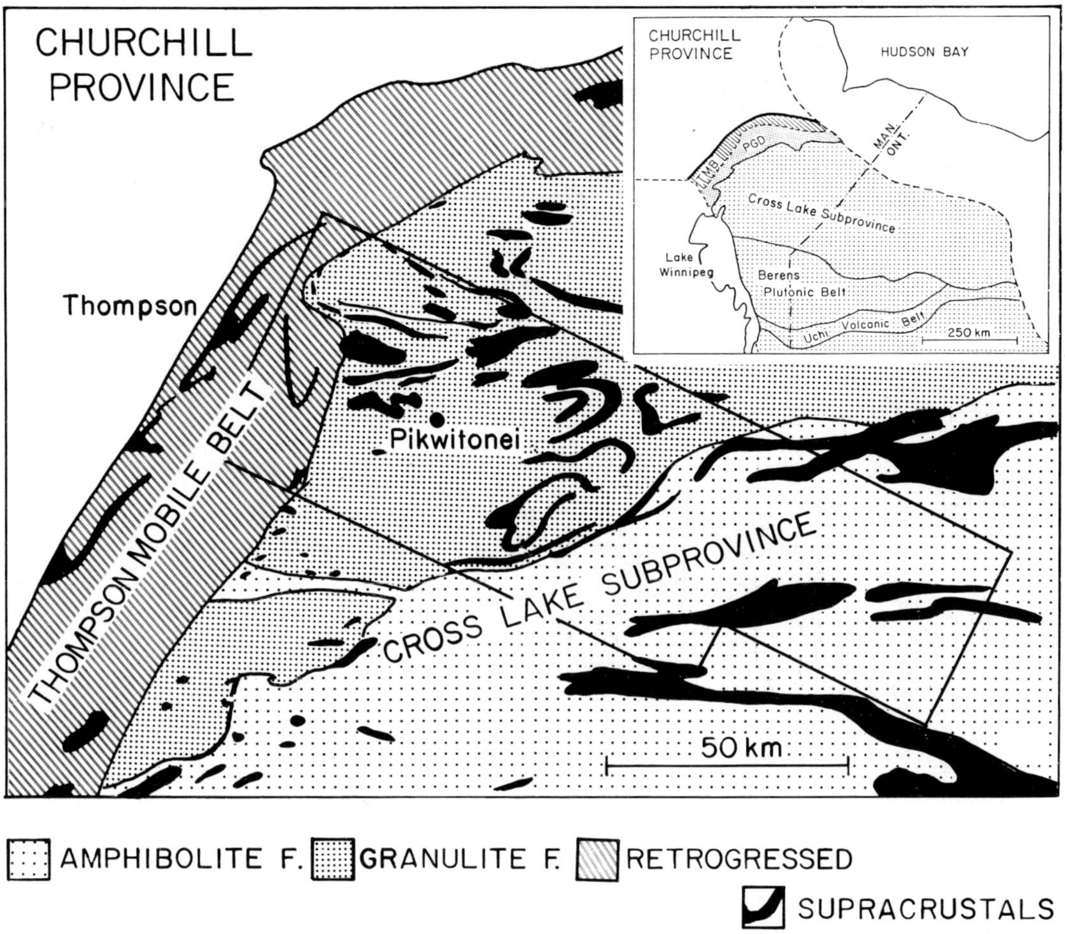

Fig. 12-3. Generalized geologic map of the NW Superior Province (after Manitoba Mineral Resources, 1980). Areas in black represent supracrustal units. Outlined area is enlarged in Fig. 12-4.

tramafic to granitic), arenaceous and argillaceous sedimentary rocks, banded iron formation, calc-silicate rocks and marbles. The greenstone belt lithologies and the surrounding tonalitic and trondhjemitic gneisses show locally abundant migmatitic features. A layered anorthosite complex forms a distinct lithologic unit in the northern part of the Cauchon Lake area.

In the Natawahunan and Burntwood River areas the greenstone belts consist of amphibolites, garnet–biotite–hornblende/pyroxene gneisses, and minor garnet–biotite–sillimanite gneisses and iron formations. The greenstone belts are surrounded by tonalitic gneisses with mafic and ultramafic inclusions. Toward the west the granulite facies mineral assemblages are retrogressed to greenschist to amphibolite facies assemblages within the Thompson mobile belt (Weber and Scoates, 1978).

3.2.2. Geologic history

The earliest evidence for geologic activity along the northwest margin of the Superior Province comes from Nd isotopes. Nd-model ages (T_{CHUR}) for meta-igneous and metasedimentary samples from the Cauchon Lake area range from 3640 to 2720 Ma (Fig. 12-5)

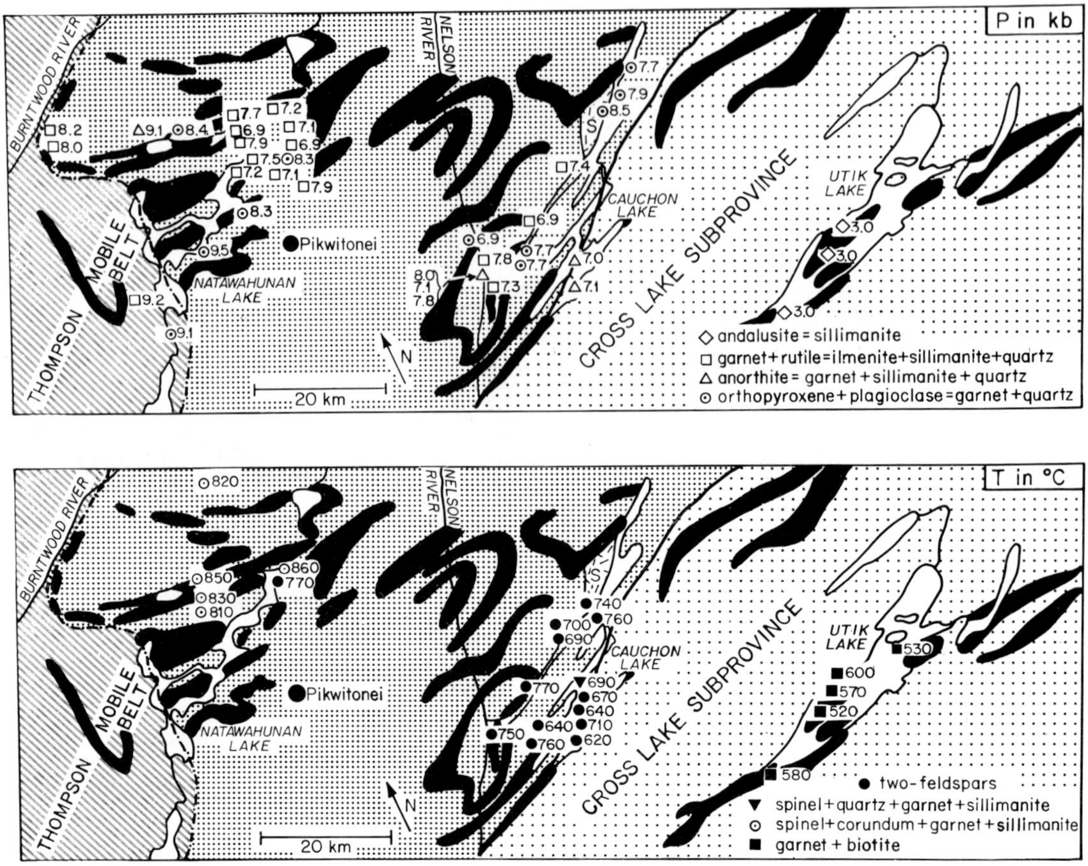

Fig. 12-4. Maps showing a section from Utik Lake in the east to the Burntwood river in the west. The metamorphic temperatures (a) increase from ca. 575°C at Utik Lake to ca. 820°C at Natawahunan Lake and the pressures (b) increase from ca. 3 kbar at Utik Lake to ca. 9 kbar close to the retrogressed Thompson mobile belt. These increasing metamorphic conditions indicate that the section shown represents an oblique cross-section through about 20 km of Archean crust.

TABLE 12-2

Geologic history of the northwestern margin of the Superior Province (after Mezger et al., 1989, 1990a)

Ages (in Ma)	Events
>2984	amphibolite facies metamorphism?
2744–2738	upper amphibolite facies to granulite facies metamorphism, T locally >800°C
2700–2687	upper amphibolite facies with partial melting, T >700°C.
2660–2637	amphibolite to granulite facies with extensive melting, T and P increasing from 600°C/3 kbar at Utik Lake to 830°C/8 kbar along the western margin
2605–2591	amphibolite facies, intrusion of hydrous granites and pegmatites
1883	intrusion of Molson Dikes
ca. 1700	collision of Superior with Churchill Province, retrogression along the collision zone

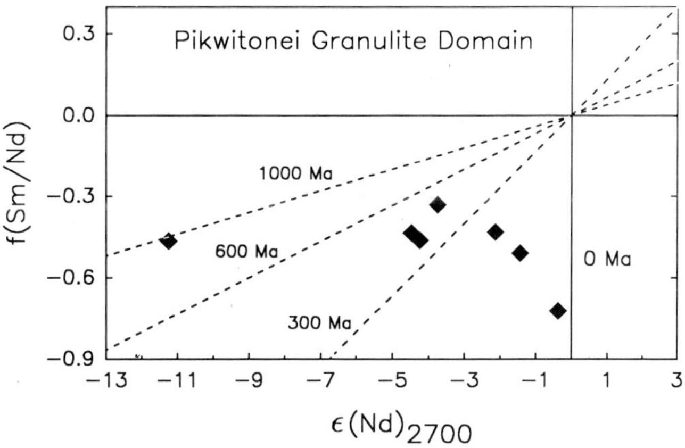

Fig. 12-5. f(Sm/Nd) versus ϵ_{Nd} plot (Shirey and Hanson, 1986) for meta-igneous (diamonds) and meta-sedimentary (circles) rocks from the Cauchon Lake area (Pikwitonei granulite domain). The values for ϵ_{Nd} are normalized to CHUR at 2700 Ma. f(Sm/Nd) is the Sm/Nd ratio normalized to the chondritic ratio. Also shown are reference isochrons for 300, 600 and 1000 Ma before 2700 Ma. The spread in the data indicates that crustal material in the Cauchon Lake area is older than 2700 Ma and some components are possibly as old as 3700 Ma.

(Mezger, unpubl. data). The model ages indicate that the majority of the currently exposed crust in the area was generated before 2900 Ma (Table 2).

Mineral ages from a large variety of rock types from Utik Lake in the east to Natawahunan Lake in the west indicate five distinct periods of pre-Hudsonian (older than 1700 Ma) mineral growth (Table 2). Not all periods of mineral growth may have occurred throughout the whole area, but may have been only of local importance. This indicates that mineral reactions did not occur continuously but were punctuated, lasting only for relatively short intervals during the >150 Ma metamorphic history of the terrane. These periods of mineral growth may indicate times of geologic activity characterized by changing physical conditions, particularly increase in temperature. These episodes were separated by relative quiescent intervals with no new mineral growth. It is possible that during these times the terrane cooled following a thermal disturbance. The episodes of mineral growth have to be considered as relative metamorphic peaks or as punctuations in the long metamorphic history.

Two early metamorphic episodes (2984 ± 3 and 2742 ± 2 Ma) that reached conditions high enough to form garnets are recorded in two xenocrystic garnets extracted from a post-kinematic granite at Cauchon Lake. Ages around 2740 Ma are also recorded from Natawahunan Lake where they date the garnet growth according to the reaction:

$$\text{staurolite} + \text{quartz} = \text{garnet} + \text{sillimanite} + H_2O$$

at 2744 ± 2 Ma and the reaction:

$$\text{staurolite} = \text{garnet} + \text{sillimanite} + \text{spinel} + H_2O$$

at 2738 ± 5 Ma.

This indicates that around 2740 Ma the metamorphic conditions at Cauchon Lake reached at least lower amphibolite facies, and upper amphibolite to granulite facies at Natawahunan Lake.

Leucosomes in metapelitic gneiss surrounding a xenolith-bearing enderbite at Cauchon Lake yielded a garnet with an age of 2700 ± 6 Ma, and leucosomes in mafic gneisses contain zircons with an age of 2695 ± 3 Ma (Heaman et al., 1986a; Krogh et al., 1986). These ages are interpreted as the time of intrusion of the xenolith bearing enderbite. This large intrusive body of tonalitic composition has a metamorphic texture and contains abundant pristine to partially digested xenoliths. Within an outcrop up to 50% of the rock may be made up of xenoliths from all the different rock types in the area except for marbles. Ternary plagioclases from an iron formation xenolith within the enderbite record temperatures >1100°C. These temperatures are considered minimum values for the magma. Because of their significant heat capacity and latent heat such hot and dry synmetamorphic intrusions have to be considered an important additional heat source to promote high-temperature metamorphism (Wells, 1980; Powers and Bohlen, 1985). The intrusion of such melts could have been a mechanism that led to an episodic increase in the geothermal gradient and thus to intervals of enhanced mineral reactions. As indicated by garnets from the Cauchon Lake area and a sphene core from Utik Lake, the area was undergoing regional metamorphism at the time of intrusion of the xenolith bearing enderbite.

From 2648 to 2641 Ma metamorphic conditions at Cauchon Lake favored partial melting that caused the formation of extensive migmatites in metapelitic and mafic gneisses as well as in the surrounding tonalitic to trondhjemitic gneisses. The formation of the last metamorphic garnets in a metarhyolite at 2637 ± 5 Ma corresponds to the time of peak metamorphic conditions for this last prograde metamorphic episode. Subsequently, the terrane cooled nearly isobarically and from 2629 to 2591 Ma garnet and sillimanite-bearing granites and pegmatites of minimum melt composition intruded. The associated hydration reset the two-feldspar temperatures to ca. 650°C and caused the alteration of orthopyroxene to amphibole (Mezger et al., 1990a). The intrusion of post-kinematic granites and pegmatites seems to have been the last major geologic activity until 1883 Ma when the ultramafic to mafic Molson Dyke Swarm intruded (Heaman et al., 1986b). These mafic to ultramafic dikes are a few centimeters to several tens of meter wide. They strike approximately N–S and do not follow any metamorphic fabric.

Geochronologic information, combined with pressure and temperature estimates, textural observations and zoning in garnets is used to construct quantitative P–T–t (pressure–temperature–time) paths. The P–T–t path for the Cauchon Lake area is depicted in Figure 12-6. The path for Natawahunan Lake is similar, with the major difference that high-grade conditions were reached there already at 2740 Ma (Mezger et al., 1989, 1990a). The peak pressures and temperatures shown in Figure 12-4 correspond to the last metamorphic peak around 2640 Ma and therefore represent conditions for only a short time interval within the long metamorphic history of the northwestern Superior Province.

The cooling rate from 2637 Ma to 2591 Ma was ca. 1.8°C/Ma. Based on rutile ages of 2430 Ma from Cauchon Lake and 2360 Ma from Natawahunan Lake, the time-integrated cooling rate between 2591 and 2360 Ma was ca. 1°C/Ma (Mezger et al., 1989, 1990a,b). The slow cooling rate is consistent with the model of Wells (1980) for the formation of granulites in a terrane where the thermal budget is controlled largely by magmatism and erosional uplift is slow. The slow cooling rate combined with the nearly isobaric cooling following the metamorphism at ca. 2640 Ma indicates that time integrated erosion-rates were <70 m/Ma following this last high-grade metamorphic episode and the terrane was probably near isostatic equilibrium. Based on ages from Fe-Ti minerals from the Burntwood River area, the western part of the terrane was at ca. 300–400°C at ca. 2180 Ma. Therefore, uplift must have occurred long after high-grade metamorphism and may have been unrelated to the tectonic processes that caused the metamorphism (Mezger et al., 1990a).

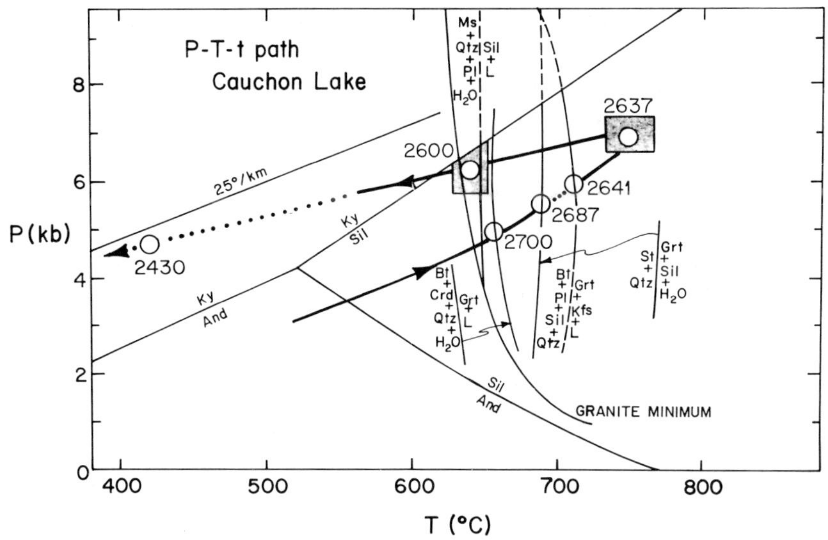

Fig. 12-6. P–T–t path for Utik Lake. For this diagram the information from reaction textures, mineral zoning, geothermometry and geobarometry was combined with geochronologic information from rutile (low temperature point) and garnet (all others). The path may not be continuous but may have excursions towards lower temperatures between the periods of mineral growth (see Table 12-2). Circles indicate ages, numbers represent ages in Ma, ruled boxes represent P–T estimates, solid lines refer to parts of the path where pressure, temperature and age constraints on the metamorphic history are available, the dotted lines are tentative.

Some uplift may have been the result of rifting and crustal heating during the time of the intrusion of the Molson dike swarm around 1883 Ma (Heaman et al., 1986b). The major part of the uplift may be related to the collision of the Superior Province with the Churchill Province during the Hudsonian orogeny at ca. 1700 Ma (Machado et al., 1987). This collisional event caused extensive retrogression of granulite grade gneisses to greenschist and amphibolite facies along the western margin of the Pikwitonei granulite domain, and thus strongly altered what might have been the deepest crustal rocks exposed in the area.

3.2.3. Possible tectonic setting

A tectonic setting for high-grade metamorphism that is consistent with the features and geologic history described above is a magmatic arc superimposed on older continental crust. In this model metamorphism is thought to be superimposed on a continental crust that was in major part assembled before the onset of high-grade metamorphism at 2740 Ma. The supracrustal belts were initially brought into the lower parts of the crust by processes that may be completely unrelated to those that caused the high-grade metamorphism.

Possible modern analogues may be volcanic arcs built on pre-existing continental crust (e.g., Andes, central America). In such a setting, metamorphism is caused by the intrusion and underplating of magmas related to subduction. The increased heat causes partial melting within the deeper parts of the continental arc. These mostly granitic melts will be displaced to higher crustal levels. Addition of mantle-derived melts will cause an increase in crustal thickness. Upward movement of crustal and mantle-derived melts will lead to increasing pressure in the lower crust. This upward movement of heat will lead to a steep geothermal gradient at lower crustal levels and a to a shallower gradient at deeper levels (Richardson, 1970).

The duration of metamorphism is intimately linked to the duration of subduction. For a given area, magmatism may not be continuous through time but rather episodic. The epeirogenic uplift rate following the end of the orogeny depends on crustal thickness as well as composition. Cooling of underplated basaltic material through the basalt/eclogite transition could significantly inhibit uplift.

3.3. Adirondack terrane (Grenville Province)

3.3.1. Geologic setting

The Adirondack terrane is an exposed outlier of the Grenville Province, continuous with the larger region of Grenvillian rocks across the Frontenac Arc. Based on the topography, lithology and metamorphic grade, the Adirondack terrane is traditionally divided into the northwest Lowlands region and central Highlands region. The Adirondack Highlands region is underlain by predominantly granulite-grade orthogneisses, including extensive anorthosites with thin infolded, pelitic and calc-silicate units (McLelland and Isachsen, 1986) (Fig. 12-7). The northwest Lowlands region is dominantly underlain by upper-amphibolite grade pelitic and calc-silicate gneisses, felsic orthogneisses, marbles and minor amphibolite. The two areas are separated by the Carthage–Colton line, a SW–NE-trending shear zone.

An extensive set of data is available for the Adirondack terrane. This includes information concerning the deformational history (McLelland and Isachsen, 1986), metamorphic conditions (Bohlen et al., 1985), stable isotopes and fluid flow during metamorphism (Valley et al., 1990) as well as geochronologic data (Silver, 1969; Heizler and Harrison, 1986; McLelland et al., 1988; Mezger et al., 1991).

3.3.2. Geologic history

Based on structural relationships and the geochronologic data, the Adirondack terrane underwent a complex igneous, metamorphic and deformational history (McLelland and Isachsen, 1986, McLelland, 1986, McLelland et al., 1988). A U-Pb zircon age of 1416 ± 6 Ma is recorded from a metasedimentary gneiss form the Lowlands. This represents the oldest age from the Adirondack terrane. Magmatic ages around 1300 Ma are found in tonalitic rocks from the Highlands as well as the Lowlands (McLelland et al., 1988) (Table 3).

TABLE 12-3

Geologic history of the Adirondacks (after McLelland et al., 1988; Mezger et al., 1991).

Ages (in Ma)	Events
1415–1300	tonalitic intrusion, formation of oldest crust
1167–1130	intrusion of AMC suite, contact metamorphic overprint on areas undergoing regional metamorphism. Subsequent cooling of the Lowlands
1100–996	intrusion of hornblende granites and alaskites, granulite facies metamorphism in Highlands (T up to 800°C, P = 6–7 kbar). High-grade conditions lasted until ca. 1040 Ma in the southern and until ca. 1000 Ma in the central Highlands

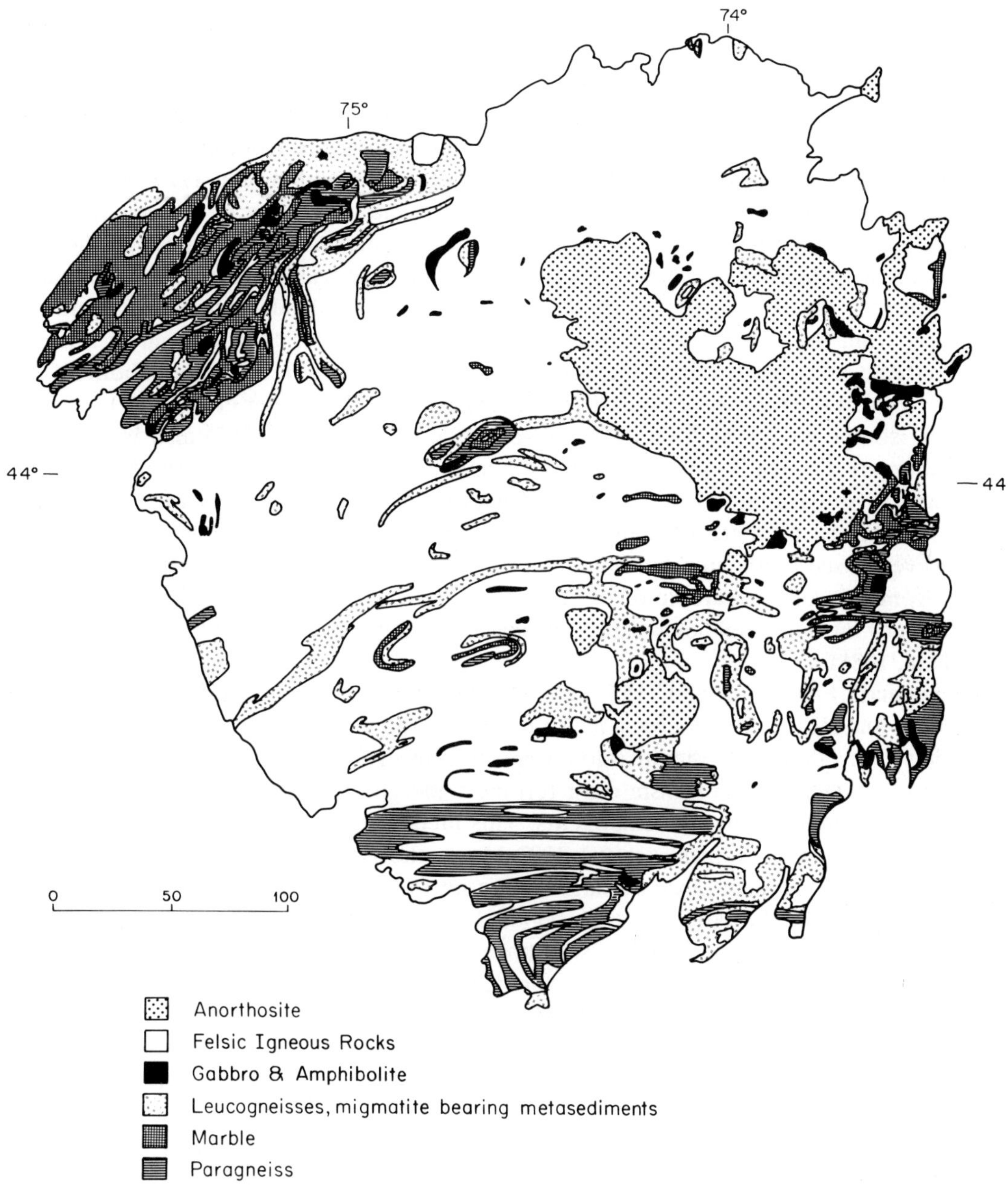

Fig. 12-7. Generalized geologic map of the Adirondack terrane (from Bohlen et al., 1985).

McLelland (1986) combined field relationships, deformation patterns and geochronologic data to infer that anorthositic and associated alaskitic, mangeritic, charnockitic and granitic rocks represent a single suite (Anorthosite–Mangerite–Charnockite (AMC) suite). Based on U-Pb zircon ages McLelland et al. (1988) further suggest that the syenitic bodies as well as hornblende–granitic gneisses may also belong to the same suite that intruded prior to the Grenville orogeny. This pervasive magmatic and metamorphic episode lasted from ca. 1170 to 1130 Ma. It is recorded in zircons extracted from intrusions as well as sphenes, monazite and garnets from marbles and metasedimentary gneisses (McLelland et al., 1988, Mezger et al., 1991).

By combining ages deduced from zircon and garnet it is evident that most of the magmatic bodies intruded within a limited time span from ca. 1155 to 1134 Ma. Field relationships combined with U-Pb ages obtained from zircons and baddeleyite from the anorthosite and cross-cutting metagabbro indicate that the anorthosites intruded between 1113 Ma and 1135 Ma (McLelland and Chiarenzelli, 1990). This supports the suggestion by McLelland (1986) that there is a close temporal relationship between the large anorthosite bodies and the other members of the AMC series. These intrusions must have provided a significant amount of heat to promote the metamorphism in the Lowlands and Highlands. Because a garnet and a monazite from the Lowlands were found to be 5–10 Ma older than the oldest intrusion of the suite, it is suggested that the terrane was undergoing at least amphibolite facies metamorphism at the time this magmatic episode was initiated (Mezger et al., 1991).

Oxygen isotope studies by Valley and O'Neil (1982) and Valley (1985) also indicate that the members of the AMC suite intruded prior to the last peak of metamorphism. The low $\delta^{18}O$ found in wollastonite skarns associated with anorthositic intrusions were interpreted as evidence for the presence of meteoric waters and thus intrusion depths of less than 10 km. However, based on the mineral assemblage and the metamorphic pressures derived from samples from the contact aureole of the Diana syenite (located along the Carthage–Colton line), this body intruded well below 10 km, probably at a depth between 20 and 25 km (Powers and Bohlen, 1985).

The magmatic episode around 1155–1130 Ma apparently occurred throughout the Adirondack terrane and must be considered a major magmatic and metamorphic event. The extensive magmatic activity, that led to the intrusion of mostly anhydrous melts and superimposed high-temperature contact metamorphism on a terrane undergoing regional metamorphism, may have caused the loss of volatile components. This metamorphism seems to have many similarities with the metamorphism around 3100 Ma in the Napier Complex (abundant felsic intrusions, high T/P gradient, extensional tectonic setting), however, the Adirondacks do not show any evidence for the extremely high regional metamorphic temperatures (>900°C) that are reported from the Napier Complex (see above).

In the Lowlands there is no geologic activity recorded subsequent to the magmato-metamorphic episode from 1170 to 1130 Ma. From 1130 to 911 Ma the Lowlands cooled at a rate of 1–2°C/Ma. In contrast, in the Highlands hornblende granites intruded around 1100 Ma and alaskitic bodies around 1070 Ma (McLelland et al., 1988). Metamorphic ages from zircon and garnet range from 1070–996 Ma (Silver, 1969; McLelland et al., 1988; McLelland and Chiarenzelli, 1990; Mezger et al., 1991). Zircon and garnet ages indicate that granulite facies metamorphism occurred around 1060 Ma throughout the Highlands. In the southern Highlands sphene and monazite ages of ca. 1030 Ma indicate cooling from high-grade metamorphism, whereas garnets and zircons from the central Highland indicate that granulite facies metamorphism occurred in this area from 1030 to 996 Ma.

Metamorphic pressures and temperatures, as recorded by a large variety of geothermometers and geobarometers, record peak metamorphic conditions for the last high-grade event of ca. 600–700°C at 7.0 kbar in the Lowlands and 700–800°C at 7.5 kbar in the Highlands [see Bohlen et al. (1985) for a summary] (Fig. 12-8a, b). These conditions were reached in the Lowlands from ca. 1150 to 1130 Ma, in the southern Highlands around 1060 Ma and the central Highlands around 1020 Ma (McLelland et al., 1988; Mezger et al., 1991).

Based on garnet zoning in rocks that contain the assemblage garnet–rutile–ilmenite–

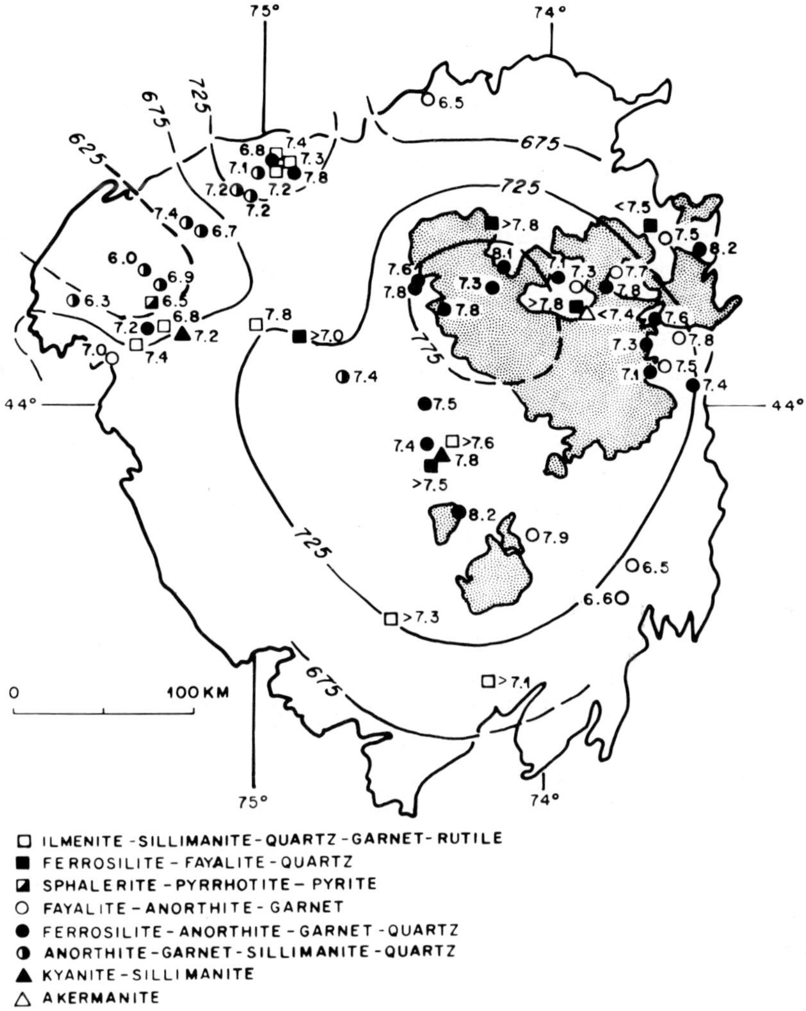

Fig. 12-8. Metamorphic temperatures and pressure for the Adirondacks (Bohlen et al., 1985). Temperatures increase from 600°C in the Lowlands to 800°C in the central Highlands, whereas pressures do not show a systematic regional variation.

sillimanite–quartz and garnet–orthopyroxene–plagioclase–quartz the terrane cooled nearly isobarically following the last episode of high-grade metamorphism (Bohlen, 1987). The cooling rate for the southern Highlands following the last peak of metamorphism is ca. 3°C/Ma at 1030 Ma, ca. 1°C/Ma at 800 Ma and ca. 0.5°C/Ma after 700 Ma (Heizler and Harrison, 1986; Mezger et al., 1991).

3.3.3. Possible tectonic setting

Comparison of the geochronologic data from the Lowlands with that of the Highlands shows that the two terranes had a common early history, but a separate younger history. There is no direct record of any geologic activity in the Lowlands after 1130 Ma, whereas the Highlands show significant magmatism and high-grade metamorphism as late as 1000 Ma. The early part of their metamorphic history is dominated by extensive magmatic intrusions and contact metamorphism in a terrane that was already undergoing regional metamorphism.

Based on the predominance of metasedimentary rocks in the Lowlands and meta-igneous rocks in the Highlands, the Carthage–Colton mylonite zone has been interpreted as a suture between two crustal blocks of different geologic history (e.g., Windley, 1986). The geochronologic data indicates, however, that the two areas had a common metamorphic episode at 1170–1130 Ma but subsequently had distinct thermal histories. Another argument against the suture model is the location of the Diana syenite. This body is structurally within the Highlands, but is sheared in its interior and on its northwestern margin (Wiener et al., 1984) indicating it is located within Carthage–Colton mylonite zone. Because the syenite body intruded at 1154 Ma and still has its contemporaneous contact aureole preserved along the northwest margin, the displacement along the Carthage–Colton mylonite zone could not have been large; the Lowlands, the Diana syenite and the Highlands are in approximately the same structural position as at ca. 1150 Ma. This is consistent with the similar metamorphic pressures obtained for the Highlands and Lowlands (Bohlen et al., 1985) that preclude significant vertical displacement along the Carthage–Colton line.

The Adirondack Highlands are characterized by anhydrous mineral assemblages and vapor-absent high-grade metamorphism (Valley et al., 1990). The terrane may have acquired a strongly volatile depleted character prior to granulite facies metamorphism. The depletion could be the result of the abundance of dry igneous rocks that intruded the terrane around 1155–1130 Ma and caused widespread high-grade contact metamorphism (Valley, 1985).

The abundance of magmas of the AMC series and their mildly to strongly alkaline composition indicates that they may have been generated in an extensional environment (McLelland, 1986). If so, this regional metamorphism was probably the result of underplating of basaltic magmas at the base of the continental crust. Heating of the lower crust caused partial melting and the rising partial melts were emplaced in the middle and upper crust where they superimposed high-grade contact metamorphism on a terrane that was undergoing regional metamorphism. These processes may have been similar to those currently active in the Basin and Range Province of the western USA.

All the rocks of the AMC suite together with their enveloping metasedimentary units are strongly deformed and folded (McLelland and Isachsen, 1986). This indicates that the tectonic regime must have changed from extensional at 1155–1130 Ma to compressional after 1130 Ma.

Bohlen (1987) proposed an Andean-type continental margin setting for the granulite grade metamorphism from ca. 1070 to 1000 Ma. The major heat source was intrusions and magmatic underplated material. The retrograde path following this high-grade metamorphism is characterized by a slow cooling rate and small retrograde dP/dT as inferred from garnet-zoning. This indicates that erosional uplift was slow. In analogy to the Pikwitonei domain this is interpreted to indicate that the terrane was nearly isostatically compensated during the last peak of metamorphism. Hence it is likely that the uplift of the Adirondack terrane was the result of significantly younger tectonic activity possibly unrelated to the processes that caused the granulite facies metamorphism.

4. Granulites and the lower crust

Results from heat flow studies indicate that the lower crust produces less heat than the upper crust (e.g., Roy et al., 1968; Heier, 1973). This requires that, in comparison with the upper crust, the radioactive and thus heat-producing elements U, Th, K and Rb are depleted in the average lower crust. The lower crust can be depleted from the beginning or it can be modified subsequently by processes that lead to the removal of heat producing elements and volatiles. In either case, lower crustal rocks must have experienced high-temperature conditions typical of the granulite facies. The crust may have obtained a depleted character during its formation by the removal of highly fractionated residual melts following the crystallization of the mantle derived melts that form the crust. If the removal of residual melts is the major depletion mechanism, the lower crust must be dominated by igneous rocks and was produced by a one-cycle process. However, xenoliths derived from the deep continental crust indicate more complex histories suggesting later reworking (Rudnick and Williams, 1987; Rudnick, this volume). If crustal depletion is the result of some later process, the crust must have reached conditions of at least lower granulite facies. Only under these conditions is it possible to generate the anhydrous mineral assemblages characteristic for granulite terranes and xenoliths derived from the deep crust.

Under granulite facies conditions, widespread anhydrous mineral assemblages without wholesale melting can only be generated by reducing the activity of H_2O [a_{H_2O}] significantly below unity. This can be achieved by diluting the vapor phase with other volatiles, the most likely candidate being CO_2 (e.g., Newton et al., 1980). Another possibility is to remove H_2O with a granitic melt or by passage of H_2O-undersaturated magmas. Experimental evidence (e.g., Huang and Wyllie, 1975) (Fig. 12-1) and the widespread occurrence of migmatites at the amphibolite–granulite transition zone indicate that during prograde metamorphism, the first formed granitic melts form under upper-amphibolite facies conditions when a H_2O-dominated fluid phase is present. These first formed melts will be H_2O-saturated and therefore are difficult to remove (Wickham, 1987). In order to lower a_{H_2O} significantly below unity it is necessary to generate H_2O-undersaturated melts that can migrate to higher crustal levels and remove H_2O. Extraction of these melts will leave a residue depleted in volatiles and a granitic component and to a varying degree in the heat-producing elements U, Th, K and Rb. Once a_{H_2O} has dropped significantly below unity or vapor-absent conditions have been reached, partial melting is only possible by the breakdown of hydrous phases such as biotite or hornblende. Under vapor-absent conditions, significant amounts of melts will be generated by the dehydration melting reactions involving the breakdown of biotite only above ca. 850°C (Vielzeuf and Holloway, 1988). Such temperatures are recorded in only a few granulite terranes (Harley, 1989). Since granulite facies terranes that yield maximum metamorphic temperatures of <850°C show a depletion in volatiles and in some cases also in heat-producing elements (e.g., Newton, 1987), it seems that partial melting under upper-amphibolite to granulite grade conditions combined with devolatilization by the breakdown of hydrous phases (biotite, cordierite, hornblende) and carbonates may be sufficient to cause this observed depletion. The suggestion that extraction of CO_2-rich fluids may lead to a depletion in LIL elements (e.g., Collerson and Fryer, 1978) must await the experimental study of reliable distribution coefficients between silicate minerals and CO_2-rich fluids.

Most of the stable continental shield areas expose low-grade metamorphic rocks, and in many cases these rocks are covered by a young thin veneer of sedimentary rocks. This suggests that large amounts of uplift are atypical for the major part of the stable continental crust and that the lower continental crust typically shows long-term geologic stability. Rapid uplift (200–1000 m/Ma) seems to be restricted to young tectonically active orogens (Wagner et al., 1977; Hollister, 1982; Zeitler et al., 1982; Koons, 1987). In contrast uplift rates in terranes that expose middle and lower crustal rocks seem to be typically smaller than 100 m/Ma (e.g., Cosca 1989; Mezger et al., 1990a). In young orogens, rapid uplift may be a consequence of thrusting in the upper parts of the crust or extension, possibly due to gravitational collapse (e.g., Dewey, 1988). The lower parts of the crust in these tectonicaly active terranes will respond by ductile deformation that results in only small pressure changes during orogenesis. In addition, rapid uplift can be only of short duration and is most likely restricted to individual thrust sheets and the time of active orogenesis. The lower crust in these terranes would show rapid post-orogenic erosional uplift only if the orogenesis led to the formation of thick felsic crust (i.e. >50 km). If such strongly thickened crustal was generated during an orogeny this would lead to the exposure of regional granulites within a few tens of Ma following the peak of metamorphism. Since vast parts of the stable continental crust exposes only low-grade metamorphic rocks rapid and extensive uplift immediately following the cessation of an orogeny may be atypical for the major parts of the stabilized crust.

Granulites represent some of the deepest crustal material accessible. On a large scale, only regional granulite terranes show the depletion in heat-producing elements and volatiles and possibly the extraction of a granitic component. These terranes also show the high-grade thermal history and long-term tectonic stability that seem to be characteristic for the lower crust. For these reasons, it is generally assumed that granulites can be used as models for the evolution of the lowermost continental crust. However, not all exposed granulites are suitable since some of them do not show a depletion in radioactive elements or a granitic component (Weaver, 1980; Sighinolfi et al., 1981; Weaver and Tarney, 1984; Rudnick et al., 1985; Newton, 1987) that seem to be characteristic for the deep continental crust. In addition mineral assemblages from most regional granulite terranes record metamorphic pressures that suggest they do not represent exhumed lowermost crust (Fig. 12-1) (Perkins and Newton, 1981; Newton and Perkins; 1982; Newton, 1987; Bohlen and Mezger, 1989; Harley, 1989) and, therefore, not all granulites may be suitable as models for the lowermost continental crust. If their composition is not typical for the composition of the lowermost crust, the processes that formed these now exposed terranes may be similar to those operating in the deep crust or may at least give an indication of how these processes work. Although, it is likely that exposed regional granulites and the lower crust are not identical (Bohlen and Mezger, 1989), it is anticipated that they will have at least some common characteristics, particularly with respect to their chemistry and history, and that a study of regional granulites can give some important insight into the evolution of the lowermost continental crust.

5. Discussion

From the detailed studies of the three regional granulite terranes it is evident that in each area stable lower continental crust evolved over an extended period of time (several hundred Ma) and experienced several major tectonothermal episodes. During these episodes magmatism played an important role in generating new crust and modifying it to develop stable crust.

The isobaric cooling paths for the terranes indicate that the terranes did not obtain a very thick crust during high-grade metamorphism. The time-integrated uplift rates of less than ca.

100 m/Ma (Harley, 1987a; Mezger et al., 1990a) following the last peak of metamorphism are significantly slower than those observed in recent tectonically active orogenic belts with over-thickened crust (Wagner et al., 1977; Zeitler et al., 1982). The slow time-integrated cooling rates of <3°C/Ma indicate that the granulite terranes cooled at great depth. The slow cooling is also consistent with slow erosional uplift, although uplift may have been caused by later tectonic activities that were unrelated to the processes that led to the formation of the granulites.

In order to achieve the high temperatures recorded in granulite terranes without leading to a double-thickened crust it was necessary to add heat to the crust (e.g., Wells, 1980). A likely mechanism is underplating of mantle-derived melts and intrusion of magmas of deep crustal or mantle origin. Such syn-metamorphic intermediate to felsic intrusions are identified in the Adirondack and Pikwitonei granulites and the Napier Complex (Powers and Bohlen, 1985; Mezger et al., 1990a; Harley and Black, 1987; Ellis, 1987) but may be typical for most regional granulite terranes. Geothermometry on such synmetamorphic intrusions and their contact aureoles from the Pikwitonei granulite domain and the Adirondack terrane indicate that the minimum temperatures of these dry melts were 1100°C (Powers and Bohlen, 1985; Mezger et al., 1990a). Some of these melts may be derived directly from the mantle, but others may be derived from the lowermost crust by partial melting and may have formed as a consequence of underplating of mantle-derived melts. The vertical movement of primary mantle-derived melts and secondary crustal melts leads to an advective heat flow during metamorphism. This results in buffered temperatures in crustal regions where partial melting occurs and leads to a transient geothermal gradient that decreases with depth. Such thermal conditions have been suggested for the Napier Complex, the Grenville terrane and Pikwitonei granulite domain (Harley, 1987b; Anovitz and Essene, 1990; Mezger et al., 1990a).

Large amounts of partial melts would be produced if the geothermal gradient recorded in granulite terranes is stable, and is the result of dominantly conductive heat transfer. In this case, temperature increases almost linearly with depth as it is characteristic of tectonic environments where convective heat transport by upward-moving melts is insignificant (Chapman, 1986). In this situation even rocks of intermediate composition would undergo extensive partial melting. If the geothermal gradient is dominated by advective heat transport, the gradient will be steep at lower crustal levels and decrease strongly with depth, where the temperature may be buffered by partial melting over a significant depth range (Harley, 1987b; Vielzeuf and Holloway, 1988). In the latter case, rocks of intermediate composition will undergo only minor partial melting in the lowermost crust, provided $a_{H_2O} \ll 1$. This allows for the presence of mafic and also some intermediate rocks in the lowermost continental crust. Indeed, suites of deep crustal xenoliths tend to be magmatic in origin and intermediate to ultramafic in composition. Xenoliths from suites that are dominated by mafic rocks yield equilibration pressures typically in excess of 10 kbar whereas regional granulites yield pressures typically <9 kbar (Bohlen and Mezger, 1989). Xenoliths of metasedimentary composition that may be derived from the lowermost crust are extremely rare (Hayob et al., 1989; Rudnick, this volume).

All three regional granulite terranes described above are characterized by the dominance of magmatic over sedimentary rocks, suggesting an increasing importance of magmatism with crustal depth. As indicated by the geochronologic data, a large proportion of these igneous rocks may be synmetamorphic. This is consistent with the observation that during high-grade metamorphism heat appears to be added to the crust by underplating and intrusion of mafic to intermediate magmas. As a consequence it is expected that the lowermost crust is dominantly made up of mafic magmatic rocks. This would also imply that vertical crustal growth by underplating is a process of major importance for generating and modifying continental crust. Lateral accretion may be restricted to the early stages of crust formation, but does not seem to lead to stable continental crust, unless the lower crust is later overprinted by a high-T metamorphism. Such magmaticaly underplated terranes seem not to get exposed during the uplift of regional granulite terranes but may be represented by the deepest level rocks exposed in some crustal cross-sections like the Ivrea Zone or Doubtful Sound (Fountain and Salisbury, 1981; Percival et al., this volume).

If this model is correct, the lowermost continental crust is made up of dominantly juvenile mantle-derived material mixed with older crustal material. In terms of Nd-isotope systematics, the lower crustal rock will exhibit a significant spread on a f(Sm/Nd) versus ϵ_{Nd} diagram (Fig. 12-5) and thus in model ages. The nature of the crustal material will be a function of the prehistory of the crust (e.g., Dewey, 1986) and may involve dominantly rocks of plutonic and only rarely rocks of supracrustal origin.

One place where the results of magmatic underplating of mantle-derived magmas under pre-existing continental crust may be observed in situ, is the Ivrea zone (northern Italy) (Fountain, 1989; Voshage et al., 1990; Percival et al., this volume). In this area, mafic mantle-derived melts intruded along the contact between peridotite and metasediments. Repeated injection of mafic melts led to the formation of a long-lived magma chamber and continuous partial melting in the metapelitic country rock. Mixing of the crustal melts with the mafic melt and concomitant fractional crystallization led to a continuously differentiating magma chamber and a homogeneous isotopic signature in the evolving mafic layered intrusion. The isotopic values of the layered intrusion are between the values of the metasediments and uncontaminated mantle melts.

The widespread supracrustal rocks in the three granulite terranes discussed above seem to be significantly older than the orogeny that led to the major stabilization of the crust (Tables 12-1 to 12-3). The prograde metamorphic paths that could give an indication about the processes by which the supracrustal rocks were brought to their deep crustal levels are almost completely obscured by the high metamorphic grade. Only for the Pikwitonei granulite domain are there some traces preserved in the form of reaction textures and pseudomorphs that give some information on the prograde path (Mezger et al., 1990b).

An underplated mafic magmatic root may be a major reason for the difficulty in exposing lower crustal rocks, but also may account for its long-term stability. During cooling parts of this underplated crust may transform to eclogite and prevent further uplift (Furlong and Fountain, 1986; Griffin and O'Reilly, 1987). In the terranes discussed above, uplift seems to be unrelated to the tectonic processes that generated and stabilized the crust. The Napier Complex may have been exposed as a result of additional magmatic underplating at ca. 1100–900 Ma or underthrusting of the Proterozoic Rayner Complex (Sandiford, 1985; Harley, 1987b); the Pikwitonei domain may have been exposed around 1900–1700 Ma as a result rifting or the collision of the Superior with the Churchill Province; the mechanism for the uplift of the Adirondack granulites following the Grenville orogeny is not known.

Magmatic underplating and the resulting metamorphism imprints a layered character on the continental crust. The layering in the upper crust is dominated by the change in metamorphic grade rather than composition (see also Fountain and Salisbury, 1981). In contrast, the laminar structure in the lowermost crust may be more the result of changing composition and multiple magmatic intrusions at the base of the crust. Such a model is consistent with the less pronounced layering in the upper crust and the apparent laminar structures observed in the lowermost crust as revealed by seismic studies (e.g., Fountain, 1986).

The anhydrous character of the granulite terranes and the lower crust can be due to the predominance of magmatic rocks that were emplaced as dry melts. Water may have been expelled as a result of low pressure contact metamorphism at shallow levels during an earlier high-grade event as is suggested for the Adirondack granulites (e.g., Valley et al., 1990; Mezger et al., 1991). Significant dehydration that also leads to low a_{H_2O}, may be due to partial melting and removal of these melts, rather than large scale infiltration of CO_2-rich fluids. The dehydration and partial melting may also be responsible for the depletion of the lower crust in heat-producing elements particularly K. However, no satisfactory explanation has been proposed yet for the variation in the U/Th ratios observed in some granulite facies rocks (Vidal et al., 1988; Heier, 1973). Strongly variable U/Th ratios seem to be restricted to granulite-grade rocks, but do not occur in all high-grade terranes. Since U and Th have identical charge and similar ionic radii, partial melting may not be an efficient mechanism for the fractionation of U from Th. The observed variation in U/Th ratios in some granulite terranes may require the removal of a metamorphic fluid at upper-amphibolite to granulite grade or dehydration melting (Clemens and Vielzeuf, 1987). The mafic underplated rocks may have had low concentrations in LIL elements and some LIL elements may also have been removed to higher crustal levels with late differentiates leaving behind a depleted residue. It should also be noted that although depletion in heat-producing elements is a characteristic of the average lower crust (e.g., Taylor and McLennan, 1985) not all regional granulite terranes show this depletion [for a compilation see Newton (1987)]. Although the Pikwitonei is strongly depleted in U and Th as indicated by the Pb-isotope compositions of whole rocks (Mezger et al., 1989), such information is lacking for the Napier Complex and the Adirondack granulites. In a study of the heat production and thermal conductivity of rocks from the Pikwitonei granulite domain and the adjacent Cross Lake subprovince (Manitoba), Fountain et al. (1987) found that heat production in tonalitic gneisses decreases by a factor of two from the amphibolite to granulite terrane. The heat production in the mafic gneisses is unchanged. This observation strongly suggests that in the Pikwitonei granulite domain the depletion in the heat-producing elements is the result of partial melting. Currently, there is no satisfactory explanation for the depletion of the lower crust and the varying degrees of depletion observed in granulite terranes. So far no explanation has found acceptance for the extreme depletion of the Lewisian granulite-facies gneisses of northern Scotland, despite the fact that this is, in terms of geochemical studies, the best studied granulite terrane in the world. It is still not clear whether the depleted character of these granulites is due to the extraction of a melt, a fluid or a primary feature (Pride and Muecke, 1980; Weaver and Tarney, 1981, 1983; Tarney and Weaver, 1987).

It is evident from this discussion that regional granulites tend to be polymetamorphic and can have been part of different tectonic settings at different times [see also Fountain and Salisbury (1981)]. The major requirement for them to end up in the lower parts of stable continental crust seems to be the extraction of a granitic component, the reduction of a_{H_2O} as well as some depletion in the heat-producing elements. This variety in tectonic histories implies that there is no unique way of forming stable lower continental crust, but intrusion of mafic and extraction of more felsic magmas seems to play a major role. Although the terranes discussed above can have certain attributes that are thought to be typical for the lowermost crust, none of them should be considered to be representative for the total lowermost continental crust. In addition, it is necessary to evaluate whether the composition and temporal evolution of granulites can be reconciled with the properties inferred for the lower crust. Therefore, it is not advisable to make broad generalizations before several terranes are studied in detail and comparisons are possible. Unfortunately, no complete data set that includes *P–T* estimates, major- and trace-element data on a large variety of rocks and detailed geochronological studies, is available for any regional granulite terrane. For most of the terranes *P–T* estimates are published, but for only a few there are sufficient geochemical data available to evaluate their state of depletion in LIL elements and even fewer have detailed geochronological data to allow the reconstruction of the major part of their geologic history. This makes it particularly difficult to compare granulite terranes and to derive information that is generally valid for the lower crust.

It is probably safe to consider most regional granulite terranes to represent the transition between the upper crust and the lowermost crust. Therefore, the processes they record can be used to elucidate the temporal evolution of the lowermost crust, whereas their composition does not directly reflect the composition of the lowermost crust. The regional granulite terranes discussed in this chapter show that the lower and middle continental crust has been formed by prolonged geologic processes involving magmatism, metamorphism and tectonism. Many of these processes occurred at high temperatures and, thus, later events may have eradicated information of the early processes. Only careful investigation of selected components can give information of the early processes. In comparison with the upper crust, the lower crust has had an equally complex geologic history and cannot be seen as an uniform component or reservoir that formed in a specific way. Because the history of the crust has significant influence on its petrogenesis and thus composition, it will be essential for future research to address the question of absolute timing of geologic events in the studies of regional granulite terranes. Since the geologic histories for only a few high-grade terranes are reconstructed so far, it is also not known whether processes that led to the generation of the lower crust change over geologic time.

Acknowledgments

This study was supported by NSF grants EAR 84-16250 and EAR 86-15714 to S.R. Bohlen, EAR 86-07973 to G.N. Hanson and EAR 88-05083 to E.J. Essene. S.R. Bohlen, D.M. Davis, C.P. DeWolf, E.J. Essene, G.N. Hanson, K.V. Hodges, E.J. Krogstad, and S.M. McLennan read various versions of this paper and provided helpful suggestions. The review by D.M. Fountain lead to the clarification of several important statements. B.R. Frost also reviewed the paper.

References

Anovitz, L.A. and Essene, E.J., 1990. Thermobarometry and pressure–temperature paths in the Grenville Province of Ontario. J. Petrol., 31: 197–241.

Barnicoat, A.C., 1987. The causes of high-grade metamorphism of the Scourian complex, NW Scotland. In: R.G Park and J. Tarney (Editors), Evolution of the Lewisian and Comparable Precambrian High Grade Terrains. Geol. Soc. London, Spec. Publ., 27: 73–79.

Black, L.P., Harley, S.L., Sun, S.S. and McCulloch, M.T., 1987. The Rayner Complex of East Antarctica: complex isotopic systematics within a Proterozoic mobile belt. J. Metamorph. Geol., 4: 1–26.

Black, L.P., James, P.R. and Harley, S.L., 1983. The geochronology, structure and metamorphism of early Archaean rocks at Fyfe Hills, Enderby Land, Antarctica. Precambrian Res., 21: 197–222.

Black, L.P. and McCulloch, M.T., 1987. Evidence for isotopic equilibration of Sm-Nd whole-rock systems in early Archaean crust of Enderby Land, Antarctica. Earth Planet. Sci. Lett., 82: 15–24.

Black, L.P., Sheraton, J.W. and James, P.R., 1986a. Late Archean granites of the Napier Complex, Enderby Land, Antarctica: A comparison of Rb-Sr, Sm-Nd and U-Pb isotopic systematics in a complex terrain. Precambrian Res., 32: 343–368.

Black, L.P., Williams, I.S. and Compston, W., 1986b Four zircon age from one rock: the history of a 3930 Ma-old granulite from Mount Sones, Enderby Land, Antarctica. Contrib. Mineral. Petrol., 94: 427–437.

Bohlen, S.R., 1987. Pressure–temperature–time paths and a tectonic model for the evolution of granulites. J. Geol., 95: 617–632.

Bohlen, S.R., 1990. On the formation of granulites. J. Metamorph. Geol., 9: 223–229.

Bohlen, S.R. and Mezger, K., 1989. Origin of granulite terranes and the formation of the lowermost continental crust. Science, 244: 326–329.

Bohlen, S.R., Valley, J.W. and Essene E.J., 1985. Metamorphism in the Adirondacks. I. Petrology, pressure and temperature. J. Petrol., 26: 971–992.

Brady, J.B., 1988. The role of volatiles in the history of metamorphic terranes. J. Petrol., 29: 1187–1213.

Chacko, T., Ravindra Kumar, G.R. and Newton, R.C., 1987. Metamorphic $P–T$ conditions of the Kerala, South India Khondalite Belt, a granulite facies supracrustal terrain. J. Geol., 95: 343–358.

Chamberlain, C.P. and Rumble, D., 1988. Thermal anomalies in a regional metamorphic terrane. An isotopic study of the role of fluids. J. Petrol., 29: 1215–1232.

Chapman, D.S., 1986. Thermal gradients in the continental crust. In: J.B. Dawson, D.A. Carswell, J. Hall and K.H. Wedepohl (Editors), The Nature of the Lower Continental Crust. Geol. Soc. London, Spec. Publ., 24: 63–70.

Clemens, J.C. and Vielzeuf, D., 1987. Constraints on melting and magma production in the crust. Earth Planet. Sci. Lett., 86: 287–306.

Collerson, K.D. and Fryer, B.J, 1978. The role of fluids in the formation and subsequent development of early continental crust. Contrib. Mineral. Petrol., 67: 151–167.

Coolen, J.J.M., 1980. Chemical petrology of the Furua granulite complex, southern Tanzania. GUA (Amsterdam), 13: 1–258.

Corfu, F. and Ayres, L.D., 1984. U-Pb age and genetic significance of heterogeneous zircon populations in rocks from the Favourable Lake area northwestern Ontario. Contrib. Mineral. Petrol., 88: 86–101.

Cosca, M.A., 1989. Cooling and inferred uplift/erosion history of the Grenville orogen, Ontario: Constraints from $^{40}Ar/^{39}Ar$ thermochronology. PhD thesis, Univ. Michigan, 223 pp.

Davidson, P.M. and Lindsley, D.H., 1985. Thermodynamic analysis of quadrilateral pyroxenes. Part II. Model calibration from experiments and application to geothermometry. Contrib. Mineral. Petrol., 91: 390–404.

Dewey, J.F., 1986. Diversity in the lower continental crust. In: J.R. Dawson, D.A. Carswell, J. Hall and K.H. Wedepohl (Editors), The Nature of the Lower Crust. Geol. Soc. London, Spec. Publ., 24: 71–78.

Dewey, J.F., 1988. Extensional collapse of orogens. Tectonics, 7: 1123–1139.

Edwards, R.L. and Essene, E.J., 1988: Pressure, temperature, and C-O-H fluid fugacities across the amphibolite–granulite transition, N.W. Adirondack Mtn., NY. J. Petrol., 29: 39–72.

Ellis, D.J., 1980. Osumilite-sapphirine granulites from Enderby Land, Antarctica: P–T conditions and metamorphism, implications for garnet–cordierite equilibria and the evolution of the deep crust. Contrib. Mineral. Petrol., 74: 201–210.

Ellis, D.J., 1983. The Napier and Rayner Complexes of Enderby Land, Antarctica — Contrasting styles of metamorphism and tectonism. In: R.L. Oliver, P.R. James and J.B. Jago (Editors), Antarctic Earth Science. Aust. Acad. Sci., Canberra, pp. 20–24.

Ellis, D.J., 1987. Origin and evolution of granulites in normal and thickened crusts. Geology, 15: 167–170.

Ellis, D.J., Sheraton, J.W., England, R.N. and Dallwitz, W.B., 1980. Osumilite-sapphirine-quartz granulites from Enderby Land, Antarctica — mineral assemblages and reactions. Contrib. Mineral. Petrol., 72: 123–143.

Ermanovics, I.F. and Davison, W.L., 1976. The Pikwitonei granulites in relation to the northwestern Superior Province of the Canadian Shield. In: B.F. Windley (Editor), The Early History of the Earth. Wiley, New York, N.Y., pp. 331–347.

Essene, E.J., 1989. Thermobarometry of metamorphic rocks. In: S.R. Daly, R. Cliff and B. Yardley (Editors), Evolution of Metamorphic Belts. J. Geol. Soc., Spec. Pap., 41: 1–65.

Ferry, J.M. and Burt, D.M., 1982. Characterization of metamorphic fluid composition through mineral equilibria. In: J.M. Ferry (Editor), Characterization of Metamorphism through Mineral Equilibria. Reviews in Mineralogy,

10: 207–262.

Fountain, D.M., 1986. Implications of deep crustal evolution for seismic reflection interpretation. In: Reflection Seismology: The Continental Crust. Am. Geophys. Union, Geodyn. Ser., 14: 1–7.

Fountain, D.M., 1989. Growth and modification of lower continental crust in extended terrains: The role of extension and magmatic underplating. Am. Geophys. Union, Monogr., 51: 287–299.

Fountain, D.M. and Salisbury, M., 1981. Exposed cross-sections through the continental crust: implications for crustal structure, petrology, and evolution. Earth Planet. Sci. Lett., 56: 263–277.

Fountain, D.M., Salisbury, M. and Furlong, K.P., 1987. Heat production and thermal conductivity of rocks from the Pikwitonei–Sachigo continental cross section, central Manitoba: implications for the thermal structure of Archean crust. Can. J. Earth Sci., 24: 1583–1594.

Freer, R., 1981. Diffusion in silicate minerals and glasses: a digest and guide to the literature. Contrib. Mineral. Petrol., 76: 440–454.

Frost, B.R. and Chacko T., 1989. The granulite uncertainty principle: limitations on thermobarometry in granulites. J. Geol., 97: 435–450.

Frost, B.R. and Frost, CD., 1987. CO_2, melts and granulite metamorphism. Nature, 327: 503–506.

Fuhrman, M.L. and Lindsley, D.H., 1988. Ternary feldspar modelling and thermometry. Am. Mineral., 73: 201–215.

Furlong, K.P and Fountain, D.M., 1986: Continental crustal underplating: Thermal considerations and seismic-petrologic consequences. J. Geophys. Res., 91: 8285–8294.

Ganguly, J. and Saxena, S.K., 1984. Mixing properties of aluminosilicate garnets: constraints from natural and experimental data and applications to geothermo-barometry. Am. Mineral., 69: 88–97.

Green, A.G., Hajnal, Z. and Weber W., 1985. An evolutionary model of the western Churchill Province and the western margin of the Superior Province in Canada and north-central United States. Tectonophysics, 116: 281–322.

Grew, E.S., 1980. Sapphirine+quartz associations from Archean rocks in Enderby Land, Antarctica. Am. Mineral., 65: 821–836.

Grew, E.S., 1982. Osumilite in the sapphirine-quartz terrane of Enderby Land, Antarctica: implications for osumilite petrogenesis in the granulite facies. Am. Mineral., 67: 762–787.

Griffin, W.L. and O'Reilly, S.Y., 1987. Is the continental Moho the crust–mantle boundary. Geology, 15: 241–244.

Hansen, E.C., Newton, R.C. and Janardhan, A.S., 1984. Pressures, temperatures and metamorphic fluids across an unbroken amphibolite facies to granulite facies transition in southern Karnataka. In: A. Kröner, G.N. Hanson and A.M. Goodwin (Editors), Archean Geochemistry. Springer Verlag, Berlin, pp. 161–181.

Harley, S.L., 1985a. Garnet-orthopyroxene bearing granulites from Enderby Land, Antarctica: metamorphic pressure–temperature–time evolution of the Archean Napier Complex. J. Petrol., 26: 819–856.

Harley, S.L., 1985b. Paragenetic and mineral–chemical relationships in orthoamphibole-bearing gneisses from Enderby Land, east Antarctica: a record of Proterozoic uplift. J. Metamorph. Geol., 3: 179–200.

Harley, S.L., 1987a. A pyroxene-bearing meta-ironstone and other pyroxene-granulites from Tonagh Island, Enderby Land, Antarctica: Further evidence for very high temperature (>980°C) Archaean regional metamorphism in the Napier Complex. J. Metamorph. Geol., 5: 341–356.

Harley, S.L., 1987b. Long-lived continental stability in east Antarctica: Evidence and implications from metamorphic P–T–t paths. Terra Cognita, 7: 615.

Harley, S.L., 1989. The origins of granulites: a metamorphic perspective. Geol. Mag. 126: 215–247.

Harley, S.L. and Black, L.P., 1987. The Archean geological evolution of Enderby Land, Antarctica. In: R.G. Park and J. Tarney (Editors), Evolution of the Lewisian and Comparable Precambrian High Grade Terrains. Geol. Soc. London, Spec. Publ., 27: 285–296.

Hayob, J.L., Essene, E.J., Ruiz, J. and Ortega-Gutierrez F., 1989. Young high-temperature granulites from the base of the crust in central Mexico. Nature, 342: 265–268.

Heaman. L., Machado, N., Krogh, T.E. and Weber, W., 1986a. Preliminary U-Pb zircon results from the Pikwitonei granulite domain, Manitoba. Geol. Assoc. Can.–Mineral. Assoc. Can. Joint Annu. Meet. 1986, Ottawa, Ont., Program Abstr., 11: 79.

Heaman, L., Machado, N., Krogh, T.E. and, Weber W., 1986b. Precise U-Pb zircon ages for the Molson dyke swarm and the Fox River sill. Implications for Early Proterozoic crustal evolution in NE Manitoba, Canada. Contrib. Mineral. Petrol., 94: 82–89.

Heier, K.S., 1973. Geochemistry of granulite facies rocks and problems of their origin. Philos. Trans. R. Soc. London, Ser. A, 273: 429–442.

Heizler, M. and Harrison, T.M., 1986. Chronology of the Adirondacks epeirogeny. EOS, Trans. Am. Geophys. Union, 67: 400.

Hodges, K.V., Bartley, J.M. and Burchfiel, B.C., 1982. Structural evolution of an A-type subduction zone, Lofoten–Rombak area, Northern Scandinavian Caledonides. Tectonics, 1: 441–462.

Hollister, L.S., 1982. Metamorphic evidence for rapid (2mm/yr) uplift of a portion of the Central Gneiss Complex, Coast Mountains, B.C. Can. Mineral., 20: 319–332.

Hörmann, P.K., Raith, M., Raase, P., Ackermand, D. and Seifert, F., 1980. The granulite complex of Finnish Lapland: petrology and metamorphic conditions in the Ivalojoki–Inarjärvi area. Geol. Surv. Finl. Bull., 308: 1–95.

Huang, W.-L. and Wyllie, P.J., 1975. Melting reactions in the system $NaAlSi_3O_8$–$KAlSi_3O_8$–SiO_2 to 35 kilobars, dry and with excess water. J. Geol., 83: 737–748.

Hubregtse, J.J.M.W., 1980. The Archean Pikwitonei granulite domain and its position at the margin of the northwestern Superior Province (central Manitoba). Dept. Energy Mines Manitoba, Geol. Pap., GP80-3.

Janardhan, A.S., Newton, R.C. and Smith, J.V., 1979. Ancient crustal metamorphism at low P_{H_2O}; charnockite formation at Kabbaldurga, S. India. Nature, 278: 511–514.

Janardhan, A.S., Newton, R.C. and Hansen, E.C., 1982. The transformation of amphibolite facies gneiss to charnockite in southern Karnataka and northern Tamil Nadu, India. Contrib. Mineral. Petrol., 79: 130–149.

Kay, R.W. and Kay, S.M., 1981. The nature of the lower continental crust. Rev. Geophys. Space Phys., 19: 271–297.

Kay, S.M. and Kay, R.W., 1983. Thermal history of the deep crust inferred from granulite xenoliths, Queensland, Australia. Am. J. Sci., 283a: 486–513.

Koons, P.O., 1987. Some thermal and mechanical consequences of rapid uplift: an example from the southern Alps, New Zealand. Earth Planet. Sci. Lett., 86: 307–319.

Krogh, T.E., Heaman, L., Machado. N., Davis, D. and Weber W., 1986. U-Pb geochronology program, Northern Superior Province. Dept. Energy Mines Manitoba, Rep. Field Act., 1986: 178–180.

Kröner, A., Williams, I.S., Compston. W., Baur, N., Vitanage, P.W. and Perera, L.R.L., 1987. Zircon ion microprobe dating of high-grade rocks in Sri Lanka. J. Geol., 95: 775–791.

Kuehner, S.M. and Green, D.H., 1987. Uplift history of the East Antarctic Shield: constraints imposed by high pressure experimental studies of Proterozoic mafic dykes. 5th Int. Symp. Antarctic Earth Sci. (abstracts), Cambridge, p. 84.

Lamb, W. and Valley J.W., 1984. Metamorphism of reduced granulites in low-CO_2, vapor-free environments. Nature, 312: 56–58.

Machado, N., Heaman, L., Krogh, T.E. and Weber, W., 1987. U-Pb Geochronology Program: Thompson belt — northern Superior Province. Dept. Energy Mines Manitoba, Rep. Field Act., 1987: 145–147.

McCulloch, M.T. and Black, L.P., 1984. Sm-Nd isotopic systematics of Enderby Land granulites and evidence for the redistribution of Sm and Nd during metamorphism. Earth Planet. Sci. Lett., 71: 46–58.

McLelland, J.M., 1986. Pre-Grenvillian history of the Adirondacks as an orogenic bimodal caldera complex of mid-Proterozoic age. Geology, 14: 229–233.

McLelland, J.M. and Chiarenzelli, J., 1990. Isotopic constraints on the emplacement age of anorthositic rocks of the Marcy Massif, Adirondack Mts. New York. J. Geol., 98: 19–41.

McLelland, J.M., Chiarenzelli, J., Whitney, P. and Isachsen, Y., 1988. U-Pb zircon geochronology of the Adirondack Mountains and implications for their tectonic evolution. Geology, 16: 920–924.

McLelland, J.M. and Isachsen, Y., 1986. Geological synthesis of the Adirondack Mts. and their tectonic setting within the SW Grenville Province. In: J. Moore, A. Davidson and A. Baer (Editors), The Grenville Province: New Perspectives. Geol. Assoc. Can., Spec. Pap., 31: 75–94.

Mezger, K., Hanson, G.N. and Bohlen, S.R., 1989. U-Pb systematics of garnet: Dating the growth of garnet in the late Archean Pikwitonei granulite domain at Cauchon and Natawahunan Lakes, Manitoba, Canada. Contrib. Mineral. Petrol. 101: 136–148.

Mezger, K., Bohlen, S.R. and Hanson, G.N., 1990a. Metamorphic history of the Archean Pikwitonei granulite domain and the Cross Lake subprovince, Superior Pronvince, Manitoba, Canada. J. Petrol., 31: 483–517.

Mezger, K., Hanson, G.N. and Bohlen, S.R., 1990b. U-Pb ages of metamorphic rutiles: Application to the cooling history of high grade terranes. Earth Planet. Sci. Lett., 96: 106–118.

Mezger, K., Rawnsley, C.M., Bohlen, S.R. and Hanson, G.N., 1991. U-Pb garnet, sphene, monazite and rutile ages: Implications for the duration of high-grade metamorphism and cooling histories, Adirondack Mts., New York. J. Geol., 99: 415–428.

Moorbath, S., Taylor, P.N. and Jones, N.W., 1986. Dating the oldest terrestrial rocks — fact and fiction. Chem. Geol., 57: 63–86.

Nelson, B.K. and DePaolo, D.J., 1985. Rapid production of continental crust 1.7 to 2.9 b.y. ago: Nd isotopic evidence from the basement of the North American mid-continent. Geol. Soc. Am. Bull., 96: 746–754.

Newton, R.C., 1985. Temperature, pressure and metamorphic fluid regimes in the amphibolite facies to granulite facies transition zone. In: A.C. Tobi and J. Touret (Editors), The Deep Proterozoic Crust in the North Atlantic Provinces. Reidel, Dordrecht, pp. 75–104.

Newton R,C., 1987. Petrologic aspects of Precambrian granulite facies terrains bearing on their origins. In: A. Kröner (Editor), Proterozoic Lithospheric Evolution. Am. Geophys. Union, Geodyn. Ser., 17: 11–26.

Newton, R.C., 1989. Metamorphic fluids in the deep crust. Annu. Rev. Earth Planet. Sci., 17: 385–412.

Newton, R.C. and Perkins D., 1982. Thermodynamic calibration of geobarometers based on the assemblages garnet–plagioclase–orthopyroxene (clinopyroxene)–quartz. Am. Mineral., 67: 203–222.

Newton, R.C., Smith, J.V. and Windley B.F., 1980. Carbonic metamorphism, granulites, and crustal growth. Nature, 288: 45–50.

Peacock, S.M., 1987. The thermal effects of metamorphic fluids in subduction zones. Geology, 15: 1057–1060

Perchuck, L.L., Aranovich, L.Ya., Podlesskii, K.K., Lavrent'eva, I.V., Gerasimov, V.Yu., Fed'kin, V.V., Kitsul, V.I., Karsakov, L.P. and Berdnikov, N,V., 1985. Precambrian granulites of the Aldan Shield, eastern Siberia, USSR. J. Metamorph. Geol., 3: 265–310.

Perkins. D. and Newton, R.C., 1981. Charnockite geobarometers based on coexisting garnet–pyroxene–plagioclase–quartz. Nature, 292: 144–146.

Peterson, J.W. and Newton, R.C., 1989. CO_2-enhanced melting of biotite-bearing rocks at deep-crustal pressure–temperature conditions. Nature, 340: 378–580.

Powell, R., 1983. Processes in granulite-facies metamorphism. In: M.P. Atherton and C.D. Gribble (Editors), Migmatites, Melting and Metamorphism. Shiva Publ., Nantwich, Cheshire, pp. 127–139.

Powell, R. and Holland, T.J.B., 1988. An internally consistent data set with uncertainties and correlations. 3. Applications to geobarometry, worked examples and computer program. J. Metamorph. Geol., 6: 173–204.

Powers, R.E. and Bohlen, S.R., 1985. The role of synmetamorphic igneous rocks in the metamorphism and partial melting of meta-sediments, northwest Adirondacks. Contrib. Mineral. Petrol., 90: 401–409.

Pride, C. and Muecke, G.K., 1980. Rare earth element geochemistry of the Scourian complex NW Scotland — evidence for the granite–granulite link. Contrib. Mineral. Petrol., 73: 403–412.

Raith, M., Hengst, C., Nagel, B., Bhattacharya, A. and Srikantappa, C., 1988. Metamorphic conditions in the Nilgiri granulite terrane and the adjacent Moyar shear zones: A reevaluation. J. Geol. Soc. India, 31: 112–113.

Richardson, S.W., 1970. The relationship between petrogenetic grid, facies series and geothermal gradient in metamorphism. Fortschr. Mineral., 47: 65–76

Roy, R.F., Blackwell, D.D. and Birch, F., 1968. Heat generation of plutonic rocks and continental heat flow provinces. Earth Planet. Sci. Lett., 5: 1–12.

Rudnick, R.L., McLennan, S.M. and Taylor, S.R., 1985. Large ion lithophile elements in rocks from high pressure granulite facies terranes. Geochim. Cosmochim. Acta, 49: 1645–1655.

Rudnick, R.L. and Taylor, S.R., 1987. The composition and petrogenesis of the lower crust: A xenolith study. J. Geophys. Res., 92: 13981–14005.

Rudnick, R.L. and Williams, I.S., 1987. Dating the lower crust by ion microprobe. Earth Planet. Sci. Lett., 85: 145–161.

Sandiford, M., 1985. The metamorphic evolution of granulites at Fyfe Hills, implications for Archean crustal thickness in Enderby Land, Antarctica. J. Metamorph. Geol., 3: 155–178.

Sandiford, M. and Powell, R., 1986a. Deep crustal metamorphism during continental extension: modern and ancient examples. Earth Planet. Sci. Lett., 79: 151–158.

Sandiford, M. and Powell, R., 1986b. Pyroxene exsolution in granulites from Fyfe Hills, Enderby Land, Antarctica: Evidence for 1000°C metamorphic temperatures in Archean continental crust. Am. Mineral., 71: 946–954..

Sandiford, M., Powell, R., Martin, S.F. and Perera, L.R.K., 1988. Thermal and baric evolution of granulites from Sri Lanka. J. Metamorph. Geol., 6: 351–364.

Schenk, V., Hölzl, S., Köhler, H., Raase, P. and Schumacher, R., 1990. Metamorphic zonation and P–T history of the Highland Series in Sri Lanka. Terra Abstr., 2: 48.

Schenk, V., Raase, P. and Schumacher, R., 1988. Very high temperatures and isobaric cooling before tectonic uplift in the Highland Series in Sri Lanka. Terra Cognita, 8: 264.

Sheraton, J.W. and Black, L.P., 1983. Geochemistry of Precambrian gneisses: relevance for the evolution of the East Antarctic shield. Lithos, 16: 273–296.

Sheraton, J.W., Offe, L.A., Tingey, R.J. and Ellis, D.J., 1980. Enderby Land, Antarctica — an unusual Precambrian high-grade metamorphic terrain. J. Geol. Soc. Aust., 27: 1–18.

Shirey, S.B. and Hanson, G.N., 1986. Mantle heterogeneity and crustal recycling in Archean granite–greenstone belts: evidence from Nd isotopes and trace elements in the Rainy Lake area, Ontario. Geochim. Cosmochim. Acta, 50: 2631–2651.

Sighinolfi, G.P., Figueredo, M.C.H., Fyfe, W.S., Kronberg, B.I. and Tanner Oliveria, M.A.F., 1981. Geochemistry and petrology of the Jequie granulite complex (Brazil). Contrib. Mineral. Petrol., 78: 263–271.

Silver, L., 1969. A geochronologic investigation of the anorthosite complex, Adirondack Mts., NY. In: Y. Isachsen (Editor), Origin of Anorthosites and Related Rocks. New York State Mus. Sci. Serv. Mem., 18: 233–252.

Stähle, H.J., Raith, M., Hoernes, S. and Delfs, A., 1987. Elemental mobility during incipient granulite formation at Kabbaldurga, southern India. J. Petrol., 28: 803–830.

Stosch, H.-G. and Lugmair, G.W., 1984. Evolution of the lower continental crust: granulite facies xenoliths from the Eifel, West Germany. Nature, 311: 368–370.

Tarney, J. and Weaver, B.L., 1987. Geochemistry of the Scourian complex: petrogenesis and tectonic models. In: Park, R.G. and Tarney, J., (Editors), Evolution of the Lewisian and Comparable Precambrian High Grade Terrains. Geol. Soc. London Spec. Publ., 27: 45–56.

Taylor, S.R. and McLennan, S.M., 1985. The Continental Crust: Its Composition and Evolution. Blackwell, Oxford, 312 pp.

Touret, J., 1971. Le facies granulite et Norvege meridionale. Association minerales et inclusions fluides. Lithos, 4: 239–249 and 423–435

Touret, J., 1987. Fluid distribution in the continental crust. In: A. Kröner (Editor), Proterozoic Lithospheric Evolution. Am. Geophys. Union, Geodyn. Ser., 17: 27–33.

Valley, J.W., 1985. Polymetamorphism in the Adirondacks: Wollastonite at contacts of shallowly intruded anorthosite. In: A.C. Tobi and J. Touret (Editors), The Deep Proterozoic Crust in the North Atlantic Provinces. Reidel, Dordrecht, pp. 217–236.

Valley, J.W., Bohlen, S.R., Essene, E.J., and Lamb, W., 1990. Metamorphism in the Adirondacks: II. The role of fluids. J. Petrol., 31: 555–596.

Valley, J.W., McLelland, J., Essene, E.J., and Lamb, W., 1983. Metamorphic fluids in the deep crust: evidence from the Adirondacks. Nature, 301: 226–228.

Valley, J.W. and O'Neil, J.R., 1982. Oxygen isotope evidence for shallow emplacement of Adirondack anorthosite. Nature, 300: 497–500.

Valley, J.W. and O'Neil JR., 1984. Fluid heterogeneity during granulite facies metamorphism in the Adirondacks: stable isotope evidence. Contrib. Mineral. Petrol., 85: 158–173.

Vidal, Ph., Peucat, J.J., Bernard-Griffiths, J. and Condie, K.C., 1988. Sr, Nd and Pb isotope systematics of Archean low- to high-grade transition zone of southern India. Terra Cognita, 8: 262.

Vielzeuf, D. and Holloway, J.R., 1988. Experimental determination of the fluid-absent melting relations in the pelitic system. Consequences for crustal differentiation. Contrib. Mineral. Petrol., 98: 257–276.

Voshage, H., Hofmann, A.W., Raczek, I., Feldmann, H., Mazzucchelli, Rivalenti, G. and Sinigoi, S., 1990. Evolution of a lower crustal magma chamber, Ivrea Zone, Varallo, Italy. Terra Abstr., 2: 49.

Vry, J., Brown, P.E., Valley, J.W. and Morrison, J., 1988. Constraints on granulite genesis from carbon isotope compositions of cordierite and graphite. Nature, 332: 66–68.

Wagner, G.A., Reimer, G.M. and Jäger E., 1977. Cooling ages derived by apatite fission track, mica Rb-Sr and K-Ar dating: the uplift and cooling history of the Central Alps. Mem. Ist. Geol. Petrol. Univ. Padova., 30: 1–28.

Waters, D.J., 1985. Metamorphic zonation and thermal history of pelitic gneisses from Western Namaqualand, South Africa. Trans. Geol. Soc. S. Afr., 88: 323–335.

Waters, D.J., 1986. Metamorphic history of sapphirine-bearing and related magnesian gneisses from Namaqualand, South Africa. J. Petrol., 27: 541–565.

Waters, D.J. and Whales, C.J., 1984. Dehydration melting and the granulite transition in metapelites from southern Namaqualand, S. Africa. Contrib. Mineral. Petrol., 88: 269–265.

Weaver, B.L., 1980. Rare-earth element geochemistry of Madras granulites. Contrib. Mineral. Petrol., 71: 271–279.

Weaver, B.L. and Tarney, J., 1981. Lewisian gneiss geochemistry and Archean crustal development models. Earth Planet. Sci. Lett., 55: 171–180.

Weaver, B.L. and Tarney, J., 1983. Elemental depletion in Archean granulite facies rocks. In: M.P. Atherton and C.D. Gribble (Editors), Migmatites, Melting and Metamorphism. Shiva Publ., Nantwich, Cheshire, pp. 250–263.

Weaver, B.L. and Tarney, J., 1984. Major and trace element composition of the continental lithosphere. Phys. Chem. Earth, 15: 39–68.

Weber, W., 1980. The Pikwitonei granulite domain, edge of the Superior Province craton. EOS, Trans. Am. Geophys. Union, 61: 386–387.

Weber, W. and Loveridge, W.D., 1981. Geological relationships and Rb-Sr age studies of the Cauchon Lake tonalite and Pikwitonei enderbite in the northwestern Superior Province, central Manitoba. Geol. Surv. Can. Pap., 81-1C: 115–121.

Weber, W. and Scoates, R.F.J., 1978. Archean and Proterozoic metamorphism in the northwestern Superior Province and along the Churchill–Superior boundary, Manitoba. Geol. Surv. Can. Pap., 78-10: 5–16.

Wells, P.R.A., 1979. Chemical and thermal evolution of Archean sialic crust, southern eastern Greenland. J. Petrol.. 20: 187–226.

Wells, P.R.A., 1980. Thermal models for the magmatic accretion and subsequent metamorphism of continental crust. Earth Planet. Sci. Lett., 46: 253–265.

Wickham, S.M., 1987. The segregation and emplacement of granitic magmas. J. Geol. Soc. London, 144: 281–297.

Wickham, S.M. and Oxburgh, E.R., 1985. Continental rifts as a setting for regional metamorphism. Nature, 318: 330–333.

Wiener, R.W., McLelland, J.M. and Isachsen, Y.W., 1984. Stratigraphy and structural geology of the Adirondack Mountains, N.Y.: Review and synthesis. In: M.J. Bartholomew (Editor), The Grenville Event in the Appalachians and Related Topics. Geol. Soc. Am., Spec. Pap., 194: 1–56.

Windley, B.F., 1986. Comparative tectonics of the western Grenville and the Western Himalaya. In: J. Moore, A. Davidson and A. Baer (Editors), The Grenville Province: New Perspectives. Geol. Assoc. Can., Spec. Pap., 31: 342–348.

Zeitler, P.K., Johnson, N.M., Naeser, C.W. and Tahirkheli, R.A.K., 1982. Fission-track evidence for Quarternary uplift of the Nanga Parbat region, Pakistan. Nature, 298: 255–257.

Subject index